ANNUAL REVIEW OF ECOLOGY AND SYSTEMATICS

Annual Review of Ecology and Systematics

VOLUME 30, 1999

DAPHNE GAIL FAUTIN, *Editor*
University of Kansas

DOUGLAS J. FUTUYMA, *Associate Editor*
State University of New York at Stony Brook

FRANCES C. JAMES, *Associate Editor*
Florida State University

www.AnnualReviews.org   science@annurev.org   650-493-4400

ANNUAL REVIEWS
4139 El Camino Way • P.O. Box 10139 • Palo Alto, California 94303-0139

ANNUAL REVIEWS
Palo Alto, California, USA

International Standard Serial Number: 0066-4167
International Standard Book Number: 0-8243-1430-1
Library of Congress Catalog Card Number: 71-135616

∞ The paper used in this publication meets the minimum requirements of American National Standards for Information Sciences—Permanence of Paper for Printed Library Materials. ANSI Z39.48-1992.

TYPESET BY TECHBOOKS, FAIRFAX, VA
PRINTED AND BOUND IN THE UNITED STATES OF AMERICA

Contents

RELATED ARTICLES

From the ***Annual Review of Earth and Planetary Sciences***, Volume 27 (1999)

Stromatolites in Precambrian Carbonates: Evolutionary Mileposts or Environmental Dipsticks? John P. Grotzinger and Andrew H. Knoll

Major Events in the History of Carnivorous Mammals, Blaire Van Valkenburgh

From the ***Annual Review of Energy and the Environment***, Volume 23 (1998)

Science and Nonscience Concerning Human-Caused Climate Warming, J. D. Mahlman

The O_2 Balance of the Atmosphere: A Tool for Studying the Fate of Fossil Fuel CO_2, Michael L. Bender, Mark Battle, and Ralph F. Keeling

From the ***Annual Review of Entomology***, Volume 44 (1999)

Mites in Forest Canopies: Filling the Size Distribution Shortfall? David Evans Walter and Valerie Behan-Pelletier

The Evolution and Development of Dipteran Wing Veins: A Systematic Approach, Julian Stark, James Bonacum, James Remsen, and Rob DeSalle

Pathogens and Predators of Ticks and Their Potential in Biological Control, M. Samish and J. Rehacek

The Role of Stingless Bees in Crop Pollination, Tim A. Heard

Hyperparasitism: Multitrophic Ecology and Behavior, Daniel J. Sullivan and Wolfgang Völkl

Congruence and Controversy: Toward a Higher-Level Phylogeny of Diptera, D. K. Yeates and B. M. Wiegmann

From the ***Annual Review of Microbiology***, Volume 53 (1999)

Wolbachia pipientis: Microbial Manipulator of Arthropod Reproduction, R. Stouthamer, J. A. J. Breeuwer and G. D. D. Hurst

Poles Apart: Biodiversity and Biogeography of Sea Ice Bacteria, James T. Staley and John J. Gosink

Circadian Rhythms in Cyanobacteria: Adaptiveness and Mechanism, Carl Hirschie Johnson and Susan S. Golden

From the ***Annual Review of Phytopathology***, Volume 37 (1999)

Effects of Plants on Nematode Community Structure, G. W. Yeates

The Evolution of Asexual Fungi: Reproduction, Speciation, and Classification, John W. Taylor, D. J. Jacobson, and Matthew C. Fisher

Host Variation for Interactions with Beneficial Plant-Associated Microbes, Kevin P. Smith and Robert M. Goodman

From the ***Annual Review of Plant Physiology and Plant Molecular Biology***, Volume 50 (1999)

Roots in Soil: Unearthing the Complexities of Roots and Their Rhizospheres, Margaret E. McCully

ANNUAL REVIEWS is a nonprofit scientific publisher established to promote the advancement of the sciences. Beginning in 1932 with the *Annual Review of Biochemistry*, the Company has pursued as its principal function the publication of high-quality, reasonably priced *Annual Review* volumes. The volumes are organized by Editors and Editorial Committees who invite qualified authors to contribute critical articles reviewing significant developments within each major discipline. The Editor-in-Chief invites those interested in serving as future Editorial Committee members to communicate directly with him. Annual Reviews is administered by a Board of Directors, whose members serve without compensation.

Annual Reviews of

Anthropology
Astronomy and Astrophysics
Biochemistry
Biomedical Engineering
Biophysics and Biomolecular Structure
Cell and Developmental Biology
Earth and Planetary Sciences
Ecology and Systematics
Energy and the Environment
Entomology
Fluid Mechanics
Genetics
Immunology
Materials Science
Medicine
Microbiology
Neuroscience
Nuclear and Particle Science
Nutrition
Pharmacology and Toxicology
Physical Chemistry
Physiology
Phytopathology
Plant Physiology and Plant Molecular Biology
Political Science
Psychology
Public Health
Sociology

SPECIAL PUBLICATIONS
Excitement and Fascination of Science, Vols. 1, 2, 3, and 4

For the convenience of readers, a detachable order form/envelope is bound into the back of this volume.

Annu. Rev. Ecol. Syst. 1999. 30:1–22

The Origin and Early Evolution of Turtles

Olivier Rieppel
Department of Geology, The Field Museum, Chicago, Illinois 60605-2496; e-mail: rieppel@fmppr.fmnh.org

Robert R. Reisz
Department of Zoology, University of Toronto, Erindale Campus, Mississagua, Ontario L5L 1C6, Canada; e-mail: rreisz@credit.erin.utoronto.ca

Key Words turtles, Triassic, phylogeny, paleobiology

■ **Abstract** A critical reexamination of turtle relationships continues to support a sister-group relationship of turtles with a clade of marine reptiles, Sauropterygia, within crown-group Diapsida (Sauria). The high Homoplasy Index raises concerns about the phylogenetic information content of various morphological characters in broad-scale phylogenetic analyses. Such analyses may also suffer from inadequate statements of primary homology. Several such statements that have played an important role in the analysis of turtle relationships (dermal armor, acromion, astragalo-calcaneal complex, hooked fifth metatarsal) are reviewed in detail. An evolutionary scenario for the origin of the turtle *bauplan* suggests an aquatic origin of turtles, which is supported not only by their sauropterygian relationships, but also by paleobiogeographic and stratigraphic considerations. However, turtle relationships remain labile, and further investigations of their relationships are required, involving molecular and physiological data.

INTRODUCTION

In a comprehensive evaluation of turtle relationships, Gregory (26) compared living and fossil turtles with placodonts, "cotylosaurs" (captorhinids, pareiasaurs, procolophonoids, and diadectomorphs, all considered amniotes at the time), and with seymouriamorphs. He concluded that Testudines were derived from Paleozoic "cotylosaurs," and that among those, pareiasaurs approached Triassic turtles more closely than the geologically older diadectids. Although placodonts, especially *Henodus*, had evolved an amazingly turtle-like appearance, Gregory concluded that they were not related to turtles and that convergent evolution, especially related to dermal armor, causes a serious problem in recognizing testudine relationships.

Olson (52) reconsidered the origin of turtles as part of a reevaluation of Paleozoic and Mesozoic amniotes. He argued for a basic division of amniotes into Parareptilia and Eureptilia, and he suggested a derivation of turtles from basal parareptiles, i.e., the diadectids. Much later, Carroll (9) proposed an origin of turtles from among basal captorhinids. The advent of cladistics caused a major shift in ways of looking at turtle origins. The first large-scale computer-assisted phylogenetic analysis of amniote relationships (20) supported turtle relationships with captorhinids. Two other important results of this analysis were the exclusion of diadectomorphs from amniotes, and the recognition of a clade of parareptiles including mesosaurs, millerosaurs, pareiasaurs, and procolophonoids. Monophyly of this clade was poorly supported, however, mainly because of lack of detailed anatomical information about its members.

Largely as a response to this paper, Reisz & Laurin (58) proposed an alternative hypothesis of relationships, i.e., that turtles were nested within parareptiles, closely related to procolophonoids. This result was based, in part, on the study of *Owenetta*, a basal procolophonian that shares a suite of synapomorphies with turtles. By contrast, the study of pareiasaurs led Lee (39) to conclude that they, rather than procolophonoids, are the closest known relatives of turtles. Subsequently, Laurin & Reisz (38) incorporated the new anatomical data on pareiasaurs presented by Lee (39) into their own analysis, but they continued to find support for the hypothesis of procolophonoid-turtle relationships. However, possible turtle relationships were constrained in this analysis because, other than parareptiles and captorhinids, it included only basal diapsids and basal synapsids for possible comparison. A second, slightly later publication by Lee (41) included previously withheld anatomical data that again turned turtle relationships to pareiasaurs. However, Lee's (41) analysis constrained turtle relationships even more seriously because the choice of terminal taxa was based on the a priori assumption that turtles are, indeed, parareptiles.

Continuing controversy over turtle relationships (42, 66) culminated in two adjoining articles in a single issue of the *Zoological Journal of the Linnean Society* (43, 11). Lee (43) now included most pareiasaurs as terminal taxa, with the result that turtles were found to be nested within pareiasaurs, as sister-taxon to the poorly known yet derived genus *Anthodon*. DeBraga & Rieppel (11) pursued a global approach instead by including representatives of most Paleozoic and early Mesozoic amniote taxa in order to test for patterns of turtle relationships among a broad array of amniotes. As a result of their analysis, deBraga & Rieppel (11) proposed a highly controversial hypothesis, i.e., that turtles are nested within diapsids as sister-group of a clade of Mesozoic marine reptiles, the Sauropterygia. It is the latter hypothesis that we propose to reexamine in this review by modifying the data set in accordance with recent criticisms (44) and with other recent increases in our knowledge of the relevant taxa, and by reanalyzing the data using the software packages PAUP version 3.1.1. (74) and McClade version 3 (47).

CHANGES TO THE GLOBAL DATA MATRIX

Lee (44) examined the data matrix of deBraga & Rieppel (11, 66) and argued that many of the characters were incorrectly coded. We have examined the proposed corrections and agree with the majority (the characters are numbered as in the above two papers: 44–46, 70, 73, 77, 87, synapsid part of 120, 124, 126, 150, 160, and 164). Changes were incorporated in the present study in accordance with Lee's (44) interpretations. However, we disagree with several of the proposed corrections, and for these retain the original coding. They include: [65] distinct basal tubera are absent in pareiasaurs, which show a secondarily derived condition, i.e., basal tubera on the parasphenoid; [82] the position of the mandibular joint relative to occiput is polymorphic in turtles; [103] cervical centra of pareiasaurs are ridged but not keeled; [120, 121] one coracoid ossification is present, and the coracoid foramen is enclosed by the coracoid and scapula, in pareiasaurs; [127] the presence of the ectepicondylar foramen in the humerus is polymorphic for turtles; [140] a weak 4th trochanter is present in pareiasaurs and shifted to the edge of the femur; [152] the first distal tarsal is retained as polymorphic in turtles.

We also reexamined the data matrix published by deBraga & Rieppel (11) as part of our attempt to identify possible biological causes for apparent patterns of character conflicts in their phylogeny. This effort resulted in the following changes of character coding: Character 69 (occipital flange of parietal) has been almost entirely recoded, taxa 1–14 and 20–22 all having this feature (1), which is sometimes covered posteriorly by other elements such as the postparietal and tabular. *Paleothyris*: 139/0 → 1; Millerettidae: 125/? → 1; *Bradysaurus*: 80/1 → 0; 125/? → 0; 144/1 → 0; *Scutosaurus*: 80/1 → 0; 125/? → 0; 144/1 → 0; *Anthodon*: 80/1 → 0; 125/? → 0; 144/1 → ?; *Owenetta*: 52/0 → 0&2; Kuehneosauridae: 29/? → 0; 78/2 → 0; 125/0 → 1; Testudines: 83/0 → 1;142/0 → 1; 167/1 → 0&1; Rhynchocephalia: 68/0 → 0&1; *Placodus*: 41/? → 1; 64/1 → ?; 77/0 → 1; 83/1 → 1&2; 89/1 → 0; 112/? → 1; 140/0 → 1; 165/2 → 1; Eosauropterygia: 6/0 → 1; 31/1 → 0&1; 37/0 → 1; 41/? → 1; 59/2 → 1&2; 87/0 → 1; 97/? → 1, 112/0 → 0&1; 140/0&1 → 1; 142/0&1 → 1&2.

In addition, we decided to code the placodont *Cyamodus* to ensure that we can provide a proper test of the possibility that turtles are either nested within placodonts or that the striking similarities of cyamodontoid placodonts and turtles are, indeed, convergent. The following list represents the coding, in groupings of five, for all the characters listed in deBraga & Rieppel (1997): *Cyamodus* 10000 101(0&1)0 10000 0212? 00(0&1)0(0&1) 00001 00001 11110 1?00? ?0121 32212 01021 00??1 01101 01201 21201 21111 01000 ?1101 00?10 ??111 1111? 1111? 00010 11111 10121 ??1?0 01011 0?111 ??1?? 0???? ????? ?0002 000.

The resulting data matrix was reanalyzed using PAUP version 3.1.1. (74), implementing the same search procedures as in the original analysis (11). The results duplicate the tree topology of the original study (11, 66), with turtles nested within diapsids as the sister-group of sauropterygians (Figure 1). To ensure proper comparison of the phylogenetic analyses of Lee (43) and deBraga & Rieppel (11), we refrained from adding new characters to the data matrix used by the latter authors. Instead we checked in the data matrix of Lee (43) for characters that would appear relevant to a more extensive, global analysis but that were not included by deBraga & Rieppel (11). Three characters are of potential importance (numbers in accordance with Ref. 43): [5] exoccipital lateral flange absent (0), or present (1); [104] lateral pubic process absent (0), or present (1); [105] median pubic process absent

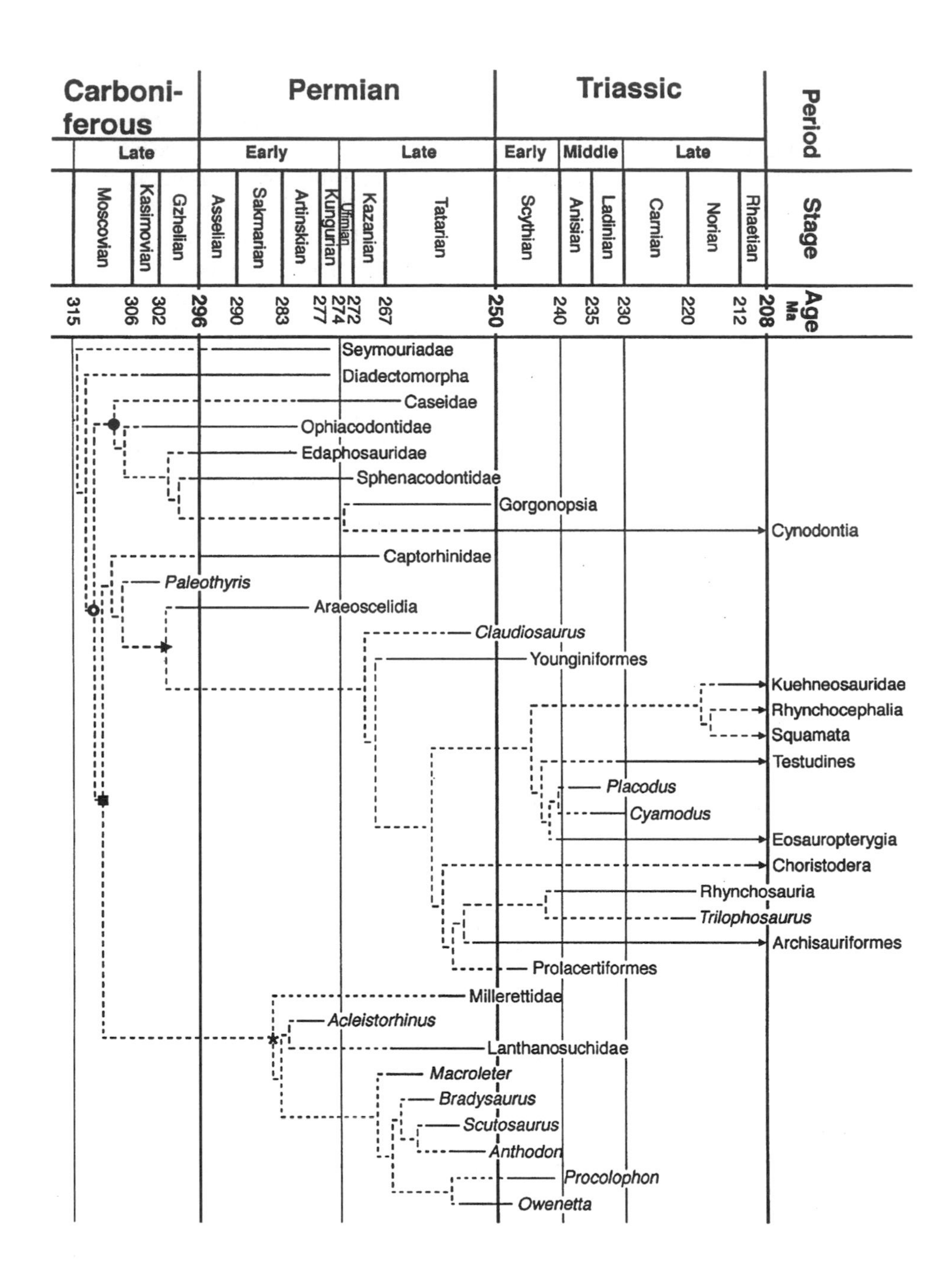
Carboniferous
Permian
Triassic
Period
Late
Early
Late
Early
Middle
Late
Moscovian
Kasimovian
Gzhelian
Asselian
Sakmarian
Artinskian
Kungurian
Ufimian
Kazanian
Tatarian
Scythian
Anisian
Ladinian
Carnian
Norian
Rhaetian
Stage
315 306 302 296 290 283 277 274 272 267 250 240 235 230 220 212 208
Age Ma
Seymouriadae
Diadectomorpha
Caseidae
Ophiacodontidae
Edaphosauridae
Sphenacodontidae
Gorgonopsia
Cynodontia
Captorhinidae
Paleothyris
Araeoscelidia
Claudiosaurus
Younginiformes
Kuehneosauridae
Rhynchocephalia
Squamata
Testudines
Placodus
Cyamodus
Eosauropterygia
Choristodera
Rhynchosauria
Trilophosaurus
Archisauriformes
Prolacertiformes
Millerettidae
Acleistorhinus
Lanthanosuchidae
Macroleter
Bradysaurus
Scutosaurus
Anthodon
Procolophon
Owenetta

(0), or present (1). However, these characters are expected not to change tree topology, but only to increase the homoplasy index because all of them are found convergently in pareiasaurs and saurians. For example, the derived state for character 5 is found not only in turtles and pareiasaurs, but also in crocodilians, eosauropterygians, and in lizards where the exoccipital is not (yet) fused to the opisthotic.

CLADISTIC ANALYSIS

As in the original analysis (11), two most parsimonious trees (MPTs) were found (a single unresolved trichotomy within archosauromorphs) with a *T*ree *L*ength (TL) of 793 steps, a consistency index (CI) of 0.503, and a *H*omoplasy *I*ndex (HI) of 0.731 (search procedures as in the original analysis; characters 2, 54, 166 are uninformative; if ignored, the two MPTs have a TL = 789, CI = 0.501, and HI = 0.735). Pruning of Testudines from its position among diapsids and grafting it into parareptiles as the sister-taxon of pareiasaurs increases *T*ree *L*ength by five steps (not significant at $p \leq 0.05$ based on the Templeton test [$T_S = 562.5$, $n = 55$]. Decay analyses yielded similar results, as five extra steps were required to collapse the reptilian node that would allow turtles to shift into parareptiles. Constraining the overall pattern of relationships in PAUP, yet forcing turtles to be the sister-group of pareiasaurs (Testudines (*Bradysaurus* (*Scutosaurus, Anthodon*))) results in a single tree with a TL of 798 (TL = 794 if uninformative characters are ignored), i.e., a tree again five steps longer. If turtles are constrained to be the sister-group of pareiasaurs, but tree topology is allowed to change in all other aspects, two trees (lack of resolution confined to Lepidosauriformes) are found with a TL = 796, i.e., three steps longer than the original MPTs. However, tree topology has changed: Procolophonoids become paraphyletic in these trees, and the interrelationships of pareiasaurs change to (Testudines (*Anthodon* (*Bradysaurus, Scutosaurus*))). Finally, allowing PAUP to search for all trees that are five steps longer than the most parsimonious reconstruction (TL = 798) results in more than 1000 trees with little or no resolution.

The Consistency Index is low, although not out of line for such a large-scale analysis. More importantly, the high Homoplasy Index suggests rampant convergence. This is also indicated by relatively low bootstrap values. The bootstrap support for the node linking turtles and sauropterygians is only 53% (2000 replications). We reexamined the list of characters and concluded that many of the proposed

←

Figure 1 The phylogeny and temporal occurrence of Reptilia, based on the analysis discussed in the text. The tree for Reptilia is rooted on Seymouriadae, Diadectomorpha, and the synapsids. In the phylogeny, dashed lines represent ghost lineages, solid lines represent the fossil record. Arrows at the top indicate continuation of the fossil record beyond the Triassic and may lead up to extant taxa. *Open circle* at the base of the phylogeny denotes Amniota, *solid circle* denotes Synapsida, *solid square* denotes Reptilia, *solid triangle* denotes Diapsida, *asterisk* denotes Parareptilia.

primary homologies appear to be either flawed or uninformative in an analysis covering such a broad variety of taxa. We would like, therefore, to argue that rigorous evaluation of statements of primary homology is an effective way both to minimize evolutionary steps and to maximize parsimony.

STATEMENTS OF PRIMARY HOMOLOGY

The essence of all phylogenetic reconstructions of evolutionary relationships is the character data base, i.e., the identification of "characters," or "primary homologies" (13), which are then subjected to the test of congruence (54). Although the test of congruence represents the ultimate arbiter on homology (similarity due to common ancestry) versus homoplasy (independently acquired similarity), it is itself not rooted in anatomical investigation, but is designed merely to maximize congruence and hence to minimize homoplasy. A high degree of character congruence (resulting in a relatively short TL and a relatively high CI) therefore says nothing about the quality of the character data base. Similarly, the sheer number of characters is not going to improve the quality of a phylogenetic analysis unless every one of these characters is founded in careful anatomical comparison.

The test of congruence critically depends on the rigorous application of the "test of similarity" (54) in the identification of primary homologies. The establishment of "similarity" (in terms of equivalence of topographic relations, or connectivity) may be difficult if organisms of highly derived anatomy such as turtles have to be dealt with. In such cases, the analysis of ontogenetic development, as well as comparison of organisms (fossil and extant) in terms of hypothetical transformation series in search for intermediate conditions of form, may play a crucial role in the identification of primary homologies (59). In this section we explore the importance of detailed anatomical comparison for four selected character complexes that have played a prominent role in recent discussions of turtle relationships: the dermal armor, the acromion process on the scapula, the astragalus-calcaneum complex, and the hooked fifth metatarsal. These examples are chosen for the following reasons: dermal armor and the acromion process on the scapula have been used as synapomorphies uniting pareiasaurs and turtles, but closer anatomical comparison reveals a flawed basis for the identification of primary homologies. Conversely, detailed anatomical comparison supports the astragalus-calcaneal complex and the hooked fifth metatarsal as potential homologies of turtles and diapsids on the basis of close developmental and structural correspondence.

Dermal Armor

Lee (41–43) considered the presence of a heavy dermal armor a synapomorphy of turtles and pareiasaurs, some of which show interlocking osteoderms lying closely above the vertebral column and ribs. As he believed developmental evidence indicated a primarily dermal nature of the carapace in turtles, Lee suggested that the turtle carapace developed by fusion of ancestral osteoderms. Turtles are unique among amniotes in that the scapular blade lies at a morphological level deep (ventral) to the ribs, which according to Lee (39, 40) would have resulted from a backward shift of the shoulder girdle.

The primary homology and potential synapomorphy of pareiasaur osteoderms and the turtle "dermal" armor is problematic because of the morphological complexity of the turtle carapace. The latter is recognized as a composite structure, involving "thecal" as well as "epithecal" ossifications (77, 81, 34). The thecal ossifications comprise a central longitudinal row of neurals overlying and fused to the neural arches of dorsal vertebrae, a lateral row of costal plates associated with the dorsal ribs, a marginal row of marginals, an anterior nuchal, and a posterior pygal. Epithecal ossifications are osteoderms superimposed on thecal ossifications (81, 82). It had been claimed (29, 51) that epithecal ossifications are primitive for turtles and would have covered the body of the ancestral turtle prior to the development of a theca. If so, the turtle shell could easily be derived from fused ancestral osteoderms. However, epithecal elements ossify later than thecal components during ontogeny in living turtles, and mapping the occurrence of epithecal ossifications on a cladogram of Testudines indicates their derived nature. Whereas the genuinely dermal nature of the marginal, supramarginal (*Proganochelys*—15), nuchal, and pygal plates is generally accepted (34, 75–77), the nature of the neural and costal plates remains controversial.

The basic distinction of turtles from other amniotes is not a (posterior) shift of the pectoral girdle (scapula) to a level medial (ventral) of the ribs, but a "deflection" of the ribs to a position dorsal (lateral) to the scapula (71). The carapacial ridge redirects the migration of those somitic cells that will eventually form the ribs (7), such that the ribs chondrify in a position dorsal to the scapula and within the dermal carapacial disc. Completion of perichondral ossification of the ribs shows that these are not expanded in turtles (34) at the cartilaginous stage (Figure 2). Ossification of the costal plates proceeds by the formation of trabecular bone starting from and remaining in continuity with the periost of the rib (34).

The fact that the developing ribs and neural arches pierce the dermal carapacial disc renders the identification of neurals and costals as endoskeletal versus exoskeletal elements difficult. However, endo-versus exoskeletal cannot be distinguished on the basis of histogenesis but must be defined with reference to a phylogenetic framework (53, 73). Exoskeletal elements are elements homologous to structures, which in the ancestral condition combine bone, dentine, and enamel, i.e., develop at the ectoderm-mesoderm interface. Endoskeletal elements are elements that in the ancestral condition are preformed in cartilage, while the cartilaginous stage may be deleted in the descendant (membrane bone). The neural and costal plates ossify from, and in continuity with, the periost of their endoskeletal component. This pattern of ossification corresponds to the definition of *Zuwachsknochen* given by Starck (73:13). As such, neural and costal plates are endoskeletal components of the turtle carapace and cannot be derived from a hypothetical ancestral condition by fusion of exoskeletal osteoderms.

Following this analysis of primary homology, the turtle carapace is unique, i.e., autapomorphic for turtles, and morphogenetically very distinct even from its closest counterpart among other amniotes, which is the carapace of cyamodontoid

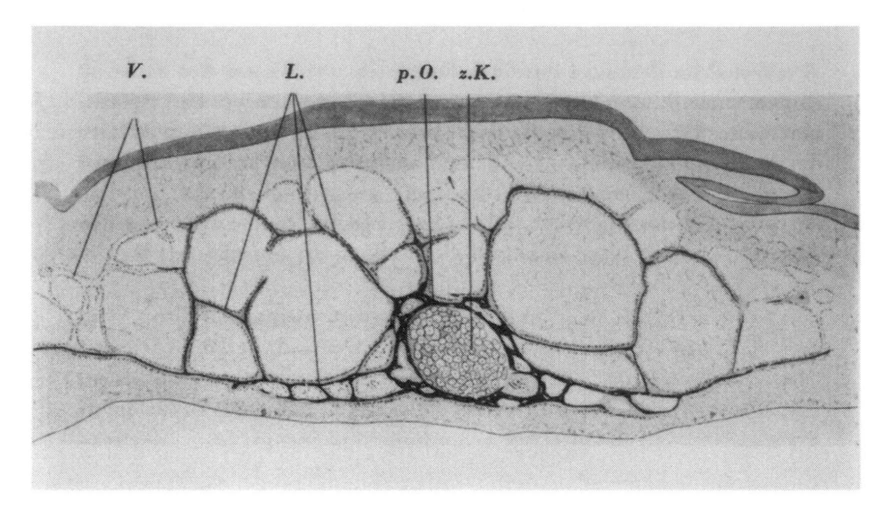

Figure 2 Cross-section through the sixth rib of an embryo of *Podocnemis unifilis*, showing the trabecular bone of the costal plate developing in continuity with the periost of the rib (from Ref. 34, with permission of S. Karger Publ.).

placodonts. The latter is composed of a mosaic of (exoskeletal) osteoderms, which may or may not be closely associated with the underlying endoskeleton (vertebrae and ribs), but which never incorporates endoskeletal elements (80).

The Acromion

The endoskeletal pectoral girdle of turtles is a triradiate structure of a highly specialized nature. The scapula forms a dorsal process with a ligamentous attachment to the anterior border of the first rib and the adjacent carapace. The coracoid forms a posteroventral process, while a medioventrally directed process of the scapula establishes a ligamentous connection to the entoplastron (78, 79). Because the ventromedial scapular process is shorter in *Proganochelys* (15) than in extant turtles, Goodrich (24) considered it an acromion. Lee (41, 43) considered the acromion a primary homology synapomorph in pareiasaurs and turtles, a conjecture of similarity rejected by deBraga & Rieppel (11).

Although it ossifies from a separate ossification center (61), it is in continuity with the scapula at the cartilaginous as well as at the fully ossified stage of its development, which suggests that the "acromion" of turtles is a process of the scapula. However, the medioventral scapular process of *Proganochelys*, homologous to the "acromial" process of modern turtles (Figure 3), fails the test of similarity with the acromion of pareiasaurs and therapsids. Both in therapsids and in pareiasaurs, the acromion is located at the dorsal tip of the clavicle, whereas the "acromion" of turtles is near the base of the clavicle and medial to it (*Proganochelys*). The "acromion" process in turtles is a medioventral extension of the scapula, not an anterolateral extension as is the acromion of pareiasaurs and nonmammalian therapsids. This difference of orientation also results in different muscle attachments

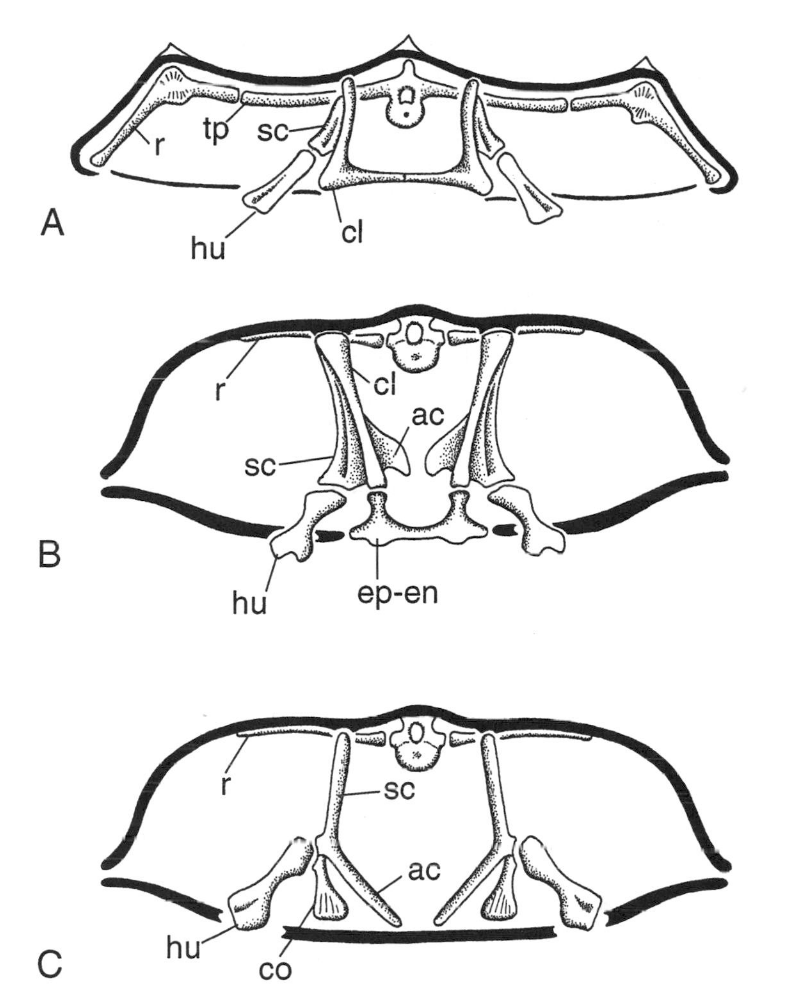

Figure 3 Schematic representation of the relation of the pectoral girdle to the dermal armor in the cyamodontoid placodont *Henodus* (A; coracoid not known), in *Proganochelys* (B; coracoid not shown as it is positioned perpendicular to the plane of the drawing), and in modern turtles (C). Abbreviations: ac, "acromion" process; cl, clavicle; co, coracoid; ep-en, epiplastron-entoplastron; hu, humerus; r, rib; sc, scapula; tp, transverse process.

(78). The fact that the "acromial" process is shorter in *Proganochelys* than in other turtles may indicate that it evolved within the clade. The ligamentous attachment of the "acromion" process to the entoplastron suggests a function in locomotion (78).

The sprawling gait primitively found in reptiles results in two principal reaction forces at the shoulder joint: a vertical component that is absorbed by the scapula and its muscular suspension from the axial skeleton, and a medially directed component that is absorbed by the clavicular-interclavicular complex. In turtles, the clavicles and interclavicle are incorporated in the plastron, and while the vertical component is absorbed by the scapula and the carapace, the medially directed force component is absorbed by the "acromion" process and the plastron (79). In *Proganochelys*, the complex epiplastra carry elongate dorsal (clavicular) processes that articulate with the dorsal process of the scapula. It is likely that medially directed forces were partly deflected from the scapula to the epiplastron-entoplastron complex via the clavicular process.

The Astragalo-Calcaneal Complex

Amniota reduce the proximal tarsal ossifications to an astragalus and a calcaneum. The astragalus is believed to have evolved by fusion of the ancestral intermedium, tibiale, and proximal centrale, whereas the calcaneum represents the ossified fibulare (55, 62). Lee (41, 43) considered the ontogenetic fusion of the astragalus and calcaneum a synapomorphy of pareiasaurs and turtles. However, an ontogenetic fusion of the astragalus and calcaneum is also observed in lepidosaurian diapsids. Secondarily marine clades, subject to skeletal paedomorphosis, may retain separate astragalus and calcaneum, as do marine turtles (77), and sauropterygians. The use of fusion of the astragalus and calcaneum in adult turtles and pareiasaurs as a synapomorphy of the two groups does not account for morphological complexity, which results from an ontogenetic repatterning of the development of the tarsus in turtles and those diapsids that develop a mesotarsal joint.

The development of the tarsus of turtles is subject to a truncation of the preaxial series (6), such that no true tibiale, connected to the tibia, forms in this group. Originally separate precartilaginous precursors in the proximal tarsus (fibulare, intermedium, and centrale), all fuse into a single block of cartilage, the "*astragaloscaphoideum*" of Gegenbaur (22), or "*tritibiale*" of Rabl (57), or "*tarsale proximale*" of Sewertzoff (72). The astragalus and calcaneum each ossify from a single ossification center within the proximal tarsal cartilage; the ossification of the astragalus precedes that of the calcaneum, and its first appearance is in an intermedium position. A third center of ossification may variably appear distal to the astragalus within the proximal tarsal cartilage of turtles, giving rise to a centrale, an element absent in *Proganochelys* (15). The development of a proximal block of cartilage in the tarsus makes it impossible for the perforating artery (84) to pass between astragalus and calcaneum, as it does in the primitive condition (62), and it accordingly shifts to a position proximal to the astragalus, passing between the distal heads of the tibia and fibula. A developmental pattern similar to that of extant turtles may be inferred for *Proganochelys*, not only because of the detailed similarity of the astragalo-calcaneal complex, but also because *Proganochelys* lacks a perforating foramen between astragalus and calcaneum (15).

The tarsus of lepidosaurian diapsids (*Sphenodon*, squamates) develops in an identical manner (although the proximal centrale is always absent in this clade). The separate precartilaginous precursors of the proximal tarsal elements again fuse to form a single proximal tarsal cartilage, within which the astragalus and calcaneum ossify from separate ossification centers (1, 22, 31, 72), and the astragalus and calcaneum fuse in the adults. The perforating artery passes proximal to the tarsal cartilage, between the distal ends of tibia and fibula (50). No fossil lepidosaur is known that has separate astragalus and calcaneum ossifications enclosing between them a foramen for the perforating artery.

Ontogenetic repatterning of the proximal tarsus in turtles and lepidosaur reptiles results in the cartilaginous preformation of the mesotarsal joint in cartilage, between the proximal tarsal cartilage and the distal tarsal 4. That way, the mesotarsal joint is fully functional in hatchlings and later growth stages. But whereas the astragalus and calcaneum also fuse in adult pareiasaurs, there is no indication of ontogenetic repatterning of the tarsus. A foramen for the perforating artery is retained within the astragalo-calcaneal complex, there is no indication of a mesotarsal joint in pareiasaurs, and the astragalo-calcaneal complex retains the primitive amniote configuration (except for its fusion) in that it is much more plate like, much broader proximo-distally, and more horizontally positioned than that of turtles and lepidosaurs.

The Hooked Fifth Metatarsal

A hooked fifth metatarsal is present in turtles (including *Proganochelys*—15), in the diapsid *Youngina* (25), as well as in the lepidosauromorph and archosauromorph clades, collectively referred to as Sauria, but it is absent in pareiasaurs or procolophonoids. Lee (45) uses arguments based on parsimony to treat the hooked fifth metatarsal of turtles and diapsids as convergent and, based on the (presumed) absence of a hooked fifth metatarsal in basal lepidosauriforms (kuehneosaurs), also considers the lepidosaurian and archosauromorph hooked fifth metatarsal as convergent (21).

The development of the hooked fifth metatarsal is identical in all three extant groups where it occurs (60, 61, 63), in that it is severely delayed in ossification in comparison to other metatarsals. If well developed (reduced in *Alligator*), perichondral ossification starts at the angulation on the medial (preaxial) side, where the hooked fifth metatarsal articulates with the proximal head of the fourth metatarsal (in contrast to other metatarsals, which show the normal mid-diaphyseal ossification). The only difference in the development of the fifth metatarsal in turtles and archosauromorphs on the one hand, and lepidosaurs on the other, is the ossification of separate epiphyses or apophyses in the latter group, adding to the complexity of the fully formed hooked fifth metatarsal in lepidosaurs (69). Ossification of epiphyses is an autapomorphy of Lepidosauria (28), and hence it is uninformative for phylogenetic inference.

Among nonlepidosaur lepidosauriforms, i.e., kuehneosaurs, the hind feet remain unknown for *Icarosaurus* (10). *Kuehneosaurus* is known from fully disarticulated material only (68), and the presence or absence of a hooked fifth metatarsal must remain conjectural. Sauropterygians, nested within saurians (64, 48), lack a hooked fifth metatarsal. Given its presence in the outgroup of saurians, *Youngina*

(25), the absence of a hooked fifth metatarsal in sauropterygians is likely due to skeletal paedomorphosis, as is also the case in marine turtles (*Dermochelys*; 77). A hooked fifth metatarsal thus represents a synapomorphy of *Youngina* plus saurians (37), including turtles, unless congruence of other characters would indicate that turtles are nested within parareptiles.

The foot of a generalized lepidosaur is strongly asymmetrical, digit length increasing from digit I through digit IV. During the initial phase of pedal plantarflexion, the metatarsus rotates around its long axis at the mesotarsal joint, such that subsequent extension of the ankle joint moves the proximal end of the metatarsus anteriorly and laterally. The fourth digit rolls off the substrate first during extension of the ankle, but digits I through III remain in contact with the substrate until the end of the propulsive phase. During this locomotor cycle, the hooked fifth metatarsal acts as a dual lever for different muscle action, once with respect to the rotation of the metatarsus around its long axis, then with respect to the extension of the ankle joint (4, 5).

In turtles, including *Proganochelys* (15), the digits are much more symmetrical in their relative length. The excursion range of (humerus and) femur retraction is somewhat restricted as compared to squamates. Initially, in turtles, during retraction of the femur, all toes point anteriorly. Toward the end of the retracting cycle, the foot performs a distinct "heel swing" that turns the heel medially around the toes, which at this time act as a pivot and which point laterally after completion of the heel swing (78, 85). In contrast to squamates, there is little plantar flexion at the end of the retractive phase. The foot is moved forward with its sole facing the substrate; its heel as well as longer claws may drag during initial phases of protraction (78). Locomotion is slower and less sophisticated in turtles compared to squamates, yet the pes of turtles shares with that of squamates a well-developed mesotarsal joint as well as a hooked fifth metatarsal. This suggests that these latter features were inherited by turtles from their common ancestor, rather than developed within turtles in convergence to squamates, a conclusion supported by our phylogenetic analysis.

THE PHYLOGENETIC INFORMATION CONTENT OF MORPHOLOGICAL CHARACTERS

The modified global analysis used to evaluate turtle relationships is based on a data matrix characterized by an unusually large number of homoplastic characters that are not useful in resolving relative relationships, whereas a relatively small and different set of characters defines the observed topology. We expected that those characters that cause the high Homoplasy Index would be clustered in certain parts of the anatomy, such as the locomotory system, and could therefore be related to some functional trends seen in amniotes. For example, the evolution of Paleozoic amniotes appears to involve the gradual change from a locomotor system that relies primarily on lateral undulation of the vertebral column for propulsion, to one that can be described as a combination of limb-driven locomotion with lateral undulation. This evolutionary innovation has been achieved independently within several amniote clades and necessitates changes in the shoulder girdle, propodials, metapodials, and phalanges to allow for increasing stride length. This phenomenon could be viewed as an evolutionary arms race at the predator-prey level, involving forms that coexisted but were not closely related (pareiasaurs and carnivorous therapsids, for example). However, in our analysis the homoplastic (convergent) characters are distributed randomly throughout the skeleton.

Convergent characters affect the skull roof pattern: 6, 11, 17(1), 20, 38, 55; braincase: 66, 68; palate: 75, 76(1), 77, 78, 79; mandible: 83(1), 85, 86, 87, 91, 92; axial skeleton: 110, 111; appendicular skeleton: 115, 130, 144, 149, 150, 153, ?158, 160.

Examination of convergent cranial features (the derived condition appearing convergently in parareptiles, diapsids, turtles, and often also in derived synapsids) suggests that certain morphological transformations cannot convey a phylogenetic signal at the taxonomic level addressed by the evaluation of turtle relationships. The shape and orientation of nares [6], the shape and size of the lacrimal bone [17], the loss of bones such as the tabular [55], and the shape of particular palatal bones [75, 76, 77, 79] evolved independently several times within amniotes. However, the same characters may still be useful in evaluating patterns of relationships within more restricted clades of amniotes. The shape and size of the lacrimal bone, for example, provides important phylogenetic information within basal synapsids or within basal parareptiles (38).

Similarly, palatal characters such as the shape and size of the pterygoid [76, 77, 78, 79] may be associated with reduction and/or loss of palatal dentition, an evolutionary event that occurred independently in several clades, including derived synapsids, lepidosaurs, and some parareptiles. Several convergent mandibular characters may be associated with changes of insertion of jaw closing [83, 85, 86, 87, 91] and jaw opening [92] muscles.

In addition, larger analyses generally tend to have a progressively smaller Consistency Index, a progressively higher Homoplasy Index, and the resulting pattern of relationships tends to be rather weakly supported. The resolution of the above pattern of relationships is based on a relatively small number of characters within a much larger data set. Ultimately, the number of synapomorphies supporting a diapsid-turtle relationship outnumbers those that support a parareptile-turtle pattern. Yet a significant number of characters (listed above) appear to remain uninformative as relative relationships are analyzed across a broad range of taxa that have dramatically different body designs. We would like to argue for the need to reconsider the phylogenetic signal provided by morphological features at different levels of inclusiveness, and we would also like to point out that larger is not necessarily better when considering the size of data matrices in large-scale phylogenetic analyses.

EVOLUTIONARY SCENARIOS, PHYLOGENIES, AND CONVERGENCE

A case has recently been made that evolutionary scenarios, if based on independent evidence, should be allowed to provide the basis for choice among several equally parsimonious phylogenetic hypotheses, or even to overturn weakly supported cladograms (46). Independent evidence for evolutionary scenarios is to be derived from genetic, developmental, and/or functional correlation. We believe this claim to be flawed, because hypotheses of character correlation can only be tested by reference to a phylogeny. To import evolutionary scenarios into phylogeny reconstruction results in an empirically empty research program. By contrast, in a case such as turtle interrelationships, with competing yet vastly different phylogenetic hypotheses, the question can be asked, which one of the alternative hypotheses of turtle relationships better supports a plausible and parsimonious evolutionary scenario of turtle origins.

Lee (39, 40, 42) argued in favor of a gradual transition from terrestrial and herbivorous pareiasaurs to a turtle body plan. Shortening of the trunk, a shift from axial to paraxial locomotion, broadening of the ribs, fusion of osteoderms above the ribs to form a carapace, ankylosis of the vertebrae underlying the carapace, and a posterior migration of the shoulder girdle are considered essential steps in that transition. Here we argue that Lee's (39, 40, 42) scenario of turtle origins by "correlated progression" greatly oversimplifies the evolutionary changes that took place in the origin of the turtle body plan, and that functional constraints discussed below render it unlikely that turtles had a terrestrial origin.

Our phylogenetic hypothesis cannot resolve the question of whether the hypothetical common ancestor of turtles and sauropterygians was aquatic or terrestrial. If turtles are coded polypmorphic for terrestrial and/or aquatic habits, optimization shows this character to be equivocal at the node uniting turtles and sauropterygians. If turtles are coded terrestrial, their origin becomes unequivocally terrestrial; if they are coded aquatic, their origin becomes unequivocally aquatic. *Proganochelys* was reconstructed as an aquatic bottom walker on the basis of limb proportions, but some characters such as the elephantine feet, the high domed shell, and the heavy armor suggest terrestrial habits as well (15). By contrast, functional constraints implicit in the evolution of the turtle body plan would seem to suggest an aquatic origin for the clade.

There is no broadening of the ribs in turtles, and the carapace cannot be derived from a simple fusion of osteoderms. Although a slight posterior shift of the shoulder girdle sems to have occurred in turtles (8), the position of the scapula inside the rib cage is primarily the result of a deflection of rib growth to a more superficial position (7, 71). As the ribs become part of costal plates, and the tips of the neural arches expand to form neurals, complete ossification of the carapace fully immobilizes the dorsal vertebral column. The neural arches of turtles move forward by half a segment in the dorsal region, carrying the ribs with them (23). As a result, the proximal heads of the ribs are positioned intrasegmentally (with respect to the primary segmentation of the paraxial mesoderm), i.e., on the boundary between successive centra (assuming these form by resegmentation as in other amniotes). Likewise, the neural arches bridge the boundaries between successive centra, and in marine turtles subject to skeletal paedomorphosis, the neural arches can be seen to meet each other in a suture located above the midpoint of the dorsal centra. At this level, the two successive neural arches together with the centrum enclose the opening for the exit of the spinal nerve, which hence comes to lie intersegmentally. In other turtles, the dorsal neural arches fuse with one another to form a vertical blade, the "neural spine" (30), pierced by the spinal nerves at an intersegmental level. The myomeric and neuromeric segmentation is secondarily established in the dorsal region of turtles, as was already the case in *Proganochelys* (15). Intercostal muscles atrophy during embryonic development (23).

As a consequence of dramatic restructuring in the evolution of the turtle body plan, the ribs lose their function in support of respiration and locomotion. In a generalized tetrapod reptile, aspiration of air is effected by an expansion of the body cavity through muscular action exerted on the ribs. Exhalation is effected either by passive recoil of the body walls, and/or by compression of the lungs as a result of active compression of the rib cage. By contrast, respiration in turtles depends on volume changes of the thoracico-peritoneal cavity inside the rigid dermal armor, which is achieved by altering the position of the limb flanks through the activity of anterior and posterior muscles (18). A comparison of respiration in an aquatic (*Chelydra*) and a terrestrial (*Testudo*) turtle resulted in the identification of three forces that influence the relative volume of the thoracico-peritoneal cavity: These forces are muscular activity, gravity (supporting inhalation), and, in aquatic turtles, hydrostatic pressure (supporting exhalation) (19). Experimental analysis of muscle activity during breathing in *Chelydra* as compared to *Testudo* showed an adjustment of inhalatory and exhalatory muscle output to hydrostatic pressure in the aquatic turtle, the generated forces generally approaching a minimum. At the limit, both inhalation and exhalation may be passive in the aquatic environment, powered by gravity and hydrostatic pressure, respectively (19). Furthermore, both terrestrial and aquatic turtles show a much higher degree of tolerance of anoxia as compared to crocodiles or squamates (2). Collectively, these data suggest that the transition to the mode of respiration characteristic of turtles would be easier to achieve in an aquatic than in a terrestrial environment.

In terrestrial reptiles, body weight is transferred from the shoulder girdle (scapula) to the axial skeleton by means of a muscular sling, principally the *serratus ventralis*. Its probable homologue in turtles is the *testocoracoideus* muscle, which participates in the expansion of the thoracico-peritoneal cavity and hence supports inhalation (18, 19, 78). It would seem difficult to simultaneously derive the turtle mode of respiration and locomotion in a terrestrial environment that requires continuous body support. By contrast, the aquatic environment provides buoyancy, which greatly facilitates both body support and locomotion. The initial integration of the clavicular-interclavicular complex into the plastron, and the concomitant development of an acromion process on the scapula to absorb medially directed forces generated by limb movements, would appear to be easier to achieve in an aquatic environment for a bottom-walking turtle supported by some degree of buoyancy, more than in the terrestrial environment.

The development of a plastron in itself indicates an aquatic origin of turtles, as it provides dermal protection for the ventral body surface, which in terrestrial tetrapods is not immediately exposed to predatory impact. Indeed, the only other amniotes that evolved a plastron joined to a carapace by a bridge covering the lateral body wall are derived members of marine cyamodontoid placodonts (27). Although superficially very similar, the dermal armor of cyamodontoids is convergent upon that of turtles (see discussion above). However, such a striking case of convergence could be considered to lend additional support, by analogy, to the hypothesis of an aquatic origin of turtles. In addition to protection, both groups may have developed extensive dermal armor as an osmotic barrier. Experimental studies have shown a significantly smaller rate of gain of water (in fresh water) or loss of water (in sea water) in a slider turtle (*Trachemys scripta*) with a well-ossified carapace and plastron, as opposed to a soft-shelled turtle (*Apalone spiniferus*) or a caiman (*Caiman crocodilus*) (3). The function of the dermal armor as an osmotic barrier may explain why cyamodontoid placodonts were the last stem-group sauropterygians to go extinct at the end of the Triassic, as shallow epicontinental and nearshore marine habitats deteriorated in the wake of an eustatic sea-level drop. *Henodus* for example persisted under extreme environmental conditions in a lagoonal lake environment that oscillated between rain flooding and hypersalinity (14). The function of the dermal armor as an osmotic barrier may also explain the diversity and wide geographic distribution in a broad range of habitats of early fossil occurrences of turtles as detailed below.

The scenario of an aquatic origin of turtles is in accordance with the occurrence of the earliest known fossil turtle in the shallow epicontinental marine deposits of the Germanic Muschelkalk (Middle Triassic, lower Ladinian, 235 mys; R Wild, personal communication). The occurrence of a proganochelyid turtle in marine deposits indicates that this clade of early turtles, if not including marine members, at least pursued an amphibious mode of life, living in delta systems or lake systems close to the sea shore; fully terrestrial animals are not expected in Muschelkalk deposits. The same general type of habitat is indicated for the geologically younger occurrences of *Proganochelys quenstedti* in the Upper Triassic (middle Norian) Stubensandstein (15).

The occurrence of the earliest fossil turtle, as well as of *Proganochelys*, in the Germanic Triassic (i.e., in the western Tethyan faunal province) also matches paleobiogeographic patterns. Triassic Sauropterygia, sister-group of Testudines, comprise two major clades, the Placodontia and the Eosauropterygia (64). Of these, the Placodontia are of strictly western Tethyan distribution. Eosauropterygia have their earliest occurrences in the uppermost Lower Triassic (Scythian) of the Germanic Basin (western Tethyan faunal province), of southern China (western Pacific faunal province), and in the western United States (eastern Pacific faunal province). Throughout the Middle Triassic, the western Tethyan faunal province was a center of diversification for stem-group Sauropterygia (67). By Upper Triassic times, the diversity of stem-group sauropterygians declines, while turtles are found in the Upper Triassic of Thailand (12), of Greenland (33), and of Argentina (70). Lower Jurassic turtles are known from northeastern Arizona (17) and from South Africa (16). This shows that during Late Triassic and Early Jurassic times, a substantial radiation of turtles took place at a cosmopolitan scale, some of the clades adopting a fully terrestrial mode of life in an arid environment (16).

The hypothesis of a sauropterygian sister-group relationship of turtles furthermore closes an important gap in the fossil record. Large amounts of tetrapod material have been collected from the Permian of Russia, South Africa, and North America, but not one single turtle carapace fragment. Given that pareiasaurs arise in the lower-most part of the Upper Permian (*Tapinocephalus* zone—36), the origin of turtles would have to date back to the Lower Permian if turtles and pareiasaurs are sister taxa (41), leaving a gap of over 30 million years between the origin and the first fossil occurrence of turtles. However, if Lee (43) is correct and turtles are nested within Pareiasauria as the sister-group to his Fleocyclopsia (including the late Permian pareiasaur *Anthodon*), turtles might have originated at the end of the Upper Permian, which still leaves a gap of about 15–20 million years. Molecular data, which also support diapsid affinities of turtles, indicate an age of 207 ± 20.5 Ma for the turtle lineage (29a) bringing them closer to the Middle Triassic diversification of Sauropterygia. Admittedly, this argument is based on negative evidence, i.e., the absence of turtles in the late Paleozoic fossil record, but the cosmopolitan distribution of turtles by Upper Triassic times does indicate that once turtles are present in a fauna, their carapace, or fragments thereof, stand a reasonably good chance to fossilize.

CONCLUSIONS AND FUTURE PERSPECTIVES

It is intuitively difficult to accept a sister-group relationship of turtles with Sauropterygia among crown-group diapsids, even for the authors of this paper. Yet this is what the osteological data at hand do indicate at this time. As a cautionary note, we would like to emphasize that removal of Sauropterygia from the phylogenetic analysis discussed above results in the movement of turtles into parareptiles, as the sister-group of pareiasaurs (with paraphyletic procolophonians). This indicates that inclusion of Sauropterygia did have a significant effect on the placement of turtles among diapsids. Bootstrap support for the node linking turtle to sauropterygia is weak (53%). The question could be raised whether the turtle-sauropterygian relationship picks up aquatic adaptations in those two groups, rather than a phylogenetic signal? However, the synapomorphies nesting turtles within crown-group Diapsida are distributed throughout the skeleton (11) and cannot readily be linked to aquatic adaptations. The addition of ichthyosaurs, a group of dolphin-like Mesozoic aquatic reptiles, to the data matrix of deBraga & Rieppel (11) again appears to switch turtles back into parareptiles (49). Although we do not anticipate the same result if ichthyosaurs were added to the data matrix in its current corrected form, it certainly is a hypothesis that has to be tested. Thalattosaurs is another group of marine reptiles that may represent the sister-group of Sauropterygia (48), with turtles being the sister-taxon of these two clades (65), but knowledge of thalattosaur anatomy is presently very incomplete, and relative relationships may change once thalattosaurs become more fully known.

Beyond the addition of new taxa, and also of new osteological data to the analysis, there is an obvious need for further testing the three-taxon problem involving Lepidosauria, Archosauria, and Testudines on the basis of other, perhaps also less conventional, characters of molecular and/or physiological nature. Recent contributions (29a, 35, 56, 83) have provided interesting support for turtles as diapsids. It is striking how little comparative physiological work has been done on turtles in a phylogenetic context compared to lizards (32). Although hampered by the relative lack of outgroups when compared to paleontological sources, molecular and physiological studies have the potential to contribute significantly to our understanding of turtle origins and relationships.

We remain convinced that phylogenetic analyses with incorporation and integration of new data not only from paleontological sources, but also neontological sources when relevant (gross morphology, ontogenetic, molecular, physiological), provide the best line of investigation for evaluation of the evolutionary history of amniotes and the origin of turtles.

ACKNOWLEDGMENTS

We thank Diane Scott for drafting Figure 1. ES Gaffney, AG Kluge, M Laurin, GC Mayer, and B Shaffer kindly read the manuscript, offering helpful advice and criticism while not always agreeing with our conclusions. RR Reisz's contribution to this paper was supported by the Borg-Warner Robert O. Bass Visiting Scientist Fund.

LITERATURE CITED

1. Baur G. 1885. Zur Morphologie des Carpus und Tarsus der Reptilien. *Zool. Anz.* 8:631–38
2. Bennett AF, Dawson WR. 1976. Metabolism. In *Biology of the Reptilia*, ed. C Gans, WR Dawson, 5:127–223. London: Academic
3. Bentley PJ. 1976. Osmoregulation. In *Biology of the Reptilia*, ed. C Gans, WR Dawson, 5:365–412. London: Academic
4. Brinkman D. 1980. Structural correlates of tarsal and metatarsal functioning in *Iguana* (Lacertilia; Iguanidae) and other lizards. *Can. J. Zool.* 58:277–89
5. Brinkman D. 1981. The hind limb step cycle of *Iguana* and primitive reptiles. *J. Zool.* 181:91–103
6. Burke AC, Alberch P. 1985. The development and homologies of the chelonian carpus and tarsus. *J. Morph.* 186:119–31

7. Burke AC. 1989. Development of the turtle carapace: implications for the evolution of a novel bauplan. *J. Morph.* 199:363–78
8. Burke AC. 1991. The development and evolution of the turtle body plan: inferring intrinsic aspects of the evolutionary process from experimental embryology. *Am. Zool.* 31:616–27
9. Carroll RL. 1969. A Middle Pennsylvanian captorhinomorph, and the interrelationships of primitive reptiles. *J. Paleontol.* 43:151–70
10. Colbert EH. 1970. The Triassic gliding reptile *Icarosaurus*. *Bull. Am. Mus. Nat. Hist.* 143:85–142
11. deBraga M, Rieppel O. 1997. Reptile phylogeny and the interrelationships of turtles. *Zool. J. Linn. Soc.* 120:281–354
12. deBroin F. 1984. *Proganochelys ruchae* n. sp., chélonien du Trias supérieur de Thailande. *Stud. Geol. Salamanticensia-Stud. Palaeocheloniológica* 1:87–97
13. dePinna MCC. 1991. Concepts and tests of homology in the cladistic paradigm. *Cladistics* 7:367–394
14. Fischer W. 1959. Neue Funde von *Henodus chelyops* v. Huene im Tübinger Gipskeuper. *N. Jb. Geol. Paläontol., Mh.* 1959:241–47
15. Gaffney ES. 1990. The comparative osteology of the Triassic turtle *Proganochelys*. *Bull. Am. Mus. Nat. Hist.* 194:1–263
16. Gaffney ES, Kitching JW. 1994. The most ancient African turtle. *Nature* 369:55–58
17. Gaffney ES, Hutchison JH, Jenkins FA Jr, Meeker L. 1987. Modern turtle origins: the oldest known cryptodire. *Science* 237:289–91
18. Gans C, Hughes GM. 1997. The mechanism of lung ventilation in the tortoise *Testudo graeca* Linné. *J. Exp. Biol.* 47:1–20
19. Gaunt AS, Gans C. 1969. Mechanics of respiration in the snapping turtle, *Chelydra serpentina* (Linné). *J. Morph.* 128:195–228
20. Gauthier J, Kluge AG, Rowe T. 1988. Amniote phylogeny and the importance of fossils. *Cladistics* 4:105–209
21. Gauthier JR, Estes R, deQueiroz K. 1988. A phylogenetic analysis of Lepidosauromorpha. In *Phylogenetic Relationships of the Lizard Families*, ed. R Estes, G Pregill, pp. 15–98. Stanford, CA: Stanford Univ. Press
22. Gegenbaur C. 1864. *Untersuchungen zur vergleichenden Anatomie der Wirbelthiere. Erstes Heft. Carpus und Tarsus*. Leipzig: Wilhelm Engelman
23. Goette A. 1899. Über die Entwicklung des knöchernen Rückenschildes (Carapax) der Schildkröten. *Z. wiss. Zool.* 66:407–34
24. Goodrich ES. 1930. *Studies on the Structure and Development of Vertebrates*. London: Macmillan
25. Goodrich ES. 1942. The hind foot of *Youngina* and the fifth metatarsal in Reptilia. *J. Anat.* 76:308–12
26. Gregory WK. 1946. Pareiasaurs versus placodonts as near ancestors to turtles. *Bull. Am. Mus. Nat. Hist.* 86:275–326
27. Haas G. 1969. The armor of placodonts from the Muschelkalk of Wadi Ramon (Israel). *Israel J. Zool.* 18:135–47
28. Haines RW. 1969. Epiphyses and sesamoids. In *Biology of the Reptilia*, ed. C Gans, TS Parsons, AdA Bellairs, 1:81–115. London: Academic
29. Hay OP. 1898. On *Protostega*, the systematic position of *Dermochelys*, and the morphogeny of the chelonian carapace and plastron. *Am. Nat.* 32:929–48
29a. Hedges SB, Poling LL. 1999. A molecular phylogeny of reptiles. *Science* 283:998–1001
30. Hoffstetter R, Rage J-C. 1969. Vertebrae and ribs of modern reptiles. In *Biology of the Reptilia*, ed. C Gans, TS Parsons, AdA Bellairs, 1:201–310. London: Academic Press
31. Howes GB, Swinnerton HH. 1901. On the development of the skeleton of the Tuatara,

Sphenodon punctatus; with comments on the egg, the hatching, and the hatched young. *Trans. Zool. Soc. Lond.* 16:1–86
32. Huey RB. 1987. Phylogeny, history, and the comparative method. In *New Direction in Physiological Ecology*, ed. ME Fered, AF Bennett, WW Burggren, RB Huey, pp. 76–98. New York: Cambridge Univ. Press
33. Jenkins FA Jr. 1994. Late Triassic continental vertebrates and depositional environments of the Fleming Fjord Formation, Jameson Land, East Greenland. *Medd. Gronl. Geosci.* 32:1–25
34. Kaelin J. 1945. Zur Morphogenese des Panzers bei den Schildkröten. *Acta Anat.* 1:144–76
35. Kirsch JAW, Mayer GC. 1998. The platypus is not a rodent: DNA hybridization, amniote phylogeny and the palimpsest theory. *Philos. Trans. R. Soc. Lond. B* 353: 1221–37
36. Kitching JW. 1977. *The Distribution of the Karroo Vertebrate Fauna*. Bernard Price Inst. for Palaeontol. Res., Memoir 1. Johannesburg: Univ. Witwatersrand
37. Laurin M. 1991. The osteology of a Lower Permian eosuchian from Texas and a review of diapsid phylogeny. *Zool. J. Linn. Soc.* 101:59–95
38. Laurin M, Reisz RR. 1995. A reevaluation of early amniote phylogeny. *Biol. J. Linn. Soc.* 113:165–223
39. Lee MSY. 1993. The origin of the turtle body plan: bridging a famous morphological gap. *Science* 261:1716–20
40. Lee MSY. 1994. The turtle's long lost relatives. *Nat. Hist.* 6:63–65
41. Lee MSY. 1995. Historical burden in systematics and the interrelationships of 'parareptiles'. *Biol. Rev.* 70:459–547
42. Lee MSY. 1996. Correlated progression and the origin of turtles. *Nature* 379:811–15
43. Lee MSY. 1997. Pareiasaur phylogeny and the origin of turtles. *Zool. J. Linn. Soc.* 120:197–280
44. Lee MSY. 1997. Reptile relationships turn turtle. *Nature* 389:245–46
45. Lee MSY. 1997. The evolution of the reptilian hindfoot and the homology of the hooked fifth metatarsal. *J. Evol. Biol.* 10:253–63
46. Lee MSY, Doughty P. 1997. The relationship between evolutionary theory and phylogenetic analysis. *Biol. Rev.* 72:471–95
47. Maddison WP, Maddison R. 1992. *MacClade, Version 3*. Sunderland: Sinauer
48. Merck JW. 1997. A phylogenetic analysis of the euryapsid reptiles. *J. Vert. Paleontol.* 17:65A
49. Motani R, Minoura, Ando T. 1998. Ichthyosaurian relationships illuminated by new primitive skeletons from Japan. *Nature* 393:255–57
50. O'Donoghue CH. 1921. The blood vascular system of the Tuatara, *Sphenodon punctatus*. *Philos. Trans. R. Soc. Lond.* 210: 175–252
51. Oguschi K. 1911. Anatomische Studien an der japanischen dreikralligen Lippenschildkröte (*Trionyx japanicus*). *Morph. Jb.* 43:1–106
52. Olson EC. 1947. The family Diadectidae and its bearing on the classification of reptiles. *Fieldiana (Geol.)* 11:1–53
53. Patterson C. 1977. Cartilage bones, dermal bones and membrane bones, or the exoskeleton versus the endoskeleton. In *Problems in Vertebrate Evolution*, ed. SM Andrews, RS Miles, AD Walker, pp. 77–121. London: Academic
54. Patterson C. 1982. Morphological characters and homology. In *Problems of Phylogenetic Reconstruction*, ed. KA Joysey, AE Friday, pp. 21–74. London: Academic
55. Peabody FE. 1951. The origin of the astragalus of reptiles. *Evolution* 5:339–44
56. Platz JE, Conlon JM. 1997. Reptile relationships turn turtle ... and turn back again. *Nature* 389:245–46
57. Rabl C. 1910. *Bausteine zu einer Theorie*

der Extremitäten der Wirbeltiere. 1. Theil. Leipzig: Wilhelm Engelmann
58. Reisz RR, Laurin M. 1991. *Owenetta* and the origin of turtles. *Nature* 349:324–26
59. Remane A. 1952. *Die Grundlagen des natürlichen Systems, der vergleichenden Anatomie und der Phylogenetik.* Leipzig: Akademische Verlagsgesellschaft
60. Rieppel O. 1992. Studies on skeleton formation in reptiles. III. Patterns of ossification in the skeleton of *Lacerta vivipara* Jacquin (Reptilia, Squamata). *Fieldiana (Zool.) n.s.* 68:1–25
61. Rieppel O. 1993. Studies on skeleton formation in reptiles: patterns of ossification in the skeleton of *Chelydra serpentina. J. Zool.* 231:487–509
62. Rieppel O. 1993. Studies on skeleton formation in reptiles. IV. The homology of the reptilian (amniote) astragalus revisited. *J. Vert. Paleontol.* 13:31–47
63. Rieppel O. 1993. Studies on skeleton formation in reptiles. V. Patterns of ossification in the skeleton of *Alligator mississippiensis* Daudin (Reptilia, Crocodylia). *Zool. J. Linn. Soc.* 109:301–25
64. Rieppel O. 1994. Osteology of *Simosaurus gaillardoti*, and the phylogenetic interrelationships of stem-group Sauropterygia. *Fieldiana (Geol.) n.s.* 28:1–85
65. Rieppel O. 1998. The systematic status of *Hanosaurus hupehensis* (Reptilia, Sauropterygia) from the Triassic of China. *J. Vert. Paleont.* 18:545–76
66. Rieppel O, deBraga M. 1996. Turtles as diapsid reptiles. *Nature* 384:453–55
67. Rieppel O, Hagdorn H. 1997. Paleobiogeography of Middle Triassic Sauropterygia in Central and Western Europe. In *Ancient Marine Reptiles*, ed. JM Callaway, EL Nicholls, pp. 121–44. San Diego: Academic
68. Robinson PL. 1962. Gliding lizards from the Upper Keuper of Great Britain. *Proc. Geol. Soc. Lond.* 1601:137–46
69. Robinson PL. 1975. The functions of the hooked fifth metatarsal in lepidosaurian reptiles. *Coll. Int. CNRS* 218:461–83
70. Rougier GW, Fuente MS, Arcucci AB. 1995. Late Triassic turtles from South America. *Science* 268:855–58
71. Ruckes H. 1929. Studies in chelonian osteology. Part II. The morphological relationships between girdles, ribs and carapace. *Ann. N. Y. Acad. Sci.* 31:81–120
72. Sewertzoff AN. 1908. Studien über die Entwicklung der Muskeln, Nerven und des Skeletts der Extremitäten der niederen Tetrapoda. *Bull. Soc. Imp. Nat. Mosc. N.S.* 21:1–430
73. Starck D. 1979. *Vergleichende Anatomie der Wirbeltiere auf evolutionsbiologischer Grundlage,* Vol. 2. Berlin: Springer Verlag
74. Swofford DL, Begle DP. 1993. *PAUP–Phylogenetic Analysis Using Parsimony, Version 3.1.* Washington, DC: Smithsonian Inst.
75. Vallén E. 1942. Beiträge zur Kenntnis der Ontogenie und der vergleichenden Anatomie des Schildkrötenpanzers. *Acta Zool. Stockholm,* 23:1–127
76. Versluys J. 1914. Über die Phylogenie des Panzers der Schildkröten und über die Verwandtschaft der Lederschildkröte (*Dermochelys coriacea*). *Paläontol. Zeit.* 1:321–47
77. Völker H. 1913. Über das Stamm, Gliedmassen-, und Hautskelett von *Dermochelys coriacea* L. *Zool. Jb., Anat.* 33:431–552
78. Walker WF. 1971. A structural and functional analysis of walking in the turtle, *Chrysemis picta marginata. J. Morph.* 134:195–214
79. Walker WF. 1973. The locomotor apparatus in turtles. In *Biology of the Reptilia*, ed. C Gans, TS Parsons, 4:1–100. London: Academic
80. Westphal F. 1975. Bauprinzipien im Panzer der Placodonten (Reptilia triadica). *Paläont. Z.* 49:97–125
81. Zangerl R. 1939. The homology of the shell elements in turtles. *J. Morph.* 65:383–406
82. Zangerl R. 1969. The turtle shell. In *Biol-*

ogy of the Reptilia, ed. C Gans, TS Parsons, AdA Bellairs, 1:311–339. London: Academic

83. Zardoya R, Meyer A. 1998. Complete mitochondrial genome suggests diapsid affinities of turtles. *Proc. Natl. Acad. Sci. USA* 95:14226–31
84. Zuckerkandl E. 1895. Zur Anatomie und Entwicklungsgeschichte der Arterien des Unterschenkels und des Fusses. *Anat. Hefte* 5:207–291
85. Zug GR. 1971. Buoyancy, locomotion, morphology of the pelvic girdle and hindlimb, and systematics of cryptodiran turtles. *Misc. Publ. Mus. Zool. Univ. Mich.* 142:1–98

Annu. Rev. Ecol. Syst. 1999. 30:23–49

USES OF EVOLUTIONARY THEORY IN THE HUMAN GENOME PROJECT

Alan R. Templeton
Department of Biology, Washington University, St. Louis, Missouri 63130-4899; e-mail: temple_a@biodec.wustl.edu

Key Words complex traits, gene trees, genetic disease, genetic architecture, population structure

■ **Abstract** The Human Genome Project (HGP) originally sought to sequence the human genome but excluded studies on genetic diversity. Now genetic diversity is a major focus, and evolutionary theory provides needed analytical tools. One type of diversity research focuses on complex traits. This is often done by screening genetic variation at candidate loci functionally related to a trait followed by gene/phenotype association tests. Linkage disequilibrium creates difficulties for association tests, but evolutionary analyses using haplotype trees can circumvent these problems and result in greater statistical power, better disease risk prediction, the elimination of some polymorphisms as causative, and physical localization of causative variation when combined with an analysis of recombination. The HGP also now proposes to map over 100,000 single nucleotide polymorphisms to test for gene/phenotype associations through linkage disequilibrium in isolated human populations affected by past founder or bottleneck events. This strategy requires prior knowledge of recent human evolutionary history and current population structure, but other evolutionary considerations dealing with disequilibrium and nonrandom mutation pose difficulties for this approach. Studies on population structure also focus upon traits of medical relevance, and an understanding of the evolutionary ultimate cause for the predisposition of some populations to certain diseases is a useful predictor for shaping public health policies. Studies on the genetic architecture of common traits reveal much epistasis and variation in norms of reaction, including drug response. Because of these interactions, context dependency and sampling bias exist in disease association studies that require population information for effective use. Overall, the population thinking of evolutionary biology is an important counterweight to naive genetic determinism in applying the results of the HGP to issues of human health and well-being.

INTRODUCTION

The Human Genome Project (HGP) began in 1990 with the goal of sequencing an entire human genome by the year 2005. This project also included the goal of obtaining genome sequences of several other organisms, many of which have

0066-4162/99/1120-0023$08.00

already been completely sequenced (17). The sequencing of genomes from several species obviously creates opportunities for evolutionary studies in comparative genomics. However, these macroevolutionary studies are outside the scope of this review, which is restricted to studies solely on humans. At first glance, this restriction may seem to preclude any relevance of evolutionary biology to the HGP. This perception is reinforced by the fact that the initial HGP excluded studies on human genetic diversity (16, 17)—the raw material of all evolutionary phenomena at the intraspecific level. Partly to remedy this perceived deficiency of the HGP, a Human Genome Diversity Project was first proposed in 1992, but it never became an integrated, focused project (118). Meanwhile, sequencing technology advanced such that in May 1998, private initiatives were announced to sequence much of the human genome de novo within just three years and at a tenth of the cost of the federally funded HGP (105). Following the announcements of these private initiatives, a major reassessment of the goals of the HGP was undertaken (17). The HGP now has a "new focus on genetic variation" because such variation is the "fundamental raw material for evolution" and "is also the basis for variations in risk among individuals for numerous medically important, genetically complex human diseases" (17). The HGP has finally opened the door to evolutionary biology.

Because of the deliberate exclusion of studies on human genetic diversity until 1998, few of the papers cited in this review were funded by the HGP. Nevertheless, many non-HGP studies have examined human genetic diversity at the molecular level, often with a clinical or medical objective. This review includes such projects because they now fall within the new goals of the HGP and show the contribution of evolutionary principles to such research.

DISEASE ASSOCIATIONS

General Considerations

The primary justification of the HGP is its promise to yield new insights into and possibly cures for a variety of human diseases. Mapping and understanding the "simple Mendelian diseases" of humanity is part of this goal. However, only about 1.25% of human births are affected by such Mendelian diseases (25). In contrast, 65.41% of live births are affected at some time in their life by a disease that has a heritability greater than 0.3 (25). Accordingly, the new goals of the HGP include a major focus upon the "genetically complex human diseases" (17). This review is limited to genetically complex diseases because evolutionary theory and practice can provide critical analytical tools for their study.

The primary impact of molecular genetics upon studies of complex diseases has been a strong shift from unmeasured to measured genotype approaches (126). In the unmeasured approach, individual phenotypes are scored and pedigree information is obtained to define the genetic relationships among the individuals. The phenotypic and pedigree information is then used to obtain estimates of classical quantitative genetic parameters such as heritabilities. In contrast, measured genotype approaches start with some assay of genetic variation (usually at the molecular level) and test for associations between measured genetic variation and phenotypic variation. Measured genotype approaches predate the HGP (e.g. 45, 127), but the ready availability of molecular markers throughout the human genome has now made measured genotype approaches increasingly common. There are three major measured genotype approaches in human genetics: candidate loci, quantitative trait loci (QTL) mapping through linkage associations within families, and QTL mapping through linkage disequilibrium in populations.

Candidate Loci

With the candidate locus approach, prior information about the basis of the phenotype and the function of known genes is used to identify loci that are likely to contribute to the phenotype of interest (126). For example, after age and sex, various lipid traits are the major risk factors for coronary artery disease (CAD), which accounts for about a third of the total human mortality in developed nations. The etiology of CAD involves the deposition of cholesterol-laden plaque onto the interior walls of the coronary arteries. Many genes have been identified that play critical roles in lipid metabolism, so these are all candidate genes for CAD (6, 126). Our knowledge of the etiology of many diseases and the function of specific genes is increasing at a rapid rate, so the candidate approach will play an increasingly important role in human genome studies.

With this approach, a population is screened for variation at the candidate loci followed by tests for associations between genetic and phenotypic variation. With low-resolution genetic screens such as protein electrophoresis, such tests of association were relatively straightforward by measuring mean phenotypes of genotypes. For example, one of the candidate genes involved in lipid metabolism is *Apoprotein E* (*ApoE*), and protein electrophoresis reveals three alleles that are common in most human populations, the *ε2*, *ε3*, and *ε4* alleles. In most human populations, the *ε2* allele is associated with lowered total serum and LDL (low density lipoprotein) cholesterol levels, and *ε4* is associated with elevated levels (24, 43, 120, 121, 126). Prospective studies confirm that individuals bearing the *ε4* allele die disproportionately from CAD (130). The marginal effects of *ApoE* alone account for about 7% of the total phenotypic variation in cholesterol levels and about 14% of the total genetic variance (121). Hence, *ApoE* is considered to be a major locus for CAD risk prediction.

As genetic resolution increases, certain difficulties arise. This is illustrated by studies on another candidate gene for CAD, the *lipoprotein lipase* locus (*LPL*): 71 individuals from three populations were sequenced for a 9.7-kb region within this locus (98). Of the 88 polymorphic sites were discovered, and 69 of these sites had their phases determined to define 88 distinct haplotypes (14). Thus, using only a subset of the known polymorphic sites in just a third of a single gene, a sample of 142 chromosomes reveals 88 "alleles" or haplotypes that in turn define 3916 possible genotypes—a number considerably larger than the sample size. Obviously, the increased genetic resolution now poses some daunting statistical challenges.

One simple and common approach for dealing with such vast arrays of genetic variation is to analyze each polymorphic site separately. However, polymorphic sites within a candidate locus are, virtually by definition, tightly linked and often show strong linkage disequilibrium, as is the case for *LPL* (14, 98). As a consequence, the multiple single-site tests are not independent from one another, making statistical and biological interpretation difficult. Most commonly, such single-site approaches are interpreted as a minilinkage analysis; each polymorphic site is treated as a "marker," and those markers that show the strongest phenotypic associations are regarded as being physically closest to the causative site or being the causative site. For example, *ApoE* is also a candidate gene for Alzheimer's disease, and the $\varepsilon 4$ allele often shows a significant marginal association with sporadic Alzheimer's disease (1, 8, 10, 13, 19, 56, 78, 104, 107), although not in all cases (68, 70, 75). This discovery has led to a plethora of papers trying to explain how this specific allele "causes" Alzheimer's disease (80, 93, 112, 132, 157). However, three other apoprotein loci are tightly linked to the *ApoE* locus (2), and the genetic markers throughout this chromosomal region show strong linkage disequilibria, apparently due to low rates of recombination (135). The question therefore arises: Which of these functionally related loci harbors causative variation?

At this point, it is necessary to consider the evolutionary forces that determine the amount and pattern of linkage disequilibrium. A standard equation from population genetics is:

$$D_t = D_0(1 - r)^t \qquad 1.$$

where D_t is the linkage disequilibrium between two sites at generation t, r is the recombination frequency between the sites, and D_0 is the linkage disequilibrium at the initial generation 0. Attention is often focused upon r in Equation 1, leading to the general belief that the magnitude of linkage disequilibrium can be used as a proxy for recombination frequency. However, when dealing with small regions of DNA, recombination is often extremely rare (but not always; 14). When recombination is rare, the magnitude of disequilibrium is determined primarily by the chromosomal background upon which a mutation initially occurred and its subsequent evolutionary fate (137, 139). Therefore, the magnitude of disequilibrium in many candidate regions reflects the temporal positioning of mutational events over evolutionary time, and not the spatial positioning over a physical region of DNA. Indeed, many studies reveal little to no correlation between linkage disequilibrium and physical distance within small regions of DNA (48, 50, 61, 74, 76, 145, 165, 166). This lack of correlation means that there is no reliable positional information even when a strong association is found only with a single marker in a candidate region. Hence, it is premature to infer that *ApoE* is the only plausible locus for "causing" Alzheimer's disease in this chromosome region and that the mutation leading to the electrophoretic $\varepsilon 4$ allele is a causative mutation. Multiple single-marker association studies in a candidate region are statistically indefensible because of non-independence and easily misinterpretated biologically unless the evolutionary basis of disequilibrium is taken into account.

One way to eliminate the problems caused by disequilibrium is to organize all the genetic variation in the DNA region into haplotypes and then use haplotypes as the units of analysis (149, 150). However, as noted above for *LPL*, the number of haplotypes is often still so large that even extremely large samples will only provide sparse coverage of the possible genotypic space defined by haplotypes. Evolutionary history can be used to solve this problem. Given a candidate gene region in which there has been either no or little recombination, it is possible to estimate a haplotype tree that reflects the evolutionary history that generated the disequilibrium and the haplotypes (139). The estimation of such haplotype trees can be accomplished by standard phylogenetic inference procedures, but many of these procedures were originally worked out for interspecific phylogenies and make assumptions that are inappropriate for haplotype trees (22, 140). Accordingly, new procedures have been developed for estimating and testing haplotype trees (21, 22, 139, 141).

A simple example of the use of a haplotype tree is provided by a study of the *ApoAI-CIII-AIV* gene cluster (49). Seven haplotypes (numbered *1* through *7*) defined by restriction site polymorphisms were found in a population of 147 French Canadians scored for total serum cholesterol and other lipid phenotypes. Of these individuals 140 bore at least one copy of haplotype *6* and therefore only one haplotype state varied among these 140 individuals. Thus, the genotypes in this subset could be analyzed as if they were haploid. The haplotype tree in this case is a simple star in which the six rarer haplotypes radiate from the central and common haplotype *6*, with all six of the tip haplotypes differing from haplotype *6* by only a single restriction site. Note that there are 21 different pairwise contrasts among these 7 haplotypes, but there are only 6 degrees of freedom. In general, there are $n-1$ degrees of freedom in a haploid analysis of n haplotypes and $\frac{1}{2}n(n-1)$ pairwise contrasts, so this discrepancy gets worse with increasing haplotype diversity. One of the basic principles of the comparative method of evolutionary biology is that the most meaningful contrasts are between evolutionary neighbors. Applying this principle, the evolutionarily relevant contrasts are those between haplotype *6* and the six others. Accordingly, only these six pairwise contrasts were performed, each using one of the available degrees of freedom, and several significant phenotypic associations were discovered (49). In contrast, no significant phenotypic associations were discovered with a traditional, one-way ANOVA that does not use evolutionary principles to concentrate statistical power upon the most relevant comparisons (49).

Most haplotype trees deviate from the pure star topology. Still, any haplotype tree of n haplotypes will have $n - 1$ connections in a fully resolved tree, so in theory all haplotype trees could define contrasts across their branches that would fully use all available degrees of freedom and no more. Statistical problems can still exist because many haplotypes, particularly those on the tips of the haplotype tree, are expected to be quite rare in the population (22), resulting in little power. This problem can be diminished by pooling haplotypes into larger categories, particularly tip haplotypes. One method of pooling is to start with tip haplotypes (in general the rarest) and move one mutational step toward the interior of the haplotype tree, pooling together all the tip and interior haplotypes that are reached by such a movement into a "1-step clade" (139). The resulting initial set of 1-step clades on the tips of the haplotype tree are then pruned off, and the same procedure is then repeated upon the remainder of the haplotype tree until all haplotypes have been placed into 1-step clades. One then applies the same nesting algorithm to the tree of 1-step clades to create 2-step clades. This procedure is iterated until all clades are nested into a single category. Special rules of nesting are needed to deal with symmetries in the tree and regions of topological ambiguity (141). This nesting procedure has several advantages. First, nesting categories are determined exclusively by the evolutionary history of the haplotypes and not by a phenotypic pre-analysis, thereby eliminating a major source of potential bias. Second, the clades define a nested design that makes full and efficient use of the available degrees of freedom. Third, statistical power has been enhanced by contrasting pooled clades across many of these branches instead of individual haplotypes (124, 125, 136, 139).

Most human studies involve diploid genotypes. The nested clade statistics were therefore extended to deal with diploid populations through the quantitative genetic device of the average excess of a haplotype (134, 143). The statistical significance of these average excesses can be determined by a variety of methods. For example, a nonparametric nested permutational analysis of the *ApoAI-CIII-AIV* region revealed a significant association with the log of the serum triglyceride level, another risk factor for CAD that was undetectable with a standard nonevolutionary analysis (143). Alternatively, likelihood-ratio tests of linear models of parameterized haplotype effects were used to identify clades of haplotypes in the *Apoprotein B* (*ApoB*) region that explained about 10% of the genetic variance and 5% of the total variance in HDL-cholesterol and triglyceride levels (44). None of the sites used to define these *ApoB* haplotypes showed any significant phenotypic associations when tested individually (44). As these examples show, evolutionary analyses provide greater statistical power than nonevolutionary alternatives.

The evolutionary approach also results in better individual level risk prediction over the single-marker approach. Analyzing one marker at a time is inherently a bivariate analysis, but phenotypes typically come in more than two categories. When this occurs, single-marker analyses are inherently incapable of detecting the full range of phenotypic heterogeneity. For example, a single marker analysis of case-control data concluded that the *ApoE* $\varepsilon 4$ allele and a restriction site marker in the tightly linked *ApoCI* locus can each divide people into higher and lower Alzheimer's disease risk categories (10). The nested evolutionary analysis of the same data (Figure 1) identifies the *ApoCI* marker as being associated with a significant change in Alzheimer's risk, as is *ε2* (but not *ε4*) at the *ApoE* locus, with the *ε4* marginal effect being due to linkage disequilibrium (135). Thus, both approaches identified the *ApoCI* marker as providing information about Alzheimer's risk that is independent from the *ApoE* locus markers, but the evolutionary analysis results in three risk categories (Figure 1), not two (135). The "high-risk" category in the marginal analysis of the *ApoCI* marker contains those individuals with both lowest risk (clade 1-4 in Figure 1) and highest risk (clades 1-1 and 1-2). Hence, using the *ApoCI* marker by itself would mean that many people with the lowest risk for Alzheimer's disease would incorrectly be advised that they are in a high-risk category. Thus, evolutionary analyses provide greater precision of risk prediction for individuals than nonevolutionary analyses of the same data.

Another advantage of the evolutionary approach is that it provides guidance for studies attempting to find causative variation. For example, a nested clade analysis of restriction-site haplotypes at the *low-density lipoprotein receptor* locus (*LDLR*) localized a significant phenotypic change in CAD risk factors to a branch between two haplotypes (designated *H1* and *H5*) defined by a single *Taq*I restriction site change (46, 47, 124). This *Taq*I site also had a significant marginal association, so a non-evolutionary analysis detects this association as well. However, because there is no reliable relationship between physical distance and degree of association in such small gene regions, the non-evolutionary analysis gives no guidance as to the nature or location of the causative site or sites. Similarly, in the nested clade analysis, the mutational change/s that define the branch in the haplotype tree that is associated with a phenotypic change are not necessarily causative (139). Therefore, neither analysis justifies equating the *Taq*I site to the causative site. However, a comparison of individuals with and without the *Taq*I site would include much more background genetic variation than the more refined comparison of bearers of *H1* versus *H5* (47). Hence, the evolutionary analysis minimizes the differences in background variation in favor of candidate functional variations (47). More detailed genetic surveys can then be performed upon just these two haplotypes to execute a more refined nested analysis to eliminate additional sites as being causative (47, 123). For example, a nested clade analysis of haplotypes in the *ApoB* region revealed a clade of haplotypes defined by an *Xba*I restriction site polymorphism that explained 6.5% of the phenotypic variance in reduction of total cholesterol and 22.3% of the phenotypic variance in reduction of LDL-cholesterol (the "bad" cholesterol for CAD risk) in response to placing 63 male students upon a controlled low-cholesterol diet (38). It was hypothesized that the putative LDL receptor binding region of apo B might determine how individuals respond to low-cholesterol diets, so this region was sequenced in individuals homozygous for the two haplotype categories on either side of the *Xba*I branch in the *ApoB* haplotype tree. The sequence analysis revealed no differences in this region despite these haplotypes being associated with marked differences in

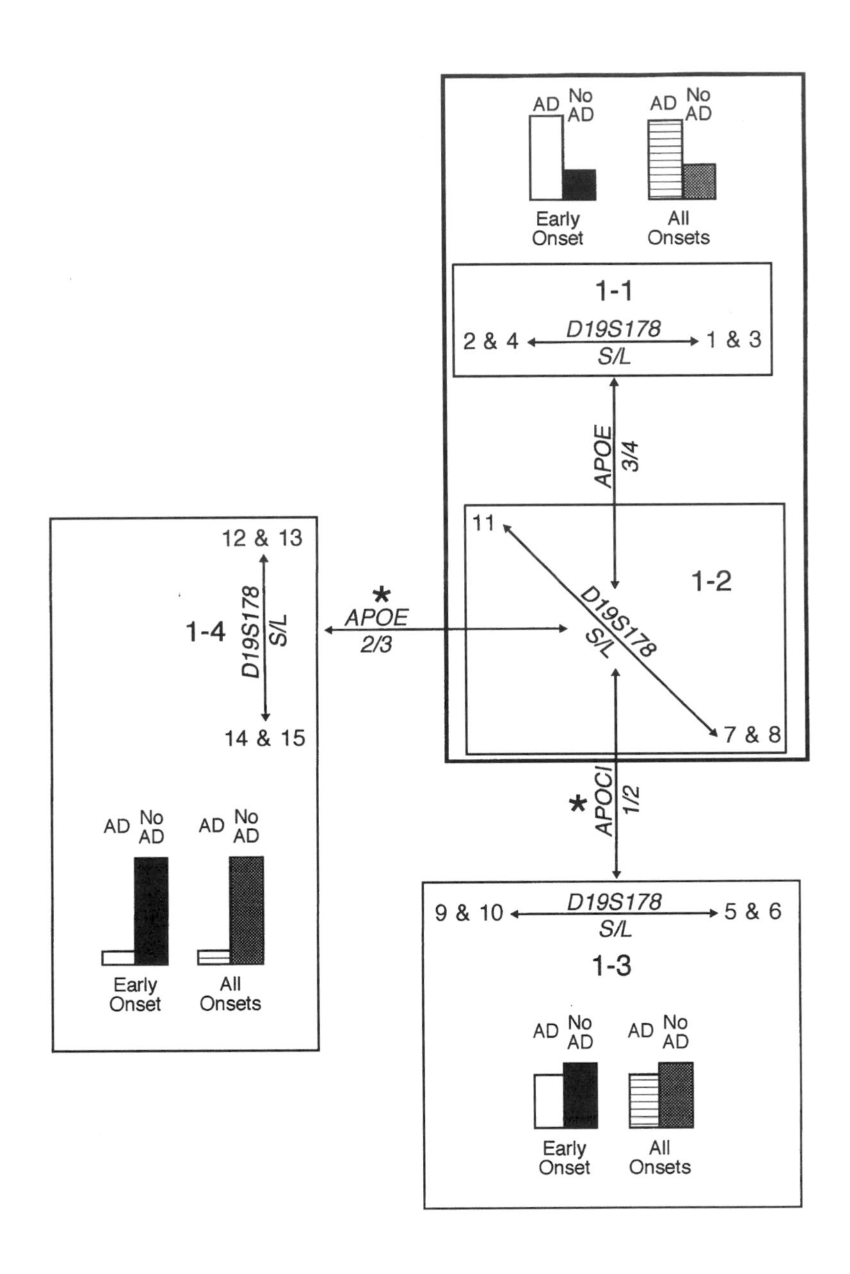

AD
No AD
AD
No AD
Early Onset
All Onsets
1-1
2 & 4
D19S178
S/L
1 & 3
APOE
3/4
11
1-2
D19S178
S/L
7 & 8
12 & 13
1-4
D19S178
S/L
14 & 15
*
APOE
2/3
*
APOCI
1/2
9 & 10
D19S178
S/L
5 & 6
1-3
AD
No AD
AD
No AD
Early Onset
All Onsets
AD
No AD
AD
No AD
Early Onset
All Onsets

dietary response (38). Thus, the causative sites are not in the putative binding region.

The study on dietary response demonstrates that the candidate locus approach can be applied to many types of traits, including responses to environmental variables. The idea that a single genotype can give rise to a variety of phenotypes depending upon environmental conditions (phenotypic plasticity) and the idea that different genotypes will display different phenotypic responses to the same environmental change (norm of reaction) have long played an important role in evolutionary biology (115). Although it has long been known that humans display different genetically determined reactions to environmental variables [e.g., the genetic polymorphisms for lactose intolerance (35) or drug response (72, 87, 90, 159)], a widespread appreciation for the importance of these evolutionary concepts is only recent in human genetics. This appreciation was slow in developing because drug and food metabolism are sufficiently redundant and complex as to mask the effects of single-locus variants with classic Mendelian studies (83). However, there are a plethora of candidate loci in this area, so the increased feasibility and power of the candidate locus approach is revolutionizing this field (60, 83). Indeed, "pharmacogenomic" companies have been founded in recent years with the expressed purpose of developing haplotype data at candidate loci that would be amenable to evolutionary lineage analysis in predicting drug response (26).

Any evolutionary analysis of haplotype variation is based upon the presumption that a haplotype tree can be constructed in the first place. If recombination is frequent and uniform in a candidate region, a nested clade analysis would be impossible. However, the presence of a few recombinants can actually enhance biological inference by providing information about physical location of causative sites (141). For example, the *angiotensin-I converting enzyme* (*ACE*) locus is a candidate gene for hypertension and CAD. Two major clades (*A* and *B*) explain 36% of the phenotypic variance for the amount of circulating ace levels (59). A third clade (*C*) was inferred to have arisen from a recombination event between clades *A* and *B*, and this recombinant clade had phenotypic effects indistinguishable from those of clade *B* (59). Given that clade *C* received its 3′ end from clade *B*, the causative sites or sites must be 3′ to the inferred crossover event, thereby

Figure 1 Nested haplotype analysis of frequency of Alzheimer's disease in a case (AD)-control (no AD) study. Haplotypes are indicated by the numbers 1–15. Each arrow in the tree indicates a single mutational change, with the nature of the change indicated by the arrow. Microsatellite variation at the marker *D19S178* was simply characterized as short (*S*) or long (*L*) alleles, and transitions between these two allelic classes could not be ordered in the tree. Four 1-step clades are defined by nesting, indicated by 1-1 through 1-4. Two significant transitions were identified in the tree, as indicated by asterisks, that subdivide people into three categories. The relative incidences of early onset and all onset Alzheimer's disease is indicated for each of these three categories. Modified from Templeton (135).

excluding the 5′ region and the *ACE* promoter from harboring the causative variation (59).

A nested clade analysis is also possible in DNA regions with frequent recombination as long as the recombination is concentrated into a hotspot. For example, 30 statistically significant recombination/gene conversion events were inferred to have occurred in the 9.7-kb region of the *LPL* gene sequenced in 71 individuals from three populations, but 24 of these 30 events were concentrated into intron 6 in the *LPL* gene (AR Templeton, AG Clark, KM Weiss, DA Nickerson, E Boerwinkle, CF Sing, —139a). As a consequence, there is much phylogenetic structure in the regions flanking this recombinational hotspot.

Given the improvements in sequencing technology, data sets such as *LPL* should become more common in the future. Such data will provide the abundance of markers that are necessary to infer both recombination events and highly resolved haplotype trees. Although rigorous statistical analyses of haplotype trees are still relatively rare, more informal analyses of haplotype trees that make use of evolutionary relatedness are becoming increasingly common in human disease association studies (3, 53, 54, 71, 84, 151). Hence, the new goals of the HGP should be a boon to evolutionary analysis of candidate gene regions.

QTL Mapping Through Linkage Associations Within Families

Under this approach, the measured genotypes are used as markers of physical location within the genome, and phenotypic associations are detected through linkage. This methodology is not discussed in this review because it is an application of quantitative genetics and not evolutionary genetics per se, although it certainly interfaces with evolutionary genetics. However, such QTL studies do have implications for genetic architecture, as is discussed later.

QTL Mapping Through Linkage Disequilibrium in Populations

One of the major new goals of the HGP is to develop a map of over 100,000 single nucleotide polymorphisms (SNPs) distributed over the entire human genome. Although the new HGP goals include developing technologies for scoring any type of polymorphism (17), SNPs are the initial focus because they are abundant and amenable to rapid and efficient screening of large samples (16, 17, 156). The justification for creating such a map is the expectation that linkage disequilibrium, at least in relatively isolated, homogeneous human populations, should exist between at least one of these SNPs and the common alleles at any locus in the genome.

As pointed out earlier, linkage disequilibrium is commonly found among closely linked polymorphisms in regions with infrequent recombination. The magnitude and physical extent of likely disequilibria values can be augmented by founder events that establish relatively isolated populations. Indeed, the first success in positional cloning of a human genetic disease gene, Huntington's chorea, depended critically upon first identifying a relatively isolated founder population that was polymorphic for this disease (42). Such relatively isolated populations also diminish the importance of genetic background effects, making it more likely to detect marginal effects of alleles linked to a SNP. However, linkage disequilibrium can be induced by population subdivision even between unlinked loci (73). Hence, if a sample includes individuals drawn from more than one subpopulation, disequilibria could exist with many markers regardless of their linkage relationships, undermining the SNP strategy. However, if the population subdivisions are known a priori and admixture is occurring, this information can be used in disequilibrium mapping (131). It is therefore critical in SNP studies to either sample within a relatively panmictic and homogeneous population or to use a known admixture situation. In either case, analyses of the evolutionary history and genetic structure of human populations are necessary prerequisites in designing SNP studies. Accordingly, human populations thought to have undergone founder events within the last several centuries and that have been relatively isolated since are now extremely valuable resources for association studies of both classical genetic diseases and the more common complex diseases (28, 41, 119). Indeed, such relatively isolated human populations are now valuable commercial entities. For example, a private company, deCODE, is seeking legislation that will give the company a 12-year license to operate a database containing the medical records of the entire population of Iceland and sell access to third parties (32, 41).

Although evolutionary considerations have gone into designing the SNP strategy, other evolutionary considerations undermine the claim that "this strategy should in theory reveal the identity of the gene or genes underlying any phenotype not due to a rare allele" (16). Given that the current strategy has about one SNP per gene, this claim requires that a single SNP marker near or in the locus of interest would show significant disequilibrium with nearly all common alleles at this locus. As pointed out earlier, not all marker pairs display significant disequilibrium within a gene region, and the disequilibrium that exists is not well correlated with physical distance. Therefore, it is doubtful if a single SNP would display disequilibrium with all common polymorphic sites within a gene. Moreover, a significant phenotypic association with a SNP is not a reliable guide to the actual location of the gene causing the phenotypic variation because of the physical clustering of functionally related genes in the genome. This problem is only accentuated when dealing with founder events because chance events (genetic drift) play an important role in creating disequilibrium even between markers that are well separated physically.

Another problem with randomly chosen SNP's is the danger of selecting a highly mutable site. Such sites will show complicated patterns of association due to the fact that identity by state does not reflect identity by descent. It is commonplace in much of the human genetic literature to assume that the infinite sites model (which does not allow multiple hits) is appropriate for nuclear DNA (e.g. 155). At first the case for the infinite sites model seems compelling. For example, only 88 sites out of 9734 are polymorphic in the sequenced region of *LPL* (98)—a figure that seems to be well below saturation. Moreover, almost all these 88 polymorphic sites are biallelic, which seemingly further bolsters the argument against multiple hits (14). However, both of these arguments are based upon the premise that mutations are equally likely to occur at all sites and, given a mutation, that any of the three nucleotide states are likely to arise. Neither of these premises is justified by the human genetic literature. For example, about a third of all mutations in human nuclear DNA are transitions from 5-methylcytosine to thymine that occur exclusively at CpG dinucleotides, a combination markedly underrepresented in human DNA (55, 64, 81, 110, 116, 164). Mutational hotspots have also been reported for mononucleotide-repeat regions, DNA polymerase α arrest sites, and other rarely occurring sequence motifs in human DNA (63, 91, 147, 148). The pattern of site polymorphism in the *LPL* gene parallels the results of these mutation studies, with 9.6% of the nucleotides in CpG sites being polymorphic, 3.3% of the nucleotides in mononucleotide runs of length 5 or greater, 3.0% of the nucleotides within three base pairs of the polymerase α arrest site motif of TG(A/G)(A/G)GA, and 0.5% at all other sites (AR Templeton, AG Clark, KM Weiss, DA Nickerson, E Boerwinkle, CF Sing, unpublished observations). Altogether, almost half of the polymorphic sites in the sequenced portion of the *LPL* gene were from one of these three highly mutable classes and therefore would be less than ideal choices for a SNP marker. This consideration further dims the likelihood of a randomly chosen SNP providing adequate coverage of a gene.

The nonrandom nature of mutation means that the infinite sites model is not applicable to the human nuclear genome. If the patterns observed at the *LPL* locus hold true for other loci, a model that is a mixture of the finite and infinite sites model would be more appropriate, but such a mixture is qualitatively more similar to a finite sites model (which allows multiple hits) than to an infinite sites model (which forbids multiple hits). This is a serious implication because many of the standard descriptors and test statistics for looking at genetic diversity, effective sizes, recombination, etc, are based upon the infinite sites model, primarily because of its greater mathematical simplicity. Consequently, these human genome studies constitute a major challenge to population geneticists. New analytical and statistical models need to be developed to advance our understanding of the nature of genetic diversity within the human genome.

HUMAN POPULATION STRUCTURE

As discussed above, the genetic structuring of human populations is now medically and commercially valuable knowledge that is essential for meeting the new goals of the HGP. Obviously, the SNP surveys themselves will greatly augment our knowledge of human population structure. Therefore, the study of human population structure and recent evolutionary history is and will remain a major focus of the HGP. There is also a great need to understand the patterns of differentiation among human populations for clinically relevant phenotypes and the candidate loci or chromosomal regions associated with them. For example, the incidence of CAD varies tenfold among populations (129). As previously noted, the $\varepsilon 4$ allele of the *ApoE* locus is a major predictor for individual CAD-risk and cholesterol levels within several human populations despite much variation in the population incidence of CAD (43). The $\varepsilon 4$ allele frequency varies from 0.07 to 0.20 and explains 75% of the variation in CAD mortality rate among the populations, with higher $\varepsilon 4$ allele frequencies being found in those populations with higher CAD (129). However, discrepancies exist between the inter- versus intrapopulational estimates of the effects of the $\varepsilon 4$ allele (129), and studies that integrate the intra- and interpopulation components of the sort already performed in evolutionary genetics (152–154, 160) are obviously going to be needed in examining the dual roles of genetic variation as sources of phenotypic variability within and among populations.

As the above studies illustrate, human populations differ greatly in the incidence of common diseases that have a strong underlying genetic component (158). This component of human population structure has obvious public health importance (52, 117). Evolutionary biology is an important tool for predicting which human populations are at greatest risk for certain diseases because it can provide insights into the ultimate reason why a particular disease is common (158). For example, the insight that sickle-cell anemia (associated with the *S* allele at the *β-globin* locus in humans) is an adaptation to malaria immediately leads to the prediction that *S* alleles should be found in high frequencies in populations that live now or in the recent past in malarial regions. Although sickle cell is commonly portrayed as a disease associated with one "race" of humanity, the evolutionary prediction is far more accurate because the *S* allele is found in high frequency in some, but not all, sub-Saharan African populations, populations in the Mediterranean and Middle East, and populations in India (7, 31, 100, 109, 114). Sickle cell is not cleanly associated with any "race" but rather is more strongly associated with the presence of malaria, its selective agent. Similar considerations hold for the more common diseases. For example, certain human populations, scattered widely throughout the world, have extremely high incidences of adult-onset diabetes (158). Neel (94) proposed an evolutionary explanation for this complex pattern; namely, that populations exposed to prolonged low-calorie diets or periodic famines would be selected for "thrifty genotypes," and these same thrifty genotypes when placed into an environment with a high-calorie diet would be predisposed to diabetes. Recent studies have supported the thrifty genotype hypothesis (5, 29, 94a, 106); with increasing knowledge about the genes responsible for adult-onset diabetes (27, 88, 106, 146, 149), this hypothesis can be tested with greater rigor than hitherto possible. Evolutionary explanations have been proposed for many other diseases in humans (51, 94a, 96, 117, 158), and studies testing these evolutionary hypotheses are needed in order to exploit more fully the public health predictions possible through an evolutionary explanation of ultimate causation.

GENETIC ARCHITECTURE

The genetic architecture of a trait refers in part to the number of loci and their genomic positions, and the number of functional alleles per locus that influence the trait. Genetic architecture also includes the patterns of dominance, epistasis, pleiotropy, and gene-by-environment interactions that characterize the transition from genotype to phenotype. The relative frequency of different types of genetic architecture constitutes one of the most longstanding debates within evolutionary genetics. Fisher (34) argued that mutations of large phenotypic effect would mostly be deleterious and thereby would be rapidly eliminated from populations. Fisher also argued that the underlying genetic architecture of most traits should consist of a large number of alleles and loci of small phenotypic effect with epistasis playing an unimportant role. Wright (161) also regarded genetic architecture as generally involving multiple loci, but he argued that the effect of any single allele is highly dependent upon its genetic and environmental context, thereby making the distinction between minor polygenes and "major" genes (66) a false one. Wright also regarded epistasis and pleiotropy as "universal" phenomena and emphasized the importance of gene-by-environment interactions and the dynamics of gene expression throughout development and the aging process (162). This debate has continued to the present day (33), in part because the nature of genetic architecture plays an important role in evolutionary models of adaptation (34, 161) and speciation (133).

The new HGP research focus on complex, common traits contributes directly to the debate about genetic architecture. In turn, the resolution of this evolutionary debate has profound implications for the HGP. All of the approaches discussed in the previous section can provide information on genetic architecture. The linkage and linkage-disequilibrium marker approaches can identify some of the loci that influence the trait of interest. These studies clearly show that not all loci are minor in their phenotypic contributions (9, 12, 15, 18, 20, 27, 39, 65, 86, 88, 146) and that the marginal effects of specific chromosomal regions are highly context dependent (to be discussed shortly).

Although epistasis can create difficulties for the linkage marker approach (30), epistasis can be tested in QTL studies (11, 67, 111, 163). However, several biases exist against the detection of epistasis. First, many QTL studies only search for epistasis among markers or chromosomal regions that first show significant marginal effects, but significant epistasis can exist between marked chromosomal regions with no marginal phenotypic effects (142). Second, most searches for epistasis only deal with pairwise interactions, but additional epistasis can exist at higher levels (142). Third, and most importantly, epistasis is not detected because no effort is made to discover it (37). Despite these biases, there are already several QTL studies that report significant epistasis for some common human phenotypes (12, 15, 20).

In principle, the biases against discovering epistasis are less severe when dealing with the candidate locus approach because the loci are chosen on the basis of prior knowledge rather than a significant marginal effect. However, in practice most candidate loci studies do not look for epistasis (37), and those that do are generally limited to pairwise interactions among candidate loci having significant marginal effects. Nevertheless, much epistasis has been detected with the candidate loci approach for many common human phenotypes (4, 40, 57, 58, 62, 69, 77, 79, 89, 92, 99, 101–103, 108, 113, 122, 126, 144).

The study that has best used the potential of the candidate loci approach to avoid bias in the detection of epistasis is that of Nelson et al (95), although even this study retains the bias of only examining pairwise interactions. Nelson et al (95) surveyed genetic variation at 18 markers in six candidate gene regions in a human population scored for the trait of log-transformed triglyceride level, a risk predictor of CAD. Nelson et al (95) used an extensive computer search called the combinatorial partitioning method to identify the pairwise combinations of markers that explained a significant proportion ($\alpha = 0.01$) of the observed phenotypic variance. Because of the immense number of tests, the significant effects identified by the initial search were subjected to tenfold cross-validation. This process was carried out with no a priori model of gene action. The pairwise markers that best explained phenotypic variation in females and males are shown in Figure 2, along with the percent of the phenotypic variation explained by the pair jointly and their single-locus marginal effects. Obviously, epistasis is of critical importance for this lipid trait. The included loci would clearly be overlooked in a single-locus analysis in both females and males.

As more studies of this nature are performed under the HGP, this long-standing debate over genetic architecture should become increasingly resolved. It is already clear that interactions among loci are important in many clinically relevant

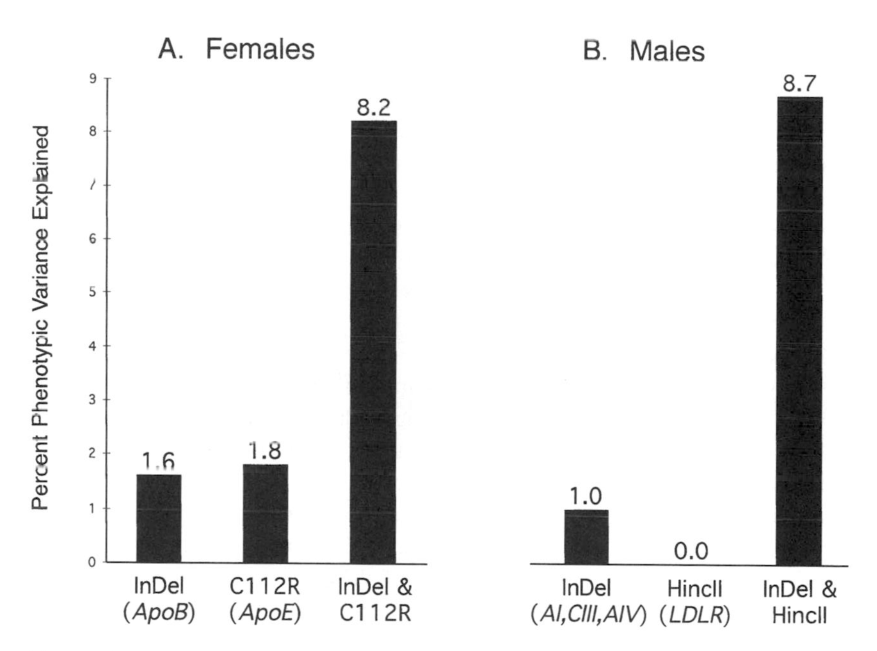

Figure 2 Percent of phenotypic variance in the natural logarithm of serum triglyceride levels in females (*a*) and males (*b*) that explained best by pairs of markers in candidate genes as well as the single marker contributions. Modified from Nelson et al (95).

phenotypes (37). Also, as pointed out before, interactions between genes and environments are important. Together, these interactions imply that the effects of any single locus could display extreme context dependency. This context dependency has profound implications for the study of disease and health under the HGP.

There are now many cases in which one research group identified a particular locus or DNA region as having a strong association with a medically important phenotype only to have another group find no association with that locus or region (23). Such "disappearing" genes occur too frequently to be explained by multiple false positives (23). Context dependency, due to either epistasis or gene-by-environment interactions, provides an explanation for this phenomenon. To illustrate this, consider an example based upon the strong documented epistasis between the *ApoE* and *LDLR* loci such that individuals tend to have elevated serum cholesterol levels only if they have the $\varepsilon 4$ allele at *ApoE* and are homozygous A_2A_2 at the *LDLR* locus (102, 103). Figure 3(*a*) shows a quantitative genetic partitioning of the phenotypic variance in this case using allele frequencies typical of Western European populations: 0.152 for the $\varepsilon 4$ allele and 0.78 for A_2 (138). Note that most of the epistasis in this system is converted to "additive variance" at *ApoE*, a type of conversion expected in epistatic systems (11). Figure 3(*b*) shows the results from the same calculations, but now using allele frequencies of 0.95 for $\varepsilon 4$ and 0.5 for A_2. Although the genotype/phenotype relationships are identical in 3(*a*) and 3(*b*), *ApoE* emerges as a "major" gene and *LDLR* is a "minor" gene in 3(*a*), but the reverse is true in 3(*b*). This illustrates a well-known but little appreciated property of complex systems in which interactions are the true agents of causation and not individual components; namely, there is a confoundment between the frequency of an element and apparent marginal causation at the population level. In particular, the rarer components of the interactive system appear to have the stronger marginal effects, as illustrated in Figure 3. Rareness and commonness are properties of populations, not genes or genotypes. Hence, no matter how elegant the molecular biology, disease associations must always be interpreteted at the population level.

Context dependency is particularly important in medical research because the populations studied are often highly nonrandom samples from the general population. For example, in families with four or more cases of breast cancer, certain mutations at the *BRCA 1* locus are predictive of 52% of the cases of breast cancer, and this increases to 76% of the cases in those families that have both male and female breast cancer (36). When one ascertains families by their having only two cases of breast cancer before the age of 45, *BRCA 1* mutants are only predictive of 7.2% of the cases (82), and when one looks at sporadic cases (the bulk of the disease in the general population), only 1.4% of women with breast cancer have *BRCA 1* mutants (97). Obviously, recommendations for prophylactic mastectomies based on *BRCA 1* testing need to be made in a highly context-dependent fashion (see also 108a).

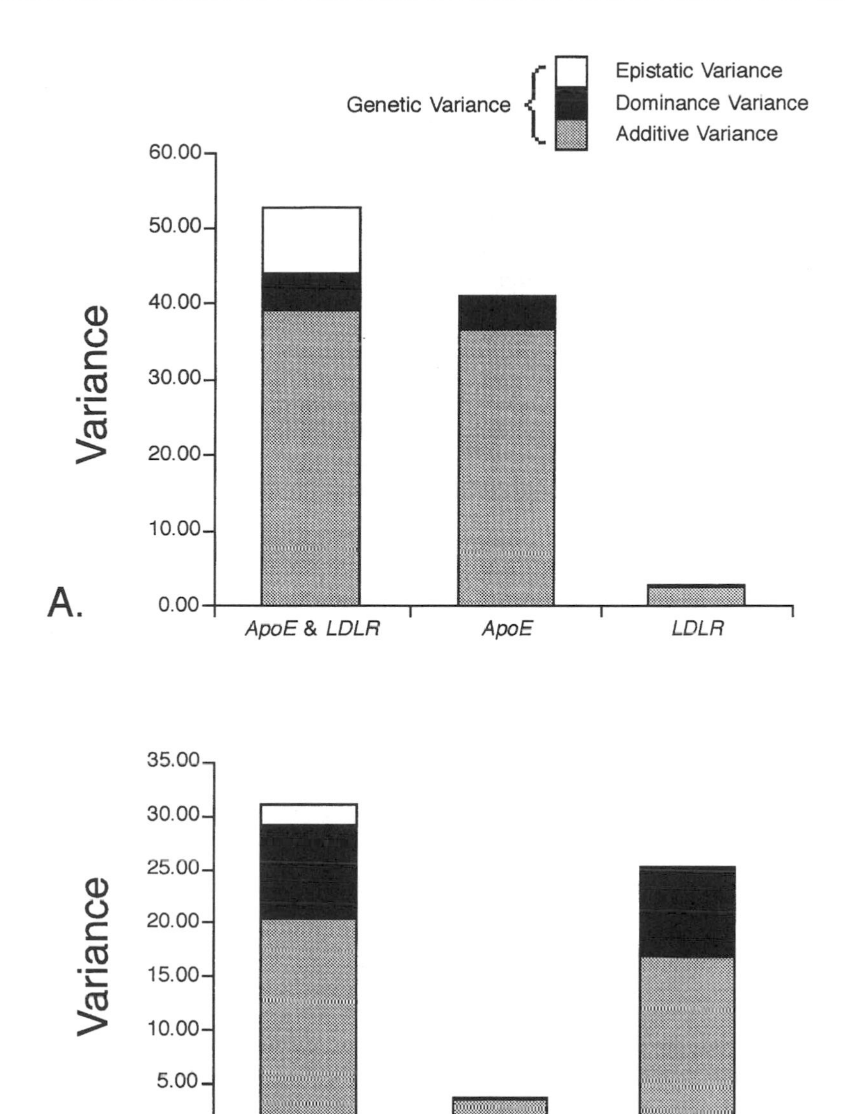

Figure 3 Partitioning of genetic variance for the phenotype of total serum cholesterol level using a constant genotype-to-phenotype mapping of the *ApoE* and *LDLR* loci in two hypothetical populations with differing allele frequencies, as described in the text. The first column shows the decomposition into additive, dominance, and epistatic variance for the two-locus system, and the next two columns shows the marginal decomposition into additive and dominance variance for the two loci considered separately. [From Templeton (138).]

OVERVIEW

The *BRCA 1* story illustrates the practical importance of population thinking in applying the results of the HGP to human health and well-being. Indeed, population thinking (85) may represent the single most important contribution of evolutionary biology to the HGP. Population thinking is central to the interpretation and application of virtually all phases of the diversity research proposed under the new goals of the HGP. Population thinking draws a clear distinction between association and causation, a distinction that is often ignored in human research as illustrated by the earlier discussion of Alzheimer's disease. This distinction is difficult to make because of the evolutionary implications of context-dependent sampling, linkage disequilibrium, and the clustering of functionally related genes within the genome.

The distinction between association and causation is even more critical in light of claims that we will soon be able to "cure" many diseases by altering "defective genes"—a "new eugenics" that "could, at least in principle, be implemented on a quite individual basis" (128). Obviously, such treatments would depend upon an absolute inference of causation, but evolutionary considerations indicate such an absolute inference will be exceedingly difficult. Moreover, population thinking undercuts the idea of a "defective" gene for common diseases. For example, is the $\varepsilon 4$ allele at the *ApoE* locus a "defective gene" and $\epsilon 2$ a "good gene" for CAD because of their marginal associations? Population thinking makes it clear that a marginal association does not always translate well to the individual level—the proposed treatment level of the "new eugenics." For example, considering only the pairwise interactions between *ApoE* genotype and serum cholesterol level upon the odds of CAD, the highest incidence of CAD is found in bearers of the "good" $\varepsilon 2$ allele who have high cholesterol (123). The "defective" $\varepsilon 4$ allele is only associated with the highest incidence of CAD among genotypes when cholesterol levels are closer to average (123). Moreover, given that cholesterol levels can be altered by life style changes (diet, exercise, etc) or drugs, a genetic treatment that takes these interactions into account may still not benefit a particular individual because the individual's phenotypes is temporally dynamic. Indeed, the extensive interactions with environmental factors referred to earlier indicates that the most probable and practical medical applications of genetic knowledge will not be through gene therapy but rather, ironically, by augmenting the importance of the environment; treatment of patients will be individualized by taking into account their norms of reaction to diet, drugs, and other life style variables.

Population thinking therefore serves as an important counterweight to the naive genetic determinism inherent in the almost daily proclamations of the discovery of some new important "defective" gene. Perhaps the greatest use of evolutionary theory in the HGP will be to foster an appreciation of human diversity and its clinical implication that treatment must focus upon an individual with a disease and not upon the disease of the individual.

Visit the Annual Reviews home page at http://www.AnnualReviews.org

LITERATURE CITED

1. Albert SM, Gurland B, Maestre G, Jacobs DM, Stern Y, et al. 1995. *ApoE* genotype influences functional status among elderly without dementia. *Am. J. Med. Genet.* 60:583–87
2. Allan CM, Walker D, Segrest JP, Taylor JM. 1995. Identification and characterization of a new human gene (*ApoC4*) in the *apolipoprotein E, C-I*, and *C-II* gene locus. *Genomics* 28:291–300
3. Angelicheva D, Calafell F, Savov A, Jordanova A, Kufardjieva A, et al. 1997. Cystic fibrosis mutations and associated haplotypes in Bulgaria: a comparative population genetic study. *Hum. Genet.* 99: 513–20
4. Berg K. 1990. Risk factor variability and coronary heart-disease. *Acta Geneticae Medicae Et Gemellologiae* 39:15–24
5. Bindon JR, Baker PT. 1997. Bergmann's rule and the thrifty genotype. *Am. J. Phys. Anthropol.* 104:201–10
6. Boerwinkle E, Ellsworth DL, Hallman DM, Biddinger A. 1996. Genetic analysis of atherosclerosis: a research paradigm for the common chronic diseases. *Hum. Mol. Genet.* 5:1405–10
7. Boletini E, Svobodova M, Divoky V, Baysal E, Curuk MA, et al. 1994. Sickle-cell-anemia, sickle-cell beta-thalassemia, and thalassemia major in Albania–characterization of mutations. *Hum. Genet.* 93:182–87
8. Cai XG, Fallin D, Stanton J, Scibelli P, Duara R, et al. 1997. *ApoE* is linked to Alzheimer's disease in a large pedigree. *Am. J. Med. Genet.* 74:365–69
9. Cardon LR, Smith SD, Fulker DW, Kimberling WJ, Pennington BF, et al. 1994. Quantitative trait locus for reading disability on chromosome 6. *Science* 266:276–79
10. Chartier-Harlin M, Parfitt M, Legrain S, Pérez-Tur J, Brousseau T, et al. 1994. *Apolipoprotein E, ε4* allele as a major risk factor for sporadic early and late-onset forms of Alzheimer's disease: analysis of the 19q13.2 chromosomal region. *Human Mol. Genet.* 3:569–74
11. Cheverud JM, Routman EJ. 1995. Epistasis and its contribution to genetic variance components. *Genetics* 139:1455–61
12. Cho JH, Nicolae DL, Gold LH, Fields CT, Labuda MC, et al. 1998. Identification of novel susceptibility loci for inflammatory bowel disease on chromosomes 1q, 3q, and 4q: evidence for epistasis between 1p and Ibd1. *Proc. Natl. Acad. Sci. USA* 95:7502–7
13. Clair DS, Rennie M, Slorach E, Norrman J, Yates C, et al. 1995. *Apolipoprotein E epsilon-4* allele is a risk factor for familial and sporadic presenile Alzheimer's disease in both homozygote and heterozygote carriers. *J. Med. Genet.* 32:642–44
14. Clark AG, Weiss KM, Nickerson DA, Taylor SL, Buchanan A, et al. 1998. Haplotype structure and population genetic inferences from nucleotide sequence variation in human *lipoprotein lipase*. *Am. J. Hum. Genet.* 63:595–612
15. Cloninger CR, Vaneerdewegh P, Goate A, Edenberg HJ, Blangero J, et al. 1998. Anxiety proneness linked to epistatic loci in genome scan of human personality traits. *Am. J. Med. Genet.* 81:313–17
16. Collins FS, Guyer MS, Chakravarti A. 1997. Variations on a theme: cataloging human DNA sequence variation. *Science* 278:1580–81
17. Collins FS, Patrinos A, Jordan E, Chakravarti A, Gesteland R, et al. 1998. New goals for the US human genome project—1998–2003. *Science* 282:682–89
18. Comuzzie AG, Allison DB. 1998. The search for human obesity genes. *Science* 280:1374–77

19. Corder EH, Saunders AM, Strittmatter WJ, Schmechel DE, Gaskell PC, et al. 1993. Gene dose of *Apoliprotein E* type *4* allele and the risk of Alzheimer's disease in late onset families. *Science* 261:921–23
20. Cornelis F, Faure S, Martinez M, Prudhomme JF, Fritz P, et al. 1998. New susceptibility locus for rheumatoid arthritis suggested by a genome-wide linkage study. *Proc. Natl. Acad. Sci. USA* 95:10746–50
21. Crandall KA. 1994. Intraspecific cladogram estimation: accuracy at higher levels of divergence. *Syst. Biol.* 43:222–35
22. Crandall KA, Templeton AR. 1993. Empirical tests of some predictions from coalescent theory with applications to intraspecific phylogeny reconstruction. *Genetics* 134:959–69
23. Crow TJ. 1997. Current status of linkage for schizophrenia—polygenes of vanishingly small effect or multiple false positives. *Am. J. Med. Genet.* 74:99–103
24. Cummings AM, Robertson FW. 1984. Polymorphism at the *Apoprotein-E* locus in relation to risk of coronary disease. *Clinical Genet.* 25:310–13
25. Czeizel A. 1989. Application of DNA analysis in diagnosis and control of human diseases. *Biol. Zentrabl.* 108:295–301
26. Davidson B. 1998. Corporate profile Genaissance Pharmaceuticals. *Genet. Engin. News* 18:41,62
27. Davies JL, Kawaguchi Y, Bennett ST, Copeman JB, Cordell HJ, et al. 1994. A genome-wide search for human type 1 diabetes susceptibility genes. *Nature* 371:130–36
28. de la Chapelle A, Wright FA. 1998. Linkage disequilibrium mapping in isolated populations: the example of Finland revisited. *Proc. Natl. Acad. Sci. USA* 95:12416–23
29. Diamond JM. 1992. Diabetes running wild. *Nature* 357:362–63
30. Eaves LJ. 1994. Effect of genetic architecture on the power of human linkage studies to resolve the contribution of quantitative traitloci. *Heredity* 72:175–92
31. El-Hazmi M 1990. *Beta globin* gene haplotypes in the Saudi sickle-cell-anemia patients. *Hum. Hered.* 40:177– 86
32. Enserink M. 1998. Human genetics—opponents criticize Iceland's database. *Science* 282:859
33. Fenster CB, Galloway LF, Chao L. 1997. Epistasis and its consequences for the evolution of natural populations. *Trends Evol. Ecol.* 12:282–86
34. Fisher RA. 1930. *The Genetical Theory of Natural Selection*. Oxford: Clarendon
35. Flatz G. 1992. Lactase deficiency biologic and medical aspects of the adult human lactase polymorphism. In *Genetic Basis of Common Diseases*, ed. RA King, JI Rotter, AG Motulsky, pp. 305–25. Oxford: Oxford Univ. Press
36. Ford D, Easton DF, Stratton M, Narod S, Goldgar D, et al. 1998. Genetic heterogeneity and penetrance analysis of the *BRCA1* and *BRCA2* genes in breast cancer families. *Am. J. Hum. Genet.* 62:676–89
37. Frankel WN, Schork NJ. 1996. Who's afraid of epistasis? *Nat. Genet.* 14:371–73
38. Friedlander Y, Berry EM, Eisenberg S, Stein Y, Leitersdorf E. 1995. Plasma lipids and lipoproteins response to a dietary challenge:analysis of four candidate genes. *Clin. Genet.* 47:1–12
39. Ginns EI, Jean PS, Philibert RA, Galdzicka M, Damschroder-Williams P, et al. 1998. A genome-wide search for chromosomal loci linked to mental health wellness in relatives at high risk for bipolar affective disorder among the Old Order Amish. *Proc. Natl. Acad. Sci. USA* 95:15531–36
40. Gloria-Bottini F, Borgiani P, Amante A, Lucarelli P, Bottini E. 1995. Genetic interactions and the environment: a study of *ADA* and *ACP1* systems in the Sardinian population. *Hum. Hered.* 45:129–34
41. Gulcher J, Stefansson K. 1998. Population genomics laying the groundwork for

genetic disease modeling and targeting. *Clin. Chem. Lab. Med.* 36:523–27

42. Gusella JF, Wexler NS, Conneally PM, Naylor SL, Anderson MA, et al. 1983. A polymorphic DNA marker genetically linked to Huntington's disease. *Nature* 306:324–28
43. Hallman DM, Boerwinkle E, Saha N, Sandholzer S, Menzel HJ, et al. 1991. The *apolipoprotein E* polymorphism: a comparison of allele frequencies and effects in nine populations. *Am. J. Hum. Genet.* 49:338–49
44. Hallman DM, Visvikis S, Steinmetz J, Boerwinkle E. 1994. The effect of variation in the *apolipoprotein B* gene on plasmid lipid and apolipoprotein B levels. I. A likelihood-based approach to cladistic analysis. *Ann. Hum. Genet.* 58:35–64
45. Harris H. 1966. Enzyme polymorphisms in man. *Proc. R. Soc. Lond. B* 164:298–310
46. Haviland MB, Betard C, Kessling AM, Davignon J, Sing CF. 1991. An application of cladistics to the low density lipoprotein receptor (*LDLR*) gene to identify functional mutations that influence lipid and lipoprotein variation among familial hypercholesterolemia (FH) patients with a 10 kb deletion. *Am. J. Hum. Genet.* 49 (Suppl):471
47. Haviland MB, Ferrell RE, Sing CF. 1997. Association between common alleles of the *low-density lipoprotein receptor* gene region and interindividual variation in plasma lipid and apolipoprotein levels in a population-based sample from Rochester, Minnesota. *Hum. Genet.* 99:108–14
48. Haviland MB, Kessling AM, Davignon J, Sing CF. 1991. Estimation of Hardy-Weinberg and pairwise disequilibrium in the *apolipoprotein AI-CIII-AIV* gene cluster [published erratum appears in *Am. J. Hum. Genet.* 1992 Dec; 51(6):1457]. *Am. J. Hum. Genet.* 49:350–65
49. Haviland MB, Kessling AM, Davignon J, Sing CF. 1995. Cladistic analysis of the *apolipoprotein AI-CIII-AIV* gene cluster using a healthy French Canadian sample. I. Haploid analysis. *Ann. Hum. Genet.* 59:211–31
50. Heizmann C, Kirchgessner T, Kwiterovich PO, Ladias JA, Derby C, et al. 1991. DNA polymorphism haplotypes of the human *lipoprotein-lipase* gene: possible association with high-density-lipoprotein levels. *Hum. Genet.* 86:578–84
51. Hill AVS, Motulsky AG. 1999. Genetic variation and human disease: the role of natural selection. In *Evolution in Health and Disease*, ed. SC Stearns, pp. 50–61. Oxford: Oxford Univ. Press
52. Hill AVS, Sanchez-Mazas A, Barbujani G, Dunston G, Excoffier L, et al. 1999. Human genetic variation and its impact on public health and medicine. In *Evolution in Health and Disease*, ed. SC Stearns, pp. 50–61. Oxford:Oxford Univ. Press
53. Hofmann S, Jaksch M, Bezold R, Mertens S, Aholt S, et al. 1997. Population genetics and disease susceptibility: characterization of Central European haplogroups by mtDNA gene mutations, correlation with D loop variants and association with disease. *Human Mol. Genet.* 6:1835–46
54. Jeunemaitre X, Inoue I, Williams C, Charru A, Tichet J, et al. 1997. Haplotypes of *angiotensinogen* in essential hypertension. *Am. J. Hum. Genet.* 60:1448–60
55. Jones PA, Rideout WMd, Shen JC, Spruck CH, Tsai YC. 1992. Methylation, mutation and cancer. *BioEssays* 14:33–36
56. Kamboh MI. 1995. *Apolipoprotein E* polymorphism and susceptibility to Alzheimer's disease. *Hum. Biol.* 67:195–215
57. Kamboh MI, Sanghera DK, Aston CE, Bunker CH, Hamman RF, et al. 1997. Gender-specific nonrandom association between the *alpha-1-antichymotrypsin* and *apolipoprotein E* polymorphisms in the general population and its implication for the risk of Alzheimer's-disease. *Gen. Epidemiol.* 14:169–80
58. Kamboh MI, Sanghera DK, Ferrell RE, Dekosky ST. 1995. *ApoE*-4* associated

Alzheimer's disease risk is modified by *alpha-1-antichymotrypsin* polymorphism. *Nat. Genet.* 10:486–88
59. Keavney B, McKenzie CA, Connell JMC, Julier C, Ratcliffe PJ, et al. 1998. Measured haplotype analysis of the *angiotensin-I converting enzyme* gene. *Human Mol. Genet.* 7:1745–51
60. Kleyn PW, Vesell ES. 1998. Genetic variation as a guide to drug development. *Science* 281:1820–21
61. Klitz W, Stephens JC, Grote M, Carrington M. 1995. Discordant patterns of linkage disequilibrium of the *peptide-transporter* loci within the *HLA class II* region. *Am. J. Hum. Genet.* 57:1436–44
62. Ko YL, Ko YS, Wu SM, Teng MS, Chen FR, et al. 1997. Interaction between obesity and genetic polymorphisms in the *apolipoprotein CIII* gene and *lipoprotein lipase* gene on the risk of hypertriglyceridemia in Chinese. *Hum. Genet.* 100:327–33
63. Krawczak M, Cooper DN. 1991. Gene deletions causing human genetic disease: mechanisms of mutagenesis and the role of the local DNA sequence environment. *Hum. Genet.* 86:425–41
64. Krawczak M, Reitsma PH, Cooper DN. 1995. The mutational demography of protein C deficiency. *Hum. Genet.* 96:142–46
65. Krushkal J, Xiong MM, Ferrell R, Sing CF, Turner ST, et al. 1998. Linkage and association of adrenergic and dopamine receptor genes in the distal portion of the long arm of chromosome 5 with systolic blood pressure variation. *Human Mol. Genet.* 7:1379–83
66. Lande R. 1983. The response to selection on major and minor mutations affecting a metrical trait. *Heredity* 50:47–65
67. Lander ES, Schork NJ. 1994. Genetic dissestion of complex traits. *Science* 265:2037–48
68. Lannfelt L, Lilius L, Nastase M, Viitanen M, Fratiglioni L, et al. 1994. Lack of association between *apolipoprotein-E* allele *epsilon-4* and sporadic Alzheimer's disease. *NeuroScience Letters* 169:175–78
69. Lehmann DJ, Johnston C, Smith AD. 1997. Synergy between the genes for *butyrylcholinesterase K* variant and *apolipoprotein* $\epsilon 4$ in late-onset confirmed Alzheimer's-disease. *Hum. Mol. Genet.* 6:1933–36
70. Lendon CL, Martinez A, Behrens IM, Kosik KS, Madrigal L, et al. 1997. *E280a Ps-1* mutation causes Alzheimer's-disease but age of onset is not modified by *ApoE* alleles. *Hum. Mutation* 10:186–95
71. Levo A, Partanen J. 1997. Mutation-haplotype analysis of *steroid 21-hydroxylase* (*Cyp21*) deficiency in Finland: implications for the population history of defective alleles. *Hum. Genet.* 99:488–97
72. Levy G. 1998. Predicting effective drug concentrations for individual patients. Determinants of pharmacodynamic variability *Clin. Pharmacokinetics* 34:323–33
73. Li W, Nei M. 1974. Stable linkage disequilibrium without epistasis in subdivided populations. *Theor. Pop. Biol.* 6:173–83
74. Litt M, Jorde LB. 1986. Linkage disequilibria between pairs of loci within a highly polymorphic region of chromosome 2q. *Am. J. Hum. Genet.* 39:166–78
75. Liu L, Forsell C, Lilius L, Axelman K, Corder EH, et al. 1996. Allelic association but only weak evidence for linkage to the *apolipoprotein E* locus in late-onset Swedish Alzheimer families. *Am. J. Med. Genet.* 67:306–11
76. Lobos EA, Rudnick CH, Watson MS, Isenberg KE. 1989. Linkage disequilibrium study of RFLPs detected at the human *muscle nicotinic acetylcholine receptor subunit* genes. *Am. J. Hum. Genet.* 44:522–33
77. Lucarini N, Nicotra M, Gloria-Bottini F, Borgiani P, Amante A, et al. 1995. Interaction between ABO blood groups and *ADA* genetic polymorphism during intrauterine life. A comparative analysis of couples with habitual abortion and normal

puerperae delivering a live-born infant. *Hum. Genet.* 96:527–31
78. Lucotte G, Visvikis S, Leiningermuler B, David F, Berriche S, et al. 1994. Association of *apolipoprotein E* allele *epsilon-4* with late-onset sporadic Alzheimer's disease. *Am. J. Med. Genet.* 54:286–88
79. Ludwig EH, Borecki IB, Ellison RC, Folsom AR, Heiss G, et al. 1997. Associations between candidate loci *angiotensin-converting enzyme* and *angiotensinogen* with coronary heart disease and myocardial infarction—the NHLBI Family Heart Study. *Ann. Epidemiol.* 7:3–12
80. Ma J, Yee A, H. B. Brewer J, Das S, Potter H. 1994. Amyloid-associated proteins α_1-antichymotrysin and apolipoprotein E promote assembly of Alzheimer β-protein into filaments. *Nature* 372:92–94
81. Magewu AN, Jones PA. 1994. Ubiquitous and tenacious methylation of the CpG site in codon 248 of the p53 gene may explain its frequent appearance as a mutational hot spot in human cancer. *Mol. Cell. Biol.* 14:4225–32
82. Malone KE, Daling JR, Thompson JD, Obrien CA, Francisco LV, et al. 1998. *BRCA1* mutations and breast cancer in the general population:analyses in women before age 35 years and in women before age 45 years with first-degree family history. *JAMA: J. Am. Med. Assoc.* 279:922–29
83. Marshall A. 1997. Laying the foundations for personalized medicines. *Nat. Biotech.* 15:954–57
84. Matsumoto T, Imamura O, Yamabe Y, Kuromitsu J, Tokutake Y, et al. 1997. Mutation and haplotype analyses of the Werner's-syndrome gene based on its genomic structure: genetic epidemiology in the Japanese population. *Hum. Genet.* 100:123–30
85. Mayr E. 1974. The challenge of diversity. *Taxon* 23:3–9
86. McClearn GE, Johansson B, Berg S, Pedersen NL, Ahern F, et al. 1997. Substantial genetic influence on cognitive abilities in twins 80 or more years old. *Science* 276:1560–63
87. Meyer UA. 1999. Medically relevant genetic variation of drug effects. In *Evolution in Health and Disease*, ed. SC Stearns, pp. 41–49. Oxford: Oxford Univ. Press
88. Morahan G, Huang D, Tait BD, Colman PG, Harrison LC. 1996. Markers on distal chromosome 2q linked to insulin-dependent diabetes mellitus. *Science* 272:1811–13
89. Morgan K, Morgan L, Carpenter K, Lowe J, Lam L, et al. 1997. Microsatellite polymorphism of the *alpha(1)-antichymotrypsin* gene locus associated with sporadic Alzheimer's disease. *Hum. Genet.* 99:27–31
90. Motolsky AG. 1957. Drug reactions enzymes, and biochemical genetics. *J. Am. Med. Assoc.* 165:835–37
91. Nakagawa H, Koyama K, Miyoshi Y, Ando H, Baba S, et al. 1998. Nine novel germline mutations of *STK11* in ten families with Peutz-Jeghers-Syndrome. *Hum. Genet.* 103:168–72
92. Nakauchi Y, Suehiro T, Yamamoto M, Yasuoka N, Arii K, et al. 1996. Significance of *angiotensin I-converting enzyme* and *angiotensin II type 1 receptor* gene polymorphisms as risk factors for coronary heart disease. *Atherosclerosis* 125:161–69
93. Nathan BP, Bellosta S, Sanan DA, Weisgraber KH, Mahley RW, et al. 1994. Differential effects of apolipoproteins E3 and E4 on neuronal growth *in vitro*. *Science* 264:850–52
94. Neel JV. 1962. Diabetes mellitus: a "thrifty genotype" rendered detrimental by "progress". *Am. J. Hum. Genet.* 14:353–62
94a. Neel JV, Weder AB, Julius S. 1998. Type II diabetes, essential hypertension, and obesity as "syndromes of impaired genetic homeostasis": the "thrifty genotype" enters the 21st century. *Perspect. Biol. Med.* 42:44–74
95. Nelson MR, Kardia SLR, Ferrell RE, Sing CF. 1998. *A combinatorial optimization*

approach to identify multilocus genotypic classes associated with quantitative trait variability. Presented at Annu. Meet. Am. Soc. Hum. Genet., 48th, Denver
96. Nesse RM, Williams GC. 1998. Evolution and the origins of disease. *Sci. Am.* Nov.:86–93
97. Newman B, Mu H, Butler LM, Millikan RC, Moorman PG, et al. 1998. Frequency of breast cancer attributable to *BRCA1* in a population-based series of American women. *J. Am. Med. Assoc.* 279:915–21
98. Nickerson DA, Taylor SL, Weiss KM, Clark AG, Hutchinson RG, et al. 1998. DNA sequence diversity in a 9.7-kb region of the human *lipoprotein lipase* gene. *Nat. Genet.* 19:233–40
99. Okuizumi K, Onodera O, Namba Y, Ikeda K, Yamamoto T, et al. 1995. Genetic association of the *very low density lipoprotein (VLDL) receptor* gene with sporadic Alzheimer's disease. *Nat. Genet.* 11:207–9
100. Oner C, Dimovski AJ, Olivieri NF, Schiliro G, Codrington JF, et al. 1992. *Beta-S* haplotypes in various world populations. *Hum. Genet.* 89:99-104
101. Peacock RE, Temple A, Gudnason V, Rosseneu M, Humphries SE. 1997. Variation at the *lipoprotein lipase* and *apolipoprotein AI-CIII* gene loci are associated with fasting lipid and lipoprotein traits in a population sample from Iceland: interaction between genotype, gender, and smoking status. *Gen. Epidemiol.* 14:265–82
102. Pedersen JC, Berg K. 1990. Gene-gene interaction between the *low-density-lipoprotein receptor* and *apolipoprotein-E* loci affects lipid levels. *Clinical Genet.* 38:287–94
103. Pedersen JD, Berg K. 1989. Interaction between *Low-Density Lipoprotein Receptor (LDLR)* and *Apolipoprotein-E (ApoE)* alleles contributes to normal variation in lipid level. *Clinical Genet.* 35:331–37
104. Pereztur J, Campion D, Martinez M, Brice A, Tardieu S, et al. 1995. Evidence for *apolipoprotein E epsilon-4* association in early-onset Alzheimer's patients with late-onset relatives. *Am. J. Med. Genet.* 60:550–53
105. Pickering L. 1998. Perkin-Elmer and Incyte to map the human genome. *Genet. Engin. News* 18:18, 41, 51
106. Pillay TS, Langlois WJ, Olefsky JM. 1995. The genetics of non-insulin-dependent diabetes mellitus. *Adv. Genet.* 32: 51–98
107. Poduslo SE, Riggs D, Schwankhaus J, Osborne A, Crawford F, et al. 1995. Association of apolipoprotein E but not B with Alzheimer's disease. *Hum. Genet.* 96:597–600
108. Porkka KV, Taimela S, Kontula K, Lehtimaki T, Aaltosetala K, et al. 1994. Variability gene effects of DNA polymorphisms at the *ApoB, ApoAI/CIII* and *ApoE* loci on serum-lipids:the cardiovascular risk in young Finns study. *Clin. Genet.* 45:113–21
108a. Rahman N, Stratton MR. The Genetics of breast cancer susceptibility. *Annu. Rev. Genet.* 32:95–121
109. Reddy PH, Modell B. 1996. Reproductive behaviour and natural selection for the sickle gene in the Baiga tribe of Central India–the role of social parenting. *Ann. Hum. Genet.* 60:231–36
110. Rideout WMd, Coetzee GA, Olumi AF, Jones PA. 1990. 5-Methylcytosine as an endogenous mutagen in the human *LDL receptor* and *p53* gene. *Science* 249:1288–90
111. Routman EJ, Cheverud JM. 1997. Gene effects on a quantitative trait: two-locus epistatic effects measured at microsatellite markers and at estimated QTL. *Evolution.* 51:1654–62
112. Sandbrink R, Hartmann T, Masters CL, Beyreuther K. 1996. Genes contributing to Alzheimer's disease. *Mol. Psychol.* 1:27–40
113. Sanghera DK, Aston CE, Saha N,

Kamboh MI. 1999. DNA polymorphisms in two paraoxonase genes (*PON1* and *PON2*) are associated with the risk of coronary heart disease. *Am. J. Hum. Genet.* 62:36–44
114. Schiliro G, Spena M, Giambelluca E, Maggio A. 1990. Sickle haemoglobinpathies in Sicily. *Am. J. Hematol.* 33:81–85
115. Schmalhausen II. 1949. *Factors of Evolution*. Philadelphia: Blakiston Company. 327 pp.
116. Schmutte C, Jones PA. 1998. Involvement of DNA methylation in human carcinogenesis. *Biol. Chem.* 379:377–88
117. Schork NJ, Cardon LR, Xu X. 1998. The future of genetic epidemiology. *Trends Genet.* 14:266–72
118. Schull WJ, Annas GJ, Arnheim N, Blangero J, Chakravarti A, et al. 1997. *Evaluating Human Genetic Diversity.* Washington, DC: Natl. Acad. 91 pp.
119. Sheffield VC, Stone EM, Carmi R. 1998. Use of isolated inbred human populations for identification of disease genes. *Trends Genet.* 14:391–96
120. Sing CF, Boerwinkle EA. 1987. Genetic architecture of inter-individual variability in apolipoprotein, lipoprotein and lipid phenotypes. *Ciba Found. Symp.* 130:99–127
121. Sing CF, Davignon J. 1985. Role of the *Apolipoprotein E* polymorphism in determining normal plasma lipid and lipoprotein variation. *Am. J. Hum. Genet.* 37:268–85
122. Sing CF, Haviland MB, Reilly SL. 1996. Genetic architecture of common multifactorial diseases. In *Variation in the Human Genome*, ed. G Cardew, pp. 211–32. Chichester: Wiley
123. Sing CF, Haviland MB, Templeton AR, Reilly SL. 1995. Alternative genetic strategies for predicting risk of atherosclerosis. In *Atherosclerosis X. Excerpta Medica International Congress Series*, ed. FP Woodford, J Davignon, AD Sniderman, pp. 638–44. Amsterdam: Elsevier Sci. B.V.
124. Sing CF, Haviland MB, Templeton AR, Zerba KE, Reilly SL. 1992. Biological complexity and strategies for finding DNA variations responsible for inter-individual variation in risk of a common chronic disease, coronary artery disease. *Ann. Med.* 24:539–47
125. Sing CF, Haviland MB, Zerba KE, Templeton AR. 1992. Application of cladistics to the analysis of genotype-phenotype relationships. *Eur. J. Epidemiol.* 8:3–9
126. Sing CF, Moll PP. 1990. Genetics of atherosclerosis. *Annu. Rev. Genet.* 24:171–88
127. Sing CF, Orr JD. 1976. Analysis of genetic and environmental sources of variation in serum cholesterol in Tecumseh, Michigan. III. Identification of genetic effects using 12 polymorphic genetic blood marker systems. *Am. J. Hum. Genet.* 28:453–64
128. Sinsheimer RG. 1969. The prospect of designed genetic change. *Engin. Sci.* 32:8–13
129. Stengard JH, Weiss KM, Sing CF. 1998. An ecological study of association between coronary heart disease mortality rates in men and the relative frequencies of common allelic variations in the gene coding for Apolipoprotein E. *Hum. Genet.* 103:234–41
130. Stengard JH, Zerba KE, Pekkanen J, Ehnholm C, Nissinen A, et al. 1995. *Apolipoprotein E* polymorphism predicts death from coronary heart disease in a longitudinal study of elderly Finnish men. *Circulation* 91:265–69
131. Stephens JC, Briscoe D, O'Brien SJ. 1994. Mapping by admixture linkage disequilibrium in human populations: limits and guidelines. *Am. J. Hum. Genet.* 55:809–24
132. Strittmatter WJ, Weisgraber KH, Goedert M, Saunders AM, Huang D, et al. 1994. Hypothesis–microtubule instability and

paired helical filament formation in the Alzheimer disease brain are related to apolipoprotein E genotype. *Exp. Neuro.* 125:163–71

133. Templeton AR. 1982. Genetic architectures of speciation. In *Mechanisms of Speciation*, ed. C Barigozzi, pp. 105–21. New York: Liss
134. Templeton AR. 1987. The general relationship between average effect and average excess. *Genet. Res.* 49:69–70
135. Templeton AR. 1995. A cladistic analysis of phenotypic associations with haplotypes inferred from restriction endonuclease mapping or DNA sequencing. V. Analysis of case/control sampling designs: Alzheimer's disease and the *Apoprotein E* locus. *Genetics* 140:403–9
136. Templeton AR. 1996. Cladistic approaches to identifying determinants of variability in multifactorial phenotypes and the evolutionary significance of variation in the human genome. In *Variation in the Human Genome*, ed. G Cardew, pp. 259–83. Chichester: Wiley
137. Templeton AR. 1998. Linkage mapping versus the candidate gene approach. *Proceedings of the 6th World Congress on Genetics Applied to Livestock Production* 26:175–82
138. Templeton AR. 1999. Epistasis and complex traits. In *Epistasis and the Evolutionary Process*, ed. M Wade, E Brodie III, J Wolf. Oxford: Oxford Univ. Press. In press
139. Templeton AR, Boerwinkle E, Sing CF. 1987. A cladistic analysis of phenotypic associations with haplotypes inferred from restriction endonuclease mapping. I. Basic theory and an analysis of Alcohol Dehydrogenase activity in *Drosophilaz. Genetics* 117:343–51

139a. Templeton AR, Clark AG, Weiss KM, Nickerson DA, Boerwinkle E, Sing CF. 1999. Recombinational and mutational hotspots within the human *Lipoprotein Lipase* gene. *Am. J. Hum. Genet.* In press

140. Templeton AR, Crandall KA, Sing CF. 1992. A cladistic analysis of phenotypic associations with haplotypes inferred from restriction endonuclease mapping and DNA sequence data. III. Cladogram estimation. *Genetics* 132:619–33
141. Templeton AR, Sing CF. 1993. A cladistic analysis of phenotypic associations with haplotypes inferred from restriction endonuclease mapping. IV. Nested analyses with cladogram uncertainty and recombination. *Genetics* 134:659–69
142. Templeton AR, Sing CF, Brokaw B. 1976. The unit of selection in *Drosophila mercatorum*. I. The interaction of selection and meiosis in parthenogenetic strains. *Genetics* 82:349–76
143. Templeton AR, Sing CF, Kessling A, Humphries S. 1988. A cladistic analysis of phenotypic associations with haplotypes inferred from restriction endonuclease mapping. II. The analysis of natural populations. *Genetics* 120:1145–54
144. Thomas AE, Green FR, Lamlum H, Humphries SE. 1995. The association of combined *alpha* and *beta fibrinogen* genotype on plasma fibrinogen levels in smokers and non-smokers. *J. Med. Genet.* 32:585–89
145. Thompson EA, Deeb S, Walker D, Motulsky AG. 1988. The detection of linkage disequilibrium between closely linked markers:RFLPs at the *AI-CIII apolipoprotein* genes. *Am. J. Hum. Genet.* 42:113–24
146. Todd JA. 1995. Genetic analysis of type I diabetes using whole genome approaches. *Proc. Natl. Acad. Sci. USA* 92:8560–65
147. Todorova A, Danieli GA. 1997. Large majority of single-nucleotide mutations along the *dystrophin* gene can be explained by more than one mechanism of mutagenesis. *Hum. Mut.* 9:537–47
148. Tvrdik T, Marcus S, Hou SM, Falt S,

Noori P, et al. 1998. Molecular characterization of two deletion events involving ALU-sequences, one novel base substitution and two tentative hotspot mutations in the *Hypoxanthine Phosphoribosyltransferase* (*HPRT*) gene in five patients with Lesch-Nyhan-Syndrome. *Hum. Genet.* 103:311–18

149. Valdes AM, McWeeney S, Thomson G. 1997. HLA class II DR-DQ amino acids and insulin-dependent diabetes mellitus:application of the haplotype method. *Am. J. Hum. Genet.* 60:717–28
150. Valdes AM, Thomson G. 1997. Detecting disease predisposing variants: the haplotype method. *Am. J. Hum. Genet.* 60:703–16
151. Virtaneva K, Damato E, Miao IM, Koskiniemi M, Norio R, et al. 1997. Unstable minisatellite expansion causing recessively inherited Myoclonus Epilepsy, EPM1. *Nat. Genet.* 15:393–96
152. Wade MJ. 1996. Adaptation in subdivided populations: kin selection and interdemic selection. In *Adaptation*, pp. 381–405. New York: Academic Press
153. Wade MJ, Goodnight CJ. 1991. Wright's shifting balance theory: an experimental study. *Science* 253:1015–18
154. Wade MJ, Griesemer JR. 1998. Populational heritability empirical studies of evolution in metapopulations. *Am. Nat.* 151:135-47
155. Wakeley J, Hey J. 1997. Estimating ancestral population parameters. *Genetics* 145:847–55
156. Wang DG, Fan JB, Siao CJ, Berno A, Young P, et al. 1998. Large-scale identification, mapping, and genotyping of single-nucleotide polymorphisms in the human genome. *Science* 280:1077–82
157. Weisgraber KH, Mahley RW. 1996. Human apolipoprotein E: the Alzheimer's disease connection. *FASEB J.* 10:1485–94
158. Weiss KM. 1993. *Genetic Variation and Human Disease:Principles and Evolutionary Approaches.* Cambridge, UK: Cambridge Univ. Press. 354 pp.
159. West WL, Knight EM, Pradhan S, Hinds TS. 1997. Interpatient variability:genetic predisposition and other genetic factors. *J. Clin. Pharmacol.* 37:635-48
160. Whitlock MC, Phillips PC, Wade MJ. 1993. Gene interaction affects the additive genetic variance in subdivided populations with migration and extinction. *Evol.* 47:1758–69
161. Wright S. 1932. The roles of mutation, inbreeding, crossbreeding, and selection in evolution. *Proc. Sixth Int. Cong. Genet* 1:356–66
162. Wright S. 1934. Physiological and evolutionary theories of dominance. *Am. Nat.* 68:25–53
163. Wu WR, Li WM. 1994. A new approach for mapping quantitative trait loci using complete genetic marker linkage maps. *Theor. Appl. Genet.* 89:535–39
164. Yang AS, Gonzalgo ML, Zingg JM, Millar RP, Buckley JD, et al. 1996. The rate of CpG mutation in Alu repetitive elements within the *p53 tumor suppressor* gene in the primate germline. *J. Mol. Biol.* 258:240–50
165. Zerba KE, Ferrell RE, Sing CF. 1998. Genetic structure of five susceptibility gene regions for coronary artery disease:disequilibria within and among regions. *Hum. Genet.* 103:346–54
166. Zerba KE, Kessling AM, Davignon J, Sing CF. 1991. Genetic structure and the search for genotype-phenotype relationships: an example from disequilibrium in the *Apo B* gene region. *Genetics* 129:525–33

Annu. Rev. Ecol. Syst. 1999. 30:51–81

Streams in Mediterranean Climate Regions: Abiotic Influences and Biotic Responses to Predictable Seasonal Events

Avital Gasith[1] and Vincent H. Resh[2]

[1]*Institute for Nature Conservation Research, Tel Aviv University, Tel Aviv 69978, Israel; e-mail: avigas@post.tau.ac.il* [2]*Department of Environmental Science, Policy & Management, University of California, Berkeley, California 94720; e-mail: vresh@nature.berkeley.edu*

Key Words flooding, drying, human impact

■ **Abstract** Streams in mediterranean-climate regions (areas surrounding the Mediterranean Sea, parts of western North America, parts of west and south Australia, southwestern South Africa and parts of central Chile) are physically, chemically, and biologically shaped by sequential, predictable, seasonal events of flooding and drying over an annual cycle. Correspondingly, aquatic communities undergo a yearly cycle whereby abiotic (environmental) controls that dominate during floods are reduced when the discharge declines, which is also a time when biotic controls (e.g. predation, competition) can become important. As the dry season progresses, habitat conditions become harsher; environmental pressures may again become the more important regulators of stream populations and community structure. In contrast to the synchronous input of autumn litterfall in forested temperate streams, riparian input to mediterranean-type streams is more protracted, with fall and possibly spring peaks occurring in streams in the Northern Hemisphere and a summer peak existing in their Southern Hemisphere counterparts. We present 25 testable hypotheses that relate to the influence of the stream hydrograph on faunal richness, abundance, and diversity; species coexistence; seasonal changes in the relative importance of abiotic and biotic controls on the biotic structure; riparian inputs and the relative importance of heterotrophy compared to autotrophy; and the impact of human activities on these seasonally water-stressed streams. Population increases in mediterranean-climate regions (particularly in fertile regions) result in an intensification of the competition for water among different users; consequently, water abstraction, flow regulation, increased salinity, and pollution severely limit the ability of the streams to survive as sustainable, self-regulated systems.

0066-4162/99/1120-0051$08.00

INTRODUCTION

Climatic and geomorphic setting strongly influence structural and functional features of rivers and streams (2, 45, 88). Consequently, lotic ecosystems within a large geographic area may exhibit greater similarity to those in other regions that have similar climates and geomorphology than to those occurring in the same region (58, 93, 134). Streams and rivers in mediterranean-climate areas throughout the world are excellent examples of such convergence. Although confined to relatively small regions (<1% or up to 4% of the continental area according to different definitions, e.g. 5 and 50, respectively) and widely separated on all continents, mediterranean ecosystems support similar types of sclerophyllous vegetation (50, 51, 156, 157). Thus, because of the strong influence of the climate and catchment vegetation on stream structure and function, streams of coastal California, for example, may better be compared with streams in other mediterranean-type climates in Australia, Europe, South America, or Africa than, for example, with coastal streams of the adjoining Pacific Northwest.

In a recent review of the history of research on mediterranean-type ecosystems, 89 comparative studies (in 13 subject categories) among regions with a mediterranean-type climate were listed (84); none of these studies involve aquatic ecosystems. The effect of human-influenced disturbance on the integrity of aquatic habitats in these regions and the need to understand the ecological basis for their rehabilitation prompted us to assess the extent of scientific understanding about the organization and functioning of such habitats and their responses to key environmental factors. Some topics that we examined in mediterranean-type streams are better understood (e.g. hydrology) than others (e.g. riparian inputs and in-stream dynamics, biotic structure and interactions). As a result, we often had to draw on principles and examples from studies on streams in more mesic and xeric regions.

What are the defining characteristics of streams in mediterranean climates? Our central thesis is that because these ecosystems have a distinct cool and wet season followed by a warm and dry season, they are influenced by a sequence of regular and often extreme flooding and drying periods. In this review we describe how (*a*) these streams are sometimes distinctly different from, and other times the same as, streams in either temperate mesic or xeric areas; (*b*) the annual, predictable floods determine the temporal and spatial dynamics of these systems; (*c*) the drying that follows the rainy season results in gradual increases in biotic (e.g. competition and predation) and abiotic (e.g. low oxygen, high temperatures, loss of habitat) controls in structuring biological communities; (*d*) annual variability in rainfall can lead to extreme conditions of flooding and drying, and populations and communities respond differently over a multiannual scale; (*e*) the timing and dynamics of riparian inputs in mediterranean-climate regions differ from those of temperate, forested streams; (*f*) human activities may impact these lotic systems more than their counterparts in more humid or arid regions because of the severe competition for water that occurs in mediterranean-climate regions. Following a description of the environmental setting and the riparian vegetation of mediterranean regions, we organize our discussion of mediterranean-climate streams around a series of testable hypotheses about the response of the aquatic biota to the flooding and drying sequence. This first review of mediterranean-type streams is selective; aspects that deserve further treatment are mentioned in the conclusion.

CLIMATIC SETTING

On all continents, certain coastal regions (often westerly positioned) in the middle latitudes, most extending between 30° and 40° north and south of the equator, are governed by a symmetrical atmospheric circulation that produces a climate characterized by cool, wet winters and hot, dry summers (5). The moderating ocean influence keeps winter temperatures mild, with mean monthly minima ranging from about 8° to 12°C, and frosts infrequent except at high elevations or well inland; summer mean monthly maxima usually vary between 18° and 30°C (49).

Over one half of the area worldwide with this mediterranean-type climate is located in the Mediterranean Sea basin (latitude 30°–45° N) and embraces parts of three continents–Europe, Asia, Africa (see map in 102); hence, the commonly used name for the climate type. This climate and the associated biome occur in four other limited and widely scattered areas of the world (see Figure 1 in 93): the Pacific Coast of North America from southern Oregon to northern Baja California (latitude 31°–41° N); parts of West and of South Australia (latitude 32°–38° S); the south-western Cape Region of South Africa (latitude 32°–35° S); and the central Chilean coast (latitude 32°–41° S) (for further details see 51, 84). Except for Australia, where the topography is characterized by moderate relief, the mediterranean lands are often rugged with a marked change in elevation and, therefore, also in climatic conditions along relatively short horizontal distances (5). Consequently, headwaters of some streams may be in high elevation areas where the climate is too wet and cold to be mediterranean (e.g. Ter River, Spain; 145).

Seasonality and variability in rainfall is the principal attribute of the mediterranean-type climate. At least 65% and often 80% or more of the rain falls in the three months of winter, with most of the precipitation often falling during a few major storm events that may produce flooding (100, 102, 119). The strong seasonality in rainfall (during the cool wet season) and the associated seasonal flooding clearly distinguish the mediterranean climate from most mesic or xeric temperate climates; in the latter climate regions, storms can occur during the colder wet season but also at other times of the year. Although the seasonal precipitation pattern is highly predictable in mediterranean-climate areas, annual rainfall can vary markedly in some regions from year to year; a deviation of 30% or more from a multiannual average is not uncommon. Such precipitation patterns are best described as having low constancy (sensu 40) and high predictability (36, 40, 126). Relatively long-term dry and wet cycles (>10 years) have been detected in mediterranean-climate regions (e.g. 48, 145), but the pattern is far from being consistent and underscores that short-term and year-to-year variations in rainfall are unpredictable.

Because annual precipitation usually ranges between 275 and 900 mm (5, 102), certain mediterranean-climate regions fall into the category of semi-arid regions (200 and 500 mm annual rainfall; 73, 170); however, others may encompass regions with annual rainfall exceeding 1000 mm, including small amounts of summer rain (e.g. in Australia and South Africa—50, 148, 155). Similar to arid regions, mediterranean-climate regions are naturally water stressed because of high annual water losses by evapotranspiration relative to inputs from precipitation (71, 96, 170).

HYDROLOGIC REGIME

The discharge regime of mediterranean-type streams generally follows that of the rainfall pattern, and consequently exhibits both strong seasonal and annual variability (48, 140, 145). In mediterranean-type streams, the high flows abruptly commence in fall or early winter and floods occur during a few months in late fall, winter, and early spring. Subsequent drying and declining flow are gradual over a period of several months in summer, ending abruptly in fall or early winter when the next year's rains commence (see also "extreme-winter" category in 77, and the "perennial" and "intermittent" seasonal winter categories in 171). This seasonal pattern of high discharge coinciding with cool temperatures, which is followed by low discharges with warm or hot temperatures, greatly differs from that of temperate desert streams, which can be disturbed repeatedly and unpredictably by wetting and drying over an annual cycle (e.g. 59).

Deviations from the above-described pattern can occur in mediterranean-type streams. In regions of karst geology (e.g. Spain, Israel), large amounts of water may be stored in subterranean aquifers, which may moderate seasonal fluctuations in flow in mediterranean-type streams (e.g. 4). Furthermore, mediterranean-type streams that have part of their catchment in elevated areas with subfreezing winter temperatures and snow accumulation may exhibit a bimodal mean flow pattern, with highest discharges following the onset of rain (e.g. in fall) and after snow melt in spring (e.g. 4, 145).

Although the occurrence of floods during fall, winter, or spring is predictable in mediterranean-type streams, the intensity and frequency of the floods vary greatly from year-to-year depending on the frequency and intensity of rainfall. Low rainfall may eliminate significant floods and extend the drying period; we refer to these years as drought years. Fluctuations of the mean monthly discharge ranging from 100% above to 50% below the multiannual average have been reported (145), and this is not atypical of other mediterranean-type streams (e.g. 140).

The often brief and sometimes intense rainstorms that occur in mediterranean climates produce flashy hydrographs. The streams flood and attain peak discharges shortly after the rainstorm begins (possibly within hours in situations of a saturated soil and/or a high gradient topography) and decline to normal seasonal flow shortly after the storm ends (e.g. Figures 2 and 5 in 43).

Local geomorphology may play a major role in hydrograph dynamics. Brief, violent floods (characterized by rapid onset and short duration) typically occur in high gradient constrained channels. Mediterranean-type streams are expected to be less "flashy" than desert streams, which are often restricted by steep bedrock canyons or alluvium, and more "flashy" than streams in mesic drainages of low relief that have extensive tributary networks and floodplains and where the floods build in intensity over periods of hours to days (e.g. 77, 109, 126).

In the wetter areas of mediterranean regions, streams usually maintain permanent flow or at least hold surface water throughout the year. In the drier ranges, mediterranean-type streams often have sections of the stream that lack continuous surface water and are composed of a series of isolated pools (e.g. 43, 56); we refer to this mosaic of isolated pools and dry areas as intermittent reaches. In Israel, for example, drying of the headwater tributaries is characteristic of many of the coastal streams, and the duration of intermittency and proportion of the channel that dries generally increases from north to south as rainfall decreases. Seepage of groundwater, return of agricultural water, and effluent discharge may maintain channel wetness in the lower reaches of some of these streams during the dry, summer period. Streams in areas that are intermediate in the wetness gradient of the mediterranean climate may alternate between perennial flow in rainy years and intermittency with surface water limited to isolated pools in drought years (e.g. some coastal mountain streams in California, 43, 140).

THE STRUCTURING ROLE OF FLOODING AND DRYING

Floods are characteristic disturbance events in streams in both mediterranean- and nonmediterranean-type rivers (139). Streams throughout the world also undergo a drying process (i.e. a gradual reduction in discharge and wetted area); however, drying in most mesic and temperate desert streams over an annual cycle is interrupted by rain, whereas drying in mediterranean-type streams is a continuous, gradual process over the dry summer period.

Floods have a variety of effects on the stream ecosystem, depending on their frequency, intensity, timing, and duration. These effects include scouring of accumulated sediment and debris, and redistribution of streambed substrate and organic matter in the channel; changing channel morphology and forming new erosional (riffles) and depositional (point and mid-channel bars, pools) zones; washing away in-channel and encroaching riparian vegetation; restoring channel connectivity; and homogenizing water quality conditions along the stream channel (48, 78, 90, 93, 145).

Watershed-stream interactions in mediterranean-type streams are most pronounced during the relatively short wet season. The first floods usually flush the tributaries of accumulated debris and carry a high load of dissolved and suspended matter from the drainage basin into the stream. Following this and continuing throughout the wet season, the concentration of total dissolved solids is usually relatively low and that of suspended solids relatively high (e.g. 145), water quality conditions along the stream channel are least variable, and diurnal ranges in variables such as temperature and dissolved oxygen are usually lowest (e.g. 32).

Drying involves a gradient of events from reductions in flow, to formation of isolated pools, to complete drying of substrate; a temporal sequence of this is illustrated for a desert stream (160). The physical effect of drying involves habitat contraction, which is essentially the opposite of the habitat expansion that occurs during flooding. Depending on the severity of drying, the following conditions may occur in streams (55) and typically occur in mediterranean-type streams during the dry season: highly fluctuating oxygen concentrations caused by a combination of factors (elevated temperatures, high daytime photosynthesis and nighttime respiration, high biochemical oxygen demand associated with increased algal production and enhanced microbial respiration, reduced dilution of effluents in streams, and reduced turbulence); reduced dilution of nutrients; increased salinity; increased deposition of fine sediments; loss of connectivity with the hyporheic zone; encroachment of vegetation into stream channels (that further enhances siltation); and increased channel erosion because of the prolonged drying of banks.

Droughts can exaggerate the average drying sequence and not only result in the elimination or reduction of scouring flows but also result in an alteration of the temporal and spatial dynamics of habitat structure. Consecutive drought years may have an effect similar to that of water diversion in that accrual of fine sediments, expansion of deposition zones, bank erosion, and vegetation encroachment may produce major changes in channel morphology, including a complete choking of the channel (e.g. 78). There also may be an indirect response of flow to riparian vegetation through changes in geomorphology; this may be of greater consequence to the vegetation than the direct effects. In regions where beaches build, coastal mediterranean-type streams may become seasonally or annually disconnected from the sea during drought years because there is insufficient water to wash away the deposited sand (e.g. 85); this turns the lowermost reach into an elongated, closed lagoon (67).

An increase in evapotranspiration in summer may result in an increase in concentration of dissolved minerals (16, 32, 145). The diurnal fluctuations in water quality condition are highest during this period (32), and a spatial gradient in water quality conditions can develop along the stream, particularly in streams where the surface flow is fragmented into isolated pools. The changes in water quality conditions that occur during the drying process in mediterranean-type streams extend over a period of several months and thus have a more gradual impact on the biota than is typical of more arid streams. As in other streams, water quality conditions in mediterranean-type streams are influenced by stream-specific attributes such as local geomorphology, edaphic features, and human impact, and thus may vary spatially irrespective of climatic conditions (e.g. 139).

RIPARIAN VEGETATION AND INPUT OF ALLOCHTHONOUS ORGANIC MATTER

Riparian Vegetation

The vegetation in mediterranean climates is typically sclerophyllous and evergreen, adapted to water stress during the dry summer period, and able to grow on infertile soils (22, 25, 50, 83, 91, 103, 156, 163). The availability of year-round moisture near streams enables deciduous woody vegetation to occur in the riparian zone as seen in mediterranean-type streams in the Northern Hemisphere (86), with equivalent species pairs occurring in different mediterranean regions (e.g. Israel and California, 150).

With increasing aridity, the riparian vegetation of mediterranean-type streams becomes shorter, more scattered, more restricted to the side of the active channel, and markedly different from the upland regions (58, 103). Because the riparian vegetation is also related to site-specific attributes such as elevation, slope, and lithology (e.g. 56, 111), it may exhibit spatial variability in species richness, composition, and density from the headwaters at high elevation to the lowland reaches (56). Shrubs and herbaceous plants often predominate on exposed banks, particularly in areas impacted by humans, and their richness increases in exposed streambeds of intermittent streams (56). Although riparian vegetation of humid regions cannot withstand even mild fires (121), vegetation along mediterranean-type streams is frequently exposed to natural and human-made fires (22, 23, 158, 168) and therefore may be more fire-adapted and show more rapid recovery after fire.

Inputs of Allochthonous Organic Matter

Riparian input may be the primary energy source to consumers in low-order streams worldwide (e.g., 110). Its role in stream ecosystems is well recognized, based mostly on studies in temperate forests (e.g. 3, 110, 174). Although far less is known about riparian inputs to mediterranean-type streams, canopy cover and organic matter input in low-order streams in the wetter range of the mediterranean climate are apparently as important as they are in temperate, forested streams (e.g. Table III in 103).

mediterranean-type streams differ from temperate deciduous-forest streams in the timing and dynamics of allochthonous detritus input. In the latter, the bulk of litter fall occurs in a short pulse in autumn (ca. 70%–80% in 1–2 months, e.g. 60), with additional material entering the streams by lateral movement over the remainder of the year (e.g. 10, 42, 112). The few studies that have measured organic matter input to mediterranean-type streams suggest a more protracted period of litter fall than in non-mediterranean-type streams, with peaks occurring at different times of the year in the Northern and Southern Hemispheres (autumn and spring cf. summer, respectively, 22, 91, 103, 163). In the Northern Hemisphere, the presence of fall-winter or summer deciduous trees along with sclerophyllous evergreens in the riparian community of mediterranean-type streams (e.g. 43, 86, 103) can partially explain the autumn peak and the extended period of litter production. In the Southern Hemisphere, protracted litter fall results from evergreen phenology whereby new leaf growth is balanced by almost simultaneous leaf fall (157); perhaps the summer peak results from the relic phenology of the tropical-subtropical vegetation that occurs there (155, 157). In Northern Hemisphere mediterranean-type streams, most input occurs when discharge is high and the temperature is relatively low. In contrast, most leaf fall enters mediterranean-type streams in the Southern Hemisphere when discharge is low and water temperature is usually increasing.

Riparian litter that enters streams by lateral movement, i.e. blown by the wind, sliding in from the stream banks, or being carried in by receding flood water (10, 136), may vary considerably in quantity and quality, but it may be especially important in areas of low plant cover, such as in higher-order streams or mediterranean-type streams in drier regions. The lateral movement of the riparian litter, however, is expected to vary locally depending on geomorphology and the nature and density of the understory vegetation (34), and whether the litter fall is wind blown or floodwater transported.

Retention and Transport of Allochthonous Organic Matter

The availability of allochthonous detritus to consumers depends not only on inputs but also on the instream distribution of these inputs. The latter is a function of the retentive capacity of the stream, which is generally inversely related to discharge and positively influenced by retentive structures (i.e. rocks and debris dams) where coarse particulate organic matter (CPOM–> 1 mm in diameter) accumulates (e.g. 133, 136, 149, 153, 154). During the wet season, spates scour the streambed and transport particulate matter downstream; therefore, low-order mediterranean-type streams in both hemispheres are likely to exhibit lowest standing stock of CPOM during the wet season. In mediterranean-type streams in the Southern Hemisphere, most of the allochthonous detritus is expected to be transported downstream several months after it enters the streams. In Northern Hemisphere mediterranean-type streams, the residence time of the allochthonous matter may be much shorter because it enters the streams shortly before or at the time of flooding. In streams with broad valleys, litter may either enter the stream or be deposited in the floodplain during the rise and fall of the hydrograph (44, 136, 151). Thus, fluctuating discharges during the wet period may sequentially accumulate, replace, or distribute CPOM in streams (47); however, the organic material is rearranged only during high discharge periods.

The retentive capacity of forested streams is markedly enhanced by accumulations of large woody material originating in the riparian zone (120), which forms debris dams (12). Because of sparser woody riparian vegetation (except for restricted headwater reaches in the more mesic mediterranean range), mediterranean-type streams are likely to have fewer debris dams (on an areal basis), lower retention capacity, and less allochthonous detritus available than their temperate forest counterparts with similar physical structure (e.g. 91). Human interference that typically occurs in mediterranean regions (deforestation, grazing, fire, channelization, urbanization) further reduces riparian inputs of woody material and the likelihood of natural formation of debris dams and backwater habitats in mediterranean-type streams.

In the Northern Hemisphere where the largest amount of litter fall apparently enters mediterranean-type streams in autumn and winter (e.g. ca. 60% in a Moroccan stream; 103), CPOM is expected to be retained in low-order reaches for only a short period in autumn (1–2 months) until the rainy period commences and the seasonal successive floods occur. In short and steep gradient mediterranean-type streams with low retentive capacity, a significant portion of the CPOM that enters the stream during fall may be exported downstream and possibly even leave the system incompletely processed (energy "leakage" sensu 172). This loss may be partly compensated for later in the year by the influx of litter in spring and summer (e.g. ca. 40% of the annual input in a Moroccan stream, 103). In contrast, in the Southern Hemisphere where litter fall production peaks in late spring and summer at the time when the discharge in mediterranean-type streams is decreasing, CPOM can be effectively retained in low-order reaches for a period of 4–6 months (22, 92), but this CPOM may be reduced by high numbers of shredders that occur there (162, 163).

ECOSYSTEM AND AQUATIC BIOTA RESPONSE TO FLOODING, DRYING, AND HUMAN IMPACT

Expected Response

Based on the characteristics of the hydrograph of mediterranean-type streams, associated seasonal changes in habitat conditions and resource availability, the patterns of input of allochthonous organic matter, and the seasonal scarcity of water, we would expect that the biota of mediterranean-type streams should follow certain predictable trends. We propose 25 testable hypotheses (sequentially numbered in brackets) organized around (*a*) abundance and diversity, (*b*) richness, (*c*) abiotic and biotic pressures, (*d*) species coexistence, (*e*) riparian inputs and the importance of allochthonous organic matter compared to autochthonous production, and (*f*) competition for water and the impact of human activities. After presenting the hypotheses, we examine what the literature pertaining to flooding and drying, and the status of stream ecosystems in mediterranean areas, can tell us about actual influence of environmental and human factors.

[1.1] Unless the stream dries completely, the abundance of the biota is expected to be lowest during the wet season when flooding occurs. [1.2] In spring, following the cessation of flooding, the biota will rapidly increase in abundance from individuals recolonizing from refugia (e.g. fish) or from individuals migrating from nearby aquatic habitats (e.g. insects, amphibians). [1.3] The most flood-resistant species will remain, and the resilient species (the "pioneer species" that are effective colonizers after disturbance) will be the first to colonize after floods cease. [1.4] These are gradually augmented with other species that are less resistant to flooding and slower to recover but that can use the window of opportunity present during late spring and early summer, when habitat and resource availability are high and habitat conditions are relatively moderate, to become established, grow, and reproduce. [1.5] Correspondingly, the abundance of the biota in mediterranean-type streams is expected to be highest at the end of the intermediate period between flooding and extreme drying.

[1.6] With the progression of drying in late summer and fall, and the associated decline in habitat availability and deterioration of water quality, the biota gradually becomes dominated by species that tolerate low discharge, warmer water, and poor water quality. [1.7] Correspondingly, diversity (i.e. the interaction of richness and evenness) should differ between winter and summer assemblages, reflecting temporal changes in species dominance that result from changes in resource availability and habitat conditions; evenness should be lowest at the end of the drying period when extreme physical and physicochemical conditions prevail. [1.8] Depending on the extent of drying, overall abundance can remain high, increase even further when drying is slow and isolated pools maintain water, or decline when drying is rapid and habitat conditions deteriorate or pools dry. [1.9] Furthermore, the species composition and abundance of the biota in mediterranean-type streams is expected to shift rapidly from the late summer-fall assemblage to that of the winter assemblage with the onset of floods; this is in marked contrast to the gradual transition that occurs from the end-of-flood winter faunal assemblage to the summer assemblage. [1.10] The change from the winter assemblage to that of the summer assemblage occurs earlier in drought years as a result of the earlier decline of the hydrograph. [1.11] Correspondingly, the community succession for macroinvertebrates in mediterranean-type streams occurs over the scale of one year, whereas that of temperate humid streams can be longer and that of temperate desert streams can be shorter.

[2.1] Although faunal composition changes occur, species richness is expected to be relatively persistent, reflecting the evolutionary adaptation of the biota in mediterranean-type streams to flooding and drying. [2.2] Moreover, species that evolved in streams that are disturbed predictably, as mediterranean-type streams are, may be expected to have life-history adaptations that maximize growth and reproduction during the period of moderate habitat conditions and minimize exposure of vulnerable stages to extreme conditions of flood or drought.

[3.1] Within an annual cycle, the regulation of assemblage structure in mediterranean-type streams is expected to temporally alternate between mostly abiotic controls during flooding ("reset" periods) and toward the end of the drying period (e.g. extreme physical-chemical conditions, such as high temperature, low dissolved oxygen, loss of flow), and biotic controls (e.g. predation, competition) that occur when isolated pools form. This pattern correlates with the hydrograph of mediterranean-type streams, where extreme forces of mechanical stress and abrasion drastically alter the physical and biological nature of the stream during flooding (high abiotic pressures), which is followed by a period of high resource availability combined with moderate habitat conditions (low abiotic pressures) and then the gradual transformation of the lotic into a lentic habitat. [3.2] During the latter period, the biotic interactions peak (high biotic pressures) but then may decline when extreme conditions prevail.

[4.1] Species in mediterranean-type streams differ in their resistance and resiliency in coping with the sequential abiotic and biotic pressures that occur in these systems. Consequently, the competitive advantage of different species varies with the temporal changes in intensity of abiotic conditions and biotic interactions, and so facilitates species coexistence.

[5.1] The importance of allochthonous organic matter in mediterranean-type streams is generally expected to decrease along increasing gradient of aridity; in contrast, the importance of autochthonous organic matter increases along this gradient. [5.2] Although in both hemispheres the input of leaf litter to mediterranean-type streams is temporally protracted, in the Northern Hemisphere it enters the streams shortly before or at the time of flooding when discharge is high, and temperature and autotrophic activity are relatively low; in the Southern Hemisphere, more of the litter fall enters streams in late spring and summer when discharge is low, temperature is increasing, and autotrophic activity is high. [5.3] As a result of the short contact time between shredders and other decomposers with the leaf litter, a relatively larger portion of the CPOM is likely to be washed downstream (and possibly out of the system), not completely processed in mediterranean-type streams in the Northern Hemisphere than in their counterparts in the Southern Hemisphere (assuming streams with similar retention capacity). [5.4] However, in drought years when extreme habitat conditions develop early in the drying season, shredder activity may be inhibited, and the remaining incompletely processed CPOM in mediterranean-type streams in the Southern Hemisphere may be washed downstream by the next year's floods. [5.5] Overall, shredders may be expected to play a smaller role and filter feeders and scrapers a larger role in mediterranean-type streams relative to the role they play in streams in more humid regions. [5.6] Riparian vegetation of mediterranean-type streams should be more adapted to fire (i.e. able to recover rapidly after fire) than riparian vegetation along temperate, forested streams.

[6.1] The seasonal availability of water in mediterranean-climate regions leads to a competition between human needs for water and the needs of water in the normal functioning of mediterranean-type streams. The impact of this competition is strongest in drought years when discharge is naturally reduced; then even a small diversion of water may have a disproportional large adverse effect. [6.2] Moreover, increases in diversion of water for human use exacerbates the impact by the combined effect of reduced dilution capacity and increased pollution of return water. [6.3] Given the present trend of population increase and increasing demand for fresh water, the competition for water among different users will intensify and result in most mediterranean-type streams having little chance of surviving as sustainable, self-regulating ecosystems.

Assemblage Response to Flooding and Drying

The role of flow in determining the distribution and abundance of aquatic organisms has been of interest to ecologists since early this century. Numerous examples of behavioral (161) and morphological (Table 8A in 141) adaptations and responses of stream organisms to flooding have been proposed. The survival of many organisms depends on finding a refuge from severe hydraulic stresses (e.g. 98); however, even the most effective instream refuge can become ineffective when flood intensity is high (e.g. in constrained channels).

For benthic macroinvertebrates, floods in all stream types consistently reduce population densities (because of reduced abundance and/or an increase in wetted area), but floods have a less consistent effect on reducing richness (for mediterranean-type streams: e.g. 6, 32, 93, 107); lower densities can result from drift (catastrophic drift in the sense of 177) and substrate disturbance (e.g. 39). Macroinvertebrates typically occurring in the wet season can persist into the dry season following wet winters, whereas typical dry-season taxa may appear earlier following drier winters (43). Severe floods greatly reduce population size of organisms that are entirely aquatic (e.g. snails, ostracods, some hemipterans), while those with terrestrial aerial stages (e.g. most aquatic insects) rapidly recolonize these streams (43). Density reductions in the populations with terrestrial aerial stages are not reflected in densities of the following generation (e.g. 57, 107), but in entirely aquatic organisms they are (43). Typically, benthic macroinvertebrate faunal assemblages differ before and after the flood period in mediterranean-type streams (e.g. 32, 107, 140); in part this is because flow-dependent taxa (e.g. filter feeders) are affected when flow is eliminated in isolated pools.

Relative to flooding, the effects of drying on biota have been little studied, and most studies have been descriptive accounts of physiological or behavioral responses, or lists of taxa collected (14, 20). For biota of temporary streams in both mediterranean and other climates, drying is a key environmental factor influencing their distribution, abundance, and life histories (159, 164, 173). While flooding reduces number of individuals, drying may increase numbers locally (from concentration of individuals into smaller areas), decrease (from predation or oxygen stress), or leave densities of individuals unchanged (159).

Although flooding affects stream organisms through a direct abiotic effect (e.g. the scouring of substrate) that results in their displacement, drying can produce a combination of abiotic and biotic effects. Flow and consequent oxygen reductions may result when isolated pools form, and these severe abiotic conditions can exclude some taxa. Moreover, biotic interactions related to food availability may occur as well. Predation has been shown to increase in isolated pools (17, 38, 109, 159), and intraspecific competition has been shown to result in reduced fecundity among macroinvertebrates (e.g. 57). In the latter study of a caddisfly population in a northern California mediterranean-climate stream, the authors used long-term weather records to suggest that dry years that result in food limitation occur about every ten years, droughts that result in reductions of population fitness occur about 4/100 years, but during normal rainfall and wet years (which include most years), population densities are sufficiently reduced by floods (97) so that neither food limitation nor reductions in fecundity typically occur (57).

The effect of drying is most severe when the stream bed becomes completely dry. Under these conditions, taxa lacking desiccation-resistant stages or those unable to find refugia are eliminated until recolonization occurs. In a coastal California springbrook, a severe drought resulted in loss of habitat for over 3 months; the caddisfly population present shifted from one with an age structure typical of a multi-cohort population to one typical of a single-cohort population (137); perhaps even more striking was that age structure did not return to the pre–loss-of-habitat condition until 10 years later (138).

As the temporal sequence of drying in mediterranean-type streams progresses, a gradual shift occurs from macroinvertebrates typical of lotic systems to those that are typical of lentic systems in the isolated pools that form (e.g. 1, 38), to semi-aquatic taxa (e.g. higher Diptera) that colonize when only moist substrates remain, to completely terrestrial taxa (e.g. carabid beetles) that occur on the dry substrate surface as in any other cleared terrestrial habitat. Complete drying of stream bed channels occurs with some regularity in the drier parts of mediterranean-climate areas and less frequently in more moist areas; drying will eliminate all species that lack either resistant stages or the ability to find refugia (e.g. the hyporheic zone, 108). With the initiation of floods a rapid shift occurs from summer macroinvertebrate assemblage to a winter one.

The responses of fish communities in mediterranean-type streams to the sequential pattern of flooding and drying are less well known than the responses of macroinvertebrates to these stresses. Studies in streams with contrasting flow regimes suggest that variability in flow can temporally change the structure of stream fish assemblages by affecting mortality and recruitment rates of species differently (e.g. 72, 124). Moreover, the impact of flow variability on both assemblage structure and habitat use is much stronger than the impact of resource limitation and biotic interactions (e.g. 72). However, the relatively long flood-free period during the dry season that is typical in mediterranean-type streams suggests that the significance of density-dependent biotic interactions may play a greater role in fish communities in mediterranean-type streams than in temperate desert streams affected by repetitive, unpredictable disturbances of flooding and drying. For example, the characteristics of the fish assemblage in a small California stream with a hydrologic regime similar to that of mediterranean-type streams place this stream toward the deterministic end (i.e. mainly competition and predation regulation) of the deterministic-stochastic (i.e. flow-variability regulation) continuum of community structure (118; see also 165).

Fish can avoid the impact of high velocities by moving toward the stream bottom or edge where the shear stress is minimal (101), or they can escape and assume a proper orientation to high flows, which allows them to persist during floods (109). Fish may also migrate into tributaries prior to or during the flood to avoid drifting downstream (e.g., 54) where they may get stranded in isolated reaches and markedly reduce their chance for recovery.

Floods have variable effects on fish populations. Several studies have shown a decline in abundance following intense floods (e.g. 36, 54, 109), but physically (and thus hydraulically) complex reaches lost proportionally fewer fish than hydraulically simple stream reaches (e.g. 124). Studies show that fish densities can rapidly recover after intense and even catastrophic flood events (e.g. 38, 101). Others found persistence of the fish assemblage (e.g. 101, 118), which may be attributed to the timing of flooding with respect to life cycles (i.e. absence of early life stages that are more vulnerable to mechanical stress; 106, 124), or the different intervals of recovery after flooding that were considered (106). Likewise, site-specific features such as pool size and depth, availability of floodplain habitat, and presence of stable substrate have been shown to significantly influence the ability of fishes to survive flood events (e.g. 106). Differences in these features may account for contrasting observations. Persistence of fish populations after flooding could also result from reduction in number of early stages (e.g. eggs or embryos), which could reduce competition and increase the survivorship of young fish (e.g. 101).

Fish in mediterranean-type streams are expected to be highly adapted to drying (e.g. 117, 135). Cessation or reduction of high flows in drought years might increase survivorship of pool-dwelling fish by reducing direct mortality, lowering energy expenditure, and increasing foraging success (72), but this response has not been consistently reported (e.g. 62). Fish may avoid drought effects by migrating upstream to inhabit remaining suitable habitats (72). Drying may differentially reduce survivorship depending on the species and, like with macroinvertebrates, intraspecific competition may occur within populations of tolerant species. This may be particularly important in fish populations that are stranded in isolated pools.

Species Traits Related to Flooding and Drying

In geological terms, the climate of the mediterranean basin is very young, appearing in the Pliocene approximately 3.2 million years ago (167). Many of the extant plant taxa in the mediterranean existed prior to the appearance of this climatic type (83). Certainly, many inhabitants of mediterranean-type streams are widespread in their distribution in different climatic regions, but the levels of endemism found for lotic fauna in mediterranean-type streams can be extremely high. For example, the highest rates of endemism in Europe occur in the mediterranean-climate areas of Italy and Greece for the insect orders Ephemeroptera, Plecoptera, and Trichoptera, all of which occur predominantly in lotic environments (data from Tables in 89), and for fish as well (113). Others have also noted that ecologically distinct taxa occur in mediterranean-type streams throughout the world (e.g. 21). However, at a broad taxonomic level and after removing endemic taxa, the macroinvertebrate fauna collected when flow resumes after drying was similar among desert streams and intermittent streams (including mediterranean-type streams) in Africa, Australia, and North America (20). The species richness of biota that occur in mediterranean-type streams is also related to a variety of historical (e.g. climatic history; 30) and local factors such as vegetation type (e.g. 104), recent land use (e.g. 80), salinity (e.g. 70), and of course permanence of water.

Disturbance regimes are a strong component in the evolutionary history of biota (59, 139). Although both flooding and drying occur in streams in almost all climate types, the high seasonal predictability of the hydrograph of mediterranean-type streams should select for life-history features that favor resistance from being removed by floods, and survival during periods of no flow and high temperatures (or even loss of habitat from drying during droughts). Accompanying or in lieu of these solutions, organisms may have evolved responses that allow them to recover rapidly from disturbance, i.e. that emphasize resilience.

From the perspective of an organism with less than a one-year life cycle (e.g. most benthic macroinvertebrates), the annual predictability of a wet season with floods and a dry season with gradual reductions in (or cessation of) flow in mediterranean-type streams is very high. For an organism with a multiple-year life cycle (e.g. fish and amphibians), the year-to-year variation in the timing, intensity, and frequency of floods, and the rate of drying, are of critical importance to survival but are much less predictable. In the long term (more than 10 years), the likelihood of wet years with extremely high flows and drought years with extreme drying are somewhat predictable, but the timing within that cycle is not. As a result, short-lived and long-lived organisms would be expected to have evolved different adaptive strategies to these conditions. The short-lived organisms (e.g. macroinvertebrates) will maximize recovery rates through short life cycles and high reproduction; the long-lived organisms (e.g. fish) will maximize resistance by behavioral, morphological, and physiological adaptations.

The survival of benthic macroinvertebrates during the drying period in a mediterranean-type stream may involve resistance mechanisms, such as tolerance to abiotic stresses in isolated pools (e.g. 64, 65), having desiccation resistent stages (e.g. 15, 20), finding refugia such as entering moist substrate or the hyporheic zone (e.g. 15), or recolonizing from nearby permanent sources (e.g. 21, 137, 138). Life cycles may also be tied to the sequential predictability in mediterranean-type streams by seasonal breeding patterns (31), or flexible life cycles (21, 142) such as alternating short and long generation times to produce three generations in two years (29). Macroinvertebrates of highly seasonal waters such as some mediterranean-type streams and desert streams exhibit unique adaptations in that: (*a*) egg-hatching coincides with the historically optimum time that water reappears; (*b*) there are predictable, seasonally adjusted growth rates; (*c*) there is some specialization of feeding on predictable food types; and (*d*) there is a high correlation between life history stages and environmental factors (179).

Macroinvertebrates living in some mediterranean-type streams show adaptations that represent trade-offs between resistance to flooding and drying, and avoidance of predators (129). Mobile invertebrates are exposed to predators but may have a better chance of finding refugia that enable them to avoid being washed away or stranded in a drying reach and may be more efficient in locating and gathering food, thus growing and recovering faster. Sessile, armored, and sedentary organisms may be less vulnerable to predation than mobile species but may grow slower and be at a greater risk in areas of scour or desiccation. Therefore, more mobile species that are less vulnerable to high discharges are expected to dominate in spring, shortly after the cessation of floods, and predator-resistant species will do best under the low flows that occur in summer (129).

The recovery rate of fish (i.e. their resilience) is constrained by relatively long life cycles, but they can have greater resistance to flooding and drying than macroinvertebrates because of their greater mobility compared to the passive mobility (i.e. via drift) of most macroinvertebrates. Given the high seasonal predictability of floods and drying, we would expect fish inhabiting mediterranean-type streams to exhibit life cycles that are synchronized with times of moderate flows, increasing temperature, and increasing food availability in order to maximize the growing season and population recruitment, and to minimize exposure of vulnerable stages to extreme conditions. Indeed, most of the fish in mediterranean-type streams reproduce in spring during the declining stage of the hydrograph when temperatures are increasing (e.g. 144). In contrast, in temperate streams where flooding may occur at almost any time of the year, the spawning seasons of most species are protracted or staggered (118). Consequently, the composition of the fish assemblages in temperate streams may vary considerably from year to year, whereas that of mediterranean-type streams may be relatively persistent (118), except where exotic species have invaded disturbed streams (7, 115).

Abiotic versus Biotic Regulation of Assemblage Structure and Species Coexistence

Intensely disturbed streams (i.e. those with violent spates and/or a high degree of intermittency associated with harsh physicochemical conditions) are mostly regulated by abiotic factors (e.g. 38, 61, 87, 126). Biotic controls such as predation and competition may be more important under relatively more stable conditions (e.g. 147, 171). Biotic interactions generally correlate with density (e.g. 152); however, little is known about the importance of density in terms of some biotic interactions (131), as for example in mutualism and other types of symbiosis. The expected increase of populations in mediterranean-type streams during late spring to the middle of summer suggests that biotic interactions will peak at the end of this period.

As the dry season progresses, abiotic pressures increase, perhaps even culminating in a loss of habitat (e.g. 137). Biotic interactions can occur during drying as well. For example, increases in density of macroinvertebrates in isolated pools attract invertebrate predators (e.g. 17, 159) and increase the predation rate by fish (99), oftentimes on nectonic species (e.g. mosquito larvae and dytiscid beetles, 38).

Several authors have suggested that coexistence and high resource overlap for assemblages inhabiting a disturbed ecosystem are possible because intense environmental variation reduces species abundance to levels below which resource limitation occurs, or variation may shift competitive advantage from one species to another (e.g. 41, 72, 109, 176). Observations in an unregulated Northern California mediterranean-type stream have shown that invertebrates are the primary consumers in three- or four-level food chains that result because scouring floods reduce grazers; algae recover before grazers increase and provide the most important colonizable substrate. If scouring floods do not occur, a two-level food chain results, with high densities of grazers and low algal growth (129, 130). In a southern California mediterranean-type stream, spate-cleared areas were rapidly colonized by black fly larvae; in contrast, the substrate was increasingly dominated by a competitively superior caddisfly in the absence of high flow (81).

Heterotrophy versus Autotrophy

The late spring and summer peak of litter input to mediterranean-type streams in the Southern Hemisphere occurs when discharge is reduced, which allows relatively long contact (at least four months) of decomposers and detritivores with CPOM (unless the stream dries). Combined with elevated water temperature in summer, this should promote rapid and efficient processing of the organic matter (79, 166). However, the biotic response observed in Australian streams (including mediterranean-type streams) with bank vegetation of certain eucalyptus trees may contradict this prediction (28). Fewer shredders, poor colonization by aquatic hyphomycetes, and the subsequent slow breakdown of eucalyptus leaves in summer were attributed to the low nitrogen and high polyphenol concentrations that make senescent leaves of this species a low-quality food source (8, 9, 25–28, 35). Inhibited growth rate was also found for a detritivore feeding on eucalyptus leaves in a mediterranean-type stream (35). The inhibitory effects of the polyphenols of eucalyptus leaves dissipate with time and, thus, it is less apparent in leaves that are washed downstream (9).

Low flow, possibly higher concentrations of leached polyphenol compounds, and high water temperature that result in low oxygen level could also restrict microbial conditioning and eliminate some of the detritivores in summer, contributing to slow breakdown rates of CPOM observed in Australian mediterranean-type streams (25–27). This, however, appears to be a special case for Australian streams and possibly more so for those in a mediterranean climate because highly refractory, sclerophyllous species of eucalyptus dominate in the riparian zone and environmental conditions in summer are less conducive to metabolism by consumers (25).

Microbial processing of CPOM on land is directly influenced by the extent of soil wetness; as a result, different loads of dissolved organic carbon (DOC) are expected to enter mediterranean-type streams in wet and drought years (33, 105). Moreover, because semi-arid land leaches more refractory DOC than does soil in more humid regions, lower metabolic rates of biofilms result in the former (94).

The importance of autochthonous production of organic matter in streams is generally expected to increase downstream from the headwaters because of the widening of the stream channel, which allows more light to reach the water, production is expected to decline in high order reaches where turbidity increases (e.g. 172). This effect will be less pronounced in regions with infertile soils and low nutrient loading in streams (e.g. Australia). Because of sparser vegetation and thus greater light penetration, the importance of autochthonous production is expected to be higher in lower-order mediterranean-type streams than in similar order streams in more humid regions. Autochthonous production will also increase in importance further upstream in mediterranean-type streams that are in regions with increasing aridity and will be similar to that in desert streams. Environmental conditions typical to mediterranean-type streams, such as long periods of clear sky, large sunlit instream areas, mild winter temperatures, and low flows, promote algal growth such as that of the attached filamentous alga *Cladophora glomerata* (13, 43, 127–129, 178). In disturbance-controlled systems such as lakes undergoing water level fluctuations (63) and in a spate-scoured mediterranean-type stream, *Cladophora* has been shown to produce large biomass after the cessation of the flooding period, which provides substrate for colonization of epiphytes and invertebrates and serves as an important food source (11, 53, 127–129). Rapid colonization of spate-scoured substrate in riffles, first by diatoms and then by colonial algae, occurred as the time increased since the last disturbance in a southern California mediterranean-type stream (43). The establishment of colonial algae was thereafter significantly affected by macroinvertebrate grazing and perhaps sedimentation.

The prolonged low flow and high temperature during summer in mediterranean-type streams enhance primary production by allowing a massive biofilm matrix (composed of algae and heterotrophs) to develop on solid surfaces (e.g. 74, 75, 146). A different biofilm formed by microorganisms and invertebrates in the hyporheos during the period of low flow may be an alternative energy and nutrient source to mediterranean-type streams, in addition to the above mentioned allochthonous and autochthonous sources (47). High discharges in winter and upwelling of groundwater wash the biofilm into the stream where it becomes a utilizable resource for surface dwellers. Furthermore, this rich biofilm can enhance leaf litter conditioning and processing (47).

Competition for Water and the Associated Human Impact

Mesic regions generally have a surplus of river water, a predictable and relatively stable supply, and a per-capita water renewable volume that exceeds the demand. In contrast, mediterranean-climate regions (like arid regions) have a deficit of water resources (71, 96), which is typical of sub-humid zones (sensu 170) where the ratio of mean annual precipitation to mean potential evapotranspiration is <0.75 (76, 170). However, unlike arid regions where the shortage of water and harsh climatic conditions limit population growth and development, an abundance of sunshine and mild winters make mediterranean-climate regions particularly suitable for human settlement and intensive agricultural production (e.g. 122). This results in a high demand for freshwater, particularly for irrigation, which often consumes water more than all other sectors of the economy combined (e.g., ca. 70% in California and Israel; 180). Consequently, mediterranean-type streams are particularly susceptible to human impact (46, 67, 145). Moreover, the seasonal availability of water in mediterranean-climate regions is a strong catalyst for flow regulation through water diversions and reservoir construction (e.g. 46, 47, 66, 129, 169).

When the ratio of renewable water to withdrawal volume is <1 (e.g., Israel; 180), the excess water demand for freshwater must be complemented by the use of reclaimed wastewater, desalinized water, or water imported from elsewhere. This produces a strong competition for water among consumer sectors. This competition is accentuated in drought years when the increased water deficit imposes further reductions on water consumption (e.g. 67). In practice, a greater proportion of water is abstracted from streams for human use during drought years, which leaves less for the biota at times when they need water the most (114).

Mediterranean-type streams, like other streams in semi-arid regions, are particularly susceptible to water diversion (directly or via groundwater withdrawal) and flow regulation. Diversions, like flow regulation in mediterranean-type streams, interfere with the fundamental mechanisms that structure the stream habitats: They reduce the intensity and frequency of scouring floods, alter the normal stream-floodplain interaction (18, 19), and change water quality conditions (e.g. 169).

Macroinvertebrates and fish are directly affected by dam construction in mediterranean-type streams (e.g. 113, 132, 169, 175). The elimination of eels from central Spain (66) and the drastic decline of Chinook salmon in California (181) are extreme examples of such effects on fish.

In summer, when flow in mediterranean-type streams is naturally reduced, even a relatively small diversion of water from the stream may drastically reduce the water surface and have an effect similar to that of extended drought. Long periods of reduced flow may also result in extinction of threatened species (e.g. 114). Furthermore, the dilution capacity of the stream is lowest during this period and, consequently, even small discharges of poor quality water into streams may have disproportionally large, adverse effects.

Although diversion of water from the stream may change a perennial stream to an intermittent one, the return of agriculture water, and the discharge of sewage effluents into streams may lengthen inundation time (e.g. 125) or even transform intermittent reaches to perennial ones (e.g. 24, 52). This supplementation may alter the distribution of the biota (e.g. favoring pollution resistant species; 69), life history cues, and the structure of the intermittency-adapted communities that occur there.

Riparian vegetation is especially sensitive to changes in minimum and maximum flows (123). Floods facilitate the recruitment of new plants and maintain heterogeneity within the riparian zone (120). In contrast, hydrologic alterations can result in shifts in riparian plant community composition as well as senescence of woody communities (120). Reduced channel wetness for prolonged periods may also result in loss of riparian vegetation and in extensive bank erosion, particularly when combined with steep topography (78). A loss of vegetation and a ten-fold channel width increase were reported in a California mediterranean-type stream in response to local underground water withdrawal that lowered the water table and dried the river bed (95). Elimination of woody vegetation may result in overall reduction in shading, and inputs of allochthonus detritus and large woody debris.

The sensitivity of the riparian vegetation to water diversions is expected to be highest in streams that are within the intermediate wetness range of the mediterranean climate, where a relatively small reduction in the water balance can have a disproportionally large effect, transforming perennial reaches to intermittent ones. In contrast, loss of scouring flow combined with low but constant summer release from upstream dams may enable in-channel intrusion of riparian vegetation that reduces conveyance and increases shading and input of allochthonous detritus.

One major but rarely discussed consequence of water diversions is the change in salinity that puts the biota of mediterranean-type streams at risk. Reduced flow followed by an increase in salt content is expected in mediterranean-type streams during drought periods (e.g. 32). Moreover, reduced discharge also enables greater surface or subsurface intrusion of seawater into coastal streams, transforming freshwater habitats into brackish ones. Agricultural practices such as irrigation-water return and salt flushing increase salinity in mediterranean-type streams (e.g. 122). Lowering of the watertable can result in intrusion of seawater into coastal aquifers or seepage of salt water from inland salt deposits historically covered by sea (e.g. mediterranean Basin). Salinity increases in streams may also result from selective diversion of freshwater sources, whereby the proportion of water contributed by the brackish springs increases (e.g. 67), and by removal of native vegetation in areas where evapotranspiration exceeds precipitation and salts accumulate in the soil (e.g. Australia, SE Bunn, personal communication). mediterranean-type streams can therefore be subjected to salinity pulses in response to fluctuations in annual rainfall and human exploitation of freshwater, which in turn may significantly change the assemblage structure (e.g. from insects to crustacean predominance; 82) or riparian composition (e.g. an increase in salt-tolerant species such as tamarisk).

The competition for water in mediterranean-climate regions is often compounded by water pollution, which threatens the existing supply of water and exacerbates the damage to stream ecosystem. Israel, one of the most water stressed countries (180), has extensive water reclamation and reuse projects; nevertheless, less than 50% of the wastewater is reused and the remaining effluents end up in streams and the sea (67).

Pollution impacts of streams in mesic regions are often ameliorated by the high dilution capacity of receiving waters that have high stream discharges; however, mediterranean-type streams experience a relatively long period of natural low flow, resulting in reduced conveyance and lower dilution capacity. Consequently, effluent release criteria developed for receiving waters in humid regions are unsuitable for mediterranean type streams (68) and can result in extreme deterioration of water quality and elimination of most of the natural biota (67).

The above-mentioned factors put mediterranean-type streams at a greater risk of adverse human impact and thus make them inherently more difficult to rehabilitate than most other stream ecosystems. Can we "have our water and drink it too"? This question may describe a rehabilitation option in water stressed regions. To alleviate the effect of stream water abstraction, the release of high quality reclaimed wastewater is presently being considered for rehabilitation of streams in Israel. However, this solution is not without potential problems: The allocation of high quality effluent for stream rehabilitation is not guaranteed because in drought years the agriculture sector will strongly compete for the higher quality effluents; public health reasons may restrict recreational uses, such as swimming and fishing; and the addition of effluents of any kind may setback stream rehabilitation efforts anytime the wastewater treatment operation fails. Successful rehabilitation has occurred in mediterranean-type streams, particularly in urban environments (e.g. 37, 143), but streams chosen may have unique reasons for selection and the costs of rehabilitation may be high (e.g. 37, 116). Allocation of water for rehabilitation of aquatic habitats in the Sacramento River Basin (California's largest river system) is being considered as part of the implementation of the Central Valley Project Improvement Act (78). In addition to reversing the increasing trend of water pollution as required by the federal Water Quality Act, rehabilitation projects in California also focused on channel stabilization, flood management, fishery enhancement, and riparian floodplain revegetation (78).

CONCLUSIONS

Although streams in mediterranean-climate regions are less studied than their counterparts in temperate, forested regions, ecological information is available for four of the five mediterranean regions worldwide (i.e. not Chile). In Australia and South Africa, however, most information is available for headwater reaches and mountain streams. Important aspects of stream ecology such as nutrient loading and spiraling, algal dynamics and primary production, secondary production, and hyporheic communities were not included in this review and deserve special attention in future overviews of mediterranean-type streams.

First and foremost, mediterranean-type stream ecosystems are ecological reflections of the unique mediterranean climate; they are fluvial systems that are physically, chemically, and biologically shaped by sequential, seasonally predictable events of flooding (late fall-winter) and drying (late summer-early fall) that vary markedly in intensity on a multi-annual scale. Although the biota is under abiotic pressure from floods, there is a period that may last for several months (spring-early summer) during which moderate ecological conditions and high resource availability allow the biota to recover from floods. However, as densities increase, biotic pressures such as competition and predation also increase. If there is extreme drying or desiccation, abiotic regulation returns. This seasonal sequence of abiotic, biotic, and abiotic regulation is a unique characteristic of mediterranean-type streams.

Second, in this review we have presented 25 testable hypotheses about the structure and function of mediterranean-type stream ecosystems, including predictions about structural responses of biotic assemblages, the long debated importance of abiotic versus biotic regulation in stream ecosystems, riparian inputs and the importance of allochthonous matter, and the ecological impact of water scarcity and the resulting severe competition for water. mediterranean-type streams are particularly well suited for experimental tests of the relative importance of biotic and abiotic controls on populations and communities.

Third, the significance of the protracted leaf litter input in terms of the timing of flooding and autotrophic production peaks requires further study. More measurements of allochthonous organic-matter input in streams of different sizes and in different locations are needed, as is an assessment of its effect on community structure and function. Special attention should also be given to the effect of fires, which are common in mediterranean regions, on nutrient and organic-matter input dynamics.

Fourth, the relatively sparse riparian vegetation and abundance of sunshine in mediterranean regions suggest that autotrophic processes may play a more important role in mediterranean-type streams than heterotrophic processes. Reductions in riparian vegetation or increases in nutrient loading further enhance autotrophic processes in mediterranean-type streams.

Finally, unlike the transitory characteristic of the natural disturbances, human disturbances in streams (e.g. water diversion, flow regulation, pollution) are often permanent and tend to increase as population size increases. Population increases in mediterranean-climate regions result in an intensification of the competition for water among different users (e.g. agriculture, domestic consumption, industry, nature conservation). Consequently, mediterranean-type streams (particularly in fertile regions) face major challenges in surviving as sustainable, self-regulated systems. Attempts to rehabilitate these streams must address the inherent problem of how to compensate users for the loss of water that is needed to reinstate the typical hydrological regime of mediterranean-type streams. Rehabilitation efforts in mediterranean-type streams, therefore, may best succeed in urban or unique conservation areas where citizen pressure can successfully outcompete agricultural withdrawal.

ACKNOWLEDGMENTS

We thank our colleagues studying mediterranean-type streams for supplying us with reprints and reports. We are especially in debt to AJ Boulton, SE Bunn, R del Rosario, TL Dudley, GM Kondolf, PB Moyle, RJ Naiman, ME Power, N Prat, F Sabater, and EH Stanley for their reviews, and to N Kobzina, H Lee, EP McElravy,

K Rogers, M. Sommerhäusser, and A van Coller for assistance in gathering information.

Visit the Annual Reviews home page at www.AnnualReviews.org

LITERATURE CITED

1. Alba-Tercedor J, Gonzales G, Puig MA. 1992. Present level of knowledge regarding fluvial macroinvertebrate communities in Spain. *Limnetica* 8:231–41
2. Allan JD. 1995. *Stream Ecology*. London: Chapman & Hall. 388 pp.
3. Anderson NH, Sedell JR. 1979. Detritus processing by macroinvertebrates in stream ecosystems. *Annu. Rev. Entomol.* 24:351–77
4. Armengol J, Sabater S, Vidal A, Sabater F. 1991. Using the rescaled range analysis for the study of hydrological records: the River Ter as an example. *Oecologia Aquat.* 10:21–33
5. Aschmann H. 1973. Distribution and peculiarity of mediterranean ecosystems. See Ref. 51, pp. 11–20. New York: Springer-Verlag. 405 pp.
6. Badri A, Giudicelli J, Prevot G. 1987. Effets d'une crue sur la communauté d'invertébrés benthiques d'une rivière mediterranénne, Le Rdat (Maroc). *Acta Oecologia/Oecologia Gener.* 8:481–500
7. Baltz DM, Moyle PB. 1993. Invasion resistance to introduced species by a native assemblage of California stream fishes. *Ecol. Appl.* 3:246–55
8. Barlocher F, Canhoto C, Graca MAS. 1995. Fungal colonization of alder and eucalypt leaves in two streams in Central Portugal. *Arch. Hydrobiol.* 133:457–70
9. Basaguren A, Pozo J. 1994. Leaf litter processing of alder and eucalyptus in the Agêera stream system (Northern Spain) II. Macroinvertebrates associated. *Arch. Hydrobiol.* 132:57–68
10. Benfield EF. 1997. Comparison of litterfall input to streams. *J. North. Am. Benth. Soc.* 16:104–8
11. Bergey EA, Boettiger CA, Resh VH. 1995. Effects of water velocity on the architecture and epiphytes of *Cladophora glomerata* (Chlorophyta). *J. Phycol.* 31:264–71
12. Bilby RE. 1981. Role of organic debris dams in regulating the export of dissolved and particulate matter from a forested watershed. *Ecology* 62:1234–43
13. Blum LJ. 1956. The ecology of river algae. *Bot. Rev.* 22:291–341
14. Boulton AJ, Lake PS. 1988. Australian temporary streams–some ecological characteristics. *Verh. Internat. Verein Limnol.* 23:1380–83
15. Boulton. AJ. 1989. Over-summering refuges of aquatic macroinvertebrates in two intermittent streams in Central Victoria. *Trans. R. Soc. Aust.* 113:23–24
16. Boulton AJ, Lake PS. 1990. The ecology of two intermittent streams in Victoria, Australia. I. Multivariate analyses of physicochemical features. *Freshwater Biol.* 24:123–41
17. Boulton AJ, Lake PS. 1992. The ecology of two intermittent streams in Victoria, Australia. III. Temporal changes in faunal composition. *Freshwater Biol.* 27:123–38
18. Boulton AJ, Lloyd LN. 1992. Flood frequency and invertebrate emergence from dry floodplain sediments of the River Murray, Australia. *Reg. Riv.* 7:137–51
19. Boulton AJ, Sheldon F, Thoms MC, Stanley EH. 1999. Problems and constraints in managing rivers with contrasting flow regimes. In *Global Perspectives on River Conservation: Science, Policy and Practice*, ed. PJ Boon, BR Davies, GE Petts. London: Wiley. In press
20. Boulton AJ, Stanley EH, Fisher SG, Lake PS. 1992. Over-summering strategies of

macroinvertebrates in intermittent streams in Australia and Arizona. In *Aquatic Ecosystems in Semi-Arid Regions: Implications for Resource Management*, ed. RD Roberts, ML Bothwell, pp. 227–37. Natl. Hydrology Res. Inst., Symp. Ser. 7, Saskatoon, (Canada): Environment Canada. 375 pp.

21. Boulton AJ, Suter PJ. 1986. Ecology of temporary streams–an Australian perspective. In *Limnology in Australia*, ed. P De Decker, WD Williams, pp. 313–27. Melbourne: CSIRO. 671 pp.
22. Britton DL. 1990. Fire and the dynamics of allochthonous detritus in a South African mountain stream. *Freshwater Biol.* 24:347–60
23. Britton DL, Day JA, Henshall-Howard M-P. 1993. Hydrochemical response during storm events in a South African mountain catchment: the influence of antecedent conditions. *Hydrobiologia* 250:143–57
24. Bromley HJ, Por FD. 1975. The metazoan fauna of a sewage-carrying wadi, Nahal Soreq (Judean Hills, Israel). *Freshwater Biol.* 5:121–33
25. Bunn SE. 1986. Origin and fate of organic matter in Australian upland streams. In *Limnology in Australia*, ed. P De Decker, WD Williams, pp. 277–91. Melbourne: CSIRO. 671 pp.
26. Bunn SE. 1986. Spatial and temporal variation in the macroinvertebrate fauna of streams in the northern jarah forest, Western Australia: functional organization. *Freshwater Biol.* 16:621–632
27. Bunn SE. 1988. Processing of leaf litter in a northern jarrah forest stream, Western Australia: I. Seasonal differences. *Hydrobiologia* 162:201–10
28. Bunn SE. 1988. Processing of leaf litter in two northern jarrah forest streams, Western Australia: II. The role of macroinvertebrates and the influence of soluble polyphenols and inorganic sediment. *Hydrobiologia* 162:211–23
29. Bunn SE. 1988c. Life histories of some benthic invertebrates from streams of the northern jarrah forest, Western Australia. *Aust. J. Mar. Freshwater Res.* 39:785–804
30. Bunn SE, Davies PM. 1990. Why is the stream fauna of south-western Australia so impoverished? *Hydrobiologia* 194:169–76
31. Bunn SE, Davies PM, Edwards DH. 1989. The association of *Glacidorbis occidentalis* Bunn and Stoddart 1983 (Gastropoda: Glacidorbidae) with intermittently-flowing, forest streams in south-western Australia. *J. Malac. Soc. Aust.* 10:25–34
32. Bunn SE, Edward DH, Lonegran NR. 1986. Spatial and temporal variation in the macroinvertebrate fauna of streams of the northern jarrah forest, Western Australia: community structure. *Freshwater Biol.* 16:67–91
33. Butturini A, Sabater F. 1998. Ammonium and phosphate retention in a Mediterranean stream: hydrological versus temperature control. *Can. J. Fish. Aquat. Sci.* 55:1938–45
34. Campbell IC, James KR, Hart BT, Devereaux A. 1992. Allochthonous coarse particulate organic material in forest and pasture reaches of two south-eastern Australian streams. I. Litter accession. *Freshwater Biol.* 27:341–52
35. Canhoto C, Graca MAS. 1995. Food value of introduced eucalypt leaves for a Mediterranean stream detritivore: *Tipula lateralis*. *Freshwater Biol.* 34:209–14
36. Chapman LJ, Kramer DL. 1991. The consequences of flooding for the dispersal and fate of poeciliid fish in an intermittent tropical stream. *Oecologia (Berlin)* 87:299–306
37. Charbonneau R, Resh VH. 1992. Strawberry Creek on the University of California, Berkeley campus: a case history of urban stream restoration. *Aquat. Conserv.: Mar. Freshwater Ecosyst.* 2:293–307
38. Closs GP, Lake PS. 1996. Drought differential mortality and the coexistence of a native and an introduced fish species in a

south east Australian intermittent stream. *Environ. Biol. Fishes* 47:17–26
39. Cobb DG, Galloway TD, Flannagan JF. 1992. Effects of discharge and substrate stability on density and species composition of stream insects. *Can. J. Fish. Aquat. Sci.* 49:1788–95
40. Colwell RK. 1974. Predictability, constancy, and contingency of periodic phenomena. *Ecology* 55:1148–53
41. Connell JH. 1978. Diversity in tropical rainforests and coral reefs. *Science* 199: 1302–10
42. Conners ME, Naiman RJ. 1984. Particulate allochthonous inputs: relationships with stream size in an undisturbed watershed. *Can. J. Fish. Aquat. Sci.* 41:1473–84
43. Cooper SD, Dudley TL, Hemphill N. 1986. The biology of chaparral streams in southern California. In *Proceedings of the Chaparral Ecosystems Research Conference*, ed. JJ DeVries, pp. 139–51. *California Water Resources Center Rep. 62*. Univ. Calif., Davis. 155 pp.
44. Cuffney TF. 1988. Input, movement and exchange of organic matter within a subtropical coastal blackwater river-floodplain system. *Freshwater Biol.* 19:305–20
45. Cushing CE, Cummins KW, Minshall GW, eds. 1995. *River and Stream Ecosystems*. Amsterdam: Elsevier. 817 pp.
46. Davies BR, O'Keeffe JH, Snaddon CD. 1993. A synthesis of the ecological functioning, conservation and management of South African river ecosystems. *Water Res. Commission Rep. TT62/93*. South Africa: Pretoria
47. Davies BR, O'Keeffe JH, Snaddon CD. 1995. River and stream ecosystems in Southern Africa: predictably unpredictable. See Ref. 45, pp. 537–99
48. Davies BR, Thoms MC, Walker KF, O'Keeffe JH, Gore JA. 1994. Dryland rivers: their ecology, conservation and management. In *The Rivers Handbook*, Vol. 21, ed. P. Calow, GE Petts, pp. 484–511. Oxford: Blackwell Sci. 523 pp.
49. Dell B, Havel JJ, Malajczuk N, eds. 1989. *The Jarrah Forest: A Complex Mediterranean Ecosystem*. London: Kluwer. 408 pp.
50. diCastri F. 1981. Mediterranean-type shrublands of the world. See Ref. 50a, pp. 1–52.
50a. diCastri F, Goodall DW, Specht RL, eds. 1981. *Mediterranean-Type Shrublands*. Amsterdam: Elsevier. 643 pp.
51. diCastri F, Mooney HA, eds. 1973. *Mediterranean Type Ecosystems. Origin and Structure*. New York: Springer–Verlag. 405 pp.
52. Dor I, Schechter H, Shuval HI. 1976. Biological and chemical succession in Nahal Soreq: a free-flowing wastewater stream. *J. Appl. Ecol.* 13:475–89
53. Dudley TL. 1992. Beneficial effects of herbivores on stream macroalgae via epiphyte removal. *Oikos* 65:121–7
54. Erman DC, Andrew ED, Yoder-Williams M. 1988. Effects of winter floods on fishes in the Sierra Nevada. *Can. J. Fish. Aquat. Sci.* 45:2195–200
55. Everard M. 1996. The importance of periodic droughts for maintaining diversity in the freshwater environment. *Freshwater Forum* 7:33–50
56. Faber PM, Keller E, Sands A, Massey BM. 1989. The ecology of riparian habitats of the Southern California coastal region: a community profile. *US Fish Wildlife Serv. Biol. Rep. 85(7.27)*. Washington: US Dept. Interior. 178 pp.
57. Feminella JW, Resh VH. 1990. Hydrologic influences, disturbance, and intraspecific competition in a stream caddisfly population. *Ecology* 71:2083–94
58. Fisher SG. 1995. Stream ecosystems of the western United States. See Ref. 45, pp. 61–87
59. Fisher SG, Grimm NB. 1988. Disturbance as a determinant of structure in a Sonaran desert stream ecosystem. *Verh. Int. Verein. Limnol.* 23:1183–89
60. Fisher SG, Likens GE. 1973. Energy flow

in Bear Brook, New Hampshire: an integrated approach to stream ecosystem metabolism. *Ecol. Monogr.* 43:421–39
61. Fisher SG, Minckley WL. 1978. Chemical characteristics of a desert stream in flash flood. *J. Arid Environ.* 1:25–33
62. Frenette M, Caron M, Julien P, Gibson RJ. 1984. Interaction entre le débit et les populations de tacons (*Salmo salar*) de la riviére Matamec, Québec. *Can. J. Fish. Aquat. Sci.* 41:954–63
63. Gafny S, Gasith A. 1999. Spatial distribution and temporal dynamics of the epilithon community in the littoral zone of Lake Kinneret, Israel. *Verh. Int. Verein. Limnol.* 27: In press
64. Gallardo A, Prenda J. 1994. Influence of some environmental factors on the freshwater macroinvertebrates distribution in two adjacent river basins under Mediterranean climate. I. Dipteran larvae (excluding chironomids and simuliids) as ecological indicators. *Arch. Hydrobiol.* 131:435–47
65. Gallardo A, Prenda J, Pujante A. 1994. Influence of some environmental factors on the freshwater macroinvertebrates distribution in two adjacent river basins under Mediterranean climate. II. Molluscs. *Arch. Hydrobiol.* 131:449–63
66. Garcia de Jalón D, Gonzalez del Tanago M, Casado C. 1992. Ecology of regulated streams in Spain: an overview. *Limnetica* 8:161–66
67. Gasith A. 1992. Conservation and management of the coastal streams of Israel: an assessment of stream status and prospect of rehabilitation. In *River Conservation and Management*, ed. PJ Boon, P Calow, GE Petts, pp. 51–64. New York: Wiley. 470 pp.
68. Gasith A, Bing M, Raz Y, Goren M. 1998. Fish community parameters as indicators of habitat conditions: the case of the Yarqon, a lowland, polluted stream in a semi-arid region (Israel). *Verh. Int. Verein. Limnol.* 26:1023–26
69. Gasith A, Sidis I. 1984. Polluted water bodies, the main habitat of the Caspian terrapin (*Mauremys caspica rivulata*) in Israel. *Copeia* 1984:216–19
70. Gerecke R. 1991. Taxonimische, faunistische und ekologische Untersuchungen an Wassermilben (Acari, Actinedida) aus Sizilien unter Beruecksichtigung anderer aquatischer. *Lauterbornia* 7:1–303
71. Gleick PH, ed. 1993. *Water in Crisis: A Guide to the World's Freshwater Resources*. New York: Oxford Univ. Press. 473 pp.
72. Grossman GD, Ratajczak RE Jr, Crawford M, Freeman MC. 1998. Assemblage organization in stream fishes: effects of environmental variation and interspecific interactions. *Ecol. Monogr.* 68:395–420
73. Grove AT. 1977. The geography of semiarid lands. *Philos. Trans. R. Soc. Lond, Series B* 278:457–75
74. Guasch H, Sabater S. 1995. Seasonal variations in photosynthesis-irradiance responses by biofilms in Mediterranean streams. *J. Phycol.* 31:727–35
75. Guasch H, Marti E, Sabater S. 1995. Nutrient enrichment effects on biofilm metabolism in a Mediterranean stream. *Freshwater Biol.* 33:373–83
76. Gustard A. 1992. Analysis of river regimes. In *The Rivers Handbook*, Vol. 1, ed. P Calow, GE Petts, pp. 29–47. London: Blackwell Sci. 526 pp.
77. Haines AT, Finlayson BL, McMahon TA. 1988. A global classification of river regimes. *Appl. Geogr.* 8:255–72
78. Haltiner JP, Kondolf GM, Williams PB. 1996. Restoration approaches in California. In *River Channel Restoration: Guiding Principles for Sustainable Projects*, ed. A Brookes, FD Shields Jr., pp. 291–329. London: Wiley. 433 pp.
79. Hanson BJ, Cummins KW, Barnes JR, Carter MW. 1984. Leaf litter processing in aquatic systems: a two variable model. *Hydrobiologia* 111:21–9
80. Haslam SM. 1997. Deterioration and

fragmentation of rivers in Malta. *Freshwater Forum* 9:55–61

81. Hemphill N, Cooper SD. 1983. The effect of physical disturbance on the relative abundances of two filter-feeding insects in a small stream. *Oecologia (Berlin)* 58:378–82
82. Herbst GN, Mienis HK. 1985. Aquatic invertebrate distribution in Nahal Tanninim, Israel. *Isr. J. Zool.* 33:51–62
83. Herrera CM. 1995. Plant-vertebrate seed dispersal systems in the Mediterranean: ecological, evolutionary, and historical determinants. *Annu. Rev. Ecol. Syst.* 26:705–27
84. Hobbs RJ, Richardson DM, Davis GW. 1995. Mediterranean-type ecosystems: opportunities and constraints for studying the function of biodiversity. In *Mediterranean-Type Ecosystems, The Function of Biodiversity*, ed. GW Davis, DM Richardson, pp. 1–42. Berlin: Springer–Verlag. 366 pp.
85. Hodgkin EP, Lenanton RCJ. 1981. Estuaries and coastal lagoons of southwestern Australia. In *Estuaries and Nutrients*, ed. BJ Nielson, LE Cronin, pp. 307–21. New Jersey: Humana. 643 pp
86. Holstein G. 1984. California riparian forests: deciduous islands in an evergreen sea. In *California Riparian Systems*, ed. RE Warner, KM Hendrix, pp. 2–22. Berkeley: Univ. Calif. Press. 1035 pp.
87. Horowitz RJ. 1978. Temporal variability patterns and the distributional patterns of stream fishes. *Ecol. Monogr.* 48:307–21
88. Hynes HBN. 1970. *The Ecology of Running Waters*. Toronto: Univ. Toronto Press. 555 pp.
89. Illies J, ed. 1978. *Limnofauna Europea*. Amsterdam: Swets en Zeitlinger. 532 pp.
90. Keller EA. 1971. Areal sorting of bed-load material: the hypothesis of velocity reversal. *Geol. Soc. Am. Bull.* 82:753–56
91. King JM, Day JA, Davies BR, Henshall-Howard M-P. 1987. Particulate organic matter in a mountain stream in the south-western Cape, South Africa. *Hydrobiologia* 154:165–87
92. King JM, Henshall-Howard MP, Day JA, Davies BR. 1987. Leaf-pack dynamics in a Southern African mountain stream. *Freshwater Biol.* 18:325–40
93. King JM, Day JA, Hurly PR, Henshall-Howard M-P, Davies BR. 1988. Macroinvertebrate communities and environment in a Southern African mountain stream. *Can. J. Fish. Aquat. Sci.* 45:2168–81
94. Koetsier P III, McArthur JV, Leff LG. 1997. Spatial and temporal response of stream bacteria to sources of dissolved organic carbon in a blackwater stream system. *Freshwater Biol.* 37:79–89
95. Kondolf GM, Curry RR. 1986. Channel erosion along the Carmel River, Monterey County, California. *Earth Surf. Proc. Landforms* 11:307–19
96. Korzoun VI, ed. 1977. *Atlas of World Water Balance*. Leningrad: Hydrometeorological Pub. House. 34 pp.
97. Lamberti GA, Feminella JW, Resh VH. 1987. Herbivory and intraspecific competition in a stream caddisfly population. *Oecologia* 73:75–81
98. Lancaster J, Hildrew AG. 1993. Flow refugia and microdistribution of lotic macroinvertebrates. *J. North Am. Benth. Soc.* 12: 385–93
99. Larimore RW, Childers WF, Heckrotte C. 1959. Destruction and re-establishment of stream fish and invertebrates affected by drought. *Trans. Am. Fish. Soc.* 88:261–85
100. LeHouórou HN. 1990. Global change: vegetation, ecosystems, and land use in the southern Mediterranean Basin by the mid twenty-first century. *Isr. J. Bot.* 39:481–508
101. Lobón-Cerviaé J. 1996. Response of a stream fish assemblage to a severe spate in Northern Spain. *Trans. Am. Fish. Soc.* 125:913–19
102. Lulla K. 1987. Mediterranean climate. In *Encyclopedia of Climatology*, ed.

JE Oliver, RW Fairbridge, pp. 569–71. New York: Van Nostrand Reinhold. 986 pp.

103. Maamri A, Chergui H, Pattee E. 1994. Allochthonous input of coarse particulate organic matter to a Moroccan mountain stream. *Acta Oecologia* 15:495–508
104. Malicky H. 1987. The miraculous island of Serifos–one possible key to understand the evolution of Mediterranean stream ecosystems. *Biologia Gallo-hellenica* 13: 43–46
105. Marti E, Sabater F. 1996. High variability in temporal and spatial nutrient retention in Mediterranean streams. *Ecology* 77:854–69
106. Matthews WJ. 1986. Fish faunal structure in an Ozark stream: stability, persistence and a catastrophic flood. *Copeia* 1986:388–97
107. McElravy EP, Lamberti GA, Resh VH. 1989. Year-to-year variation in the aquatic macroinvertebrate fauna of a northern California stream. *J. North Am. Benth. Soc.* 8:51–63
108. McElravy EP, Resh VH. 1991. Distribution and seasonal occurrence of the hyporheic fauna in a northern California stream. *Hydrobiologia* 220:233–46
109. Meffe GK. 1984. Effects of abiotic disturbance on coexistence of predator-prey fish species. *Ecology* 65:1525–34
110. Minshall GW, Cummins KW, Petersen RC, Cushing CE, Bruns DA, Sedell JR, Vannote RL. 1985. Developments in stream ecosystem theory. *Can. J. Fish. Aquat. Sci.* 42:1045–55
111. Minshall GW, Petersen RC, Cummins KW, Bott TL, Sedell JR, et al. 1983. Interbiome comparison of stream ecosystem dynamics. *Ecol. Monogr.* 53:1–25
112. Moser H. 1991. Input of organic matter (OM) in a low order stream (Ritrodat-Lunz study area, Austria). *Verh. Int. Verein. Limnol.* 24:1913–16
113. Moyle PB. 1995. Conservation of native freshwater fishes in the mediterranean-type climate of California, USA: a review. *Biol. Conserv.* 72:271–79
114. Moyle PB, Herbold B, Stevens DE, Miller LW. 1992. Life history and status of delta smelt in the Sacramento-San Joaquin estuary, California. *Trans. Am. Fish. Soc.* 121:67–77
115. Moyle PB, Light T. 1996. Fish invasions in California: Do abiotic factors determine success? *Ecology* 77:1666–70
116. Moyle PB, Marchetti MP, Baldrige J, Taylor TL. 1998. Fish health and diversity: justifying flows for a California stream. *Fisheries Manage.* 23:7:6–15
117. Moyle PB, Smith JJ, Daniels RA, Taylor TL, Price DG, Baltz DM. 1982. Distribution and ecology of stream fishes of the Sacramento-San Joaquin drainage system, California. *Univ. Calif. Publ. in Zool.* 115:1–256
118. Moyle PB, Vondracek B. 1985. Persistence and structure of the fish assemblage in a small California stream. *Ecology* 66:1–13
119. Nahal I. 1981. The mediterranean climate from a biological viewpoint. See Ref. 50a, pp. 63–86
120. Naiman RJ, Dócamps H. 1997. The ecology of interfaces: riparian zones. *Annu. Rev. Ecol. Syst.* 28:621–58.
121. Naiman RJ, Fetherston KL, McKay S, Chen J. 1998. Riparian forests. In *River Ecology and Management: Lessons from the Pacific Coastal Ecoregion*, ed. RJ Naiman, RE Bilby, pp. 289–323. New York: Springer-Verlag. 705 pp.
122. Narasimhan TN, Quinn NWT. 1996. Agriculture, irrigation and drainage on the west side of the San Joaquin Valley, California: unified perspective on hydrogeology, geochemistry and management. *Lawrence Berkeley Lab., Univ. Calif. LBL Rep. 38498.* 86 pp.
123. National Research Council (NRC). 1992. Rivers and streams. In *Restoration of Aquatic Ecosystems*, pp. 165–260. Washington, DC: Natl. Acad. Press. 552 pp.

124. Pearsons TN, Lee HW, Lamberti GA. 1992. Influence of habitat complexity on resistance to flooding and resilience of stream fish assemblages. *Trans. Am. Fish. Soc.* 121:427–36
125. Poff NL, Allan JD, Bain MB, Karr JR, Prestegaard KL, et al. 1997. The natural flow regime. *BioScience* 47:769–84
126. Poff NL, Ward JV. 1989. Implications of streamflow variability and predictability for lotic community structure: a regional analysis of streamflow patterns. *Can. J. Fish. Aquat. Sci.* 46:1805–17
127. Power ME. 1990a. Benthic turfs vs floating mats of algae in river food webs. *Oikos* 58:67–79
128. Power ME. 1990b. Effects of fish in river food webs. *Science* 250:811–4
129. Power ME. 1992. Hydrologic and trophic controls of seasonal algal blooms in northern California rivers. *Arch. Hydrobiol.* 125:385–410
130. Power ME. 1995. Floods, food chains, and ecosystem processes in rivers. In *Linking Species and Ecosystems*, ed. CG Jones, JH Lawton, pp. 52–60. New York: Chapman & Hall. 387 pp.
131. Power ME, Stout RJ, Cushing CE, Harper PP, Hauer FR. et al. 1988. Biotic and abiotic controls in river and stream communities. *J. North Am. Benth. Soc.* 7:456–79
132. Prat N, Ward JV. 1994. The tamed river. In *Liminology Now: A Paradigm of Planetary Problems*, ed. R. Margalef, pp. 219–36. Amsterdam: Elsevier. 553 pp.
133. Prochazka K, Stewart BA, Davies BR. 1991. Leaf litter retention and its implications for shredder distribution in two headwater streams. *Arch. Hydrobiol.* 120:315–25
134. Puckridge JT, Sheldon F, Walker KF, Boulton AJ. 1998. Flow variability and the ecology of large rivers. *Mar. Freshwater Res.* 49:55–72
135. Pusey BJ. 1990. Seasonality, aestivation and the life history of the salamanderfish *Lepidogalaxias salamandroides* (Pisces: Lepidogalaxiidae). *Environ. Biol. Fishes* 29:15–26
136. Ractliffe G, Davies BR, Stewart BA, Snaddon CD. 1995. The influence of discharge on entrainment of bank litter in a headwater stream. *Arch. Hydrobiol.* 134:103–17
137. Resh VH. 1982. Age structure alteration in a caddisfly population after habitat loss and recovery. *Oikos* 38:280–84
138. Resh VH. 1992. Year-to-year changes in the age structure of a caddisfly population following loss and recovery of a springbrook habitat. *Ecography* 15:314–17
139. Resh VH, Brown AV, Covich AP, Gurtz ME, Li HW, et al. 1988. The role of disturbance in stream ecology. *J. North Am. Benth. Soc.* 7:433–55
140. Resh VH, Jackson JK, McElravy EP. 1990. Disturbance, annual variability, and lotic benthos: examples from a California stream influenced by a mediterranean climate. *Mem. Ist. Ital. Idrobiol.* 47:309–29
141. Resh VH, Solem JO. 1996. Phylogenetic relationships and evolutionary adaptations of aquatic insects. In *An Introduction to the Aquatic Insects of North America*, ed. RW Merritt, KW Cummins, pp. 98–107. Dubuque, IA: Kendall/Hunt. 862 pp.
142. Resh VH, Wood JR, Bergey EA, Feminella JW, Jackson JK, McElravy EP. 1997. Biology of *Gumaga nigricula* (McL.) in a northern California stream. In *Proc. 8th Int. Symp. Trichoptera*, ed. RW Holzenthal, OS Flint Jr, pp. 401–10. Columbus: Ohio Biol. Survey. 496 pp.
143. Riley AL. 1998. *Restoring Streams in Cities*. Washington, DC: Island Press. 423 pp.
144. Rodriquez-Ruiz A, Granada-Lorencho C. 1992. Spawning period and migration of three species of cyprinids in a stream with Mediterranean regimen (SW Spain). *J. Fish Biol.* 41:545–56
145. Sabater F, Guasch H, Marti E, Armengol J, Sabater S. 1995. The River Ter: a

Mediterranean river case-study in Spain. See Ref. 45, pp. 419–38

146. Sabater S, Sabater F. 1992. Longitudinal changes of benthic algal biomass in a Mediterranean river during two high production periods. *Arch. Hydrobiol.* 124:475–87
147. Schlosser IJ. 1987. A conceptual framework for fish communities in small warmwater streams. In *Community Evolutionary Ecology of North American Fishes*, ed. WJ Mathews, DC Heins, pp. 17–24. Norman: Univ. Okla. Press. 310 pp.
148. Schofield NJ, Stoneman GL, Loh IC. 1989. Hydrology of the jarrah forest. In *The Jarrah Forest, A Complex Mediterranean Ecosystem*, ed. B Dell, JJ Havel, N Malajczuk, pp. 179–201. London: Kluwer. 408 pp.
149. Sedell JR, Naiman RJ, Cummins KW, Minshall GW, Vannote RL. 1978. Transport of particulate organic material in streams as a function of physical processes. *Verh. Int. Verein. Limnol.* 20:1366–75
150. Shmida A. 1981. Mediterranean vegetation in California and Israel: similarities and differences. *Isr. J. Bot.* 30:105–23
151. Shure DJ, Gottschalk MR. 1985. Litterfall patterns within a floodplain forest. *Am. Midl. Nat.* 114:98–111
152. Smith RW, Pearson RG. 1987. The macroinvertebrate communities of temporary pools in an intermittent stream in tropical Queensland. *Hydrobiologia* 150:45–61
153. Snaddon CD, Stewart BA, Davies BR. 1992. The effect of discharge on leaf retention in two headwater streams. *Arch. Hydrobiol.* 125:109–20
154. Speaker RW, Moore K, Gregory S. 1984. Analysis of the process of retention of organic matter in stream ecosystems. *Verh. Int. Verein. Limnol.* 22:1835–41
155. Specht RL. 1973. Structure and functional response of ecosystems in the mediterranean climate of Australia. See Ref. 51, pp. 113–120
156. Specht RL, ed. 1979. *Heathlands and Related Shrublands*, Part A. Amsterdam: Elsevier. 497 pp.
157. Specht RL, ed. 1981. *Heathlands and Related Shrublands*, Part B. Amsterdam: Elsevier. 383 pp.
158. Specht RL. 1981. Primary production in mediterranean-climate ecosystems regenerating after fire. See Ref. 50a, pp. 257–68
159. Stanley EH, Buschman DL, Boulton AJ, Grimm NB, Fisher SG. 1994. Invertebrate resistance and resilience to intermittency in a desert stream. *Am. Midl. Nat.* 131:288–300
160. Stanley EH, Fisher SG, Grimm NB. 1997. Ecosystem expansion and contraction in streams. *BioScience* 47:427–35
161. Statzner B, Gore GA, Resh VH. 1988. Hydraulic stream ecology: observed patterns and potential applications. *J. North Am. Benth. Soc.* 7:307–60
162. Stewart BA. 1992. The effect of invertebrates on leaf decomposition rates in two small woodland streams in southern Africa. *Arch. Hydrobiol.* 124:19–33
163. Stewart BA, Davies BR. 1990. Allochthonous input and retention in a small mountain stream, South Africa. *Hydrobiologia* 202:135–46
164. Storey AW, Bunn SE, Davies PM, Edward DH. 1990. Classification of the macroinvertebrate fauna of two river systems in southwestern Australia in relation to physical and chemical parameters. *Reg. Riv.* 5:217–32
165. Strange EM, Moyle PB, Foin TC. 1992. Interactions between stochastic and deterministic processes in stream fish community assembly. *Environ. Biol. Fishes* 36: 1–15
166. Suberkropp KM, Klug MJ, Cummins KW. 1975. Community processing of leaf litter in woodland streams. *Verh. Int. Verein. Limnol.* 19:1653–58

167. Suc JP. 1984. Origin and evolution of the Mediterranean vegetation and climate in Europe. *Nature* 307:429–32
168. Trabaud L. 1981. Man and fires: impact on Mediterranean vegetation. See Ref. 50a, pp. 523–38
169. Tuch A, Gasith A. 1989. Effects of an upland impoundment on structural and functional properties of a small stream in a basaltic plateau (Golan Heights, Israel). *Reg. Riv.* 3:153–67
170. Unesco. 1979. *Map of the World Distribution of Arid Regions*. MAB Technical Note 7. Paris: Unesco. 56 pp.
171. Uys MC, O'Keeffe JH. 1997. Simple words and fuzzy zones: early directions for temporary river research in South Africa. *Environ. Manage.* 4:517–31
172. Vannote RL, Minshall GW, Cummins KW, Sedell JR, Cushing CE. 1980. The river continuum concept. *Can. J. Fish. Aquat. Sci.* 37:130–37
173. Vidal-Abarca MR, Suarez ML, Ramírez-Díaz L. 1992. Ecology of spanish semiarid streams. *Limnetica* 8:151–60
174. Wallace JB, Eggert SL, Meyer JL, Webster JR. 1997. Multiple trophic levels of a forest stream linked to terrestrial litter input. *Science* 277:102–47:253–72
175. Ward JV, Garcia de Jalon D. 1991. Ephemeroptera of regulated mountain streams in Spain and Colorado. In *Overview and Strategies of Ephemeroptera and Plecoptera*, ed. J. Alba-Tercedor, A. Sanchez-Ortega, pp. 567–78. Gainesville, FL: Sandhill. 588 pp.
176. Ward JV, Stanford JA. 1983. The serial discontinuity concept of lotic ecosystems. In *Dynamics of Lotic Ecosystems*, ed. TD Fontaine, SM Bartell, pp. 29–43. Ann Arbor, MI: Ann Arbor Sci. 494 pp.
177. Waters TF. 1972. The drift of stream insects. *Annu. Rev. Entomol.* 17:253–72
178. Whitton BA. 1970. Biology of *Cladophora* in freshwaters. *Water Res.* 4:457–76
179. Williams DD. 1996. Environmental constraints in temporary fresh waters and their consequences for the insect fauna. *J. North. Am. Benth. Soc.* 15:634–50
180. World Resources Institute (WRI). 1996–1997. *World Resources*. New York: Oxford Univ. Press. 365 pp.
181. Yoshiyama RM, Fisher FW, Moyle PB. 1998. Historical abundance and decline of Chinook Salmon in the Central Valley region of California. *N. Am. J. Fish Manage.* 18:487–521

Annu. Rev. Ecol. Syst. 1999. 30:83–108

Choosing the Appropriate Scale of Reserves for Conservation

Mark W. Schwartz
Department of Environmental Science and Policy, University of California, Davis, CA 95616; e-mail: mwschwartz@ucdavis.edu

Key Words Endangered Species Act, ecosystem management, reserve selection, fine-filter, coarse-filter

■ **Abstract** Over the past ten years the scientific basis for reserve selection and design have rapidly developed. This period has also been characterized by a shift in emphasis toward large spatial and organizational scales of conservation efforts. I discuss the evidence in support of this shift toward larger scale conservation by contrasting the success of fine-filter (genes, populations, species) conservation and coarse-filter (communities, habitats, ecosystems, landscapes) conservation. Conservation at both organizational scales has been successful and merits continued support, although fine-filter conservation is more straightforward. Ecological theory suggests that conservation at large scales is preferred. Despite this preference, both fine- and coarse-filter conservation objectives have been met by small reserves. In many landscapes there are no opportunities for the conservation of native species diversity that encompass a large spatial scale. Thus, reserve selection at any organizational scale may include conservation at a variety of spatial scales. A variety of methods have been suggested that integrate across scales of conservation. Some, such as umbrella, flagship, and indicator species, remain very problematic. Reserve selection algorithms and gap analyses, in contrast, offer promising opportunities to increase the efficiency of conservation at all scales.

INTRODUCTION

Conservation is limited by a small pool of resources directed toward a large and not entirely attainable goal: saving global representation of all unique populations, species, communities, and ecosystems within their natural context. Maximizing efficiency in the protection of biological diversity is critical. Conservation actions are constrained by past losses of biotic resources and prioritized by threats to remaining resources. Exactly how to prioritize threatened natural resources for protection is an area of particular interest to conservation biologists. A large body of recent conservation literature has attempted to center protection strategies on large scales (19, 27, 30, 35, 102, 116). This emphasis on large-scale conservation is exemplified by the recent adoption of ecosystem-based management policies by

0066-4162/99/1120-0083$08.00

18 US federal agencies (19, 26, 44). The US Fish and Wildlife Service, the agency principally responsible for implementing the Endangered Species Act, has typified this shift in emphasis by adopting an approach of ecosystem-based management (5, 6). My goal is to review issues of organizational (i.e., species, ecosystems) and spatial scale (reserve size) with respect to the protection of biological diversity. In particular, I assess the degree to which the emphasis on large scales is supported by scientific evidence.

Targets of conservation efforts range from genes, populations, and species to communities, habitats, ecosystems, and landscapes (60, 87). I use the term "fine-filter" to refer to conservation efforts directed at genes, populations, or species (sensu The Nature Conservancy; 97, 102). The term "coarse-filter" (sensu The Nature Conservancy) is used to refer to conservation efforts aimed at communities, ecosystems, or landscapes. This terminology is useful because it obviates the often confusing distinction between communities and ecosystems, both of which are often described by their dominant vegetation. For example, Wisconsin (152) and Illinois (58) have focused on describing communities for conservation purposes using dominant vegetation, while Noss et al (104) uses many of the same descriptors to describe endangered ecosystems.

During this review I highlight several points regarding the contrast between fine- and coarse-filter conservation. First, coarse-filter conservation, although scientifically appealing for a variety of reasons, is made difficult by a lack of general and objective measures of success or failure. Resolving this difficulty may be the greatest challenge for conservation biology during the coming decade. Second, conservation planning often inappropriately equates priorities for coarse-filter conservation of ecosystems and those for fine-filter conservation of large vertebrate species. Conservation programs ought to adopt approaches that distinguish, but also incorporate, concerns at both scales. Third, focusing protection on large sites, to exclusion of small sites, entails considerable sacrifices with respect to capturing diversity. A large share of diversity is restricted to sites where large reserves are not an option. If a conservation target is solely embedded within a human-dominated landscape, which they frequently are, then small reserves may be the only protection option. Finally, I argue that conservation actions at any organizational level may appropriately entail small or large reserves.

FINE-FILTER CONSERVATION

Any conservation target that is protected must have attributes that can be quantified in order to measure the success or failure of management. Although there are multiple rationales for fine-filter conservation, the simplest is this: Populations and species must be saved from extinction in order to preserve biotic diversity. The simple objective of preventing extinction makes evaluation of success straightforward. Fine-filter targets are identifiable: If a species can be documented as rare within a state or country, then it is an appropriate conservation target. Formal programs exist for identifying species at risk. International conservation organizations identify species as vulnerable, threatened, or endangered (62, 171). Mace & Lande (78) have incorporated quantitative measures of threat into these categories. Additional work continues to fine-tune these generally accepted threat categories (67, 149). Using these widely accepted criteria, Red Data books provide international lists of vulnerable, threatened, and endangered taxa (24, 62, 171). Within the United States and Canada, state-based Natural Heritage programs classify threat by the number of extant populations and the degree of threat to those populations to track threatened species (83).

This is not to assert that all target identifications are unambiguous. Subspecies, isolated populations, and hybrids, in particular, present problems. For the purposes of federal listing of endangered species (93), the rule for specifying targets of fine-filter conservation is that they be evolutionary significant units (ESUs). An ESU is a lineage that is evolutionarily isolated (93). With recent advances in molecular genetics, techniques exist with which to measure the degree of genetic isolation of proposed targets (93). For example, the Florida panther (*Felis concolor coryi*) has evidence of introgression from South American genotypes (105). Despite this introgression, the Florida panther retains its designation as a distinct ESU because of its isolation from remaining other *Felis concolor*. Molecular taxonomy has confirmed that the red wolf (*Canis rufus*) is a hybrid between the gray wolf (*C. lupus*) and coyote (*C. latrans*). With unclear directives regarding hybrids, it is not clear how this evidence will affect listing for the red wolf (13).

Another advantage of fine-filter conservation is that evaluation is clear: Persistence is success, extinction is failure. Populations can be monitored from year to year to estimate critical demographic parameters. Demographic data can be used to predict population trajectories and estimate population viability (150). Although a thorough viability analysis frequently requires more demographic data than are available (45), the methods are well developed (9, 146, 166). Despite sophisticated techniques for estimating risk, setting the threshold level of acceptable extinction risk (e.g., 90% probability of survival for 100 years) is subjective (93). Although scientists may debate the choice of threshold levels, assigning an acceptable level of extinction risk for recovery remains an agency decision that may negotiated upon by stakeholders.

Finally, fine-filter programs have a history of legislated protection. Within the United States, the Endangered Species Act (ESA), designed to protect species and their critical habitat, has been called the strongest conservation legislation ever enacted (4, 111). Passed in 1973 and later amended, the ESA provides for the protection and recovery of endangered species and their habitats (93). Nonetheless, the ESA is a political tool and subject to variable application. A recent spate of lawsuits has found that the US Fish and Wildlife Service (USFWS) and Department of Interior have acted in an arbitrary and capricious manner in not taking listing action on behalf of certain endangered species (144).

Despite the considerable number of potential advantages that fine-filter conservation programs hold, they have often been cited as fatally flawed (e.g., 127, 165,

170). One reason is that the problem of conserving biodiversity species by species is simply too large. As of August 1998 a total of 1143 species of vertebrates, plants, and invertebrates appeared on the ESA list (168), yet the IUCN (171) lists over 4500 species of plants alone that are at risk of extinction within the United States. Given that arthropod diversity far outstrips plant diversity, and that plants at risk outnumber all listed species 4 to 1, the ESA can only be considered a partial listing based on extinction risk and policy priorities.

Once a species is listed, funds are often insufficient to implement protection measures. Expenditures of the US Fish and Wildlife Service on behalf of endangered species in 1995 were distributed such that most species received no discernible funding for recovery actions, while the top ten funded species received more than 90% of all funds (S Johnson, USFWS—168a).

Since its inception, the ESA has resulted in the recovery and delisting of five species within the United States (168). Past recoveries are principally a result of two measures: alleviating predation pressure and reducing pollutants that cause reproductive failure. It is much more difficult to recover species where habitat degradation and loss are the primary causes of endangerment. Yet, habitat loss and degradation are the most frequent causes of species endangerment (39, 178).

The lack of recovery, however, only serves to raise the question of whether recovery ought to be the benchmark of success or failure of the ESA (21). There are clear weaknesses in both legislation and implementation of the ESA (21, 47, 63, 93, 127, 128). These include: 1. undue attention to high profile species (47); 2. insufficient protection of critical habitat (21, 127, 128); 3. a lack of critical data with which to construct a recovery plan (127, 132); 4. insufficient funding to implement recovery actions (93); 5. insufficient attention to interagency cooperation (21); 6. inappropriate attention to future uncertainty in population size (21, 47, 93, 127); and 7. delaying of listing actions until populations are at critically low numbers (21). An example of this final problem is *Pritchardia munroi*, a Hawaiian palm listed in 1992. At the time of listing this species was known from a single individual whose entire seed crop is devastated each year by nonnative predators (37, 156). Extinction of this species in the wild seems inevitable.

If species are not listed until they face a palpable likelihood of extinction, then the abeyance of extinction may be a better benchmark of success. Since 1973 only seven listed species within the United States have gone extinct (168; tecopa pupfish *Cyprinidon nevadensis calidae*, longjaw cisco *Coregonus alpenae*, blue pike *Stizostedion vitreum glaucum*, Santa Barbara song sparrow *Melospiza melodia graminae*, Sampson's pearly mussel *Epioblasma sampsoni*, Amistad gambusia *Gambusia amistadensis*, dusky seaside sparrow *Ammodramus maritimus nigrescens*). Given an average population size among listed species of 1075 for vertebrates, 999 for plants, and 120 for invertebrates (177), we might expect that more species would have gone extinct during this time. Without quantitative measures of extinction risk for each endangered species, it is impossible to calculate an exact number of expected extinctions, but consider the following: Mace (77) analyzed quantitative measures of extinction risk to predict that 100% of "critical" and 67% of "endangered" species are likely to go extinct within the next 100 years. Although Mace's criteria for "critical" and "endangered" species are not the same as under the ESA, quantification of extinction risk has been shown to increase, not decrease, the number of species in these high risk categories (76, 137). Using this estimate, we can adopt a few simplifying assumptions to predict how many species ought to have gone extinct between the inception of the ESA in 1973 and 1998. I make the conservative assumption that all endangered species fit Mace's endangered category (none are critical). I also assume that extinction is a stochastic process and that the suite of 1143 currently listed endangered species have actually been endangered since 1973. Under these conditions, one expects that during the past 25 years (1143 species $\times$ .67 species $\times$ 25/100 years =) 192 listed species should have gone extinct. Despite the simplicity of these assumptions, I assert that the seven observed extinctions represents a significant benchmark of success of the ESA.

The difficulty with fine-filter protection, however, rests in the fact that organisms require habitat. Habitat protection may involve multiple species, stakeholders, and habitats. Habitat Conservation Plans (HCP's) are currently the principal tool for both protection planning and issuing take permits for listed species under the ESA (93). The HCP process grew out of 1982 amendments to the ESA (93) and is an integrated plan intended to provide sufficient habitat for the long-term persistence of endangered species within a region or ecosystem. Early reviews of HCPs are mixed. The National Research Council (93) claims that the San Bruno Mountains (California) HCP has successfully protected resources, while Shilling (141) finds that the Yolo County (California) HCP principally provides a mechanism to avoid protection of critical species. Mann & Plummer (79) assert that HCPs are costly and inefficient and should be avoided. Further reviews of HCPs are needed to assess their success.

COARSE-FILTER CONSERVATION

Coarse-filter conservation has several distinct conceptual advantages over fine-filter conservation. First, coarse-filter conservation seeks to preserve not just targeted species and their immediate habitats, but also potentially important ecosystem linkages and processes (e.g., 116). These processes include attributes that vary over the short-term, such as nutrient flux and primary productivity, as well as longer term processes, such as natural disturbance regimes, soil development, or natural selection and evolution (88). Second, coarse-filter conservation can preemptively protect resources before they become critically endangered (19, 40, 102). In this sense, coarse-filter conservation actions may be more efficient by capturing more diversity for each action. Third, conservation arguably ought to extend beyond the preservation of species diversity and natural processes to include a broader array of environmental concerns, such as sustainable harvest of species and the sustainability of ecosystems within human-altered landscapes (19).

Despite the appealing logical motivation for coarse-filter conservation, implementation is not a simple task. A variety of attributes of ecosystem value fall under the banner of coarse-filter conservation. The relative conservation value of ecosystems may be assessed by (*a*) endangerment of ecosystems (104); (*b*) hotspots of species diversity within ecoregions (2, 33, 121, 124, 169); (*c*) representation among habitat types (46, 58); or (*d*) critical ecosystem functions (30, 88). It is the multitude of valued attributes of ecosystems that make coarse-filter conservation difficult. There is no a priori basis upon which to prioritize one set of attributes over another. For example, do we value endangered habitats more or less than diverse areas? Is function more or less important than diversity? With unclear priorities, it follows that measures of success for coarse-filter conservation are also unclear.

A healthy, well-maintained ecosystem changes through time (19, 51, 88). For example, Meyer (88) defines natural ecosystems as open, temporally varying, spatially heterogeneous systems with flux and linkages across boundaries. These ecosystems are dependent upon both direct and indirect effects among species and have functions that depend on species diversity (88). A number of studies, recognizing the dynamic nature of ecosystems, attempt to identify critical attributes with respect to ecosystem management. For example, Christensen et al (19) recommend that the focus of ecosystem management goals should be on ecosystem trajectories and behavior. Similarly, Norton (95) suggested five axioms of environmental management for ecosystem health: 1. ecosystems are dynamic objects with fluxes; 2. ecosystem processes are interrelated; 3. ecosystems are hierarchically arranged with respect to time and space; 4. ecosystems are self-organizing and self-maintaining; and 5. some ecosystems are more fragile than others.

I use a "parking lot" analogy for evaluating suggested rules for ecosystem management. Without exception, a parking lot fits the aforementioned descriptions of functioning ecosystems: there is flux, species diversity, hierarchical arrangement, temporal and spatial variation, stability and variable fragility. If one can describe a parking lot as a healthy ecosystem without violating any major premise of ecosystem rules, then I argue that the rules are not very helpful. The obvious difference between parking lots and natural ecosystems is in the magnitude of the attributes. Specifying appropriate magnitude, however, may be difficult. For example, natural processes of allochthonous and autochthonous sedimentation cause lakes to become marshes or peatlands through time. If this process takes thousands of years, we call it succession. If this process takes decades as a result of anthropogenic increases in nutrient and sediment load, we call it environmental degradation. Given that we have quantitatively studied most natural processes for less than a century, we are not equipped to provide exact estimates of natural rates of variability or acceptable rates of change for many ecosystem processes affected by humans. Thus, we use a combination of observations, models, and theory to make predictions of natural rates. Nonetheless, criteria for success or failure are sometimes a matter of uncomfortably broad interpretation.

Therein lies the principal challenge of all coarse-filter strategies. The magnitude of critical variables depends on the community, habitat, or ecosystem in question. A desert will have different critical attributes than a lake, forest, prairie, or ocean. Further, every particular geographic location carries unique attributes of size, initial condition, and context within its landscape such that no two deserts are likely to have the same expected attribute values. Since ecosystems vary, their evaluation criteria pose a multivariate problem with few objective guidelines (19).

Lacking general objective criteria, evaluation is an issue with ecosystem management. Brunner & Clark (14) evaluated three approaches to improving ecosystem-based management and concluded that neither clarification of general goals nor creating a better scientific foundation were necessary for making better decisions regarding ecosystem management. Instead, they advocate a practice-based approach that stresses societal and practical considerations along with science when making management decisions. Similarly, Grumbine (48) stresses the need to integrate social and scientific needs into management. Buzzwords of ecosystem management include "contextual thinking," "adaptive management," and "ecological integrity" (48). These are all relative concepts that enhance management flexibility and potential but also inhibit accountability of management actions. With no general rules we risk the criticism that every project is claimed a success even if it results in environmental degradation or diversity losses. If all management constitutes successful management, then ecosystem management is destined for a short period of utility. Despite the need for flexibility, specific benchmarks are required to evaluate specific management actions.

One way to alleviate uncertainty in defining appropriate goals is to use reference ecosystems as benchmarks for management (e.g., 107, 175). Most ecosystems provide some sort of background historical descriptions from which to benchmark. Nonetheless, these are likely to be inadequate in most cases. For example, longleaf pine (*Pinus palustris*) ecosystems of the southeast are a conservation priority because they are rich in endangered species and are a critically endangered ecosystem (104). In particular, the federally listed red-cockaded woodpecker (*Picoides borealis*) is dependent upon mature pine trees for nesting sites, and longleaf pine is a preferred species (36, 72). Perhaps as a result of this conservation concern, longleaf pine savannah is a particularly well-studied habitat. Longleaf pine dynamics in response to fire have been described using the old growth Wade Tract in southern Georgia (42, 118). Noel et al (94) compared longleaf pine forest structure across the Gulf Coast with the Wade Tract and found the latter to be structurally different from other stands. Although structural differences can be largely attributed to stand age, we cannot predict what proportion of these differences are driven by location. We know something, albeit considerably less, about arthropod diversity (50) and nutrient dynamics in response to fire (180). Less than 2% of the original 33 million ha of longleaf forest remain, and less than 0.5% of what is left is old growth (86, 104). As a result, longleaf pine conservation and restoration from Virginia to Texas will likely draw upon experiences gained from the Wade Tract. With few intact old-growth stands, we cannot estimate the variation in vital processes among complete longleaf pine savanna ecosystems, and we know little regarding the appropriateness of this model for longleaf pine conservation.

Despite these many uncertainties, the abstract description of the problem may overstate actual difficulties. General rules have already been enunciated (88, 95). The chief objective for conservation biologists may be to fit attribute values of specific cases into these guidelines. Aplet & Keeton (1) define the "historic range of variability" (HRV) as the bounds of ecosystem structure and functions prior to disturbance by post-Columbian humans. Recognizing the unique nature of ecosystems, Aplet & Keeton suggest establishing ecosystem-specific HRV values as the principal guideline for ecosystem management. Historic range of variability, and similar constructs, encourage parameterization of critical ecosystem values and suggest management guidelines.

SPATIAL SCALE

Increasing emphasis on larger reserves has been argued from the perspective of both coarse- and fine-filter approaches. With respect to coarse-filter targets, large reserves have the ability to capture a mosaic landscape of different habitats as well as habitats in different stages of maturity or succession (102, 117, 143). Biological reasons to support large reserve size on behalf of fine-filter targets are that large reserves (*a*) have the unique ability to protect species with large habitat requirements (22, 65, 173); (*b*) minimize negative impacts of reserve borders and edge effects (71, 81, 84, 182, 159); and (*c*) typically support large populations with lower extinction probabilities (15, 112). There may also be simple pragmatic reasons to support a preference for large reserves. Large reserves may be preferred over small reserves because they are logistically easier to manage. Alternatively, large reserves may also be easier to acquire when located in regions where real estate values are low compared to urbanizing landscapes (AP Dobson, unpublished data).

Small Reserves Can Be Effective for Some Targets

Diamond (31) proposed six reserve design guidelines based on island biogeographic theory. While not explicitly stated, these guidelines prioritize the objective of minimizing the likelihood of extinction of a fine-filter target. These guidelines, adopted by the International Union for the Conservation of Nature and Natural Resources (61), are commonly cited regarding reserve selection (102, 115) and continue to be elaborated upon (139, 140).

Diamond's (31) first guideline asserts that a large reserve is better than a small one. A principal argument for increasing size is that larger reserves decrease extinction risk. Most empirical data support this contention (15, 112), and, all things being equal, securing large areas is an appropriate planning objective. Noss &

Cooperrider (102) consider any reserve less than 1000 ha as "tiny," with a clear implication that tiny reserves are a low conservation priority (101). The corollary to the "more is better" paradigm, however, is not "small is bad." Small size, by itself, is not cause to dismiss potential reserves. For example, Cowling & Bond (28) compared species-area relationships on mainland and island patches of Fynbos in South Africa and conclude that isolated reserves of 4–15 ha are sufficient to maintain floristic diversity for long periods of time.

The high degree of habitat loss in many regions leaves no alternative for conservation. Prioritization of large sites discourages consideration of many critically endangered species restricted to small potential reserves. For example, MacDougall et al (75) found that seven of nine habitat types in southeastern New Brunswick have a small total area and that 57% of all rare plants were found in these spatially restricted habitats. Turner & Corlett (167) review the evidence of species losses in small (<100 ha) tropical forest remnants. They conclude that while these remnant patches fail as reserves for most large mammals, they are likely to retain a considerable proportion of their plant diversity for decades. Turner & Corlett (167) summarize their findings by concluding that something is better than nothing in an environment where tropical forest is being lost at an alarming rate. Similarly, the largest numbers of California's rare plants are found in mixed chaparral, valley grassland, and coastal scrub (113). Nearly 80% of the over 1700 rare plant taxa are found in these three habitats. Very little habitat remains of either valley grassland or coastal scrub, and most of what is left is found in small patches (54, 55, 90). Although mixed chaparral is a widely distributed habitat type (49), the largest number of rare chaparral taxa are found on serpentine (113), which is frequently isolated to small patches.

Over 60% of the flora of Mauritius is endemic to the Mascarene Islands (Mauritius, Reunion, and Rodrigues), and approximately 80% of this flora is endangered (157). Less than 5% of the area of Mauritius remains in native vegetation. Median patch size of the 21 largest natural habitats remaining on Mauritius is 280 ha, with only a single site exceeding 1000 ha (129). Four of seven biotic regions, and three of five general vegetation types (e.g., rain forest), are not represented in any patch larger than 500 ha (129). More than 80 plant species of Mauritius have gone extinct; another 20 species are known from fewer than 10 individuals (157). Although there is concern regarding the persistence of diversity in small sites (130), it seems more of an error to fail to try than to try but fail. Further, if preserving diversity is envisioned through restoration of larger habitat patches, the process must begin by protecting representatives of this diversity.

A second reason for not abandoning small sites is that reserve size does not necessarily predict population size nor diversity. Owing to the positive correlation between site size and degradation, habitat quality must be considered in tandem with size. For example, large tallgrass prairie sites are frequently low-quality habitats that have a history of grazing and lack disturbance-sensitive, prairie-dependent plants and butterflies (110). Species-area curves of plant diversity on tallgrass prairies suggest that a 2-ha area of tallgrass prairie is often more diverse than much larger sites (126). Similarly, butterfly and leafhopper diversity levels off at sizes of isolated reserves much less than 100 ha (110). Most Midwestern states have lost an excess of 99% of their original tallgrass habitat (126, 131). As a result of this habitat loss, small reserves are the norm. The Nature Conservancy owns more than 100 tallgrass prairie reserves, 97% of which are less than 1000 ha (153). Nearly a third (70 of 236) of the dedicated nature reserves in Illinois are less than 10 ha in size (85), and an estimated 75% of remaining high-quality prairie sites in Illinois are less than 2 ha (126). Nonetheless, the existing suite of small reserves in the Midwest has been effective at capturing prairie diversity. Despite near total habitat loss, and a century of isolation in small fragments, there are few endangered species (10 plants, 2 insects) in the tallgrass prairie ecosystem (126). With a legacy of selecting prairie reserves on the basis of plant diversity, reserves in northern Illinois and southern Wisconsin have also effectively captured prairie-dependent insect diversity (110). In Illinois, restricting protection programs to large grasslands would have resulted in protection of fewer populations of many of the states' sensitive plant and insect species. Klein (69) argues that small prairie conservation has been so successful that restoration of sites such as schoolyards is a valuable exercise not just for education, but for conservation of ecosystem attributes.

A third reason not to abandon small reserves is that many species are naturally restricted to small patches. Diamond's second guideline states that single large reserves are better than several small ones. This guideline triggered a long and often heated debate over the ability of single large or several small (SLOSS) reserves to capture biological diversity (151). The SLOSS debate has been thoroughly reviewed elsewhere (87, 102, 151). Suffice it to say that protection is a prerequisite for maintaining diversity. If diversity is finely distributed over numerous small sites, then protecting small sites is warranted. For example, small isolated granitic outcrops of the southern Appalachians contain a number of endemic species (64). The chief threat to these species is trampling by visitors (64). Rather than focus on creating large reserves, appropriate actions involve increased protection of existing small patches. Similarly, serpentine outcrops of California have an evolutionary history of isolation. Semlistch & Bodie (138) note that 87% of Carolina bay wetlands along the southeast Atlantic coast of the United States are less than 4 ha in size, the US Corps of Engineers cutoff for regulatory protection. They also note that these small and often isolated Carolina bays are vitally important for maintaining breeding populations of amphibians. Conservation of diversity on serpentine, southern Appalachian outcrops, and Carolina bays will rely more on protecting a large number of patches, rather than protecting a few large sites.

The previous examples notwithstanding, not all conservation targets are well protected by small sites. Several studies have focused on assessing isolation and fragmentation effects on a suite of native scrub vegetation of San Diego. These studies generally indicate strong size-related effects and rapid degradation of small patches of this community type (11, 12, 158). Conservation planning needs methods for predicting when small reserves can effectively protect and maintain diversity. The answer to this question is likely to be habitat specific. Nonetheless, we may gain some clues from certain ecosystems. The degradation of the aforementioned scrub habitats is strongly influenced by nonnative species. Sites surrounded and invaded by aggressive nonnatives are likely to be unsuccessful. Polis et al (119) identify ecosystems that depend on the influx of energy across landscape boundaries through subsidized food webs. It is likely that ecosystems that are heavily augmented by energy flow from adjacent ecosystems are not likely to succeed in maintaining diversity in isolation.

The Case for Holistic Reserves

The importance of large reserves is often argued on the basis of completeness. Large sites can support the full suite of regional natural diversity and the interactions that support this diversity. This concept has been enunciated by a number of authors. Pickett & Thompson (117) define a "minimum dynamic area" as the smallest area that contains patches unaffected by the largest expected disturbances. Large size is required to allow recolonization from undisturbed patches within the reserve. Shugart & West (143) argue that in order to maintain a landscape in dynamic equilibrium, a reserve ought to be 50–100 times larger than a typical large disturbance. Some versions of these arguments, however, appear practically untenable. Recalling that the recent Yellowstone fires were larger in size than the National Park, Grumbine (47) wonders whether the fact that Yellowstone is too small to be in dynamic equilibrium (143) is relevant to its functioning as a reserve. Nonetheless, how big is big enough remains an important question for conservation biologists.

Arguments for large reserves using minimum area concepts assume that only large sites can sustain a disturbance regime supporting the full array of habitats at different stages of succession; that large sites containing many patches support a more complete array of species than a suite of isolated sites that lack interactions among patches; and that disturbance results in local extirpation of species such that recolonization is a necessity. Empirical data do not universally support these assumptions. Most disturbances can dramatically reduce populations, but extirpation is not a general expectation. For example, there is concern that managed fire in small prairie remnants may jeopardize insect populations (108, 160, 161). Panzer (109) studied the potential of managed fire to cause the extirpation of vulnerable, prairie-restricted butterflies and leafhoppers on small (<100 ha) isolated prairie remnants. Panzer found no evidence that managed fires caused insect population extirpations over a six-year study. While populations of most species observed declined as a result of fire, they also tended to rebound quickly. Siemann et al (145) observed a similar response of insects to fire in a Minnesota prairie: a strong negative short-term effect, rapid rebounding, and no evidence of extirpations. These sorts of data suggest that tallgrass prairie reserves in the Midwest protect the historic range of variability despite being distributed in many small and isolated reserves. This example does not argue for a preference toward small reserves. Nonetheless, I argue that size and isolation of coarse-filter targets alone is not sufficient to dismiss them as valuable conservation targets.

Large Size Is Required for Some Targets

There is no substitute for large reserves for certain conservation objectives. For example, large size is a defining attribute of wilderness (99, 102). Similarly, a common concern of humanity is to assure that ecosystems continue to provide a supply of clean water (30). It would be ludicrous to assert that conserving a small reserve within a large watershed protects the hydrology of that watershed. On the other hand, this raises an important issue regarding ecosystem conservation: Degraded ecosystems may support many types of vital ecosystem functions. Research on the relationship between diversity and ecosystem function predicts that relatively low diversity ecosystems retain nearly full capacity of ecosystem functions such as nutrient flux and primary productivity (e.g., 134, 164). Thus, large degraded ecosystems provide benefits to humanity independent of their diversity. Whether these benefits are endangered, and hence represent a conservation benefit, remains a matter of interpretation.

Conservation has often focused efforts toward vertebrates with large habitat requirements. This emphasis is justified on the basis that society places more value on larger vertebrate species (174). Despite a bias toward funding conservation of larger species, we still often lack sufficient habitat for viable populations of large vertebrates (22, 65, 173). Problems relating to providing habitats for lions, tigers, and bears are well known, but these problems extend to many species in many regions. The existing suite of Midwestern tallgrass reserves has failed to protect species with large habitat requirements. Grassland birds in Illinois have declined throughout the twentieth century (56, 57). Grassland birds, formerly dependent upon pasture lands, are currently restricted to a few large but low-quality grasslands (56, 57). In this case, however, the fact that they thrive on low-quality sites benefits the birds. The potential exists to increase protection for grassland by focusing on large low-quality sites. There may even be a potential to create bison reserves within this context (154). The result, however, is that the negative correlation between site size and habitat quality in tallgrass prairie has resulted in a tension between balancing the needs of the few species that require large habitats versus the many species that require high-quality habitats (133).

INTEGRATING ACROSS SCALES

Reconciling the need for continued fine-filter conservation while accommodating coarse-filter conservation is a central issue of conservation planning (46, 53, 91, 123, 162, 163). Several approaches are being explored to assess the extent to which fine- and coarse-filter conservation objectives can be simultaneously met. I detail several common approaches below.

Fine-Filter Targets for Coarse-Filter Objectives: Flagships, Umbrellas

Fine-filter conservation targets have been used to augment coarse-filter conservation under the guise of flagship and umbrella species concepts (142, 147). A flagship is a species that can garner resources in support of conservation as a result of its popularity. It has been shown that vertebrates, the larger the better, make effective flagship species owing to their public appeal (174). Fine-filter efforts on behalf of flagship species provide collateral protection of the habitats upon which they depend. There are, however, problems with using flagship species as a surrogate for coarse-filter conservation (147). First and foremost is the cost of dishonesty. If the flagship species is not located in a particular reserve or potential reserve, then there arises a problem justifying conservation actions on the coarse-filter target (147). If an ecosystem is the target of conservation, then reserve design ought to rely on a coarse-filter approach and not a fine-filter surrogate.

The umbrella species concept provides another expression of collateral value in conservation (102). Umbrellas are species with large area requirements such that other species are protected through conservation actions directed at the umbrella species. As with flagships, the use of umbrella species is logically appealing, but often problematic when considering specific cases (147). In particular, species with large habitat requirements may act as leaky, or partial umbrellas. For example, black rhinos in Namibia have been considered for use as umbrellas for other large herbivores because they have large home ranges. Berger (10) found that rhino behavior during extreme wet and dry cycles differed significantly from that of six other large herbivores: Other herbivores moved between habitats, while rhinos did not. As a result, protection programs aimed specifically at rhinos may fail to protect other species adequately by virtue of partially nonoverlapping habitat requirements (10).

Hierarchical Reserve Design

A variant of the umbrella approach is to use vertebrates to set coarse-filter reserve priorities (102, 176). Noss & Harris (103) proposed a tiered strategy for reserve design. Multiple use module (MUM) strategies include core reserve areas with a high degree of protection surrounded by buffer habitats with less protection. Core areas may then be connected through corridors (102, 115). Hierarchical reserve design plans such as MUMs are recommended in several recent conservation texts (87, 102, 115). Hierarchical reserve designs have been published for Florida (29, 98) and western Oregon (100) and are in development elsewhere (102, 115).

An assumption of hierarchical reserve design is that buffers enhance the ability of the core reserve to protect critical resources. Chief supporting evidence for the utility of buffer areas comes from a large number of studies on detrimental edge effects (71, 81, 84, 182, 159). A keyword search of only two conservation journals (*Biological Conservation, Conservation Biology*) found in just the past six years 38 studies that addressed edge effects (1993–1998). There can be no doubt that a multitude of deleterious edge effects can diminish the effectiveness of reserves. The problem with mitigating edge effects is that different species have vastly differing effective edge widths. Typically edges are narrow (<100 m) with respect to plants (e.g., 84) and wide (>1 km) with respect to animals (159). Thus, an appropriate buffer for one species may be insufficient for another species and overkill for another. Reserve buffers are defined with respect to particular threats and the species that benefit from this protection.

Another key assumption of hierarchical reserve design is that dispersal corridors decrease extinction likelihoods (103). Several studies have attempted to document the utility of dispersal corridors, but supporting evidence has been found in fewer than half of these empirical studies (8). Studies that document corridor usage by wildlife typically do not take the additional step of addressing whether corridor usage affects demographic rates (8). In contrast, corridors may carry serious disadvantages such as providing avenues for exotic species, disturbance, or disease (59, 148), the same negative effects that make habitat edges poor reserves. Clinchy (23) further questions the utility of corridors by modeling alternative hypotheses for why adjacent reserves may have similar fates independent of corridors or their usage. After reviewing the evidence and recognizing the equivocal support for corridors, Beier & Noss (8) suggest that the burden of proving that connecting reserves lacks utility remains on those who would destroy the connections. To turn this around, I suggest that the burden of proof ought to rest on those who would divert needed conservation resources away from core habitats in favor of establishing dispersal corridors.

Despite these constraints, hierarchical reserve design may be the appropriate strategy for the protection of target species. Inherent in the idea of hierarchically designed core areas, buffers, and corridors is a target species to which these units are scaled. As such, hierarchical reserve design is an elaborate fine-filter program and not a coarse-filter strategy. We must not assume that a hierarchical plan is a coarse-filter program simply by virtue of large area.

Predicting Conservation Value: Indicators, Hotspots

Indicator species are those that predict the presence or diversity of other taxa (142). It is logically appealing to posit that attributes that make a habitat rich in one suite of species may also give rise to high diversity in other groups. The arduous task of identifying high priority conservation sites may be eased by using indicator taxa and not an exhaustive survey of diversity. An obvious first step in using indicator species is specifying the intent of the indicator. Nonetheless, vertebrates have often been used as indicator species without clear objectives as to what they were supposed to indicate (70, 147). Caro & O'Doherty (17) identify three types of indicator species in use those that predict attributes of: (*a*) ecosystem condition, (*b*) biodiversity, or (*c*) population trends in species that are difficult to sample.

Unfortunately, empirical studies where the objectives are clear have also yielded equivocal results. Supportive results have typically used narrowly constrained indicators. For example, high correlations have been documented between tiger beetle (Cicindelidae) diversity and Lepidoptera in both North America (18) and Amazonia (114). Martikainen et al (82) found that white-backed woodpeckers (*Dendrocopos leucotos*) co-occur with a high number of threatened saproxylic beetles with which they share a common resource: decaying wood. Dufrene & Legendre (34) successfully apply phytosociological methods to discern patterns of covariation among plant species in order to better detect which species are the best indicators of overall diversity.

In contrast, many studies fail to support indicators as shortcuts, perhaps because they use a scale of measure that is too coarse. For example, Prendergast et al (121) and Prendergast & Eversham (120) lump diversity within 10 km^2 grid cells in England and then find low covariation among groups that may be using different habitats within those grid cells. Dobson et al (33) and Flather et al (38) failed to find strong covariation in taxon diversity but used all of North America as a study region. Weaver (172) used coarsely lumped recognizable taxonomic units (RTUs) of arthropods (e.g., mites, thrips, beetles) to assess covariation, but they did not find strong patterns. Caro & O'Doherty (17) suggest selection criteria that may help to identify successful indicators. Given the equivocal support, the current use of indicators is debated (2, 125) and requires empirical data to justify utility in each case.

Identification of biodiversity hotspots (92) in order to prioritize reserve selection is another method whereby fine-filters are used to predict coarse-filter value. Several studies have used biodiversity hotspots to identify conservation priorities (32, 33, 89). As with indicator taxa, there are important issues of scale to consider in the utility of biodiversity hotspots. When faced with reserve selection, however, hotspot information may not be sufficient, as one would also like to know the degree to which hotspots of diversity overlap in constituent species (179).

Selecting Reserves: Minimum Sets, Gap Analysis, and Gap Analogs

A variety of approaches has been developed to prioritize the selection of reserves for the protection of biological diversity. One approach uses algorithms that maximize selection efficiency in order to capture each conservation target in a potential reserve (7, 20, 43, 68, 80). Any attribute, fine- or coarse-filter, may be used as input data for reserve prioritization (122). Nonetheless, minimum set algorithms prioritize the representation of rarity in selection and are most frequently applied to fine-filter targets (e.g., species occurrences). These searching algorithms are complex, and a number of different methods have been proposed (20, 28, 41, 74). The scale of the units, not surprisingly, is important to the efficiency of reserve selection algorithms. Smaller selection units are observed in some cases to be more than an order of magnitude more efficient at identifying required reserves (122). Smaller scales, however, require more specific data. In addition, cells for the selection algorithms are abstractions on the physical landscape that may not reflect realistic reserve boundaries. For example, Lombard et al (73) map the density of endemic species and vegetation types in 9 km^2 grid cells, and they prioritize them for reserve selection for a portion of South Africa. Yet, it is not clear from this study whether 9 km^2 grid cells are realistic reserve units.

Gap analysis is a way to plan protection programs by assessing the degree to which existing reserves have already captured existing biotic diversity. Similar to minimum-set algorithm methods, gap analysis seeks to maximize the coverage of the full representation of all biotic resources. Using a geographical information system (GIS), gap analysis builds regional coverages of biotic and abiotic attributes such as vegetation cover, animal distributions, rare species occurrences, and land ownership (136). Overlays of information are used to distinguish resources that are protected in the existing reserve network from those that are not (gaps). Gap analysis is agnostic on whether to prioritize fine- or coarse-filter targets and is typically used to develop a mixed strategy that incorporates both scales. Nonetheless, since the emphasis in gap analysis is on landscape units, the bias is typically toward coarse-filter targets.

Less formal programs aimed at targeting gaps in protection abound. For example, during the 1970s, Illinois developed a program to prioritize reserve acquisition toward unprotected habitats and rare species (58). Similar programs exist in many other states. Haufler et al (53) advocate identifying reserves based on coarse-filter targets, and then overlaying potential reserve areas with rare species occurrences in order to identify gaps. Priorities can then be adjusted in order to capture the full spectrum of biological diversity. Regardless of the specifics, separately identifying fine- and coarse-filter targets and then designing methods to protect both are appealing strategies.

There are two principal limitations to gap analysis and its many less formal analogs. First, location data for species typically lack specificity. As a result, data layers on species distributions often rely on habitat descriptors to predict species occurrences (16, 136). For example, if there is no way to determine which habitat patches a particular species occupies, then all appropriate habitats within the region may be assumed to represent occurrences. Second, habitat descriptors are typically limited to attributes described with satellite imagery (136). Thus, habitat classification may be crude relative to the biotic variability observed on the ground (16). Despite these limitations, gap analysis has gained considerable popularity. Regional analyses are being published (16, 66, 155), and a national gap analysis program (GAP) is currently underway (115).

CONCLUSION

It has been posited that the era of species conservation is over (147) and that ecosystems integrated within landscapes will be the conservation unit of the future (106). Although these statements may be prophetic, we currently operate in an environment where knowledge of species endangerment far exceeds that of threats to ecosystem health and stability. Yaffee (181) suggested that the Endangered Species Act is important as a benchmark for societal values. Human society, along with scientists, is wrestling with issues of what it is about the natural world that merits protection and at what cost (95, 96, 124). Species protection legislation represents a line in the sand defining unacceptable negative impacts of humans in the natural environment. It remains a challenge to ecologists to formally delineate the line in the sand within the context of coarse-filter conservation.

The emerging emphasis of conservation on larger organizational and spatial scales has been punctuated by extremes. For example, Barrett & Barrett (3) characterize early conservation efforts as focused on: 1. bounded natural areas often small in size; 2. management for stability and persistence in systems assumed to be at equilibrium; and 3. diversity as objects of natural heritage value. In contrast, the "new" conservation focuses on processes and context in open systems that are generally large and interconnected (3). Within this context, management is active and assumes non-equilibrium dynamics. Although there is some truth to this description, abstract caricatures often fail in specific examples. For example, Midwestern tallgrass prairie conservation has long since adopted the view that non-equilibrium dynamics are the norm and active management is a necessity (25). Fire management of tallgrass prairie has been common practice for at least twenty years (126, 135).

Conservation managers have been both blessed and cursed by the attention of academic biologists. In an effort to find tools to maximize the efficiency of conservation, scientists rapidly suggest and then sometimes test new ideas for conservation. The immediacy of conservation encourages management agencies to adopt these new ideas before empirical knowledge is able to verify their efficacy. This has led to overenthusiasm for ideas that were later found to be only weakly supported (e.g., 52, 150). Academicians are principally rewarded for novelty. Research support is more readily available to work on a novel approach to conservation than it is for increasing our biological understanding of a particular conservation target. Society expects resource management agencies to use the best possible scientific information in decision making, yet there is ambiguity as to what the best scientific information is. There are many competing protection strategies available. Within this context there is one clear message: Theory is nice, but empirical data are required before abandoning tried methods in favor of novel ideas. Nonetheless, a body of evidence is now building around several strategic conservation planning tools. While shortcuts such as indicators, flagships, and umbrellas are not gaining as much support as we would like, minimum set algorithms and gap analysis seem to hold great promise for conservation planning.

Through this review I have made several general points with respect to the organizational and spatial scale of conservation. First, no matter what we view to be a minimum area for long-term viability of a conservation target, we cannot sustain biological resources until we protect them. If this entails protecting smaller units than we would like, then we ought to prefer the risk of losing diversity in small reserves over the guaranteed loss of diversity by neglect. Second, accepting non-equilibrium dynamics, disturbance ecology, and patch dynamics as paradigms of ecology does not require a preference for coarse-filter reserves. Embracing coarse-filter conservation is not linked to the abandonment of fine-filter conservation. Fine-filter actions remain some of our most successful programs, and we should expect them to continue to succeed. Similarly, embracing ecosystem-based management does not entail restricting conservation to large reserves. Choosing the appropriate spatial scale for reserves entails careful consideration of the alternatives. If a fine- or coarse-filter target is contained within a small parcel of land, then the appropriate action may entail a small reserve. Even with a hierarchical design, many good conservation projects encompass significantly less than the 1000s hectares required for conserving integrated ecosystems in a complex landscape. In aggregate, these points are in sharp contrast to the trend in conservation to restrict protection actions toward spatially large reserves in order to protect coarse-filter objectives and fine-filter targets with large habitat requirements.

Early conservation efforts may have inordinately focused attention on capturing fine- and coarse-filter targets of diversity as static objects of natural history, much as one would assemble a museum. Conservation actions are now redressing this by appropriately focusing efforts on conserving interactions among species and processes within ecosystems. Ecologists have argued the need to move away from the crisis-driven approach of species conservation (40, 170). While I agree with the sentiment, avoiding conservation crisis by adopting coarse-filter priorities to the exclusion of fine-filter conservation all but guarantees the crisis to be resolved through extinction. The effort to increase emphasis on coarse-filter approaches for large reserves must abandon neither fine-filter conservation nor conservation at small spatial scales. Conservation, in order to be effective, must strive to balance the protection of countable objects of diversity and the protection of natural processes. This balance will entail a broad array of programs and strategies on a variety of spatial and organizational scales.

LITERATURE CITED

1. Aplet GH, Keeton WS. 1999. Application of historic range of variability concepts to biodiversity conservation. In *Practical Approaches to the Conservation of Biological Diversity*, ed. RK Baydack, H Campa III, JB Haufler, pp. 71–86. Washington, DC: Island. 313 pp.
2. Balmford A. 1998. On hotspots and the use of indicators for reserve selection. *Trends Evol. Ecol.* 13:409
2a. Barbour MG, Major J, ed. 1988. *Terrestrial Vegetation of California*. Spec. Publ. #9, California Native Plant Society, Sacramento. 1030 pp.
3. Barrett NE, Barrett JP. 1997. Reserve design and the new conservation theory. See Ref. 116, pp. 236–51
4. Bean M. 1983. *The Evolution of National Wildlife Law*. New York: Praeger. 449 pp.
5. Beattie M. 1996. Biodiversity policy and ecosystem management. In *Biodiversity and the Law*, ed. WJ Snape III, pp. 11–15. Washington, DC: Island Press. 259 pp.
6. Beattie M. 1996. An ecosystem approach

to fish and wildlife conservation. *Ecol. Appl.* 6:696–99
7. Bedward M, Pressey RL, Keith DA. 1992. A new approach for selecting fully representative reserve networks: addressing efficiency, reserve design and land suitability with an iterative analysis. *Biol. Conserv.* 62:115–25
8. Beier P, Noss RF. 1998. Do habitat corridors provide connectivity? *Conserv. Biol.* 12:1241–52
9. Beissinger SR, Westphal MI. 1998. On the use of demographic models of population viability in endangered species management. *J. Wild. Manage.* 62:821–41
10. Berger J. 1997. Population constraints associated with the use of black rhinos as an umbrella species for desert herbivores. *Conserv. Biol.* 11:69–78
11. Bolger DT, Scott TA, Rotenberry JT. 1997. Breeding bird abundance in an urbanizing landscape in coastal Southern California. *Conserv. Biol.* 11:406–21
12. Bolger DT, Alberts AC, Sauvajot RM, Potenza P, McCalvin C, et al. 1997. Response of rodents to habitat fragmentation in coastal southern California. *Ecol. Appl.* 7:552–63
13. Brownlow CA. 1996. Molecular taxonomy and the conservation of the red wolf and other endangered carnivores. *Conserv. Biol.* 10:390–96
14. Brunner RD, Clark TW. 1997. practice-based approach to ecosystem management. *Conserv. Biol.* 11:48–58
15. Burkey TV. 1995. Extinction rates in archipelagoes: implications for populations in fragmented habitats. *Conserv. Biol.* 9:527–41
16. Caicco SL, Scott JM, Butterfield B, Csuti B. 1995. A gap analysis of the management status of the vegetation of Idaho (U.S.A.). *Conserv. Biol.* 9:498–511
17. Caro TM, O'Doherty G. 1999. On the use of surrogate species in conservation biology. *Conserv. Biol.* In press
18. Carroll SS, Pearson DL. 1998. Spatial modeling of butterfly species richness using tiger beetles (Cicindelidae) as a bioindicator taxon. *Ecol. Appl.* 8:531–43
19. Christensen NL, Bartuska AM, Brown JH, Carpenter S, D'Antonio C, et al. 1996. The report of the Ecological Society of America Committee on the Scientific Basis for Ecosystem Management. *Ecol. Appl.* 6:665–91
20. Church RL, Stoms DM, Davis FW. 1996. Reserve selection as a maximal covering location problem. *Biol. Conserv.* 76:105–12
21. Clark JA. 1994. The Endangered Species Act: its history, provisions, and effectiveness. In *Endangered Species Recovery: Finding the Lessons, Improving the Process*, ed. TW Clark, RP Reading, AL Clarke, pp. 19–43. Washington, DC: Island. 450 pp.
22. Clark TW, Paquet PC, Curlee AP. 1996. Introduction: special section: large carnivore conservation in the Rocky Mountains of the United States. *Conserv. Biol.* 10:936–39
23. Clinchy M. 1997. Does immigration "rescue" populations from extinction? Implications regarding movement corridors and the conservation of mammals. *Oikos* 80:618–22
24. Collar NJ. 1996. The reasons for Red Data Books. *Oryx* 30:121–30
25. Collins SL, Wallace LL, ed. 1990. *Fire in North American Tallgrass Prairies*. Norman: Univ. Okla. Press. 175 pp.
26. Congressional Research Service. 1994. *Ecosystem Management: Federal Agency Activities*. CRS report for Congress 94-339 ENR. Washington, DC: Library of Congress
27. Costanza R, Norton BG, Haskell BD, ed. 1992. *Ecosystem Health: New Goals for Environmental Management*. Washington, DC: Island. 269 pp.
28. Cowling RM, Bond WJ. 1991. How small can reserves be? An empirical approach in Cape Fynbos, South Africa. *Biol. Conserv.* 58:243–56

29. Cox J, Kautz R, MacLaughlin M, Gilbert T. 1994. *Closing the Gaps in Florida's Wildlife Habitat Conservation System*. Tallahassee, FL: Fla. Game and Fresh Water Fish Com.
30. Daily GC. 1997. Introduction: what are ecosystem services? In *Nature's Services: Societal Dependence on Natural Ecosystems*, ed GC Daily. pp 1–10. Washington, DC: Island. 392 pp.
31. Diamond J. 1975. The island dilemma: lessons of modern biogeographic studies for the design of natural reserves. *Biol. Conserv.* 7:129–46
32. Dinerstein E, Wikramanayake ED. 1993. Beyond "hotspots": how to prioritize investments in biodiversity in the Indo–Pacific region. *Conserv. Biol.* 7:53–65
33. Dobson AP, Rodriguez JP, Roberts WM, Wilcove DS. 1997. Geographic distribution of endangered species in the United States. *Science* 275:750–52
34. Dufrene M, Legendre P. 1997. Species assemblages and indicator species: the need for a flexible asymmetrical approach. *Ecol. Monogr.* 67:345–66
35. Edwards PJ, May RM, Webb NR, ed. 1994. *Large-Scale Ecology and Conservation Biology*. Oxford, UK: Blackwell Sci. 375 pp.
36. Engstrom RT, Mikusinski G. 1998. Ecological neighborhoods in red-cockaded woodpecker populations. *Auk* 115:473–78
37. Falk DA. 1992. From conservation biology to conservation practice: strategies for protecting plant diversity. In *Conservation Biology: The Theory and Practice of Nature Conservation Preservation and Management*, ed. PL Fiedler, SK Jain, pp 397–431. New York: Chapman & Hall. 507 pp.
38. Flather CH, Wilson KR, Dean DJ, McComb WC. 1997. Identifying gaps in conservation networks: of indicators and uncertainty in geographic-based analyses. *Ecol. Appl.* 7:531–42
39. Foin TC, Riley SPD, Pawley AL, Ayres DR, Carlsen TM, et al. 1998. Improving recovery planning for threatened and endangered species: comparative analysis of recovery plans can contribute to more effective recovery planning. *Bioscience* 48:177–84
40. Franklin JF. 1993. Preserving biodiversity: species, ecosystems, or landscapes? *Ecol. Appl.* 3:202–5
41. Freitag S, Nicholls AO, van Jaarsveld AS. 1996. Nature reserve selection in the Transvaal, South Africa: What data should we be using? *Biodiv. Conserv.* 5:685–98
42. Glitzenstein JS, Platt WJ, Streng DR. 1995. Effects of fire regime and habitat on tree dynamics in North Florida longleaf pine savannas. *Ecol. Monogr.* 65:441–76
43. Goldsmith FB. 1987. Selection procedures for forest nature reserves in Nova Scotia, Canada. *Biol. Conserv.* 41:185–201
44. Gordon DR, Provencher L, Hardesty JL. 1997. Measurement scales and ecosystem management. See Ref. 116, pp. 262–73
45. Groom MJ, Pascual MA. 1998. The analysis of population persistence: an outlook on the practice of viability analysis. In *Conservation Biology for the Coming Decade*, ed. PL Fiedler, PM Kareiva, pp 4–27. New York: Chapman & Hall. 533 pp.
46. Grossman DH, Faber–Langendoen D, Weakley AW, Anderson M, Bourgeron P, et al. 1998. *International Classification of Ecological Communities: Terrestrial Vegetation of the United States*. Vol. I: *The National Vegetation Classification Standard*. Arlington, VA: Nature Conservancy
47. Grumbine RE. 1992. *Ghost Bears: Exploring the Biodiversity Crisis*. Washington, DC: Island. 290 pp.
48. Grumbine RE. 1997. Reflections on "What is Ecosystem Management?" *Conserv. Biol.* 11:41–47
49. Hanes TL. 1988. Chaparral. See Ref. 2a, pp. 417–70
50. Hanula JL, Franzreb K. 1998. Source distribution and abundance of macroarthro-

pods on the bark of longleaf pine: potential prey of the red-cockaded woodpecker. *For. Ecol. Manage.* 102:89–102

51. Haskell BD, Norton BG, Constanza R. 1992. What is ecosystem health and why should we worry about it? In *Ecosystem Health: New Goals for Environmental Management*, ed. R Costanza, BG Norton, BD Haskell, pp 3–20. Washington, DC: Island. 269 pp.
52. Hastings A, Harrison S. 1994. Metapopulation dynamics and genetics. *Annu. Rev. Ecol. Syst.* 25:167–188
53. Haufler JB, Mehl CA, Roloff GJ. 1999. Conserving biological diversity using a coarse-filter approach with a species assessment. In *Practical Approaches to the Conservation of Biological Diversity*, ed. RK Baydack, H Campa III, JB Haufler, pp 107–26. Washington, DC: Island. 313 pp.
54. Heady HF. 1988. Valley grassland. See Ref. 2a, pp 491–514
55. Heady HF, Foin TC, Hektner MM, Taylor DW, Barbour MG, et al. 1988. Coastal prairie and northern coastal scrub. See Ref. 2a, pp 733–60
56. Herkert JR. 1994. Breeding bird communities of Midwestern prairie fragments: the effects of prescribed burning and habitat-area. *Nat. Areas J.* 14:128–35
57. Herkert JR. 1995. An analysis of Midwestern breeding bird trends: 1966–1993. *Am Midl. Nat.* 134:41–50
58. Herkert JR. 1997. Nature preserves, natural areas, and the conservation of endangered and threatened species in Illinois. In *Conservation in Highly Fragmented Landscapes*, ed. MW Schwartz, pp 395–406. New York: Chapman & Hall. 436 pp.
59. Hess GR. 1994. Conservation corridors and contagious disease: a cautionary note. *Conserv. Biol.* 8:256–62
60. Heywood VH. 1994. The measurement of biodiversity and the politics of implementation. In *Systematics and Conservation Evaluation*, ed. PF Forey, CJ Humphries, RI Vane-Wright, pp. 15–22. Oxford, UK: Clarendon. 438 pp.
61. International Union for Conservation of Nature and Natural Resources (IUCN). 1980. *World Conservation Strategy*. Gland, Switzerland: IUCN
62. International Union for Conservation of Nature and Natural Resources (IUCN). 1996. *IUCN Red List of Threatened Animals*. Gland, Switzerland: IUCN. 368 pp.
63. Irvin WR. 1992. The Endangered Species Act: prospects for reauthorization. In *Transactions of the Fifty-Seventh North American Wildlife and Natural Resources Conference*, ed. RE McCabe. Washington, DC: Wildlife Manage. Inst.
64. Johnson BR. 1996. Southern Appalachian rare plant reintroductions on granitic outcrops. In *Restoring Diversity: Strategies for Reintroduction of Endangered Plants*, ed. DA Falk, CI Millar, M. Olwell, pp 433–43. Washington, DC: Island. 505 pp.
65. Keiter RB, Locke H. 1996. Law and large carnivore conservation in the Rocky Mountains of the U.S. and Canada. *Conserv. Biol.* 10:1003–12
66. Keister AR, Scott JM, Csuti B, Noss RF, Butterfield B, et al. 1996. Conservation prioritization using GAP data. *Conserv. Biol.* 10:1332–42
67. Keith DA. 1998. An evaluation and modification of World Conservation Union red list criteria for classification of extinction risk in vascular plants. *Conserv. Biol.* 12:1076–1090
68. Kirkpatrick JB. 1983. An iterative method for establishing priorities for the selection of nature reserves: an example from Tasmania. *Biol. Conserv.* 25:127–34
69. Klein VM. 1997. Planning a restoration. In *The Tallgrass Restoration Handbook: For Prairies, Savannas and Woodlands*, ed. S. Packard, CF Mutel, pp. 31–46. Washington, DC: Island. 463 pp.
70. Landres PB, Verner J, Thomas JW. 1988. Ecological uses of vertebrate indicator species: a critique. *Conserv. Biol.* 2:316–28

71. Laurance WF, Yensen E. 1991. Predicting the impacts of edge effects in fragmented habitats. *Biol Conserv.* 55:77–92
72. Letcher BH, Priddy JA, Walters JR, Crowder LB. 1998. An individual-based, spatially-explicit simulation model of the populations dynamics of the endangered red-cockaded woodpecker, *Picoides borealis*. *Biol. Conserv.* 86:1–14
73. Lombard AT, Cowling RM, Pressey RL, Mustart PJ. 1997. Reserve selection in a species-rich and fragmented landscape on the Agulhas plain, South Africa. *Conserv. Biol.* 11:1101–16
74. Lomolino MV. 1994. An evaluation of alternative strategies for building networks of nature reserves. *Biol. Conserv.* 69:243–49
75. MacDougall AS, Loo JA, Clayden SR, Goltz JR, Hinds HR. 1998. Defining conservation priorities for plant taxa in southeastern New Brunswick, Canada using herbarium records. *Biol. Conserv.* 86:325–39
76. Mace GM. 1994. An investigation into methods for categorising the conservation status of species. In *Large-Scale Ecology and Conservation Biology*, ed. PJ Edwards, R May, NR Webb, pp. 295–314. Oxford, UK: Blackwell
77. Mace GM. 1995. Classification of threatened species and its role in conservation planning. In *Extinction Rates*, ed. JH Lawton, RM May, pp. 197–213. Oxford, UK: Oxford Univ. Press. 233 pp.
78. Mace GM, Lande R. 1991. Assessing extinction threats: toward reevaluation of IUCN threatened species categories. *Conserv. Biol.* 5:148–57
79. Mann CC, Plummer ML. 1995. *Noah's Choice: The Future of Endangered Species*. New York: Harper & Row. 302 pp.
80. Margules CR, Nicholls AO, Pressey RL. 1988. Selecting networks of reserves to maximize biological diversity. *Biol. Conserv.* 43:63–76
81. Marini MA, Robinson SK, Heske EJ. 1995. Edge effects of nest predation in the Shawnee National Forest, southern Illinois. *Biol. Conserv.* 74:203–13
82. Martikainen P, Kaila L, Haila Y. 1998. Threatened beetles in white-backed woodpecker habitats. *Conserv. Biol.* 12:293–301
83. Master LL. 1991. Assessing threats and setting priorities for conservation. *Conserv. Biol.* 5:555–63
84. Matlack GR. 1993. Microenvironment variation within and among forest edge sites in the eastern United States. *Biol. Conserv.* 66:185–94
85. McFall D, Kearns J, ed. 1995. *A Directory of Illinois Nature Preserves*. Volumes I & II. Springfield, IL: Illinois Dept. Nat. Resources
86. Means DB. 1996. Longleaf pine forest, going, going... In *Eastern Old-growth Forests: Prospects for Rediscovery and Recovery*, ed. MB Davis, pp. 210–29. Washington, DC: Island. 383 p.
87. Meffe GK, Carroll CR. 1997. *Principles of Conservation Biology*. Sunderland, MA: Sinauer. 729 pp. 2nd ed.
88. Meyer JL. 1997. Conserving ecosystem function. See Ref. 116, pp. 136–45
89. Mittermeier RA, Myers N, Thomsen JB, Da Fonseca GAB, Olivieri S. 1998. Biodiversity hotspots and major tropical wilderness areas: approaches to setting conservation priorities. *Conserv. Biol.* 12:516–20
90. Mooney HA. 1988. Southern coastal scrub. See Ref. 2a, pp 471–90
91. Moss MR, Milne RJ. 1998. Biophysical processes and bioregional planning: the Niagara Escarpment of southern Ontario, Canada. *Landscape Urban Planning* 40:251–68
92. Myers N. 1988. Threatened biotas: hotspots in tropical forests. *Environmentalist* 8:178–208
93. National Research Council. 1995. *Science and the Endangered Species Act*. Washington, DC: Natl. Acad. Press. 271 pp.
94. Noel JM, Platt WJ, Moser EB. 1998. Structural characteristics of old- and second-growth stands of longleaf pine (*Pinus*

palustris) in the Gulf Coastal Region of the U.S.A. *Conserv. Biol.* 12:533–48
95. Norton BG. 1991. *Toward Unity Among Environmentalists*. Oxford, UK: Oxford Univ. Press. 287 pp.
96. Norton BG. 1994. On what we should save: the role of culture in determining conservation targets. In *Systematics and Conservation Evaluation*, ed. PF Forey, CJ Humphries, RI Vane-Wright, pp. 23–40. Oxford, UK: Clarendon. 438 pp.
97. Noss R. 1987. From plant communities to landscapes in conservation inventories: a look at The Nature Conservancy (USA). *Biol. Conserv.* 41:11–37
98. Noss R. 1987. Protecting natural areas in fragmented landscapes. *Nat. Areas J.* 7:2–13
99. Noss R. 1992. The Wildlands Project: land conservation strategy. *Wild Earth* (Special Issue) 10–25
100. Noss R. 1993. A bioregional conservation plan for the Oregon Coast Range. *Nat. Areas J.* 13:276–90
101. Noss R. 1996. Protected areas: How much is enough? In *National Parks and Protected Areas*, ed. R.G. Wright, pp 91–120. Oxford, UK: Blackwell Sci. 470 pp.
102. Noss RF, Cooperrider AY. 1994. *Saving Nature's Legacy, Protecting and Restoring Biodiversity*. Washington, DC: Island. 416 pp.
103. Noss RF, Harris LD. 1986. Nodes, networks, and MUMs: preserving diversity at all scales. *Environ. Manage.* 10:299–309
104. Noss RF, LaRoe ET III, Scott JM. 1995. *Endangered Ecosystems of the United States: A Preliminary Assessment of Loss and Degradation. Biol. Rep. 28.* Washington, DC. Natl. Biol. Serv., Dept. Interior. 58 pp.
105. O'Brien SJ, Roelke ME, Yuhki N, Richards KW, Johnson WE, et al. 1990. Genetic introgression with the Florida panther *Felis concolor coryi*. *Nat. Geog. Res.* 6:485–94
106. Ostfeld RS, Pickett STA, Shachak M, Likens GE. 1997. Defining the scientific issues. See Ref. 116, pp. 3–10
107. Packard S. 1997. Restoration options. In *The Tallgrass Restoration Handbook: For Prairies, Savannas and Woodlands*, ed. S Packard, CF Mutel, pp. 47–62. Washington, DC: Island. 463 pp.
108. Panzer R. 1988. Managing prairie remnants for insect conservation. *Nat. Areas J.* 8:83–90.
109. Panzer R. 1998. *Insect Conservation Within the Severely Fragmented Eastern Tallgrass Prairie Landscape*. Phd dissertation, Univ. Illinois. 108 pp.
110. Panzer R, Schwartz MW. 1998. Effectiveness of a vegetation-based approach to insect conservation. *Conserv. Biol.* 12:693–702
111. Patlis J. 1996. Biodiversity, ecosystems and endangered species. In *Biodiversity and the Law*, ed. WJ Snape III, pp. 43–58. Washington, DC: Island. 259 pp.
112. Patterson BD. 1987. The principle of nested subsets and its implication for biological conservation. *Conserv. Biol.* 1:323–34
113. Pavlik BM, Skinner MW. 1994. Ecological characteristics of California's rare plants. In *Inventory of Rare and Endangered Vascular Plants of California*, ed. MW Skinner, BM Pavlik, pp. 4–6. Spec. Pub. #1. Calif. Native Plant Soc., Sacramento. 5th ed
114. Pearson DL. 1992. Tiger beetles as indicators for biodiversity patterns in Amazonia *Nat. Geog. Res. Expl.* 8:116–17
115. Peck S. 1998. *Planning for Biodiversity*. Washington, DC: Island. 221 pp.
116. Pickett STA, Ostfeld RS, Shachak M, Likens GE, ed. 1997. *The Ecological Basis of Conservation: Heterogeneity, Ecosystems and Biodiversity*. New York: Chapman & Hall. 466 pp.
117. Pickett STA, Thompson J. 1978. Patch dynamics and the design of nature reserves. *Biol. Conserv.* 13:27–37

118. Platt WJ, Evans GW, Rathbun SL. 1988. The population dynamics of a long-lived conifer (*Pinus palustris*). *Am. Nat.* 131:491–525
119. Polis GA, Anderson WB, Holt RD. 1997. Toward an integration of landscape and food web ecology: The dynamics of spatially subsidized food webs. *Annu. Rev. Ecol. Syst.* 28:289–316
120. Prendergast JR, Eversham BC. 1997. Species richness covariance in higher taxa: empirical tests of the biodiversity indicator concept. *Ecography* 20:210–16
121. Prendergast JR, Quinn RM, Lawton JH, Eversham BC, Gibbons DW. 1993. Rare species, the coincidence of diversity hotspots and conservation strategies. *Nature* 365:335–37
122. Pressey RL, Logan VS. 1998. Size of selection units for future reserves and its influence on actual vs targeted representation of features: a case study in western New South Wales. *Biol. Conserv.* 85:305–19
123. Redford KH. et al. 1997. *Geography of Hope: Guidelines for Ecoregion-Based Conservation.* Arlington VA: Nature Conservancy
124. Reid WV. 1994. Setting objectives for conservation evaluation. In *Systematics and Conservation Evaluation*, ed. PF Forey, CJ Humphries, RI Vane-Wright, pp. 1–14. Oxford, UK: Clarendon. 438 pp.
125. Reid WV. 1998. Biodiversity hotspots. *Trends Evol. Ecol.* 13:275–80
126. Robertson KR, Anderson RC, Schwartz MW. 1997. The tallgrass prairie mosaic. In *Conservation in Highly Fragmented Landscapes*, ed. MW Schwartz, pp. 55–87. New York: Chapman & Hall. 436 pp.
127. Rohlf DJ. 1991. Six biological reasons the Endangered Species Act doesn't work–and what to do about it. *Conserv. Biol.* 5:275–82
128. Rohlf DJ. 1992. Response to O'Connell. *Conserv. Biol.* 6:144–45
129. Safford RJ. 1997. A survey of the occurrence of native vegetation remnants on Mauritius in 1993. *Biol. Conserv.* 80:181–88
130. Safford RJ, Jones CG. 1998. Strategies of land-bird conservation on Mauritius. *Conserv. Biol.* 12:169–76
131. Samson FB, Knopf FL. 1994. Prairie conservation in North America. *Bioscience* 44:418–21
132. Schemske DW, Husband BC, Ruckelshaus MH, Goodwillie C, Parker IM, et al. 1994. Evaluating approaches to the conservation of rare and endangered plants. *Ecology* 75:584–600
133. Schwartz MW. 1994. Conflicting goals for conserving biodiversity: issues of scale and value. *Nat. Areas J.* 14:213–16
134. Schwartz MW, Brigham CA, Hoeksema JD, Lyons KG, Mills MH, et al. Is biodiversity-for-ecosystem-function an appropriate conservation paradigm? Ms in review
135. Schwartz MW, Hermann SM. 1997. Midwestern fire management: prescribing a natural process in an unnatural landscape. In *Conservation in Highly Fragmented Landscapes*, ed. MW Schwartz, pp. 213–33. New York: Chapman & Hall. 436 pp.
136. Scott JM, Anderson H, Davis F, Caicco S, Csuti B, et al. 1993. Gap analysis: a geographic approach to protection of biological diversity. *Wildlife Monogr.* 123:1–41
137. Seal US, Foose TJ, Ellis–Joseph S. 1993. Conservation assessment and management plans (CAMPs) and global action plans (GCAPs). In *Creative Conservation: The Interactive Management of Wild and Captive Animals*, ed. PJ Olney, GM Mace, ATC Feistner, pp. London: Chapman & Hall
138. Semlitsch RD, Bodie JR. 1998. Are small, isolated wetlands expendable? *Conserv. Biol.* 12:1129–33
139. Shafer C. 1994. Beyond park boundaries. In *Landscape Planning and Ecological Networks*, ed. EA Cook, HN van Lier, pp 201–223. Amsterdam: Elsevier

140. Shafer C. 1997. Terrestrial nature reserve design at the rural/urban interface. In *Conservation in Highly Fragmented Landscapes*, ed. MW Schwartz, pp. 345–78. New York: Chapman & Hall. 436 pp.
141. Shilling F. 1998. Do habitat conservation plans protect endangered species? *Science* 276:1662–63
142. Shrader-Frechette KS, McCoy ED. 1993. *Method in Ecology: Strategies for Conservation*. Cambridge, UK: Cambridge Univ. Press. XXX pp.
143. Shugart H, West D. 1981. Long-term dynamics of forest ecosystems. *Am. Sci.* 69:647–52
144. Sidle JG. 1998. Arbitrary and capricious species conservation. *Conserv. Biol.* 12:248–49
145. Siemann E, Haarstad J, Tilman D. 1997. Short-term and long-term effects of burning on oak savanna arthropods. *Am. Midl. Nat.* 137:349–61
146. Silvertown J, Franco M, Menges E. 1996. Interpretation of elasticity matrices as an aid to the management of plant populations for conservation. *Conserv. Biol.* 10:591–97
147. Simberloff D. 1998. Flagships, umbrellas, and keystones: is single-species management passe in the landscape era? *Biol. Conserv.* 83:247–57
148. Simberloff D, Farr JA, Cox J, Mehlman DW. 1992. Movement corridors conservation bargains or poor investments? *Conserv. Biol.* 6:493–504
149. Skinner MW, BM Pavlik, ed. 1994. *Inventory of Rare and Endangered Vascular Plants of California*. Spec. Publ. #1 Calif. Native Plant Society, Sacramento. 5th ed.
150. Soule ME, ed. 1987. *Viable Populations for Conservation*. Cambridge, UK: Cambridge Univ. Press
151. Soule ME, Simberloff D. 1986. What do genetics and ecology tell us about the design of nature reserves? *Biol. Conserv.* 35:19–40
152. Stearns F, Matthiae P. 1997. The history of natural areas programs in Wisconsin. In *Conservation in Highly Fragmented Landscapes*, ed. MW Schwartz, pp. 407–17. New York: Chapman & Hall. 436 pp.
153. Steinauer EM, Collins SL. 1996. Prairie ecology–the tallgrass prairie. In *Prairie Conservation: Preserving North America's Most Endangered Ecosystem*, ed. FB Samson, FL Knopf, pp. 39–52. Washington, DC: Island. 339 pp.
154. Steuter AA. 1997. Bison. In *The Tallgrass Restoration Handbook: for Prairies, Savannas, and Woodlands*, ed. S Packard, CF Mutel, pp. 339–47. Washington, DC: Island. 463 pp.
155. Stoms DM, Davis FW, Driese KL, Cassidy KM, Murray MP. 1998. Gap analysis of the vegetation of the intermountain semi-desert ecoregion. *Great Basin Naturalist* 58:199–216
156. Stone CP, Scott JM ed. 1985. *Hawaii's Terrestrial Ecosystems: Preservation and Management*. Cooperative National Park Resources Study Unit, Honolulu: Univ. Hawaii.
157. Strahm W. 1996. Conservation of the flora of the Mascarene Islands. *Curtis' Botanical Magazine* 13:228–37
158. Suarez AV, Bogler DT, Case TJ. 1998. Effects of fragmentation and invasion on native ant communities on coastal southern California. *Ecology* 79:2041–56
159. Suarez AV, Pfennig KS, Robinson SK. 1997. Nesting success of a disturbance-dependent songbird on different kinds of edges. *Conserv. Biol.* 11:928–35
160. Swengel AB. 1996. Effects of fire and hay management on abundance of prairie butterflies. *Biol. Conserv.* 76:73–85
161. Swengel AB. 1998. Effects of management on butterfly abundance in tallgrass prairie and pine barrens. *Biol. Conserv.* 83:77–89
162. Tartowski SL, Allen EB, Barrett NE, Berkowitz AR, Colwell RK, et al. 1997. Integration of species and ecosystem

approaches to conservation. See Ref. 116, pp 187–92
163. Thackway R, Cresswell ID. 1997. A bioregional framework for planning the national system of protected areas in Australia. *Nat. Areas J.* 17:241–47
164. Tilman D. 1997. Biodiversity and ecosystem functioning. In *Nature's Services: Societal Dependence on Natural Ecosystems*, ed. GC Daily, pp 93–112. Washington, DC: Island. 392 pp.
165. Tobin RJ. 1990. *The Expendable Future: U.S. Politics and the Protection of Biological Diversity*. Durham, NC: Duke Univ. Press
166. Tuljapurkar S, Caswell H, ed. 1996. *Structured-Population Models in Marine, Terrestrial and Freshwater Systems*. New York: Chapman & Hall. 643 pp.
167. Turner IM, Corlett RT. 1996. The conservation value of small, isolated fragments of lowland tropical rain forest. *Trends Evol. Ecol.* 11:330–33
168. United States Fish and Wildlife Service. 1996. *Endangered and Threatened Wildlife and Plants. 50 Cfr 17.11 & 17.12*. Washington, DC: USGPO
168a. United States Fish and Wildlife Service. 1998. *Federal and State Endangered Species Expenditures. Fiscal Year 1995.* Washington, DC: USGPO
169. van Jaarsveld AS, Freitag S, Chown SL, Muller C, Koch S, et al. 1998. Biodiversity assessment and conservation strategies. *Science* 279:2106–08
170. Walker BH. 1992. Biodiversity and ecological redundancy. *Conserv. Biol.* 6:18–23
171. Walter KS, Gillett HJ, ed. 1998. *1997 UCN Red List of Threatened Plants*. Compiled by the World Conservation Monitoring Centre. IUCN–The World Conservation Union, Gland, Switzerland and Cambridge, UK. lxiv + 862 pp.
172. Weaver JC. 1995. Indicator species and scale of observation. *Conserv. Biol.* 9:939–42
173. Weber W, Rabinowitz A. 1996. A global perspective on large carnivore conservation. *Conserv. Biol.* 10:1046–54
174. White PCL, Gregory KW, Lindley PJ, Richards G. 1997. Economic values of threatened mammals in Britain: a case study of the otter *Lutra lutra* and the water vole *Arvicola terrestris*. *Biol. Conserv.* 82:345– 54
175. White PS, Walker JL. 1997. Approximating nature's variation: selecting and using reference information in restoration ecology. *Restor. Ecol.* 5:338–49
176. Wilcove D. 1993. Getting ahead of the extinction curve. *Ecol. Appl.* 3:218–20
177. Wilcove D, McMillan M, Winston KC. 1993. What exactly is an endangered species? An analysis of the endangered species list, 1985–1991. *Conserv. Biol.* 7:87–93
178. Wilcove DS, Rothstein D, Dubow J, Phillips A, Losos E. 1998. Quantifying threats to imperiled species in the United States. *Bioscience* 48:607–15
179. Williams P, Gibbons D, Margules C, Rebelo A, Humphries C, et al. 1996. A comparison of richness hotspots, rarity hotspots, and complementary areas conserving diversity of British birds. *Conserv. Biol.* 10:155–74
180. Wilson CA, Houseal GA, Mitchell RJ, Hendricks JJ, Boring LR. 1996. Complex ecological gradients in longleaf pine ecosystems: VII. Annual patterns of nitrogen and phosphorus availability as related to site productivity. *Bull. Ecol. Soc. Am.* 77(Suppl. Pt 2):485
181. Yaffee SL. 1994. *The Wisdom of the Spotted Owl: Policy Lessons for a New Century*. Washington, DC: Island. 430 pp.
182. Young A, Mitchell N. 1994. Microclimate and vegetation edge effects in a fragmented podocarp-broadleaf forest in New Zealand. *Biol. Conserv.* 67:63–72

Annu. Rev. Ecol. Syst. 1999. 30:109–32

CONSPECIFIC SPERM AND POLLEN PRECEDENCE AND SPECIATION

Daniel J. Howard

Department of Biology, New Mexico State University, Las Cruces, New Mexico 88003-8001; e-mail: dahoward@nmsu.edu

Key Words fertilization, gametic incompatibility, reproductive barrier, reproductive isolation, sperm competition

■ **Abstract** The evolution of reproductive isolation is perhaps the most significant stage in the process of species formation, and the study of reproductive barriers currently dominates investigations of speciation. The discovery that conspecific sperm and pollen precedence play an important role in the reproductive isolation of some closely related animals and plants is one of the real surprises to emerge from this field in recent years. This review begins with a brief history of the study of reproductive isolation with the aim of understanding why conspecific sperm and pollen precedence were generally overlooked in early work on reproductive barriers. It then examines: case studies, the prevalence of conspecific sperm and pollen precedence, the isolating potential of this class of reproductive barriers, the mechanisms that account for the operation of these barriers, and potential explanations for the rapid divergence of populations in traits related to fertilization. Conspecific sperm and pollen precedence appear to be quite effective in limiting gene exchange; these barriers are widespread although not universal in animals and plants, and they operate through a number of different mechanisms. Much more work remains to be done on a number of fronts to elucidate the processes responsible for the evolution of these reproductive barriers.

Introduction

Speciation, one of the great themes of evolution, is usually an extended process involving the splitting and genetic divergence of two lineages, the evolution of trait differences that reduce potential or actual gene flow between the lineages, the irreversible separation of the lineages, and the evolution of exclusive genealogical relationships among the genes carried by each lineage (46, 69, 77, 143). Although the evolution of trait differences that isolate two lineages represents but one aspect of this process, it is seen as an especially important one by evolutionists; it is the event that irrevocably separates two lineages and assures their future independence. As such, the study of reproductive barriers has assumed a central role in investigations of speciation. Among the questions under active investigation are: In what geographic circumstances do reproductive barriers evolve (19, 56, 57);

0066-4162/99/1120-0109$08.00

what role does divergent natural selection play in the evolution of reproductively isolating trait differences (141); does reinforcement (selection against hybridization) strengthen barriers to gene flow (76, 118, 119); and, what number and kind of genes control reproductive barriers (39, 43, 115, 156, 167)? Pivotal to all of these investigations is the identification of the trait differences responsible for reproductive isolation between two closely related taxa.

Perhaps the greatest surprise to emerge from recent studies of reproductive barriers among terrestrial organisms has been evidence suggesting that conspecific sperm and pollen precedence can play important roles in isolating closely related taxa (2, 8, 24, 25, 65, 79, 89, 114, 135, 137, 169). Postinsemination barriers to fertilization were not unknown to earlier evolutionists; indeed, Dobzhansky (49) included gametic or gametophytic isolation in his list of the principal reproductive isolating mechanisms. However, from the 1940s through the 1980s such barriers were rarely invoked to explain reproductive isolation in terrestrial organisms. In this review, I provide a brief history of the study of reproductive isolation, and I discuss the factors that have, until recently, hidden postinsemination barriers to fertilization from the view of evolutionists. I then go on to examine case studies of conspecific sperm and pollen precedence, what is known of the mechanisms by which these barriers operate, their effectiveness at isolating closely related taxa, and hypotheses that might explain their rapid evolution. But before I begin, a digression into terminology is necessary.

Terminology

In 1951, Dobzhansky described gametic or gametophytic isolation as resulting when "Spermatozoa, or pollen tubes, of one species are not attracted to the eggs or ovules, or are poorly viable in the sexual ducts of another species" (50, p. 181). This is a definition in keeping with his description of fertilization problems in species crosses in 1937, and it is one that emphasizes a breakdown in the highly intricate interplay between male and female gametes or between male gametes and the female reproductive tract. Earlier authors, such as Darwin (42) and Romanes (138), used the term "prepotency" to denote the advantage in fertilization enjoyed by conspecific pollen over heterospecific pollen. More recently, terms such as "homogamy," "barrier to fertilization," "postinsemination barrier to fertilization," "concentration-dependent barrier to fertilization," "conspecific sperm precedence," "interspecific precedence," and "postpollination, prefertilization isolating mechanisms" have been used to describe reproductive barriers that come into play after insemination or pollination (8, 25, 65, 79, 81, 137). For the most part, the terms are self-explanatory, and several may be used interchangeably.

I have adopted the term "conspecific sperm precedence" when dealing with animals because the term "sperm precedence" has a clear meaning. Sperm precedence is defined as "the nonrandom utilization of sperm from one of several males to mate with a female" (145, p. 342). Conspecific sperm precedence, then, is the favored utilization of sperm from conspecific males in fertilization when both conspecific and heterospecific males have inseminated a female. The term says little about mechanism of action. The precedence may occur because the sperm of conspecific males outcompetes the sperm of heterospecific males or because heterospecific sperm generally fail to fertilize eggs. In either case, females produce fewer hybrid progeny than expected on the basis of the frequency of heterospecific mating. For plants, I use the term "conspecific pollen precedence." Although plants produce sperm and eggs, much of the research on differential fertilization in this group has focused on pollen tube growth rates (24, 25, 135). Moreover, pollen precedence has essentially the same meaning for botanists that sperm precedence has for zoologists. However, it should be kept in mind that conspecific pollen precedence may result from interactions that occur at the final stage of fertilization—the fusion of egg and spermatozoon.

A Brief History of the Study of Reproductive Barriers

Although the study of reproductive barriers is a major component of current investigations of species formation (69), such has not always been the case. Darwin did not consider species to be well-defined entities, clearly separated from one another by barriers to reproduction. Instead, he emphasized the lack of discontinuity in nature and the arbitrary character of species boundaries.

> From these remarks it will be seen that I look at the term species, as one arbitrarily given for the sake of convenience to a set of individuals closely resembling each other, and that it does not essentially differ from the term variety, which is given to less distinct and more fluctuating forms. (41, p. 52)

Darwin's point of view was to dominate evolutionary biology for the next 78 years. His influence was clearly evident in a symposium held by the Botanical Society of America on "Aspects of the Species Question," reports of which were published in 1908 in *The American Naturalist*. The prevalent points of view expressed by contributors were that species are created by men rather than by nature (9, 34, 35) and that the major criterion for drawing species boundaries is morphological distinctness (3, 18, 34, 173). No mention of reproductive isolation occurs in any of the papers.

This is not to say that the idea that species are real entities that are isolated from one another by a number of factors had not occurred to evolutionists. Gulick (67, 68) not only emphasized the importance of isolation in species formation, he also outlined various means by which species remain isolated when in spatial contact with one another. Similarly, Romanes (138) wrote of the importance of mutual infertility in setting species apart from one another. Even Darwin (41) noted the prepotency of conspecific over heterospecific pollen in plant crosses, influenced no doubt by the pioneering pollen competition studies of Kölreuter (104). However, these ideas did not attract a large number of followers, perhaps because of the influence of Darwin's clear statement on the nature of species, or perhaps because biologists found the vast new terminology introduced by Gulick daunting.

Whatever the merits of treating species as human-made constructs, this point of view did little to encourage research into species formation. Coherent research programs devoted to the study of speciation were rare early in the twentieth century. The ideas that reproductive isolation exists between species and that the evolution of this isolation is worthy of investigation were virtually absent from major journals such as *The American Naturalist* and *Genetics* through the 1930s. Nevertheless, enough data were gathered on the ecological, behavioral, and crossing relationships of closely related species during this period (e.g., 47, 97, 105, 150) to allow Dobzhansky (49) to list a variety of barriers that isolate such taxa. His clear thinking and language, together with his conclusion that an adequate understanding of the process of speciation could not be achieved without understanding the genetics of reproductive barriers, galvanized the study of reproductive isolation.

Dobzhansky's influence is plainly evident in the evolution literature of the 1940s and 1950s, a literature filled with studies of the factors that reproductively isolate closely related species. Among the noteworthy investigations were those of Blair (16, 17) on anurans; Moore (112) on races of *Rana pipiens*; Spieth (147, 148) on sexual isolation in *Drosophila*; Fouquette (59) on tree frogs; Fulton (60) on field crickets; Grant (63) on angiosperms; and Wasserman (172) on spadefoot toads. Despite intensive crossing efforts in many of the above and other studies (16, 59, 60, 72, 168, 172), gametic isolation was rarely implicated as a reproductive barrier, the insemination reaction in some closely related *Drosophila* (127, 128) serving as the major exception to this trend.

Enthusiasm for studies of isolating barriers waned during the 1960s and 1970s as the conclusion emerged that species were typically isolated by multiple barriers (51, 103). This finding raised two troubling questions: Which trait should be investigated in detail, and what could be learned about speciation from a narrow focus on a single reproductive barrier. Moreover, behavioral traits seemed to be of special importance in isolating closely related animals (103), and such traits were perceived as intransigent to genetic analysis. Many evolutionists turned their attention away from the evolution of reproductive barriers to a related problem, one for which tools were available—the amount of genetic divergence between closely related taxa. Protein electrophoresis took center stage in studies of speciation, a place it would hold until the 1980s, when the hope that new insights into speciation would emerge from studies of genetic differentiation finally foundered on the bewildering diversity of results reported by various laboratories (4, 40, 87, 88, 139, 179). Stung by the failure of the genetic divergence approach, biologists redoubled their efforts to develop model systems of two or more closely related taxa in which traits responsible for reproductive isolation could be identified and studied (36, 37, 44, 71, 100, 101, 113). These efforts were spurred on by a growing conviction that Dobzhansky (49, 50) was correct when he argued that an understanding of speciation required an understanding of the genetics of traits responsible for reproductive isolation (19, 154).

It has been during this second round of interest in reproductive barriers that conspecific sperm and pollen precedence have come to the attention of evolutionists.

For the most part, these barriers are not detectable in typical species crossing experiments in which males and females are placed together for extended periods of time, or in which the pollen from only one species is deposited on the stigma of a plant. To detect the advantage of conspecific sperm or pollen in fertilization usually requires sperm or pollen competition experiments. The need for sperm and pollen competition studies explains why the significance of barriers to fertilization was not recognized by earlier workers—they rarely carried out such experiments. These investigators expected species crosses to fail if gametic isolation, as described by Dobzhansky (49, 50), was in place. The often successful production of hybrids in the laboratory and the greenhouse was seen as an argument against the importance of postinsemination, prefertilization factors in the isolation of closely related species. The incorporation of gamete competition studies into investigations of reproductive isolation has taken place piecemeal, and even now, many evolutionists examining reproductive isolation fail to recognize the importance of including such analyses in their research programs.

Case Studies

Epilachna vigintioctomaculata and *E. pustulosa* are closely related phytophagous ladybirds that are widely sympatric in northern Japan. Although they tend to occur on different host plants in the field, heterospecific matings are common (114). In contrast to most of the other studies to be discussed, heterospecific matings result in very low egg hatch rates, about 5% of those of conspecific matings. However, females mated both to a conspecific and to a heterospecific male have egg hatch rates equivalent to those of females mated to two conspecifics. This result indicates that conspecific sperm fertilizes most of the eggs of doubly-mated females. The conspecific sperm precedence largely eliminates the negative impact of a heterospecific mating and lessens the possibility that other isolating barriers will arise as a result of selection against hybridization (114).

The grasshopper subspecies *Chorthippus parallelus erythropus* and *C. p. parallelus* meet in a narrow hybrid zone that runs along the Pyrenees (74). In multiple choice mating tests, pure species individuals from outside the zone demonstrate a tendency to mate assortatively (136). The subspecies also display song differences and differences in mate finding strategies (20). Postmating barriers to gene exchange also occur, in the form of F1 hybrid male sterility (8). Supplementing all of these barriers is consubspecific sperm precedence. Regardless of the order of mating, when mated with two males, one consubspecific and the other heterosubspecific, females produce fewer hybrid offspring than expected from random fertilization (8). The effect is asymmetrical, being stronger for *parallelus* females than for *erythropus* females. Despite the occurrence of multiple reproductive barriers between the two subspecies, extensive genome mixing has occurred within the zone (20).

The ground crickets *Allonemobius fasciatus* and *A. socius*, are sister species that meet and interact in a mosaic hybrid zone in the eastern United States.

Within mixed populations (populations containing individuals of both species), pure species individuals usually predominate, and individuals classified as hybrid typically possess genotypes characteristic of backcrosses (82). Thus, reproductive isolation between the two species in areas of contact appears strong, although incomplete. This isolation cannot be attributed to premating factors. Although calling songs differ between the species (166), females are not sensitive to the song differences (52), and both males and females show little tendency toward assortative mating in mate choice experiments (66). The two species occur in the same habitats at the same time of the year, eliminating habitat and phenology as potential reproductive barriers (83). Postzygotic barriers are weak, as well. Hybrid individuals are viable in the field and laboratory (64, 83), and interspecific crosses result in the formation of many fertile offspring (64). The only reproductive barrier that has yet been uncovered is conspecific sperm precedence. When females of either species are mated sequentially to two males, a conspecific and a heterospecific, regardless of the order of mating, the vast majority of progeny are sired by the conspecific male (65, 79). The fertilization advantage of the conspecific male persists even when a female is mated only once to a conspecific but multiple times to a heterospecific (81).

The three main species of the Louisiana iris complex are *Iris fulva, I. hexagona*, and *I. brevicaulis*. Although hybrid populations exist, F1 hybrids are rarely found in nature, suggesting that the three taxa are strongly isolated from one another (55). A variety of factors appear to contribute to the isolation, including conspecific pollen precedence. In the case of *I. fulva* and *I. hexagona*, habitat differences exist, but phenological and pollinator differences are slight, suggesting that pollen transfer between the two species is common in nature (2, 25). When applied by itself to stigmas, heterospecific pollen is quite effective in siring seeds in both species (25). However, when conspecific pollen and heterospecific pollen are both applied to receptive stigmas, fewer hybrid seeds are produced than expected based on the composition of the pollen load. The drop in hybrid seed production is especially great when *I. fulva*, rather than *I. hexagona*, is the seed parent. For both types of seed parents, hybrid seed production increases when heterospecific pollen is applied to a stigma prior to the application of conspecific pollen, a finding that suggests pollen tube growth rate plays an important role in mediating conspecific pollen precedence (25). In the case of *I. brevicaulis*, conspecific pollen precedence occurs but is probably relatively unimportant in isolating this species from its congeners (55). *I. brevicaulis* occurs in habitats distinctly different from those that harbor *I. fulva* and *I. hexagona*, and its flowering time does not overlap that of the other two.

The sunflowers *Helianthus annuus* and *H. petiolaris* are morphologically distinct plants that form a mosaic hybrid zone over most of the central region of the United States (135). Although reproductive isolation is incomplete, several barriers limit gene flow between the two species, including habitat differences and poor viability of hybrid pollen and seeds. However, conspecific pollen precedence may represent the strongest reproductive barrier. In mixed pollinations of *H. annuus*, fewer than 10% of the progeny are hybrid, even when heterospecific pollen outnumbers conspecific pollen 9:1. Precedence is even stronger in *H. petiolaris*. A maximum of 2% of progeny are hybrid when heterospecific pollen outnumbers conspecific pollen by a factor of nine (135). In common with many of the detailed studies of conspecific gamete precedence, heterospecific pollen fertilizes both *H. annuus* and *H. petiolaris* quite effectively when not competing with conspecific pollen (135).

Tribolium castaneum is a human commensal with a worldwide distribution, and *T. freemani* is a close relative found in Kashmir (75). The two species do not exhibit premating reproductive isolation in the laboratory (169), and certain strains of *T. castaneum* produce large numbers of sterile hybrids in reciprocal crosses with *T. freemani*. Conspecific sperm precedence is very strong in both species. When females of either species are mated with both conspecific and heterospecific males, the conspecific male sires virtually all of the offspring (137, 169). The strong conspecific sperm precedence coupled with the high level of sterility in hybrids severely constrains the potential for gene exchange between the two species.

Drosophila simulans is a human commensal with a worldwide distribution. Its two closest relatives, *D. mauritiana* and *D. sechellia* (130), are island endemics found on Mauritius and the Seychelles, respectively (158). Some premating isolation exists between the species; females of *D. mauritiana* and *D. sechellia* are reluctant to mate with males of *D. simulans* (131). Nevertheless, laboratory crosses of *D. simulans* with the other two species produce numerous offspring, although hybrid males are sterile. However, when a female mates sequentially with a conspecific and a heterospecific male, the conspecific male sires the vast majority of offspring, regardless of the order of mating. Among the factors that contribute to the conspecific sperm precedence are poor performance by heterospecific sperm in the reproductive tract of females and the defensive and offensive effects of the seminal fluid of conspecific males (131). It is clear from this example and those that have preceded it, that conspecific sperm precedence can evolve as quickly as, if not more quickly than, behavioral isolation, hybrid sterility, and hybrid inviability as populations diverge from one another.

Marine invertebrates are one group in which the importance of barriers to fertilization in reproductive isolation has come as no surprise to evolutionists. These are often creatures with relatively simple nervous systems, and in many species males and females do not interact behaviorally prior to gamete release. Organisms such as sea urchins and abalone can be induced to release a great number of gametes in a laboratory setting, and they have become models for the study of fertilization. Species specificity of interactions between sperm and eggs of sympatric sea urchins was recognized as early as 1915 by Loeb (98), who noted that when sperm of *Strongylocentrotus purpuratus* is added to a mixture of *Strongylocentrotus franciscanus* and *S. purpuratus* eggs in seawater, the eggs of *S. purpuratus* are fertilized more quickly than the eggs of *S. franciscanus*. The reverse is true if sperm of *S. franciscanus* is added to the mixture. The level of specificity was measured quantitatively by Lillie (97), who found that the concentration of

S. franciscanus sperm that would fertilize from 73% to 100% of *S. franciscanus* eggs would fertilize only from 0 to 1.5% of *S. purpuratus* eggs. *S. purpuratus* sperm display a similar, although somewhat reduced, level of species specificity.

Gametic incompatibility in sea urchins occurs, in large measure, because heterospecific sperm are less likely than conspecific sperm to attach to the egg vitelline layer (107, 124). The molecule on the surface of sperm that mediates species-specific attachment is bindin. Bindin has been intensively studied in a number of closely related sea urchins, and variation in this protein is clearly related, albeit in a complex manner, to variation in egg recognition by sperm (106, 108, 111, 121). Small pieces of the bindin protein can inhibit fertilization in a species-specific manner, but single amino acid substitutions do not appear to have a major effect on bindin-egg interactions (122). Within species, many bindin alleles exist and intraspecific variation resembles interspecific variation. These results suggest that there is variation in bindin function within a species and that the differences between species represent the end of a continuum rather than an abrupt discontinuity (122). It should be noted that the bindin gene represents a "speciation" gene of the most elegant sort. The protein coded for by the gene plays a direct role in the reproductive isolation of closely related species, and the relationship between genetic variation and phenotypic effect can be easily assessed. It is a gene worthy of special attention when assessing hypotheses that might account for the evolution of conspecific sperm precedence.

Another gamete recognition protein worthy of close attention is lysin, a protein of the acrosomal granule of the sperm of abalone that creates a hole in the egg vitelline envelope, allowing sperm to fuse with the plasma membrane of eggs (96). Species-specificity is not absolute, but in studies of sympatric California abalone, lysin appears to operate more effectively on conspecific than on heterospecific vitelline envelopes (142). Lysin is well understood from a molecular perspective. The gene has been sequenced from seven species of California abalone, and the resulting amino acid sequences deduced (93, 163). In contrast to bindin, lysin is monomorphic within species and highly divergent between species.

The Effectiveness of Conspecific Sperm and Pollen Precedence as Barriers to Reproduction

Understanding the isolating potential of a reproductive barrier is critical to assessing its importance in nature. Assuming random mating between two taxa, Gregory & Howard (65) argued that strong conspecific sperm precedence would serve most effectively as a reproductive barrier when two species occur with equal abundance and females mate many times. Under such circumstances, females of both species are likely to mate at least once with a conspecific male, and the advantage of conspecific sperm in fertilization should reduce the production of hybrid offspring. As the abundance of one species decreases, so too does the chance that females of the less abundant species will encounter and mate with conspecific males. Thus, at some threshold of relative abundance, a threshold that will vary from case to case, conspecific sperm precedence will not, by itself, serve as an effective barrier to reproduction. The level of relative abundance at which isolation breaks down will depend on the strength of conspecific sperm precedence and the number of times a female mates. Arnold et al (2) noted that the effectiveness of conspecific pollen precedence as a reproductive barrier is subject to similar constraints.

The most rigorous examinations of the isolating potential of conspecific sperm or pollen precedence have occurred in the ground crickets *Allonemobius fasciatus* and *A. socius* and in the irises *I. fulva* and *I. hexagona*. In the case of the ground crickets, Howard and colleagues (80) established three sets of population cages; in one set, the two species occurred in equal abundance, and in the other two sets one species was in the minority (20% of the population). The number of hybrids produced in each cage was compared to the number predicted in the absence of conspecific sperm precedence, using a stochastic simulation model. The stochastic simulation model was calibrated by the observed pattern of mating in each cage (which, in general, was random), and by the mean and variance of both the number of offspring produced in laboratory crosses and their survivorship to maturity (data contained in 64). In all cages, the numbers of hybrid progeny were low (usually less than 6% of the offspring produced) and differed significantly from the numbers expected in the absence of conspecific sperm precedence. Of great interest was the finding that the proportion of hybrid progeny did not differ between cages with equal and unequal numbers of the two species. These results demonstrate that conspecific sperm precedence can severely limit gene flow between closely related species, even when one species is less abundant than the other.

Arnold et al (2) tested the level of reproductive isolation between *I. fulva* and *I. hexagona* by introducing 200 individual *I. hexagona* rhizomes into a natural population of approximately 600 *I. fulva* plants. The next year, they analyzed 710 seeds from the mixed population and found that only seven were hybrids. The lack of hybrids could not be explained by differences in flowering time between the two species or by visitation by different pollinators (although pollinator behavior was not quantified). Very strong conspecific pollen precedence appears to provide the best explanation for the lack of hybrid seeds. Consistent with the expectation that conspecific pollen precedence is less effective as a reproductive barrier for the species that is in the minority, the hybrid progeny all came from *I. hexagona* maternal plants.

Mechanisms of Conspecific Sperm and Pollen Precedence

Evolutionists are relative late-comers to fertilization biology, a field that has attracted the attention of plant and animal breeders and molecular biologists for many years. Hence, much more is known about the physiology and molecular biology of fertilization than can be covered here (for reviews see 53, 160, 175). Vacquier (162) provides a generalized scheme of sperm-egg interactions in internally and externally fertilized animals that will be used to organize the discussion that follows. For animals that use internal fertilization, sperm passage through a female reproductive tract must be added to Vacquier's scheme to provide a complete picture of fertilization. In the first step of sperm-egg interactions, sperm move toward an unfertilized egg as a result of attraction by egg-released molecules. Evidence that this attraction can, in some cases, be species-specific has been found in cnidarians (23), ophiuroid echinoderms (110), and rabbits (48). In the second step of fertilization, sperm binds to the egg envelope. In some groups, such as amphibians and most mammals, the binding occurs before the acrosomal reaction, whereas in other groups, such as annelids and sea urchins, the acrosomal reaction occurs before binding (162). In all cases studied to date, binding occurs due to the interactions between proteins on the surface of sperm, and sugar moieties of glycoproteins associated with eggs (28, 91, 157, 176). As noted earlier, the best studied binding protein is the bindin of sea urchin, and there is strong evidence of species-specificity of sperm-egg binding in several closely related sea urchins (95, 121). Other evidence of species-specificity of binding between eggs and sperm has been found in oysters (6).

After sperm-egg binding, the spermatozoon produces a hole in the egg envelope and reaches the plasma membrane of the egg. Although proteases seem to be important in this step in ascidians and mice (162), the best-known dissolving agent is the lysin of abalone, which is not an enzyme. As noted earlier, lysin exhibits strong species-specificity in its action (142). The fusion of the plasma membrane of the egg with the plasma membrane of the spermatozoon represents the final stage of fertilization. This has proven to be a difficult step to study, but among the molecules implicated in mediating fusion are the bindin of sea urchin (161), which appears to be important in both sperm-egg binding and sperm-egg fusion, and an 18-kDa protein that coats the plasma membrane of abalone sperm following the acrosomal reaction (151, 152).

For species with internal fertilization, interactions between sperm and the female reproductive tract are as important a component of fertilization as interactions between eggs and sperm. The female reproductive tract, although providing a means of access to eggs, is by no measure a friendly highway. A typical male mammal inseminates a female with millions of sperm with each copulation, but only a very small fraction make it to the vicinity of the egg. The remainder are ejected from the vagina, fail to negotiate the passages through the cervix and the utero-tubal junction, are attacked by phagocytic cells, or are destroyed by anti-sperm antibodies (15, 53). Evidence that conspecific sperm fare better than heterospecific sperm (of relatively distantly related species) in passing through the female reproductive tract has been found in golden hamsters (146).

The components of fertilization in flowering plants were recently summarized by Delph & Havens (45). Fertilization begins with pollen germination on a stigma, formation of a pollen tube that transports sperm to the egg, and pollen tube growth through the style, ovary, and ovule. Upon reaching the embryo sac, the pollen tube bursts, releasing two sperm, one of which fuses with the nucleus of the egg, and the other of which fuses with two polar nuclei to form a triploid cell from which endosperm develops. Most studies of differential fertilization in plants have focused on a single component of this sequence—pollen tube growth (73, 132, 170).

The outlines of fertilization given above suggest a number of mechanisms by which conspecific sperm or pollen precedence can operate. Conspecific sperm may: have an advantage in negotiating the female reproductive tract, be better attracted to the egg, bind faster and more tightly to the egg envelope, penetrate the egg envelope more quickly, and be more likely to fuse with the egg plasma membrane. Conspecific pollen may: be more likely to germinate on the stigma, have faster pollen tube growth, and have pollen tubes that are less likely to burst prematurely and more likely to negotiate the micropyle of the egg. In the aforementioned scenarios, interactions between different ejaculates or between the pollen of different donors play no role in precedence. However, it is also possible for interference competition to mediate precedence. Evidence that the seminal fluid products of one male can interfere with fertilization by another male has been found in *Drosophila* (27, 70, 131). In a similar vein, mixed donor pollinations result in lower amounts of pollen germination than single donor pollinations in wild radish, suggesting that some sort of interference is occurring (102).

Although mechanisms underlying conspecific sperm and pollen precedence are incompletely understood in most cases studied to date, it is clear that a variety of factors mediate this precedence. The best understood cases from a mechanistic viewpoint are sea urchins and California abalone, where an advantage in sperm-egg binding and membrane fusion for conspecific sperm has been well-documented, and some of the molecules involved have been identified (96, 111, 153, 163, 164). Interestingly, the bindin of sea urchins is not related to the lysin of abalone or to gamete recognition proteins found in any other group outside of the echinoids (111). In the less-well-understood cases, such as that of *Drosophila simulans*, *D. sechellia*, and *D. mauritiana*, interference competition clearly occurs, but the components of the seminal fluid responsible for the interference have not been identified (131). In the ladybirds *Epilachna vigintioctomaculata* and *E. pustulosa*, incompatibility between heterospecific sperm and the female reproductive tract probably accounts for conspecific sperm precedence. Although there are no differences in the amount of semen ejaculated in heterospecific matings and conspecific matings, much less sperm makes it to the sperm storage organ of females mated heterospecifically (89). Similarly, in the ground crickets *Allonemobius fasciatus* and *A. socius*, preliminary evidence indicates that heterospecific sperm stored in the spermatheca are less motile than conspecific sperm (65). In *Iris fulva* and *I. hexagona*, the results of sequential pollinations suggest that pollen tube growth rate and pollen tube attrition are major factors contributing to conspecific pollen precedence (25). In contrast, initial observations have failed to implicate differences in pollen tube growth rate in the conspecific pollen precedence exhibited by *Helianthus annuus* and *H. petiolaris* (135).

How Prevalent are Conspecific Sperm and Pollen Precedence?

Conspecific sperm and pollen precedence may be common, but they are not universal, a finding that should surprise few evolutionists given the idiosyncratic nature of reproductive barriers (77). Gómez & Cabada (62) reported that the eggs of *Bufo arenum* are easily fertilized by the sperm of both *B. paracnemis* and *Leptodactylus chaquensis*. Moreover, because the fast block to polyspermy in *B. arenum* is not effective in blocking the sperm of the other species, heterospecific sperm have an advantage in fertilizing *B. arenum* eggs. In a surprising early study, Chang et al (26) reported that when domesticated rabbits (*Oryctolagus cuniculus*) were inseminated with a mixture containing equal numbers of rabbit sperm and snowshoe hare (*Lepus americanus*) sperm, the snowshoe hare sperm fertilized 66% of the females' eggs. In the case of the teleosts *Oryzias latipes* and *Oryzias melastigma*, the sperm of *O. melastigma* are thinner, longer, and swim faster than the sperm of *O. latipes*. The smaller size and greater speed of the *O. melastigma* sperm apparently give them an advantage over *O. latipes* sperm in fertilization. In mixed inseminations, almost twice as many *O. latipes* eggs are fertilized by *O. melastigma* sperm as by conspecific sperm (84).

Even in free-spawning marine invertebrates, the archetype group for barriers to fertilization, gametic incompatibility between closely related species has not been a universal finding. Byrne & Anderson (21) reported that the gametes of *Patiriella calcar* and *P. gunnii*, two asteroids found sympatrically, are reciprocally compatible. In sea urchins of the genus *Arbacia*, the amino acid sequence of bindin varies little among species, and fertilization experiments between two divergent species detected no blocks to gamete recognition (106). Similarly, an extensive series of experiments provided no evidence of barriers to fertilization among seven species of the scleractinian coral genus *Platygyra* (109).

Based on the *Patiriella* results and the results of other hybridization studies on echinoderms in which crosses between species proceeded easily (22, 99, 149), Byrne & Anderson (21) went so far as to suggest that gametic incompatibility, although clearly important in some instances, may not be the most important reproductive barrier separating closely related echinoderms and other broadcast spawning marine invertebrates. Such a conclusion is certainly premature. It is clear that gametic incompatibility need not be complete in order to play a role in reproductive isolation. None of the investigations of marine invertebrates that have failed to implicate gametic incompatibility as a reproductive barrier have incorporated sperm competition studies. In the absence of such studies, one cannot rule out the possibility that conspecific sperm will easily outcompete heterospecific sperm for fertilization of eggs. Even in groups with very strong conspecific sperm precedence, heterospecific crosses between individual males and females can produce many offspring (64, 169).

The Evolution of Conspecific Sperm and Pollen Precedence

The discovery of the widespread existence of conspecific sperm and pollen precedence suggests that barriers to fertilization evolve rapidly between diverging taxa. This suggestion is supported by a number of molecular studies. Singh and his colleagues (29, 31, 155) documented that proteins specific to the reproductive tracts of male and female *Drosophila melanogaster* and its sibling species diverge more rapidly than nongonadal proteins. Moreover, comparisons of DNA sequences coding for a variety of proteins in three species of *Drosophila* and two species of nematode demonstrate that genes involved in mating behavior, fertilization, spermatogenesis, and sex determination not only evolve quickly, they also have a high ratio of nonsynonymous-to-synonymous substitutions (32). The propensity to acquire nonsynonymous substitutions more rapidly than synonymous substitutions characterizes many of the fertilization genes studied to date, from the lysin genes of California abalone (93) and the bindin genes of sea urchins of the genus *Echinometra* (108) to the accessory protein genes of *Drosophila* (1, 30, 159). Some form of selection appears to be driving the divergence between species in these genes, but the nature of the selection is poorly understood.

A number of hypotheses have been put forward to explain the rapid evolution of traits and proteins related to fertilization (80, 165, 123, 133). The hypotheses can be summarized as: (*a*) response to pathogens; (*b*) sexual selection by male-male competition; (*c*) sexual conflict with female choice playing a pivotal role; (*d*) sexual conflict with avoidance of polyspermy playing a crucial role; (*e*) selection against hybridization (reinforcement), and (*f*) concerted evolution. The response-to-pathogens hypothesis has been advocated most forcefully by Vacquier and his colleagues (163, 165), who note that benthic invertebrates must contend with a host of microbes and that gamete recognition proteins on the surface of eggs may directly or indirectly defend the egg. Therefore, these proteins may evolve rapidly to keep pace with the evolution of the microbes. Similarly, the female reproductive tract appears to offer an easily accessible port of entry to disease organisms, so the hostility of this tract may have evolved at least in part not to ward off male gametes but to counter the entry of pathogens (144). One of the characteristics of proteins known to evolve in response to pathogens, such as the MHC complex, is extensive intraspecific variation (58, 92). This leads to the prediction that the glycoproteins mediating fertilization of broadcast spawning marine invertebrates will exhibit considerable intraspecific variation, either in amino acid sequence or in patterns of assembly on the surface of the vitelline envelope, if their evolution is being driven by the need to thwart pathogens. Unfortunately, the data necessary to test this hypothesis are not yet available (157).

The second hypothesis, sexual selection by male-male competition with the female serving as a passive, relatively unchanging playing field emanates from the field of intraspecific sperm competition, which early in its inception tended to focus on male interests (e.g., 125). This viewpoint changed as it became evident that female genotypes can influence the outcome of sperm competition between males (33, 120, 174, 178). Currently, most ecologists and evolutionists emphasize the importance of considering female interests in studies of sperm and pollen competition (12, 14, 54, 61, 171). Thus, the male-male competition hypothesis seems an unlikely explanation for the rapid evolution of conspecific sperm and pollen precedence. Moreover, if the female reproductive tract or egg surface were to remain relatively stable over evolutionary time, an optimal solution to transport and attachment would be expected to evolve in male gametes. In order for competition between males to produce variation between species and an advantage for conspecific gametes in fertilization, the conditions under which the competition occurs would need to vary (81).

The sexual conflict hypotheses note that the interests of males and females are different and that this difference establishes an antagonistic coevolution between the two sexes in traits related to fertilization. The two forms of the sexual conflict hypotheses, the female-choice hypothesis and the avoidance-of-polyspermy hypothesis, differ in the interests attributed to females. In the female-choice hypothesis, the major interest of the female is seen as control of paternity (10, 12, 14, 15, 54). According to this hypothesis, the hostile reproductive tract of females and protective vestments around the egg evolved to ensure fertilization by a vigorous male gamete (90). In the avoidance-of-polyspermy hypothesis, the major interest of the female is seen as preventing more than one sperm from fertilizing an egg (7, 81, 134, 126). The potentially lethal effects associated with polyspermy (86, 129) are presumed to lead to the evolution of hostile female reproductive tracts and the evolution of envelopes around eggs that either destroy or slow down enough male gametes that only a single spermatozoon fuses with the plasma membrane of the egg. In common with the female-choice hypothesis, males are assumed to be under constant pressure to adapt to the female reproductive tract and eggs, and females are assumed to be under constant pressure to adapt to semen and sperm or to pollen, pollen tubes, and sperm. The result of sexual conflict, whether attributable to female choice or to avoidance of polyspermy, is endless coevolution between males and females, and divergence between isolated populations in traits related to fertilization because the coevolutionary process is likely to follow different trajectories in different populations (81).

Distinguishing between the avoidance-of-polyspermy and the female-choice hypotheses will be difficult (81). Not only are many of the predictions of the hypotheses similar, but both factors may act in concert. For example, males may threaten females with polyspermy as they adapt to a hostile reproductive tract that itself evolved to promote fertilization by a vigorous male gamete. Females, in turn, become effectively more choosy as they protect themselves against this threat. It has been suggested (81) that one area where the predictions of the two hypotheses differ is with regard to monandrous animals. The reasoning goes that if gene expression in sperm is generally lacking (an assumption that is increasingly questionable; see 116, 177), female choice based on genetic differences among sperm can operate only if a female mates with more than one male. The female-choice hypothesis therefore predicts that the female reproductive tract should be less hostile, and conspecific sperm precedence less evident, in monandrous species. Polyspermy avoidance, on the other hand, remains as important in monandrous as in polyandrous species. Therefore, it is argued (81), the polyspermy-avoidance-hypothesis predicts no difference in the reproductive tracts of monandrous and polyandrous species. Unfortunately, this argument fails to consider that the interests of males and females are less antagonistic under monandry (released from sperm competition with other males, the interest of males now coincide with that of females—to ensure that each egg is fertilized by only one spermatozoon). Thus, the sexes should cooperate in avoiding polyspermy, which leads to the same prediction as the female-choice hypothesis—a less hostile female reproductive tract in monandrous species.

Regardless of the difficulty of differentiating between them, an important objective of future work should be the testing of a prediction central to both the avoidance-of-polyspermy and the female-choice hypotheses; namely, that males from local populations of a species will exhibit sperm or pollen precedence over males from allopatric populations. This prediction has received some initial support from work on the weedy plant *Turnera ulmifolia* (5), but many more systems must be examined before a general conclusion can be reached.

Before moving on to the next hypothesis, a comment on polyspermy is in order. Polyspermy appears to occur routinely in birds (10) and in salamanders and newts (85). In these organisms, polyspermy does not have obvious negative effects, an observation that seems to run counter to the polyspermy avoidance hypothesis. However, even in birds, the female reproductive tract appears to be hostile to sperm, and factors that operate within the egg prevent more than one spermatozoon nucleus from fusing with the female pronucleus (11, 12). Thus, organisms that exhibit polyspermy still possess a number of mechanisms that limit access to the female pronucleus, and dealing with excess sperm remains a problem for females.

The fifth hypothesis mentioned in conjunction with conspecific sperm and precedence is reinforcement (165), the evolution of reproductive barriers in areas of sympatry as a consequence of selection against hybridization (76). Although reinforcement fell out of favor with evolutionists during the 1970s and 1980s, it has enjoyed a comeback in the 1990s as an increasing number of empirical studies have provided support for its operation in nature (38, 118, 140). A number of findings suggest that reinforcement may play a role in the evolution of conspecific sperm precedence, most notably: the hypervariable, species-specific amino-terminal domain of lysin, which is presumed to mediate species-specific binding, is truncated in the single abalone species, *Haliotis tuberculata*, that occurs in a region free of other abalone (94); and bindin amino acid sequences of allopatric species of sea urchin demonstrate much less divergence than do the bindin sequences of sympatric species (106, 108). However, the real signature of reinforcement–stronger conspecific sperm or pollen precedence between sympatric than between allopatric populations of two closely related species–has not yet been reported.

The final hypothesis, concerted evolution, emerged from recent studies of VERL, the egg receptor for lysin in abalone, which is a large protein consisting of a repeated amino acid sequence motif. In an important paper, Swanson & Vacquier (153) presented results suggesting that concerted evolution accounts for the homogenization of the repeated sequence within species; they hypothesized that the need to respond to this process might explain the rapid divergence of lysin between species. However, as noted by Nei & Zhang (117), while concerted evolution can explain the similarity of the repeat sequences within a species, it cannot account for rapid divergence between species. Thus, some other process is probably responsible for the differences in lysin between species.

Conclusion

The study of conspecific sperm and pollen precedence is in its infancy, and much remains to be learned about the prevalence of these phenomena, their mode of action, their genetic control, and the forces that lead to their evolution. A complete understanding will require multifaceted studies that combine the techniques and approaches of ecologists, evolutionists, molecular geneticists, and fertilization biologists. Although the tasks appear daunting, determined effort promises to shed light, not only on an important group of reproductive barriers, but on the very process by which new species are formed.

ACKNOWLEDGMENTS

I am grateful to Nick Waser and Mary Price for a memorable morning spent discussing sperm and pollen competition studies, and to Mike Draney and Mohamed Noor for their comments on an earlier draft of this paper. Supported by grants from the National Science Foundation.

LITERATURE CITED

1. Aguadé M, Miyashita N, Langley CH. 1992. Polymorphism and divergence in the *Mst26A* male accessory gland gene region in *Drosophila*. *Genetics* 132:755–70
2. Arnold ML, Hamrick JL, Bennett BD. 1993. Interspecific pollen competition and reproductive isolation in *Iris*. *J. Hered.* 84:13–16
3. Arthur JC. 1908. The physiologic aspect of the species question. *Am. Nat.* 42:243–48
4. Ayala FJ, Tracey ML, Hedgecock D, Richmond RC. 1974. Genetic differentiation during the speciation process in *Drosophila*. *Evolution* 28:576–92
5. Baker AM, Shore JS. 1995. Pollen competition in *Turnera ulmifolia* (Turneraceae). *Am. J. Bot.* 82:717–25
6. Banks MA, McColdrick DJ, Borgeson W, Hedgecock D. 1994. Gametic incompatibility and genetic divergence of Pacific and Kumamoto oysters, *Crassostrea gigas* and *C. sikamea*. *Mar. Biol.* 121:127–35
7. Bedford JM. 1983. Form and function of eutherian spermatozoa in relation to the nature of egg vestments. *In Fertilization of the Human Egg In Vitro*, eds. HM Beier, HR Lindner, pp. 133–46. Berlin: Springer-Verlag
8. Bella JL, Butlin RK, Ferris C, Hewitt GM. 1992. Asymmetrical homogamy and unequal sex ratio from reciprocal mating-order crosses between *Chorthippus parallelus* subspecies. *Heredity* 68:345–52
9. Bessey CE. 1908. The taxonomic aspect of the species question. *Am. Nat.* 42:218–24
10. Birkhead TR. 1995. Sperm competition: evolutionary causes and consequences. *Reprod. Fertil. Dev.* 7:755–75
11. Birkhead TR. 1996. Mechanisms of sperm competition in birds. *Am. Sci.* 84:254–62
12. Birkhead TR. 1998. Sperm competition in birds: mechanisms and function. See Ref. 13, pp. 579–622
13. Birkhead TR, Møller AP, eds. 1998. *Sperm*

Competition and Sexual Selection. San Diego, Academic. 826 pp.

14. Birkhead TR, Møller AP. 1998. Sperm competition, sexual selection and different routes to fitness. See Ref. 13, pp. 757–81
15. Birkhead TR, Møller AP, Sutherland WJ. 1993. Why do females make it so difficult for males to fertilize their eggs? *J. Theor. Biol.* 161:51–60
16. Blair AP. 1941. Variation, isolating mechanisms, and hybridization in certain toads. *Genetics* 26:398–417
17. Blair WF, Littlejohn MJ. 1960. Stage of speciation of two allopatric populations of chorus frogs (*Pseudacris*). *Evolution* 14:82–87
18. Britton NL. 1908. The taxonomic aspect of the species question. *Am. Nat.* 42:225–42
19. Bush GL, Howard DJ. 1986. Allopatric and non-allopatric speciation: assumptions and evidence. In *Evolutionary Processes and Theory*, eds. S Karlin, E Nevo, pp. 411–38. Orlando: Academic
20. Butlin RK. 1998. What do hybrid zones in general, and the *Chorthippus parallelus* zone in particular, tell us about speciation? See Ref. 78, pp. 367–78
21. Byrne M, Anderson MJ. 1994. Hybridization of sympatric *Patiriella* species (Echinodermata: Asteroidea) in New South Wales. *Evolution* 48:564–76
22. Cameron RA. 1984. Two species of *Lytechinus* (Toxpneustidae: Echinoidea: Echinodermata) are completely cross-fertile. *Bull. South. Calif. Acad. Sci.* 83:154–57
23. Campbell RD. 1974. Cnidaria. *In Reproduction of Marine Invertebrates*, Vol. 1, ed. AC Giese, JS Pearse, pp. 133–99. New York: Academic
24. Carney SE, Cruzan MB, Arnold ML. 1994. Reproductive interactions between hybridizing Irises: analyses of pollen-tube growth and fertilization success. *Am. J. Bot.* 81:1169–75
25. Carney SE, Hodges SA, Arnold ML. 1996. Effects of differential pollen-tube growth on hybridization in the Louisiana Irises. *Evolution* 50:1871–78
26. Chang MC, Marston JH, Hunt DM. 1964. Reciprocal fertilization between the domesticated rabbit and the snowshoe hare with special reference to insemination of rabbits with an equal number of hare and rabbit spermatozoa. *Exp. Zool.* 155:437–46
27. Chapman T, Liddle LF, Kalb JM. Wolfner MF, Partridge L. 1995. Cost of mating in *Drosophila melanogaster* females is mediated by male accessory gland products. *Nature* 373:241–44
28. Chen J, Litscher E, Wassarman PM. 1998. Inactivation of the mouse sperm receptor, mZP3, by site-directed mutagenesis of individual serine residues located at the combining site for sperm. *Proc. Natl. Acad. Sci. USA* 95:6193–97
29. Choudhary M, Coulthart MB, Singh RS. 1992. A comprehensive study of genic variation in natural populations of *Drosophila melanogaster*. VI. Patterns and processes of genic divergence between *D. melanogaster* and its sibling species, *Drosophila simulans*. *Genetics* 130:843–53
30. Cirera S, Aguadé M. 1998. Molecular evolution of a duplication: the sex-peptide (*Acp70A*) gene region of *Drosophila subobscura* and *Drosophila madeirensis*. *Mol. Biol. Evol.* 15:988–96
31. Civetta A, Singh RS. 1995. High divergence of reproductive tract proteins and their association with postzygotic reproductive isolation in *Drosophila melanogaster* and *Drosophila virilis* group species. *J. Mol. Evol.* 41:1085–95
32. Civetta A, Singh RS. 1998. Sex and speciation: genetic architecture and evolutionary potential of sexual versus nonsexual traits in the sibling species of the *Drosophila melanogaster* complex. *Evolution* 52: 1080–92
33. Clark AG, Begun DJ. 1998. Female

genotypes affect sperm displacement in *Drosophila*. *Genetics* 149:1487–93

34. Clements FE. 1908. An ecologic view of the species conception. *Am. Nat.* 42:253–64
35. Cowles HC. 1908. An ecological aspect of the conception of species. *Am. Nat.* 42:265–71
36. Coyne JA. 1984. Genetic basis of male sterility in hybrids between two closely related species of *Drosophila*. *Proc. Natl. Acad. Sci. USA* 81:4444–47
37. Coyne JA. 1985. Genetic studies of three sibling species of *Drosophila* with relationship to theories of speciation. *Genet. Res. Camb.* 46:169–92
38. Coyne JA, Orr HA. 1997. Patterns of speciation in *Drosophila* revisited. *Evolution* 51:295–303
39. Coyne JA, Simeonidis S, Rooney P. 1998. Relative paucity of genes causing inviability in hybrids between *Drosophila melanogaster* and *Drosophila simulans*. *Genetics* 150:1091–1103
40. Craddock EM, Johnson WE. 1979. Genetic variation in Hawaiian *Drosophila*. V. Chromosomal and allozymic diversity in *Drosophila silvestris* and its homosequential species. *Evolution* 33:137–55
41. Darwin C. 1859. *On the Origin of Species*. London: John Murray. 490 pp.
42. Darwin C. 1898. *The Effects of Cross and Self Fertilisation in the Vegetable Kingdom*. New York: D. Appleton. 482 pp.
43. Davis AW, Wu C-I. 1996. The broom of the sorcerers apprentice: the fine structure of a chromosomal region causing reproductive isolation between 2 sibling species of *Drosophila*. *Genetics* 143:1287–98
44. Dawley EM. 1986. Behavioral isolating mechanisms in sympatric terrestrial salamanders. *Herpetologica* 42:156–64
45. Delph LF, Havens K. 1998. Pollen competition in flowering plants. See Ref. 13, pp. 149–73
46. de Queiroz K. 1998. The general lineage concept of species, species criteria, and the process of speciation. See Ref. 78, pp. 57–75
47. Dice LC. 1933. Fertility relationships between some of the species and subspecies of mice in the genus *Peromyscus*. *J. Mamm.* 14:298–305
48. Dickmann Z. 1963. Chemotaxis of rabbit spermatozoa. *J. Exp. Biol.* 40:1–5
49. Dobzhansky T. 1937. *Genetics and the Origin of Species*. New York: Columbia Univ. Press. 364 pp.
50. Dobzhansky T. 1951. *Genetics and the Origin of Species*. New York: Columbia Univ. Press. 364 pp. 3rd ed.
51. Dobzhansky T. 1970. *Genetics of the Evolutionary Process*. New York: Columbia Univ. Press
52. Doherty JA, Howard DJ. 1996. Lack of preference for conspecific calling songs in female crickets. *Anim. Behav.* 51:981–90
53. Drobnis EZ, Overstreet JW. 1992. Natural history of mammalian spermatozoa in the female reproductive tract. *Oxford Rev. Reprod. Biol.* 14:1–45
54. Eberhard WG. 1998. Female roles in sperm competition. See Ref. 13, pp. 91–116
55. Emms SK, Hodges SA, Arnold ML. 1996. Pollen-tube competition, siring success, and consistent asymmetric hybridization in Louisiana Irises. *Evolution* 50:2201–6
56. Feder JL, Reynolds K, Go W, Wang EC. 1995. Intra- and interspecific competition and host race formation in the apple maggot fly, *Rhagoletis pomonella* (Diptera: Tephritidae). *Oecologia* 101:416–25
57. Feder JL, Roethele JB, Wlazlo B, Berlocher SH. 1997. Selective maintenance of allozyme differences among sympatric host races of the apple maggot fly. *Proc. Natl. Acad. Sci. USA* 94:11417–21
58. Figueroa F, Gunther E, Klein J. 1988. MHC polymorphism pre-dating speciation. *Nature* 335:167–70
59. Fouquette MJ Jr. 1960. Isolating mechanisms in three sympatric treefrogs in the canal zone. *Evolution* 14:484–97

60. Fulton BB. 1952. Speciation in the field cricket. *Evolution* 6:283–95
61. Gomendio M, Harcourt AH, Roldán ERS. 1998. Sperm competition in mammals. See Ref. 13, pp. 667–51
62. Gómez MI, Cabada MO. 1994. Amphibian cross-fertilization and polyspermy. *J. Exp. Zool.* 269:560–65
63. Grant V. 1949. Pollination systems as isolating mechanisms in angiosperms. *Evolution* 3:82–97
64. Gregory PG, Howard DJ. 1993. Laboratory hybridization studies of *Allonemobius fasciatus* and *A. socius* (Orthoptera: Gryllidae). *Ann. Entomol. Soc. Am.* 86:694–701
65. Gregory PG, Howard DJ. 1994. A post-insemination barrier to fertilization isolates two closely related ground crickets. *Evolution* 48:705–10
66. Gregory PG, Remmenga MD, Howard DJ. 1998. Patterns of mating between two closely related ground crickets are not influenced by sympatry. *Entomol. Exp. Applic.* 87:263–70
67. Gulick JT. 1887. Divergent evolution through cumulative segregation. *Linn. Soc. J., Zool.* 20:189–274
68. Gulick JT. 1905. *Evolution, Racial and Habitudinal.* Washington, DC: Carnegie Inst. 269 pp.
69. Harrison RG. 1998. Linking evolutionary pattern and process: the relevance of species concepts for the study of speciation. See Ref. 78, pp. 19–31
70. Harshman LG, Prout T. 1994. Sperm displacement without sperm transfer in *Drosophila melanogaster*. *Evolution* 48:758–66
71. Heady SE, Denno RF. 1991. Reproductive isolation in *Prokelisia* planthoppers (Homoptera: Delphacidae): acoustic differentiation and hybridization failure. *J. Insect Behav.* 4:367–90
72. Heiser CB. 1951. Hybridization in the annual sunflowers: *Helianthus annuus* x *H. argophyllus*. *Am. Nat.* 85:65–72
73. Heslop-Harrison J. 1975. Male gametophyte selection and the pollen-stigma interaction. In *Gamete Competition in Plants and Animals*, ed. DL Mulcahy, pp. 177–89. Amsterdam: North-Holland
74. Hewitt GM. 1990. Divergence and speciation as viewed from an insect hybrid zone. *Can. J. Zool.* 68:1701–15
75. Hinton HE. 1948. A synopsis of the genus *Tribolium* MacLeay, with some remarks on the evolution of its species-groups (Coleoptera, Tenebrionidae). *Bull. Entomol. Res.* 39:13–55
76. Howard DJ. 1993. Reinforcement: origin, dynamics, and fate of an evolutionary hypothesis. In *Hybrid Zones and the Evolutionary Process*, ed. RG Harrison, pp. 46–69. New York: Oxford Univ. Press
77. Howard DJ. 1998. Unanswered questions and future directions in the study of speciation. See Ref. 78, pp. 439–48
78. Howard DJ, Berlocher SH, eds. 1998. *Endless Forms: Species and Speciation.* New York, Oxford Univ. Press. 470 pp.
79. Howard DJ, Gregory PG. 1993. Post-insemination signalling systems and reinforcement. *Philos. Trans. R. Soc. Lond. B* 340:231–36
80. Howard DJ, Gregory PG, Chu J, Cain ML. 1998. Conspecific sperm precedence is an effective barrier to hybridization between closely related species. *Evolution* 52:511–16
81. Howard DJ, Reece M, Gregory PG, Chu J, Cain ML. 1998. The evolution of barriers to fertilization between closely related organisms. See Ref. 78, pp. 279–88
82. Howard DJ, Waring GL. 1991. Topographic diversity, zone width, and the strength of reproductive isolation in a zone of overlap and hybridization. *Evolution* 45:1120–35
83. Howard DJ, Waring GL, Tibbets CA, Gregory PG. 1993. Survival of hybrids in a mosaic hybrid zone. *Evolution* 47:789–800
84. Iwamatsu T, Onitake K, Matsuyama K, Satoh M, Yukawa S. 1997. Effect of

micropylar morphology and size on rapid sperm entry into the eggs of the Medaka. *Zool. Sci.* 14:623–28

85. Iwao Y, Yasumitsu K, Narihira M, Jiang J, Nagahama Y. 1997. Changes in microtubule structures during the cell cycle of physiologically polyspermic newt eggs. *Mol. Reprod. Dev.* 47:210–21
86. Jaffe LA, Gould M. 1985. Polyspermy-preventing mechanisms. In *Biology of Fertilization*, ed. CB Metz, A Monroy, pp. 223–50. New York: Academic
87. Johnson MS, Clarke B, Murray J. 1977. Genetic variation and reproductive isolation in *Partula. Evolution* 31:116–26
88. Johnson WE, Selander RK. 1971. Protein variation and systematics in kangaroo rats (genus *Dipodomys*). *Syst. Zool.* 20:377–405.
89. Katakura H. 1986. Evidence for the incapacitation of heterospecific sperm in the female genital tract in a pair of closely related ladybirds (Insecta, Coleoptera, Coccinellidae). *Zool. Sci.* 3:115–21
90. Keller L, Reeve HK. 1995. Why do females mate with multiple males? The sexually selected sperm hypothesis. *Adv. Study Behav.* 24:291–315
91. Kitazume-Kawaguchi S, Inoue S, Inoue Y, Lennarz WJ. 1997. Identification of sulfated oligosialic acid units in the O-linked glycan of the sea urchin egg receptor for sperm. *Proc. Natl. Acad. Sci. USA* 94:3650–55
92. Lawlor DA, Ward FE, Ennis PD, Jackson AP, Parham P. 1988. HLA-A and B polymorphism predate the divergence of humans and chimpanzees. *Nature* 335: 268–71
93. Lee Y-H, Vacquier VD. 1992. The divergence of species-specific abalone sperm lysins is promoted by positive Darwinian selection. *Biol. Bull.* 182:97–104
94. Lee Y-H, Ota T, Vacquier VD. 1995. Positive selection is a general phenomenon in the evolution of abalone sperm lysin. *Mol. Biol. Evol.* 12:231–38
95. Lessios HA, Cunningham CW. 1990. Gametic incompatibility between species of the sea urchin *Echinometra* on the two sides of the Isthmus of Panama. *Evolution* 44:933–41
96. Lewis CA, Talbot CF, Vacquier VD. 1982. A protein from abalone sperm dissolves the egg vitelline envelope by a nonenzymatic mechanism. *Dev. Biol.* 92:227–39
97. Lillie FR. 1921. Studies on fertilization. VIII. On the measure of specificity in fertilization between two associated species of the sea-urchin genus *Strongylocentrotus*. *Biol. Bull.* 40:1–22
98. Loeb J. 1915. On the nature of the conditions which determine or prevent the entrance of the spermatozoon into the egg. *Am. Nat.* 49:257–85
99. Lucas JS, Jones MM. 1976. Hybrid crown-of-thorns starfish (*Acanthaster planci* x *A. bervispinus*) reared to maturity in the laboratory. *Nature* 263:409–12
100. Markow TA. 1991. Sexual isolation among populations of *Drosophila mojavensis*. *Evolution* 45:1525–29
101. Markow TA, Fogelman JC, Heed WB. 1983. Reproductive isolation in Sonoran Desert *Drosophila*. *Evolution* 37:649–52
102. Marshall DL, Folsom MW, Hatfield C, Bennett T. 1996. Does interference competition among pollen grains occur in wild radish? *Evolution* 50:1842–48
103. Mayr E. 1963. *Animal Species and Evolution*. Cambridge: Belknap. 797 pp.
104. Mayr E. 1986. Joseph Gottlieb Kölreuter's contributions to biology. *Osiris* (2nd ser.) 2:135–76
105. McCray FA. 1933. Embryo development in *Nicotiana* species hybrids. *Genetics* 13:95–110
106. Metz EC, Gómez-Gutiérrez G, Vacquier VD. 1998. Mitochondrial DNA and bindin gene sequence evolution among allopatric species of the sea urchin genus *Arbacia*. *Mol. Biol. Evol.* 15:185–95
107. Metz EC, Kane RE, Yanagimachi H, Palumbi SR. 1994. Fertilization between

closely related sea urchins is blocked by incompatibilities during sperm-egg attachment and early stages of fusion. *Biol. Bull.* 187:23–34
108. Metz EC, Palumbi SR. 1996. Positive selection and sequence rearrangements generate extensive polymorphism in the gamete recognition protein bindin. *Mol. Biol. Evol.* 13:397–406
109. Miller K, Babcock R. 1997. Conflicting morphological and reproductive species boundaries in the coral genus *Platygyra*. *Biol. Bull.* 192:98–110
110. Miller RL. 1997. Specificity of sperm chemotaxis among Great Barrier Reef shallow-water Holothurians and Ophiuroids. *J. Exp. Zool.* 279:189–200
111. Minor JE, Fromson DR, Britten RJ, Davidson EH. 1991. Comparison of the bindin proteins of *Strongylocentrotus franciscanus*, *S. purpuratus*, and *Lytechinus variegatus*: sequences involved in the species specificity of fertilization. *Mol. Biol. Evol.* 8:781–95
112. Moore JA. 1950. Further studies on *Rana pipiens* racial hybrids. *Am. Nat.* 84:247–54
113. Morrison DA, McDonald M, Bankoff P, Quirico P, Mackay D. 1994. Reproductive isolation mechanisms among four closely-related species of Conospermum (Proteaceae). *Bot. J. Linn. Soc.* 116:13–31
114. Nakano S. 1985. Effect of interspecific mating on female fitness in two closely related ladybirds (*Henosepilachna*). *Kontyû* 53:112–19
115. Naveira HF, Maside XR. 1998. The genetics of hybrid male sterility in *Drosophila*. See Ref. 78, pp. 330–38
116. Nayernia K, Adham I, Kremling H, Reim K, Schlicker M, et al. 1996. Stage and developmental specific gene expression during mammalian spermatogenesis. *Int. J. Dev. Biol.* 40:379–83
117. Nei M, Zhang J. 1998. Molecular origin of species. *Science* 282:1428–29
118. Noor M. 1995. Speciation driven by natural selection in *Drosophila. Nature* 375:674–75
119. Noor M. 1997. How often does sympatry affect sexual isolation in *Drosophila*? *Am. Nat.* 149:1156–63
120. Overstreet JW, Katz DE. 1977. Sperm transport and selection in the female reproductive tract. *Dev. Mamm.* 2:31–65
121. Palumbi SR. 1992. Marine speciation on a small planet. *Trend Ecol. Evol.* 7:114–17
122. Palumbi SR. 1994. Genetic divergence, reproductive isolation, and marine speciation. *Annu. Rev. Ecol. Syst.* 25:547–72
123. Palumbi SR. 1998. Species formation and the evolution of gamete recognition loci. See Ref. 78, pp. 271–78
124. Palumbi SR, Metz EC. 1991. Strong reproductive isolation between closely related tropical sea urchins (genus *Echinometra*). *Mol. Biol. Evol.* 8:227–39
125. Parker GA. 1970. Sperm competition and its evolutionary consequences in the insects. *Biol. Rev.* 45:525–67
126. Partridge L, Hurst LD. 1998. Sex and conflict. *Science* 281:2003–8
127. Patterson JT. 1946. A new type of isolating mechanism in *Drosophila. Proc. Natl. Acad. Sci. USA* 32:202–8
128. Patterson JT. 1947. The insemination reaction and its bearing on the problem of speciation in the *mulleri* subgroup. In *Studies in the Genetics of Drosophila V. Isolating Mechanisms. Univ. Tex. Publ.* 4720:41–77
129. Payne D, Warnes GM, Flaherty SP, Matthews CD. 1994. Local experience with zona drilling, zona cutting, and sperm microinjection. *Reprod. Fertility Dev.* 6:45–50
130. Powell JR. 1997. *Progress and Prospects in Evolutionary Biology: The Drosophila Model*. New York: Oxford Univ. Press. 562 pp.
131. Price CSC. 1997. Conspecific sperm pre-

cedence in *Drosophila*. *Nature* 388: 663–66
132. Radha MR, Vasudeva R, Hegde SG, Ganeshaiah KN, Shaanker RU. 1993. Components of male gametophytic competition in *Vigna unguiculata* L. (Walp). *Evol. Trends Plants* 7:29–36
133. Rice WR. 1998. Intergenomic conflict, interlocus antagonistic coevolution, and the evolution of reproductive isolation. See Ref. 78, pp. 261–70
134. Rice WR, Holland B. 1997. The enemies within: intergenomic conflict, interlocus contest evolution (ICE), and the intraspecific Red Queen. *Behav. Ecol. Sociobiol.* 41:1–10
135. Rieseberg LH, Desrochers AM, Youn SJ. 1995. Interspecific pollen competition as a reproductive barrier between sympatric species of *Helianthus* (Asteraceae). *Am. J. Bot.* 82:515–19
136. Ritchie MG, Butlin RK, Hewitt GM. 1989. Assortative mating across a hybrid zone in *Chorthippus parallelus* (Orthoptera: Acrididae). *J. Evol. Biol.* 2: 339–52
137. Robinson T, Johnson NA, Wade MJ. 1994. Postcopulatory, prezygotic isolation: intraspecific and interspecific sperm precedence in *Tribolium* spp., flour beetles. *Heredity* 73:155–59
138. Romanes GJ. 1897. *Darwin, and After Darwin*. Chicago: Open Court
139. Ryman N, Allendorf FW, Stahl G. 1979. Reproductive isolation with little genetic divergence in sympatric populations of brown trout (*Salmo trutta*). *Genetics* 92:247–62
140. Sætre G-P, Moum T, Bures S, Král M, Adamjan M, Moreno J. 1997. A sexually selected character displacement in flycatchers reinforces premating isolation. *Nature* 387:589–92
141. Schluter D. 1998. Ecological causes of speciation. See Ref. 78, pp. 114–29
142. Shaw A, Lee Y-H, Stout CD, Vacquier VD. 1994. The species-specificity and structure of abalone sperm lysin. *Dev. Biol.* 5:209–15
143. Shaw KL. 1998. Species and the diversity of natural groups. See Ref. 78, pp. 44–56
144. Sheldon BC. 1993. Sexually transmitted disease in birds—occurrence and evolutionary significance. *Philos. Trans. R. Soc. Lond. B* 339:491–97
145. Simmons LW, Siva-Jothy MT. 1998. Sperm competition in insects: mechanisms and the potential for selection. See Ref. 13, pp. 341–34
146. Smith TT, Koyanagi F, Yanagimachi R. 1988. Quantitative comparison of the passage of homologous and heterologous spermatozoa through the uterotubal junction of the golden hamster. *Gamete Res.* 19:227–34
147. Spieth HT. 1947. Sexual behavior and isolation in *Drosophila*. I. The mating behavior of species of the *willistoni* group. *Evolution* 1:17–31
148. Spieth HT. 1949. Sexual behavior and isolation in *Drosophila*. II. The interspecific mating behavior of species of the *willistoni* group. *Evolution* 3:67–81
149. Strathmann RR. 1981. On barriers to hybridization between *Strongylocentrotus droebachiensis* (O. F. Muller) and *S. pallidus* (G. O. Sars). *J. Exp. Mar. Biol. Ecol.* 55:39–47
150. Sturtevant AH. 1915. Experiments on sex recognition and the problem of sexual selection in *Drosophila*. *J. Anim. Behav.* 5:351–66
151. Swanson WJ, Vacquier VD. 1995. Liposome fusion induced by a Mr 18,000 protein localized to the acrosomal region of acrosome-reacted abalone spermatozoa. *Biochemistry* 34:14202–8
152. Swanson WJ, Vacquier VD. 1995. Extraordinary divergence and positive Darwinian selection in a fusagenic protein coating the acrosomal process of abalone spermatozoa. *Proc. Natl. Acad. Sci. USA* 92:4957–61
153. Swanson WJ, Vacquier VD. 1998. Con-

certed evolution in an egg receptor for a rapidly evolving abalone sperm protein. *Science* 281:710–12

154. Templeton AR. 1981. Mechanisms of speciation; a population genetic approach. *Annu. Rev. Ecol. Syst.* 12:23–48
155. Thomas S, Singh RS. 1992. A comprehensive study of genic variation in natural populations of *Drosophila melanogaster*. VII. Varying rates of genic divergence as revealed by two-dimensional electrophoresis. *Mol. Biol. Evol.* 9:507–25
156. Ting CT, Tsaur SC, Wu ML, Wu C-I. 1998. A rapidly evolving homeobox at the site of a hybrid sterility gene. *Science* 282:1501–4
157. Töpfer-Petersen E, Dostàlovà Z, Calvete JJ. 1997. The role of carbohydrates in sperm-egg interaction. *Adv. Exp. Med. Biol.* 424:301–10
158. Tsacas L, Lachaise D, David JR. 1981. Composition and biogeography of the afrotropical drosophilid fauna. In *The Genetics and Biology of Drosophila*, Vol. 3a, ed. M Ashburner, HL Carson, JN Thompson Jr, pp. 197–59. London: Academic
159. Tsaur S-C, Tinb C-T, Wu C-I. 1998. Positive selection driving the evolution of a gene of male reproduction, *Acp26Aa*, of *Drosophila*: II. Divergence versus polymorphism. *Mol. Biol. Evol.* 15:1040–46
160. Tulsiani DRP, Yoshida-Komiya H, Araki Y. 1997. Mammalian fertilization: a carbohydrate-mediated event. *Biol. Reprod.* 57:487–94
161. Ulrich AS, Otter M, Glabe CG, Hoekstra D. 1998. Membrane fusion is induced by a distinct peptide sequence of the sea urchin fertilization protein bindin. *J. Biol. Chem.* 273:16748–55
162. Vacquier VD. 1998. Evolution of gamete recognition proteins. *Science* 281:1995–98
163. Vacquier VD, Lee Y-H. 1993. Abalone sperm lysin: unusual mode of evolution of a gamete recognition protein. *Zygote* 1:181–96
164. Vacquier VD, Moy GW. 1997. The fucose sulfate polymer of egg jelly binds to sperm REJ and is the inducer of the sea urchin sperm acrosome reaction. *Dev. Biol.* 192:125–35
165. Vacquier VD, Swanson WJ, Lee Y-H. 1997. Positive Darwinian selection on two homologous fertilization proteins: What is the selective pressure driving their divergence? *J. Mol. Evol.* 44(Suppl 1):S15–S22
166. Veech JA, Benedix JH Jr., Howard DJ. 1996. Lack of calling song displacement between two closely related ground crickets. *Evolution* 50:1982–89
167. Via S, Hawthorne DJ. 1998. The genetics of speciation: promises and prospects of quantitative trait locus mapping. See Ref. 78, pp. 352–64
168. Volpe EP. 1952. Physiological evidence for natural hybridization of *Bufo americanus* and *Bufo fowleri*. *Evolution* 6:393–406
169. Wade MJ, Patterson H, Chang N, Johnson NA. 1993. Postcopulatory, prezygotic isolation in flour beetles. *Heredity* 71:163–67
170. Walsh NE, Charlesworth D. 1992. Evolutionary interpretation of differences in pollen tube growth rates. *Q. Rev. Biol.* 67:19–37
171. Waser NM, Price MV. 1993. Crossing distance effects on prezygotic performance in plants: an argument for female choice. *Oikos* 68:303–8
172. Wasserman AO. 1957. Factors affecting interbreeding in sympatric species of spadefoots (genus *Scaphiopus*). *Evolution* 11:320–38
173. Williston SW. 1908. What is a species? *Am. Nat.* 42:184–94
174. Wilson N, Tubman SC, Eady PE, Robertson GW. 1997. Female genotype affects male success in sperm competition. *Proc. R. Soc. Lond. B* 264:1491–95
175. Yanagimachi R. 1988. Mammalian fertilization. In *The Physiology of Repro-*

duction ed. E Knobil, J Neill et al., pp. 135–85. New York: Raven

176. Yonezawa N, Mitsui S, Kudo K, Nakano M. 1997. Identification of an N-glycosylated region of pig zona pellucida glycoprotein ZPB that is involved in sperm binding. *Eur. J. Biochem.* 248:86–92
177. Zeh JA, Zeh DW. 1997. The evolution of polyandry II: post-copulatory defences against genetic incompatibility. *Proc. R. Soc. Lond. B* 264:69–75
178. Zimmering S, Fowler GL. 1968. Progeny: sperm ratios and non-functional sperm in *Drosophila melanogaster*. *Genet. Res. Camb.* 12:359–63
179. Zimmerman EG, Kilpatrick CW, Hart BJ. 1978. The genetics of speciation in the rodent genus *Peromyscus*. *Evolution* 32:565–79

Annu. Rev. Ecol. Syst. 1999. 30:133–65

GLOBAL AMPHIBIAN DECLINES: A PROBLEM IN APPLIED ECOLOGY

Ross A. Alford and Stephen J. Richards
School of Tropical Biology and Cooperative Research Centre for Tropical Rainforest Ecology and Management, James Cook University, Townsville, Queensland 4811, Australia; e-mail: rossalford@jcu.edu.an

Key Words conservation, frog, salamander, null hypothesis, metapopulation

■ **Abstract** Declines and losses of amphibian populations are a global problem with complex local causes. These may include ultraviolet radiation, predation, habitat modification, environmental acidity and toxicants, diseases, changes in climate or weather patterns, and interactions among these factors. Understanding the extent of the problem and its nature requires an understanding of how local factors affect the dynamics of local populations. Hypotheses about population behavior must be tested against appropriate null hypotheses. We generated null hypotheses for the behavior of amphibian populations using a model, and we used them to test hypotheses about the behavior of 85 time series taken from the literature. Our results suggest that most amphibian populations should decrease more often than they increase, due to highly variable recruitment and less variable adult mortality. During the period covered by our data (1951–1997), more amphibian populations decreased than our model predicted. However, there was no indication that the proportion of populations decreasing changed over time. In addition, our review of the literature suggests that many if not most amphibians exist in metapopulations. Understanding the dynamics of amphibian populations will require an integration of studies on and within local populations and at the metapopulation level.

INTRODUCTION

The current wave of interest in amphibian population biology and in the possibility that there is a global pattern of decline and loss began in 1989 at the First World Congress of Herpetology (10). By 1993 more than 500 populations of frogs and salamanders on five continents were listed as declining or of conservation concern (189). There is now a consensus that alarming declines of amphibians have occurred (30, 51, 125, 147, 192). Because most amphibians are exposed to terrestrial and aquatic habitats at different stages of their life cycles, and because they have highly permeable skins, they may be more sensitive to environmental toxins or to changes in patterns of temperature or rainfall than are other terrestrial vertebrate

TABLE 1 Techniques used in 46 studies[1] to quantify populations of frogs salamanders. Many studies used more than one technique or studied both taxa, so the total number of techniques used does not equal the number of studies cited.

Technique	Habitat[2]	Frogs	Salamanders
Egg mass counts	B	4	1
Counts of individuals	B	12	1
Drift fence/pit trap counts	B	7	8
Mark-recapture estimates[3]	B	7	2
Calling male counts	B	3	N/A
Dipnet samples for larvae	B	2	2
Counts of individuals	N	1	6
Mark-recapture estimates	N	0	1
Aquatic traps	N	0	1

[1]Sources used in compiling table: 15, 19, 20, 21, 41, 49, 53, 54, 63, 64, 73, 83, 84, 86, 88, 90, 91, 92, 96, 102, 107, 111, 113, 117, 131, 136, 137, 142, 157, 158, 159, 160, 161, 165, 169, 170, 171, 174, 177, 178, 179, 180, 181, 194, 196, 202.

[2]B = Breeding, N = Non-breeding.

[3]Includes, in descending order of frequency used, Jolly-Seber, Petersen, Manly-Parr, Schnabel, and Zippin techniques.

groups (29, 190). The best-documented declines have occurred in Europe and North America, are usually associated with habitat modification (87, 116), and are often attributed to interactions among causal factors (114, 125, 147). The factors associated with population declines in relatively undisturbed habitats such as montane tropical rainforests have been more difficult to elucidate (131, 163).

Although they have been the subject of many experimental and monitoring studies, the autecology of amphibians in nature is poorly understood (87). The majority of studies of ecology and population biology of amphibians (Table 1) have been conducted on aggregations at reproductive sites. Relatively little is known of their movements or activities away from breeding sites, or of rates of exchange between populations.

Many authors have suggested that there is a need for long-term studies directed toward a combination of understanding ecological theory and increasing knowledge of the autecology of amphibians (17, 39, 79, 80, 190). Simple long-term programs that monitor the fluctuations of single populations and associated environmental factors, and then apply standard population models, are unlikely to be useful for understanding the dynamics of amphibian populations as they have not worked for that purpose when applied to other terrestrial vertebrates (164). It appears likely that understanding the problem of amphibian declines will require much more information on the ecology of the metapopulations in which many species live (87).

Our goals in this review are to summarize and synthesize the literature on potential causes of amphibian declines, and to use the literature on amphibian population dynamics to develop a null hypothesis for the behavior of amphibian populations. We then use our null hypothesis and data from the literature to determine whether the incidence of declines has recently increased. Finally, we place the dynamics of amphibian populations and their declines in the context of metapopulation dynamics.

POTENTIAL CAUSES OF AMPHIBIAN DECLINES

Ultraviolet Radiation

Depletion of stratospheric ozone and resultant seasonal increases in ultraviolet B (UV-B) radiation at the Earth's surface (119) have stimulated interest in the possible relationship between resistance of amphibian embryos to UV-B damage and population declines. Significant variation among species in levels of photolyase, a photoreactivating DNA repair enzyme that repairs UV-B damage, is correlated with exposure of natural egg deposition sites to sunlight (25, 98). In a survey of 10 Oregon amphibian species, photolyase activity, and hence ability to repair UV damage, was lowest in declining species and highest in nondeclining species (25). Field experiments demonstrated that embryos of *Hyla regilla*, a nondeclining species with high photolyase activity, had significantly higher hatching success than did two declining species (*Rana cascadae* and *Bufo boreas*) with low photolyase levels (25).

A number of other studies have demonstrated that ambient (6, 24, 28, 132) or enhanced (144) UV-B radiation reduces survival or hatching success of amphibian embryos. Synergistic interactions between UV-B and other environmental stresses such as pathogens (120) and low pH (133) may also significantly increase embryonic mortality. *Rana pipiens* embryos that are unaffected when exposed to UV-B and low pH separately have significantly reduced survival when exposed to these factors simultaneously (133).

Other studies have produced more equivocal results. *Rana aurora* is a declining species with high levels of photolyase (98), and experimental hatching success is unaffected by exposure to UV-B (26). The declining frog *Litoria aurea* from eastern Australia has a lower photolyase activity than two sympatric nondeclining species, *L. dentata* and *L. peroni*, but there is no significant difference among the three species in hatching success under UV-B exposure (187). In many aquatic habitats UV-B radiation is largely absorbed in the first few centimeters of the water column (138), so increased UV-B may only affect species breeding in habitats with a narrow range of chemical and physical parameters. Ecologically relevant levels of UV-B had no effect on embryos of several Canadian amphibians, and experimental protocols used to test UV impacts have been questioned (85, 130).

Most studies that have examined the relationship between UV-B and population declines have focused their attention on species that breed in shallow, clear water, where exposure to UV-B is expected to be greatest (6, 25, 28, 132). Exposure to intense UV-B in shallow high-altitude ponds may exclude amphibians from these habitats (138).

Even when UV-B causes higher embryonic mortality in declining species, the ecological significance of this at the population level is difficult to assess. More needs to be understood about the basic natural history of amphibian species that might be at risk. For example, information is needed on variation in oviposition site characteristics (depth, vegetation) within local populations. Experiments at "natural oviposition sites" using embryos of *Ambystoma macrodactylum* were conducted in shallow water although this species lays eggs in a variety of microhabitats (28). Loss of a large proportion of near-surface clutches to UV-B damage may have negligible impacts on populations if even a small number of deeper clutches survive, as the survivors are released from density-dependent effects (1).

Even fewer data are available to assess the indirect effects of increasing UV-B on amphibian populations. Potential indirect effects include changes in water chemistry and food supplies, and shifts in competitive and predator-prey relationships with other UV-B affected species (143). Exposure to increased UV-B may reduce survival rates of adult amphibians through damage to eyes (77), increased frequency of cancers or tumors (143), and immunosuppression (143).

Predation

Biotic interactions among amphibians, and between amphibians and other organisms, can play a significant role in determining their distribution and population dynamics (1). Larval amphibians are extremely vulnerable to vertebrate and invertebrate predators (1), and the diversity of aquatic amphibian assemblages is frequently reduced in habitats containing predatory fish (1, 100).

Larval amphibians that coexist with aquatic predators have evolved a range of antipredator mechanisms (4, 48, 118). However, widespread introductions of predatory fish have increasingly exposed native amphibians to predators with which they have not previously interacted. Inappropriate responses to novel predators may increase mortality of native amphibians (82, 121), leading to significant effects on populations.

Colonization of normally fish-free water bodies by predatory fish can result in rapid extinction of amphibian assemblages (76). The allotopic distributions of native frogs and introduced fishes in many high-elevation (>2500 m asl) Sierra Nevada lakes indicate that introduced predatory fishes have caused the extinction of local frog populations there (31). Sixty percent of lakes that frogs could formerly occupy now contain introduced fishes and no frogs. Fish introductions have had a particularly severe impact on *Rana muscosa*, which breeds in the deep lakes inhabited by fishes (31). A similar pattern of allotopic distributions has been recorded for larval newts, *Taricha torosa*, and an introduced fish (*Gambusia affinis*) and crayfish (*Procambarus clarki*) (both predators of newt eggs or larvae) in Californian mountain streams (82).

Introduced predators may also have more subtle effects. Some *Rana muscosa* populations persisting in fish-free environments have become isolated from other populations by surrounding aquatic habitats containing introduced fishes. This may eventually lead to regional extinction by preventing migration among local populations (35).

North American bullfrogs (*Rana catesbeiana*) that have become established outside their natural range have been implicated in declines of native frogs (76, 102, 127, but see 97). Adult bullfrogs consume native frogs and reach densities at which they are likely to have a severe impact on local amphibian populations (166). Experimental studies have shown that *Rana aurora* larvae exposed to adult or larval bullfrogs have increased larval periods, smaller mass, and, when exposed to both, lower survival (122).

Humans have devastated frog populations in several countries for the frog-leg trade. Before 1995, about two hundred million frogs were exported annually from Asia. By 1990 India was still illegally exporting approximately seventy million frogs each year, resulting in serious population declines (145).

Habitat Modification

Habitat modification is the best documented cause of amphibian population declines. Habitat loss certainly reduces amphibian abundance and diversity in the areas directly affected (99, 101). Removal or modification of vegetation during forestry operations has a rapid and severe impact on some amphibian populations (8). Clearcutting of mature forests in the southern Appalachians has reduced salamander populations by almost 9%, or more than a quarter of a billion salamanders, below the numbers that could be sustained in unlogged forests (149). Logging exposes terrestrial amphibians to drastically altered microclimatic regimes (9), soil compaction and desiccation, and reduction in habitat complexity (197). It exposes aquatic amphibians to stream environments with increased siltation (52) and reduced woody debris (43). Although populations may recover as regenerating forests mature, recovery to predisturbance levels can take many years (9) and may not occur at all if mixed forests are replaced with monocultures (108).

Draining wetlands directly affects frog populations by removing breeding sites (116), and by fragmenting populations (74, 168), which increases the regional probability of extinction (e.g. 53). Modification of terrestrial and aquatic habitats for urban development can reduce or eliminate amphibian populations. Populations of some amphibians in urban Florida declined after degradation of upland, dry season refuges and modification of wetlands used for breeding (62). Protection of aquatic breeding sites may be of little value if adjacent terrestrial habitats used by amphibians for feeding and shelter are destroyed (167).

More subtle alterations to habitat structure can have severe impacts on amphibian populations. *Bufo calamita* populations in Britain declined over a 40-year period due to shifts in land use practices that altered vegetation characteristics (13). Changing vegetation structure and an associated increase in shading were detrimental to *B. calamita* and provided conditions under which the common toad *Bufo bufo* became a successful competitor.

Although habitat alterations can reduce amphibian populations, in some cases even severe habitat modifications can have little effect. The response of a savanna woodland frog assemblage at Weipa, Queensland, Australia to strip mining appears in Figure 1 (200). The structure and floristics of the plant assemblage at 60 revegetated sites vary widely; none strongly resemble the original native woodland.

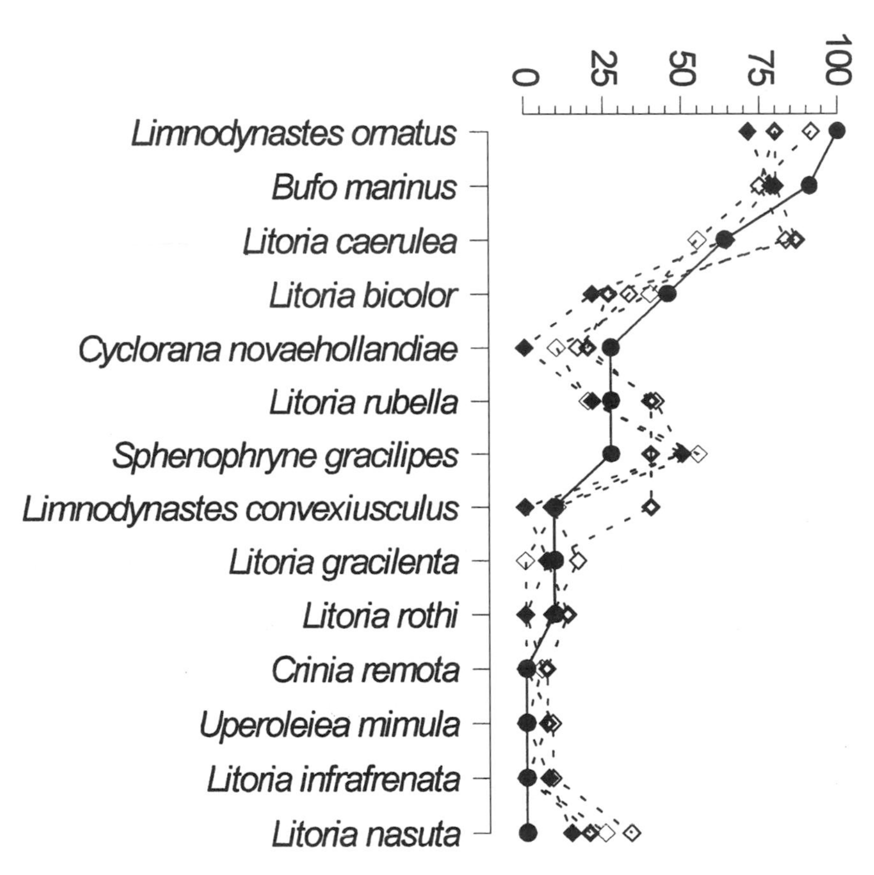

Figure 1 Profiles of percent of sites of five habitat types at Weipa, Queensland, Australia at which frogs of 14 species occurred. Species sorted in order of frequency of occurrence in native woodland. Dots and solid lines indicate native woodland habitat (13 sites). Diamonds and dashed lines indicate sites revegetated following strip mining. Density of diamonds reflects age of revegetation; from least to most dense this is: age ≤ 7 years, 7 years $<$ age $\leq$ 10.5 years, 10.5 years $<$ age $\leq$ 16 years, age $>$16 years. There were 15 revegetation sites in each age group in the survey.

The distance from revegetation sites to native woodland varies from a few meters to over 1 km. Despite this, within 7 years, the majority of frog species occur at revegetation sites at frequencies very similar to the frequencies at which they are found at sites in native woodland. This indicates that the frogs in this assemblage are insensitive to radical alterations in the soil characteristics, flora, and structure of their terrestrial habitat, and they can recolonize rapidly following disturbances that have eliminated them over a relatively wide area. It is possible that other amphibian species assemblages behave similarly.

Acidity and Toxicants

The acidity of aquatic habitats has major impacts on amphibian distribution, reproduction, and egg and larval growth and mortality (78, 79). Sensitivity to low pH varies among (79) and within (150) species and is influenced by complex chemical interactions among pH and other factors, particularly aluminum concentration (71, 110, 151). Mortality occurs in both the embryonic and larval stages via several mechanisms including incomplete absorption of the yolk plug, arrested development, and deformation of larvae (11, 79, 109). Sublethal effects of acidification include delayed (109) or early (36) hatching, reduced larval body size (36), disturbed swimming behavior (5), and slower growth rates resulting from reduced response to, and capture of, prey (155). Indirect sublethal effects include changes to tadpole food sources through impacts on algal communities (188), and shifting predator-prey relationships resulting from differential mortality of predatory fish and invertebrates in acidified habitats (104).

The population-level effects of acidity are less well understood. It is possible that the effects of low pH, in combination with other abiotic factors, lead to decreased recruitment into adult populations (12). Acidic breeding sites often contain less diverse amphibian assemblages, at lower densities, than do less acidic sites (205). Long-term acidification of ponds in Britain has excluded *Bufo calamita* from many sites (16). Reduced pH and increased metal concentrations in an Appalachian stream eliminated virtually all salamander larvae, causing severe long-term declines in populations of *Desmognathus quadramaculatus* and *Eurycea wilderae* (124). Low soil pH also influences the distribution, abundance, and diversity of terrestrial amphibians (204, 205).

Despite the well-documented effects of low pH on amphibians, there are few data to implicate acidification in recent, unexplained catastrophic population declines. Acid deposition was proposed as a factor in the decline of tiger salamanders, *Ambystoma tigrinum*, in the Rocky Mountains (96), but subsequent field studies demonstrated that mortality due to pond drying was equally likely to be the cause of this decline (201). Acid deposition is unlikely to be involved in population declines of frogs and salamanders at high altitudes in the Sierra Nevada Mountains (33, 34, 36) and Rocky Mountains (54, 55, 188, 201). There may be no rigorously documented cases where acidification of natural habitat has led to the extinction of an amphibian population (71). However, studies of acid tolerance have been biased toward species that are likely to have evolved tolerance to low pH (195).

Similarly, although there is an extensive literature on the toxic effects on larval amphibians of metals and chemicals used in insecticides and herbicides (154), insufficient data exist to determine their long-term impacts on amphibian population dynamics (22). Environmental toxicants act directly to kill animals, or indirectly by impairing reproduction, reducing growth rates, disrupting normal development and reproduction (endocrine disruption), or increasing susceptibility to disease by immunosuppression or inhibition of immune system development (22, 46).

Diseases

Little is known about the diseases of wild amphibians. Many disease agents are present in healthy animals, and disease occurs when immune systems are compromised (56, 57). Declines in populations of *Bufo boreas boreas* between 1974 and 1982 were associated with *Aeromonas hydrophila* infection, but Carey (47) suggested that environmental factor(s) caused sublethal stress in these populations, directly or indirectly suppressing their immune systems. A pathogenic fungus largely responsible for egg mortality in one population of *Bufo boreas* in Oregon may have been more virulent to embryos under environmental stress (27).

Epidemics can cause mass mortality of amphibians (123). In 1981 *Aeromonas hydrophila* killed all larval *Rana sylvatica* in a Rhode Island pond, and three years later few adult frogs were breeding at this site (141). The same bacterium was implicated in a well-documented decline to local extinction of a population of *Rana muscosa* in California (32). A chytridomycete fungus found on moribund anurans in Australia and Panama during mass mortality is fatal to healthy frogs under experimental conditions (18). This fungus was proposed as the proximate cause of declines in these two regions (18), but this hypothesis is untested at present.

Viruses have been isolated from dead and dying frogs during mass mortality events (59, 60) and may be the primary cause of mortality in animals where other infections such as bacteria have been identified (57). Laurance et al (129) argued that the pattern of population declines among Australian rainforest frogs was indicative of a "wave" of epidemic disease caused by an unidentified waterborne virus. This interpretation was challenged (3, 106) on statistical grounds and because numerous other explanations for the observed patterns were equally parsimonious. The involvement of a virus in these declines has now been largely discounted, but the possible involvement of a disease has not. The pattern of population declines in Central America has also led to the suggestion that a wave of epidemic disease might be responsible (131).

Climate/Weather

Immediately prior to the disappearance of golden toads, *Bufo periglenes*, the rainforests of Monteverde, Costa Rica, had the lowest twelve-month rainfall in 20 years. Toads were forced to shift their habitat use, and the dry conditions may have interacted with an unidentified factor such as disease or a pulse of contaminants in cloud water to eliminate toad populations (152). Unusual weather conditions were dismissed as a cause of declines of Australian rainforest frogs (128). This result relies on seasonal rainfall totals calculated for fixed groups of months, while in northern Australia the date of onset of the wet season is highly variable, occurring between October and February (140). The analysis is therefore likely to have missed the extremes and underestimated the variances of rainfall, and a reexamination of the data seems warranted. Severe, short-term climatic events such as violent storms can alter the dynamics of amphibian populations. Hurricane Hugo caused extensive damage to the forests of Puerto Rico in 1989. In the short term, populations of the terrestrial frog *Eleutherodactylus richmondi* decreased by 83%, but increased availability of ground cover due to disturbance led to a six-fold increase in the densities of *E. coqui*, followed by a long period of gradual population decrease (202, 203).

Alterations in local weather conditions caused by global climate change will influence the ecology of amphibians in a number of ways. The onset of spawning in *Rana temporaria* in Finland between 1846 and 1986 shifted earlier by 2–13 days, following shifts in air and water temperature and dates of snow cover loss (182). At some sites in Britain there has been a statistically significant trend toward earlier first sighting and spawning of *Bufo calamita*, *Rana esculenta*, and *R. temporaria* between 1978 and 1994, correlated with changing patterns of spring temperatures (14).

Amphibians in Canada are affected by decreases in summer precipitation and increased temperatures and winter rainfall (105, 143). In the neotropics, increased temperatures, extended dry seasons, and increasing inter-year rainfall variability may affect litter species by reducing prey populations and altering amphibian distributions on increasingly dry soil (67). Shifting rainfall patterns will affect the reproductive phenologies of pond-breeding species. Ponds will fill later and persist for shorter periods, leading to increased competition and predation as amphibians are concentrated at increasingly limited aquatic sites (67). Frogs exposed to these stresses may also become more vulnerable to parasites and disease (67).

Interactions Among Environmental Factors

Most studies invoke multiple causes or interactions among factors. Increased UV-B exposure may alter species interactions or vulnerability to pathogens or changes in pH. Predation may eliminate local populations and have larger-scale effects by altering rates of migration between populations. Outbreaks of disease may only occur when other stresses reduce immune function. Pesticides, pollutants, and environmental acidity may interact to produce unforeseen effects. All local effects may interact with global climate change. Proving the existence of these complex effects in natural populations will require well-planned programs of observation and experimentation. To plan such studies, and to determine how stresses affect population behavior, requires an understanding of the nature of the populations being studied and the limitations of study techniques. It also requires the development of null hypotheses regarding how amphibian populations behave in the absence of external pressures.

DEFINING AND STUDYING AMPHIBIAN POPULATIONS

Monitoring and Censusing Techniques

We have summarized the techniques used in 46 long-term population studies of frogs and salamanders in Table 1. All but one of the studies of frogs were carried out at breeding sites, while about half of the studies of salamanders included at least some data collected on densities of animals in nonbreeding habitat. The most commonly used technique is direct counts, where animals are located by intensive searching, localization of calls, or by drift fences with pitfall traps. Many studies that reported direct counts also reported the results of mark-recapture estimates. Frequently, however, the standard errors of mark-recapture estimates are very large, so that counts are regarded as being better estimates. The high standard errors typically obtained in mark-recapture estimates of animals at breeding sites probably reflect the fact that the degree of attraction of breeding sites varies widely over time, as does the activity of individual animals near them.

Problems with Studying Breeding Aggregations as Populations

Most estimates of frog populations are expressed in units such as total numbers of frogs attending a pond, numbers of frogs per m^2 of pond area, and maximum numbers of frogs at a pond on a single night during the breeding season (70). The frog groups for which data are available on densities in nonreproductive habitat tend to be species that do not aggregate at breeding sites (70, 115, 203). A few studies have examined aquatic-breeding species in nonreproductive habitats (156, 206). The use of breeding aggregations in population studies can cause problems in data interpretation.

A simple illustration of why censusing amphibians at breeding sites can cause problems in the interpretation of population dynamics appears in Figure 2. The number of frogs per unit of nonreproductive habitat (the entire rectangle) remains constant at 50. In Figure 2*a*, neither of the temporary pools in the habitat contains water. If censuses are carried out by visiting pool 1, a census at the time of Figure 2*a* would detect only 7–10 frogs in the habitat. A census carried out after pool 1 filled (Figure 2*b*) would detect 43 frogs, while one carried out after both pools filled (Figure 2*c*) would detect only 23 frogs at pool 1. Similar variations in measured density could occur over time even if the pools remain filled, because of changes in the reproductive behavior of frogs. Changes in habitat availability and in the attractiveness of bodies of water are both likely to affect the numbers of amphibians detected (191), complicating the use of data of this type as an indication of the size or density of the population occupying the surrounding habitat. Because behavior and site attributes vary seasonally and temporally with weather

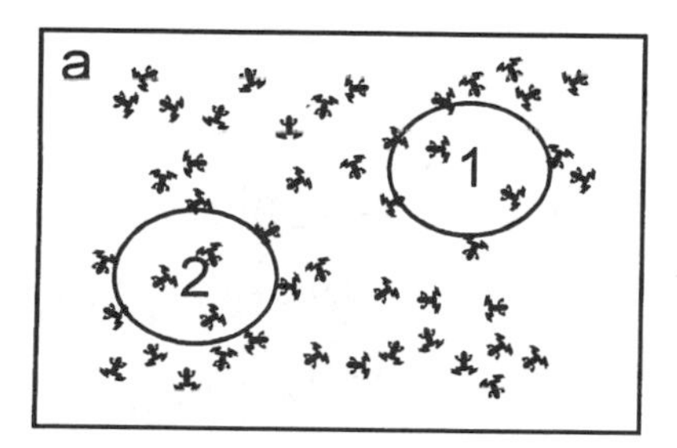

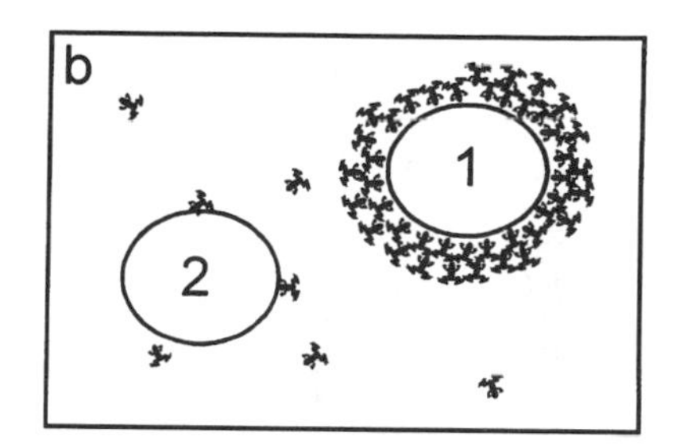

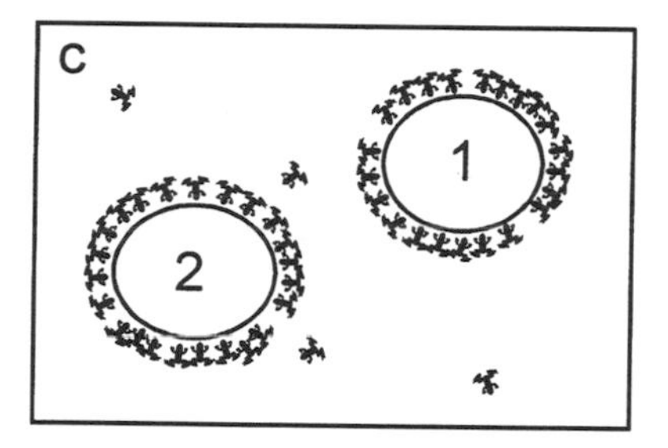

Figure 2 An illustration of the difficulties for data interpretation created by sampling animals in breeding aggregations. An area of habitat contains a fixed number of animals and two potential breeding sites. Surveys are carried out at site 1. In (*a*), animals are not attracted to breeding sites, so a census at site 1 would encounter only 9–10 individuals. In (*b*), animals are attracted to site 1, and a census would find 43 animals. In (*c*), animals are attracted to both sites, and a census at site 1 would encounter 22 animals. All of these situations may occur within a few days in many habitats, making counts at breeding sites highly variable.

and internal rhythms of the animals, the best hope for obtaining accurate estimates using censuses at breeding sites is to perform censuses frequently.

Effects of Sampling Intensity on Population Size Estimates

We used data collected by one of us (SJ Richards, unpublished data) on the number of adult males of the stream-breeding hylid *Litoria genimaculata* to investigate the effects of sampling intensity on the accuracy of count data taken at a breeding aggregation. A 60-m transect along a rainforest stream site was visited an average of 19 times each year over seven years. All frogs on the transect were counted, marked, and released. We used a resampling technique to explore how less intense sampling might have affected the results. We set the probability that a visit would occur to 0.75, 0.5, and 0.25, and resampled the data set 5000 times using each of these probabilities. It is apparent (Figure 3) that decreasing the intensity of sampling below about 0.75 times the number of visits actually made would greatly decrease the accuracy of the annual means obtained. At intensities of 0.5, 0.25,

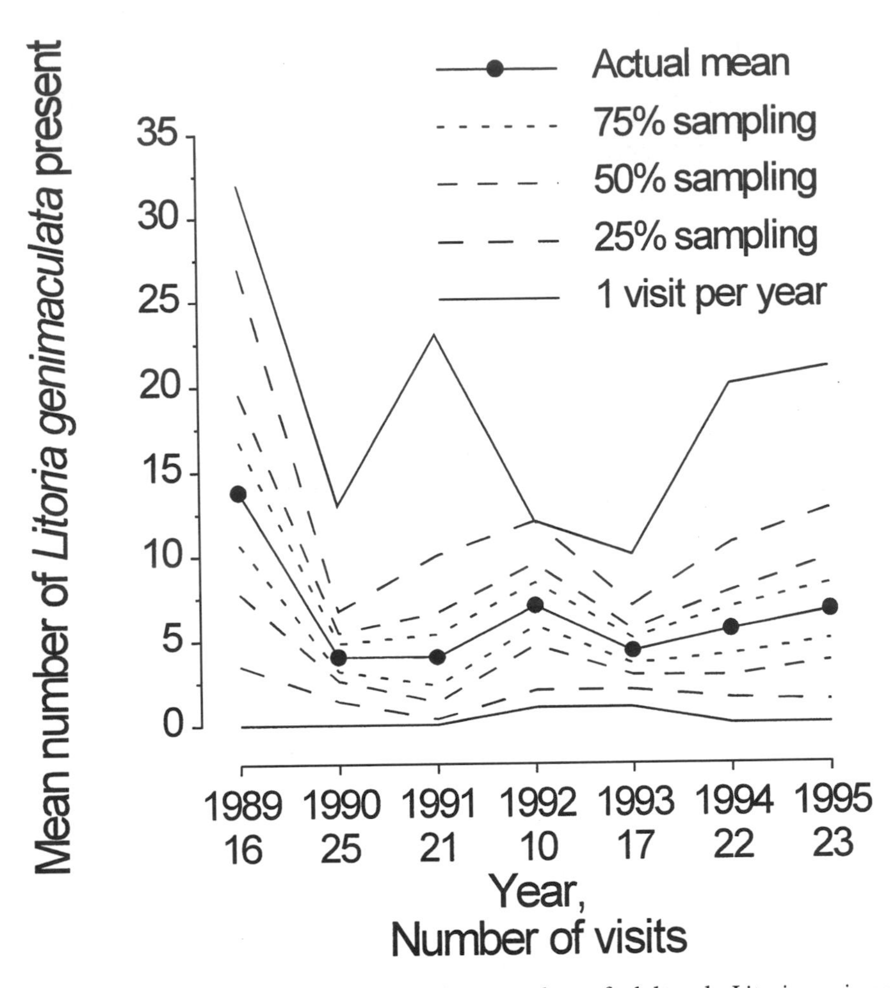

Figure 3 Results of resampling analysis on numbers of adult male *Litoria genimaculata* present on a 60-m transect on Birthday Creek, Mt. Spec, Queensland, Australia. Points indicate the mean number of individuals present taken over all visits in each year. The lines connecting the points are for illustration only. The dashed lines at increasing distances from the mean are the upper and lower 95% confidence limits for the location of the mean for each year at the given sampling intensity, as determined from 5000 resamplings of the data. Upper and lower solid lines are the 95% confidence limits for numbers obtained if the site was visited once annually; they are the minimum and maximum numbers encountered each year.

or one visit per year, the observed pattern of change through time would probably have differed from the pattern that appears in the full data set, and the coefficient of variation of the annual mean would have increased substantially.

Problems created by the changing attractiveness of breeding sites can also occur when animals are censused in nonreproductive habitat, because the attractiveness of patches changes. A population of *Rana arvalis* in a 2000-m^2 sampling area increased more than tenfold over the period 1984–1988 (89); this might have been due to a decrease in the local availability of water, causing animals to concentrate where it was available (89). Densities of the neotropical litter frogs *Bufo typhonius* and *Colostethus nubicola* in nonreproductive habitat varied more than tenfold over the course of a one-year study (183). This variation was probably caused by differences in the activity level and catchability of frogs, rather than by changes in population size (183).

Once a reasonable set of population estimates at a site is available, the next step in an analysis of the behavior of the population is to examine how it changes over time. It is unrealistic to expect perfect stability from any natural population; even if the population is truly constant, sampling error will introduce some variation. Given that we expect populations to fluctuate, we should try to determine how they will behave over time in the absence of directional pressures toward expansion or decline.

A Null Hypothesis for the Behavior of Amphibian Populations over Time

There has been substantial debate in recent years (23, 146, 148, 169) regarding how "normal" amphibian populations should change through time, but little theoretical work on this question. The problem has been defined clearly: Fluctuating populations of amphibians will be either increasing or decreasing at any time (146, 148). There is a very large opportunity for bias if populations are declared to be "declining" on the basis of short time series indicating that numbers are decreasing.

Amphibian Populations Fluctuate It is almost universally agreed that most local populations of amphibians are likely to fluctuate considerably in size. This occurs because recruitment is highly variable (7, 15, 20, 44, 65, 81, 185, 186, 194). Survival rates of terrestrial stages often appear to be relatively constant, with some degree of fluctuation (7, 44, 73, 185, 186). The survival rates of eggs, hatchlings, and larvae often vary over several orders of magnitude (1, 20, 199). Pechmann & Wilbur (148) provided a useful review of approaches that have been used to examine population behavior.

Examining Changes of Numbers with Time One approach that has been used to examine trends within populations of amphibians is simple correlations of numbers with time. Fifteen-to-twenty-year time series for six species of salamanders in the Appalachian mountains of the United States showed no evidence for consistent trends in numbers over time (91). A lack of significant correlations of numbers with years over a long-term study may indicate that variation among years represents fluctuations in numbers rather than declines (146, 169). Significant negative correlations between population size and time have been interpreted as indicating that populations are in decline (115, 117). Of 16 studies that monitored amphibian populations for four or more years, five reported that populations were declining, one that six populations had become extinct, seven that populations were fluctuating, and three that they were stable (30). Most of the evaluations of population change were based on correlations of population size with time, and some statements that populations were not in decline were based on failure to find a significant correlation. Failure to find a significant correlation does not necessarily imply that none exists (162); it can also result from a lack of statistical power. In eight studies of amphibian populations that reported no trend in population size over time, the power of a product-moment correlation was insufficient to allow acceptance of the null hypothesis (162). Reed & Blaustein (162) suggested that studies should not conclude that populations are not in decline unless that conclusion can be supported statistically. Unfortunately, the power analysis they used, relying on simple Pearson correlations, is not statistically valid, as the adjacent annual values in time series on a single population are not independent random samples. We develop an alternative approach below.

A Simple Model of Population Behavior In order to suggest how more appropriate tests might be carried out, we explored how time series data on amphibian populations should be analyzed and interpreted, and what the "normal" patterns in such data might be. We constructed a simple verbal model of frog population dynamics and compared its predictions with the behavior of 57 time series of frog abundances and 28 salamander time series (all studies cited in Table 1; details are available on the World Wide Web in the Supplemental Materials section of the main Annual Reviews site [www.AnnualReviews.org]). Finally we compared the behavior of those 85 time series with that of a simulated population. The simple verbal model is:

> Assumptions: 1. amphibian populations often persist for many generations, neither decreasing to extinction nor exploding to infinity; 2. survival rates of terrestrial stages may vary, but that variation is typically over less than a single order of magnitude; 3. survival rates of aquatic stages often vary over several orders of magnitude. Deductions: (A) Assumption 1 implies that recruitment must on average be sufficient for replacement to occur. (B) Assumptions 1 and 2 together imply that most variation in the size of terrestrial populations must be due to fluctuations in recruitment from the aquatic stages. (C) Assumptions 1, 2, and 3 together imply that when aquatic survival is high, populations must rapidly increase, but that aquatic survival must frequently be below replacement level, so that the increases do not tend to lead to sustained explosions. Deductions (A–C) taken together

suggest that populations should decrease more often than they increase, because increases can be very rapid and must be counterbalanced by slower decreases.

This model suggests that populations of species with highly variable recruitment from the aquatic to the terrestrial stage might be expected to decrease during more than 50% of time intervals. This is a potentially important point in the context of the problem of amphibian declines: If populations naturally decrease in numbers more often than they increase, relatively short-term population studies may often find that the population or populations studied appear to be in decline.

Comparing the Predictions of the Model with Data To examine this problem in more detail, we extracted data from a database of information on time series of population sizes of frogs and salamanders collated from the literature and available from the United States Geological Survey (USGS) (72). Most of the populations studied were not regarded as being in decline. We modified the database from its raw form in several ways. The database includes a number of time series collected using different techniques on the same species at the same sites and times. We included only one time series of data on any species for any site and set of dates. We did not use data on larvae or juveniles in our analysis because they reflect reproductive input modified by highly variable aquatic mortality rates and so are least likely to reflect the dynamics of adult populations. When data were available for both male and female adults, we combined them. When count data and other estimates such as mark-recapture values were available, we included the count data if they had been collected in a manner that was likely to be comparable among years; otherwise we used the estimated numbers. When only data on numbers of egg masses were available, we included them in our analyses, as they should be highly correlated with numbers of adult females. We included only time series taken over four or more years. We also included data from Cohen (50) that were not present in the USGS database. A summary of the original sources of the data we used appears in Table 1, and details of the data for each time series appear on the World Wide Web in the Supplemental Materials section of the main Annual Reviews site (www.AnnualReviews.org).

We classified each yearly population estimate as either the initial number in a series, an increase, a decrease, or no change. Three families of frogs (Bufonidae, Hylidae, and Ranidae) and two families of salamanders (Ambystomatidae and Plethodontidae) contained sufficient numbers of time series. We used nonparametric statistics (Table 2) to determine whether the percentage of the time in which populations decreased from one year to the next varied among families. In both orders, families differed in the percentage of the time that populations declined between years. In the Caudata, populations of species in the family Ambystomatidae, which are pool-breeding species with relatively large clutch sizes, decreased between years an average of 59.2% of the time. Species in the Plethodontidae, primarily terrestrial egg layers and stream-breeders with smaller average clutch

TABLE 2 Means, standard deviations, and numbers of time series analyzed for the percentage of year-to-year changes in population size that are decreases, for amphibians of five families, and tests for significant differences in the mean among families within orders.

	Percent of population changes that are decreases		
Family	**Mean**	**Standard Deviation**	***N***
Salamanders. Wilcoxon 2-sample test, $Z = 2.43$, $P = 0.015$			
Ambystomatidae	59.2	18.1	12
Plethodontidae	42.0	16.1	14
Frogs. Kruskal-Wallis test, $\chi^2 = 8.82$, 2 d.f., $P = 0.012$			
Bufonidae	55.5	19.8	19
Hylidae	47.3	6.7	15
Ranidae	60.6	13.0	23

sizes (see references in Table 1), declined between years only 42% of the time. In the Anura, species in the families Ranidae and Bufonidae decreased in more than 50% of intervals, while species in the family Hylidae decreased in slightly less than 50% of intervals. (Table 2). The ranids and bufonids included in the database generally produce larger clutches of eggs and often have highly variable offspring survival, while hylids produce smaller clutches and may have less variable rates of offspring survival (references in Table 1).

The results of this analysis suggest strongly that the expected behavior of populations over time varies among families of frogs and salamanders, and that the adult populations of species that have more variable survival of premetamorphic offspring tend to decrease between years more often than they increase. This result has implications for the ability to draw conclusions about population trends from simple time series of numbers over years. A species in the family Plethodontidae, in which the mean population behavior is to decrease between 40% of years, has a probability of four successive decreases of only 0.031, and finding four successive decreases between years in a censused population might be cause for alarm. However, a ranid species would need to decrease in numbers six times in succession before the probability of that sequence of decreases was less than 0.05.

Our data set included populations from many times and places, some of which might have been undergoing declines caused by external pressures. This would bias the results of our analysis. To erect a null hypothesis independent of the data from natural populations, we examined the behavior of a simulated population with known characteristics.

A Numerical Model of Population Behavior For comparison with our verbal model and the results of our data analysis, we used a simulation model based on a long-term study of the population dynamics of *Bufo marinus* in northern Australia (2). Our model incorporated simple density-dependence in the adult stage and lognormal variation in recruitment success. Mean recruitment rates were adjusted so that populations tended to fluctuate about a mean size rather than explode or decline rapidly to extinction. The population arbitrarily started with 100 adult female toads. Year-to-year survival of adult females was set at the lower of 50% or 50 total. Each female produced 9000 eggs, which survived to reach maturity as females in one year at variable rates equal to 0.00047^{ea+1}, where a is a normally distributed random variable with mean 0, standard deviation 1. The number of adults was truncated to an integer following survival and recruitment in each year.

We ran the simulation 5000 times, for 1000 "years" each time. On average, the population persisted for 423 years before extinction and contained 296 adult females while extant. Within a run, recruitment differed on average approximately 300-fold between the lowest and highest years, while adult survival differed 15-fold as a consequence of density-dependence. While the population persisted, it decreased in 56.32% of intervals between years, a result that is in very close agreement with the observation (Table 2) that populations of bufonids declined between 55.5% of years. This suggests that our initial verbal model was correct: When population fluctuations are driven by highly variable recruitment, it is likely that population dynamics will be characterized by occasional outbreaks with longer intervening periods of decrease, so that they are "in decline" more than 50% of the time. This result is similar to the "storage" effect in open populations (45, 193).

It is clear that a population decreasing in more years than it increases is not necessarily in decline. However, if there has recently been an increase in the general tendency of amphibian populations to decline, there should be a correlation between the year in which a study ended and the frequency of decreases in that study. We tested this hypothesis using the 85 time series in our database. We correlated the final year of each study with the percentage of intervals across which the studied population declined. We analyzed the data for each family of frogs and salamanders separately, as combining them might have confounded real effects of time with the effects of changes in the proportions of studies that were carried out on each taxon. We calculated Spearman rank correlations because these straighten nonlinear relationships and decrease the effects of outliers and data that are not normally distributed. We found no evidence for any significant correlation of proportion decreases with time in any family (maximum $r_s = 0.327$, minimum $P = 0.234$), an outcome supported by examination of the data (Figure 4), which do not suggest any strong trend for either frogs or salamanders.

The proportion of years in which a population declines may be a weak indicator of trends and is potentially subject to difficulties in deciding how large a change in population size represents a real decrease rather than noise in the data. Correlations between population size and time might be a better indicator of population status (30, 148).

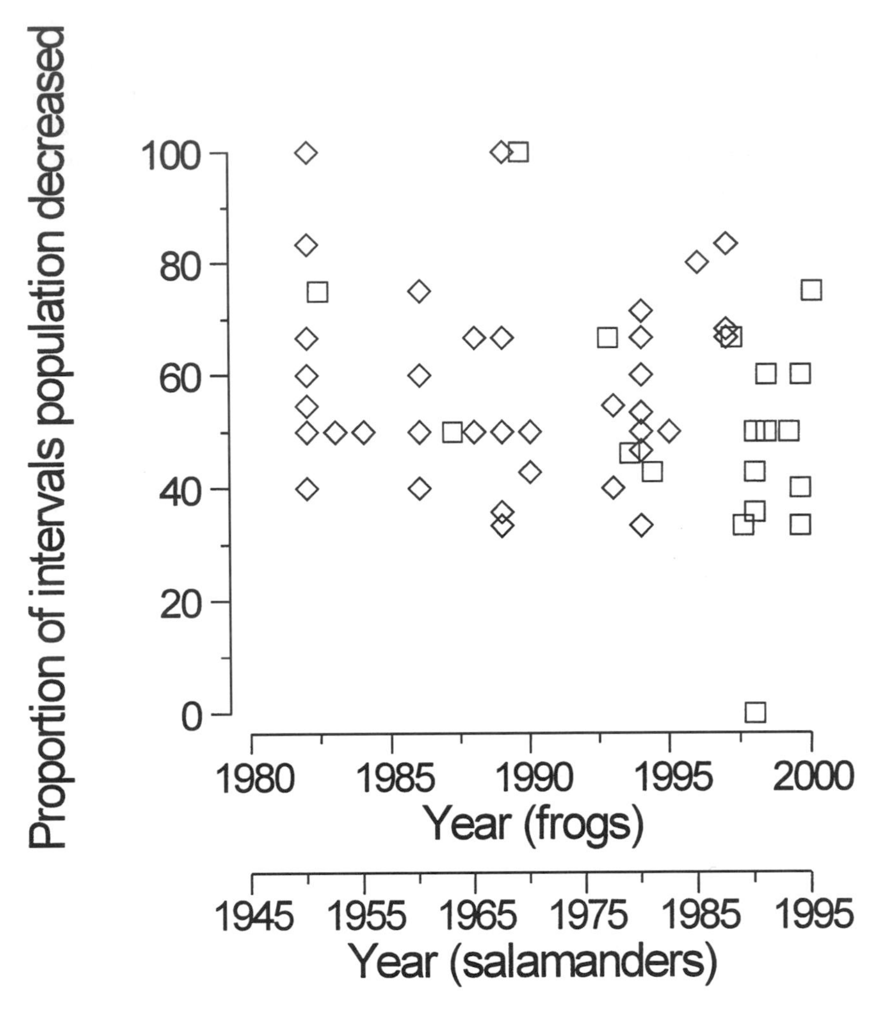

Figure 4 Proportion of the intervals from one year to the next over which populations decreased during long-term studies (4 or more years) of 85 local amphibian populations (28 of salamanders, shown as squares, 57 of frogs, shown as diamonds). The abscissa is the final year of each study.

CORRELATIONS OF POPULATION SIZE WITH TIME

If there is a general global phenomenon acting on all amphibians and increasing their tendency to decrease in numbers, we would expect that recently completed studies would show more or stronger negative relationships between population size and time than would older ones. For each of the 85 time series in our database, we regressed $\log_{10}(N+1)$ on year of the series. We used $\log_{10}(N+1)$ because this stabilizes variances and because a population increasing or decreasing at a constant proportional rate will show a linear relationship of $\log_{10}(N+1)$ with time. Years were adjusted within each time series so that the first year was year zero, to reduce possible influences of rounding errors on the regressions. We used the correlations and slopes of these regressions in further analyses to determine whether there was any evidence for an overall trend toward an increase in the incidence or intensity of negative relationships between population size and time. In these analyses we examined the data for frogs and salamanders separately because trends might be present in one but not in the other. We first looked for correlations of either the slope or the correlation coefficient with the year in which each time series ended, using Spearman rank correlations because of the unknown sampling properties of these two measures. None of the correlations of the correlation coefficient or the slope with final year were significant (salamanders: $r_s = -0.021$, $P = 0.917$ and $r_s = -0.024$, $P = 0.905$; frogs: $r_s = 0.175$, $P = 0.174$ and $r_s = 0.161$, $P = 0.232$; respectively).

Although there was no trend in either order for changes over time in the relationship of population size to year of study, it is still possible that a general declining trend was present throughout the period. To examine this possibility, we plotted the correlations of population size with year and the slopes of the regressions of population size on year against final year of the study (Figure 5*a* and 5*b*). Initial examination of these figures could be a source of alarm because there are many more negative than positive correlations and because most of the apparently significant correlations and slopes are negative. Further analysis shows that both of these effects are probably artifacts of the population dynamics of amphibians and the fact that standard assessments of significance should not be applied to time series data. In order to more rigorously assess the significance of these correlations and slopes, we returned to the results of our population simulation. On each of the 5000 iterations of the time series, we calculated correlations and slopes for regressions of $\log_{10}(N+1)$ on year for time series containing 4 through 9, 11, 12, 14, 15, 16, 23, and 28 years, starting arbitrarily at year 20 of the simulated time series to allow the effects of initial conditions to disappear. This resulted in at least 4904 coefficients for each combination of parameter and series length (less than 5000 because a few populations went extinct before the final year of each simulated series). We sorted the vector of coefficients for each combination of parameter and series length into ascending order, and we took the coefficients at 0.025 from the bottom and top of the series as the upper and lower 95% confidence limits for the correlation and slope of regressions of an amphibian population on year. These confidence limits appear in Figure 5, *c* and *d*, with plots of the correlations and slopes obtained from the time series in our database. Using the criterion that to be significant a parameter must fall outside these empirical 95% confidence limits, only a single correlation and four slopes are significant, well within the number that would be expected due to Type I error when 85 comparisons are made.

Although we must conclude that there is no evidence in the 85 time series we analyzed to suggest that correlations or slopes of regressions of $\log_{10}(N+1)$ against

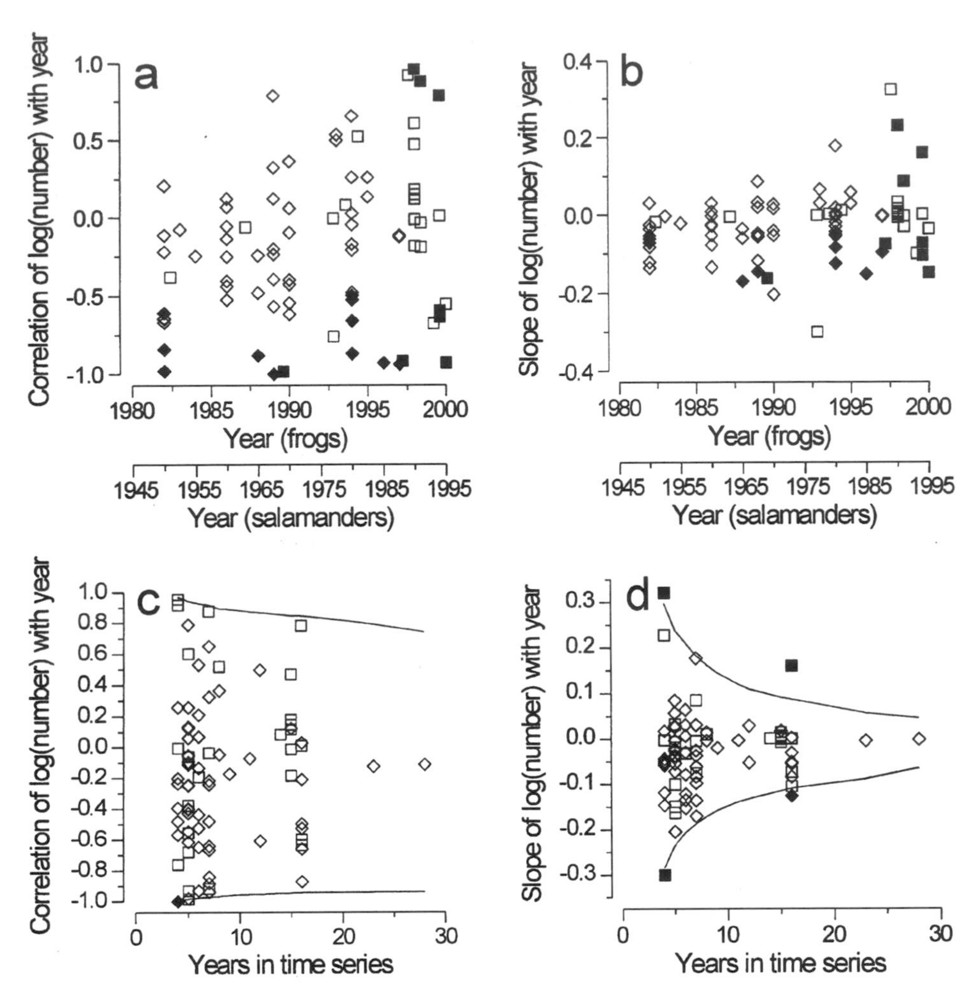

Figure 5 (*a*) Product-moment correlation of $\log_{10}(N + 1)$ with year of study for 85 time series of four or more years of data on local populations of amphibians, plotted against final year of the study. Data for salamanders shown as squares, frogs as diamonds. Correlations that would be significant at $\alpha = 0.05$ using standard parametric criteria are indicated by filled shapes. (*b*) Slopes of regressions of $\log_{10}(N + 1)$ on year of study, other details the same as in (*a*). (*c*) The same correlation coefficients as in (*a*), plotted against number of years in the time series from which the correlation was derived. Lines indicate upper and lower 95% confidence limits derived from correlations calculated on 5000 simulated time series of each length. Only the single correlation outside these confidence limits should be considered significant (indicated by the filled shape). (*d*) The same slopes as in (*b*), plotted against number of years in the time series from which they were derived. Lines indicate 95% confidence limits, derived as in (*c*). Only four of the 85 slopes should be regarded as significant at the 0.05 level (*filled shapes*).

time have changed in recent years, we might still be concerned over the apparent excess of negative relationships of population size with time (Figure 5, *a* and *b*). Fifty-seven of the 85 relationships of population size with time (67%) are negative. The proportion of negative correlations from our simulation increased with length of the time series from 54.8% negative with a series length of 4 years to 57.5% with a series length of 28 years. We used these expected proportions, weighted by the number of each of the real time series that were of each length, to calculate the expected numbers of correlations that should have been greater and less than 0 (37.6 and 47.4, respectively), and compared these with the observed numbers (28 and 57) using a chi-squared goodness-of-fit test (chi-squared $= 4.42$, $P = 0.037$). This significant result indicates that the amphibian populations we examined had a greater number of negative correlations with time than would be expected, as compared with our simulated populations. This could reflect a general tendency toward decline, but it could also reflect the fact that our simulation model, while it probably provides a more realistic null hypothesis than the simple assumption that effects in both directions should be equal, does not perfectly reflect the population behavior of all amphibians. Using a greater variety of models to generate null hypotheses more appropriate for each family, genus, or even species would obviously be preferable. It would also be useful to examine the sensitivity of our conclusions to variations in model parameters and form.

The use of appropriate null hypotheses will allow more rigorous examination of the behavior of individual populations. However, many amphibians appear to live in local populations that interact strongly with other populations, so that understanding the implications of local population dynamics for species persistence requires a knowledge of their metapopulation biology.

AMPHIBIAN METAPOPULATION BIOLOGY

A metapopulation consists of a group of local populations inhabiting more or less discrete patches of habitat (94). A metapopulation differs from a collection of independent populations in that there is substantial migration between local populations, so that no local population is likely to remain extinct for any length of time. Migration rates may be high enough to affect rates of local population increase and decrease (95, 139). A metapopulation differs from a single subdivided population by having sufficiently low rates of migration between local populations that they exhibit some degree of independence in their dynamics, including the possibility of declining to extinction (93).

Metapopulation Studies on Amphibians

Although it has been suggested that amphibians are generally highly philopatric (30, 61, 172), many species depart from this pattern. One of the problems that plagues mark-recapture studies of amphibian populations at single sites is the high rates at which animals disappear from local populations (2, 31, 40, 111, 134, 156). Substantial rates of dispersal among local populations have been documented in many species (2, 21, 37, 38, 42, 58, 66, 81, 112, 161, 173, 198)). Additional evidence that many amphibians live in metapopulations comes from explicit metapopulation studies (175). Breeding populations of the newt *Notophthalmus viridescens* act as cells in a regional metapopulation (83). The European pool frog *Rana lessonae* lives in spatially complex metapopulations; the quality of potential breeding sites and their degree of isolation from other sites determines their probability of occupancy and the probability of local extinction (176). Increases in the isolation of habitat patches due to natural succession or habitat destruction decrease the persistence of local populations (176). Surveys of *Rana clamitans* occupancy at 160 ponds in three distinct regions (103) demonstrated the existence of regional metapopulations. Colonization rates varied from 0 to 0.25 ponds $\cdot$ (pond occupied $\cdot$ year)$^{-1}$, while local extinction rates were between 0 and 0.5 ponds $\cdot$ (pond occupied $\cdot$ year)$^{-1}$. Small populations were more prone to local extinction than were large ones, and there was no overall trend in occupancy rates when all three metapopulations were considered (103). Ten other species examined at 97 ponds in the same regions (101) also exhibited metapopulation dynamics, with rates of turnover from 0.07 to 0.30 species $\cdot$ (pond $\cdot$ year)$^{-1}$. Pool size and isolation both affected species richness in 77 pools in the southern Netherlands (126) and 332 habitats in Bavaria (68).

Models of Amphibian Metapopulations and Their Implications

Because amphibians often live in metapopulations, declines and extinctions of local populations may be common events. Detailed studies of local populations may give useful insights into the autecology of species, but they are of limited use in evaluating the status of regional metapopulations. One approach to examining the behavior of metapopulations is to examine simple probabilistic models for the frequency with which local populations might change in status (184). A probabilistic null model for population declines and disappearances was used to examine whether the declines and disappearances of frogs that occurred in the 1980s at Monteverde, Costa Rica, might be due to chance (153). Pounds et al (153) used long-term studies to estimate probabilities of disappearance. They then compared the numbers of species disappearances at their study sites to the numbers of disappearances predicted by the cumulative binomial distribution, and they found that far more species had disappeared than would have been predicted.

The probabilistic approach (153) seems to be a useful way to quantify the idea that, when certain regions or taxa are considered, species are disappearing at a rate "too great to be coincidental." Several studies have produced data that could be examined using this technique. It seems likely that the disappearances of seven species from all sites at elevations above 400 m throughout the Australian Wet Tropics (135, 163) would be shown to be extremely improbable, as would the disappearances of three species and large declines in site occupancy of four others in the Yosemite area of California (69), the disappearance of *Rana cascadae* from the southern end of its range (75), and the disappearances of many species at Las Tablas, Costa Rica (131).

A more complex analytical model was used to predict the rates of extinction of local populations of the common toad *Bufo bufo* and the crested newt *Triturus cristatus* in Europe, and to examine how those rates should respond to the size of the local habitat patch and its distance from a source of migrants (92). The persistence of populations of both species should increase with the carrying capacity of the local habitat and should decrease with increasing distance to a source of migrants. Over a wide range of carrying capacities, the critical distances are approximately 10 times as great (~5 km) for toads as they are for newts (~500 m) (92). This study suggested that both patch size and spatial distribution must be taken into account when managing amphibian metapopulations. Information on patch occupancy from a geographically referenced database (168) indicates that small, relatively isolated wetlands are important in the metapopulation dynamics of amphibians in South Carolina, USA. Loss of these habitats might lead to disproportionately large rates of extinction in regional metapopulations that depend upon them as stepping stones in colonization and as refuges from local extinctions (168).

Delineating and monitoring the status of metapopulations requires extensive sampling, but because metapopulation dynamics are concerned mostly with the presence or absence of species in local populations, sampling of local populations does not need to be intense (87). Fully understanding the dynamics of amphibian metapopulations will require much more information on movements and dispersal among local populations than is presently available.

CONCLUSIONS

It is clear that local populations of many amphibian species have declined in recent years, and there are several well-documented cases of declines at and above the level of regional metapopulations. Although many environmental factors can adversely affect the growth, survival, and reproduction of amphibians, few studies have convincingly demonstrated that these effects alter their population dynamics. Studies linking factors that negatively affect amphibians in the laboratory or in artificial field trials with effects on population dynamics in more natural settings are urgently needed.

Local populations of amphibians tend to fluctuate, and our results show it is likely that many normally decrease more often than they increase. It is therefore important to develop realistic null hypotheses for their behavior. If we had not based our null hypothesis on a simulation of frog population dynamics, we would have reached very different conclusions in our analysis of population behavior. Additional data on an ecologically diverse range of species will allow the development of more sophisticated and specific null hypotheses for a greater range of populations. This is necessary to make rigorous tests of the responses of local populations to environmental factors possible.

Many amphibian species occur as metapopulations, so the dynamics of local populations may be poor indicators of their status. Declines and extinctions of metapopulations are likely to result from interactions between changes in the dynamics of local populations and habitat modification or loss (93, 94, 168). For many species, understanding the factors affecting the status and dynamics of metapopulations should therefore be the ultimate goal of studies aiming to prevent or reverse declines. Monitoring metapopulations requires different data collection techniques than monitoring isolated populations, so the first step in designing any monitoring program should be to determine whether the species of interest forms a metapopulation. Studies integrating research within local populations with investigations at the metapopulation level are most likely to discover the causes of amphibian declines and provide a basis for the conservation of amphibian diversity.

ACKNOWLEDGMENTS

The original data included in this review were collected with the support of funding from the Australian Research Council and the Cooperative Research Centre for Tropical Rainforest Ecology and Management. We thank J Winter and the Commonwealth Aluminium Corporation for permission to include data from Reference 200, and S Droege for permission to use data extracted from the USGS amphibian count database (72). The manuscript was improved with the help of MJ Caley, CN Johnson, and L Schwarzkopf.

LITERATURE CITED

1. Alford RA. 1999. Ecology: resource use, competition, and predation. In *Tadpoles: The Biology of Anuran Larvae*, ed. RW McDiarmid, R Altig. Chicago: Univ. Chicago Press. In press
2. Alford RA, Cohen MP, Crossland MR, Hearnden MN, Schwarzkopf L. 1995. Population biology of *Bufo marinus* in northern Australia. In *Wetland Research in the Wet-Dry Tropics of Australia*, ed. CM Finlayson, pp. 173–181. Canberra: Commonwealth of Australia
3. Alford RA, Richards SJ. 1997. Lack of evidence for epidemic disease as an agent in the catastrophic decline of Australian rain forest frogs. *Conserv. Biol.* 11:1026–29
4. Altig R, Channing A. 1993. Hypothesis: functional significance of colour and pattern of anuran tadpoles. *Herpetol. J.* 3:73–75
5. Andren C, Henrikson L, Olsson M, Nilson G. 1988. Effects of pH and aluminium on embryonic and early larval stages of Swedish brown frogs *Rana arvalis*, *R. temporaria* and *R. dalmatina*. *Holarctic Ecol.* 11:127–35
6. Anzalone CR, Kats LB, Gordon MS. 1998. Effects of solar UV-B radiation on embryonic development in *Hyla cadaverina*, *Hyla regilla*, and *Taricha torosa*. *Conserv. Biol.* 12:646–53
7. Arntzen JW, Teunis SFM. 1993. A six-year study on the population dynamics of the

crested newt (*Triturus cristatus*) following the colonization of a newly created pond. *Herpetol. J.* 3:99–110

8. Ash AN. 1988. Disappearance of salamanders from clearcut plots. *J. Elisha Mitchell Sci. Soc.* 104:116–22
9. Ash AN. 1997. Disappearance and return of Plethodontid salamanders to clearcut plots in the southern Blue Ridge Mountains. *Conserv. Biol.* 11:983–89
10. Barinaga M. 1990. Where have all the froggies gone? *Science* 247:1033–34
11. Beattie RC, Tyler-Jones R. 1992. The effects of low pH and aluminum on breeding success in the frog *Rana temporaria*. *J. Herpetol.* 26:353–60
12. Beattie RC, Tyler-Jones R, Baxter MJ. 1992. The effects of pH, aluminium concentration and temperature on the embryonic development of the European common frog, *Rana temporaria*. *J. Zool., London* 228:557–70
13. Beebee TJ. 1977. Environmental change as a cause of Natterjack Toad *Bufo calamita* declines in Britain. *Biol. Cons.* 11:87–102
14. Beebee TJC. 1995. Amphibian breeding and climate. *Nature* 374:219–20
15. Beebee TJC, Denton JS, Buckley J. 1996. Factors affecting population densities of adult natterjack toads *Bufo calamita* in Britain. *J. Appl. Ecol.* 33:263–68
16. Beebee TJC, Flower RJ, Stevenson AC, Patrick ST, Appleby PG, et al. 1990. Decline of the Natterjack Toad *Bufo calamita* in Britain: paleoecological, documentary, and experimental evidence for breeding site acidification. *Biol. Conserv.* 53:1–20
17. Beiswenger RE. 1986. An endangered species, the Wyoming toad *Bufo hemiophrys baxteri*–the importance of an early warning system. *Biol. Cons.* 37:59–71
18. Berger L, Speare R, Daszak P, Green DE, Cunningham AA, et al. 1998. Chytridiomycosis causes amphibian mortality associated with population declines in the rain forests of Australia and Central America. *Proc. Natl. Acad. Sci., USA* 95:9031–36
19. Bertram S, Berrill M. 1997. Fluctuations in a northern population of gray treefrogs, *Hyla versicolor*. See Ref. 87, pp. 57–63
20. Berven KA. 1990. Factors affecting population fluctuations in larval and adult stages of the wood frog (*Rana sylvatica*). *Ecology* 71:1599–1608
21. Berven KA, Grudzien TA. 1990. Dispersal in the wood frog (*Rana sylvatica*): implications for genetic population structure. *Evolution* 44:2047–56
22. Bishop CA. 1992. The effects of pesticides on amphibians and the implications for determining causes of declines in amphibian populations. In *Declines in Canadian Amphibian Populations: Designing a National Monitoring Strategy*, ed. CA Bishop, KE Pettit, pp. 67–70. *Occas. Pap. No. 76, Can. Wildlife Serv.*
23. Blaustein AR. 1994. Chicken Little or Nero's fiddle? A perspective on declining amphibian populations. *Herpetologica* 50: 85–97
24. Blaustein AR, Edmund B, Kiesecker JM, Beatty JJ, Hokit DG. 1995. Ambient ultraviolet radiation causes mortality in salamander eggs. *Ecol. Appl.* 5:740–43
25. Blaustein AR, Hoffman PD, Hokit DG, Kiesecker JM, Walls SD, Hays JB. 1994. UV repair and resistance to solar UV-B in amphibian eggs: a link to population declines? *Proc. Natl. Acad. Sci., USA* 91: 1791–95
26. Blaustein AR, Hoffman PD, Kiesecker JM, Hays JB. 1996. DNA repair activity and resistance to solar UV-B radiation in eggs of the red-legged frog. *Conserv. Biol.* 10: 1398–1402
27. Blaustein AR, Hokit DG, O'Hara RK, Holt RA. 1994. Pathogenic fungus contributes to amphibian losses in the Pacific Northwest. *Biol. Conserv.* 67:251–54
28. Blaustein AR, Kiesecker JM, Chivers DP, Anthony RG. 1997. Ambient UV-B radiation causes deformities in amphibian

embryos. *Proc. Natl. Acad. Sci., USA* 94: 13735–37
29. Blaustein AR, Wake DB. 1990. Declining amphibian populations: a global phenomenon? *Trends Ecol. Evol.* 5:203–4
30. Blaustein AR, Wake DB, Sousa WP. 1994. Amphibian declines: judging stability, persistence, and susceptibility of populations to local and global extinctions. *Conserv. Biol.* 8:60–71
31. Bradford DF. 1989. Allotopic distributions of native frogs and introduced fishes in high Sierra Nevada lakes of California: implications of the negative effect of fish introductions. *Copeia* 1989:775–78
32. Bradford DF. 1991. Mass mortality and extinction in a high-elevation population of *Rana muscosa. J. Herpetol.* 25:174–77
33. Bradford DF, Gordon MS, Johnson DF, Andrews RD, Jennings WB. 1994. Acidic deposition as an unlikely cause for amphibian population declines in the Sierra Nevada, California. *Biol. Conserv.* 69:155–61
34. Bradford DF, Swanson C, Gordon MS. 1994. Effects of low pH and aluminum on amphibians at high elevation in the Sierra Nevada, California. *Can. J. Zool.* 72:1272–79
35. Bradford DF, Tabatabai F, Graber DM. 1993. Isolation of remaining populations of the native frog, *Rana muscosa*, by introduced fishes in Sequoia and King's Canyon National Parks, California. *Conserv. Biol.* 7:882–88
36. Bradford DF, Swanson C, Gordon MS. 1992. Effects of low pH and aluminum on two declining species of amphibians in the Sierra Nevada, California. *J. Herpetol.* 26:369–77
37. Breckenridge WJ, Tester JR. 1961. Growth, local movements and hibernation of the Manitoba toad, *Bufo hemiophrys. Ecology* 42:637–46
38. Breden F. 1987. The effect of post-metamorphic dispersal on the population genetic structure of Fowler's toad, *Bufo woodhousei fowleri. Copeia* 1987:386–95
39. Brooks RJ. 1992. Monitoring wildlife populations in long-term studies. See Ref. 22, pp. 94–97
40. Brown WC, Alcala AC. 1970. Population ecology of the frog *Rana erythraea* in southern Negros, Philippines. *Copeia* 1970:611–22
41. Bruce RC. 1995. The use of temporary removal sampling in a study of population dynamics of the salamander *Desmognathus monticola. Aust. J. Ecol.* 20:403–12
42. Buckley D, Arano B, Herrero P, Llorente G. 1996. Population structure of Moroccan water frogs: genetic cohesion despite a fragmented distribution. *J. Zool. Syst. Evol. Res.* 34:173–79
43. Bury RB, Corn PS. 1988. Responses of aquatic and streamside amphibians to timber harvest: A review. In *Streamside Management: Riparian Wildlife and Forestry Interactions*, ed. KJ Raedeke, pp. 165–81. Seattle: Univ. Was. Press
44. Caldwell JP. 1987. Demography and life history of two species of chorus frogs (Anura:Hylidae) in South Carolina. *Copeia* 1987:114–27
45. Caley MJ, Carr MH, Hixon MA, Hughes TP, Jones GP, Menge BA. 1996. Recruitment and the local dynamics of open marine populations. *Annu. Rev. Ecol. Syst.* 27:477–500
46. Carey C, Bryant CJ. 1995. Possible interrelationships among environmental toxicants, amphibian development, and decline of amphibian populations. *Environ. Health Perspec.* 103(Suppl. 4):13–17
47. Carey CL. 1993. Hypothesis concerning the causes of the disappearance of boreal toads from the mountains of Colorado. *Conserv. Biol.* 7:355–62
48. Chovanec A. 1992. The influence of tadpole swimming behaviour on predation by dragonfly nymphs. *Amphibia-Reptilia* 13: 341–49
49. Clay D. 1997. The effects of temperature and acidity on spawning of the spotted

salamander, *Ambystoma maculatum*, in Fundy National Park. See Ref. 87, pp. 226–32

50. Cohen MP. 1995. *Ecology of two populations of* Bufo marinus *in north-eastern Australia*. PhD thesis, James Cook Univ., Townsville, Australia. 203 pp.
51. Corn PS. 1994. What we know and don't know about amphibian declines in the west. In *Sustainable Ecological Systems: Implementing an Ecological Approach to Land Management*, ed. LF DeBano, WW Covington, pp. 59–67. Fort Collins, CO: USDA For. Serv., Rocky Mountain For. Range Exp. Station
52. Corn PS, Bury RB. 1989. Logging in western Oregon: responses of headwater habitats and stream amphibians. *For. Ecol. Manage.* 29:39–57
53. Corn PS, Fogleman JC. 1984. Extinction of montane populations of the Northern Leopard Frog (*Rana pipiens*) in Colorado. *J. Herpetol.* 18:147–52
54. Corn PS, Stolzenburg W, Bury RB. 1989. *Acid Precipitation Studies in Colorado and Wyoming: Interim Report of Surveys of Montane Amphibians and Water Chemistry. Biol. Rep. 80, US Fish & Wildlife Serv.* pp. 1–56
55. Corn PS, Vertucci FA. 1992. Descriptive risk assessment of the effects of acidic deposition on Rocky Mountain amphibians. *J. Herpetol.* 26:361–69
56. Crawshaw GJ. 1992. The role of disease in amphibian decline. See Ref. 22, pp. 60–62
57. Crawshaw GJ. 1997. Disease in Canadian amphibian populations. See Ref. 87, pp. 258–70
58. Crump ML. 1986. Homing and site fidelity in a neotropical frog, *Atelopus varius* (Bufonidae). *Copeia* 1986:438–44
59. Cunningham AA, Langton TES, Bennett PM, Lewin JF, Drury SEN, et al. 1993. Unusual mortality associated with poxvirus-like particles in frogs (*Rana temporaria*). *Vet. Rec.* 133:141–42
60. Cunningham AA, Langton TES, Bennett PM, Lewin JF, Drury SEN, et al. 1996. Pathological and microbiological findings from incidents of unusual mortality of the common frog (*Rana temporaria*). *Philos. Trans. R. Soc. Lond. Ser. B.* 351:1539–57
61. Daugherty CH, Sheldon AL. 1982. Age-specific movement patterns of the frog *Ascaphus truei*. *Herpetologica* 38:468–74
62. Delis PR, Mushinsky HR, McCoy ED. 1996. Decline of some west-central Florida anuran populations in response to habitat degradation. *Biodiv. Conserv.* 5:1579–95
63. Dodd CK Jr. 1991. Drift fence-associated sampling bias of amphibians at a Florida sandhills temporary pond. *J. Herpetol.* 25: 296–301
64. Dodd CK Jr. 1992. Biological diversity of a temporary pond herpetofauna in north Florida sandhills. *Biodiv. Conserv.* 1:125–142
65. Dodd CK Jr. 1994. The effects of drought on population structure, activity, and orientation of toads (*Bufo quercicus* and *B terrestris*) at a temporary pond. *Ecol. Ethol. Evol.* 6:331–49
66. Dole WJ. 1965. Summer movements of adult leopard frogs, *Rana pipiens* Schreber, in northern Michigan. *Ecology* 46:236–55
67. Donnelly MA, Crump ML. 1998. Potential effects of climate change on two neotropical amphibian assemblages. *Climate Change* 39:541–61
68. Dorn WMP, Brandl R. 1991. Local distribution of amphibians: the importance of habitat fragmentation. *Global Ecol. Biogeogr. Lett.* 1:36–41
69. Drost CA, Fellers GM. 1996. Collapse of a regional frog fauna in the Yosemite area of the California Sierra Nevada, USA. *Conserv. Biol.* 10:414–25
70. Duellman WE, Trueb L. 1986. *Biology of Amphibians*. New York: McGraw-Hill. 670 pp.
71. Dunson WA, Wyman RL, Corbett ES. 1992. A symposium on amphibian declines and acidification. *J. Herpetol.* 26:349–52
72. Eagle P. 1999. Amphibian Count Database.

United States Geological Survey, http://www.mp2-pwrc.usgs.gov/ampCV/ampdb.cfm

73. Elmberg J. 1990. Long-term survival, length of breeding season, and operational sex ratio in a boreal population of common frogs, *Rana temporaria* L. *Can. J. Zool.* 68: 121–27
74. Elmberg J. 1993. Threats to boreal frogs. *Ambio* 22:254–255
75. Fellers GM, Drost CA. 1993. Disappearance of the cascades frog *Rana cascadae* at the southern end of its range, California, USA. *Biol. Conserv.* 65:177–81
76. Fisher RN, Shaffer HB. 1996. The decline of amphibians in California's Great Central Valley. *Conserv. Biol.* 10:1387–97
77. Fite KV, Blaustein A, Bengston L, Hewitt HE. 1998. Evidence of retinal light damage in *Rana cascadae*: a declining amphibian species. *Copeia* 1998:906–14
78. Freda J, Dunson WA. 1986. Effects of low pH and other chemical variables on the local distribution of amphibians. *Copeia* 1986:454–66
79. Freda J, Sadinski WJ, Dunson WA. 1991. Long term monitoring of amphibian populations with respect to the effects of acidic deposition. *Water, Air, Soil Pollut.* 55:445–62
80. Freedman B, Shackell NL. 1992. Amphibians in the context of a national environmental monitoring program. See Ref. 22, pp. 101–104
81. Friedl TWP, Klump GM. 1997. Some aspects of population biology in the European treefrog, *Hyla arborea*. *Herpetologica* 53:321–30
82. Gamradt SC, Kats LB. 1996. Effect of introduced crayfish and mosquitofish on California Newts. *Conserv. Biol.* 10:1155–62
83. Gill DE. 1978. The metapopulation ecology of the red-spotted newt, *Notophthalmus viridescens* (Rafinesque). *Ecol. Monogr.* 48:145–66
84. Gittins SP. 1983. Population dynamics of the common toad (*Bufo bufo*) at a lake in mid-Wales. *J. Anim. Ecol.* 52:981–88
85. Grant KP, Licht LE. 1995. Effects of ultraviolet radiation on life-history stages of anurans from Ontario, Canada. *Can. J. Zool.* 73:2292–306
86. Green DM. 1992. Fowler's toads (*Bufo woodhousei fowleri*) at Long Point, Ontario: changing abundance and implications for conservation. See Ref. 22, pp. 37–43
87. Green DM. 1997. Perspectives on amphibian population declines: defining the problem and searching for answers. In *Amphibians in Decline. Canadian Studies of a Global Problem*, ed. DM Green, pp. 291–308. *Herpetological Conserv.*, Vol 1.
88. Green DM. 1997. Temporal variation in abundance and age structure in Fowler's toads, *Bufo fowleri*, at Long Point, Ontario. See Ref 87, pp. 45–56
89. Gyovai F. 1989. Demographic analysis of the moor frog (*Rana arvalis olterstorffi* Feje'rva'ry 1919) population in *Fraxino pannonicae-Alnetum* of the Tisza basin. *Tiscia (Szeged)* 24:107–21
90. Hairston NG. 1983. Growth, survival, and reproduction of *Plethodon jordani*: trade-offs between selective pressures. *Copeia* 1983:1024–35
91. Hairston NG Sr, Wiley RH. 1993. No decline in salamander (Amphibia:Caudata) populations: a twenty year study in the southern Appalachians. *Brimleyana* 18: 59–64
92. Halley JM, Oldham RS, Arntzen JW. 1996. Predicting the persistence of amphibian populations with the help of a spatial model. *J. Appl. Ecol.* 33:455–70
93. Hanski I. 1997. Metapopulation dynamics: from concepts and observations to predictive models. In *Metapopulation Biology: Ecology, Genetics, and Evolution*, ed. I Hanski, M. Gilpin, pp. 69–91. San Diego, CA: Academic Press. 512 pp.
94. Hanski I. 1998. Metapopulation dynamics. *Nature* 396:41–49

95. Hanski I, Pakkala T, Kuussaari M, Lei G. 1995. Metapopulation persistence of an endangered butterfly in a fragmented landscape. *Oikos* 72:21–28
96. Harte J, Hoffman E. 1989. Possible effects of acidic deposition on a Rocky Mountain population of the tiger salamander *Ambystoma tigrinum*. *Conserv. Biol.* 3:149–58
97. Hayes MP, Jennings MR. 1986. Decline of frog species in western North America: are bullfrogs (*Rana catesbeiana*) responsible? *J. Herpetol.* 20:490–509
98. Hays JB, Blaustein AR, Kiesecker JM, Hoffman PD, Pandelova I, Coyle D, Richardson T. 1996. Developmental responses of amphibians to solar and artificial UVB sources: a comparative study. *Photochem. Photobiol.* 64:449–55
99. Hecnar SJ. 1997. Amphibian pond communities in southwestern Ontario. See Ref. 87, pp. 1–15
100. Hecnar SJ, M'Closkey RT. 1996. The effects of predatory fish on amphibian species richness and distribution. *Biol. Conserv.* 79:123–31
101. Hecnar SJ, M'Closkey RT. 1996. Regional dynamics and the status of amphibians. *Ecology* 77:2091–97
102. Hecnar SJ, M'Closkey RT. 1997. Changes in the composition of a ranid frog community following bullfrog extinction. *Am. Midl. Nat.* 137:145–50
103. Hecnar SJ, M'Closkey RT. 1997. Spatial scale and determination of species status of the green frog. *Conserv. Biol.* 11:670–82
104. Henrikson B-I. 1990. Predation on amphibian eggs and tadpoles by common predators in acidified lakes. *Holarctic Ecol.* 13:201–6
105. Herman TB, Scott FW. 1992. Assessing the vulnerability of amphibians to climatic warming. See Ref. 22, pp. 46–49
106. Hero J-M, Gillespie GR. 1997. Epidemic disease and amphibian declines in Australia. *Conserv. Biol.* 11:1023–25
107. Heyer WR. 1979. Annual variation in larval amphibian populations within a temperate pond. *J. Wash. Acad. Sci.* 69:65–74
108. Homolka M, Kokes J. 1994. Effect of air pollution and forestry practice on the range and abundance of *Salamandra salamandra*. *Folia Zoologica* 43:49–56
109. Horne MT, Dunson WA. 1994. Exclusion of the Jefferson salamander, *Ambystoma jeffersonianum*, from some potential breeding ponds in Pennsylvania: effects of pH, temperature, and metals on embryonic development. *Arch. Environ. Contamination Toxicol.* 27:323–30
110. Horne MT, Dunson WA. 1995. Effects of low pH, metals, and water hardness on larval amphibians. *Arch. Environ. Contamination Toxicol.* 29:500–5
111. Husting EL. 1965. Survival and breeding structure in a population of *Ambystoma maculatum*. *Copeia* 1965:352–59
112. Ishchenko VG. 1989. Population biology of amphibians. *Sov. Sci. Rev. F Physiol. Gen. Biol.* 3:119–55
113. Jaeger RG. 1980. Density-dependent and density-independent causes of extinction of a salamander population. *Evolution* 34:617–21
114. Jennings MR, Hayes MP. 1994. Decline of native ranid frogs in the desert southwest. In *Southwestern Herpetol. Soc., Spec. Publ. No. 5*, ed. PR Brown, JW Wright, pp. 183–211
115. Joglar RL, Burrowes PA. 1996. Declining amphibian populations in Puerto Rico. In *Contributions to West Indian Herpetology: A Tribute to Albert Schwartz*, ed. R Powell, RW Henderson, pp. 371–80. *SSAR Contrib. to Herpetol.*, Volume 12.
116. Johnson B. 1992. Habitat loss and declining amphibian populations. See Ref. 22, pp. 71–75
117. Kagarise Sherman C, Morton ML. 1993. Population declines of Yosemite toads in the eastern Sierra Nevada of California. *J. Herpetol.* 27:186–98
118. Kats LB, Petranka JW, Sih A. 1988.

Antipredator defenses and the persistence of amphibian larvae with fishes. *Ecology* 69:1865–70
119. Kerr JB, McElroy CT. 1993. Evidence for large upward trends of ultraviolet-B radiation linked to ozone depletion. *Science* 262:1032–34
120. Kiesecker JM, Blaustein AR. 1995. Synergism between UV-B radiation and a pathogen magnifies amphibian embryo mortality in nature. *Proc. Natl. Acad. Sci., USA* 92:11049–52
121. Kiesecker JM, Blaustein AR. 1997. Population differences in responses of red-legged frogs (*Rana aurora*) to introduced bullfrogs. *Ecology* 78:1752–60
122. Kiesecker JM, Blaustein AR. 1998. Effects of introduced bullfrogs and smallmouth bass on microhabitat use, growth and survival of native red-legged frogs (*Rana aurora*). *Conserv. Biol.* 12:776–87
123. Koonz W. 1992. Amphibians in Manitoba. See Ref. 22, pp.19–20
124. Kucken DJ, Davis JS, Petranka JW, Smith KC. 1994. Anakeesta stream acidification and metal contamination: effects on a salamander community. *J. Environ. Qual.* 23:1311–17
125. Kuzmin S. 1994. The problem of declining amphibian populations in the Commonwealth of Independent States and adjacent territories. *Alytes* 12:123–34
126. Laan R, Verboom B. 1990. Effects of pool size and isolation on amphibian communities. *Biol. Conserv.* 54:251–62
127. Lannoo MJ, Lang K, Waltz T, Phillips GS. 1994. An altered amphibian assemblage: Dickinson County, Iowa, 70 years after Frank Blanchard's survey. *Am. Midl. Nat.* 131:311–19
128. Laurance WF. 1996. Catastrophic declines of Australian rainforest frogs: Is unusual weather responsible? *Biol. Conserv.* 77:203–12
129. Laurance WF, McDonald KR, Speare R. 1996. Epidemic disease and the catastrophic decline of Australian rain forest frogs. *Conserv. Biol.* 10:406–13
130. Licht LE, Grant KP. 1997. The effects of ultraviolet radiation on the biology of amphibians. *Am. Zool.* 37:137–45
131. Lips KR. 1998. Decline of a tropical montane amphibian fauna. *Conserv. Biol.* 12:106–17
132. Lizana M, Pedraza EM. 1998. The effects of UV-B radiation on toad mortality in mountainous areas of central Spain. *Conserv. Biol.* 12:703–7
133. Long LE, Saylor LS, Soule ME. 1995. A pH/UV-B synergism in amphibians. *Conserv. Biol.* 9:1301–3
134. Martof B. 1956. Factors influencing the size and composition of populations of *Rana clamitans*. *Am. Midl. Nat.* 56:224–45
135. McDonald KR, Alford RA. 1999. A review of declining frogs in northern Queensland. In *Declines and Disappearances of Australian Frogs, National Threatened Frog Workshop 1997*, ed. A Campbell. Canberra: Environment Australia. In press
136. Meyer AH, Schmidt BR, Grossenbacher K. 1998. Analysis of three amphibian populations with quarter-century long time series. *Proc. R. Soc. Lond., Ser. B* :523–28
137. Mierzwa KS. 1998. Status of northeastern Illinois amphibians. In *Status and Conservation of Midwestern Amphibians*, ed. MJ Lannoo, pp. 115–24. Iowa City: Univ. Iowa Press
138. Nagle MN, Hofer R. 1997. Effects of ultraviolet radiation on early larval stages of the Alpine newt, *Triturus alpestris*, under natural and laboratory conditions. *Oecologia* 110:514–19
139. Nee S, May RM, Hassell MP. 1997. Two-species metapopulation models. In *Metapopulation Biology: Ecology, Genetics, and Evolution*, ed. I Hanski, M Gilpin, pp. 123–47. San Diego, CA: Academic Press. 512 pp.

140. Nicholls N. 1984. A system for predicting the onset of the north Australian wet-season. *J. Climatol.* 4:425–35
141. Nyman S. 1986. Mass mortality in larval *Rana sylvatica* attributable to the bacterium, *Aeromonas hydrophilum. J. Herpetol.* 20:196–201
142. Olson DH. 1989. Predation on breeding western toads (*Bufo boreas*). *Copeia* 1989:391–97
143. Ovaska K. 1997. The vulnerability of amphibians in Canada to global warming and increased solar radiation. See Ref. 87, pp. 206–25
144. Ovaska K, Davis TM, Flamarique IN. 1997. Hatching success and larval survival of the frogs *Hyla regilla* and *Rana aurora* under ambient and artificially enhanced solar ultraviolet radiation. *Can. J. Zool.* 75:1081–88
145. Oza GM. 1990. Ecological effects of the frog's legs trade. *Environmentalist* 10:39–41
146. Pechmann JHK, Scott DE, Semlitsch RE, Caldwell JP, Vitt LJ, Gibbons JW. 1991. Declining amphibian populations: the problem of separating human impacts from natural fluctuations. *Science* 253:892–95
147. Pechmann JHK, Wake DB. 1997. Declines and disappearances of amphibian populations. In *Principles of Conservation Biology*, ed. GK Meffe, CR Carroll, pp. 135–37. Sunderland, MA: Sinauer. 2nd ed.
148. Pechmann JHK, Wilbur HM. 1994. Putting declining amphibian populations in perspective: natural fluctuations and human impacts. *Herpetologica* 50:65–84
149. Petranka JW, Eldridge ME, Haley KE. 1993. Effects of timber harvesting on southern Appalachian salamanders. *Conserv. Biol.* 7:363–70
150. Pierce BA, Wooten DK. 1992. Genetic variation in tolerance of amphibians to low pH. *J. Herpetol.* 26:422–29
151. Portnoy JW. 1990. Breeding biology of the spotted salamander *Ambystoma maculatum* (Shaw) in acidic temporary ponds at Cape Cod, USA. *Biol. Conserv.* 53:61–75
152. Pounds JA, Crump ML. 1994. Amphibian declines and climate disturbance: the case of the golden toad and the harlequin frog. *Conserv. Biol.* 8:72–85
153. Pounds JA, Fogden MPL, Savage JM, Gorman GC. 1997. Tests of null models for amphibian declines on a tropical mountain. *Conserv. Biol.* 11:1307–22
154. Power T, Clark KL, Harfenist A, Peakall DB. 1989. A review and evaluation of the amphibian toxicological literature. *Tech. Rep. 61, Can. Wildlife Serv.*
155. Preest MR. 1993. Mechanisms of growth rate reduction in acid-exposed larval salamanders, *Ambystoma maculatum. Physiol. Zool.* 66:686–707
156. Ramirez J, Vogt RC, Villareal-Benitez J-L. 1998. Population biology of a neotropical frog (*Rana vaillanti*). *J. Herpetol.* 32:338–44
157. Ramotnik CA. 1997. *Conservation Assessment of the Sacramento Mountain Salamander. Gen. Tech. Rep. RM-GTR-293.* Fort Collins, CO: USDAFS, Rocky Mountain For. Range Exp. Stat.
158. Raymond LR. 1991. Seasonal activity of *Siren intermedia* in northwestern Louisiana (Amphibia: Sirenidae). *Southwestern Nat.* 36:144–47
159. Raymond LR, Hardy LM. 1990. Demography of a population of *Ambystoma talpoideum* (Caudata: Ambystomatidae) in northwestern Louisiana. *Herpetologica* 46:371–82
160. Raymond LR, Hardy LM. 1991. Effects of a clearcut on a population of the mole salamander, *Ambystoma talpoideum*, in an adjacent unaltered forest. *J. Herpetol.* 25:509–12
161. Reading CJ, Loman J, Madsen T. 1991. Breeding pond fidelity in the common toad, *Bufo bufo*. *J. Zool. Lond.* 225:201–11

162. Reed JM, Blaustein AR. 1995. Assessment of "nondeclining" amphibian populations using power analysis. *Conserv. Biol.* 9:1299–1300
163. Richards SJ, McDonald KR, Alford RA. 1993. Declines in populations of Australia's endemic tropical rainforest frogs. *Pacific Conserv. Biol.* 1:66–77
164. Sarkar S. 1996. Ecological theory and anuran declines. *BioScience* 46:199–207
165. Schlupp I, Podloucky R. 1994. Changes in breeding site fidelity: a combined study of conservation and behaviour in the common toad *Bufo bufo*. *Biol. Conserv.* 69:285–91
166. Schwalbe CR, Rosen PC. 1988. Preliminary report on effect of bullfrogs on wetland herpetofaunas in southeastern Arizona. In *Management of Amphibians, Reptiles and Small Mammals in North America*, pp. 166–73. US Dep. Agric.
167. Semlitsch RD. 1998. Biological delineation of terrestrial buffer zones for pond-breeding salamanders. *Conserv. Biol.* 12: 1113–19
168. Semlitsch RD, Bodie JR. 1998. Are small, isolated wetlands expendable? *Conserv. Biol.* 12:1129–33
169. Semlitsch RD, Scott DE, Pechmann JHK, Gibbons JW. 1996. Structure and dynamics of an amphibian community. In *Long-Term Studies of Vertebrate Communities*, ed. ML Cody, JA Smallwood, pp. 217–50. San Francisco: Academic Press
170. Shirose LJ, Brooks RJ. 1997. Fluctuations in abundance and age structure in three species of frogs (Anura: Ranidae) in Algonquin Park, Canada, from 1985 to 1993. See Ref. 87, pp. 16–26
171. Shoop CR. 1974. Yearly variation in larval survival of *Ambystoma maculatum*. *Ecology* 55:440–44
172. Sinsch U. 1991. Mini-review: the orientation behaviour of amphibians. *Herpetol. J.* 1:541–44
173. Sinsch U. 1992. Structure and dynamics of a natterjack toad metapopulation (*Bufo calamita*). *Oecologia* 90:489–99
174. Sinsch U. 1996. Population dynamics of natterjack toads (*Bufo calamita*) in the Rhineland: a nine-years study. *Verh. Deutschen Zool. Ges.* 89:1–127
175. Sjögren P. 1991. Extinction and isolation gradient in metapopulations: the case of the pool frog (*Rana lessonae*). *Biol. J. Linn. Soc.* 42:135–47
176. Sjögren Gulve P. 1994. Distribution and extinction patterns within a northern metapopulation of the pool frog, *Rana lessonae*. *Ecology* 75:1357–67
177. Sredl MJ, Collins EP, Howland JM. 1997. Mark-recapture of Arizona leopard frogs. In *Ranid Frog Conservation and Management. Nongame and Endangered Wildl. Prog. Tech. Rept 121. Ariz Game Fish Dept*, ed. MJ Sredl, pp. 1–20.
178. Stebbins RC. 1954. Natural history of the salamanders of the Plethodontid genus *Ensatina*. *Univ. Calif. Publ. in Zool.* 54: 47–124
179. Stromberg G. 1995. The yearly cycle of the jumping frog (*Rana dalmatina*) in Sweden. A 12 year study. *Scientia Herpetologica* 1995:185–86
180. Stumpel AHP. 1987. Distribution and present numbers of the tree frog *Hyla arborea* in Zealand Flanders, The Netherlands (Amphibia, Hylidae). *Bijdragen tot de Dierkunde* 57:151–63
181. Taub FB. 1961. The distribution of the red-backed salamander, *Plethodon c. cinereus*, within the soil. *Ecology* 42:681–98
182. Terhivuo J. 1988. Phenology of spawning of the common frog (*Rana temporaria*) in Finland from 1846 to 1986. *Ann. Zool. Fennici* 25:165–75
183. Toft CA, Rand AS, Clark M. 1982. Population dynamics and seasonal recruitment in *Bufo typhonius* and *Colostethus nubicola* (Anura). In *The Ecology of a Tropical Forest: Seasonal Rhythms and Long-Term Changes*, ed. EG Leigh Jr, AS Rand,

DM Windsor, pp. 397–403. Washington, DC: Smithsonian Inst. Press

184. Travis J. 1994. Calibrating our expectations in studying amphibian populations. *Herpetologica* 50:104–8
185. Turner FB. 1960. Population structure and dynamics of the western spotted frog, *Rana p. pretiosa* Baird & Girard, in Yellowstone Park, Wyoming. *Ecol. Monogr.* 30:251–78
186. Turner FB. 1962. The demography of frogs and toads. *Q. Rev. Biol.* 37:303–14
187. van de Mortel T, Buttemer W, Hoffman P, Hays J, Blaustein A. 1998. A comparison of photolyase activity in three Australian tree frogs. *Oecologia* 115:366–69
188. Vertucci FA, Corn PS. 1996. Evaluation of episodic acidification and amphibian declines in the Rocky Mountains. *Ecol. Appl.* 6:449–57
189. Vial JL, Saylor L. 1993. *The Status of Amphibian Populations: a Compilation and Analysis. IUCN/SSC Declining Amphibian Populations Taskforce. Work. Doc. No. 1*
190. Vitt LJ, Caldwell JP, Wilbur HM, Smith DC. 1990. Amphibians as harbingers of decay. *BioScience* 40:418–18
191. Voris HK, Inger RF. 1995. Frog abundance along streams in Bornean forests. *Conserv. Biol.* 9:679–83
192. Waldman B, Tocher M. 1998. Behavioral ecology, genetic diversity, and declining amphibian populations. In *Behavioural Ecology and Conservation Biology*, ed. T Caro, pp. 394–443. New York: Oxford Univ. Press
193. Warner RR, Chesson PL. 1985. Coexistence mediated by recruitment fluctuations: a field guide to the storage effect. *Am. Nat.* 125:769–87
194. Waringer-Löschenkohl A. 1991. Breeding ecology of *Rana dalmatina* in lower Austria: a 7-years study. *Alytes* 9:121–34
195. Wassersug R. 1992. On assessing environmental factors affecting survivorship of premetamorphic amphibians. See Ref. 22, pp. 53–59
196. Weitzel NH, Panik HR. 1993. Long-term fluctuations of an isolated population of the Pacific chorus frog (*Pseudacris regilla*) in northwestern Nevada. *Great Basin Nat.* 53:379–84
197. Welsh HW Jr. 1990. Relictual amphibians and old-growth forests. *Conserv. Biol.* 4: 309–319
198. Werner JK. 1991. A radiotelemetry implant technique for use with *Bufo americanus. Herpetol. Rev.* 22:94–95
199. Wilbur HM. 1980. Complex life cycles. *Annu. Rev. Ecol. Syst.* 11:67–93
200. Winter JE, Alford RA. 1999. *Terrestrial Vertebrate Fauna in Regenerated Mine Areas at Weipa*. Rep. to Commonwealth Aluminium Co. 150 pp.
201. Wissinger SA, Whiteman HH. 1992. Fluctuation in a Rocky Mountain population of salamanders: anthropogenic acidification or natural variation? *J. Herpetol.* 26:377–91
202. Woolbright LL. 1991. The impact of Hurricane Hugo on forest frogs in Puerto Rico. *Biotropica* 23:462–67
203. Woolbright LL. 1996. Disturbance influences long-term population patterns in the Puerto Rican frog, *Eleutherodactylus coqui* (Anura: Leptodactylidae). *Biotropica* 28:493–501
204. Wyman RL, Hawksley-Lescault DS. 1987. Soil acidity affects distribution, behavior and physiology of the salamander *Plethodon cinereus. Ecology* 68:1819–27
205. Wyman RL, Jancola J. 1992. Degree and scale of terrestrial acidification and amphibian community structure. *J. Herpetol.* 26:392–401
206. Zug GR, Zug PB. 1979. The marine toad, *Bufo marinus*: a natural history resume of native populations. *Smithson. Contrib. Zool.* 284:1–58

Annu. Rev. Ecol. Syst. 1999. 30:167–99

USING PHYLOGENETIC APPROACHES FOR THE ANALYSIS OF PLANT BREEDING SYSTEM EVOLUTION

Stephen G. Weller and Ann K. Sakai
Department of Ecology and Evolutionary Biology, University of California, Irvine, California 92697 sgweller@uci.edu and aksakai@uci.edu

Key Words plant reproductive systems, phylogeny, breeding systems, pollination systems, dioecy, self-incompatibility

■ **Abstract** Until recently, studies of plant reproductive systems have been at the population level, using microevolutionary approaches. The development of cladistic approaches, combined with the emergence of molecular systematics, has resulted in an explosion of phylogenetic studies and an increase in interdisciplinary approaches combining ecological and systematic methodology. These new approaches offer the possibility of testing explicit hypotheses about the number of evolutionary transitions in reproductive characters and the evolutionary relationship of these characters to changes in the environment. Character mapping may be especially useful for detecting convergent evolution. In a number of cases, character mapping has provided new insights into the evolution of plant breeding systems and pollination biology, especially in suggesting the number of times evolutionary transitions have taken place, indicating where there have been reversals and suggesting when preadaptation has been important. The insights provided by character mapping are determined by a number of factors, including the degree of confidence in phylogenies underlying these studies and the identification of appropriate outgroups. Assumptions about character coding, character ordering, inclusion vs. exclusion of characters that are mapped on trees in the data matrix, and weighting of characters will have profound effects on interpretation of character evolution. Highly labile characters that evolve frequently and have the potential to undergo reversals may make it difficult to detect the pattern of character evolution. Characters that are very strongly correlated with each other or with ecological shifts may make prediction of cause and effect using phylogenetic approaches difficult because changes in characters and ecological shifts will occur, apparently simultaneously, on the same branches. Results from microevolutionary studies have been used in several cases to weight transitions, suggesting that results of phylogenetic studies may not provide fully independent assessments of character evolution. While not a simple cure to understanding problems that have been studied only in the realm of microevolutionary studies, phylogenetic approaches offer clear potential for providing new insights for evolutionary studies.

0066-4162/99/1120-0167$08.00

INTRODUCTION

It is only recently that phylogenetic approaches have been used to address questions about the evolution of plant reproductive systems and plant-pollinator interactions. Although many population-level studies have been comparative in nature, fewer attempts have been made to incorporate detailed historical analysis of characters important to population biologists. Conversely, systematic treatments and monographs have rarely treated information on plant reproductive systems in a manner permitting explicit historical analysis.

The development of cladistic approaches that define monophyletic groups on the basis of shared, derived character states, combined with the emergence of molecular systematics, has resulted in an explosion of phylogenetic studies and an increased awareness of the need for interdisciplinary approaches combining ecological and systematic methodology (6, 7, 9, 19, 39, 52, 66). These phylogenetic approaches offer the hope of understanding the historical context for the evolution of character evolution, including plant reproductive systems. For example, it may be possible to determine the number of times a character state evolves and whether the character state evolves in response to ecological shifts hypothesized as causative in character evolution. The potential for understanding both the number of times a character state has evolved and the timing of the evolution of that character state relative to environmental shifts is a powerful complement to population-level studies, which provide detailed information but are necessarily limited to one or a few populations, often of a single species. The purpose of this review is to determine how character mapping using phylogenetic trees has been used for the analysis of plant breeding system evolution and plant-pollinator interactions. The assumptions and difficulties associated with the use of character mapping are discussed, using examples from plant reproductive systems, including pollination mechanisms (Table 1). A second goal is to determine how these macroevolutionary approaches have extended our understanding of plant reproductive biology beyond the insights derived from studies of microevolutionary processes. The potential for using phylogenetic studies to guide population level studies is also examined. Several examples of insights from these approaches are addressed in detail, and factors contributing to successful use of character mapping are summarized.

THE NECESSITY FOR HISTORICAL APPROACHES

Felsenstein (26) recognized that analyses correlating traits of organisms with each other or with ecological features suffered from a potential lack of independence. A single evolutionary transition in the ancestor of several species would produce a pattern of similarity among the descendent species that would confound attempts to understand the evolutionary significance of relationships. For example, if a shift to a novel environment occurred in a lineage at the same time a new

TABLE 1 Case studies where phylogenetic approaches have been used to reconstruct the evolutionary histories of diverse reproductive systems and plant-pollinator interactions. Case studies are organized by the type of reproductive system or plant-pollinator interaction and are listed alphabetically within categories by family

Genus or lineage	Type of data (morphological, molecular, or combination of both)	Were characters that were mapped included or excluded from the data matrix?	Conclusions based on character mapping	Reference
Androdioecy				
Datisca (Datiscaceae)	Molecular data	Characters not included	Conclusions about the evolution of androdioecy (occurrence of staminate and hermaphroditic individuals in populations) varied greatly depending on molecular data set used to reconstruction phylogeny.	56, 70
Dioecy:				
Schiedea and *Alsinidendron* (Caryophyllaceae)	Morphological and molecular data	Both	Dimorphism (gynodioecy, subdioecy and dioecy) hypothesized to evolve 1–6 times depending on character coding and whether reproductive system was ordered. Inclusion of reproductive system in data matrix had little effect on results, although inclusion of characters related to dioecy resulted in many more equally parsimonious trees.	60, 67, 75, 81
Lepechinia (Lamiaceae)	Morphological	Both	When breeding system character was included in the data matrix, dioecy hypothesized to evolve on two occasions. When the character was excluded, dioecy hypothesized to evolve on four occasions.	37, 38

(*continued*)

TABLE 1 (*continued*)

Genus or lineage	Type of data (morphological, molecular, or combination of both)	Were characters that were mapped included or excluded from the data matrix?	Conclusions based on character mapping	Reference
Freycinetia, Pandanus, Saranga (Pandanaceae)	Morphological	Included	Dioecy and vertebrate pollination ancestral, broader geographic range found in species with incomplete or wind pollination and facultative apomixis.	15
Siparuna (Siparunaceae)	Molecular	Characters not included	Dioecy has evolved from monoecy.	57
Heterostyly and multi-allelic self-incompatibility				
Lythraceae	Morphological	Characters included	Five transitions to tristyly hypothesized.	34
Pontederiaceae	Morphological and molecular	Characters excluded in morphological analysis	Morphological analysis produced many equally parsimonious trees, preferred tree did not split homostylous and heterostylous species of *Eichhornia*. In molecular analysis, with equal weighting of characters and assuming characters were unordered, four gains of tristyly were possible. With weighting favoring gains of tristyly, a single gain of tristyly occurred.	24, 45
Multi-allelic self-incompatibility in angiosperms	Morphological	Characters excluded	Using different phylogenetic hypotheses and different systems of coding self-incompatibility produced little evidence for basal self-incompatibility in angiosperms. Self-incompatibility is very unlikely to be homologous in divergent angiosperm lineages.	76

Evolution of self-fertilization vs. outcrossing				
Amsinckia, (Boraginaceae)	Molecular	Characters not included	When the outgroup was assumed to be distylous, and the breeding systems was treated as unordered, four separate transitions to homostyly and selfing were hypothesized. Weighting loss of distyly as more likely produced a similar result. If the outgroup was assumed to be homostylous, and characters were unweighted and unordered, homostyly was hypothesized as basal.	62
Schiedea and *Alsinidendron* (Caryophyllaceae)	Morphological and molecular	Both	Two separate transitions in lineage to facultative or obligate self-fertilization. Very wet habitats appear to favor autogamy, perhaps because enclosure of anthers within calyx protects pollen from rainfall and facilitates selfing.	60, 67, 75, 81
Scutellaria angustifolia complex (Lamiaceae)	Morphological and molecular (allozyme data)	Not included	Outcrossing has evolved on three occasions from selfing ancestors.	55
Cuphea (Lythraceae)	Morphological	Not included	*Cuphea* section *Brachyandra*, comprised of mostly self-fertilizing species, is highly polyphyletic, with species distributed to four other sections of *Cuphea*.	35
Linanthus section *Leptosiphon* (Polemoniaceae)	Molecular	Not included	Self-incompatibility basal in *Linanthus*, 3–4 transitions to self-compatibility.	33, 33a

(*continued*)

TABLE 1 *(continued)*

Genus or lineage	Type of data (morphological, molecular, or combination of both)	Were characters that were mapped included or excluded from the data matrix?	Conclusions based on character mapping	Reference
Pollination biology: Vertebrate pollination				
Aphelandra pulcherrima complex (Acanthaceae)	Morphological	Information not given	Pollination by short-billed trochiline hummingbirds derived in complex; pollination by traplining hermit hummingbirds represents the ancestral character state. Shift to use of short-billed hummingbirds as pollinators associated with shift to dry habitats. Morphology of corollas provides further evidence that short corollas have been independently derived.	53
Adansonia (Bombacaceae)	Molecular	Characters excluded	Basal pollination system could not be determined because of diversity of pollinators in closely related genera. Assuming that hawkmoth pollination is basal, mammalian pollination has arisen on two occasions in lineage.	8
Erythrina (Fabaceae)	Morphological and molecular	Characters included	Passerine pollination is ancestral, four transitions to hummingbird pollination hypothesized. Morphological information suggests that floral features associated with hummingbird pollination in these four lineages are not homologous.	10

Parkia (Fabaceae)	Morphological	Characters included	Entomophily basal in genus, with a single shift to bat pollination hypothesized.	49
Insect pollination				
Abrotanella (Asteraceae)	Morphological	Characters included	White-flowered and purple-flowered species confined to a single clade within genus; purple-flowered species found in sub-Antarctic Islands, a pattern repeated for other primarily New Zealand genera. White coloration is a possible response to fly-dominated pollinator fauna; purple pigment may protect against effects of sunlight.	69
Dalechampia (Euphorbiaceae)	Morphological	Characters not included	Pollination systems very labile, with many independent gains of similar pollination systems. In three lineages where fragrances are collected by euglossine bees, different morphological structures secrete fragrances.	2, 3
Lapeirousia (Iridaceae)	Morphological	Information not given	Mapping of pollination systems indicated a wide range of pollinators with two types of long-tongued fly pollination evolving convergently on several occasions.	32

(*continued*)

TABLE 1 *(continued)*

Genus or lineage	Type of data (morphological, molecular, or combination of both)	Were characters that were mapped included or excluded from the data matrix?	Conclusions based on character mapping	Reference
Subtribes Cyrtopodiinae and Catasetinae (Orchidaceae)	Molecular	Characters not included	Two shifts to pollination by euglossine bees hypothesized; a single shift at the base of the tree is equally parsimonious, but improbable in view of the Old and New World distribution of the Cytopodiinae and the New World distribution of the euglossine bees.	14
Disa (Orchidaceae)	Morphological	Characters included	Nineteen pollination syndromes found among 27 species included in study. Many pollination syndromes have evolved on numerous occasions, and autogamy has evolved from outcrossing on three occasions.	42
Bird and insect pollination				
Nectar spur diversification in *Aquilegia* (Ranunculaceae)	Molecular	Not included	Evolution of nectar spurs a key innovation promoting species diversification.	40, 41

Hydrophyly and wind pollination				
Cymodoceaceae	Morphological	Characters included	All taxa in Cymodoceaceae dioecious, leading to hypothesis that the ancestor was also dioecious. Hydrophily assumed to have evolved at the base of the lineage containing the Cymodoceaceae and two related families (Zosterdaceae and Posidoniaceae).	16
Marine angiosperms	Molecular	Characters excluded	Hydrophily has evolved on several occasions; conclusion of Cox and Humphries (16) resulted from exclusion of many nonmarine sister taxa. Unisexuality is basal in some marine angiosperm lineages, and has not necessarily evolved after hydrophily.	46
Schiedea and *Alsinidendron* (Caryophyllaceae)	Morphological and molecular	Both	Wind pollination strictly associated with evolution of dimorphism (gynodioecy, subdioecy, and dioecy), and has probably evolved each time there has been a shift to dimorphism (1–6 occasions).	78

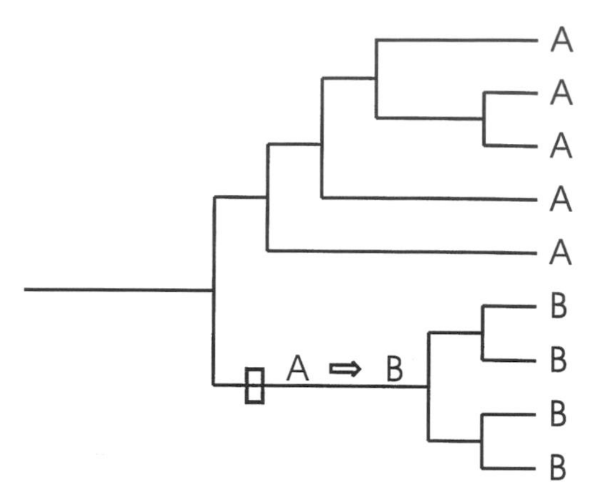

Figure 1 Hypothesized single ecological shift (*open rectangle*) in one branch of lineage, followed by shift from character state A to B. All four derivative species would show the same correlation of habitat and trait, suggesting, in the absence of phylogenetic information, a strong causal basis.

morphological trait appeared, perhaps incidentally, subsequent diversification of the lineage would give the impression of a strong correlation between the novel environment and the trait (Figure 1). Despite the strong correlation, there might not be a causal relationship. In contrast, if species in a lineage colonized the novel environment multiple times and acquired the morphological trait on each occasion, the case for a causal relationship would be much stronger (Figure 2). Subsampling, or

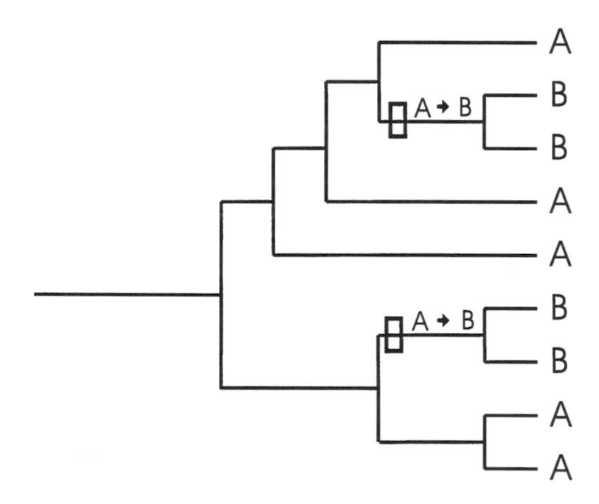

Figure 2 Ecological shifts (*open rectangles*) occur independently in different portions of the phylogeny, followed in each case by a transition from character state A to B. Evidence that the ecological shift has caused the transition is stronger than for the phylogeny shown in Figure 1.

sampling at different taxonomic levels, would not resolve the problem (26). Only when phylogenetic approaches are used, where the relationship of changes in the trait with the relative timing of shifts in ecology can be specified, can causality be understood. This approach asks whether traits evolve with greater frequency in portions of trees where other traits have changed or where shifts in ecology have been detected (19, 50). The process involves first the construction of a phylogenetic tree using parsimony or maximum likelihood methods (72). The phylogenetic tree is viewed as a hypothesis of evolutionary relationships. Morphological or molecular characters, or a combination of the two types of data, are used to construct the phylogeny. Using the principle of parsimony, traits are mapped (or optimized) on the tree in a manner that minimizes the degree of convergent evolution. Parsimony is the most significant assumption underlying this approach. As emphasized by Felsenstein and others (17a, 26), evolution may occur nonparsimoniously, and transitions in character states could be radically underestimated because of undetected convergent evolution. The possibility that character state changes may be underestimated must be recognized in all studies. Use of maximum likelihood for the reconstruction of character states (17a, 61a) circumvents some of the problems of parsimony. Maximum likelihood methods take into account differences in evolutionary rates and in the lengths of internode branches, and may favor nonparsimonious reconstructions of character state evolution (17a). To date, these approaches have been used only rarely in studies of the evolution of reproductive systems (e.g. 33a, 61).

CONSTRUCTING PHYLOGENIES

Morphological and Molecular Data Sets

Construction of the phylogeny is the critical and obvious first step of the process of character mapping using phylogenetic approaches. Morphological data may be an important source of information (22) and often may lead to robust phylogenetic hypotheses. While molecular data sets are sometimes viewed as superior because they are less likely to show convergence, independent gains of morphological traits are often phylogenetically informative (22). This will be true so long as they are recognized as independent gains (31). The introduction of molecular approaches to phylogenetic reconstruction has led to a very rapid increase in the number of phylogenetic analyses, providing the hope that character mapping can answer large-scale questions about trait evolution and cause-effect relationships.

Molecular data are sometimes viewed as the best or only appropriate data for phylogenetic analysis of character evolution because of the difficulties of detecting convergent evolution (homoplasy) using morphological data (31, 65). Concerted convergence has been defined as environmentally induced convergence so strong that entire syndromes of morphological traits converge, with little hope for detecting convergence using morphological characters (31). Although molecular data are often thought to show little homoplasy, this is not always the case. Homoplasy may occur in older lineages where many molecular changes have accumulated, so that unrelated taxa may occur in close proximity on phylogenetic trees constructed using molecular data (22).

Combined data sets consisting of morphological and molecular data are often employed, particularly when the individual data sets give congruent results. When similar results are obtained from independent data sets, the level of confidence in the phylogenetic hypothesis increases, and often resolution of the branching patterns in the phylogenetic tree is enhanced (22). Adequate phylogenetic resolution using only molecular data may be a problem, particularly for rapidly evolving or young clades where few molecular differences have accumulated. For these groups, morphological data may provide the only phylogenetically informative characters. In contrast, basal portions of clades are often well defined using both morphological and molecular characters because of the time available for divergence for both types of characters.

Future studies will undoubtedly rely heavily on molecular approaches, especially as new regions of the genome become available for molecular analysis. It seems likely, however, that there will be renewed emphasis on morphological characters and phylogenies constructed using combinations of molecular and morphological characters.

Identification of Appropriate Outgroups

Correct identification of the outgroup, the taxon most closely related to the group under study, is critical for identification of ancestral character states within lineages and is therefore of fundamental importance to character mapping (21, 51). When outgroups cannot be identified with certainty, the outgroup substitution approach may be a satisfactory alternative (21). Using this approach, the ancestral state of the character is determined using a variety of potential outgroups, and the effects of varying the outgroup on character mapping are analyzed. The ability to precisely identify the outgroup provides a powerful means for understanding character evolution. For example, in the Hawaiian silversword alliance (*Argyroxiphium*, *Dubautia*, *Wilkesia*; Asteraceae), molecular data and artificial hybridization have shown with a high degree of probability that the ancestor of the Hawaiian silverswords was closely related to self-incompatible, perennial species of *Madia* and *Raillardia* in California (5). The current widespread self-incompatibility in the silverswords (11) is therefore likely to be the result of colonization by a self-incompatible ancestor of this lineage. Unfortunately, precise outgroup identification is often problematic. Even in remote archipelagoes like the Hawaiian Islands, where the rarity of colonization events means that most lineages are very likely to be monophyletic, it may be difficult to determine the correct outgroup because lineages in Hawaii have often diverged dramatically from continental relatives (e.g. 81).

Several studies illustrate approaches used to address difficulties in outgroup identification. Analysis of the evolution of distylous reproductive systems in *Amsinckia* using phylogenetic approaches is complicated by the difficulty in identifying the appropriate outgroups (62). To circumvent this problem, different assumptions about the breeding system of the outgroup were used. When the outgroup was assumed to be distylous and the breeding system character was treated as unordered, distyly was hypothesized to be basal in the lineage, with four separate transitions to homostyly and selfing (43, 62). When the outgroup was considered homostylous and loss of distyly was weighted as more likely than a gain (cf. 45), a single transition to distyly at the base of the lineage was most parsimonious. Homostyly was the basal condition in *Amsinckia* only when the outgroup was treated as homostylous and the breeding system character was unweighted and unordered (62).

In a molecular phylogenetic study of *Linanthus* section *Leptosiphon* (Polemoniaceae; 33a), outgroup analysis could not be used to resolve the question of whether self-incompatibility is the ancestral breeding system state. Self-incompatibility is widely distributed in the Polemoniaceae and sporophytically controlled in *Linanthus parviflorus* (33). Because of uncertainty over relationships among sections of *Linanthus*, and between *Linanthus* and other genera of Polemoniaceae, a more distant outgroup (*Navarretia*) was chosen to root the tree. Because this outgroup was not appropriate for determining the basal breeding system condition (33a), no assumptions were made about the basal breeding system condition based on outgroup analysis. When losses and gains of self-incompatibility were weighted equally, or when loss of self-incompatibility was considered more likely than a gain, self-incompatibility proved to be the ancestral character state in *Linanthus*, and selfing evolved on three to four occasions (33a). *Linanthus bicolor*, a widely distributed selfing species, was shown to be polyphyletic in origin. Alternate trees required fewer transitions to self-compatibility but broke apart a clade with 100% bootstrap support and had significantly lower likelihood values using the maximum likelihood approach for constructing phylogenies. When a loss of self-incompatibility was considered slightly less likely than a gain (weighting ratio of 1.2:1), there were five independent origins of self-incompatibility.

The basal condition for pollination systems could not be estimated in baobabs (*Adansonia*, Bombacaceae) based on outgroup comparison because of the wide diversity of pollination systems in closely related genera and the lack of phylogenetic information for these genera (8). If the basal pollination condition in *Adansonia* is assumed to be hawkmoth pollination, then mammalian pollination has arisen on two occasions in the lineage. Morphological features associated with mammalian pollination are diverse, as are the mammals that pollinate these species, and it seems likely that these differences will provide further evidence for independent transitions to mammalian pollination in *Adansonia*.

Difficulties in outgroup identification may eventually be resolved for many lineages as increased information about generic and family level relationships becomes available. Molecular variation is likely to be especially suited to this task, particularly using those sequences whose variation is too limited to provide species-level resolution in many cases, but which may be sufficient for higher level resolution.

Character Inclusion or Exclusion

Whether characters that will be mapped on trees should be included in the morphological data matrices used to produce trees has been debated extensively. A commonly held view is that inclusion of these characters is circular because taxa possessing a character state would group together and cases of multiple evolution of the character state (homoplasy) would be missed (2, 9, 65). In contrast, others (18, 22, 71) argue that traits should be included if they are phylogenetically informative. Even with substantial homoplasy, these characters may be useful in phylogenetic reconstruction, assuming that there are enough unrelated characters that homoplasy can be detected. Luckow & Bruneau (48) concluded that traits under study should always be included in analyses, and they argued that exclusion of characters may lead to weaker phylogenetic hypotheses. They also distinguished between characters coded on a functional basis that may be misleading versus characters that follow the rules of homology.

This distinction is exemplified in Bruneau's (10) investigation of passerine and hummingbird pollination in *Erythrina*. Similar phylogenies were obtained from morphological and molecular approaches. Passerine pollination is the ancestral condition in *Erythrina*, with four hypothesized transitions to hummingbird pollination. Transitions to hummingbird pollination involve shifts from horizontal to upright inflorescences, secund flowers to flowers radially arranged along the inflorescence axis, and standard petals that are open and expose the reproductive structures to standard petals that are folded to form a pseudotube.

Despite similarities among the hummingbird-pollinated species of *Erythrina*, each transition involved differences in calyx and pollen morphology indicating that transitions to hummingbird pollination were not strictly homologous. Deletion of characters associated with the pollination systems of *Erythrina* would have resulted in substantial loss of phylogenetically useful information. Luckow & Bruneau (48) argued that deletion of characters in phylogenetic analysis would always be arbitrary because it would be necessary to know which characters were associated with the adaptation in question at the outset of the analysis. The characters that one might choose to exclude would vary depending on the ecological hypothesis under investigation.

The argument of whether it is better to include or exclude particular characters in morphological data sets is unlikely to have a single answer. Armbruster (3) suggested that characters with a low consistency index (those that evolve or are lost on more than a single occasion in a tree) should be used preferentially in phylogenetic analysis because they would be less likely to interfere with phylogenetic reconstruction. A similar approach would be to determine whether characters should be excluded on a case-by-case basis. If multiple independent origins for a character state are hypothesized, inclusion of the character in the data matrix would be appropriate because inclusion would favor a single origin of the character state, and the bias is in the appropriate direction for a hypothesis of multiple transitions (18). Conversely, including a character would be inappropriate when a single origin is considered likely because this would favor the hypothesized result (18).

In the orchid genus *Disa*, multiple shifts were inferred to occur in several pollination syndromes; apparently enough characters unrelated to these floral syndromes were used in the phylogenetic analysis to prevent all species possessing similar pollination biology from grouping together (42). Johnson et al (42), like de Queiroz (18), viewed the use of a potentially functionally related character as a conservative approach reducing the likelihood of detecting multiple transitions, rather than overestimating them. Regardless of whether characters show single or multiple origins, the actual number of transitions is likely to be underestimated (48). Luckow & Bruneau (48) and de Queiroz (18) have argued that discarding characters with high consistency indices will result in less accurate phylogenies.

The effects of character exclusion will undoubtedly vary depending on the features of lineages under investigation. If the characters in question constitute a small fraction of the total number of characters in the matrix, then character exclusion may have little effect on phylogenetic reconstruction, assuming that the remaining characters contain phylogenetically useful information. Armbruster (3) was able to obtain a highly resolved tree, despite elimination of all floral traits from his analysis. If characters that are phylogenetically informative are few in number, then inclusion of the character will exert undue influence on the tree, and cases of convergence might not be detected. One solution is the inclusion of additional, presumably unrelated characters (10, 42, 48), assuming that they are available (10).

The minor role of floral characters in establishing phylogenetic relationships in *Schiedea* (Caryophyllaceae) may explain why deletion of the breeding system trait had no effect on estimates of the number of transitions to dimorphism (81). In *Schiedea*, flowers are small and have relatively few characters of phylogenetic significance. In contrast, vegetative characters, which are quite variable, delineate four major clades in the genus. Addition or subtraction of the breeding system character (presence of hermaphroditism, gynodioecy, subdioecy, or dioecy) had relatively little effect on phylogenetic reconstruction (81), presumably because of the overwhelming effect of the vegetative characters. Addition of characters thought to be associated with breeding system, such as the degree of condensation of the inflorescence (a trait associated with wind pollination and dimorphism) resulted in increasing numbers of equally parsimonious, less-resolved trees. The significant impact of adding traits associated with dimorphism indicates that one or several of these traits had a more significant effect on phylogenetic reconstruction than the breeding system character alone (81). Including all characters thought to be related to the evolution of dimorphism, especially inflorescence structure, resulted in changes in tree topology that implied basal positions for hermaphroditic species within clades. This occurred because outgroups were coded as having diffuse inflorescences, in contrast to the highly condensed inflorescences of several dimorphic *Schiedea* species.

One limitation in *Schiedea* for assessing potential convergence in the evolution of dimorphism is the small number of characters directly associated with the shift to dimorphism. Although the evolution of dimorphism might have occurred on several occasions in *Schiedea*, there are few morphological features that might be used to detect whether changes have occurred independently (are not truly homologous), because evolution of dioecy involves loss or reduction of floral parts rather than modification of parts. In contrast, the complex shifts that occur in *Erythrina* species adapted to hummingbird pollination have taken several routes, permitting detection of different shifts to the derived pollination syndrome (10). In this study of *Erythrina*, the question of failure to detect convergence is somewhat moot with respect to the pollination features since detailed morphological studies demonstrated that with each origin of hummingbird pollination different patterns of floral modification could be detected. Bruneau's study (10) approaches the case where knowing the "true homology" of the character means that each character state appears at a single point on the tree (48). Cases where an in-depth analysis produces this type of result may be rare, but the analysis of *Erythrina* indicates how morphological analyses provide insights into shifts in a complex trait.

Phylogenetic analysis was used to determine whether bat pollination in *Parkia* (Fabaceae) has evolved on separate occasions in the New and Old World (49). This question was motivated by the diverse inflorescence architecture of species from the two geographic areas, and the absence of a close relationship between the New and Old World bat species that pollinate *Parkia*. Another question of interest was whether insect pollination in this genus is the basal condition or secondarily derived from bat pollination. All characters were used in the phylogenetic analysis; exclusion of evidence was projected to weaken the phylogenetic hypothesis, and there was no a priori basis for judging which characters should be excluded because of their potential association with bat pollination (49). In this study, inclusion of characters associated with pollination biology biases results in the opposite direction from the prediction of more than one transition to bat pollination.

In contrast to the expectation that bat pollination might have evolved on several occasions in *Parkia*, phylogenetic analysis indicated that a single transition to bat pollination has occurred, presumably in the New World. Only one character, a change in anther attachment, was perfectly correlated with the shift to bat pollination, although the functional significance of this character was unclear. Few characters in *Parkia* provide obvious clues for understanding why bat pollination evolved because many of the features associated with bat pollination are found in outgroup species. Entomophily is basal in the genus, and a species pollinated by nocturnal bees may represent a preadaptation for bat pollination (49).

With only a single shift to bat pollination hypothesized in *Parkia*, understanding general ecological conditions and morphological features favoring the evolution of this type of pollination will depend on additional phylogenetic studies of other lineages where bat pollination has evolved, as well as microevolutionary studies focused on the functional significance of characters potentially associated with bat pollination.

Despite the different conclusions about the number of transitions to derived pollination states in *Erythrina* and *Parkia*, the studies are similar in that a large number of characters are associated with the floral syndromes under consideration.

In *Erythrina*, these characters indicated slightly different syndromes of features associated with transitions to hummingbird pollination (a lack of homology in the feature), while in *Parkia*, despite the diversity of bat-pollinated flowers, a single origin of the shift in pollination biology was likely. The rich diversity of characters makes these conclusions robust, and the issue of character inclusions vs. exclusion is less relevant than in those studies where reproductive systems or pollination biology have few characters for analysis.

Hart's (37, 38) study of the evolution of dioecy in *Lepechinia* (Lamiaceae) was the first use of phylogenetic methods for mapping a reproductive system trait. He found that dioecy has evolved from gynodioecy on two occasions, if the reproductive system was included as a character. Deletion of the breeding system character (dioecy vs. gynodioecy) in *Lepechinia* resulted in an increase in the number of transitions to dioecy from two to four. This result demonstrates the substantial impact of excluding the reproductive character in *Lepechinia*. Although Hart determined the effect of deleting the single breeding system character from his matrix, he did not remove characters such as flower size that are likely to be related to breeding system.

In a phylogenetic analysis of the Pontederiaceae, Eckenwalder & Barrett (24) excluded characters known to be part of the tristylous breeding system from their data matrix. Many equally parsimonious trees resulted, and the authors chose as their preferred tree one that did not split homostylous and heterostylous species of *Eichhornia* because of "the clear relationships among these species." Based on microevolutionary studies, the authors could pair homostylous and heterostylous species of *Eichhornia*, suggesting that homostyly has been derived on several occasions from tristyly. These relationships could not be detected using phylogenetic analysis of morphological characters. The selfing syndrome may have provided enough characters to unite homostylous species in a single clade. This tendency was probably heightened by deliberate exclusion of traits associated with tristyly, as traits associated with tristyly appeared useful for recognizing species pairs (24). Some features of the tristylous reproductive system appear to provide useful data for assessing phylogenetic relationships; the difficulty is in distinguishing these characters from those that result in grouping of species with the same breeding system that may not be closely related.

In summary, approaches to inclusion versus exclusion of characters have been extremely diverse, and there is probably less agreement on the best approach to this problem than any other aspect of character mapping. When inclusion of characters in data matrices appears to result in misleading conclusions about phylogenetic relationships (cf. 71), investigators are likely to delete these characters from matrices, despite the difficulty in determining a priori which characters are problematic. The complexity of the traits that are mapped on trees and the total number of characters used to construct trees will undoubtedly influence approaches taken by investigators. When deletion of a single or small number of characters from a tree has a pronounced effect on the topology of the tree and conclusions about the evolution of a characters, it seems clear that the phylogenetic hypothesis is poorly supported, and character mapping is unlikely to yield significant insights about character evolution.

Character Coding, Ordering, and Weighting

How characters are coded and whether characters are ordered may have profound effects on interpretation of character evolution. Morphological characters that are coded as a number of different states will more accurately reflect the diversity within the group in question but usually result in more transitions to derived character states. When characters with multiple states are mapped on branches of a tree, derived character states occur nearer to the terminal branches of the tree. In contrast, when the character state is binary, derived character states may occur at deeper levels of the tree, ultimately resulting in fewer transitions to the derived character state and more reversals near the branch tips. The effects of differences in character coding can be seen in *Schiedea*, where coding the breeding system as four states (hermaphroditic, gynodioecious, subdioecious, and dioecious) rather than two states (hermaphroditic, dimorphic) resulted in a greater number of hypothesized transitions from hermaphroditism to dimorphism (81).

Ordering of character states and binary character coding have similar effects on the estimation of numbers of transitions to derived character states. In *Schiedea*, ordering of the character states implies that dioecy evolves from hermaphroditism through intermediate gynodioecious and subdioecious character states. Given that the character states are defined in large part by the frequency of females in populations, ordering of the character states seems very reasonable. Such an approach may be unrealistic, however, if changes in breeding systems occur rapidly relative to branching events in trees (81). In *Schiedea*, ordering of characters resulted in fewer transitions to gynodioecy and more reversals to hermaphroditism.

Character weighting, where transitions between character states are more likely in one direction than the other, will have major effects on interpretation of character evolution. Using a molecular approach, Kohn et al (45) reconstructed the phylogeny of the Pontederiaceae. As predicted from earlier studies (24), tristylous species of *Eichhornia* were most closely related to monomorphic species. When monomorphic species were forced to form a single clade, trees were 23–26 steps longer. Two optimization schemes were used to map tristyly and related characters, including self-incompatibility. In the first scheme, all character states were unordered and weighted equally. In the second approach, loss of tristyly (or enantiostyly, a second breeding system found within the Pontederiaceae) was favored by a two-fold margin over a gain of either breeding system or an interconversion between them. Not surprisingly, conclusions about breeding system evolution varied substantially, depending on the weighting scheme. With equal weighting and the characters unordered, up to four gains of tristyly were possible, and self-incompatibility evolved twice in the family. With unequal weighting, a single gain in tristyly occurs, and one to two gains of self-incompatibility are possible. The conclusion from the molecular study that tristyly has originated a single time is based in large part on the assumption that the weighting scheme is appropriate. If a heavy bias in favor of a particular outcome is introduced at the outset of the study, one might argue that there is little point in the phylogenetic analysis, at least for the purpose of character mapping. Kohn et al (45) argue, however, that character weighting is perhaps more reasonable than leaving the characters unweighted, in view of the complexity of tristyly and the low probability that this breeding system evolves very frequently. Using the weighting scheme that favors the loss over the gain of tristyly, self-incompatibility appears to evolve after the evolution of tristyly (45). The evolution of self-incompatibility after tristyly is obtained when losses and gains are equally weighted or when a loss of self-incompatibility is weighted more heavily than a gain. Using different rootings of the tree and different coding schemes for self-incompatibility could change this interpretation (45).

Using morphological phylogenetic data, Graham et al (34) concluded that tristyly in the Lythraceae has evolved on at least five occasions. In this analysis, heterostyly was included in the data matrix. Because of the difficulty in finding a sufficient number of characters, the phylogeny based on these characters was very weakly supported. A weighting scheme that favored loss of heterostyly over gains would presumably have resulted in substantial modifications of the phylogeny, especially in view of the limited number of characters used for the analysis, and fewer transitions to heterostyly. Self-compatibility in the Lythraceae is characteristic of all monomorphic genera, and families related to the Lythraceae are also self-compatible, suggesting that acquisition of tristyly in the Lythraceae has been accompanied by the evolution of self-incompatibility (76). If heterostyly has evolved on separate occasions, then self-incompatibility may also have evolved independently.

These examples illustrate the complexity of issues related to character delineation and assumptions about these characters. Although the use of phylogenetic trees for the analysis of character evolution may appear to be an objective means of obtaining additional insights into evolutionary processes, it seems clear that in many cases, prior views of character evolution may strongly influence results.

Limitations of Phylogenetic Hypotheses

The use of phylogenetic approaches for understanding the evolution of plant reproductive biology is clearly limited by the phylogenies underlying the studies. Difficulties in the use of phylogenies have diverse causes (19). Phylogenies may be poorly supported or have topologies that do not reflect the evolutionary history of the lineage. Even when phylogenies are well supported, attempts to map characters may lead to ambiguity in interpretation because of the topology of the tree.

There are many causes for weakly supported phylogenetic hypotheses. Recent evolution of a lineage and reduced phylogenetically informative variation lead to poorly resolved phylogenies. Lineages within the Hawaiian Islands illustrate difficulties resulting from reduced variation. Area cladograms (phylogenies mapped on the geographic distribution of species within a monophyletic lineage) demonstrate for many lineages that basal portions of radiations occur on the older high islands. The older portions of these lineages are more likely to be resolved than those occurring on the younger islands because of the accumulation of derived character states (74). Alternatively, lineages that colonized the younger islands initially (e.g. *Geranium* [Geraniaceae] and *Tetramolopium* [Asteraceae]) are likely to have less morphological diversity than lineages that have migrated from older to younger islands. Even in cases where there is substantial morphological diversity, molecular differentiation may be limited, as in the case of *Schiedea* (67).

Lineages in the Hawaiian Islands are particularly suitable for illustrating the difficulties resulting from extinction of lineages. Species found on the older islands may well represent the remnants of lineages that were once extensive but that have largely disappeared due to the erosion and subsidence of islands (75). As species become extinct, clades will become compressed, and species that would not have been viewed as closely related and may be very different in morphology will eventually appear as sister taxa. Outgroup analysis could be misleading in these circumstances and could predict unlikely combinations of character states. For example, *Schiedea membranacea*, a species occurring in the basal lineage of *Schiedea* and *Alsinidendron*, is placed in a clade of two *Schiedea* species (*S. membranacea* and *S. verticillata*) sister to highly selfing *Alsinidendron* (*Schiedea* is paraphyletic). On that basis, the breeding system of *S. membranacea* was predicted to be highly selfing (17), a prediction not borne out by studies of outcrossing rates and allozyme variation (17, 79). *Schiedea membranacea* and *S. verticillata* may represent the sole surviving species in a much larger clade that is now largely extinct. Additional information on the breeding system of *S. verticillata* would be useful. If this species is outcrossing, then the prediction that *S. membranacea* would be a selfer is clearly not warranted. *Schiedea verticillata* occurs on Nihoa, a small, largely eroded island occurring 200 km northwest of Kauai. Geological evidence indicates that Nihoa was once as large as Kauai; at one point there may have been many other species related to *S. membranacea* and *S. verticillata* on older Hawaiian Islands before erosion and subsidence, and most of these now extinct taxa may have been outcrossers.

Although the effect of extinction on the process of character mapping is particularly easy to conceptualize for the Hawaiian Islands, where the islands are progressively older to the northwest and extinction rates higher, the same arguments are likely to apply to many clades occurring in continental areas. If basal portions of these clades have suffered from disproportionate extinction, character mapping might result in misleading predictions for expected character states. The occurrence of long branches due to reduced branching could have a similar effect on the reliability of character state assessment, assuming that the probability of change in character states is constant over time and over the phylogeny. Maximum likelihood approaches to character-state reconstruction would minimize some of these problems.

Portions of trees where species have evolved recently may present equally difficult problems for character mapping if characters states are labile (29). For these characters, there will be considerable ambiguity in predicting points at which character states undergo modification. When predicted character states are equivocal, delayed transitions (DELTRAN) and accelerated transitions (ACCTRAN) mapping procedures will give very different results. With accelerated transitions, equivocal character state changes will occur at deeper branching points of the phylogeny. Delayed transitions of equivocal character states occur closer to the tips of the tree. When character mapping produces equivocal results, character state changes can only be bracketed on the tree, rather than pinpointed with accuracy.

Characters that are likely to be labile are those that may present the greatest difficulties for character mapping (26, 29), although these characters might be of considerable interest. Modeling studies indicate, however, that unless clades are very small and transition probabilities are very high, ancestral character states should be determined with high reliability (64). Paradoxically, in portions of clades where species have evolved recently and extinction is unlikely to interfere with interpretation of character evolution, difficulties in resolving phylogenies and the lability of characters of interest may limit the ability to infer ancestral character states. One approach to these difficulties is the use of alternate phylogenetic hypotheses that include trees with different topologies and alternate rootings to determine the range of possible character reconstructions (21).

INFERRING CAUSE AND EFFECT USING PHYLOGENETIC APPROACHES

One of the potential strengths of phylogenetic approaches is the ability to infer causality by mapping the sequence of events on phylogenetic trees. Donoghue (19), using Maddison's (50) method of asking whether gains or losses of traits are concentrated on portions of trees, investigated the relationship of the evolution of fleshy seeds and dioecy in gymnosperms. An earlier contingency analysis of gymnosperms by Givnish (30) indicated a strong relationship between fleshy seeds and dioecy. This relationship was used to suggest that accelerating fitness gains resulting from increasingly large allocation to female function have selected for separate sexes in gymnosperms. As discussed by Donoghue (19), Givnish's use of contingency analysis implies that each case of fleshy seeds and dioecy represents an independent evolutionary event, as assumption unlikely to be true. When Maddison's phylogenetic approach was used, inconclusive results were obtained because of the very strong correlation between the two traits. The tight correlation of dioecy and fleshy fruits suggests the possibility of a functional relationship, but the order of evolutionary events, critical to testing Givnish's hypothesis, could not be tested with this data set. Maddison (50) applied the same test to the larger data set from Donoghue (19) and obtained ambiguous results, depending on the interpretation of coincidental gains of traits. When all shifts to dioecy occurring on branches with animal dispersal were assumed to take place after the evolution of animal dispersal, the probability of the association occurring by chance was very low ($P = 0.018$).

The distribution of dimorphism in *Schiedea* demonstrates a similar problem. In one of the two clades in the lineage where dimorphism evolves, all character reconstructions indicate unambiguously that the transition to dimorphism occurred at the base of the clade. The transition from mesic to dry habitats, a switch hypothesized as a causative factor in the evolution of dimorphism, occurs on the same branch as the evolution of dimorphism, making it impossible to determine which change occurred first. In a second clade with dimorphism, the uncertainty over the number of transitions to dimorphism makes inferences about causality with respect to habitat difficult. One unambiguous transition to dry habitats occurs in a third clade containing only hermaphroditic species, indicating that habitat shifts do not always lead to the evolution of dimorphism in *Schiedea* (81). Armbruster (2) and Frumhoff & Reeve (29) argued that phylogenetic analyses are likely to help determine the order of historical events only if the selective associations between characters are relatively weak. Paradoxically, the method works least well when relationships are strong and likely to be of the greatest interest to evolutionary biologists.

Using a consensus tree based on morphological characters of the Pandanaceae, Cox (15) concluded that dioecy and vertebrate pollination are ancestral character states and that the distribution of the three genera in the family has been influenced strongly by subsequent modifications of both the breeding system and pollination biology. Vertebrate-pollinated *Freycinetia* is widely distributed in the Pacific region, and its distribution is attributed to increased colonization potential as a result of incomplete dioecy, a derived character state. *Saranga* has retained complete separation of sexes but evolved insect pollination, whereas *Pandanus*, with the widest geographic distribution, is dioecious, wind pollinated, and facultatively apomictic. Interpreting ancestral character states in the Pandanaceae is complicated by the lack of a readily defined outgroup (15). The hypothesis that dioecy evolved as an adaptation protecting against destructive vertebrate pollinators may be more testable once the sister group of the Pandanaceae has been determined.

The significance of presumed key innovations, behavioral or morphological traits that are thought to result in new adaptive radiations can be tested using phylogenetic approaches (e.g. 61). Using a molecular phylogeny for *Aquilegia* and related genera, Hodges & Arnold (41) demonstrated that species diversity is very likely to have increased dramatically after the evolution of nectar spurs. The increased diversity is presumed to result from greater reproductive isolation associated with the presence of nectar spurs. No other derived character states appeared to be associated with diversification. The presumed role of nectar spurs and diversification is bolstered by the occurrence of similar higher diversity in the majority of lineages where nectar spurs have appeared (41).

The rapidity of the radiation in *Aquilegia* results in some difficulties in interpretation: The position of one spurless species of *Aquilegia* is unresolved within the genus (40). Presumably, if greater phylogenetic resolution were obtained, this species would be placed in a basal position in the genus (41), although a secondary reversion to spurless flowers is also possible (40). Because of the limited molecular diversification in *Aquilegia* (41) and the limited number of morphological traits useful for phylogenetic analysis, it may be difficult to attain any further resolution of the phylogeny.

DETECTING HOMOLOGY IN REPRODUCTIVE SYSTEMS

The difficulty in detecting convergent evolution (homoplasy) is likely to vary dramatically depending on the breeding system or pollination syndrome question. As complexity in reproductive systems or pollination biology increases, ability to detect homology is likely to increase (cf. 23). Bruneau (10) was able to differentiate different combinations of morphological traits associated with transitions to hummingbird pollination in *Erythrina*, providing a strong rationale in her study for using characters associated with pollination biology in the phylogenetic analysis. Armbruster (3) was able to identify different syndromes of floral characters in *Dalechampia* associated with a shift to pollination by fragrance-collecting, male euglossine bees. Transitions in these cases are not homologous because they have occurred by different routes.

In other lineages, it may not be possible to determine whether there have been independent transitions (except through phylogenetic approaches) because the changes involved are too few in number (i.e., the characters are simple; cf. 23). Independent transitions to dioecy from hermaphroditism or from outcrossing to selfing may be difficult to distinguish because the changes result from loss of relatively small numbers of parts. Renner & Ricklefs (58) for example, have suggested that dioecy is more common among basal angiosperm groups because the greater morphological complexity in more derived lineages lowers the likelihood of modifications leading to unisexual flowers. In other cases, apparent homology may result from lack of knowledge about characters. There are likely to be cases, however, where parallel shifts in reproductive biology cannot be distinguished solely on the basis of morphology because so few characters are involved. For example, the differences obtained from morphological vs. molecular phylogenetic analyses in Pontederiaceae indicate that in the morphological analyses, monomorphic, selfing species were more likely to group together because of reduction in floral parts. In the molecular analysis these monomorphic, selfing species grouped with different tristylous species (24, 45).

For some categories of breeding systems, the assumption of homology in early studies may not have been warranted. For example, the evolution of self-incompatibility has been suggested as causative in the diversification of the angiosperms (82), an argument that assumes homology for self-incompatibility. Using Whitehouse's assumption of homology, self-incompatibility was mapped on basal lineages of angiosperms (76). If self-incompatibility spurred the diversification of the angiosperms, it should be basal in the angiosperms. Phylogenetic analysis was complicated by uncertainty over relationships of basal lineages of flowering plants and lack of information about the distribution of self-incompatibility in some of these lineages. To accommodate this uncertainty, analyses were run using trees based on different data sets and different assumptions about the distribution of self-incompatibility. Regardless of which trees were used for analysis, and which assumptions were used about the presence of self-incompatibility, there was little evidence for self-incompatibility occupying a basal position in the angiosperms. Self-incompatibility will undoubtedly be discovered in additional groups of basal angiosperms (e.g. 63), but it is unlikely that the conclusion that self-compatibility is basal in the angiosperms will change.

Recent molecular studies of self-incompatibility have now revealed that there are diverse mechanisms underlying this phenomenon (76). It may well be the case that many instances of self-incompatibility represent independent gains of nonhomologous traits. If so, there would be no reason to suggest that a single gain of self-incompatibility among early angiosperms resulted in their diversification.

In general, increased knowledge of the genetic systems and molecular biology underlying breeding systems could provide information useful for understanding evolutionary processes. Dioecy, although relatively simple at the morphological level, may show greater complexity in terms of underlying genetic systems. Nuclear-cytoplasmic systems (27, 68) are well known to govern the dynamics of male sterility in some species, while in other species (1, 25, 36, 76) only nuclear genes are implicated in the expression of male sterility. In the Caryophyllaceae, both nuclear control in *Schiedea* and nuclear-cytoplasmic control in *Silene* are known, indicating the independent evolution of dimorphism in this family. Although morphological information has been used to identify different male-sterility mutations (e.g. 73), in many cases genetic information may be useful for distinguishing potential homology from cases of independent evolution (44). Even more detailed molecular information could be used to provide evidence for the degree of homology in breeding systems. In *Schiedea*, hybridizations among different dimorphic species produce the same segregation patterns of male sterile individuals observed following intraspecific hybridization, indicating that the nuclear male sterility alleles in dimorphic species are allelic (77). Evidence from phylogenetic analysis, however, indicates the likelihood that there have been at least two separate transitions to dimorphism (81). Molecular approaches might be useful for determining whether genes controlling male sterility are identical in different clades within *Schiedea*. Similarly, Hodges (40) suggests that QTL mapping of nectar spurs in *Aquilegia* may be a promising method for determining the degree of homology for this trait in *Aquilegia*.

INSIGHTS FROM PHYLOGENETIC STUDIES

In this section several examples from Table 1 are used to address the question of what has been learned about the evolution of breeding systems and pollination biology using phylogenetic approaches. In particular, how have phylogenetic approaches advanced our understanding of reproductive and pollination biology in ways that have complemented results of micro-evolutionary studies? These studies include evolution of diverse reproductive systems (androdioecy, dioecy, heterostyly, multi-allelic self-incompatibility, and self-fertilization) and aspects of pollination biology (vertebrate, insect, and abiotic pollination). Phylogenetic analysis may also be used to guide microevolutionary studies with quantitative genetic and physiological components.

Origin of Androdioecy in Datiscaceae

The evolution of androdioecy in *Datisca* (Datiscaceae) has been investigated intensively by Rieseberg and co-workers (28, 47, 56, 70). These studies demonstrate the crucial role of phylogenetic analysis in determining the evolutionary relationship of androdioecy to other reproductive systems occurring in related taxa. Androdioecy (the presence of male and hermaphroditic individuals in populations) is a potential intermediate step in the evolution of dioecy (12). The stringent conditions for maintenance of males with hermaphrodites in populations is a plausible explanation for the rarity of androdioecy and suggests that this breeding system is unlikely to serve commonly as an intermediate step in the evolution of dioecy (13).

Datiscaceae consist of three genera, *Datisca*, *Octomeles*, and *Tetrameles*. *Datisca cannabina*, *Octomeles*, and *Tetrameles* are dioecious, while the fourth species in the family, *Datisca glomerata*, is androdioecious. Datiscaceae were assumed to be monophyletic, and chloroplast restriction site analysis (56) was used to infer that androdioecy was derived from dioecy. More recent analyses using *rbcL* and 18S ribosomal sequences were used to investigate monophyly in Datiscaceae (70). Based on a consensus of the *rbcL* and 18S data, Datiscaceae are paraphyletic, with Begoniaceae nested within Datiscaceae and sister to *Datisca*. The positions of *Octomeles* and *Tetrameles* are unresolved. Reconstruction of breeding systems using the *rbcL* tree indicates that either monoecy or dioecy may be ancestral to androdioecy in *Datisca glomerata*, while the 18S tree indicates that monoecy is basal. Using the consensus tree of the *rbcL* and 18S data, dioecy appears to be ancestral. Because the basal breeding systems in taxa used to polarize breeding systems in *Datisca* have not been studied in detail, these results could change with further study.

Phylogenetic investigations of Datiscaceae show clearly that androdioecy is not an intermediate step in the evolution of dioecy. In contrast, this breeding system is more likely to be derived from dioecy. Phylogenetic studies therefore lead to the question of conditions that would favor a transition from dioecy to androdioecy, rather than the reverse.

Evolution of Dioecy in Schiedea and Alsinidendron

Morphological and molecular phylogenetic approaches were used to address the question of the number of transitions from hermaphroditism to dimorphism in the endemic Hawaiian lineage containing *Schiedea* and *Alsinidendron* (Caryophyllaceae). This example shows the difficulties in assessing phylogenetic relationships for recently evolved groups or portions of groups (60, 67, 75, 81). Dimorphism was hypothesized to have evolved in two of the four major clades in the lineage. The estimated number of transitions varied widely because of the topology of the clade of *Schiedea* containing *S. globosa* and its relatives (81), and the very weak support for the clade. Accelerated transitions predicted a single transition to dimorphism followed by reversals to hermaphroditism in several species. Use of delayed transitions resulted in multiple shifts to dimorphism. Although the *S. globosa* clade is very poorly supported using both morphological and molecular data, even a strongly supported clade with the same topological relationships would present the same ambiguities in character mapping, because of the distribution of hermaphroditic species throughout the clade.

In the clade containing *S. adamantis* and related species, the occurrence of a reversal to hermaphroditism from a gynodioecious ancestor is strongly supported for *S. lydgatei* (54, 81). *Schiedea lydgatei* is the sole hermaphroditic species in the *S. adamantis* clade and is nested within the clade, which ensures that hermaphroditism will be viewed as a reversal. In contrast to the *S. globosa* clade, morphological and molecular data support the *S. adamantis* clade. Unless there have been multiple, nonparsimonious transitions to dimorphism in the *S. adamantis* clade, there seems little doubt that hermaphroditism in *S. lydgatei* represents a reversal. While there is no reason that reversals to hermaphroditism could not occur, especially from a gynodioecious ancestor, this reversal would not necessarily have been predicted from broad comparative data that generally suggest the derivation of dioecy from hermaphroditism.

Phylogenetic information in *Schiedea* can be used to guide population-level studies in the choice of study species as well as hypotheses. *Schiedea salicaria* and *S. adamantis* are two gynodioecious species occurring within the *Schiedea adamantis* clade. Allocation patterns in hermaphrodites vary in the direction predicted by the frequency of females in populations: in *S. salicaria* (12% females; 80), hermaphrodites and females have equivalent female function (AK Sakai, SG Weller, unpublished), while in *S. adamantis* (39% females; 59), females have substantially greater female function than hermaphrodites (59). Differences in these very closely related species have been used to suggest that the appearance of females in populations of *Schiedea* is largely a function of high inbreeding depression and high selfing rates, rather than initial shifts in patterns of resource allocation (59). Future approaches include using phylogenetic information to guide the choice of species for a quantitative genetic analysis of resource allocation patterns in *Schiedea* to determine the genetic potential for shifts in allocation patterns to male and female function (AK Sakai, SG Weller, and DR Campbell, unpublished).

Pollination Systems in Aphelandra

The evolution of plant-pollinator interactions has been investigated by McDade (53) for the *Aphelandra pulcherrima* complex (Acanthaceae). Species in this complex have red, tubular flowers and are pollinated by either short-billed trochiline hummingbirds or long-billed hermit hummingbirds. Comparative studies in the past have suggested that species pollinated by long-billed hermits, which have long, curved corollas and produce abundant nectar, are evolutionarily specialized compared to species pollinated by trochiline hummingbirds, which have short, straighter corollas and produce far less nectar. Contrary to expectations, phylogenetic analysis of the *A. pulcherrima* complex indicated that pollination by hermits is ancestral, and species with short-tubed corollas have evolved on two occasions. The use of a phylogenetic approach permitted the explicit framing of questions about the evolution of floral structure and function, and its relationship to pollination in this complex. As McDade (53) emphasized, there is no reason to expect that different patterns of trochiline and hermit pollination would not emerge in other hummingbird-pollinated lineages.

Evolution of Pollination Biology in Dalechampia

In *Dalechampia* (Euphorbiaceae), Armbruster (3) mapped pollination characteristics on a phylogenetic tree produced using morphological data. Pollination systems in *Dalechampia* proved to be evolutionarily labile. The most common pollination syndrome, pollination by resin-collecting bees, evolved on one occasion, although several different resin-collecting bee species visit different species of *Dalechampia*. In contrast, pollination by fragrance-collecting male euglossine bees is less common but has evolved three to four times. In three of the lineages in which there has been a transition to fragrance collecting by euglossine bees, different morphological structures secrete rewards (2, 3), adding to the evidence that these transitions represent independent evolutionary events. Pollen collection has evolved in one to two species. *Dalechampia* is a large genus consisting of approximately 120 species, of which 40 were studied in detail (3). Presumably, conclusions about the number of transitions to different pollination systems might change considerably if additional species were added to the analysis.

Resins may first have originated as a defense system for staminate flowers and evolved later to become a reward for pollinators (4). If this evolutionary scenario is correct, resin production was a preadaptation that allowed the eventual use of resins as a reward for pollinators. A second case of preadaptation may have occurred in another species, where a defense system appears to have resulted from modification of morphological structures associated with attraction of pollinators (4). Phylogenetic analysis of *Dalechampia* has provided a powerful tool for the investigation of both transitions in pollination systems and the functions that may have predated current use of structures associated with pollination and defense.

PREDICTING WHEN PHYLOGENETIC APPROACHES WILL BE USEFUL FOR ANALYSIS OF CHARACTER EVOLUTION

Can available studies (Table 1) be used to predict when phylogenetic approaches are likely to be most useful for analyses of character evolution? Clearly, phylogenetic approaches may be most useful in lineages with well-supported phylogenies including a large proportion of extant species. Character mapping in lineages in which there has been considerable extinction (or where there are regions of the phylogeny with less branching) may be difficult if extinct sister taxa had different character states than did extant species. Younger lineages may have fewer problems associated with extinction but often have less phylogenetic resolution, making the process of character mapping more difficult.

The analyses cited in this review indicate that breeding systems or pollination syndromes that are not highly labile are most successfully analyzed using phylogenetic approaches. When traits are extremely labile, considerable uncertainty may exist for character states at ancestral nodes, and single transitions followed by reversals are likely to have the same likelihood as multiple transitions to the derived character state (81). Large, well-sampled lineages lend themselves to analysis of character evolution, particularly when traits under study are relatively conservative, because multiple transitions are likely to be correctly interpreted (e.g. 10).

When the outgroup is clearly identified in a study, characters within lineages can be polarized correctly, and the evolutionary relationships of different characters can be assessed correctly. In many other cases, inability to specify a single outgroup means that multiple outgroup comparison is essential. Diversity among potential outgroups for the traits under study will further complicate analysis for the ingroup, unless phylogenies for the outgroups are well defined.

Using phylogenies to determine the order of evolutionary events (and thus provide inferences of causality) is likely to work best when cause and effect are not very strongly related. As the correlation between a modification of the environment, for example, and the evolution of a trait increases, traits will co-occur on branches and the order of their acquisition will be impossible to determine. Inference of a relationship will remain strong, but it will not be possible to determine causality using only phylogenetic approaches. In these cases, the stronger inference for causality may come from micro-evolutionary studies.

The evolution of complex character traits is more likely to be interpreted correctly using phylogenetic analysis, particularly when homoplasy can be reinterpreted as cases of independent gains of superficially similar character states. Complex traits can be weighted so that they are more likely to break down than to evolve independently (cf. 34, vs. 45), although these approaches are then not strictly independent of the microevolutionary studies on which the weighting schemes depend. When traits under study are relative simple (e.g. dioecy), multiple gains are less likely to be distinguishable because there are fewer characters associated with the breeding system. Increasingly detailed knowledge of traits, particularly their molecular basis, will undoubtedly increase the probability of correct assessment of homology. When lack of homology is conspicuous, independent gains of pollination syndromes or breeding systems can be identified without using phylogenetic approaches. It seems more likely, however, that phylogenetic approaches will continue to have a critical role in identifying cases where true homology is more restricted than cursory examination might suggest. While not a simple cure to understanding problems that in the past have been studied only in the realm of microevolutionary studies, phylogenetic approaches offer clear potential for providing new insights.

ACKNOWLEDGMENTS

We thank Bruce Baldwin, Lucinda McDade, Shirley Graham, Peter Hoch, Leslie Gottlieb, and Ken Sytsma for discussions, Carol Goodwillie for sharing an unpublished manuscript, and Diane Campbell and Alan Prather for commenting on the manuscript.

LITERATURE CITED

1. Ahmadi H, Bringhurst RS. 1991. Genetics of sex expression in *Fragaria* species. *Am. J. Bot.* 78:504–14
2. Armbruster WS. 1992. Phylogeny and the evolution of plant-animal interactions. *BioScience* 42:12–20
3. Armbruster WS. 1993. Evolution of plant pollination systems: hypotheses and tests with the neotropical vine *Dalechampia*. *Evolution* 47:1480–1505
4. Armbruster WS. 1997. Exaptations link evolution of plant-herbivore and plant-pollinator interactions: a phylogenetic inquiry. *Ecology* 78:1661–72
5. Baldwin BG, Robichaux RH. 1995. Historical biogeography and ecology of the Hawaiian silversword alliance. See Ref. 74, pp. 259–87
6. Barrett SCH. 1995. Mating-system evolution in flowering plants: micro- and macroevolutionary approaches. *Acta Bot. Neerl.* 44:385–402
7. Barrett SCH, Harder LD, Worley AC. 1996. The comparative biology of pollination and mating in flowering plants. *Philos. Tran. R. Soc. Lond. B* 351:1271–80
8. Baum DA, Small, RL, Wendel JF. 1998. Biogeography and floral evolution of baobabs (*Adansonia*, Bombacaceae) as inferred from multiple data sets. *Syst. Biol.* 47:181–207
9. Brooks DR, McLennan DA. 1991. *Phylogeny, Ecology and Behavior*. Chicago: Univ. Chicago Press. 434 pp.
10. Bruneau A. 1997. Evolution and homology of bird pollination syndromes in *Erythrina* (Leguminosae). *Am. J. Bot.* 84:54–71
11. Carr GD, Powell EA, Kyhos DW. 1986. Self–incompatibility in the Hawaiian Madiinae (Compositae): an exception to Baker's Rule. *Evolution* 40:430–34
12. Charlesworth B, Charlesworth D. 1978. A model for the evolution of dioecy and gynodioecy. *Am. Nat.* 112:975–997
13. Charlesworth D. 1984. Androdioecy and the evolution of dioecy. *Biol. J. Linn. Soc.* 23:333–348
14. Chase MW, Hills HG. 1992. Orchid phylogeny, flower sexuality, and fragrance-seeking. *BioScience* 42:43–49
15. Cox PA. 1990. Pollination and the evolution of breeding systems in Pandanaceae. *Ann. Miss. Bot. Gard.* 77:816–40
16. Cox PA, Humphries CJ. 1993. Hydrophilous pollination and breeding system evolution in seagrasses: a phylogenetic approach to the evolutionary ecology of Cymodoceaceae. *Bot. J. Linn. Soc.* 113:217–26
17. Culley TM, Weller SG, Sakai AK, Rankin AE. 1999. Inbreeding depression and selfing rates in a self-compatible, hermaphroditic species, *Schiedea membranacea* (Caryophyllaceae). *Am. J. Bot.* 86:980–987

17a. Cunningham CW, Omland KE, Oakley TH. 1998. Reconstructing ancestral character states: a critical reappraisal. *Trends Ecol. Evol.* 13:361–66
18. de Queiroz K. 1996. Including the characters of interest during tree reconstruction and the problems of circularity and bias in studies of character evolution. *Am. Nat.* 148:700–8
19. Donoghue MJ. 1989. Phylogenies and the analysis of evolutionary sequences, with examples from seed plants. *Evolution* 43:1137–56
20. Donoghue MJ, Ackerly DD. 1996. Phylogenetic uncertainties and sensitivity analyses in comparative biology. *Philos. Tran. R. Soc. Lond. B* 351:1241–49
21. Donoghue MJ, Cantino PD. 1984. The logic and limitations of the outgroup substitution approach in cladistic analysis. *Syst. Bot.* 9:192–202
22. Donoghue MJ, Sanderson MJ. 1992. The suitability of molecular and morphological evidence in reconstructing plant phylogeny. In *Molecular Systematics of Plants*, ed. PS Soltis, DE Soltis, JJ Doyle, pp. 340–68. New York: Chapman & Hall. 434 pp.
23. Donoghue MJ, Sanderson MJ. 1994. Complexity and homology in plants. In *Homology: the Hierarchical Basis of Comparative Biology*, ed. BK Hall, pp. 393–420. New York: Academic. 621 pp.
24. Eckenwalder JE, Barrett SCH. 1986. Phylogenetic systematics of Pontederiaceae. *Syst. Bot.* 11:373–91
25. Eckhart VM. 1992. The genetics of gender and the effects of gender on floral characters in gynodioecious *Phacelia linearis* (Hydrophyllaceae). *Am. J. Bot.* 79:792–800
26. Felsenstein J. 1985. Phylogenies and the comparative method. *Am. Nat.* 125:1–15
27. Frank SA. 1989. The evolutionary dynamics of cytoplasmic male sterility. *Am. Nat.* 133:345–76
28. Fritsch PF, Rieseberg LH. 1992. High outcrossing rates maintain male and hermaphrodite individuals in populations of the flowering plant *Datisca glomerata*. *Nature* 359:633–36
29. Frumhoff PC, Reeve HK. 1994. Using phylogenies to test hypotheses of adaptation: A critique of some current proposals. *Evolution* 48:172–80
30. Givnish TJ. 1982. Outcrossing versus ecological constraints in the evolution of dioecy. *Am. Nat.* 119:849–65
31. Givnish TJ, Sytsma KJ eds. 1997. *Molecular Evolution and Adaptive Radiation.* New York: Cambridge Univ. Press. 621 pp.
32. Goldblatt P, Manning JC, Bernhardt P. 1995. Pollination biology of *Lapeirousia* subgenus *Lapeirousia* (Iridaceae) in southern Africa; floral divergence and adaptation for long–tongued fly pollination. *Ann. Mo. Bot. Gard.* 82:517–34
33. Goodwillie C. 1997. The genetic control of self-incompatibility in *Linanthus parviflorus* (Polemoniaceae). *Heredity* 79:424–32
33a. Goodwillie C. 1999. Multiple origins of self-incompatibility in *Linanthus* section *Leptosiphon* (Polemoniaceae): phylogenetic evidence from ITS sequence data. *Evolution.* In press
34. Graham SA, Crisci JV, Hoch PC. 1993. Cladistic analysis of the Lythraceae *sensu lato* based on morphological characters. *Biol. J. Linn. Soc.* 113:1–33
35. Graham SA. 1998. Relacionamentos entre as espécies autógamas de *Cuphea* P. Browne seção *Brachyandra* Koehne (Lythraceae). *Acta bot. bras.* 12:203–14
36. Hancock JF Jr, Bringhurst RS. 1980. Sexual dimorphism in the strawberry *Fragaria chiloensis*. *Evolution* 34:762–68
37. Hart JA. 1985. Peripheral isolation and the origin of diversity in *Lepechinia* sect. *Parviflorae* (Lamiaceae). *Syst. Bot.* 10:134–46
38. Hart JA. 1985. Evolution of dioecism

in *Lepechinia* Willd. sect. *Parviflorae* (Lamiaceae). *Syst. Bot.* 10:147–54
39. Harvey PH, Pagel MD. 1991. *The Comparative Method in Evolutionary Biology.* Oxford, UK: Oxford Univ. Press. 239 pp.
40. Hodges SA. 1997. Floral nectar spurs and diversification. *Int. J. Plant Sci.* 158:S81–88
41. Hodges SA, Arnold ML. 1995. Spurring plant diversification: are floral spurs a key innovation? *Proc. R. Soc. Lond.* B 262: 343–48
42. Johnson SD, Linder HP, Steiner KE. 1998. Phylogeny and radiation of pollination systems in *Disa* (Orchidaceae). *Am. J. Bot.* 85:402–11
43. Johnston MO, Schoen DJ. 1996. Correlated evolution of self-fertilization and inbreeding depression: an experimental study of nine populations of *Amsinckia* (Boraginaceae). *Evolution* 50:1478–91
44. Koelewijn HP, Van Damme JMM. 1996. Genetics of male sterility in gynodioecious *Plantago coronopus* II nuclear genetic variation. *Genetics* 139:1759–75
45. Kohn JR, Graham SW, Morton B, Doyle JJ, Barrett SCH. 1996. Reconstruction of the evolution of reproductive characters in Pontederiaceae using phylogenetic evidence from chloroplast DNA restriction-site variation. *Evolution* 50:1454–69
46. Les DH, Cleland MA, Waycot M. 1997. Phylogenetic studies in Alismatidae, II: evolution of marine angiosperms (seagrasses) and hydrophily. *Syst. Bot.* 22:443–63
47. Liston AL, Rieseberg LH, Elias TS. 1990. *Datisca glomerata* is functionally androdioecious. *Nature* 343:641–642
48. Luckow M, Bruneau A. 1997. Circularity and independence in phylogenetic tests of ecological hypotheses. *Cladistics* 13:145–51
49. Luckow M, Hopkins HCF. 1995. A cladistic analysis of *Parkia* (Leguminosae: Mimosoideae). *Am. J. Bot.* 82:1300–20
50. Maddison WP. 1990. A method for testing the correlated evolution of two binary characters: Are gains or losses concentrated on certain branches of a phylogenetic tree? *Evolution* 44:539–57
51. Maddison WP, Donoghue MJ, Maddison DR. 1984. Outgroup analysis and parsimony. *Syst. Zool.* 33:83–103
52. Martins EP, ed. 1996. *Phylogenies and the Comparative Method in Animal Behavior.* New York: Oxford Univ. Press. 415 pp.
53. McDade LA. 1992. Pollinator relationships, biogeography, and phylogenetics. *BioScience* 42:21–26
54. Norman JK, Weller SG, Sakai AK. 1997. Pollination biology and outcrossing rates in hermaphroditic *Schiedea lydgatei* (Caryophyllaceae). *Am. J. Bot.* 84:641–48
55. Olmstead R. 1989. Phylogeny, phenotypic evolution, and biogeography of the *Scutellaria angustifolia* complex (Lamiaceae): inference from morphological and molecular data. *Syst. Bot.* 14:320–38
56. Rieseberg LH, Hanson MA, Philbrick CT. 1992. Androdioecy is derived from dioecy in Datiscaceae: evidence from restriction site mapping of PCR–amplified chloroplast DNA fragments. *Syst. Bot.* 17:324–36
57. Renner SS. 1998. Phylogenetic affinities of Monimiaceae based on cpDNA gene and spacer sequences. *Perspect. in Plant Ecol., Evol. Syst.* 1:61–67
58. Renner SS, Ricklefs RE. 1995. Dioecy and its correlates in the flowering plants. *Am. J. Bot.* 82:596–606
59. Sakai AK, Weller SG, Chen M–L, Chou S–Y, Tasanont C. 1997. Evolution of gynodioecy and maintenance of females: the role of inbreeding depression, outcrossing rates, and resource allocation in *Schiedea adamantis* (Caryophyllaceae). *Evolution* 51:724–36
60. Sakai AK, Weller SG, Wagner WL, Soltis PS, Soltis DE. 1997. Phylogenetic perspective on the evolution of dioecy: adaptive radiation in the endemic Hawaiian genera *Schiedea* and *Alsinidendron* (Caryophyllaceae: Alsinoideae). In *Molecular Evolu-*

tion and Adaptive Radiation. ed. TJ Givnish, KJ Sytsma, pp. 455–73. New York: Cambridge Univ. Press. 621 pp.

61. Sanderson MJ, Donoghue MJ. 1994. Shifts in diversification rate with the origin of angiosperms. *Science* 264:1590–93

61a. Schluter D, Price T, Mooers AO, Ludwig D. 1997. Likelihood of ancestor states in adaptive radiation. *Evolution* 51:1699–1711

62. Schoen DJ, Johnston MO, L'Heureux A–M, Marsolais JV. 1997. Evolutionary history of the mating system in *Amsinckia* (Boraginaceae). *Evolution* 51:1090–99

63. Schroeder MS, Weller SG. 1997. Self-incompatibility and clonal growth in *Anemopsis californica*. *Plant Spec. Biol.* 12:55–59

64. Schultz TJ, Cocroft RB, Churchill GA. 1996. The reconstruction of ancestral character states. *Evolution* 50:504–11

65. Silvertown J, Dodd M. 1996. Comparing plants and connecting traits. *Philos. Tran. R. Soc. Lond. B* 351:1233–39

66. Simpson BB. 1988. The need for systematic studies in reconstructing paleogeographic and ecological patterns in the South American tropics. *Acta Univ. Ups. Symb. Bot. Ups.* 28:150–158

67. Soltis PE, Soltis DE, Weller SG, Sakai AK, Wagner WL. 1996. Molecular phylogenetic analysis of the Hawaiian endemics *Schiedea* and *Alsinidendron* (Caryophyllaceae). *Syst. Bot.* 21:365–79

68. Sun M. 1987. Genetics of gynodioecy in Hawaiian *Bidens* (Asteraceae). *Heredity* 59:327–36

69. Swenson U, Bremer K. 1997. Patterns of floral evolution of four Asteraceae genera (Senecioneae, Blennospermatinae) and the origin of white flowers in New Zealand. *Syst. Biol.* 46:407–425

70. Swenson SM, Luthi JN, Rieseberg LH. 1998. Datiscaceae revisited: monophyly and the sequence of breeding system evolution. *Syst. Bot.* 23:157–169

71. Swofford DL, Maddison WP. 1992. Parsimony, character-state reconstructions, and evolutionary inferences. In *Systematics, Historical Ecology, and North American Freshwater Fishes*, ed. RL Mayden, pp. 186–222. Stanford, CA: Stanford Univ. Press. 969 pp.

72. Swofford DL, Olsen GJ. 1990. Phylogeny reconstructions. In *Molecular Systematics*, ed. DM Hillis, C Moritz, pp. 411–501. Sunderland, MA: Sinauer. 588 pp.

73. Van Damme JMM, Van Delden W. 1982. Gynodioecy in *Plantago lanceolata* L. I. Polymorphism for plasmon type. *Heredity* 49:303–18

74. Wagner WL, Funk VA, eds. 1995. *Hawaiian Biogeography: Evolution on a Hot Spot Archipelago*. Washington, DC: Smithsonian. 467 pp.

75. Wagner WL, Weller SG, Sakai AK. 1995. Phylogeny and biogeography in *Schiedea* and *Alsinidendron* (Caryophyllaceae). See Ref. 74, pp. 221–58

76. Weller SG, Donoghue MJ, Charlesworth D. 1995. The evolution of self-incompatibility in flowering plants: a phylogenetic approach. In *Experimental and Molecular Approaches to Plant Biosystematics*, ed. PC Hoch, AG Stephenson, 53:355–82. St. Louis: Mo. Bot. Gard. Monogr. in Syst. Bot.

77. Weller SG, Sakai AK. 1991. The genetic basis of male sterility in *Schiedea* (Caryophyllaceae), an endemic Hawaiian genus. *Heredity* 67:265–73

78. Weller SG, Sakai AK, Rankin AE, Golonka A, Kutcher B, Ashby KE. 1998. Dioecy and the evolution of pollination systems in *Schiedea* and *Alsinidendron* (Caryophyllaceae: Alsinoideae) in the Hawaiian Islands. *Am. J. Bot.* 85:1377–88

79. Weller SG, Sakai AK, Straub C. 1996. Allozyme diversity and genetic identity in *Schiedea* and *Alsinidendron* (Caryophyllaceae: Alsinoideae) in the Hawaiian Islands. *Evolution* 50:23–34

80. Weller SG, Sakai AK, Wagner WL, Herbst DR. 1990. Evolution of dioecy in *Schiedea* (Caryophyllaceae: Alsinoideae) in the Hawaiian Islands: biogeographical and ecological factors. *Syst. Bot.* 15:266–76
81. Weller SG, Wagner WL, Sakai AK. 1995. A phylogenetic analysis of *Schiedea* and *Alsinidendron* (Caryophyllaceae: Alsinoideae): implications for the evolution of breeding systems. *Syst. Bot.* 20:315–37
82. Whitehouse HLK. 1950. Multiple-allelomorph incompatibility of pollen and style in the evolution of the angiosperms. *Ann. Bot.* 14:199–216

Annu. Rev. Ecol. Syst. 200X . 30:201–33

Evolution of Diversity in Warning Color and Mimicry: Polymorphisms, Shifting Balance, and Speciation

James Mallet[1] and Mathieu Joron[2]
[1]*Galton Laboratory, 4 Stephenson Way, London NW1 2HE, England; e-mail: j.mallet@ucl.ac.uk or http://abacus.gene.ucl.ac.uk/jim/; and* [2]*Génétique et Environnement, CC065 ISEM, Université de Montpellier 2, Place Bataillon, F-34095 Montpellier, cedex 5, France; e-mail: joron@isem.univ-montp2.fr*

Key Words aposematism, Batesian mimicry, Müllerian mimicry, defensive coloration, predator behavior

■ **Abstract** Mimicry and warning color are highly paradoxical adaptations. Color patterns in both Müllerian and Batesian mimicry are often determined by relatively few pattern-regulating loci with major effects. Many of these loci are "supergenes," consisting of multiple, tightly linked epistatic elements. On the one hand, strong purifying selection on these genes must explain accurate resemblance (a reduction of morphological diversity between species), as well as monomorphic color patterns within species. On the other hand, mimicry has diversified at every taxonomic level; warning color has evolved from cryptic patterns, and there are mimetic polymorphisms within species, multiple color patterns in different geographic races of the same species, mimetic differences between sister species, and multiple mimicry rings within local communities. These contrasting patterns can be explained, in part, by the shape of a "number-dependent" selection function first modeled by Fritz Müller in 1879: Purifying selection against any warning-colored morph is very strong when that morph is rare, but becomes weak in a broad basin of intermediate frequencies, allowing opportunities for polymorphisms and genetic drift. This Müllerian explanation, however, makes unstated assumptions about predator learning and forgetting which have recently been challenged. Today's "receiver psychology" models predict that classical Müllerian mimicry could be much rarer than believed previously, and that "quasi-Batesian mimicry," a new type of mimicry intermediate between Müllerian and Batesian, could be common. However, the new receiver psychology theory is untested, and indeed it seems to us unlikely; alternative assumptions could easily lead to a more traditional Müllerian/Batesian mimicry divide.

0066 4162/99/1120-0201$08.00

INTRODUCTION

Since their discovery, antipredator mimicry and warning colors have been used as simple and visually appealing examples of natural selection in action. This simplicity is beguiling, and controversy has often raged behind the textbook examples. Warning color and mimicry have been discussed from three different points of view: a traditional insect *natural history* angle, which makes simplistic assumptions about both predator behavior and prey evolution (6, 103, 122, 175); an *evolutionary dynamics* angle, which virtually ignores predator behavior and individual prey/predator interactions (37, 47, 54, 91, 162); and a predator *behavior* (or "*receiver psychology*") angle (59–61, 71, 72, 84a, 115, 147a–150), which tends to be simplistic about evolutionary dynamics. Assumptions are necessary to analyze any mathematical problem, but the sensitivity of mimicry to these different simplifications remains untested.

We believe it will be necessary to combine these disparate views (for example, 111, 133, 181) in order to resolve controversy and explain paradoxical empirical observations about the evolution of mimicry. Mimicry should progressively reduce numbers of color patterns, but the actual situation is in stark contrast: There is a diversity of "mimicry rings" (a mimicry ring is a group of species with a common mimetic pattern) within any single locality; closely related species and even adjacent geographic races often differ in mimetic or warning color pattern; and there are locally stable polymorphisms. The current controversies and problems are not simply niggles with the theory of mimicry, designed to renew flagging interest in a largely solved area of evolutionary enquiry. Recent challenges and critiques cast justifiable doubt on previously unstated assumptions. A further reason for reexamining the evolution of mimicry is that its frequency-dependent selective landscapes are rugged, as in mate choice and hybrid inviability (53), so that mimicry provides a model system for the shifting balance theory (41, 178–180); mimicry may also act as a barrier to isolate species (96). While a number of interesting peculiarities of mimicry arguably have little general importance (57), mimicry excels in providing an intuitively understandable example of multiple stable equilibria and transitions between them (41, 87, 89, 97, 98, 161, 167). Mimicry and warning color are highly variable both geographically within species and also between sister species. In this, they are similar to other visual traits involved in signaling and speciation, such as sexually selected plumage morphology and color in birds (2). Sexual and mimetic coloration may therefore share some explanations.

This article covers only the evolution of diversity in mimicry systems, and we skate quickly over many issues reviewed elsewhere (19–21, 32, 40, 45, 86, 123, 125, 129, 134, 170, 176; see also a list of over 600 references in Reference 90). Our discussion mainly concerns antipredator visual mimicry, though it may apply to other kinds of mimicry and aposematism, for example, warning smells (62, 130) and mimicry of behavioral pattern (18, 152–154). In addition, most of our examples are shamelessly taken from among the insect mimics and their models, usually butterflies, that we know best; careful studies on the genetics or ecology of other systems have rarely been done.

MIMICRY AND WARNING COLOR: THE BASICS

Bates (6) noticed two curious features among a large complex of butterflies of the Amazon. First, color patterns of unrelated species were often closely similar locally; second, these "mimetic" patterns changed radically every few hundred miles, "as if by the touch of an enchanter's wand" (8). Bates argued that very abundant slow-flying Ithomiinae (related to monarch butterflies) were distasteful to predators and that palatable species, particularly dismorphiine pierids (related to cabbage whites), "mimicked" them; that is, natural selection had caused the pattern of the "mimic" to converge on that of the "model" species. This form of mimicry became known as Batesian (122). The term "mimicry" had already been used somewhat vaguely by pre-Darwinian natural philosophers for a variety of analogical resemblances (13), but Bates's discovery was undoubtedly a triumph of evolutionary thinking.

Bates also noticed that rare unpalatable species such as *Heliconius* (Heliconiinae) and *Napeogenes* (Ithomiinae) often mimicked the same common ithomiine models (such as *Melinaea, Oleria*, and *Ithomia*) copied by dismorphiines. He assumed that this "mimetic resemblance was intended" (6, p. 554) because, regardless of its palatability, a rare species should benefit from similarity to a model. However, where both mimic and model were common, as in the similarity of unpalatable *Lycorea* (Danainae) to ithomiines, he felt this was "a curious result of [adaptation to] local [environmental] conditions" (6, p. 517); in other words, convergent evolution unrelated to predation. It was left to Müller (103) to explain clearly the benefits of mimicry in pairs of unpalatable species. If a constant number of unpalatable individuals per unit time must be sacrificed to teach local predators a given color pattern, the fraction dying in each species will be reduced if they share a color pattern, leading to an advantage to mimicry. Thus, mimicry between unpalatable species became known as Müllerian.

Many mimetic species are also warning-colored, but some are not: for example, the larva of a notodontid moth (6) and the pupa of *Dynastor darius* (Brassolidae) (1) both mimic highly poisonous but cryptic pit vipers (Viperidae). The former mimics even the keeled scales of its model; the latter has eyes that mimic the snake's own eyes, even down to the slit-shaped pupils. Many small clearwing ithomiines that Bates studied in the tropical rainforest understory are also very inconspicuous but are clearly mimetic. Mimicry does not require a warning-colored model, only that potential predators develop aversions to the model's appearance. Warning color, or "aposematism" (122), was first developed as an evolutionary hypothesis by Wallace in response to a query from Darwin, four years after Bates' publication on mimicry. Darwin's sexual selection theory (42) explained much bright coloration in animals but could not explain conspicuous black, yellow, and red sphingid caterpillars found by Bates in Brazil because the adult moth could not choose mates on the basis of larval colors. Wallace in 1866 (see 42, 175) suggested that bright colors advertised the unpalatability of the larvae, in the same way that yellow and black banding advertised the defensive sting of a hornet (Vespidae). Warning color in effect must increase the efficiency with which predators learn to avoid unpalatable prey (see also 59, 61 for excellent discussion of possible advantages of warning color).

NUMBER-DEPENDENT SELECTION ON MIMICRY AND WARNING COLOR

To the natural history viewpoint of early Darwinians (6, 8, 103, 122, 175), it was apparently not clear that explaining aposematism and mimicry as adaptations could be problematic: They had not fully realized that short-term individual benefits and long-term group benefits may conflict. In fact, the selective landscape of mimetic evolution has multiple stability peaks that should often prevent the spread of ultimately beneficial unpalatability, warning color, and some mimicry. To understand why this is so, we must examine the evolutionary dynamics of mimicry.

Müller (103) was the first to formulate the benefits of mimicry explicitly, using mathematical intuition from a natural history perspective (reprinted in 78). He assumed that, while learning to avoid the color pattern of unpalatable species, a predator complex killed a fixed number of individuals per unit time (n_k). Müllerian mimicry is favored, therefore, because the per capita mortality rate decreases when another unpalatable species shares the same pattern. If this traditional naturalist's "number-dependent" (162) view of mimicry is correct, it leads to two interesting predictions, only the first of which Müller himself apparently appreciated. First, although Müllerian mimicry of this kind should always be mutualistic, a rare species ultimately gains far more from mimicry than a common one, in proportion to the square of the ratio of abundances (103). Second, a novel mimetic variant in the rarer species resembling the commoner is always favored because the common species generates greater numerical protection, while a mimetic variant of the commoner species is always disfavored because it loses the strong protection of its own kind and gains only weak protection from the rarer pattern (161). Both these effects will tend to cause rarer unpalatable species to mimic commoner models, rather than the other way around, in spite of the fact that Müllerian mimicry is a mutualism (albeit with unequal benefits) once attained.

Müller's number-dependent selection applies similarly to morphs within a single species (Figure 1). A warning-colored variant within a cryptic but unpalatable prey will suffer a twofold disadvantage: First, it is more conspicuous to predators; second, it does not gain from warning color because predators, not having learned to avoid the pattern, may attack it at higher rate than the cryptic morph. This creates a barrier to initial spread (67), even though, once evolved, warning color is beneficial because by definition it reduces the number of prey eaten during predator learning. In exactly the same way, a novel warning pattern is disfavored within an already warning-colored species, essentially because of intraspecific mimicry (27, 89, 97; also Figure 1). This selection against rarity makes it easy to understand why warning-colored races are normally fixed and sharply separated by narrow overlap zones from other races (27, 50, 87, 93), but in turn makes it hard to understand how geographic races diversified in the first place (6, 8, 87, 89, 98, 137, 167). Similarly, if energy is required to synthesize or sequester distasteful compounds, unpalatability itself may be disfavored (49, 63, 67, 160) because unpalatable individuals may sacrifice their lives in teaching predators to avoid other members of their species. Hypotheses to overcome the difficulties with this new, more sophisticated evolutionary dynamic view of aposematism are detailed below.

POPULATION STRUCTURE AND THE EVOLUTION OF WARNING COLOR AND UNPALATABILITY

The Evolution of Unpalatability

Unpalatability itself is hard to define (20, 51; see also below under *Müllerian Mimicry*), but here we use the term loosely to mean any trait that acts on predators as a punishment, and that causes learning leading to a reduction in attacks. The unpalatable individual may incur costs in synthesis or processing of distasteful chemistry and is often likely to suffer damage during predator sampling, while other members of the population mostly benefit from predator learning. The frequent aposematism of gregarious larvae, often siblings from the same brood, suggests that benefits are shared among kin, and that kin selection could have been responsible for the evolution of unpalatability (49, 63, 67, 160). These authors assumed that altruistic unpalatability was unlikely to evolve unless kin-groups already existed, so explaining the association between gregarious larvae and unpalatability. However, unpalatability may not be very costly. First, although it may be expensive to process distasteful secondary compounds, in some cases the same biochemical machinery is required to exploit available food; for instance, *Zygaena* and *Heliconius*, which both feed on cyanogenic host plants, can also synthesize their own cyanogens (70, 76, 104), presumably using enzyme systems similar to those required in detoxification. Second, because most toxic compounds also taste nasty (arguably, the sense of taste has evolved to protect eaters from toxic chemistry), and because predators taste-test their prey before devouring them, and, finally, because unpalatable insects are often tough and resilient, an unpalatable insect should often gain an individual advantage by sequestering distasteful chemicals. A good example of predator behavior showing this is possible is seen in birds feeding on monarch (*Danaus plexippus*) aggregations at their overwintering sites in Mexico: Birds repeatedly taste-reject butterflies more or less unharmed, until they find a palatable individual, which is then killed and eaten (29).

Another problem with empirical evidence for kin selection is that gregariousness, which reduces per capita detectability of the prey (14, 64, 157, 160), is expected to evolve when there is any tendency toward predator satiation; and one of the best ways of satiating predators is to be distasteful. Thus the association between gregarious larvae and unpalatability can be explained easily because gregariousness will evolve more readily after unpalatability, rather than before it as required under the kin selection hypothesis. This expected pattern of unpalatability first, gregariousness thereafter, is now well supported in Lepidoptera by phylogenetic analysis (139, 141). In conclusion, the supposed necessity for kin selection in the evolution of unpalatability is now generally disbelieved (59, 97, 139), although kin selection could, of course, help.

Evolution of Novel Warning Colors in Cryptic and Aposematic Defended Prey

Although the realization that aposematic insects may be altruistic came 70 years ago (49), it was finally some 50 years later that the evolution of warning color was explicitly disentangled from the evolution of unpalatability (66, 67). Under

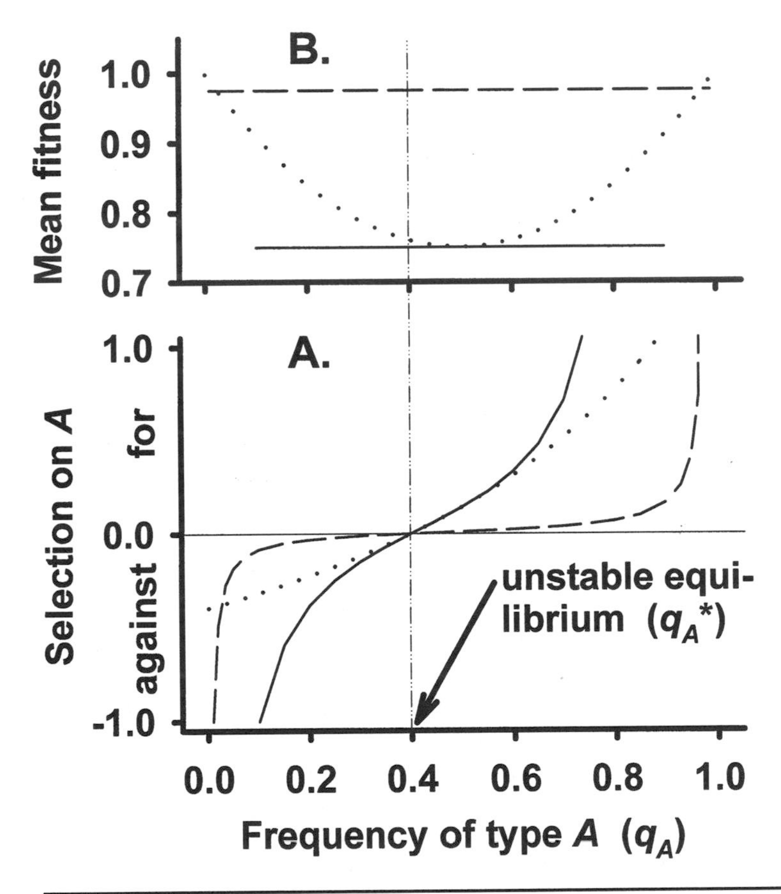

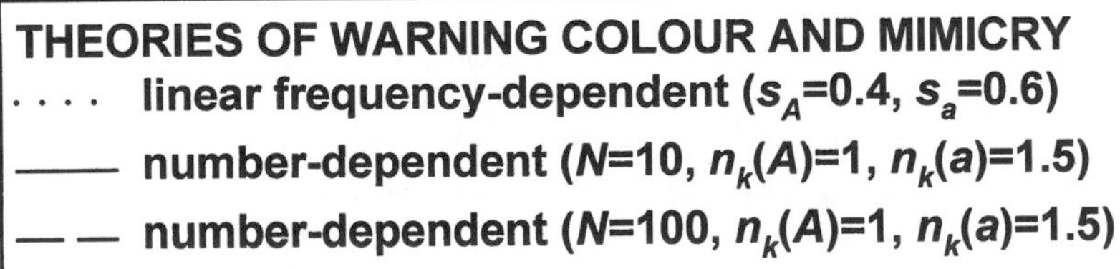

Müller's number-dependent theory, intraspecific Müllerian mimicry acting on a novel warning-colored variant A within a population would strongly favor the commonest wild-type morph a; the number-dependence gives rise to frequency-dependent selection, which is purifying, tending to prevent polymorphism (Figure 1*A*). If unpalatable prey often survive attacks, it might be argued that the problem will be surmounted (47, 73, 74, 177). However, fitness will be reduced if attacks are even potentially damaging; the "effective number killed" (n_k) may take fractional or probabilistic values, but the frequency-dependent logic applies in exactly the same way (68, 97).

A critical feature of number-dependence is the great nonlinearity of frequency-dependent selection. Many authors from the evolutionary dynamics tradition have assumed a simpler linear frequency-dependence (dotted line in Figure 1) (47, 54, 87, 91, 121). In fact, the relationship between selection and frequency becomes more sigmoidal as n_k/N decreases. When $n_k \ll N$, there are strong spikes of selection against A and a when each is rare, but much of the frequency range forms a nearly neutral polymorphic basin (e.g. $N = 100$, Figure 1). Another interesting feature of this model is that the mean fitness surface is flat: Assuming most predators learn the pattern and then avoid it, the mean fitness throughout the frequency spectrum becomes approximately constant at $[n_k(A) + n_k(a)]/N$ (Figure 1*B*). This is an extreme example of how mean fitness cannot be guaranteed to increase when selection is frequency-dependent (65). [Mimetic and warning color patterns may, of course, vary continuously, rather than as discrete patterns (82); this may also contribute to evolution of warning color and mimicry (see under *Pattern En-*

Figure 1 Number- and frequency-dependence in mimicry and aposematism. Müller's number-dependent theory supposes that, while they learn to avoid the pattern of an unpalatable insect, predators kill a constant number, $n_k(i)$ of each morph i per unit time in a given area. Assuming the local population has constant size (N) and contains a novel pattern (A) and a wild-type pattern (a), Müller's theory can give the strength of frequency-dependent selection for or against the pattern A at different frequencies (q_A) in the population. The fitness of A is $W_A = 1 - (n_k(A)/q_A N)$, while that of a is $W_a = 1 - [n_k(a)/(1 - q_A)N]$. The measure of frequency-dependent selection acting on A relative to a used here is $S_A = (W_A/W_a) - 1$; if S_A is positive, A is favored, if S_A is negative, A is disfavored. The fitnesses are shown in terms of S_A (graph A) and mean fitness (graph B). The dashed and solid lines show frequency-dependent fitnesses for a low total population size ($N = 10$) and a high total population size ($N = 100$), respectively, relative to $n_k(A)$ and $n_k(a)$ (the fractions n_k/N are more important than absolute values of n_k and N). In contrast, linear frequency-dependent selection has been more normally used to study the population genetics of warning color and mimicry (37, 47, 54, 91, 93), for example where $W_A = 1 - s_A(1 - q_A)$, and $W_a = 1 - s_a q_A$. This model gives frequency-dependent fitnesses shown in the dotted curve of the figure. In both number-dependent and frequency-dependent selection, values of s and n_k have been chosen to give an unstable equilibrium frequency of $q_A^* = 0.4$, which is the case if A has $1.5 \times$ X greater fitness than a.

hancement and Peak-Shift and *Mimetic Polymorphism and Genetic Architecture*, below).]

In nature, not only do warning colors exist, but also novel warning patterns are forever being multiplied in already warning-colored species (see also *Genetic Drift and the Shifting Balance*) in spite of barriers suspected to impede their initial evolution. Various ideas have been proposed:

1. *Novelty and Recognizability* It has been suggested that warning colors are favored because they induce predator neophobia and because they are easier to recognize and learn (59, 84a, 140, 177). Neophobia has some experimental evidence (45, 140), whereas increased memorability is part of the definition of warning color (see above). These factors, coupled with a high survival rate of attacked prey, might seem to allow warning color to increase from low frequency in spite of increased conspicuousness (140, 177). Unfortunately, the problem with fear of novelty is that this survival advantage evaporates after a time, and enhanced learning is useful only if there are enough individuals available to do the teaching. This behavior viewpoint is rarely coupled with much thought about evolutionary dynamics. Thus, an unpalatable and brightly colored sea slug that survives 100% of attacks by fish (158) seems likely to have some risk, or loss of fitness due to fish biting; any such loss of fitness will be progressively diluted as the numbers increase, leading again to frequency-dependent selection against rare the. In fact, Müllerian mimicry or monomorphic warning color would be unnecessary if this selection against rarity were not present. An increase of conspicuousness will almost always lead to an initially greater level of attack on the first few individuals with the new pattern, even if the pattern is ultimately advantageous once fixed within the population (59, 81, 97). Essentially, $n_k(A)/1 \leq n_k(a)/N$ for warning color A to spread in a population of size N—the learning advantage of first individual A variant must outweigh the population size advantage of the cryptic wild-type a. With reasonably large prey population sizes, say $N > 10$, for a reasonably unpalatable species, this seems almost impossible; given that A is more conspicuous, the possibility seems even more remote (89, 97). In any case, high rates of beak-marks on the wings of brightly colored unpalatable butterflies attest to a high frequency of potentially lethal attacks (11, 30, 31, 91, 111). Nonetheless, various possibilities allow warning colors to cheat against this apparent selective disadvantage. These are reviewed below.

2. *Preadaptation* This idea is motivated by the fact that many palatable insects, particularly butterflies, are already brightly colored. Cryptic resting postures and rapid, jinking flight allow these insects to expose conspicuous patterns in flight that may be important for intraspecific signaling in mate choice and sexual selection (42) or in territoriality and male-male interactions (138, 169), as deflection markings (45, 128, 176), or in Batesian mimicry. If these species become unpalatable, perhaps as a result of a need to process toxic secondary compounds in food, their conspicuous patterns, already adapted for signaling, could simply be reused in predator education.

3. *Pattern Enhancement and Peak Shift* The representation of a pattern in a predator's memory is likely to be a caricature of the actual pattern. Thus, an exag-

gerated pattern may be avoided by a predator more strongly than the normal pattern on which the predator originally trained, and exaggerated warning patterns will evolve to exploit this predator bias. Training an artificial neural network model can also recreate this kind of perceptual bias for supernormal stimuli (3, 48). Whether perceptual bias is produced in computer models is strongly assumption-dependent (79), but there is good evidence for exaggerated responses to supernormal stimuli in vertebrate perception (156), which seem likely to have been a cause of exaggerated male traits in sexual selection (110, 131). Similar perceptual biases in vertebrates may contribute to the gradual evolution of warning colors (82).

A related idea is "peak shift" whereby, if zones of negative and positive reinforcement are located close together along a perceptual dimension, they may each cause the perceiver to bias their responses further apart (Figure 2). Peak shift is not dissimilar to the old idea that warning colors function by appearing as different as possible from the color patterns of edible prey (49, 59–61, 164). Theory shows that peak shift can produce gradual evolution of warning colors (133, 181), and recent experiments with birds have demonstrated relevant perceptual bias (52, 84).

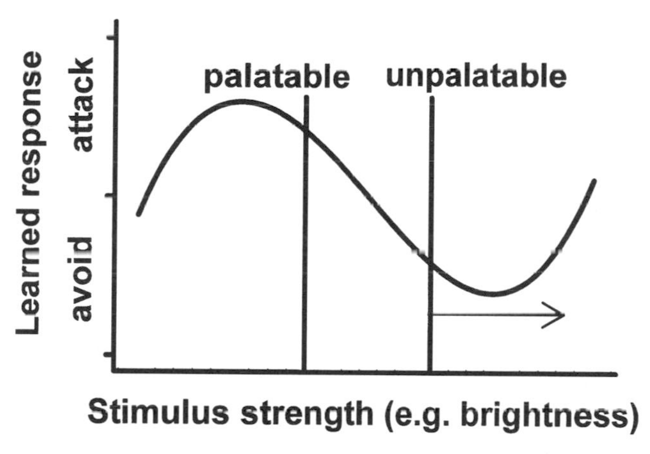

Figure 2 The theory of behavioral "peak shift." If the appearances of palatable and unpalatable species are close to each other along some stimulus dimension, such as conspicuousness, predators may develop a perceptual bias that enhances discrimination, known by behaviorial biologists as peak shift (not to be confused with evolutionary peak shift via the shifting balance). The conflicting pressures on their perceptual/learning system may lead them to avoid patterns brighter than the norm for the unpalatable species more strongly than they avoid the normal pattern; conspicuous unpalatable variants would then have an advantage over the normal pattern, allowing gradual evolution of greater and greater conspicuousness of the unpalatable species (arrow) (61). It is unclear how the palatable species will evolve; it could be selected for mimicry (to the right), or to greater inconspicuousness (to the left) to avoid detection, even though the latter may be costly due to increased predator attack rate once detected.

It seems likely that at least some warning colors evolved by pattern enhancement. For example, the patterns of conspicuous morphine butterflies *Taenaris* and *Hyantis* are clearly related to those of cryptic morphines and satyrines, such as *Morphopsis* with deflective eyespot patterns similar to many other edible members of the satyrid lineages to which they belong. *Taenaris* and *Hyantis* have evolved unpalatability, perhaps as a result of feeding on toxic Cycadaceae, both as larvae on leaves and as adults on sap and fruits. Compared with the drab *Morphopsis*, color and brightness have been enhanced, eyespot size has been increased, and eyespot number has been reduced. A variety of Batesian mimics from palatable genera such as *Elymnias agondas* (Satyrinae) and females of *Papilio aegeus* (Papilioninae) mimic *Taenaris* and *Hyantis* patterns (117, 118), showing that the latter are unpalatable. Although likely to explain some warning color evolution, it is hard to imagine that all novel warning patterns evolved by enhancement. The color patterns of related species, or even races of *Heliconius* (24, 137, 159), for example, seem so radically divergent as to preclude one being an enhancement of the other. Of course, this is a dubious anthropocentric argument, but the major gene switches in *Heliconius* suggest that radical shifts, rather than gradual enhancement of existing patterns, are responsible for much of the pattern diversity within already warning-colored lineages. If this is the case for switches between warning patterns, then the ubiquity of enhancement and predator perceptual bias, even for the initial switch, seems in doubt.

4. *Müllerian Mimicry* Another way that a newly unpalatable species might become warning-colored is via *Müllerian mimicry*. The constraints on its evolution discussed on p. 204, apply, but the widespread existence of Müllerian mimicry suggests that the idea should work both in the initial evolution of warning color and in its diversification within already unpalatable lineages. Because many species typically join in Müllerian mimicry rings (9, 10, 22–25, 116), it seems likely that, in butterflies, most warning color switches are due to Müllerian mimicry. Only the initial divergence of mimicry rings needs to be explained in some other way (24, 89, 98, 167).

5. *Density- and Apparency-dependent Warning Color* Our formulation so far of number-dependent warning color and Müllerian mimicry (see Figure 1) assumes all individuals are seen by predators, but in fact apparency as well as density per se are important for the ultimate benefits of warning colors. If more prey are killed while predators learn of a warning pattern than would be detected and killed for a cryptic population of the same size, it may pay the prey to remain cryptic. This may explain why many stationary pupae of unpalatable insects, such as *Heliconius*, are brown and resemble dead leaves, while their more apparent and mobile larvae and adults are brightly colored and classically aposematic. Density-dependent color pattern development in *Schistocerca* (desert locusts and their relatives) shows a switch from crypsis at low density to advertisement of food-induced unpalatability at high density, and predation experiments with *Anolis* lizards support this idea (155). If so, pattern enhancement (see point 3 above) of characteristics used by predators for recognition may provide a way in which this

kind of context dependent warning color evolves (133, 155, 181). Nonetheless, density-dependent facultative warning colors are unlikely in most animals, such as adult butterflies, in which color patterns are largely genetic.

6. *Kin Selection, Kin-founding, and "Green-beard" Selection* Predators attacking kin groups can kill or damage some individuals, but, after doing so, avoid others, who are relatives carrying the same pattern. A superior warning pattern may therefore increase locally under a kind of kin group selection (67). This is somewhat different to classical kin selection because benefits are transferred between individuals of like phenotype, rather than according to degree of relationship (58): The effect has therefore been called family selection (66), or kin-founding (97). Warning color is a concrete and uncheatable green-beard trait (58, 59, 166), a hypothetical type of altruism invented by Dawkins (43), whereby altruists carrying a badge (such as a green beard) recognize other altruists because they also carry the badge. More recently, the general term synergistic selection (61, 81, 82, 99, 139) has been applied to such traits. The synergism can be viewed as a behavioral explanation of the warning color trait, once evolved, but the nature of synergism does not explain its initial evolution because both the fixed absence of the trait and the fixed presence of the trait are evolutionarily stable strategies (99). The population genetic problem of frequency-dependence shown in Figure 1 still arises, and it seems clear that kin-founding could aid the initial increase of novel warning colors (59, 97). Whether kin-founding is important for the initial or subsequent evolution of novel warning colors seems hard to decide (see also under *Genetic Drift and the Shifting Balance*). However, kin-grouping and larval gregariousness in many unpalatable insects does not seem such good evidence now as formerly for kin-founding, for reasons already discussed above under *The Evolution of Unpalatability*: in most cases, gregariousness seems to have evolved after unpalatability and aposematism (139, 141).

7. *Genetic Drift and the Shifting Balance* Although kin-founding can be looked upon as a purely deterministic model similar to kin-selection (66), it is clear that, like Sewall Wright's "shifting balance" model of evolution (178–180), it requires a small local population size: The phenotypes of a small group of related individuals must dominate the learning and recognition systems of local predators, which is only possible if the total local population is low. The evolution of warning color via kin-founding is in fact a special case of phases I and II of the shifting balance (89, 97, 98, 167). In phase I, genetic drift allows a local population to explore a new adaptive peak; in phase II, local selection causes the population to adapt fully to the new adaptive peak. Although not usually treated in kin-founding models (but see 66), phase III of the shifting balance, i.e. spread of the new adaptive peak to other populations, would clearly be an important final phase in the kin-founding process. This would be equivalent to having local populations with different warning colors competing across narrow bands of polymorphism, as is actually the case in many hybrid zones between geographic races of warning-colored species today; movement of these clines for warning color would be the equivalent of Phase III (87, 98). In warning color, stable and unstable equilibria are peaks

and troughs of relative fitness, but not necessarily of mean fitness. Under purely number-dependent selection (Figure 1), mean fitness is a constant independent of frequency, $[n_k(A) + n_k(a)]/N$, and even under linear frequency-dependence, the minimum of mean fitness (at $q_A = 0.5$) is not at the unstable equilibrium ($q_A^* = 0.4$ in Figure 1). If A is more memorable than a, then $n_k(A) < n_k(a)$, but this does not increase the mean fitness when q_A is high, except very close to fixation of A when hardly any a are available to be tasted by predators.

A recent critique of the shifting balance model concluded that chromosomal evolution, warning color evolution, and more general patterns of phenotypic adaptation were almost always better explained by ordinary individual selection (41). For warning color and mimicry, the key problems are that natural selection seems too intense so that drift is unlikely, and, in common with other examples of rugged adaptive surfaces, phase III of the shifting balance seems an inefficient means of spreading better warning patterns. While these problems seem serious, key features of warning color considerably increase the chances of shifting balance occurring. First, although selection for warning color can often be extremely strong, it would be surprising if predator attacks were not sometimes reduced or suspended locally, due to temporary absence of key predators such as flycatchers or jacamars (34, 35, 119, 120). If so, populations can occasionally drift to become polymorphic because of a relaxation of selection. Provided that the prey are abundant compared with their predator, (i.e. $n_k \ll N$), the populations will quickly enter the central basin where selection is weak (e.g. for $N = 100$ in Figure 1). Here drift or mild forms of selection other than that due to warning function may cause a new pattern to rise in frequency above the unstable equilibrium (phase I), whereupon warning selection can fix and refine the new pattern (phase II). An interface between new and old patterns will form, resulting in a cline similar to hybrid zones between races observed today. If one pattern is superior at warning away predators, asymmetries of selection will drive it into the range of the other behind a narrow moving cline, as in phase III (87, 89, 98). Cline movement seems likely; with strong selection in the clines observed in nature (93), fairly rapid movement is predicted (87, 89, 91). The shifting balance proposal is speculative because we know little about the frequency, timing, and depth of episodes of selection relaxation required for phase I, the relative advantages of different warning colors across clinal boundaries required for phase III, and whether population structural constraints will prevent cline movement (4, 69, 89, 98). However, empirical evidence for all phases suggests the shifting balance is likely: (*a*) Polymorphism seems to exist regularly among Müllerian mimics (see below under *Müllerian Mimicry, Polymorphism and the Palatability Spectrum*), showing that although mimicry is sometimes strongly selected (11, 75, 80, 92), at other times, a combination of reduced selection, genetic drift, and nonmimetic selection causes polymorphism in the central basin, and therefore that events triggering phase I seem actually to occur; (*b*) the strong purifying selection that is the problem for phase I promotes phase II; and (*c*) the existence of today's narrow clines and biogeographic evidence for past cline movement and movement in historical times suggest that phase III occurs regularly.

The current disjunct distribution of genetically homologous "postman" patterns of *Heliconius erato* and its Müllerian co-mimic *Heliconius melpomene* in peripheral populations strongly suggests that some such competitive cline movement in favor of central Amazonian "dennis-ray" patterns of this nature has occurred, even if the color patterns have been sometimes restricted to Pleistocene refuges in the past (24, 27, 89, 98, 137, 164). There is some empirical evidence for movement of *Heliconius* clines this century; although slow on a historical scale, the movement of warning color clines could be very fast relative to an evolutionary time scale (89). (*d*) The shifting balance does seem to have a strong potential in explaining geographic divergence within species, the strong differences in warning color and mimicry between sister species, and also the extraordinary diversity and novelty of these patterns (98). If the shifting balance is important for current diversification, there is little reason to doubt that it could also have been important in the murky initial stages of the origins of warning color in aposematic lineages, though evidence has long since been erased by more recent color pattern evolution.

DIMORPHISM AND POLYMORPHISM IN MIMICRY

Sex-Limited Mimicry

In a minority of Batesian mimetic butterflies, females are mimetic, while males, although brightly colored, are not. Such cases can be explained if males are constrained to be nonmimetic by sexual selection, either via female choice (162, 163) or by the requirements of combat or other male-male signaling (138, 169). This topic has been reviewed excellently elsewhere (164; see also 78), so we do not treat it in detail here.

Sexual selection may explain sexually dimorphic mimicry, but there are some peculiarities of female-limited mimicry for which the answers are not known. First, female-limitation seems restricted to putative Batesian mimicry. As far as is known, Müllerian mimics lack strong sexual dimorphism. Presumably, this is explained because Müllerian mimicry is under purifying density-dependent selection: As a mimetic pattern becomes more common, its advantage increases (Figure 1). In contrast, Batesian mimicry becomes less successful as it becomes commoner; thus, sexual selection is more likely to outweigh this weakening mimetic advantage in Batesian mimics (162). Female-limited mimicry also seems virtually confined to butterflies (46, 168), whereas the sexual selection theory should apply to all examples of Batesian mimicry. Here, the explanation may be ecological. Territorial or fighting males of many butterflies fly purposefully, fast, and can escape predators easily. Female butterflies searching for oviposition sites can be particularly vulnerable to predator attacks (111) because they must at times fly slowly, like potential models; thus ecological considerations may explain why butterfly females, but not males, often mimic slow-flying models (111, 169). Ecological constraints on sexually dimorphic mimicry are well demonstrated by cases in which only the male is mimetic (168), for example, in saturniid moths with nocturnal females but diurnal males (172).

Mimetic Polymorphism and Genetic Architecture

Batesian mimetic butterflies may be polymorphic as well as sexually dimorphic. This phenomenon is best known and studied genetically among female mimetic forms of Papilionidae, particularly *Papilio dardanus* and *P. memnon*, where each morph mimics a different unpalatable model. The maintenance of this polymorphism is easily explained in common Batesian mimics because frequency-dependent selection favors rare mimics. Polymorphisms in Batesian mimics are also well-known in nonbutterfly groups: Good examples exist in hoverflies (170, 171). However, the rarity of accurate polymorphic mimicry of the kind displayed in *Papilio* suggests that special circumstances must be involved. Mimetic polymorphisms in these cases are usually determined at relatively few genomic regions with large effect ("supergenes"), often with almost complete dominance (38, 39, 134, 135). The maintenance of mimetic polymorphisms probably depends rather strongly on supergene inheritance. Without it, nonadaptive intermediates would be produced.

While it is easy to understand the maintenance of polymorphisms at mimetic supergenes, it is far from clear how these supergenes initially evolved. Early Mendelians used these genetic switches as evidence that mutations of major effect were prime movers of adaptation (123). Fisher argued forcefully that most adaptive evolution could be explained via multiple genetic changes of individually small effect being sorted by natural selection (49). Essentially, Fisher proposed that selection rather than mutation was the creative process in adaptation. Goldschmidt (56) then revived mutationist theory in more sophisticated form and proposed that mimics could exploit major ("systemic") mutations that reused the same developmental machinery originally exploited by the model. He felt it unlikely that the same genes were reused by mimics and models, proposing instead that different genes had access to the same developmental pathways. Gradualists were quick to point out cases in which development of mimicry was clearly analogous rather than homologous, such as colored spots on the head and body of models being mimicked by basal wing patches on mimetic *Papilio memnon* (135). Single gene switches in *P. memnon* were demonstrated to consist of tightly linked multiple genetic elements that could be broken apart by recombination or mutation, and it was suggested by gradualists that these "supergenes" had been gradually constructed by a process of linkage tightening to reduce the breakup of adaptive combinations by recombination (38, 39).

More recently, opinion has swung back (but only part way) toward mutationism. It has been realized that it would be hard to construct supergenes by means of natural selection alone. Separate elements of a supergene must have been tightly linked initially in order that a sufficiently high correlation between favorable traits was available for selection for tighter linkage. Thus polymorphic mimicry must to some extent have depended on the pre-existence of gene clusters (36, 37). If so, this could explain why Müllerian mimics and models such as *Heliconius* often themselves show major gene inheritance. Müllerian mimics are not expected to have polymorphisms, and usually they do not (but see below under *Müllerian Mimicry*); thus they are not expected to require supergene inheritance of their color patterns under the gradualist hypothesis. *Heliconius* patterns are inherited at multiple loci; this was interpreted as confirming a gradualist expectation for polygenic inheritance of mimicry (113, 137, 164). However, a closer look at *Heliconius* shows that many of the pattern switches are indeed major, have major fitness effects, and can also in some cases be broken down into tightly linked component parts by recombination or mutation (87a, 93, 137), again suggesting mimetic "supergenes." For example, in both Müllerian mimics *H. erato* and *H. melpomene*, a large forewing orange patch known as "dennis," and orange hindwing "ray" patterns are very tightly linked but are separable via recombination or mutation that shows up only in rare individuals from hybrid zones (87a). Probably, mutations with major effect are required even in Müllerian mimicry because, during adaptation, a Müllerian mimic loses its current warning pattern while approaching that of a model. There is thus a phenotypic fitness trough between the old pattern and the new pattern. Only if a mutation produces instant protection by the new pattern can the gene be favored, unless the two patterns are already extremely close. After approximate mimicry has been achieved by mutation, multilocus "modifiers" can improve the resemblance in the normal way (37, 105, 161, 164). This hybrid view of Müllerian mimicry, known as the Nicholson "two-step" theory, combines what is arguably a mutationist argument with a gradualist hypothesis to explain the perfection of resemblances.

This explanation fits major gene adaptations in Müllerian mimicry, especially as it is now realized that Fisher's argument for adaptation via small mutations has serious flaws (112), even without the frequency-dependent stability peaks of mimicry (Figure 1). However, two-step theory cannot explain why genes for forewing and hindwing patterns should be tightly linked in both model and mimic in *Heliconius*. Why should *H. erato* and *H. melpomene* (the former is almost certainly the model driving the divergence—see 55, 89) both diverge geographically using probable supergenes of major genetic effect? One possibility is that genetic architecture for color pattern change in *Heliconius* simply has limited flexibility (87a). We now know that there is widespread reuse of homeotic gene families throughout the animal kingdom, including some involvement in color pattern development in butterflies (16, 33). It would not be surprising if mimicry gene families were not also reused similarly (106–108, 164) in the lineages leading to *H. erato* and *H. melpomene*. This argument is similar to Goldschmidt's (56), but in one sense more extreme, since Goldschmidt thought it likely only that the same patterning control would be reused, rather than the very same genes. Others argue from similar data that the evidence is in favor of analogous rather than homologous developmental pathways and gene action (88), but a true test will be possible only when mimicry genes are characterized at the molecular level in both lineages (51, 95).

In conclusion, current opinion based on nearly a century of genetic studies and mathematical population genetic theory suggests how mimetic as well as other adaptations may often require mutations of major effect, at least initially, both because of the ruggedness of the selective landscape, and probably also because of constraints imposed by pattern genetics. Perfection of these adaptations then involves effects generated at multiple genes of increasingly small effect. The genetic architectures required, especially for polymorphic mimicry, may be rare. This would explain why some lineages involved in mimicry, such as the Papilionidae, are able to colonize multiple mimicry rings and become polymorphic (164), while others are rarely mimetic. Disruptive mimetic selection is perhaps as likely to be an agent causing an alternative, speciation, as it is to be a common cause of polymorphism (see *Mimicry and Speciation*, below).

Müllerian Mimicry, Polymorphism, and the Palatability Spectrum

Müllerian mimicry and warning color are standard textbook examples of frequency-dependent selection within species (e.g. 99, 126). Polymorphisms should be rare due to high rates of attack on rare variants (Figure 1;) (27, 47, 67, 87, 89). In general, workers in the field of mimicry assert that this is so (89, 97, 98, 161, 164, 167), but there are some very embarrassing exceptions to the rule among even the best known Müllerian mimics. The most famous case is *Danaus chrysippus* and its Müllerian mimics *Acraea encedon*, *A. encedana*, together with their Batesian mimic *Hypolimnas misippus*. While distinct color patterns are virtually fixed in the peripheries of their respective ranges, these species are highly polymorphic over an area of Central and Eastern Africa larger than Europe (57, 146). Similarly embarrassing widespread polymorphisms are found in two-spot ladybirds (15, 85) and in *Laparus doris* (Heliconiinae) (151, 159). Arguably, mimicry in many of these cases is weak: Non- or poorly mimetic morphs are common (85, 114, 143, 144, 159). However, there are equally problematic examples in which mimicry is very accurate. For instance, *Heliconius cydno* is mostly monomorphic in Central America (94, 142), but becomes polymorphic throughout much of the Andes of Colombia and Western Ecuador (80, 83); each morph can be clearly identified as an accurate mimic of other *Heliconius*, particularly *H. sapho* and *H. eleuchia*. The pinnacle of Müllerian mimetic polymorphism is found in *Heliconius numata*. This species is polymorphic throughout virtually its whole range, and some populations of the Amazon basin near the slopes of the Eastern Andes may have up to seven different morphs, each an accurate mimic of a separate species of *Melinaea* or *Mechaniti* (Ithomiinae) (23, 26). Three explanations have been proposed, and we here add a further hypothesis that may contribute to the persistence of polymorphisms once they have been established.

1. *Batesian Overload and Coevolutionary Chase* If an unpalatable species has many Batesian mimics, it may suffer from Batesian overload. According to this hypothesis, the deleterious effects of mimics may force the model to diverge from

its normal pattern to escape mimicry, leading to a coevolutionary chase of model by mimic. This idea has generated some controversy (54, 71, 72, 109, 165) but has been well reviewed recently (164, 165), and we merely summarize: It does Z not seem likely that coevolutionary chase or Batesian overload can explain polymorphisms in unpalatable models. Frequency-dependent purifying selection on the models must almost always be stronger than the diversifying selection due to mimetic load (57, 78, 109, 165).

2. *The Palatability Spectrum* Unpalatability cannot be absolute; there must be variation in unpalatability, which could lead to some interesting evolutionary effects. Müllerian and Batesian mimicry are differentiated by means of palatabilities. Models and Müllerian mimics are negatively reinforcing, while Batesian mimics positively reinforce predator attacks. Hence, the straightforward view that Batesian mimics are parasitic and hurt their models, while Müllerian mimics are mutualistic and benefit their models (103). However, a second equally straightforward idea apparently conflicts with this view: If two Müllerian mimics are not equally unpalatable, the presence of the more palatable could increase the rate of attack on the less palatable, so that weakly unpalatable mimics may harm stronger models or co-mimics, leading to a parasitic form of Müllerian mimicry. A series of behavioral modelers since the 1960s have suggested that parasitic Müllerian mimicry may explain some of the embarrassing examples of polymorphism in aposematic species. Because benefits and costs become decoupled from the Müllerian/Batesian palatability divide in this latter prediction, a new terminology must be developed. An appropriate name for the new parasitic form of Müllerian mimicry is "quasi-Batesian" (148). [There is also a category of palatability-defined Batesian mimicry that is beneficial to the model as well as the mimic—"quasi-Müllerian" mimicry (84a, 147a, 151, 165). This is possible if seeing a palatable mimic "jogs" the memory, reminding predators of unpleasant experiences with the model, thus leading to greater avoidance of the model than if there were no mimic. Quasi-Müllerian mimicry seems unlikely (151); anyway, it should not lead to polymorphism and is not discussed further.] In quasi-Batesian mimicry, the more palatable mimic may suffer increasing attacks as its numbers increase relative to the model's, even though its effect while alone would be to reduce its predation progressively as density increases (Figure 3*B*, *C*) (71, 72, 115, 148–151). It has been suggested that this leads to the evolution of polymorphism in Müllerian mimicry systems (71, 72, 147a–151).

The behavioral assumptions that lead to quasi-Batesian mimicry pose a severe threat to traditional natural history and evolutionary dynamical views of mimicry, possibly "the end of traditional Müllerian mimicry" (148). This problem never arose until behavioral biologists attempted to model memory realistically. It is apparent that Müller and subsequent naturalists and evolutionists made an unstated assumption: that the sum of learning and forgetting over all predators would cause an approximately constant number (n_k) of unpalatable individuals of each phenotype to be killed (or damaged) per unit time (Figure 1). Purifying frequency-dependent selection results from Müller's assumption because

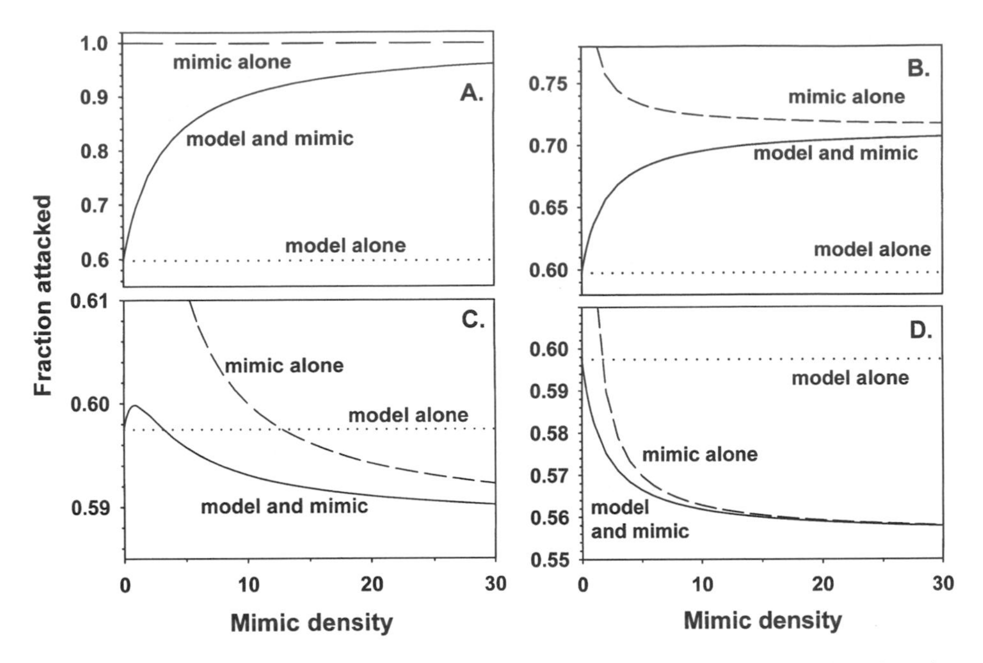

Figure 3 Mimicry and the palatability spectrum. The effect of assumptions about learning and forgetting on fitnesses of model and mimic are shown in this figure. Equilibrated attack rates at varying mimic densities are shown for model alone, mimic alone, and model-mimic pair. Comparisons of attack rate clearly demonstrate whether mimic, model, or both benefit from the association (115) (the mimic is here considered by convention to be the more palatable species). In all panels, the model density is a constant set at 1.6 (115). These assumptions (71, 115, 148, 149, 151, 166) can reproduce classical parasitic Batesian mimicry (*A*; $\lambda_{Mo} = 0.2$, $\lambda_{Mi} = 1$, $\alpha_{Mo} = 0.3$, $\alpha_{Mi} = 0.0$) and mutualistic Müllerian mimicry (*D*; $\lambda_{Mo} = 0.2$, $\lambda_{Mi} = 0.2$, $\alpha_{Mo} = 0.3$, $\alpha_{Mi} = 0.3$), but they also produce intermediate types of mimicry, including parasitic quasi-Batesian mimicry between pairs of unpalatable species (*B*; $\lambda_{Mo} = 0.2$, $\lambda_{Mi} = 0.6$, $\alpha_{Mo} = 0.3$, $\alpha_{Mi} = 0.5$), and a cusped quasi-Batesian/Müllerian combination (*C*; $\lambda_{Mo} = 0.2$, $\lambda_{Mi} = 0.3$, $\alpha_{Mo} = 0.3$, $\alpha_{Mi} = 0.5$). The curves were generated from a general equation for attack rate equilibrium at particular density (115, 149). Note that Owen & Owen's own figures are sketches only, and contain some incorrect features (149).

the average attack fraction n_k/N decreases as the total number of individuals, N, increases. The existence of quasi-Batesian mimicry, in contrast, requires that the attack fraction on a Müllerian mimic increases as N increases, implying that n_k can be a rising function of N rather than a constant. We here follow the development of these ideas and discuss why we feel the assumptions that lead to quasi-Batesian mimicry may not be met in most real situations.

The original idea for what is now called quasi-Batesian mimicry was proposed by Huheey (71 and earlier). After a single trial experience with an unpalatable individual, the predator was imagined to learn to avoid it totally; thereafter, the predator would forget after seeing, but not attacking, a fixed number of individuals with the same pattern. In this formulation, unpalatability affected only the rate of memory loss, rather than its acquisition; very unpalatable species caused slower forgetting than mildly unpalatable species. Increasing the density of less nasty mimics caused a rise in the average forgetting rate and led to an increasing fraction of models attacked. Thus, if two unpalatable species differed in palatability, only one benefited, while the other suffered, though the more palatable species on its own was still unpalatable in the sense that predators are negatively reinforced. Mimicry, even when at the point of equal palatability, was neutral; increases in density of either co-mimic caused a faster rate of both learning and forgetting, rather than a reduction in fraction attacked. The predicted absence of mutualistic mimicry in Huheey's theory was strongly attacked (12, 115, 136, 151). The problem appeared to be the event-triggered forgetting model, in which avoidance lapsed after a certain number of prey were avoided. This meant that the total number of prey in the population had no effect on the evolution; selection was assumed to depend only on relative frequency of mimics and models.

To avoid this pathology of Huheey's formulation, it was proposed that forgetting should be time-dependent (12, 115, 136, 166), rather than depending on the number of avoidances; forgetting should cause the attack fractions to decline or rise exponentially with rates α_{Mi} and α_{Mo} (for mimic and model, respectively) toward the "naive attack rate" asymptote, i.e. a naive attack fraction (115, 149). At the same time, a more flexible system of learning was proposed, in which unpalatability was represented as an asymptotic fraction of prey attacked, λ_{Mi} and λ_{Mo}; these asymptotes were again approached exponentially, with learning rates forming another set of parameters (84a, 115, 147a, 148, 151). These theories could reproduce the full spectrum of mimicry from Batesian mimicry (Figure 3*A*) to Müllerian mimicry (Figure 3*D*), including quasi-Batesian mimicry (Figure 3*B*), and also a curious form of biphasic mimicry, which is quasi-Batesian at low mimic density, but traditionally Müllerian at higher mimic densities (Figure 3*C*).

The behavior of these models is easy to explain. Learning and forgetting each result in an exponential approach to a different asymptote of attack fraction, so the combination of the two processes will itself lead to a stationary resultant attack fraction independent of density for either model or mimic on their own. The joint attack fraction on model and mimic together (assuming models and mimics are visually indistinguishable) is simply an average between the curves for model and mimic asymptotic attack fractions. When mimic density is very low, the joint response is very like that of the model; when mimic density is high, the effect of the mimic dominates, and the joint response increasingly obeys the mimic's asymptote. Because the averaging process is of the form of a harmonic mean (115) rather than an arithmetic mean, curious peaks in the density response can occur, the "Owen & Owen effect" (149) (Figure 3*C*), implying a quasi-Batesian/Müllerian transition across a density threshold. Speed & Turner (151, 167a) recently examined the behavior of a number of different formulations and combinations of these basic memory assumptions. They concluded that (*a*) many of the assumptions produce quasi-Batesian responses like that of Figure 3*B-C* and (*b*) behavioral experiments on mimicry and warning color are not usually set up to test for density responses and therefore cannot easily be used to test whether mimicry falls into quasi-Batesian categories. Well-known polymorphic Müllerian mimics often have intermediate levels of acceptance in tests both with caged and wild birds (20, 34, 35, 77, 119, 120, 132), showing that many supposedly unpalatable species may often be attacked. Therefore, the known biology of predation on unpalatable species as well as theory mesh with the possibility of a palatability spectrum that could lead to quasi-Batesian mimicry.

However, if theories like those in Figure 3 are correct, the whole basis for traditional number-dependent and frequency-dependent mimicry of Figure 1 is suspect. Our own belief is that new and incorrect assumptions lurking in the behavioral models are to blame for the conflict. Our objections, which are more fundamental than those raised in an earlier critique (84a), are as follows:

(*a*) We think it unlikely that attack rates on unpalatable species will reach an asymptotic fraction independent of density, unless that fraction is zero. To understand this, imagine that forgetting is switched off, so that all learning is perfect (see also 84a). Under this assumption, the new theories (115, 148) predict that learning should asymptote at constant frequency; number-dependence enters into memory dynamics only through time-based forgetting. With no forgetting, there is then no number-dependent selection, and mutualistic Müllerian mimicry becomes impossible (149–151). Intuitively, it seems odd that perfect memory does not lead to extremely successful Müllerian mimicry, and we here attempt to show why this intuition is correct. With no forgetting, the absence of Müllerian mimicry is due to a density-independent asymptotic attack fraction. In other words, as the density of an unpalatable mimic in Figure 3 rises, the predator is supposed to stuff itself with more and more unpalatable prey in order to maintain a constant asymptotic fraction of prey attacked. Learning to avoid prey is more likely to depend on dose received by the predator per unit time, rather than dose per individual prey. This will lead to a fraction attacked that declines with density rather than a constant asymptotic fraction. Note that this argument does not depend on "hunger levels," because unpalatable prey are unlikely to form a large component of the diet (166). The new theories in effect have the same problem in their learning module (i.e. not being time-based) as did Huheey's much- criticized forgetting module (12, 115, 136). It seems much more likely to us that for "unpalatable" prey, an asymptotic number of prey attacked per unit time would be required for learning, leading to strongly number-dependent and frequency-dependent selection like that of Figure 1, and a resultant attack fraction that declines to zero as prey densities increase.

(*b*) It is hard to justify the term "unpalatability" unless the effect is density-independent; predators should reject and increasingly avoid unpalatable prey whenever they encounter them at whatever density. However, the new theories see a species as unpalatable if it has a learning asymptote lower than the "naive attack fraction," and as palatable if it has a learning asymptote higher than the naive attack fraction (149, 151). But only when the asymptotic attack fraction is zero do we produce avoidance, whatever the attack fraction prior to experience; this was the case, for example, in an original simulation model designed to disprove Huheey's assertions, and which recovered only Batesian and Müllerian mimicry, with a sharp transition between them (166). If our argument is correct, the whole of the palatability spectrum above an asymptotic attack fraction of zero is then "palatable," and quasi-Batesian mimicry simply becomes Batesian, parasitic mimicry. The "palatability spectrum" represented by $0 <$ asymptotic attack fraction ≤ 1 is just that, a spectrum of palatability rather than of unpalatability. Under this view, levels of unpalatability may differ, but they cause changes only in rates of learning and forgetting, rather than in level of the learning asymptote itself, which must be zero.

(*c*) Another problem is that, strictly speaking, "attack fraction" is not "palatability" at all, but a transformation of palatability onto a behavioral response axis. What we mean by "unpalatability" is easiest to interpret as a simple linear, or perhaps logarithmic, function of noxious compound dosage, which can vary from zero to infinity. The behavioral effect of these compounds may be to produce an asymptotic attack fraction of 0%, 100%, or somewhere in between (Figure 3). However, the behavioral response will certainly be a sigmoidal function of dose; the majority of dosages will yield approximately 100% (palatable) or 0% (unpalatable) asymptotic attack, with only a relatively narrow intervening band of dosages giving rise to intermediate levels of attack. Thus, the behavioral "palatability spectrum" as modeled by attack fraction is a highly distorted view of the underlying palatability, or dosage, of noxious chemistry; in fact, most of the dosage spectrum is not considered by these attack rate spectrum models (115, 147a–151). In reality, intermediate asymptotic attack fractions, even if they do exist, are likely to form a small part of the palatability (dosage) spectrum.

Empirical data from caged and wild birds showing intermediate levels of attack on models are of great interest, but they do not necessarily conflict with the points made above. Attack fractions in the laboratory or in nature tell us neither how they vary with prey density (167a) nor how they asymptote. The behavioral, "receiver psychology" view, which leads to possibly novel forms of mimicry, suggests that attack fraction will reach a nonzero asymptote as density is increased; the number-dependent (natural history) view predicts that attack fractions on unpalatable insects will always decline with increasing density. Unfortunately, experiments have not clearly distinguished between these alternatives because they were designed with other ends in mind (167a). It does not seem impossible to design more appropriate experiments, however.

In conclusion, theories of palatability from a receiver psychology angle have led to a potentially major upset in traditional views of mimicry. To decide which view is correct, we need to understand memory dynamics of actual predators, and, given that many of the controversial theories are supposedly based on a standard Pavlovian learning theory (124, 148), understanding the evolutionary results of memory on mimicry could lead to advances in memory theory in general. Even if quasi-Batesian mimicry turns out, as we believe, to be unlikely, the threat posed by these new theories demonstrates the naiveté of the original natural history assumption that memory is a black box producing number-dependence.

3. *Spatial and Temporal Variation in Mimetic Selection* Geographic variation in mimetic color patterns within a mimic can obviously be maintained by geographic divergence of models. If mimicry is Müllerian, then divergence becomes self-reinforcing. Patches of habitat with different Müllerian mimetic patterns will be separated by zones of polymorphism; the width of the polymorphic region will be proportional to average dispersal distance and inversely proportional to the square root of the strength of selection (92), as for clines in general (5). Thus, if selection is weak and dispersal extensive, bands of polymorphism may be wide compared to areas of monomorphism. This situation undoubtedly pertains in many species; for example, it is often not realized how common this is within *Heliconius*. The hybrid zones between races of *Heliconius erato* or *H. melpomene* are renowned for their narrowness (e.g. 87, 93); however, zones of polymorphism between weakly differentiated races, for instance in the Amazon basin, are much broader, so that polymorphism is almost the norm (see maps in 24, 27; many other maps of *Heliconius* races oversimplify the actual distributions).

A similar situation may exist for wide bands of polymorphism in the unpalatable *Acraea encedon, A. encedana, Danaus chrysippus*, and their Batesian mimic *Hypolimnas misippus* in Central and Eastern Africa: Peripheral populations of these species are nearly monomorphic (146). Similarly, spatially varying mimetic and other selection pressures, rather than quasi-Batesian mimicry (151), may explain the polymorphisms of ladybirds such as *Adalia bipunctata* (15, 85) and butterflies such as *Laparus doris* (159).

There may also be temporal as well as spatial variation in mimetic selection. The diverse polymorphism of *Heliconius numata* may be selected because the models (ithomiines in the genus *Melinaea*) vary greatly in abundance over time and space (26). However, it would be hard to explain how polymorphism is maintained via temporal variation unless the color pattern loci have, on average, a net heterozygote advantage. Given that the supergenes affecting mimicry in *H. numata* are visually dominant (26), any heterozygous advantage must usually be nonvisual. Another example of polymorphism in a Müllerian mimic with multiple models is *Heliconius cydno*. There are strong differences across W. Ecuador in the frequency of models *Heliconius sapho* and *H. eleuchia*, causing divergent patterns of natural selection (80). In conclusion, the observed polymorphisms of many Müllerian mimics can be explained without quasi-Batesian mimicry, via spatial and possibly temporal variation in model abundance.

4. *The Shape of Frequency-Dependence* The maintenance of polymorphism in unpalatable species will be considerably aided by the shape of frequency-dependence, given number-dependent selection (Figure 1). When population sizes of prey (N) are large relative to the numbers sacrificed during predator learning, the fraction n_k/N will be small, say $1/100$ or less, and there will be little selection in the central polymorphic basin.

Although we do not know the values of n_k/N typical in the wild, a variety of experiments (11, 75, 80, 92) indicate that selection can be strong, i.e. $n_k/N \geq 1/10$. On the other hand, it seems likely that many predators will require few learning trials to avoid an aposematic insect. Models and common Müllerian mimics will often outnumber their predators considerably, and, furthermore, predators live much longer and will often be able to generalize between prey generations. Experiments by Kapan on *H. cydno* in W. Ecuador showed that selection against polymorphism was much weaker where *H. cydno* was abundant than where it was rare (80). Thus, it seems not unlikely that $n_k/N \leq 1/100$, at least some of the time.

Drift can explain the origin but not the maintenance of polymorphism in the central basin. However, polymorphisms, once attained, should be removed only slowly via mimetic selection. Second-order selective forces such as nonvisual selection (for instance, thermal selection in ladybirds), arbitrary mate choice, or other factors (15, 85,144–147) may become important and contribute to nonadaptedness of mimetic polymorphisms. Strong selection at some times and places ($n_k/N = 1/10$ or greater) is clearly required to produce near-perfect resemblance and narrow hybrid zones between races. But if a Müllerian mimic or model, perhaps by an ecological fluke, becomes abundant relative to its predators ($n_k/N \leq 1/100$), it could then be relatively free to experiment with nonadaptive and polymorphic color patterns. In short, the shape of frequency-dependence, together with varying selection for mimicry and mild selection of other types can explain Müllerian polymorphism without the need for quasi-Batesian mimicry.

MIMICRY AND SPECIATION

Bates, Wallace, and Darwin were all of the opinion that strong natural selection, which must occur sometimes to explain mimicry, could lead to speciation. The continuum between forms, races, and species of diversely patterned tropical butterflies led to this idea in the first place (6, 7, 173, 174). This view has since faded into the background, probably because of a postwar concentration on reproductive characteristics ("reproductive isolating mechanisms") thought important in speciation under the "biological species concept" (100). However, mimicry causes strong selection against nonmimetic hybrids or intermediates and should therefore contribute strongly to speciation and species maintenance, by acting as a form of ecologically mediated postmating isolation. Together with the evolution of assortative mating, mimetic shifts may have led to speciation in butterflies such as *H. himera* and *H. erato* (96, 101).

If mimicry contributes to speciation, mimetic shifts should often be associated with speciation within phylogenies. Mimicry-related speciation would explain the curious pattern of "adaptive radiation" in *Heliconius*: Müllerian co-mimics are usually unrelated, while closely related species almost always belong to different mimicry rings (164). Mimetic pattern has been switched between eight of nine pairs of terminal sister taxa in a mtDNA phylogeny of *Heliconius* (17, 95). Many sister taxa that have switched mimicry are known from other groups as well. For example, among butterflies, the viceroy (*Limenitis archippus*) mimics queen and monarch butterflies (*Danaus* spp.), while its close relative, the red-spotted purple (*Limenitis arthemis astyanax*) mimics an unpalatable papilionid, *Battus philenor*. The two *Limenitis* are very closely related and hybridize occasionally in the wild (127). Similar examples exist in the Papilionidae. Mimetic lineages do not seem to speciate more rapidly than nonmimetic lineages in the genus *Papilio* (F Sperling, in litt.); however closely related, species do often differ in their mimicry ring.

While we believe mimicry contributes to speciation, this section must remain somewhat speculative. We cannot point to any convincing case in which mimicry has been the major or only cause of speciation. But then perhaps speciation is almost always caused by multiple, rather than single, episodes of disruptive selection.

EVOLUTION AND MAINTENANCE OF MULTIPLE MIMICRY RINGS

A naive view of Müllerian mimicry would suggest that all similarly sized species should converge locally onto a single color pattern. In fact, there are often ten or more mimicry rings among ithomiine and heliconiine butterflies of the Amazon basin (6, 9, 102, 137). The reason for the lack of a single uniform mimicry ring among similarly sized butterflies is currently disputed and parallels, at an interspecific level, the debate on Müllerian polymorphisms.

Papageorgis (116) provided data from Peru showing that different heliconiine and ithomiine mimicry rings fly at different heights in the forest canopy. She suggested that dual selection for camouflage and mimicry might explain these patterns. In other words, particular mimicry rings are better camouflaged in the lighting conditions pertaining at their favored flight heights. However, heliconiine flight heights are now well-documented to overlap far more extensively than appeared from Papageorgis' data, although weak mimicry associations do exist for habitat and nocturnal roosting height (25, 28, 94). It is unclear how dual selection would work, and it is anyway hard to imagine that the garish reds, yellows, blacks, and iridescent blues of heliconiines are ever very cryptic against subdued forest backdrops.

Nonetheless, recent studies of ithomiines do demonstrate some patterning of mimicry rings in flight height as well as in horizontal (habitat-related) distribution (10, 25, 44, 102). A possible explanation for these community patterns is that different guilds of predators are found preferentially in the different habitats or microhabitats, so that, within each habitat, mimicry is tuned to local predator knowledge (10, 94). There must be some selection pressure of this sort to explain the microhabitat associations; however, it would be hard to imagine birds ignoring butterflies a meter or two higher or lower than their normal flight height in the forest understory, and it seems highly unlikely that the proposed subcommunities of predators and particular mimicry rings are very discrete. The overlap between mimicry rings is rather more noticeable to the naturalist than the somewhat statistical differences in average heights or microhabitats (10, 25, 44, 94, 102). Instead, statistical differences may exist because newly invading unpalatable species are most likely to join mimicry rings already most prevalent in their habitats. Major mimicry rings that overlap substantially may be unlikely to join together as species accumulate in each ring for the same reason that intraspecific polymorphisms have a nearly neutral central basin (Figure 1); the selection for convergence of two abundant mimicry rings will simply not be that strong.

CONCLUSIONS: MIMETIC DIVERSITY AND THE FORM OF FREQUENCY DEPENDENCE

We have shown that the shape of number-dependent selection on the color patterns of unpalatable species can help explain many mutually conflicting data of mimicry and mimetic diversity. When the attack fraction is high because of a high predator/prey ratio, selection on mimicry can clearly be extremely strong and has been measured to be so in a handful of field studies. But when predator/prey ratios are low ($n_k/N \leq 1/100$), there is a wide central basin of near-neutrality where only weak purifying selection acts on polymorphisms. Therefore, once an unpalatable butterfly becomes abundant relative to predators, n_k/N decreases hyperbolically, and its morphology becomes less constrained by selection. A temporary relaxation of selection may then result in polymorphisms, which become relatively impervious to further bouts of selection. The weakness of purifying selection in polymorphic populations can help explain why puzzling polymorphisms persist in some Müllerian mimics. Such polymorphisms enable populations to explore the selective landscape, which can increase the chances of shifting balance, one of the few ways to explain the empirical observation that utterly novel color patterns evolve continually in warning-colored and mimetic butterflies. Similarly, weak selection against multiple rings may be partially responsible for the diversity of mimicry in any one area.

But these arguments will fail if predator memory and perception do not produce number-dependent selection. If predators behave according to some current theories of "receiver psychology," these conclusions based on extensions of traditional, number-dependent Müllerian theory are in jeopardy. We do not think that this is the case; however, appropriate experimental studies are urgently required to test between these conflicting models of memory and forgetting.

ACKNOWLEDGMENTS

We are very grateful to George Beccaloni, Chris Jiggins, Gerardo Lamas, Russ Naisbit, Mike Speed, Felix Sperling, Maria Servedio, Greg Sword, John Turner, Dick Vane-Wright, Dave Williams for critiques, conversations, and comments, and to NERC, BBSRC, the British Council, and the Ministries of Higher Education and Research and of Foreign Affairs, France, for financial support.

Visit the Annual Reviews home page at www.AnnualReviews.org

LITERATURE CITED

1. Aiello A, Silberglied RE. 1978. Life history of *Dynastor darius* (Lepidoptera: Nymphalidae: Brassolinae) in Panama. *Psyche* 85: 331–45
2. Andersson M. 1994. *Sexual Selection.* Princeton, NJ: Princeton Univ. Press
3. Arak A, Enquist M. 1993. Hidden preferences and the evolution of signals. *Philos. Trans. R. Soc. London Ser. B* 340:207–14
4. Barton NH. 1979. The dynamics of hybrid zones. *Heredity* 43:341–59
5. Barton NH, Gale KS. 1993. Genetic analysis of hybrid zones. In *Hybrid Zones and the Evolutionary Process*, ed. RG Harrison, pp. 13–45. New York: Oxford Univ. Press
6. Bates HW. 1862. Contributions to an insect fauna of the Amazon valley. Lepidoptera: Heliconidae. *Trans. Linn. Soc. London* 23:495–566
7. Bates HW. 1863. *A Naturalist on the River Amazons.* London: John Murray
8. Bates HW. 1879. [commentary on Müller's paper]. *Trans. Entomol. Soc. London* 1879:xxviii–ix
9. Beccaloni GW. 1997. Ecology, natural history and behaviour of ithomiine butterflies and their mimics in Ecuador (Lepidoptera: Nymphalidae: Ithomiinae). *Trop. Lepid.* 8: 103–24
10. Beccaloni G. 1997. Vertical stratification of ithomiine butterfly (Nymphalidae: Ithomiinae) mimicry complexes: the relationship between adult flight height and larval host-plant height. *Biol. J. Linn. Soc.* 62:313–41
11. Benson WW. 1972. Natural selection for Müllerian mimicry in *Heliconius erato* in Costa Rica. *Science* 176:936–39
12. Benson WW. 1977. On the supposed spectrum between Batesian and Müllerian mimicry. *Evolution* 31:454–55
13. Blaisdell M. 1982. Natural theology and Nature's disguises. *J. Hist. Biol.* 15:163–89
14. Bradbury JW. 1981. The evolution of leks. In *Natural Selection and Social Behavior: Recent Research and New Theory*, ed. RD Alexander, DW Tinkle, pp. 138–69. New York: Chiron
15. Brakefield PM. 1985. Polymorphic Müllerian mimicry and interactions with thermal melanism in ladybirds and a soldier beetle: a hypothesis. *Biol. J. Linn. Soc.* 26: 243–67
16. Brakefield PM, Gates J, Keys D, Kesbeke F, Wijngaarden PJ, et al. 1996. Development, plasticity and evolution of butterfly eyespot patterns. *Nature* 384:236–42
17. Brower AVZ. 1994. Phylogeny of *Heliconius* butterflies inferred from mitochondrial DNA sequences (Lepidoptera: Nymphalinae). *Mol. Phylogenet. Evol.* 3: 159–74
18. Brower AVZ. 1995. Locomotor mimicry in butterflies? A critical review of the evidence. *Philos. Trans. R. Soc. London Ser. B* 347:413–25
19. Brower AVZ. 1996. Parallel race formation and the evolution of mimicry in *Heliconius* butterflies: a phylogenetic hypothesis from mitochondrial DNA sequences. *Evolution* 50:195–221
20. Brower LP. 1984. Chemical defence in butterflies. In *The Biology of Butterflies*, ed. RI Vane-Wright, PR Ackery, pp. 109–34. London: Academic
21. Brower LP. 1988. Preface. *Am. Nat.* 131(Suppl.):S1–S3
22. Brown KS. 1973. *A Portfolio of Neotropical Lepidopterology.* Rio de Janeiro, Brazil: Privately published. 28 pp.
23. Brown KS. 1976. An illustrated key to the silvaniform *Heliconius* (Lepidoptera: Nymphalidae) with descriptions of new subspecies. *Trans. Am. Entomol. Soc.* 102: 373–484
24. Brown KS. 1979. *Ecologia geográfica e*

evolução nas florestas neotropicais. Livre dc Doccncia. Campinas, Brazil: Univ. Estadual de Campinas

25. Brown KS. 1988. Mimicry, aposematism and crypsis in neotropical Lepidoptera: the importance of dual signals. *Bull. Soc. Zool. France* 113:83–101
26. Brown KS, Benson WW. 1974. Adaptive polymorphism associated with multiple Müllerian mimicry in *Heliconius numata* (Lepid.: Nymph.). *Biotropica* 6: 205–28
27. Brown KS, Sheppard PM, Turner JRG. 1974. Quaternary refugia in tropical America: evidence from race formation in *Heliconius* butterflies. *Proc. R. Soc. London Ser. B* 187:369–78
28. Burd M. 1994. Butterfly wing colour patterns and flying heights in the seasonally wet forest of Barro Colorado Island, Panama. *J. Trop. Ecol.* 10:601–10
29. Calvert WH, Hedrick LE, Brower LP. 1979. Mortality of the monarch butterfly (*Danaus plexippus* L.): avian predation at five overwintering sites in Mexico. *Science* 204:847–51
30. Carpenter GDH. 1939. Birds as enemies of butterflies, with special reference to mimicry. *Proc. VII Int. Kongr. Entomol., Berlin* 1938:1061–74
31. Carpenter GDH. 1941. The relative frequency of beakmarks on butterflies of different edibility to birds. *Proc. Zool. Soc. London Ser. A* 3:223–31
32. Carpenter GDH, Ford EB. 1933. *Mimicry*. London: Methuen
33. Carroll SB, Gates J, Keys DN, Paddock SW, Panganiban GEF, et al. 1994. Pattern formation and eyespot determination in butterfly wings. *Science* 265:109–14
34. Chai P. 1986. Field observations and feeding experiments on the responses of rufous-tailed jacamars (*Galbula ruficauda*) to free-flying butterflies in a tropical rainforest. *Biol. J. Linn. Soc.* 29:166–89
35. Chai P. 1996. Butterfly visual characteristics and ontogeny of responses to butterflies by a specialized bird. *Biol. J. Linn. Soc.* 59:37–67
36. Charlesworth B. 1994. The genetics of adaptation: lessons from mimicry. *Am. Nat.* 144:839–47
37. Charlesworth D, Charlesworth B. 1975. Theoretical genetics of Batesian mimicry. II. Evolution of supergenes. *J. Theor. Biol.* 55:305–24
38. Clarke CA, Sheppard PM. 1971. Further studies on the genetics of the mimetic butterfly *Papilio memnon*. *Philos. Trans. R. Soc. London Ser. B* 263:35–70
39. Clarke CA, Sheppard PM, Thornton IWB. 1968. The genetics of the mimetic butterfly *Papilio memnon*. *Philos. Trans. R. Soc. London Ser. B* 254:37–89
40. Cott HB. 1940. *Adaptive Coloration in Animals*. London: Methuen
41. Coyne JA, Barton NH, Turelli M. 1997. Perspective: a critique of Sewall Wright's shifting balance theory of evolution. *Evolution* 51:643–71
42. Darwin C. 1871. *The Descent of Man, and Selection in Relation to Sex*. London: John Murray. 2nd ed.
43. Dawkins R. 1976. *The Selfish Gene*. Oxford, UK: Oxford Univ. Press
44. DeVries PJ, Lande R. 1999. Associations of co-mimetic ithomiine butterflies on small spatial and temporal scales in a neotropical rainforest. *Biol. J. Linn. Soc.* In press
45. Edmunds M. 1974. *Defence in Animals*. Harlow, Essex: Longmans
46. Edmunds M, Golding YC. 1999. Diversity in mimicry. *Trends Ecol. Evol.* 14:150
47. Endler JA. 1988. Frequency-dependent predation, crypsis, and aposematic coloration. *Philos. Trans. R. Soc. London Ser. B* 319:459–72
48. Enquist M, Arak A. 1993. Selection of exaggerated male traits by female aesthetic senses. *Nature* 361:446–48
49. Fisher RA. 1930. *The Genetical Theory of Natural Selection*. Oxford: Clarendon

50. Fox RM. 1955. On subspecies. *Syst. Zool.* 4:93–95
51. French V. 1997. Pattern formation in colour on butterfly wings. *Curr. Opin. Genet. Dev.* 7:524–29
52. Gamberale G, Sillén–Tullberg B. 1996. Evidence for a peak-shift in predator generalization among aposematic prey. *Proc. R. Soc. London Ser. B* 263:1329–34
53. Gavrilets S. 1997. Evolution and speciation on holey adaptive landscapes. *Trends Ecol. Evol.* 12:307–12
54. Gavrilets S, Hastings A. 1998. Coevolutionary chase in two-species systems with applications to mimicry. *J. Theor. Biol.* 415–27
55. Gilbert LE. 1983. Coevolution and mimicry. In *Coevolution*, ed. DJ Futuyma, M Slatkin, p. 263–81. Sunderland, MA: Sinauer
56. Goldschmidt RB. 1945. Mimetic polymorphism, a controversial chapter of Darwinism. *Q. Rev. Biol.* 20:147–64; 205–30
57. Gordon IJ, Smith DAS. 1999. Diversity in mimicry. *Trends Ecol. Evol.* 14:150–51
58. Guilford T. 1985. Is kin selection involved in the evolution of warning coloration? *Oikos* 45:31–36
59. Guilford T. 1990. The evolution of aposematism. In *Insect Defenses. Adaptive Mechanisms and Strategies of Prey and Predators*, ed. DL Evans, JO Schmidt, pp. 23–61. New York: State Univ. New York Press
60. Guilford T, Dawkins MS. 1991. Receiver psychology and the evolution of animal signals. *Anim. Behav.* 42:1–14
61. Guilford T, Dawkins MS. 1993. Receiver psychology and the design of animal signals. *Trends Neurosci.* 16:430–36
62. Guilford T, Nicol C, Rothschild M, Moore B. 1987. The biological roles of pyrazines: evidence for a warning odour function. *Biol. J. Linn. Soc.* 31:113–28
63. Hamilton WD. 1964. The genetical evolution of social behaviour. *Theor. Biol.* 7:1–52
64. Hamilton WD. 1971. Geometry for the selfish herd. *J. Theor. Biol.* 31:295–311
65. Hartl DL, Clark AG. 1989. *Principles of Population Genetics*. Sunderland, MA: Sinauer. 2nd ed.
66. Harvey PH, Bull JJ, Pemberton M, Paxton RJ. 1982. The evolution of aposematic coloration in distasteful prey: a family model. *Am. Nat.* 119:710–19
67. Harvey PH, Greenwood PJ. 1978. Antipredator defence strategies: some evolutionary problems. In *Behavioural Ecology*, ed. JR Krebs, NB Davies, pp. 129–51. Oxford: Blackwell Sci.
68. Harvey PH, Paxton RJ. 1981. The evolution of aposematic coloration. *Oikos* 37: 391–93
69. Hewitt GM. 1988. Hybrid zones—natural laboratories for evolutionary studies. *Trends Ecol. Evol.* 3:158–67
70. Holzkamp G, Nahrstedt A. 1994. Biosynthesis of cyanogenic glycosides in the Lepidoptera—incorporation of [U-C-14]-2-methylpropanealdoxime, 2S-[U-C-14]-methylbutanealdoxime and D,L-[U-C-14]-N-hydroxyisoleucine into linamarin and lotaustralin by the larvae of *Zygaena trifolii*. *Insect Biochem. Mol. Biol.* 24: 161–68
71. Huheey JE. 1976. Studies in warning coloration and mimicry. VII. Evolutionary consequences of Batesian–Müllerian spectrum: a model for Müllerian mimicry. *Evolution* 30:86–93
72. Huheey JE. 1988. Mathematical models of mimicry. *Am. Nat.* 131(Suppl.):S22–41
73. Järvi T, Sillén-Tullberg B, Wiklund C. 1981. The cost of being aposematic. An experimental study of predation on larvae of *Papilio machaon* by the great tit *Parus major*. *Oikos* 36:267–72
74. Jävi T, Sillén-Tullberg B, Wiklund C. 1981. Individual versus kin selection for aposematic coloration. A reply to Harvey and Paxton. *Oikos* 37:393–95
75. Jeffords MR, Sternberg JG, Waldbauer GP. 1979. Batesian mimicry: field demonstra-

tion of the survival value of pipevine swallowtail and monarch color patterns. *Evolution* 33:275–86

76. Jones DA, Parsons J, Rothschild M. 1962. Release of hydrocyanic acid from crushed tissues of all stages in the life cycle of species of the Zygaeninae (Lepidoptera). *Nature* 193:52–63
77. Jones FM. 1932. Insect coloration and the relative acceptability of insects to birds. *Trans. R. Entomol. Soc. London* 80:345–85
78. Joron M, Mallet J. 1998. Diversity in mimicry: paradox or paradigm. *Trends Ecol. Evol.* 13:461–66
79. Kamo M, Kubo T, Iwasa Y. 1998. Neural network for female mate preference, trained by a genetic algorithm. *Philos. Trans. R. Soc. London Ser. B.* In press
80. Kapan D. 1998. *Divergent natural selection and Müllerian mimicry in polymorphic* Heliconius cydno (*Lepidoptera: Nymphalidae*). PhD diss. Univ. B-C
81. Leimar O, Enquist M, Sillén-Tullberg B. 1986. Evolutionary stability of aposematic coloration and prey unprofitability: a theoretical analysis. *Am. Nat.* 128:469–90
82. Leimar O, Tuomi J. 1998. Synergistic selection and graded traits. *Evol. Ecol.* 12: 59–71
83. Linares M. 1997. Origin of neotropical mimetic diversity from a three-way hybrid zone of *Heliconius cydno* butterflies. In *Tropical Diversity and Systematics*, ed. H Ulrich, pp. 93–108. *Proc. Int. Symp. Biodiversity Syst. Trop. Ecosyst.* Bonn, 1994. Bonn: Zool. Forsch. inst. Mus. Alex. Koenig
84. Lindström L, Alatalo RV, Mappes J, Riipi M, Vertainen L. 1999. Can aposematic signals evolve by gradual change? *Nature* 397:249–51

84a. MacDougall A, Dawkins MS. 1998. Predator discrimination error and the benefits of Müllerian mimicry. *Anim. Behav.* 55:1281–88

85. Majerus MEN. 1998. *Melanism. Evolution in Action.* Oxford: Oxford Univ. Press. xiii + 338 pp.
86. Malcolm SB. 1990. Mimicry: status of a classical evolutionary paradigm. *Trends Ecol. Evol.* 5:57–62
87. Mallet J. 1986. Hybrid zones in *Heliconius* butterflies in Panama, and the stability and movement of warning colour clines. *Heredity* 56:191–202

87a. Mallet J. 1989. The genetics of warning colour in Peruvian hybrid zones of *Heliconius erato* and *H. melpomene*. *Proc. R. Soc. London Ser. B* 236:163–85

88. Mallet J. 1991. Variations on a theme? *Nature* 354:368 (Review of HF Nijhout. 1991. *The Development and Evolution of Butterfly Wing Patterns*. Washington, DC: Smithsonian Inst.
89. Mallet J. 1993. Speciation, raciation, and color pattern evolution in *Heliconius* butterflies: evidence from hybrid zones. In *Hybrid Zones and the Evolutionary Process*, ed. RG Harrison, pp. 226–60. New York: Oxford Univ. Press
90. Mallet J. 1999. Mimicry references. http://abacus.gene.ucl.ac.uk/Jim/Mim/mimicry.htm.
91. Mallet J, Barton N. 1989. Inference from clines stabilized by frequency-dependent selection. *Genetics* 122:967–76
92. Mallet J, Barton NH. 1989. Strong natural selection in a warning color hybrid zone. *Evolution* 43:421–31
93. Mallet J, Barton N, Lamas G, Santisteban J, Muedas M, Eeley H. 1990. Estimates of selection and gene flow from measures of cline width and linkage disequilibrium in *Heliconius* hybrid zones. *Genetics* 124:921–36
94. Mallet J, Gilbert LE. 1995. Why are there so many mimicry rings? Correlations between habitat, behaviour and mimicry in *Heliconius* butterflies. *Biol. J. Linn. Soc.* 55:159–80
95. Mallet J, Jiggins CD, McMillan WO.

1996. Mimicry meets the mitochondrion. *Curr. Biol.* 6:937–40

96. Mallet J, McMillan WO, Jiggins CD. 1998. Mimicry and warning color at the boundary between races and species. In *Endless Forms: Species and Speciation*, ed. S Berlocher, D Howard, pp. 390–403. New York: Oxford Univ. Press
97. Mallet J, Singer MC. 1987. Individual selection, kin selection, and the shifting balance in the evolution of warning colours: the evidence from butterflies. *Biol. J. Linn. Soc.* 32:337–50
98. Mallet JLB, Turner JRG. 1998. Biotic drift or the shifting balance—Did forest islands drive the diversity of warningly coloured butterflies? In *Evolution on Islands*, ed. PR Grant, pp. 262–80. Oxford: Oxford Univ. Press
99. Maynard Smith J. 1998. *Evolutionary Genetics*. Oxford, UK: Oxford Univ. Press
100. Mayr E. 1963. *Animal Species and Evolution*. Cambridge, MA: Harvard Univ. Press
101. McMillan WO, Jiggins CD, Mallet J. 1997. What initiates speciation in passion-vine butterflies? *Proc. Natl. Acad. Sci. USA* 94:8628–33
102. Medina MC, Robbins RK, Lamas G. 1996. Vertical stratification of flight by ithomiine butterflies (Lepidoptera: Nymphalidae) at Pakitza, Manu National Park, Perú. In *Manu. The Biodiversity of Southeastern Peru*, ed. DE Wilson, A Sandoval, pp. 211–16. Washington, DC: Smithsonian Inst.
103. Müller F. 1879. *Ituna* and *Thyridia*; a remarkable case of mimicry in butterflies. *Trans. Entomol. Soc. London* 1879:xx–xxix
104. Nahrstedt A, Davis RH. 1983. Occurrence, variation and biosynthesis of the cyanogenic glucosides linamarin and lotaustralin in species of the Heliconiini (Insecta: Lepidoptera). *Comp. Biochem. Physiol.* 75B:65–73
105. Nicholson AJ. 1927. A new theory of mimicry in insects. *Aust. Zool.* 5:10–104
106. Nijhout HF. 1991. *The Development and Evolution of Butterfly Wing Patterns*. Washington, DC: Smithsonian Inst.
107. Nijhout HF, Wray GA. 1988. Homologies in the colour patterns of the genus *Heliconius* (Lepidoptera: Nymphalidae). *Biol. J. Linn. Soc.* 33:345–65
108. Nijhout HF, Wray GA, Gilbert LE. 1990. An analysis of the phenotypic effects of certain colour pattern genes in *Heliconius* (Lepidoptera: Nymphalidae). *Biol. J. Linn. Soc.* 40:357–72
109. Nur U. 1970. Evolutionary rates of models and mimics in Batesian mimicry. *Am. Nat.* 104:477–86
110. O'Donald P. 1980. *Genetic Models of Sexual Selection*. Cambridge, UK: Cambridge Univ. Press
111. Ohsaki N. 1995. Preferential predation of female butterflies and the evolution of Batesian mimicry. *Nature* 378:173–75
112. Orr HA. 1998. The population genetics of adaptation: the distribution of factors fixed during adaptive evolution. *Evolution* 52:935–49
113. Orr HA, Coyne JA. 1992. The genetics of adaptation: a reassessment. *Am. Nat.* 140: 725–42
114. Owen DF, Smith DAS, Gordon IJ, Owiny AM. 1994. Polymorphic Müllerian mimicry in a group of African butterflies: a reassessment of the relationship between *Danaus chrysippus, Acraea encedon* and *Acraea encedana* (Lepidoptera: Nymphalidae). *J. Zool.* 232:93–108
115. Owen RE, Owen ARG. 1984. Mathematical paradigms for mimicry: recurrent sampling. *J. Theor. Biol.* 109:217–47
116. Papageorgis C. 1975. Mimicry in neotropical butterflies. *Am. Sci.* 63:522–32
117. Parsons M. 1984. Life histories of *Taenaris* (Nymphalidae) from Papua New Guinea. *J. Lepid. Soc.* 38:69–84
118. Parsons M. 1991. Butterflies of the

Bololo–Wau Valley. *Handb. Wau Ecol. Insti.* 12. Honolulu: Bishop Mus.
119. Pinheiro CEG. 1996. Palatability and escaping ability in neotropical butterflies: tests with wild kingbirds (*Tyrannus melancholicus*). *Biol. J. Linn. Soc.* 59: 351–65
120. Pinheiro CEG. 1997. *Unpalatability, mimicry and escaping ability in neotropical butterflies: experiments with wild predators*. DPhil. thesis. Univ. Oxford
121. Plowright RC, Owen RE. 1980. The evolutionary significance of bumble bee color patterns: a mimetic interpretation. *Evolution* 34:622–37
122. Poulton EB. 1890. *The Colours of Animals*. London: Trübner
123. Punnett RC. 1915. *Mimicry in Butterflies*. Cambridge, UK: Camb. Univ. Press
124. Rescorla RA, Wagner AR. 1972. A theory of Pavlovian conditioning: variations in the effectiveness of reinforcement and non-reinforcement. In *Classical Conditioning II: Current Research and Theory*, ed. A Black, WF Prokasy, pp. 64–99. New York: Appleton-Century-Crofts
125. Rettenmeyer CW. 1970. Insect mimicry. *Annu. Rev. Entomol.* 15:43–74
126. Ridley M. 1996. *Evolution*. Oxford, UK: Blackwell Sci.
127. Ritland DB. 1990. Localized interspecific hybridization between mimetic *Limenitis* butterflies (Nymphalidae) in Florida. *J. Lepid. Soc.* 44:163–73
128. Robbins RK. 1980. The lycaenid "false head" hypothesis: historical review and quantitative analysis. *J. Lepid. Soc.* 34: 194–208
129. Rothschild M. 1985. British aposematic Lepidoptera. In *The Moths and Butterflies of Great Britain and Ireland*, ed. J Heath, AM Emmet, 2. Cossidae–Heliodinidae, pp. 9–62. Great Horkesley, Essex: Harley Books
130. Rothschild M, Moore BP, Brown WV. 1984. Pyrazines as warning odour components in the monarch butterfly, *Danaus plexippus*, and in moths of the genus *Zygaena* and *Amata* (Lepidoptera). *Biol. J. Linn. Soc.* 23:375–80
131. Ryan MJ, Rand AS. 1993. Sexual selection and signal evolution: the ghost of biases past. *Philos. Trans. R. Soc. London Ser. B* 340:187–96
132. Sargent TD. 1995. On the relative acceptabilities of local butterflies and moths to local birds. *J. Lepid. Soc.* 49:148–62
133. Servedio MR. 1998. *Preferences, signals, and evolution: theoretical studies of mate choice copying, reinforcement, and aposematic coloration*. PhD. diss. Univ. Texas at Austin
134. Sheppard PM. 1958. *Natural Selection and Heredity*. London: Hutchinson
135. Sheppard PM. 1959. The evolution of mimicry: a problem in ecology and genetics. *Cold Spring Harbor Symp. Quant. Biol.* 24:131–40
136. Sheppard PM, Turner JRG. 1977. The existence of Müllerian mimicry. *Evolution* 31:452–53
137. Sheppard PM, Turner JRG, Brown KS, Benson WW, Singer MC. 1985. Genetics and the evolution of muellerian mimicry in *Heliconius* butterflies. *Philos. Trans. R. Soc. London Ser. B* 308:433–613
138. Silberglied RE. 1984. Visual communication and sexual selection among butterflies. In *The Biology of Butterflies*, ed. RI Vane-Wright, PR Ackery, pp. 207–23. London: Academic
139. Sillén-Tullberg B. 1988. Evolution of gregariousness in aposematic butterfly larvae: a phylogenetic analysis. *Evolution* 42:293–305
140. Sillén-Tullberg B, Bryant EH. 1983. The evolution of aposematic coloration in distateful prey: an individual selection model. *Evolution* 37:993–1000
141. Sillén-Tullberg B, Hunter AF. 1996. Evolution of larval gregariousness in relation to repellant defences and warning coloration in tree-feeding Macrolepidoptera: a phylogenetic analysis based on

independent contrasts. *Biol. J. Linn. Soc.* 57:253–76
142. Smiley JT. 1978. *The host plant ecology of* Heliconius *butterflies in Northeastern Costa Rica*. PhD diss. Univ. Texas at Austin
143. Smith DAS. 1976. Phenotypic diversity, mimicry and natural selection in the African butterfly *Hypolimnas misippus* L. (Lepidoptera: Nymphalidae). *Biol. J. Linn. Soc.* 8:183–204
144. Smith DAS. 1980. Heterosis, epistasis and linkage disequilibrium in a wild population of the polymorphic butterfly *Danaus chrysippus*. *Zool. J. Linn. Soc.* 69:87–110
145. Smith DAS. 1981. Heterozygous advantage expressed through sexual selection in a polymorphic African butterfly. *Nature* 289:174–75
146. Smith DAS, Gordon IJ, Depew LA, Owen DF. 1998. Genetics of the butterfly *Danaus plexippus* (L.) in a broad hybrid zone, with special reference to sex ratio, polymorphism and intragenomic conflict. *Biol. J. Linn. Soc.* 65:1–40
147. Smith DAS, Owen DF, Gordon IJ, Lowis NK. 1997. The butterfly *Danaus chrysippus* (L.) in East Africa: polymorphism and morph–ratio clines within a complex, extensive and dynamic hybrid zone. *Zool. J. Linn. Soc.* 120:51–78
147a. Speed, MP. 1993. *Mimicry and the psychology of predation*. PhD thesis. Univ. Leeds
148. Speed MP. 1993. Muellerian mimicry and the psychology of predation. *Anim. Behav.* 45:571–80
149. Speed MP. 1999. Robot predators in virtual ecologies: the importancc of memory in mimicry studies. *Anim. Behav.* 57: 203–13
150. Speed MP. 1999. Robot predators, receiver psychology, and doubts about Mullerian mimicry: comments on MacDougall and Dawkins. *Anim. Behav.* In press
151. Speed MP, Turner JRG. 1999. Learning and memory in mimicry: II. Do we understand the mimicry spectrum? *Biol. J. Linn. Soc.* In press
152. Srygley RB. 1994. Locomotor mimicry in butterflies? The associations of positions of centres of mass among groups of mimetic, unprofitable prey. *Phil. Trans. Roy. Soc. London Ser. B* 343:145–55
153. Srygley RB. 1998. Locomotor mimicry in Heliconius butterflies: contrast analyses of flight morphology and kinematics. *Philos. Trans. R. Soc. London Ser. B* 353:1–13
154. Srygley RB, Chai P. 1990. Flight morphology of neotropical butterflies: palatability and the distribution of mass to the thorax and abdomen. *Oecologia (Berlin)* 84:491–99
155. Sword GA. 1999. Density-dependent warning coloration. *Nature* 397:217
156. Tinbergen N. 1951. *The Study of Instinct*. Oxford: Oxford Univ. Press
157. Treisman M. 1975. Predation and the evolution of gregariousness. I. Models for concealment and evasion. *Anim. Behav.* 23:779–800
158. Tullrot A, Sundberg P. 1991. The conspicuous nudibranch *Polycera quadrilineata*: aposematic coloration and individual selection. *Anim. Behav.* 41:175–76
159. Turner JRG. 1971. Studies of Müllerian mimicry and its evolution in burnet moths and heliconid butterflies. In *Ecological Genetics and Evolution*, ed. ER Creed, pp. 224–60. Oxford, UK: Blackwell Sci.
160. Turner JRG. 1975. Communal roosting in relation to warning colour in two heliconiine butterflies (Nymphalidae). *J. Lepid. Soc.* 29:221–26
161. Turner JRG. 1977. Butterfly mimicry–the genetical evolution of an adaptation. *Evol. Biol* 10:163–206
162. Turner JRG. 1978. Why male butterflies are non-mimetic: natural selection, sex-

ual selection, group selection, modification and sieving. *Biol. J. Linn. Soc.* 10: 385–432

163. Turner JRG. 1979. Oscillation of gene frequencies in Batesian mimics: a correction. *Biol J. Linn. Soc.* 11:397–98
164. Turner JRG. 1984. Mimicry: the palatability spectrum and its consequences. In *The Biology of Butterflies*, ed. RI Vane-Wright, PR Ackery, pp. 141–61. London: Academic
165. Turner JRG. 1995. Mimicry as a model for coevolution. In *Biodiversity and Evolution*, ed. R Arai, M Kato, Y Doi, pp. 131–50. Tokyo: Natl. Sci. Mus. Found.
166. Turner JRG, Kearney EP, Exton LS. 1984. Mimicry and the Monte Carlo predator: the palatability spectrum and the origins of mimicry. *Biol. J. Linn. Soc.* 23:247–68
167. Turner JRG, Mallet JLB. 1996. Did forest islands drive the diversity of warningly coloured butterflies? Biotic drift and the shifting balance. *Philos. Trans. R. Soc. London Ser. B* 351:835–45

167a. Turner JRG, Speed MP. 1996. Learning and memory in mimicry. I. Simulations of laboratory experiments. *Philos. Trans. R. Soc. London. Ser. B* 351:1157–70

168. Vane-Wright RI. 1976. A unified classification of mimetic resemblances. *Biol. J. Linn. Soc.* 8:25–56
169. Vane-Wright RI. 1984. The role of pseudosexual selection in the evolution of butterfly colour patterns. In *The Biology of Butterflies*, ed. RI Vane-Wright, PR Ackery, pp. 251–53. London: Academic
170. Waldbauer GP. 1988. Aposematism and batesian mimicry. *Evol. Biol.* 22:227–59
171. Waldbauer GP, Sheldon JK. 1971. Phenological relationships of some aculeate Hymenoptera, their dipteran mimics, and insectivorous birds. *Evolution* 25: 371–82
172. Waldbauer GP, Sternburg JG. 1975. Saturniid moths as mimics: an alternative interpretation of attempts to demonstrate mimetic advantage in nature. *Evolution* 29:650–58
173. Wallace AR. 1854. On the habits of the butterflies of the Amazon Valley. *Trans. Entomol. Soc. London* 2:253–64
174. Wallace AR. 1865. On the phenomena of variation and geographical distribution as illustrated by the Papilionidae of the Malayan region. *Trans. Linn. Soc. Lund* 25:1–71
175. Wallace AR. 1878. *Tropical Nature and Other Essays*. London: MacMillan
176. Wickler W. 1968. *Mimicry in Plants and Animals*. New York: McGraw Hill
177. Wiklund C, Järvi T. 1982. Survival of distasteful insects after being attacked by naïve birds: a reappraisal of the theory of aposematic coloration evolving through individual selection. *Evolution* 36:998–1002
178. Wright S. 1932. The roles of mutation, inbreeding, crossbreeding and selection in evolution. *Proc. 11th, Int. Congr. Genet., The Hague* 1:356–66
179. Wright S. 1982. Character change, speciation, and the higher taxa. *Evolution* 36:427–43
180. Wright S. 1982. The shifting balance theory and macroevolution. *Annu. Rev. Genet.* 16:1–19
181. Yachi S, Higashi M. 1998. How can warning signals evolve in the first place? *Nature* 394:882–84

Annu. Rev. Ecol. Syst. 1999. 30:235–56

CONSEQUENCES OF EVOLVING WITH BACTERIAL SYMBIONTS: Insights from the Squid-Vibrio Associations

Margaret J. McFall-Ngai
Kewalo Marine Laboratory, Pacific Biomedical Research Center, University of Hawaii, Honolulu, Hawaii 96813; e-mail: mcfallng@hawaii.edu

Key Words symbiosis, *Euprymna scolopes*, *Vibrio fischeri*, sepiolid

■ **Abstract** The squid-vibrio light-organ symbioses, which have been under investigation for just over 10 years, offer the opportunity to decipher aspects of the dynamics of stable associations between animals and bacteria. The two best-studied partners, the Hawaiian sepiolid squid *Euprymna scolopes* and the marine luminous bacterium *Vibrio fischeri*, engage in the most common type of animal-bacterial association, i.e., between extracellular, gram-negative bacteria and animal epithelia. Similar to most such symbioses, the squid-vibrio relationship begins anew each generation when the host animal acquires the symbiont from the surrounding environment. To establish a specific association, mechanisms have evolved to ensure recognition between the host and symbiont and the exclusion of other potential partnerships. Once the association has been established, the bacteria induce significant morphological changes in the host that result in a transition of the light organ from a form associated with initiation of the symbiosis to one characteristic of the mature, functional relationship.

INTRODUCTION

Metazoans evolved in environments rich in bacteria and, as a part of this process, formed alliances with specific subsets of these microbes that persist from generation to generation. The partnerships that resulted allowed animals to take advantage of the unique and diverse metabolic capabilities of their prokaryotic partners. Thus, animals can be accurately viewed as complex communities comprised of a principal multicellular, eukaryotic cell type and an array of microbial species. It is likely that when species arise, a specific microbiota arises with them (37); and, similarly, when an animal host becomes extinct, it is likely that some subset of the microbial community with which they associate will also not persist (73).

Despite the ubiquity of animal-microbial associations, rarely have these communities been ecologically modeled, nor has the impact of these associations been extensively studied by evolutionary biologists (37). The lack of knowledge is likely due to the complexity of these associations, which impose a sort of "Russian doll" nature to the system. A less complex biological model, while not comprehensive in scope, can provide insight into the mechanisms by which multicellular, eukaryotic organisms initiate and maintain stable associations with microbes. To date, very few such models have proven amenable to in-depth study of these questions; either they are rare or their host and symbiont are not culturable independently of one another (40). The notable exception to this trend has been in the study of the plant-bacterial associations between legumes and root-nodulating, nitrogen-fixing rhizobia, which have been the principal model for prokaryotic-eukaryotic symbioses (54, 76). Over 100 years of research with these associations, at all levels of analysis, from ecology to molecular biology, have provided the biological community with a window into the complexity of the interactions between bacteria and multicellular organisms.

With the emergence of new model systems (29, 41, 66) and the application of molecular, genetic and biochemical approaches, this frontier area of research has shown promise over the past few years (40). One model that is part of this nascent effort is the association between sepiolid squids and particular species of marine vibrios (41, 44, 65). In these relationships, the bacteria form a persistent, extracellular relationship with host epithelial tissue, the most common type of association between prokaryotes and animals. The two-partner nature of these symbioses offers opportunities for the study of animal-bacterial symbioses similar to those that have been available in plant-microbial symbioses, and it offers the opportunity to determine the similarities and differences between animal and plant associations with microbes. In addition, the study of how symbioses may interface with features unique to animals, such as cell-mediated immunity and animal developmental pathways, are possible in this system. The aim of this review is to provide an overview of progress of the study of the squid-vibrio association to date.

THE PARTNERS

Bobtail Squids

The sepiolid, or bobtail, squids are small (mantle length in sexually mature adults usually averaging between 1 and 8 cm) mollusks whose taxonomic position within the coleoid cephalopods is presently controversial (10, 11). Most systematists who study the relationships of cephalopods consider the sepiids, or cuttlefish, the sister group of this family. The family Sepiolidae contains 14 genera and between 50 and 60 species in 3 subfamilies: the Sepiolinae, the Rossiinae, and the Heteroteuthinae (53, 79). They are broadly distributed from tropical to boreal habitats. Whereas the species in the Heteroteuthinae are principally deep pelagic, the species in the Sepiolinae and Rossiinae are benthic, distributed from the shallow subtidal to deeper areas of the shelf and bathyl benthic regions of the ocean.

Over half of the 14 sepiolid genera are characterized by an internal, light-emitting organ associated with the ink sac (27, 53, 61). All members of the subfamily

Heteroteuthinae have light organs, although the luminescence is, in some cases autogenic rather than bacterial in origin (14). Some of the genera within the Sepiolinae and Rossiinae have bacterial light organs, but the taxonomic distribution within these subfamilies suggests that either this character arose independently several times or was lost several times (10).

Within the Sepiolinae, two genera have members that are readily accessible for the study of bacterial light-organ symbioses: the Indo-Pacific genus *Euprymna* and the Atlantic/Mediterranean genus *Sepiola* (44). Presently, the array of species within these two genera is being studied as a mechanism by which to understand the evolution of animal-bacterial symbioses (see below). However, the principal focus for studies of the dynamics of the symbiosis itself have focused on the Hawaiian sepiolid species *Euprymna scolopes* (41).

Euprymna scolopes is endemic to the Hawaiian archipelago (6, 70), with populations occurring from Midway to Maui; no reports of this species have been madefrom the southernmost, and youngest, island of the archipelago, Hawaii. Some evidence exists for the divergence of subspecies within *E. scolopes*; i.e., morphological and molecular sequence data suggest reproductively isolated populations exist on either side of the island of Oahu (Kimbell, MJ McFall-Ngai and Roderick, unpublished results).

Luminous Bacterial Symbionts

Most luminous bacteria are facultative anaerobes belonging to the gamma-proteobacteria group of gram-negative bacteria. Four culturable species form specific associations with light organs of marine fishes and squids: *Vibrio fischeri*, *V. logei*, *Photobacterium phosphoreum*, and *P. leiognathi* (19, 26, 45). Populations of these four species of luminous bacteria can be found in a wide variety of habitats; in addition to being the microbes in fish and squid light organs, they occur as components of the enteric microbiota of marine animals, are opportunistic saprophytes and pathogens, and are relatively common members of the bacterioplankton. These planktonic populations of the luminous bacteria provide the inoculum for the light organs of juvenile squids and fishes that have light organs. Squid species are known to harbor three of these species, *V. fischeri* and *V. logei* in the sepiolid squids with light organs (19) and *P. leiognathi* in the loliginid squid species with light organs (21).

Until recently, these light-organ associations were thought to be exclusive, i.e., one particular luminous bacterial species always found in association with a specific squid or fish host. This pattern does hold true for all *Euprymna* species studied thus far; i.e., all identified luminous bacterial isolates from *Euprymna* spp. light organs have been *V. fischeri*, although the light organs have not been sampled at all times of year. However, studies of the Mediterranean sepiolids indicate that the light organs of an individual squid can harbor mixed cultures of *V. fischeri* and *V. logei* (19). *Vibrio logei* is more psychrophilic (i.e., it grows optimally at relatively low temperatures) (4) and may be a better competitor in the light organs in colder months or in deeper populations. To determine the dynamics of these two symbionts in the host *Sepiola* spp., a more detailed study must be undertaken.

THE HOST LIFE CYCLE AND THE MAINTENANCE OF BREEDING COLONIES UNDER LABORATORY CONDITIONS

The life cycles of the sepiolid squids have been studied extensively, most likely because their life histories can be examined under laboratory conditions (3, 25). Unlike many cephalopods, which are semelparous, i.e., having a single reproductive effort (mating once, depositing their eggs, and then dying), the sepiolids are iteroparous (22). For example, the adult females of the Hawaiian sepiolid *E. scolopes* may mate numerous times and, over their life time, a given female will typically lay several clutches on hard substrates (44). Each clutch contains 50–400 eggs, each of which is about 2 mm in diameter. No direct parental care occurs during embryogenesis; immediately following the laying of a clutch, the female covers the clutch with sand and leaves the embryos to develop and hatch on their own.

Under laboratory conditions, we have found that female *E. scolopes* will not lay eggs unless they are frequently mated, although the females may be carrying many spermatophores from previous matings (MJ McFall-Ngai, personal observation). If the animals are fed to satiation and mated once per week, a breeding colony of between 10 to 12 females and 3 to 4 males produces 60,000–100,000 eggs per year. The rate of hatching success approaches 100% if the eggs are kept in clean, well-aerated seawater (44).

Following an embryonic period that varies from 17 to 25 days, the length of which depends on the environmental conditions under which the embryos develop (temperature, aeration, etc), the juvenile *E. scolopes* (averaging 1.7 mm in mantle length) hatch with internal yolk reserves that furnish nutrition over the next few days. These newly hatched juveniles have a gross morphology that is very similar to that of the adult, i.e., like most cephalopod species, *E. scolopes* does not have larval stages in its life history. Individuals of this species reach sexual maturity as early as 60 days posthatch (25), and their entire life span is thought to be about one year (70), although only fragmentary data are available to support this assumption.

The life histories of *Sepiola* spp. have also been studied extensively, and the details of their life cycle differ in several aspects from that of *E. scolopes* (10, 22). Although the adults are similar in size and gross morphology to *E. scolopes, Sepiola* spp. lay smaller clutches (between 1 and 200 eggs) of much larger eggs, each 4–7 mm in diameter (22). The embryonic period of *Sepiola* species is two to three times that of *E. scolopes*, and when the juveniles hatch, they are significantly larger, averaging 4–6 mm in mantle length (MJ McFall-Ngai, personal observation). These differences in the developmental program between *E. scolopes* and *Sepiola* spp. have proven valuable in advancing an understanding of the dynamics of squid-vibrio associations, particularly the role of symbionts in development of light organs (see below).

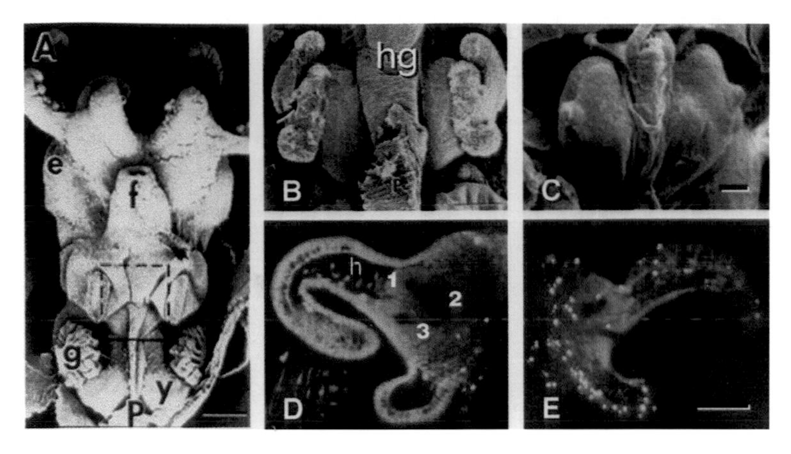

Figure 3 The juvenile *E. scolopes* light organ. (*A*) SEM of ventral dissection, showing the eye (*e*), funnel (*f*), gill (*g*), and yolk sac (*y*). The light organ is under the dotted square. Bar, 200 μm. (*B*) SEM of the ventral surface of a hatchling organ revealing the dense ciliated, microvillous field (CMS). Pores on the surface of the organ (arrow), lead to the internal crypts that will house the bacterial symbionts, hindgut (*hg*), posterior (*p*). (*C*) SEM of the light organ of a 4-d symbiotic animal showing complete regression of the CMS. Bar, 50 μm. (*D*) A confocal image of an acridine-orange (AO) stained light organ of a 14-h aposymbiotic animal revealing that the CMS is a single layer of epithelial cells over a blood sinus. Hemocytes can be seen freely floating in the sinus of the CMS appendages (arrowhead). The three pores (1, 2, 3) lead to the crypt spaces. (*E*) 14-h symbiotic organ under same conditions as (*D*). The condensed chromatin of apoptotic cells can be seen as bright spots of AO staining. Bar, 50 μm.

THE INITIATION OF THE SYMBIOSIS

Embryonic Development of the Host Organ

During embryogenesis the host squid develops an incipient light organ with a morphology that allows it to initiate a symbiosis with *V. fischeri* immediately upon hatching (39, 44, 48, 51). About half-way through embryogenesis, a lateral thickening in the hindgut-ink sac complex represents the first stages of organ development. Over the last half of embryogenesis, pairs of crypts, three on either side, invaginate from the surface epithelium in sequence. On each side one begins to invaginate early in this period of development and continues to grow throughout embryogenesis, the second pair begins several days later, and the third pair forms a few days before the juvenile host hatches. Thus, at hatching, the host has three pairs of epithelia-lined crypts of different sizes, each individual crypt connected by a pore to the surface of the organ. These crypts will house the symbiotic bacterial population after inoculation of the organ by *V. fischeri* from the environment.

Also during embryogenesis, the superficial epithelium of the developing organ, which is not directly connected with the crypt spaces, forms a complex ciliated, microvillous field (44, 48). This field spreads medially over the lateral surfaces of the organ,where it consists of three regions: an anterior appendage and a posterior appendage, which have the three crypt pores of that side at their base, and a basal field, which surrounds the pores and spreads medially. The appendages consist of a single layer of epithelium, separated from a blood sinus by a basement membrane. The medial extension of the ciliated field on each side of the developing organ terminates in a row of prominent, elongate cilia. High-speed cinematography (RB Emlet, MJ McFall-Ngai, unpublished data) has suggested that the arms form a ring-like arrangement that entrains water toward the vicinity of the pores leading to the crypt spaces. In addition, manipulation of the ciliated field with pharmacological agents has indicated that the cilia beat toward the pores (44). Studies with fluorescently labeled bacteria demonstrate that they do not adhere to this ciliated, microvillous surface but instead are entrained by the water currents into the vicinity of the pores (MJ McFall-Ngai, personal observation).

The crypt spaces are connected to the superficial pores by long cilia-lined ducts (Figure 1) (42, 48). Goblet cells embedded among the epithelial cells lining these ducts appear to release mucopolysaccharides into the duct spaces (MJ McFall-Ngai, personal observation.). The orientation of rootlets of the cilia, as viewed by TEM, suggests that they beat materials from the duct, and perhaps the crypt spaces, out through the pores (44). This ciliary activity may provide a gradient of a chemoattractant sensed by potential symbionts. However, while it is uncertain whether chemoattraction is involved in colonization of light organs, studies of the bacteria suggest that the anatomical arrangement of the duct presents a physical impediment to colonization. *Vibrio fischeri* mutants that are defective in motility, because of either a lack of flagella or an inoperative flagellar motor, are not able to enter the crypt spaces even when they are present at very high densities in the surrounding water (24). These data suggest that the bacteria must be capable of

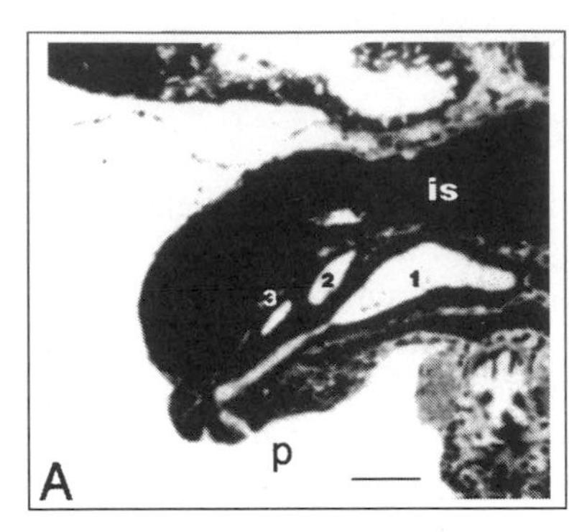

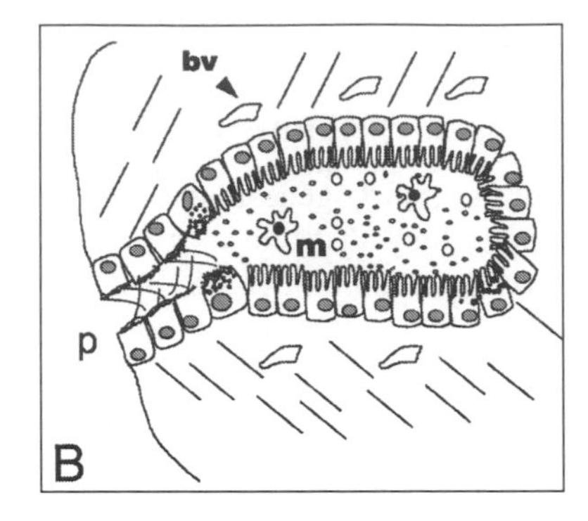

Figure 1 Cross section of the juvenile light organ of *Euprymna scolopes. A.* A histological section through the organ of a newly hatched animal. Three pores on each side of the light organ provide the site of entry into the light organ for the symbiotic bacteria. The bacteria enter the pores, travel along ducts, and enter into three separate crypt spaces [1, 2, 3]. The pore (p), the entire duct, and the largest portion of crypt 1 can be seen in this section. *is*, ink sac; bar = 50 μm. B. An illustration depicting the cells of the light organ with which the bacteria interact. Upon entering the pore, *V. fischeri* cells first encounter the ciliated duct cells and then move into the crypt spaces where they proliferate. In the crypt, they are surrounded by the microvillous epithelium that lines the crypt space. In addition, they also interact with macrophage-like cells that are likely to be delivered to the light organ through the blood vessels (*bv*) in the connective tissue matrix surrounding the crypts.

swimming to negotiate a physical barrier presented either by the mucus or by a cilia-generated current in the duct, or both. The bacteria also produce a mucinase (EG Ruby, personal communication), which may function to degrade the mucus in the duct. It has yet to be determined whether mutants defective in mucinase production are unable to infect the light organ.

The Bacterial Symbionts During the Colonization Process

Wei & Young (80) were the first to determine that the nascent light organ of juvenile *E. scolopes* was not capable of producing luminescence if the newly hatched animal was maintained in seawater from which the bacteria had been removed by filtration. Thus, it was concluded that in nature the animals obtain their initial inoculum of *V. fischeri* cells from the seawater into which they hatch. In later work under laboratory conditions, it was found that the addition of a few hundred to thousand *V. fischeri* cells per milliliter of filtered seawater is required to elicit a successful colonization of the *E. scolopes* light organ. However, methods traditionally applied to the enumeration of *V. fischeri* cells occurring naturally in the seawater [i.e., counting the number of colony-forming units (CFUs) that arise on a nutrient agar plate onto which seawater is spread] initially suggested that there existed a population density of fewer than 10 *V. fischeri* cells per milliliter in the seawater where *E. scolopes* hatchlings are typically found, and in which they become rapidly colonized. Thus, either the bacteria in natural seawater samples are far more infective than those grown under laboratory conditions, or the numbers of CFUs obtained for natural seawater populations of *V. fischeri* underestimated the actual number. Using a probe specific for *V. fischeri*—a DNA sequence complementary to the *luxA* gene of this species—Lee & Ruby (32) showed that *V. fischeri* is present at a concentration of approximately 100 to 1000 cells per milliliter of seawater in areas where adult *E. scolopes* occur. More than 99% of these cells, while infective, are in a "nonculturable" condition (35). A number of species of marine bacteria have been demonstrated to enter this cryptic state (13, 60), the nature of which has not been characterized. However, the conditions of the juvenile light organ are capable of initiating growth of *V. fischeri* in this state, so the light-organ environment causes these cryptic bacteria to return to a culturable state. The specific conditions of the light organ that promote the growth of these bacteria have not been determined.

Seawater samples taken at increasing distances from locations with known populations of adult *E. scolopes* contained increasingly fewer *V. fischeri* and were increasingly less capable of promoting colonization light organs (34). Laboratory studies of the interactions between the light-organ populations of *V. fischeri* and their populations in the surrounding seawater indicated that the host seeds the water with symbionts by a diel venting of the light-organ population into the environment. Specifically, at dawn each day, the host expels between 90% and 95% of its resident *V. fischeri* cells into the surrounding seawater through the lateral pores on the light organ (9, 23, 34). This venting appears to result in the high population densities of *V. fischeri* cells that are detected in habitats with large numbers of adult animals. While this daily expulsion behavior may serve a purpose in the maintenance of a healthy symbiosis, an additional, ecological result is an increased likelihood that subsequent generations of squids will be successfully inoculated.

Recognition and Specificity

In experiments in which juvenile *E. scolopes* are placed in seawater containing *V. fischeri* cells at a concentration typical of their natural environment, the animals become colonized with the symbiont within a few hours (43, 67). The colonization process has been monitored in the laboratory in two ways: one, by following the onset and increase in luminescence over time; and two, by the enumeration of bacteria in the light organ as a function of time. Studies of the infection process (33, 67) have shown that the initial growth rate of the bacteria results in a doubling of the cell number every 20 min. However, after about 12 h, the growth rate slows to an average doubling time of 5.5 h.

Although the squid will readily be infected with *V. fischeri* upon hatching, when *V. fischeri* is either low in number or absent, under either natural or laboratory conditions, the light organ remains uncolonized even when the numbers of nonspecific environmental bacteria are high (43). These data suggest *V. fischeri* is not simply a competitive dominant in the light organ, but rather that there is a host-imposed positive selection for *V. fischeri*. Experiments in which nonspecific, green-fluorescent protein (GFP)–labeled *Vibrio* spp.are exposed to juvenile squid have revealed that these bacteria enter the light-organ crypt spaces but do not grow and persist there (SV Nyholm, EV Stabb, personal communication). Further, having entered the host's crypt, these nonspecific bacteria are no longer culturable, indicating that they have lost viability. These data suggest that one aspect of specificity in the squid-vibrio light-organ symbiosis is the host's creation of a habitat in which only *V. fischeri* is able successfully to initiate and maintain a stable association. Data also exist that implicate receptor-ligand interactions in the positive selection of *V. fischeri* in the light organ (46).

The Organ as a Stressful Environment Studies of the cell biology, biochemistry, and molecular biology of the *E. scolopes* light organ have revealed that the crypt environment has many features typically associated with innate immune responses in mammals. Recently, mollusk macrophage-like hemocytes have been reported in the crypt spaces (Figure 1), and phagocytosed bacterial cells have been seen in transmission electron micrographs of these crypt hemocytes (58). The mechanisms by which these cells are transported to the crypt spaces and their exact function there remain to be determined. It is not yet known whether these cells engulf nonspecific bacteria, thus maintaining *V. fischeri* as the sole inhabitant of the organ, or whether they instead provide a mechanism to control symbiont number. Further analysis of the activity of these cells in adult hosts, as well as experimental manipulations of aposymbiotic and symbiotic juvenile squids, should contribute to an understanding of their precise role in the dynamics of the symbiosis.

In addition to the presence of macrophage-like cells in the light-organ crypts, a library derived from the mRNA of the light organ had unusually high concentrations of messages that encode a peroxidase with sequence similarity and biochemical affinities to mammalian myeloperoxidase (MPO) (72, 75, 82). MPO occurs abundantly in neutrophils where it catalyzes the production of hypohalous acid from halide ions and hydrogen peroxide (30). The presence of abundant hydrogen peroxide substrate is the result of a respiratory burst that occurs in response to the phagocytosis of a pathogen. The hypohalous acids that are produced by MPO are potent microbicides, and their activity provides a first line of defense in mammalian immune systems (30).

Since its discovery in the light organ, research on the squid peroxidase and the possible responses of *V. fischeri* have provided valuable insight into the nature of the "dialogue" between these symbiotic partners. Biochemical studies of the squid peroxidase revealed that it is also halide-dependent and that the light organ exhibits relatively high activity for the enzyme (72, 82). Antibodies to mammalian MPO recognize the squid peroxidase in immunocytochemical analyses and in protein-antibody hybridizations (Western blots). In addition, antibodies generated to the expressed gene product of the squid peroxidase cDNA recognize human MPO in

Western blots. Immunocytochemical localization of the squid peroxidase revealed that it is abundant along the apical surfaces of the crypt epithelia as well as in association with both host and bacterial cells in the crypt spaces.

The finding of an enzyme that catalyzes the production of a microbicide in a cooperative association presented a seeming contradiction, but its abundance suggested that it may be pivotal in the dynamics of this symbiosis. Because no bacteria have been described thus far that have mechanisms by which to resist the effects of hypohalous acids, it was reasoned that *V. fischeri* must be in some way inhibiting the activity of this enzyme, either directly or indirectly.

Two principal mechanisms have been described by which pathogens resist the activity of halide peroxidases (18, 74): (i) by inhibiting the respiratory burst activity of the cell; or (ii) by competing for the substrate of the enzyme, hydrogen peroxide. There is some evidence that *V. fischeri* may use both of these mechanisms. Reich & Schoolnik (64) determined that *V. fischeri* has two enzymes which they called halovibrin α and β (63, 64) with ADP-ribosyltransferase activity. Halovibrin α is produced by *V. fischeri* during log-phase growth, whereas the gene encoding halovibrin β is expressed primarily in stationary phase (63, 64). The activity of these enzymes is analogous to that of cholera toxin, the agent of *V. cholerae* that perturbs the physiology of mammalian host enterocytes. Among other activities, cholera toxin inhibits host cell respiratory burst activity and thus the production of superoxide anion that would have been catalyzed by superoxide dismutase to hydrogen peroxide, the MPO substrate (69). Comparisons of superoxide anion production in symbiotic and aposymbiotic light organs revealed that symbiotic light organs have significantly lower levels of this reactive oxygen species (71). Further, when aposymbiotic animals are incubated in seawater containing cholera toxin, they produce superoxide anion at the decreased levels that are characteristic of symbiotic animals. Presently, the halovibrin proteins are being isolated from *V. fischeri* to determine whether they experimentally induce similar oxidative signatures in the light organ. In addition, *V. fischeri* mutants defective in the production of their halovibrins are being generated (64; EV Stabb, EG Ruby, personal communication). As in the case of cholera toxin, which has multiple effects on host cell function, the halovibrins may participate in a variety of aspects of the squid-vibrio association, only one component of which may be the inhibition of respiratory burst activity in host cells.

Data also exist to suggest that *V. fischeri* inhibits the host peroxidase by competing for its hydrogen peroxide substrate. Visick and Ruby recently described a high level of catalase activity in the periplasmic space of *V. fischeri* (77). The periplasmic space lies between the cell wall and the cell membrane, and periplasmic catalases appear to occur most frequently in organisms that encounter high levels of environmentally imposed oxidative stress (36). Visick and Ruby found that the *V. fischeri* catalase gene is most fully induced during stationary phase. Because, with the daily venting of the light organ at dawn, the symbiont population goes through both log phase and stationary phase growth each day, the induction of the catalase late in the growth cycle of the bacterium may track fluctuations in the host's contribution to the organ's oxidative state.

In addition to the possibility that *V. fischeri* may prevent the production of and compete for the substrate of the peroxidase, preliminary data indicate that the symbiont influences the host's expression of the gene encoding the squid peroxidase. Specifically, at 48 h post inoculation, mRNA levels for the squid peroxidase in symbiotic animals are approximately half that of aposymbiotic animals, suggesting that the presence of the symbiotic *V. fischeri* results in a reduction of the expression of this gene (AL Small, AM Hirsch, MJ McFall-Ngai, unpublished results).

Additional studies of the pattern of occurrence of this peroxidase in a variety of tissues in *E. scolopes* suggest that it plays a pivotal role in a variety of interactions, and that the regulation of this protein may be crucial in defining the precise nature of the symbiosis, i.e., whether it is cooperative or pathogenic. For example, the accessory nidamental gland of the female, which contains a consortium of bacteria thought to be involved in successful development of eggs, also contains high levels of this enzyme, whereas the eye and digestive gland, tissues which do not typically interact with microbes, do not (72). In addition, as in certain insects, in which a similar peroxidase has a dual function in antimicrobial activity and melanin production, the enzyme is also present in the ink gland of *E. scolopes*, the site of melanin synthesis (38, 52).

The most compelling evidence that this enzyme is involved in modulation of pathogenic interactions with bacteria comes from studies of its production in the gills. Cephalopod gills are the site where potential pathogens that have invaded the blood stream are cleared from the circulatory system (5). At the gills, these bacteria are sequestered into cysts. In studies of adult *E. scolopes*, gills without cysts show very low activity of the peroxidase and very low levels of peroxidase gene expression (72). However, in gills with abundant cysts, both activity for this type of peroxidase and high levels of peroxidase gene expression are evident (72). These observations are supported by recent studies of the response of *E. scolopes* to elevated concentrations of *V. fischeri* in the ambient seawater. In newly hatched squid, when the light-organ symbiont is presented at levels normally experienced in the host's environment (i.e., 200 to 1000 cells $\cdot$ ml^{-1}), the light organ becomes colonized and no gill cysts form (AL Small, AM Hirsch, MJ McFall-Ngai, unpublished results). Under these conditions, *V. fischeri* appears to turn down peroxidase message expression, as mentioned above, and there are undetectable levels of peroxidase in the gills. However, in experiments in which *V. fischeri* is presented at levels of 10^5 to 10^7 cells $\cdot$ ml^{-1} in the seawater surrounding a juvenile, in addition to a light organ being colonized, gill cysts containing *V. fischeri* form in the animal (AL Small, AM Hirsch, MJ McFall-Ngai, unpublished results). Under these conditions, *V. fischeri* causes a dramatic increase in peroxidase gene expression in the gills, suggesting that the host is responding to *V. fischeri* as a pathogen.

The occurrence in the cooperative squid-vibrio symbiosis of macrophage-like cells and an enzyme heretofore associated only with antimicrobial activity present the possibility that the same genes, gene products, and cells are involved all types of

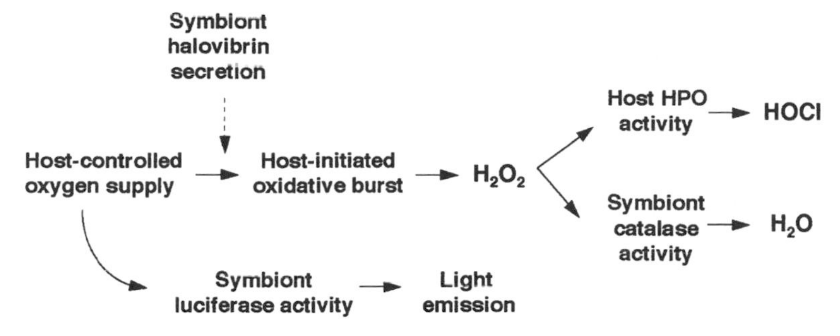

Figure 2 Model of potential oxidative-stress reactions in the squid-vibrio symbiosis. The circulatory system of the host supplies the crypt environment with oxygen, which provides the substrate for host respiratory burst activity in response to microbes. The respiratory burst leads to the eventual production of hydrogen peroxide (H_2O_2), which is the substrate for the squid halide peroxidase (HPO). The activity of the halide peroxidase creates hypochlorous acid (HOCl), which is toxic to bacteria. The symbionts can respond by lowering oxygen availability through the oxygen-requiring luminescence reaction. In addition, halovibrin secretion may inhibit respiratory burst activity. The action of either, or both, of these processes would lower the production of HOCl by inhibiting the creation of substrate for the HPO. As another defense, the bacteria may compete for existing H_2O_2 substrate of HPO with a highly active, periplasmic catalase.

animal-microbial interactions and that modulation of these components defines the outcome. Further, these data suggest that these genes may have evolved for animal-bacterial interactions, no matter what sort, rather than being strictly associated with defense against nonself.

Taken together, the data available on the oxidative environment of the host light organ, and the possible responses of the specific symbionts to this environment, suggest a model of this aspect of the squid-vibrio association (Figure 2). In this model, *V. fischeri* would enter the light organ and initially restrict the activity of the peroxidase by inhibiting a respiratory burst of host cells through the activity of the halovibrins, thus depriving the peroxidase of substrate. A second line of defense would be available to *V. fischeri* by the activity of its periplasmic catalase, which would compete with the squid peroxidase for the hydrogen peroxide substrate. The bacteria would also turn down expression of the host peroxidase gene. Some level of host peroxidase expression would remain throughout the life of the host, perhaps either to control the invasion of nonspecific bacteria into the light organ or to control the numbers of *V. fischeri*, or both. There is evidence that both these functions may be important in the biology of the symbiosis. When adult squid are incubated in water with antibiotically tagged bacteria, these bacteria are later found in the light organ, indicating that it remains open to colonization after the initial infection event. In addition, the studies with supernumerary *V. fischeri* indicate that the host can respond to them as pathogens, and that they have the potential to overgrow.

Receptor-Ligand Interactions In addition to the creation of an environment in which *V. fischeri* is able to persist, while other bacteria are not, there is evidence that specific receptor-ligand interactions may be important in the determination of specificity in the squid-vibrio symbiosis (46). Animal-bacterial cell interactions are often mediated by the recognition of sugars on the host cell membrane by bacterial surface proteins called adhesins (59). Many gram-negative bacteria have mannose-recognizing adhesins, and specificity of an interaction is conferred by variations in the bacterial adhesins that correspond to differences in the microenvironment of the mannose residues on the host receptor (59). Assays with *V. fischeri* have shown that they hemagglutinate guinea-pig red blood cells, a behavior characteristic of bacteria that have mannose-recognizing adhesins associated with their cell surfaces (46). Histochemical analyses of juvenile host light organs showed that mannose residues are abundant along the crypt brush border (46). Thus, both partners had elements of the lock-and-key mechanisms associated with this type of recognition system. Further, mannose analogs, but not other sugars and sugar analogs, introduced into the environment during the infection process, fully block colonization by *V. fischeri* (46), suggesting that these interactions are essential for successful infection of the light organ by the symbiont. Although *V. fischeri* strains mutant in mannose recognition are not yet available, a homolog of a gene encoding the mannose adhesin in *V. cholerae* (*mshA*) has recently been identified, and strains of *V. fischeri* mutant in this gene are currently being generated (EV Stabb, EG Ruby, personal communication).

These adhesin-glycan interactions in the squid-vibrio symbiosis may vary through ontogeny. In studies of the development of the mammalian intestine (29), bacteria have been shown to induce increased expression of glycans on the host cell surfaces to which they are adhering. This augmentation of host binding sites results in reinforcement of the association between the bacteria and these host cells. Mutant bacterial cells that are defective in adhesion to host cells do not induce an increase in the production of host cell glycans. This type of interaction may also occur in the squid-vibrio association. Over the first few days of the symbiosis, the bacteria induce an increase in their intimacy with host cells; i.e., the percentage of the bacterial surface in contact with the host cell increases dramatically during this time (31; see below). Studies with *V. fischeri* strains defective in adhesion to host crypt cells should reveal whether the increase in intimacy is achieved in a similar manner as demonstrated in the mammalian intestine.

Other types of receptor-ligand interactions have also been suggested as essential in the initiation of the squid-vibrio association. In some bacteria, outer membrane proteins (Omps) called porins participate in adhesion to host cells. Aeckersberg & Ruby (1) found that antibodies to a homolog (OmpL of *Photobacterium profundus*) of a *V. fischeri* porin inhibit infection, and mutants generated in the *V. fischeri* gene (*ompV*), were found to colonize the host squid less efficiently (2).

Evolution of Specificity Determinants Another mechanism by which to study specificity is to ask how specificity determinants express themselves in related host species, i.e., ask what kinds of experiments nature has done over evolutionary time that resulted in specific symbiotic associations. The phylogenetic trees derived from gene sequence data of several of the sepiolid hosts and their light-organ symbionts are congruent (55, 56). For some other symbioses, this type of data has provided evidence that the host and symbiont have coevolved (15, 28). Additional support for coevolution in the squid-vibrio symbioses has been obtained by studying the response of the newly hatched Hawaiian squid to interaction with vibrio symbionts from other sepiolid species (55). The Hawaiian squid are colonized by these heterologous symbionts when they are presented to the hatchling in the absence of any other potential symbionts. However, when the native symbiont and a strain from another host sepiolid are presented in the inoculum at equal concentrations, the native symbiont outcompetes those isolated from the light organs of all other sepiolid species. When two nonnative strains are presented to the host in such competition experiments, the strain from the host most closely related to the Hawaiian host always prevails. Thus, not only are the phylogenetic trees of the hosts and symbionts congruent, the ability of the bacterial symbionts to compete in colonization experiments with *E. scolopes* directly reflects their relative position on the phylogenetic tree.

The mechanisms underlying these differences in symbiont competitiveness remain to be determined. However, the data suggest that (i) selection of a particular symbiont occurs by processes that express themselves in the early stages of the symbiosis; and (ii) for the first time in any animal-bacterial symbiotic association, analyses of the differences between the host and their symbionts may reveal the precise nature of the specificity determinants, i.e., those biochemical and molecular features of the host and symbiont upon which selection has acted to bring about evolution of these symbioses.

DEVELOPMENT OF THE LIGHT ORGAN OF *E. SCOLOPES*

The host squid *E. scolopes* hatches with a light-organ morphology and anatomy markedly different from that of the adult (42, 51). The bacteria-containing epithelial core of the organ is surrounded by a thick reflector that serves to direct light ventrally; the ink sac has diverticula that act as an iris to control the intensity of emission; yellow filters are present over the ventral surface that may act to shift the wavelength of luminescence closer to that of downwelling moonlight and starlight; and, a thick, muscle-derived lens, with striking biochemical similarities to the squid eye lens (47, 81), covers the entire ventral surface of the organ and appears to function as a diffuser of the bacteria-produced light.

The periods of light-organ development correlate with functional stages of the symbiotic association (40, 51). In addition to the embryonic organogenesis, dramatic changes occur in the light organ over the first few hours to days posthatch that are associated with the establishment of the relationship (44, 49). This early stage is followed by late development, or maturation, of the organ, in which the components associated with modulating bacterial light emission–the reflector, lens, filters, and ink sac diverticula–are elaborated (51). Whereas specific interactions with *V. fischeri* are essential for early posthatch morphogenesis of the organ (17, 20, 31, 49, 50), the elaboration of tissues associated with the maturation of the *E. scolopes* light organ do not appear to require interaction with the bacterial symbionts. These results are supported by research on the development of the light organs of the Mediterranean sepiolids (JS Foster, SV Boletzky, and MJ McFall-Ngai, unpublished results). The embryonic period of these species is two to three times that of *E. scolopes*, and upon hatching they also have a complex, superficial, ciliated field, the loss of which is under bacterial induction. However, the shape of the light organ resembles that of *E. scolopes* at 1-month posthatching, and all of the components that modulate bacterial light (lens, reflector, and ink sac diverticula) are well developed.

Three principal changes occur in the normal, early posthatch development of the *E. scolopes* light organ. They include a loss of the superficial, ciliated, microvillous field of epithelial cells at 4-5 days following hatching (49), an increase in the microvillar density along the apical surfaces of the crypt cells (31), and a swelling of the crypt epithelial cells (49). Studies in which development of the organ is compared in symbiotic and aposymbiotic animals have shown that all of these changes are under symbiont induction; i.e., *V. fischeri* is required for the full expression of each facet of this early posthatch developmental program.

Bacteria-Induced Loss of the Superficial Field of Cells

The first experimental manipulations of the squid-vibrio developmental program showed that interaction with *V. fischeri* induces death in the cells of the superficial epithelium (Figure 3 see color insert; 43, 49). Ultrastructural, biochemical, and molecular analyses of the dying cells showed that the process of death is classic programmed cell death, or apoptosis (20). These cells die in a characteristic pattern. About 9 h following first exposure to *V. fischeri*, the cells along the medial margin of the ciliated field that bear long cilia and the cells at the tips of the appendages die. In the following hours, cells die over the entire field. By between 18 h and 24 h, the blood sinus in the center of each appendage collapses. In aposymbiotic animals, occasional cell death events sometimes occur in this field of cells, but no regression occurs and the sinus remains intact.

Experimental manipulations of the system have revealed the timing of the bacterial induction and the location in the organ where the induction takes place. Studies in which light organs were antibiotically cured of *V. fischeri* at various times following inoculation revealed that the bacteria deliver an irreversible signal at about 12 h following the initial exposure to the symbionts (17). This signaling triggers the 4-d developmental program that results in loss of the superficial field of cells. In addition, when large numbers of nonmotile mutants of *V. fischeri*, either lacking flagella or with a disabled "motor," were coincubated with the squid host, no cell death or regression was noted (17). These data suggested that the bacteria must get into the crypts and interact with cells there, rather than directly interacting with the superficial field of cells, to trigger the morphogenesis.

These findings present two questions: What is the bacterial signal(s)? How is the signal(s) transmitted to the responsive cells? The most common bacterial inducer of animal cell death is the bacterial lipopolysaccharide (LPS) (57, 62), and specifically, the lipid A portion of LPS, which is the most conserved component of the molecule. Experiments with newly hatched *E. scolopes* revealed that when either purified *V. fischeri* holo-LPS or the lipid A fraction are coincubated with the host, the typical cell death pattern is induced (JS Foster, MA Apicella, MJ McFall-Ngai, unpublished results). In addition, commercially available LPS and lipid A of other bacteria are also effective in inducing the characteristic pattern on the *E. scolopes* light organ. These responses were dose dependent, with levels of LPS in the range of pg $\cdot$ ml^{-1} of seawater causing death in only a few cells in the field, whereas levels in the range of ng to $\mu g \cdot ml^{-1}$ induced the full pattern characteristic of the infected light organ. The finding that the signal triggering cell death may be something conserved on the surface of gram-negative bacteria helped explain the observation that cell death can occasionally be seen in the superficial field of cells of aposymbiotic animals but not in animals kept axenically (personal observation). Because experiments with GFP-labeled bacteria have shown that other bacterial species get into the organ, but do not colonize, bacterial LPS as a component of the cell surface of these nonspecific bacteria is likely to occur at low levels in the light organs of aposymbiotic juveniles. Thus, the triggering of the extensive cell death pattern characteristic of the colonization of the organ by *V. fischeri* may be the result of the high doses of LPS that would be presented over a short time period by the growing culture of symbionts.

Although all of the available data suggest that symbiont LPS induces cell death, when the light organs of juveniles that have been exposed to LPS are viewed at 4 d, they do not show signs of regression of the ciliated field (JS Foster, MA Apicella, MJ McFall-Ngai, unpublished results). This finding suggested that a second bacterial signal may be required for morphogenesis of the organ. The existence of a second signal was also suggested by the fact that the full pattern of symbiont-induced and LPS-induced cell death occurs at around 9 h post-exposure (20; JS Foster, MA Apicella, MJ McFall-Ngai, unpublished results), whereas the irreversible signal for complete morphogenesis does not occur until about 12 h. One possible explanation is that the bacteria inhibit further cell proliferation, as well as inducing cell death. This hypothesis was supported by experiments with a cell proliferation inhibitor, colchicine. By itself, colchicine does not cause full regression of the superficial field, but when the animals are incubated in seawater containing both LPS and colchicine, full regression of the superficial field occurs. The nature of this second signal remains to be determined.

The mechanism by which the bacterial signals are transmitted to their site of action and the specific cell-death pathways triggered by the bacteria also remain to be determined. The bacterial cells interact with cells of two types in the crypts (Figure 1B), the epithelial cells lining the space and the macrophage-like cells free in the crypt space (58). It is likely that, following interactions with cells of one or both of these types, the signals are either carried into the circulatory system and up to the sinus of the appendages or there is a cascade of signals through the cells that connect the superficial epithelium with the deeper tissue. One interesting observation is that, if the crypt spaces are loaded with fluorescently labeled dextran, macrophage-like cells carrying dextran can be seen in the sinus of the arms a few hours later (DJ Park, MJ McFall-Ngai, personal observation). A great deal more study of the system must be undertaken before these issues can be resolved.

Bacteria-Induced Changes in Light-Organ Crypt Cells

Whereas the bacteria send an irreversible trigger that induces cell death and regression of a remote set of cells, the epithelial cells with which they directly interact in the crypt do not undergo cell death. However, significant developmental changes do occur in the crypt epithelium. Over the first four days following inoculation of the organ, the microvillar density along the apical surfaces of these cells increases fourfold (31) and the host cells swell (49). Significant differences in the number of microvilli per unit area are detectable by 12 h postinoculation with *V. fischeri*. Transmission electron micrographs of the crypt cells of adult animals reveal highly complex, lobate microvilli (42), indicating that the elaboration of the brush border continues after this initial increase. Light micrographs of the crypts have demonstrated that the crypt cells also exhibit a fourfold increase in volume over the first few days following colonization (49). A quantifiable difference in cell volume is apparent within 24 h (16). Experiments involving the antibiotic curing of the organ showed that, unlike the triggering of the regression of the superficial epithelium, the increase in microvillar density and cell swelling in the host cells in direct contact with the bacteria are reversible (16). In cured animals, the microvillar density and cell volume return to the condition characteristic of hatchling or aposymbiotic animals. The bacterial cells also become much more intimately associated with the host cells; the percent of a given bacterial cell's surface in contact with microvilli increases from less than 30% to more than 80% over the first several days (31). The increase in both microvillar density and cell swelling appears to contribute to this increase in the intimacy of the partners' cells.

Bacterial Development

While far less dramatic than those of the host, developmental changes in *V. fischeri* cells occur soon after the symbionts initiate the association. Within 6 to 12 h of entry into the light-organ crypts, the bacteria have undergone cell differentiation. Two morphological changes have been reported: (i) a loss of flagellation, and (ii) a decrease in cell volume. Neither of these phenomena occurs during the first 12 h when the inoculating cells are undergoing a period of rapid growth; however, after the light-organ population has reached its maximum level of about a million cells, the polar flagella that are typically present on *V. fischeri* cells disappear, and the cells decrease to a volume about one-seventh that of the inoculating bacteria

(67). The cause of these events remains unknown, although similar events have been reported for cells responding to a severe decrease in growth rate (65).

At the biochemical level, a large induction of bacterial luciferase becomes evident within 6 to 8 h postinoculation (67) and results in the normally dimly luminescing *V. fischeri* cells (7) producing a biologically useful amount of light for its host. This induction is due to the accumulation around the bacteria in the light-organ crypts of an acyl-homoserine lactone signal molecule called autoinducer (8), which serves as an activating cofactor when it binds to a transcriptional regulator of the *lux* operon. The synthesis of proteins besides those encoded by the *lux* genes (e.g., luciferase) are positively regulated by autoinducer accumulation (12). Discovery of the functions of these proteins may suggest the processes that change in the developing symbiont.

By applying a technology that identifies *V. fischeri* promoters that are activated specifically in cells as a response to growth in the light organ, the molecular genetics underlying the initiation of the symbiosis is being revealed (78). To date, five *V. fischeri* genes have been identified as symbiosis-induced: three are homologs of *Escherichia coli* genes that have known metabolic functions, including amino acid metabolism (23), while two others are novel and may prove to be our best route to uncovering future insights into specific events in the regulation of this association.

MAINTENANCE OF A STABLE SYMBIOSIS

Less information is presently available about the long-term maintenance of the association between *E. scolopes* and *V. fischeri* than about other aspects of the relationship. One key aspect of the dynamics of this association appears to be a complex diel rhythm (9, 23, 58). Beginning immediately after hatching and persisting throughout the life history of the host, highest levels of light are emitted from the host around dusk and through the early hours of the night, and the lowest luminescence is observed around dawn and the early hours of the morning (9). These variations in luminescence appear to be under host control. Bacteria experimentally removed from the light organ at different times of day show the same luminescence per cell, which demonstrates that expression of components the *lux* operon, including the aldehyde substrate of the luminescence reaction, do not vary in concentration within *V. fischeri* cells over the course of the day (9). Because the light organ is well vascularized and the luminescence reaction requires oxygen, the most parsimonious explanation is that the host modulates oxygen delivery to the crypt space (9).

Superimposed on this diel rhythm of luminescence is a daily release of the crypt contents into the environment at dawn (23, 58). The material exits through the lateral pores of the light organ into the mantle cavity as a thick paste, which can be collected intact for analysis. The constituents of this exudate include, on average, 95% of the bacterial symbionts (34), a population of host macrophage-like cells (58), and a dense proteinaceous matrix (23, 58). All of the bacterial cells appear to be alive and culturable (67), while not all of the host cells are viable (58). The venting of the bacteria likely serves several functions in the symbiosis, the most obvious being the control of symbiont numbers in the light organ. In addition, it has been hypothesized that this behavior also serves to seed the environment with *V. fischeri*, which must be acquired by newly hatched juveniles (34, 68).

As discussed earlier, the role of the macrophage-like cells in the crypt space is poorly understood, but some data are available on the role of the matrix material. The matrix is rich in a mixture of proteins (SV Nyholm, MJ McFall-Ngai, unpublished results). Studies with amino acid auxotroph mutants of *V. fischeri* have demonstrated that in the light organ, the symbionts obtain amino acids as a carbon and nitrogen source (23). Further analyses of the constituents of this matrix are expected to provide valuable insight into the microenvironment of the crypt space.

CONCLUSIONS

Studies of the squid-vibrio association have revealed a rich frontier for future research. The past ten years of work on this symbiosis have principally focused on describing the association and building the tools, such as microbial genetics of *V. fischeri*, for the development of this model. With this framework in place, researchers in the field are now poised to determine some of the underlying biochemical and molecular mechanisms by which bacteria form long-term, beneficial associations with animals.

ACKNOWLEDGMENTS

I thank EG Ruby for insightful comments on the manuscript and for his help in formulating the sections of this contribution that are principally microbiological. I thank the members of my laboratory, as well as the microbiologists who have studied aspects of the squid-vibrio system over the past 10 years, for their enthusiasm and dedication. This work was supported by NSF grant #IBN 96-01155 to MMN and EG Ruby and NIH grant #R01-RR12294 to EG Ruby and MMN.

LITERATURE CITED

1. Aeckersberg FT, Welch T, Ruby EG. 1997. Possible participation of an outer membrane protein in the symbiotic infection of the *Euprymna scolopes* light organ. *Abstr. Annu. Meet. Am. Soc. Microbiol.* 97: 387
2. Aeckersberg FT, Welch T, Ruby EG. 1998. Possible role of an outer membrane protein of *Vibrio fischeri* in its symbiotic infection of *Euprymna scolopes*. *Abstr. Annu. Meet. Am. Soc. Microbiol.* 98:374
3. Arnold J, Singley C, Williams-Arnold L. 1972. Embryonic development and post-hatch survival of the sepiolid squid *Eu-*

prymna scolopes under laboratory conditions. *Veliger* 14:361–64
4. Bang SS, Baumann P, Nealson KH. 1978. Phenotypic characterization of *Photobacterium logei* (sp. nov.), a species related to *P. fischeri*. *Curr. Microbiol.* 1:285–88
5. Bayne C. 1974. Molluscan immunobiology. In *The Mollusca*, Vol. 5, ed. ASM Saleuddin, EL Wilbur, pp. 408–69. New York: Academic
6. Berry S. 1912. The Cephalopoda of the Hawaiian islands. *Bull. US Bur. Fish.* 32:255–362
7. Boettcher KJ, Ruby EG. 1990. Depressed light emission by symbiotic *Vibrio fischeri* of the sepiolid squid, *Euprymna scolopes*. *J. Bacteriol.* 172:3701–6
8. Boettcher KJ, Ruby EG. 1995. Detection and quantification of *Vibrio fischeri* autoinducer from symbiotic squid light organs. *J. Bacteriol.* 177:1053–58
9. Boettcher K, Ruby EG, McFall-Ngai MJ. 1996. Bioluminescence in the symbiotic squid *Euprymna scolopes* is controlled by a daily biological rhythm. *J. Comp. Physiol.* 179:65–73
10. Boletzky SV. 1995. The systematic position of the Sepiolidae (Mollusca: Cephalopoda). In *Mediterranean Sepiolidae, Bulletin de l' Institut oceanographique, Monaco, Numero special 16*, ed. SV Boletzky, pp. 99–104. Monaco: Musee Oceanographique
11. Bonnaud L, Boucherrodoni R, Monnerot M. 1997. Phylogeny of Cephalopoda inferred from mitochondrial-DNA sequences. *Mol. Phylogenet. Evol.* 7:44–54
12. Callahan S, Dunlap PV. 1998. The autoinduction regulon of *Vibrio fischeri*: identification and partial characterization of five novel LuxR/autoinducer-regulated proteins. *Abstr. Ann. Meet. Am. Soc. Microbiol.* 98:282
13. Colwell RR, Huq A. 1994. Vibrios in the environment: viable but nonculturable *Vibrio cholerae*. In *Vibrio cholerae and Cholera*, ed. IK Wachsmuth, PA Blake, O Olsvik, pp.117–134. Washington: Am. Soc. Microbiol.
14. Dilly PN, Herring PJ. 1978. The light organ and ink sac of *Heteroteuthis dispar* (Mollusca: Celphalopoda). *J. Zool., Lond.* 172:81–100
15. Distel DL, Felbeck H, Cavanaugh CM. 1994. Evidence for phylogenetic congruence among sulfur-oxidizing chemoautotrophic bacterial endosymbionts and their bivalve hosts. *J. Mol. Evol.* 38:533–42
16. Doino JA. 1998. The role of light organ symbionts in signaling early morphological and biochemical events in the sepiolid squid *Euprymna scolopes*. PhD diss., Univ. Southern Calif. 123 pp
17. Doino JA, McFall-Ngai MJ. 1995. Transient exposure to competent bacteria initiates symbiosis-specific squid light organ morphogenesis. *Biol. Bull.* 189:347–55
18. Dukan S, Touati D. 1996. Hypochlorous acid stress in *Escherichia coli*: resistance, DNA damage, and comparison with hydrogen peroxide stress. *J. Bacteriol.* 178: 6145–50
19. Fidopiastis PM, Boletzky SV, Ruby EG. 1998. A new niche for *Vibrio logei*, the predominant light organ symbiont of squids in the genus *Sepiola*. *J. Bacteriol.* 180:59–64
20. Foster JS, McFall-Ngai MJ. 1998. Induction of apoptosis by cooperative bacteria in the morphogenesis of host epithelial tissues. *Dev. Genes Evol.* 208:295–303
21. Fukusawa ST, Dunlap PV. 1986. Identification of luminous bacteria from the light organ of the squid, *Doryteuthis kensaki*. *Agric. Bio. Chem.* 50:1645–46
22. Gabel-Deickert A. 1995. Reproductive patterns in *Sepiola affinis* and other Sepiolidae (Mollusca,Cephalopoda). In *Mediterranean Sepiolidae, Bull. de l'Institut oceanographique, Monaco, Numero special 16*, ed. SV Boletzky, pp. 73–83. Monaco: Musee Oceanographique
23. Graf J, Ruby EG. 1998. Characterization of the nutritional environment of a symbiotic light organ using bacterial mutants and bio-

chemical analyses. *Proc. Natl. Acad. Sci. USA* 95:1818–22
24. Graf J, Dunlap PV, Ruby EG. 1994. Effect of transposon-induced motility mutations on colonization of the host light organ by *Vibrio fischeri*. *J. Bacteriol.* 176:6986–91
25. Hanlon RT, Claes MF, Ashcraft SE, Dunlap PV. 1997. Laboratory culture of the sepiolid squid *Euprymna scolopes*: a model system for bacterial-animal symbioses. *Biol. Bull.* 192:364–74
26. Haygood MG. 1993. Light organ symbioses in fishes. *Crit. Rev. Microbiol.* 19:191–216
27. Herring PJ. 1988. Luminescent organs. In *The Mollusca*, Vol. 11, ed. P Hochachka, pp. 449–89. London: Academic
28. Hinkle G, Wetterer JK, Schultz TR, Sogin ML. 1994. Phylogeny of attine ant fungi based on analysis of small subunit rRNA gene sequences. *Science* 266:1695–97
29. Hooper LV, Bry L, Falk PG, Gordon JI. 1998. Host-microbial symbiosis in the mammalian intestine: exploring an internal ecosystem. *BioEssays* 20:336–43
30. Klebanoff SJ. 1991. Myeloperoxidase: occurrence and biological function. In *Peroxidases in Chemistry and Biology*, ed. J Everse, KE Everse, MB Grisham, pp. 2–35. Boca Raton, FL: CRC
31. Lamarcq LH, McFall-Ngai MJ. 1998. Induction of a gradual, reversible morphogenesis of its host's epithelial brush border by *Vibrio fischeri*. *Infect. Immun.* 66:777–85
32. Lee K-H, Ruby EG. 1992. Detection of the light organ symbiont, *Vibrio fischeri*, in Hawaiian seawater using *lux* gene probes. *Appl. Environ. Microbiol.* 58:942–47
33. Lee K-H, Ruby EG. 1994. Competition between *Vibrio fischeri* strains during the initiation and maintenance of a light organ symbiosis. *J. Bacteriol.* 176:1985–91
34. Lee K-H, Ruby EG. 1994. Effect of the squid host on the abundance and distribution of symbiotic *Vibrio fischeri* in nature. *Appl. Environ. Microbiol.* 60:1565–71
35. Lee K-H, Ruby EG. 1995. Symbiotic role of the nonculturable, but viable, state of *Vibrio fischeri* in Hawaiian seawater. *Appl. Environ. Microbiol.* 61:278–83
36. Loewen PC. 1997. Bacterial catalases. In *Oxidative Stress and the Molecular Biology of Antioxidant Defenses*, ed. JG Scandalios, pp. 273–308. New York: Cold Spring Harbor Lab.
37. Margulis L, Fester R. 1991. *Symbiosis as a Source of Evolutionary Innovation*. Cambridge, MA: Mass. Inst. Technol. Press. 454 pp
38. Marmaras VJ, Charalambidis ND, Zervas CG. 1996. Immune response in insects: the role of phenoloxidase in defense reactions in relation to melanization and sclerotization. *Arch. Insect Biochem. Physiol.* 31:119–33
39. McFall-Ngai MJ. 1994. Evolutionary morphology of a squid symbiosis. *Am. Zool.* 34:554–61
40. McFall-Ngai MJ. 1998. The development of cooperative associations between animals and bacteria: establishing détente among Domains. *Am. Zool.* 38:3–18
41. McFall-Ngai MJ. 1998. The adventure of pioneering a biological model: the squid-vibrio association. *ASM News* 64:639–45
42. McFall-Ngai MJ, Montgomery MK. 1990. The anatomy and morphology of the adult bacterial light organ of *Euprymna scolopes* Berry (Cephalopoda:Sepiolidae). *Biol. Bull.* 179:332–39
43. McFall-Ngai MJ, Ruby EG. 1991. Symbiont recognition and subsequent morphogenesis as early events in an animal-bacterial mutualism. *Science* 254:1491–94
44. McFall-Ngai MJ, Ruby EG. 1998. Bobtail squids and their luminous bacteria: when first they meet. *BioScience* 48:257–65
45. McFall-Ngai MJ, Toller WW. 1991. Frontiers in the study of the biochemistry and molecular biology of vision and luminescence in fishes. In *The Molecular Biology and Biochemistry of Fishes*, Vol. 1, ed.

P Hochachka, T Mommsen, pp. 77–107. New York: Elsevier

46. McFall-Ngai MJ, Brennan C, Weis VM, Lamarcq LH. 1998. Mannose adhesin-glycan interactions in the *Euprymna scolopes–Vibrio fischeri* symbiosis. In *New Developments in Marine Biology*, ed. Y LeGal, HO Halvorson, pp. 273–77. New York: Plenum
47. Montgomery MK, McFall-Ngai MJ. 1992. The muscle-derived lens of a squid bioluminescent organ is biochemically convergent with the ocular lens. Evidence for recruitment of ALDH as a predominant structural protein. *J. Biol. Chem.* 267:20,999–21,003
48. Montgomery MK, McFall-Ngai MJ. 1993. Embryonic development of the light organ of the sepiolid squid *Euprymna scolopes*. *Biol. Bull.* 184:296–308
49. Montgomery MK, McFall-Ngai MJ. 1994. The effect of bacterial symbionts on early post-embryonic development of a squid light organ. *Development* 120:1719–29
50. Montgomery MK, McFall-Ngai MJ. 1995. The inductive role of bacterial symbionts in the morphogenesis of a squid light organ. *Am. Zool.* 35:372–80
51. Montgomery MK, McFall-Ngai MJ. 1998. Late postembryonic development of the symbiotic light organ of *Euprymna scolopes* (Cephalopoda:Sepiolidae). *Biol. Bull.* 195:326–36
52. Nappi AJ, Vass E. 1993. Melanogenesis and the generation of cytotoxic molecules during insect cellular immune reactions. *Pigment Cell Res.* 6:117–26
53. Nesis KN. 1982. *Cephalopods of the World*. Neptune City, NJ: TFH Publ. 351 pp.
54. Niner BM, Hirsch AM. 1998. How many rhizobium genes in addition to the *nod*, *nif/fix* and *exo* are needed for nodule development and function. *Symbiosis* 24:51–102
55. Nishiguchi MK, Ruby EG, McFall-Ngai MJ. 1997. Phenotypic bioluminescence as an indicator of competitive dominance in the *Euprymna-Vibrio* symbiosis. In *Bioluminescence and Chemiluminescence: Molecular Reporting with Photons*, ed. JW Hastings, LJ Krick, PE Stanley, pp. 123–26. New York: Wiley & Sons
56. Nishiguchi MK, Ruby EG, McFall-Ngai MJ. 1998. Competitive dominance among strains of luminous bacteria provides an unusual form of evidence for parallel evolution in the sepiolid–*Vibrio fischeri* symbioses. *Appl. Environ. Microbiol.* 64:3209–13
57. Norimatsu M, Ono T, Aoki A, Ohishi K, Takahashi T, et al. 1995. Lipopolysaccharide–induced apoptosis in swine lympocytes in vivo. *Infect. Immun.* 63: 1122–1126
58. Nyholm SV, McFall-Ngai MJ. 1998. Sampling the microenvironment of the *Euprymna scolopes* light organ: description of a population of host cells with the bacterial symbiont *Vibrio fischeri*. *Biol. Bull.* 195:89–97
59. Ofek I, Doyle RJ. 1994. *Bacterial Adhesion to Cells and Tissues*. New York: Chapman & Hall
60. Oliver JD. 1995. The viable but non-culturable state in the human pathogen *Vibrio vulnificus*. *FEMS Microbiol. Lett.* 133:203–8
61. Pierantioni U. 1918. Gli organi simbiotici e la luminescenza batterica dei Cefalopodi. *Publ. Staz. Zool. Napol.* 20:15–21
62. Placido R, Mancino G, Amendola A, Mariani F, Vendetti S, et al. 1997. Apoptosis of human monocytes/macrophages in *Mycobacterium tuberculosis* infections. *J. Pathol.* 181:31–38
63. Reich KA, Schoolnik GK. 1996. Halovibrin, secreted from the light organ symbiont *Vibrio fischeri*, is a member of a new class of ADP–ribosyltransferases. *J. Bacteriol.* 178:209–15
64. Reich KA, Beigel T, Schoolnik GK. 1997. The light organ symbiont *Vibrio fischeri* possesses two distinct secre-

ted ADP-ribosyltransferases. *J. Bacteriol.* 179:1591–97
65. Ruby EG. 1996. Lessons from a cooperative, bacterial-animal association: the *Vibrio fischeri–Euprymna scolopes* light organ symbiosis. *Annu. Rev. Microbiol.* 50:591–624
66. Ruby EG. 1998. Ecology of a benign "infection": colonization of the squid luminous organ by *Vibrio fischeri*. In *Microbial Ecology and Infectious Disease*, ed. E Rosenberg, pp. 217–31. Washington: Am. Soc. Microbiol.
67. Ruby EG, Asato LM. 1993. Growth and flagellation of *Vibrio fischeri* during initiation of the sepiolid squid light organ symbiosis. *Arch. Microbiol.* 159:160–67
68. Ruby EG, Lee K-H. 1998. The *Vibrio fischeri–Euprymna scolopes* light organ association: current ecological paradigms. *Appl. Environ. Microbiol.* 64:805–12
69. Seifert R, Schultz G. 1991. *The Superoxide-Forming NADPH Oxidase in Phagocytes*. New York: Springer. 338 pp
70. Singley CT. 1983. *Euprymna scolopes*. In *Cephalopod Life Cycles*, Vol. 1, ed. PR Boyle, pp. 69–74. London: Academic
71. Small Al, McFall-Ngai MJ. 1993. Changes in the oxygen environment of a symbiotic light organ in response to infection by its luminous bacterial symbionts. *Am. Zool.* 33:61A
72. Small AL, McFall-Ngai MJ. 1998. A halide peroxidase in tissues interacting with bacteria in the squid *Euprymna scolopes*. *J. Cellul. Biochem.* 72:445–57
73. Staley JT. 1997. Biodiversity: Are microbial species threatened? *Curr. Opin. Biotechnol.* 8:340–45
74. Tartaglia LA, Storz G, Ames BN. 1989. Identification and molecular analysis of *oxyR*-regulated promoters important for the bacterial adaptation to oxidative stress. *J. Mol. Biol.* 210:709–19
75. Tomarev SI, Zinovieva RD, Weis VM, Chepelinsky AB, Piatigorsky J, McFall-Ngai MJ. 1993. Abundant mRNAs in the bacterial light organ of a squid encode a protein with high similarity to mammalian antimicrobial peroxidases: implications for mutualistic symbioses. *Gene* 132:219–26
76. Van Rhijn P, Vanderleyden J. 1995. The *Rhizobium*-plant symbiosis. *Microbiol. Rev.* 59:124–42
77. Visick KL, Ruby EG. 1998. The periplasmic, group III catalase of *Vibrio fischeri* is required for normal symbiotic competence, and is induced both by oxidative stress and by approach to stationary phase. *J. Bacteriol.* 180:2087–92
78. Visick KL, Ruby EG. 1998. Tn*luxAB* insertion mutants of *Vibrio fischeri* with symbiosis-regulated phenotypes. *Abstr. Ann. Meet. Am. Soc. Microbiol.* 98:277
79. Volpi C, Borri M, Boletzky SV. 1995. Mediterranean sepiolidae: An introduction. In *Mediterranean Sepiolidae, Bulletin de l'Institut oceanographique, Monaco, Numero special 16*, ed. SV Boletzky, pp.7–14. Monaco: Musee Oceanographique
80. Wei SL, Young RE. 1989. Development of symbiotic bacterial bioluminescence in a nearshore cephalopod, *Euprymna scolopes*. *Mar. Biol.* 103:541–546
81. Weis VM, Montgomery MK, McFall-Ngai MJ. 1993. Enhanced production of ALDH-like protein in the bacterial light organ of the sepiolid squid *Euprymna Scolopes*. *Biol. Bull.* 184:309–321
82. Weis VM, Small AL, McFall-Ngai MJ. 1996. A peroxidase related to the mammalian antimicrobial protein myeloperoxidase in the *Euprymna-Vibrio* mutualism. *Proc. Natl. Acad. Sci. USA* 93:13683–88

Annu. Rev. Ecol. Syst. 1999. 30:257–300

The Relationship Between Productivity and Species Richness

R. B. Waide[1], M. R. Willig[2], C. F. Steiner[3], G. Mittelbach[4], L. Gough[5], S. I. Dodson[6], G. P. Juday[7], and R. Parmenter[8]
[1]*LTER Network Office, Department of Biology, University of New Mexico, Albuquerque, New Mexico 87131-1091; e-mail: Rwaide@lternet.edu;* [2]*Program in Ecology and Conservation Biology, Department of Biological Sciences & The Museum, Texas Tech University, Lubbock, Texas 79409-3131; e-mail: cmmrw@ttacs.ttu.edu;* [3]*Kellogg Biological Station and the Department of Zoology, Michigan State University, Hickory Corners, Michigan 49060; e-mail: STEINER@kbs.msu.edu;* [4]*Kellogg Biological Station and the Department of Zoology, Michigan State University, Hickory Corners, Michigan 49060; e-mail: mittelbach@kbs.msu.edu;* [5]*Department of Biological Sciences, University of Alabama, Tuscaloosa, Alabama 35487-0344; e-mail: lgough@biology.as.ua.edu;* [6]*Department of Zoology, University of Wisconsin, Madison, Wisconsin 53706; e-mail: sidodson@facstaff.wisc.edu;* [7]*Forest Sciences Department, University of Alaska, Fairbanks, Alaska 99775-7200; e-mail: gjuday@mail.lter.alaska.edu;* [8]*Department of Biology, University of New Mexico, Albuquerque, New Mexico 87131-1091; e-mail: parmentr@sevilleta.unm.edu*

Key Words primary productivity, biodiversity, functional groups, ecosystem processes

■ **Abstract** Recent overviews have suggested that the relationship between species richness and productivity (rate of conversion of resources to biomass per unit area per unit time) is unimodal (hump-shaped). Most agree that productivity affects species richness at large scales, but unanimity is less regarding underlying mechanisms. Recent studies have examined the possibility that variation in species richness within communities may influence productivity, leading to an exploration of the relative effect of alterations in species number per se as contrasted to the addition of productive species. Reviews of the literature concerning deserts, boreal forests, tropical forests, lakes, and wetlands lead to the conclusion that extant data are insufficient to conclusively resolve the relationship between diversity and productivity, or that patterns are variable with mechanisms equally varied and complex. A more comprehensive survey of the ecological literature uncovered approximately 200 relationships, of which 30% were unimodal, 26% were positive linear, 12% were negative linear, and 32% were not significant. Categorization of studies with respect to geographic extent, ecological extent, taxonomic hierarchy, or energetic basis of productivity similarly yielded a heterogeneous distribution of relationships. Theoretical and empirical approaches increasingly suggest scale-dependence in the relationship between species richness and productivity;

consequently, synthetic understanding may be contingent on explicit considerations of scale in analytical studies of productivity and diversity.

INTRODUCTION

The notion that productivity (rate of conversion of resources to biomass per unit area per unit time) affects species richness can be traced to at least the mid-1960s (45, 106, 113, 153). Nonetheless, the causal mechanisms behind the patterns between productivity and species diversity, as well as the form of the relationship, have been in dispute for almost as long (53, 193). Indeed, studies of the relationship between productivity and diversity at large spatial scales have documented linear and unimodal patterns as well as no patterns at all (see review in GG Mittelbach et al, in litt. and SI Dodson et al, 51a). Experimental manipulation of productivity via fertilization of small plots long has been known to decrease plant diversity (reviews in 48, 65, 82). Importantly, both theoretical considerations (147; SM Scheiner et al, in litt.) and empirical analyses (KL Gross et al, in litt.) suggest that patterns are likely scale dependent. Some of the disparity in perceived patterns may be a consequence of variation in the spatial scale of analyses.

Efforts to determine the relationship between number of species (or number of functional types, *sensu* 41) and the properties of ecosystems have increased as global loss of biodiversity and climate change have accelerated over the past decade. One approach to this issue has been to examine the ways ecosystem processes influence species number, community composition, or trophic structure (e.g., 84, 167, 168, 191). A separate line of inquiry has focused on the importance of the number of species, the number of functional groups, and the presence or absence of particular species (or groups) on ecosystem processes (e.g., 85, 102, 133, 134, 185, 190, 192). Field manipulations and laboratory experiments have addressed the role of these aspects of biodiversity in determining rates of ecosystem processes (e.g., primary productivity and nutrient cycling). These two lines of inquiry have been largely separate in the literature despite their conceptual linkage.

In this review, we synthesize existing knowledge of the relationship between a commonly estimated property of ecosystems (primary productivity) and one aspect of biodiversity (species richness). Most theoretical studies use net primary productivity (NPP) as the driving variable, but empirical studies often use components or surrogates of NPP. Rather than introduce confusing terminology, we use primary productivity in this paper as a general term to encompass components or surrogates of NPP. We review the literature and use case studies from terrestrial, aquatic, and wetland biomes for which detailed information is available. The work we report is an extension of research initiated at the National Center for Ecological Analysis and Synthesis that focused on the influence of primary productivity on species richness. In addition, we consider how species richness may affect ecosystem function (including productivity). This is a volatile and rapidly expanding area of study (see 1, 76, 79, 85, 107, 190). Unfortunately, the database currently is too small and conflicting to draw conclusions with certainty. Nonetheless, we suggest that it is necessary to bridge these two approaches to achieve a better understanding of the relationship between primary productivity and the dynamics of populations and communities. We do not address the related and important issue of the relationship between diversity and stability (49, 91, 189), nor do we discuss in detail other possible biotic and abiotic controls of biodiversity.

How Does Productivity Affect Species Richness?

Most authors agree that productivity affects diversity (32, 45, 106, 113, 162); moreover, a plethora of mechanisms have been proposed to explain how species richness responds to variation in productivity (e.g., 84, 168, 167, 191). Nonetheless, no general consensus concerning the form of the pattern has emerged based on theoretical considerations or empirical findings. Some factors enhance richness as productivity increases, others diminish richness as productivity increases, and some, in and of themselves, produce unimodal patterns (see below for details). Rather than any one mechanism having hegemony, it may be the cumulative or interactive effect of all such factors that determines the empirical pattern within a particular study. Indeed, future research should identify the ecological context and spatial scale that predispose systems to evince one pattern rather than another (154).

Rosenzweig (167) provided a critical assessment of the mechanisms thought to affect patterns in the relationship between diversity and productivity. GG Mittelbach et al (in litt.) updated the summary and provided commentary on the ecological scale at which mechanisms likely operate. Theories that predict a positive relationship between productivity and species richness include the species-energy theory (44, 155, 156, 223) and theories invoking various forms of interspecific competition in heterogeneous environments (2). Mechanisms thought to diminish diversity with increasing productivity are more controversial and include evolutionary immaturity (especially with respect to anthropogenic emendations); habitat homogenization (*sensu* 187; 65, 88); dynamical instabilities and system infeasibilities (125, 160, 162, 164, 165, 222; JC Moore & PC de Ruiter, submitted); and predator-prey ratios (141, 142, 162, 163). Some mechanisms predict a unimodal pattern in their own right. Relevant theories include changes in environmental heterogeneity with productivity (87, 186), tradeoffs in competitive abilities and abilities to resist predation (105), effects of competitive exclusion and environmental stress (6, 71, 72), disturbance and productivity (82), productivity-dependent species-area relations (147), and changing competitive structure (167).

Understanding the productivity-diversity relationship will require the imposition of order on this apparently chaotic array of possible explanations. This can be achieved, at least in part, by careful attention to the spatial and ecological scales at which patterns are detected (124, 167), and by equally judicious consideration of the spatial and temporal scales over which likely mechanisms operate. JM Chase & MA Leibold (submitted) make significant headway in this regard. They develop simple conceptual models based on exploitative resource competition or keystone predation to show that unimodal relationships emerge at local scales, whereas monotonically positive relationships emerge at regional or global scales. Empirical data for benthic animals and vascular plants in Michigan ponds corroborate the expected scale dependence in diversity-productivity relationships.

The Effect of Species Diversity on Productivity

There is general agreement that diversity of plant species is influenced by productivity (85, 192). The converse argument, that the number and kinds of species influence productivity, has been the subject of a recent series of field and laboratory experiments (79, 133, 192). These experiments have engendered a lively debate (85, 103, 211) that has yet to reach resolution. Increasing the number of species in either field (192) or laboratory (133) experiments may have positive effects on productivity and other ecosystem processes. However, the debate concerns whether these effects are the result of increased species richness per se, or the addition of different functional groups or particular species.

Theoretical approaches to this issue occur in three categories that postulate either (*a*) a positive, linear relationship (54, 112), (*b*) a positive, nonlinear relationship (52, 209), or (*c*) no relationship between species richness and productivity (102). MacArthur (112) hypothesized that energy flow through a trophic web would increase as the number of species, and hence pathways for energy flow, increased. Elton's (54) reformulation of MacArthur's hypothesis suggested that the relationship should be linear. If, however, species have redundant functions in ecosystems, the relationship between species number and ecosystem function may be nonlinear (52, 209). Ecosystem function changes rapidly as species representing new functional groups are added, but less rapidly when new species are redundant of existing functional groups. Lawton (102) proposed a model in which species may have strong, idiosyncratic effects on ecosystems. If this is the case, there is no predictable effect of species richness per se on ecosystem function. However, if the properties or functional traits of individual species are known, then we can predict which species will have strong effects and which will not. Such hypotheses present a useful framework for the evaluation of observations and experimental results.

Some experimental studies with herbaceous plants have shown an increase in net primary productivity with an increase in the diversity of species or functional groups (e.g., 132, 133, 185, 192). However, as Huston (85) has argued (see also 1, 190), there are at least two mechanisms by which the productivity of a trophic level may increase as the diversity (species, functional groups) of that trophic level increases.

1. Increasing the number of species initially present in a system increases the probability of encountering an exceptionally productive species (e.g., a species that is proficient in converting resources to biomass). Huston (85) has labeled this the "selection probability effect." In this scenario, the productivity of the trophic level is determined by the productivity of the most productive species.

2. Increasing the number of species initially present in a system can result in complementarity in resource use, if different species use different resources. In this case, the productivity of an assemblage of species will be greater than the productivity of any single species. Also, the total resource spectrum will be more completely used in the more species-rich system.

Empirical evidence from herbaceous plant communities (e.g., 79, 80, 85, 132, 185, 192) supports hypothesis 1. No experimental study supports hypothesis 2, despite its intuitive appeal. [data from Tilman et al (190) potentially support hypothesis 2, but no information is presented on complementarity of resource use that could be used to test this mechanism]. Complementarity in resource use is the functional basis behind intercropping and polyculture (184, 203). Although not all intercropping schemes or polycultures provide higher yields, many successful examples suggest that species complementarity may be important in determining ecosystem productivity and nutrient dynamics.

The difficulty of designing and executing field experiments to determine the effect of changing species number on productivity has resulted in a scarcity of published studies. The clearest results include field experiments conducted on communities dominated by herbaceous vegetation (79, 192) and a microcosm experiment conducted under controlled laboratory conditions (133).

In the study by Naeem et al (133), conducted in a controlled environment, a series of ecosystem processes was measured in high- and low-diversity communities. The lower diversity systems were nested subsets of the higher diversity systems. Estimates of primary productivity were greater in microcosms with higher species diversity.

Tilman & Downing (188) examined the relationship between plant species diversity and primary productivity in plots fertilized with N at the Cedar Creek Long-Term Ecological Research site in Minnesota. They reported that the productivity of more diverse plots declined less and recovered more quickly after a severe drought than did the productivity of less diverse plots. They concluded that the preservation of biodiversity is important for the maintenance of productivity in fluctuating environmental conditions. This study is not a direct test of the effects of diversity on productivity because species richness was not manipulated directly but was a product of changes in nitrogen addition, and because the study measured changes in productivity in response to disturbance.

In a different experiment, Tilman et al (192) compared productivity and nutrient-use efficiency in grassland plots seeded with different numbers of native species. Productivity was greater and soil mineral nitrogen was utilized more completely in plots with greater diversity. Measurements in nearby unmanipulated grassland showed the same pattern. This study also concluded that the preservation of biodiversity is necessary to sustain ecosystem functioning.

Huston (85) criticized the conclusions of all three of these studies, claiming that each was tainted by the lack of rigorous treatment of cause and effect. In Huston's view, appropriate tests of the effect of species richness on ecosystem processes do not permit large variation in the size or function of species. Huston argued that one likely consequence of increasing species richness in an experiment involves the increased probability of introducing a productive species (the "selection probability effect"). If this happens, the effect of increasing species richness on productivity is attributed simply to the increased odds of encountering species particularly well adapted to the environment. If variation in productivity exists among species used for an experiment, the effect of increased species diversity cannot be distinguished from the effect of increased functional diversity or mean plant size leading to differences in total biomass among treatments.

Huston's position, although admirable for its insistence on rigorously designed experiments, requires studies to be circumscribed to a limited range of the variability that exists in natural ecosystems. Experiments that incorporate only species of similar size and functional status, while avoiding some of the pitfalls that Huston (85) described, may not advance substantially our understanding of natural communities. Natural communities comprise species that differ in size and function; as a result, the effect of the loss of diversity is interpreted more easily through experiments that incorporate that variability. Moreover, the question of how similar species must be to achieve experimental rigor has not been addressed. Loreau (107) and Hector (76) have recently suggested mechanisms to separate the "selection probability effect" from other effects resulting from experimental manipulations of biodiversity.

Hooper (79) approached the question of complementary resource use through an experimental design that varied the number of functional groups in experimental plots. Four functional groups (early season annual forbs, late season annual forbs, perennial bunchgrasses, and nitrogen fixers) were planted in single-group treatments, as well as in two-, three-, and four-way combinations. In this experiment, no obvious relationship existed between functional diversity and productivity of the plots. The most productive treatment included only one functional group, perennial bunchgrasses. The identity of the species in the treatments was as least as important as the number of species in affecting ecosystem processes. Competition among some combinations of functional groups reduced productivity compared to single-group treatments. These results corroborate Lawton's (102) idiosyncratic model and Huston's (85) hypothesis 1.

For practical reasons, most experiments have focused on structurally simple ecosystems with relatively few species and have manipulated only a few species from each functional group. In general, results have shown a positive, asymptotic relationship between ecosystem processes and species richness. These results suggest that once all functional groups are present, the addition of species with redundant functions has little effect on ecosystem properties.

The conclusion that diversity is important for maintaining ecosystem function (188, 192), even if justifiable based on the few studies conducted to date, has been demonstrated only for systems in which the range of richness is from 0 to about 30 species. Conclusions about the importance of the addition or loss of species in complex systems require further clarification (171). Structurally complex, species-rich ecosystems, in which much of the loss of biodiversity worldwide is occurring, require further study.

Questions of Spatial Scale

It generally is recognized that area and environmental heterogeneity have strong effects on diversity (84, 167). Equally important, their effects are intertwined (98) and produce scale-dependent relationships between productivity and diversity. For example, a unimodal pattern in the relationship between diversity and productivity can be a consequence of a correlation between productivity and the parameters of the power function ($S = CA^z$, where S is species richness, A is area, and C and z are fitted constants equivalent to the intercept and slope, respectively, of the log-form of the relationship (7). In meadow communities dominated by sedges and grasses, Pastor et al (147) document that C has a positive correlation with area, z has a negative correlation with area, and this tradeoff produces a unimodal form to the relation between diversity and productivity. In contrast, no scale dependence was detected in the relationship between species richness and latitude for New World bats or marsupials, even though latitude is often considered a broad-scale surrogate for productivity (111).

Moreover, two aspects of spatial scale—extent and focus—strongly affect the detection and form of the relation between richness and diversity (SM Scheiner et al, in litt.). Extent is the range of the independent variable, which in this context is productivity, whereas focus defines the inference space to which variable estimates apply (i.e., the area from which samples were obtained to estimate point values for productivity and richness). In particular, a series of studies, each with restricted extent along a gradient of productivity, may evince significant (positive and negative) linear relationships as well as no relationship between productivity and diversity, casting doubt on the hump-shaped pattern. If the slope of the relationship decreases with mean productivity, then a unimodal pattern emerges as a consequence of the accumulation of consecutive linear relationships (positive linear, decreasing to no relationship, decreasing to negative linear; pattern accumulation hypothesis). Guo & Berry (73) document an emergent hump-shaped relationship from a series of linear patterns based on an analysis of plant species richness along a grassland-shrubland transition in Arizona.

Nonetheless, a unimodal relationship may emerge that is not a consequence of the accumulation of patterns at smaller extents. A series of fields dominated by vascular plants in the mid-western United States exhibits a unimodal relationship across grasslands or across North America, but no or negative relationships when the extent of analyses were restricted to be within community types (KL Gross et al, in litt.). Moreover, the slopes within community types were not correlated with mean productivity of the communities, suggesting that the pattern accumulation hypothesis was not in effect. In summary, relationships between diversity and productivity have been shown to be scale dependent, with the form of the scale-dependence variable from study to study, even in situations where unimodal patterns emerge at broad spatial scales.

BIOME-SPECIFIC RELATIONSHIPS

The relationship between diversity and productivity can be mediated by different control mechanisms, depending on biome. As a consequence, patterns within biomes may differ, and emergent patterns across biomes may not be the same as those within them. We explore the range of productivity and diversity in a number of major aquatic and terrestrial biomes, and we discuss the possible control mechanisms that lead to patterns between diversity and productivity. Although we have not attempted a comprehensive coverage of all biomes, it is clear that the state of knowledge is quite variable among systems. Moreover, the form of the relationship is not always clear, and understanding of the regulatory mechanisms is only rudimentary in most cases.

Aquatic Ecosystems

Lakes

Lakes are underutilized but optimal model systems for studying the relationship between species richness and primary productivity. They are well-delineated (bounded) communities in which species can be counted relatively easily, and in which primary productivity often is measured directly using standardized ^{14}C uptake methods (208). The ^{14}C method measures productivity between gross and net on a scale of minutes to hours. Annual levels of primary productivity are estimated by summing daily productivity. In lakes, the annual level of primary productivity ranges from about 1 to about 1300 g C m^{-2} yr^{-1} (51a).

Lacustrine species richness is influenced by lake primary productivity. Pure rainwater has a primary productivity of near zero because it lacks the nutrients necessary to support life. Consequently, rainwater in rock pools supports few or no species (50). The most productive lakes, such as sewage lagoons and temple tanks, are characterized by extreme conditions, such as high temperatures, no oxygen at night, and large diel shifts in pH (e.g., 61). These conditions can be endured by only a few specialized species. Lakes between these extremes of primary productivity generally have the highest species richness. Lakes of intermediate productivity, with sufficient nutrients to support photosynthesis but without extreme conditions, support the most species in virtually all groups of aquatic organisms (51a).

The size of the body of water interacts with primary productivity to determine the number of species in a lake (11, 35, 51, 81). An increase in lake area of ten orders of magnitude is associated with an increase in zooplankton species richness of about one order of magnitude (51). Indeed, over 50% of among-lake variability in richness of crustacean zooplankton in North American lakes is the consequence of lake size. Larger lakes have more zooplankton species, regardless of other factors including primary productivity.

SI Dodson, SE Arnott, KL Cottingham (51a) investigated the relationship between the primary productivity of lake ecosystems and the number of species of lacustrine phytoplankton, rotifers, cladocerans, copepods, macrophytes and fish. In a survey of 33 well-studied lakes, species richness of all six taxa showed a significant unimodal response to annual primary productivity (^{14}C estimate, g m^{-2} yr^{-1}) after lake area was taken into account. Moreover, the relationship between richness and primary productivity for phytoplankton and fish was strongly dependent on lake area. The highest richness occurred in lakes with relatively low primary productivity (~100 g m^{-2} yr^{-1}), such as those in the northern temperate lakes area in the upper Midwest (United States) and in the Experimental Lakes Area of Ontario, Canada. When temporal and spatial scales are considered, data for lake zooplankton and macrophytes provide striking examples of unimodal relationships between species richness and primary productivity. For small lakes (<10 ha), phytoplankton species richness peaked in low-productivity lakes, whereas for larger lakes, phytoplankton species richness merely declined in more productive lakes. For lakes less than 1 ha, fish species richness peaked at low levels of primary productivity. For larger lakes, a peak was not evident, although more productive lakes had more fish species.

The relationship between species richness and productivity has been studied only for phytoplankton and zooplankton in a few other lake and marine situations. We summarize the results of those studies below.

Phytoplankton Agard et al (5) analyzed data on marine phytoplankton species richness and the primary productivity of 44 oceanographic stations in the Caribbean to test predictions of Huston's (84) dynamic equilibrium model of species richness (maximum at intermediate productivity). They argued that marine phytoplankton would likely exhibit a relationship because of the relatively large number of species and the absence of confounding factors such as spatial heterogeneity. They reported that species richness was correlated positively with primary productivity, except at high levels where the curve reached a plateau. Fishery statistics (see 84) for the region show that the diversity of harvested marine species of commercial importance mirrors the diversity of phytoplankton.

In a 4–5 year study of three productive surface mines in Pennsylvania, phytoplankton diversity was correlated inversely with primary productivity (30). These sites are within the range of productivities explored by SI Dodson, SE Arnott, KL Cottingham (51a), and the results are consistent with those of that study.

Zooplankton Microcrustacean species richness is correlated with degree days for a group of shallow Canadian and Alaskan tundra ponds (75). These low-productivity ponds represent values along the ascending portion of the unimodal relationship reported by SI Dodson, SE Arnott, KL Cottingham (51a). Patalas (148) reported data for zooplankton and July temperature (which is an indicator of lake primary productivity) in Canadian lakes. Using these same data, Rosenzweig (167) found a unimodal relationship.

Wetlands

Coastal and inland wetlands have been the subject of much of the research on the relationship between plant species diversity and productivity. The mechanisms controlling species diversity along productivity gradients in marshes may differ from those demonstrated for terrestrial communities. In particular, selection for traits allowing survival in environments with high salinity and low soil oxygen may create low-diversity, high-productivity communities without the involvement of mechanisms such as competition.

In general, coastal marshes are some of the most productive ecosystems in the world. Tidal salt marsh plant communities are structured by salinity and flooding gradients creating distinct zonation with low plant species richness (20, 89). Primary productivity is high, up to 2500 g m^{-2} yr^{-1} (123). Mangroves replace tidal salt marshes in the tropics, with low plant species richness and highly variable productivity that depends on tidal influences, runoff, and water chemistry (123). Tidal freshwater marshes associated with rivers are more diverse because of monthly flooding and lower salinity levels (139) but are equivalently productive. Inland freshwater marshes and peatlands can be highly diverse, dependent somewhat on nutrient availability. Productivity is high, often exceeding 1000 g m^{-2} yr^{-1}, and, if belowground estimates are included, can exceed 6000 g m^{-2} yr^{-1} (123).

Plant species richness in coastal marshes decreases toward the coast along natural salinity gradients (4, 37, 101, 139) and may be influenced by storm-driven salt pulses (31, 57). Flooding and soil anoxia also decrease species richness (66, 68, 117, 123). The importance of interspecific competition, disturbance, and stress tolerance in determining species distributions has been demonstrated experimentally (18, 19, 97, 151, 181). Disturbance by herbivores (14, 58, 67, 138) and wave action (178) also may affect richness.

Salt marshes may be less productive than fresh water marshes because of the metabolic cost of tolerance to salinity (70, 139). Where freshwater flows into a coastal area, bringing nutrients or reducing salinity, productivity may be higher than in areas without freshwater input (47, 123, 225). Wave exposure also may restrict productivity (97, 220, 221). In inland and coastal marshes, soil characteristics such as pH, Ca, Mg, and anoxia may correlate with productivity (e.g., 17, 62). In tidal salt marshes, sulfides rather than high salt concentrations decrease productivity (218). Mammalian and avian herbivores restrict aboveground biomass accumulation in some marshes by removing plant material (9, 14, 59, 67, 138, 202), but they may have a stimulatory effect by adding nutrients through fecal deposition (78). Based on the results of fertilization experiments in marshes, vegetation frequently is limited by nitrogen or phosphorus (67, 126, 135).

When plant species density (the number of species per unit area) and productivity (estimated by harvests of peak standing crop) are examined in concert for marshes, the relationship between them depends on the scale of measurement and other factors. When productivity is increased by fertilization, plant species richness decreases (136, 204). When herbivores are excluded, productivity usually increases, accompanied by a decrease in plant species richness (14, 58, 67). Certain abiotic variables (e.g., salinity) may have similar effects on plant species density and productivity. However, the relationship between the two is not consistent in wetlands, although it is frequently unimodal (Table 1). In most cases, data are variable, with an outer envelope of points having a peak in species density at an intermediate level of standing crop (114). In some cases, the relationship reaches an asymptote, and plant species density does not decline over an extended range of standing crops (e.g., 69, 221). Moore & Keddy (124) demonstrated a unimodal relationship across community types, although there was no relationship between plant species density and standing crop within communities.

Approximate peaks in species density are found at a range of standing crop levels from 100 to 1500 g m^{-2} (Table 1). Once biomass reaches approximately 1000 g m^{-2}, plant species density rarely exceeds 10 species per m^2 and usually remains low. However, the range in plant species numbers is quite large at levels of biomass between 0 and 1000 g m^{-2}, suggesting that other variables affect species density at a particular level of standing crop. Stress tolerance plays an important role in survival in certain marsh habitats and may control species richness independently of other factors such as biomass (68, 69). The mechanism causing consistently low species numbers above approximately 1000 g m^{-2} standing crop remains unclear but is likely a combination of abiotic stresses and biotic interactions.

Terrestrial Ecosystems

Arctic Tundra

The arctic environment restricts the presence and productivity of vascular plant species (22, 42). Because of the severity of the environment and the common origins of the flora, approximately 2200 vascular species are known in the entire arctic region (22). Many of the abiotic factors believed to control productivity also play a role in controlling diversity. In particular, low temperature, a short growing season, low rates of soil nutrient cycling, permafrost, wind exposure, and extremes of soil moisture may constrain plant productivity (reviewed in 177). Various physiological and morphological adaptations (e.g., cold hardiness, short stature, vegetative reproduction) allow arctic tundra species to survive in such an environment (22, 23). On a smaller scale, topography can dramatically influence snow cover, exposure, soil drainage, and other physical properties of the substrate that may limit or enhance accumulation of plant biomass (175, 176). Generally, the most productive arctic plant communities are those dominated by deciduous shrubs or graminoids in areas of flowing water, where nutrient availability is higher and few other vascular species are present (39, 213). Nutrient availability

TABLE 1 List of herbaceous wetlands in which the relationship between plant species density (D) and standing crop has been examined

Wetland type	Location	Standing crop range (g m^{-2})	D range	Plot size (m^2)	Peak in D[a]	N	Relationship	R^2	p-value	Reference
Fen	England	600–4800	3–26	0.25	1000	34	Negative	—[b]	—	214
Fen	England, Wales	300–4000	2–50	4	500	85	Negative	0.23	0.0001	215
Freshwater lakeshore	Nova Scotia	0.4–580.8	0–23	0.25	120	121	+, unimodal	0.40	<0.001	221
Freshwater lakeshore	Ontario	0.1–900	0–12	0.04	100	63	Unimodal	0.48	<0.01	220
Freshwater shoreline	Ontario, Quebec	0–2600	2–24	0.25	400	224	Unimodal	0.34	<0.0001	124
Freshwater shoreline	Quebec	12–1224	1–24	0.25	500	48	Unimodal	0.41	<0.05	178
Freshwater shoreline	Ottawa	10–1100	4–12	0.25	250	7[c]	None	NS	NS	47
Coastal marsh	Louisiana	100–3900	1–11	1	1500	36	Negative	0.02	0.01	68
Coastal marsh	Louisiana	100–700	4–12	1	350	32	None	NS	NS	66
Coastal marsh	Louisiana	0–700	1–17	1	150	180	+, unimodal	0.09	—	69
Salt marsh	Spain	4–1280	1–25	0.25	400	50	Unimodal	0.29	<0.001	62

+Relationship was asymptotic.

[a]The peak in D was estimated visually from figures and is approximate.

[b]—Indicates that regression statistics were not reported.

[c]Only means were reported.

consistently limits productivity in tundra ecosystems, as demonstrated by many fertilization studies (40, 77, 94, 161, 174).

In general, the regional and local species pools in the Arctic are limited by extreme temperature, short growing season, low nutrient availability, low soil moisture, and frost disturbance (21, 27, 210, 213). Local areas of enhanced resources (e.g., animal carcasses) are occupied frequently by plants found in more productive sites, suggesting nutrient limitation of species composition as well as of productivity (118). Herbivory also may be a factor affecting arctic plant communities (12, 90), but it has been studied insufficiently (for an exception, see 13).

The relationship between productivity and diversity rarely has been addressed specifically in arctic tundra (for an exception, see 60). We present a summary of mean aboveground net primary productivity [annual net primary productivity (ANPP), g m^{-2} yr^{-1}] and mean species richness (S) of vascular species for each community. In the arctic, the association between S and ANPP is weak (Figure 1). To gain insight, we divided the data into two groups (High and Low Arctic) based on floristic and ecological considerations (25).

A positive linear relationship between species richness and ANPP is obvious in the High Arctic (Alexandra Fiord, Polar Desert, Polar Semidesert, Devon Island, Russia, and Barrow; $R^2 = 0.45$, $p \ll 0.001$), but no relationship characterizes the Low Arctic (Figure 1). In the High Arctic where plant cover is sparse, light competition is rarely important. When stressful conditions are ameliorated, more species inhabit more favorable areas. This is exemplified by small sites of increased moisture or temperature that are more productive and diverse than are drier or cooler sites (26, 64, 127, 128). The lack of a relationship in the Low Arctic likely is related to greater plant cover, causing both light and nutrient competition to be important in determining species richness. Perhaps as conditions become more favorable for higher plant productivity in the Low Arctic, light competition becomes more intense, countermanding the effect of productivity on species richness. The clear relationship between productivity and species richness in the extreme environment of the High Arctic suggests similar regulation of these two parameters by abiotic factors. In the Low Arctic, the relationship becomes less clear, possibly suggesting the importance of biotic regulation of species diversity in these communities.

Hot Deserts

Desert ecosystems are typically on the low end of the productivity gradient, ranging between 0 and 600 g m^{-2} yr^{-1} (Table 2). Productivity in desert ecosystems generally is limited by moisture availability and is highly variable in space and time (104, 170, 182, 212, 216). When rainfall is abundant for extended periods, nutrient limitation (particularly nitrogen) may regulate primary production (55, 74, 109, 129). Seasonal timing of precipitation determines the period and duration of primary production, with some deserts exhibiting primarily single season pulses of productivity (e.g., Mojave Desert in early spring, and Chihuahuan Desert in mid- to late-summer), whereas other deserts have bimodal productivity peaks

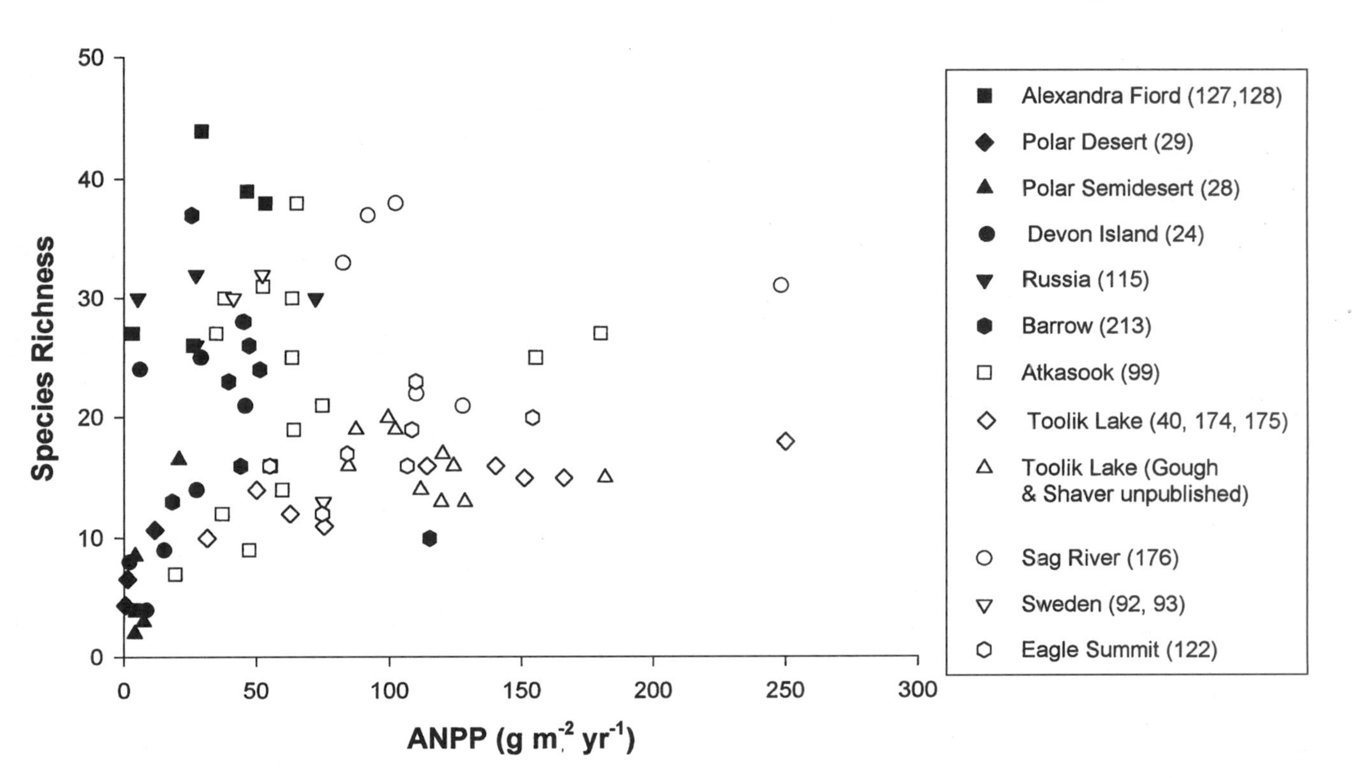

Figure 1 Scatter plot of the relationship between mean species richness and mean aboveground net primary productivity (ANPP) in arctic tundra. Site names in the insert are arranged by latitude, with Alexandra Fiord being farthest north. High Arctic sites include Alexandra Fiord through Barrow; Low Arctic sites include Atkasook through Eagle Summit.

TABLE 2 Productivity estimates of deserts

Site	Productivity/(g m^{-2} yr^{-1})	Reference
United States		
Great Basin Desert	125	150
Mojave Desert	0–62	15
	13–44	10
Sonoran Desert	9–95	149
	109	217
Chihuahuan Desert	50–100	43
Bajada: Alluvial fans	53–292	108
Bajada: Small arroyos	37–318	108
Bajada: Large arroyos	30–456	108
Basin: Slopes	51–179	108
Basin: Swales	292–592	108
Basin: Playa lake	52–258	108
Israel		
Negev Desert	4–43	144
India		
Rajastan	0–126	173
Mongolia		
Gobi Desert	0–155	96
Tunisia		
Pre-Sahara Desert	22–155	56

(e.g., the Sonoran Desert in both spring and late summer). Productivity in deserts is a consequence of slower growth of woody shrubs and succulents combined, with highly variable flushes of annual and perennial grasses and forbs. Comparisons of primary production in deserts indicate that belowground productivity may be considerably greater than aboveground (108 and references therein).

Species diversity ranges widely among deserts, depending on geographic location, biogeographic history, and extremes of moisture and temperature. However, many deserts support relatively large numbers of species (Table 3) and in many cases actually exceed the numbers of species found in other ecosystems with higher productivity. At a continental scale, the arid and semiarid ecosystems of the American Southwest support greater numbers of species of numerous taxa than do other ecosystems of North America (145, 146).

Although the relationship between productivity and species diversity in arid ecosystems has not been addressed specifically, certain relationships appear when comparing large-scale and small-scale patterns among and within deserts. Comparing large-scale patterns throughout the world, deserts that have near zero

TABLE 3 Numbers of species occurring in North American Deserts

Desert	Mammals	Birds	Reptiles	Amphibians	Plants	Reference
Great Basin	86	302	18	6	929	119
Mojave	53	332	37	3	1,836	169
Sonoran*	302	366	132	14		46
Chihuahuan	74	396	55	10		120

*Includes many island endemic species from the Sea of Cortez.

productivity (e.g., the polar desert of Antarctica, and various locations within continental deserts and dune fields) have concomitantly low species diversity, whereas deserts with hundreds of plant and thousands of animal species have comparably higher productivity. However, when comparing deserts of the same region (e.g., North America), the relationship between productivity and diversity is not as clear. For example, in North America (Table 2) the Mojave Desert has the lowest and the Chihuahuan Desert has the highest productivity, yet patterns of species numbers (Table 3) do not correspond to this pattern.

At smaller scales, various sites within a desert also differ markedly in productivity and plant species diversity. Ludwig (108) showed variations among years for productivity in different habitat types within the Chihuahuan Desert of New Mexico and found that sites that benefited from additional "run-on" moisture were most productive (Table 2). These sites typically were dominated overwhelmingly by a single grass species (*Hilaria mutica*) and exhibited high productivity with low diversity as compared to more diverse, but less productive, upland bajada slopes.

Tropical Forests

Copious data on productivity and species richness of tropical forests exist in the literature. However, detecting patterns between productivity and species richness is hampered by the great variability in environmental conditions in tropical forests and the lack of standardized sampling techniques. Ecosystems ranging from savanna to cloud forest often are grouped under the rubric of tropical forest. Soil nutrient availability, precipitation, temperature, and solar radiation exhibit strong variation, even within relatively homogeneous groupings such as lowland tropical forest. Efforts to control for these variables by selecting sites with comparable conditions reduce sample sizes to the point where statistical power is jeopardized. A search of the literature encountered 168 reported values for some measure of productivity (NPP, litter fall, leaf fall) for tropical forest sites, but only 15 were associated with any kind of measure of species richness. Joint measures of net primary productivity and species richness at the same scale are infrequent and usually restricted to woody plant species.

Reviews of studies of NPP in tropical forests have reported ranges of 6–16 t ha^{-1} yr^{-1} for tropical dry forest (131) and somewhat higher values for tropical evergreen forests [10.3–32.1 t ha^{-1} yr^{-1} (130); 11–21 t ha^{-1} yr^{-1} (34); 10.0–22.4 t ha^{-1} yr^{-1}

(121)]. A search of the recent literature did not encounter any more extreme values. Most, if not all, of the NPP measurements reported for tropical forests are based on increments of aboveground biomass and fail to include belowground productivity or losses to consumption of plant tissues. Partial measures of NPP are often used for comparison between sites. For example, there is an extensive database on litter fall (157) with values ranging from 1.0–27.0 t ha^{-1} yr^{-1}. A recent effort to estimate reasonable upper and lower bounds around the total NPP at 39 tropical forest sites resulted in ranges from 3.4–6.0 and 20.1–37.0 t ha^{-1} yr^{-1}, respectively (DA Clark, D Kicklighter, J Chambers, J Thomlinson, J Ni, E Holland, submitted).

Cumulative lists of species of plants and animals exist for diverse kinds of tropical forests, but their utility for comparison is questionable because of differences in the effort expended to construct such lists and the area upon which the lists are based. For these reasons, it is difficult to compare vertebrate species richness even in the best-studied tropical forests. Because a more standardized approach is used to count trees, it is possible to estimate the range in species richness for this group within tropical forests. For example, Phillips et al (152) reported a range of 56–283 tree species >10 cm dbh for 1 ha plots in mature tropical continental forests.

Three studies in tropical rain forest reported a positive relationship between species richness and rainfall (a surrogate for productivity). Gentry (63) interpreted positive correlations between tree species richness and annual precipitation, seasonality, and soil richness as support for a positive relationship between productivity and diversity. Huston (83) found tree-species richness positively correlated with annual precipitation and negatively correlated with soil fertility, and this was interpreted as a negative relationship between productivity and diversity. Phillips et al (152) reported positive relationships between tree species richness, climate (including increasing rainfall), and disturbance. They suggested that more dynamic systems have greater productivity, resulting in higher species richness. However, higher rainfall itself is related to increased forest disturbance and decreased soil nutrient concentrations because of leaching (83), demonstrating the tangle of cause and effect that can result when surrogates of productivity form the bases of analyses. No studies relating diversity directly to forest productivity are available to unravel this tangle.

Boreal Forests

Boreal forests occur in the coldest environments on earth in which trees survive and dominate vegetative cover. Nonetheless, a surprising diversity of climates and a wide range of ecosystem productivities are present in the boreal region. Because it controls rates of organic layer decomposition and thus the release of elements, soil temperature is a pervasive influence on productivity of boreal forests (194, 196, 197). Boreal forest productivity decreases with decreasing soil temperature, and increases with warming soil temperature (195, 197, 200), provided that moisture or other factors do not become limiting. Low soil temperatures reduce nutrient uptake particularly in higher (vascular) plants (38). In turn, soil temperatures are influenced by inherent site factors such as slope, aspect, and topographic position with respect to cold air drainage (200) and by factors that change during succession (e.g., 206). Large-scale stand disturbances warm boreal soils, especially by removing or thinning the insulating soil organic mat (198). Advancing succession rebuilds the organic layer, causing soil cooling and the build-up of high concentrations of refractory, low-quality forest litter that depresses productivity, particularly in conifer-dominated forest types (201).

On sites underlain by permafrost soil, rooting depth is restricted to the annually thawed active layer at the surface, and ground layer vegetation is dominated by mosses. Mosses filter and sequester incoming nutrients, restricting nutrient availability and productivity for rooted vegetation (140). Sphagnum moss dominance on permafrost sites produces organic soils of such high acidity that availability of particular nutrient elements is restricted for rooted vegetation. On larger river floodplains, nitrogen addition by alder (*Alnus*) shrubs promotes a substantial increase in productivity (95, 199). On low elevation sites in semi-arid central Alaska, a soil water-balance model is well correlated with basal area growth in white spruce (224), demonstrating that moisture can limit productivity on warm, dry south slopes as well. Belowground productivity in the boreal region only recently has been measured carefully, and recent progress on methodological problems suggests that most of the previous literature may not be reliable, especially because the high turnover of fine roots makes interval measurements of productivity problematic (158). Fine root production constitutes a large part of total production in boreal forests, accounting for 32% and 49% of total production in deciduous and coniferous stands, respectively, in central Alaska (159).

An often-cited value for average boreal forest aboveground net primary production is 2,700 kg ha^{-1} yr^{-1} (143). Productivity in the extensive larch forest and sparse larch taiga of Siberia typically ranges from 2500 kg ha^{-1} yr^{-1} to 1400 kg ha^{-1} yr^{-1} respectively (95). More recent multiyear measurements ofaboveground production in boreal forests of central Alaska include 9600 kg ha^{-1} yr^{-1} on a highly productive floodplain in peak alder/balsam poplar stage of development, and a range of 3600 kg ha^{-1} yr^{-1} to 4500 kg ha^{-1} yr^{-1} in 200 year-old floodplain and upland white spruce forest, respectively (159). Upland birch/aspen forest averaged 8100 kg ha^{-1} yr^{-1}, and poorly productive black spruce on permafrost averaged 680 kg ha^{-1} yr^{-1} (158). Long-term studies reveal a high degree of interannual variability in primary production.

Ecological studies and floristic surveys provide estimates of overall species richness in the boreal forest (Table 4). Total regional plant species richness is correlated positively with productivity, increasing from the less productive middle or northern boreal region to the more productive southern boreal region and boreal/temperate transition (adjusting for differences in intensity of sampling effort). Databases adequate for comparison of diversity and productivity with confidence at the local site and stand level do not exist, although some data are suggestive. Highly productive mature forests in the Bonanza Creek Long-Term Ecological Research Site have lower plant species density (e.g., 205), perhaps partly because

TABLE 4 List of studies in which estimates are provided for number of species of plants in boreal forests

Location	Vascular plants	Lichens	Mosses/ hepatics	Total plants	Number of plots	Reference
S. Ontario & Quebec	440	—	—	—	197	36
U.S. Great Lakes, Ontario	378	—	—	—	103	8
Alaska	375	107	70	552	103	110
Canada	—	—	133	—	60	100
Ontario	162	—	—	—	228	172
Finland	1,350	1,500	810	3,660	National	137
Sweden	2,000	2,100	1,000	5,100	National	137

of more complete usurpation of resources. This suggests a negative and linear relationship.

Succession plays a major role in the relationship between diversity and productivity as well. Most boreal forests are adapted to a stand-replacement disturbance regime, so they generally lack classic climax stages and an associated specialized complement of species. Total plant species richness increases during boreal forest succession, especially during primary succession (205), whereas productivity declines in later successional stages (207), suggesting a hump-shaped relationship through successional time. Many plant species of the forest understory persist (albeit at low abundance) throughout secondary succession, during which time productivity differs greatly depending on soil cooling and other influences of the stand (207).Species richness in late succession may be underestimated systematically in most of the literature because the difficult-to-identify cryptogams can constitute a large proportion, possibly even a majority, of the autotrophic plant species in boreal regions (Table 4). Vascular plant dominance is generally at a maximum during early succession, and cryptogam diversity and abundance are usually at a maximum late in succession (207).

Summary

Integration of results from the two aquatic and three terrestrial biomes discussed above suggest that there exists no universal pattern in the relationship between primary productivity and species richness. In most cases, patterns seem to change with scale, but data within biomes are inadequate for rigorous tests of this suggestion. Unimodal patterns are found in lakes, some wetlands, and through successional time in boreal forest. In terrestrial systems, a positive relationship pertains at regional or greater scales. In no case, however, are the data adequate to examine the relationship between primary productivity and species richness across scales and taxa. To address these issues, a broader survey of the literature is necessary.

SURVEY OF PATTERNS

Biome-specific consideration of studies leads to the conclusion that, one, extant data are insufficient to conclusively resolve the relationship between diversity and productivity, and two, patterns are variable, with mechanisms equally varied and complex. This is in sharp contrast to the broad claim that the unimodal pattern is among the few valid generalizations in ecology (84, 86, 166, 167). Indeed, a unimodal pattern has been heralded as the "true productivity pattern" (166) and as the "ubiquitous" pattern (87). We surveyed the published literature in ecology to assess such claims.

Data Acquisition

We conducted a literature search to examine patterns in the relationship between diversity and productivity. Using BIOSIS®, *Biological Abstracts*, and the search string: "species richness OR species diversity OR primary productivity OR production OR biomass OR rainfall OR precipitation," we searched *The American Naturalist, Oecologia, Oikos, Holarctic Ecology/Ecography, The Journal of Biogeography, The Journal of Ecology*, and *Vegetatio* for the years 1980–1997. We also manually searched these journals for the years 1968–1979 (pre-database). We combined results of this search with those from GG Mittelbach et al (in litt.). The latter study searched all issues of *Ecology* and *Ecological Monographs* to 1993 using the JSTOR® database (plus a manual search of issues between 1994–1997), and electronically searched a broad-spectrum of biological journals from 1982–1997.

In all cases, we included only studies with a sample size $\geq$4 that assessed a statistical relation (or presented data sufficient to calculate one) between species richness and productivity (or its surrogates), regardless of scale, taxon, or system. Agricultural and intensively managed systems were excluded, as were systems subject to severe anthropogenic disturbance. Systems whose potential productivities were manipulated experimentally were excluded as well.

Relationships between species diversity and productivity were classified into four types: linear positive, linear negative, unimodal, or no relationship. When possible, classifications were based on original published analyses. However, when proper statistics were not available, we used raw data to perform linear and quadratic regressions. Relationships were deemed significant if $P \leq 0.10$; regressions in which the quadratic term was significantly different from zero were classified as unimodal. Two studies produced significant "U-shaped" relationships (i.e., positive quadratic in the polynomial regressions). Because these relationships are rare, we have not included them in our figures; yet they were included when calculating percentages. Hence, histograms at some scales do not sum to 100%. More than 200 relationships between diversity and productivity were found in 154 articles. A tabulation of all studies surveyed, a summary of statistical results, and information on taxon, location, and measures of productivity are available on the World Wide Web in the Supplemental Materials section of the main Annual Reviews site (http://www.annurev.org/sup/material.htm).

We explored patterns in the relationships between species richness and productivity via five schemes of classification. We classified studies using an ecological criterion of scale as within community, across community, or continental-to-global scales. We used a shift in the structure of vegetation or plant physiognomy to define a change in community type (e.g., transitions from desert to grassland or from meadow to woodland). For most studies, we relied on descriptions of sites by the authors to generate the classifications. In a few cases, we classified studies based on knowledge of natural history. Studies whose sites were dispersed over distances greater than 4000 km were classified as continental-to-global. All studies of noncontiguous lakes, ponds, streams, or rivers were classified as across-community or as continental-or-global if the minimum distance criterion was met. Patterns at different ecological scales were explored for animals and plants separately. Our second classification scheme was based on the greatest geographic distance between sites within a study. We recognized four geographic scales: local (0–20 km), landscape (20–200 km), regional (200–4000 km), and continental-to-global (>4000 km). The third classification distinguished studies as terrestrial or aquatic, and further subdivided them into vertebrate, invertebrate, or plant. Our fourth method of classification focused on vertebrates and separately considered fish, mammals, amphibians and reptiles, and birds. We also tallied patterns for rodents separately because the literature review generated numerous studies of rodent diversity. The final classification considered whether the quantified measure of productivity was based on energy available to a trophic level or the energy assimilated by the trophic level.

The Patterns

The relationship between productivity and diversity differs with scale (Figure 2). Considering plants at the within-community scale, unimodal relationships are about as common as positive relationships (24 and 22%, respectively); however, most studies reported no pattern at all (42%). Though the proportion of studies that show no significant relationship remains large at the across-community scale, unimodal patterns are more than three times more prevalent than positive relationships (about 39% of studies compared to 11%). At the continental-to-global scale, the pattern is dominated by positive relationships (70% compared to 10% for unimodal relationships), and negative patterns are absent.

For animals, there was a less dramatic shift in the prevalence of unimodal versus positive relationships across biotic scales. Unimodal relationships predominate at the across-community scale, whereas positive relationships occur most commonly at the within-community and continental-to-global scales. As for plants, studies showing no relationship are numerous at the within- and across-community scales, but negative relationships are a clear minority, regardless of scale of classification.

Results of a geographic scale of classification (Figure 3) contrasted with those based on ecological scale. At the local scale (<20 km), studies of plant communities exhibited mostly unimodal relationships or no relationship at all. The dominance

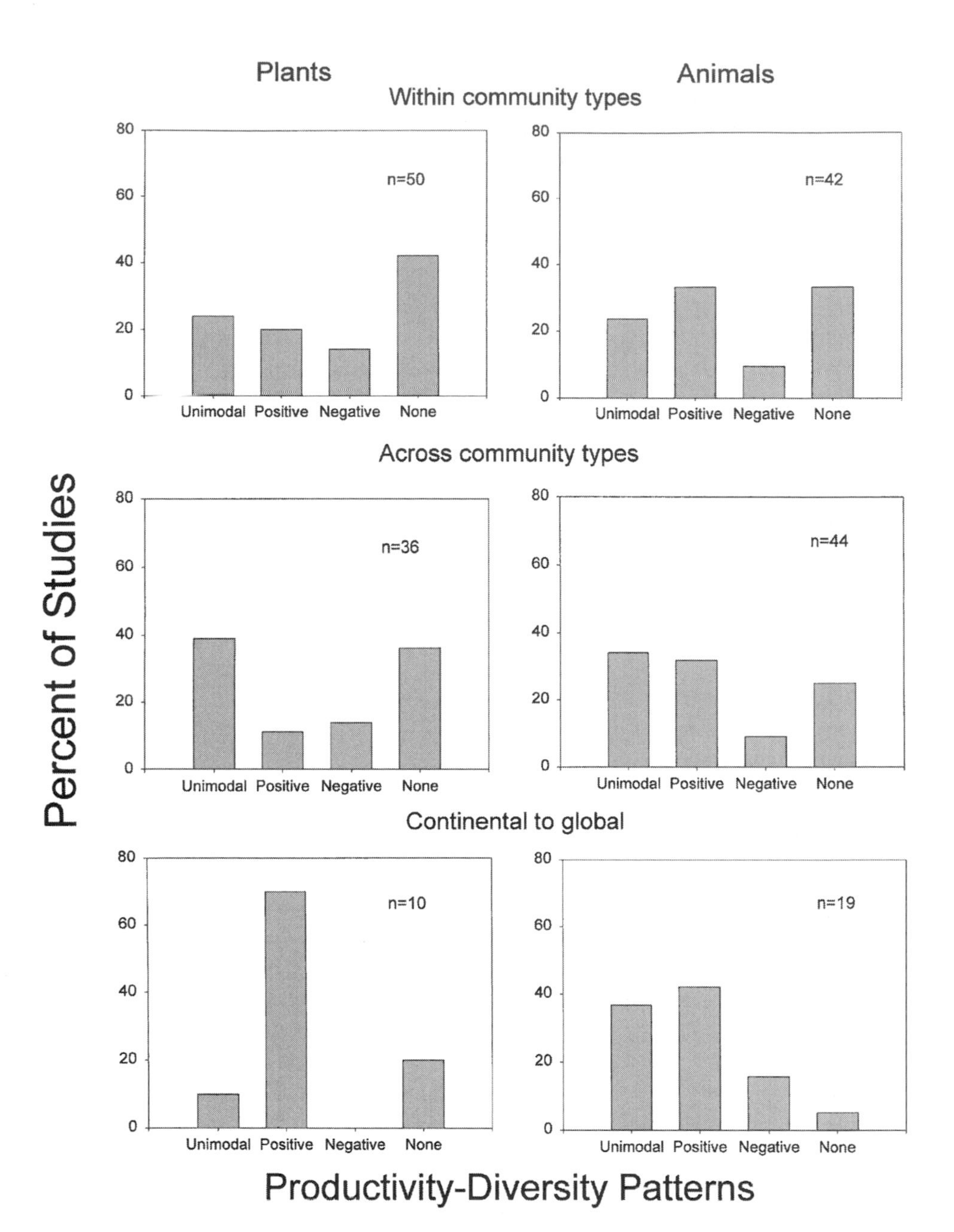

Figure 2 Percentages of published studies exhibiting particular relationships (positive linear, negative linear, unimodal, or no relationship) between species richness and productivity (or its surrogates) at each of three scales of ecological organization: within community types, among community types, and continental to global. Patterns are illustrated separately for plants and animals. Sample sizes refer to the number of analyzed data sets in each classification.

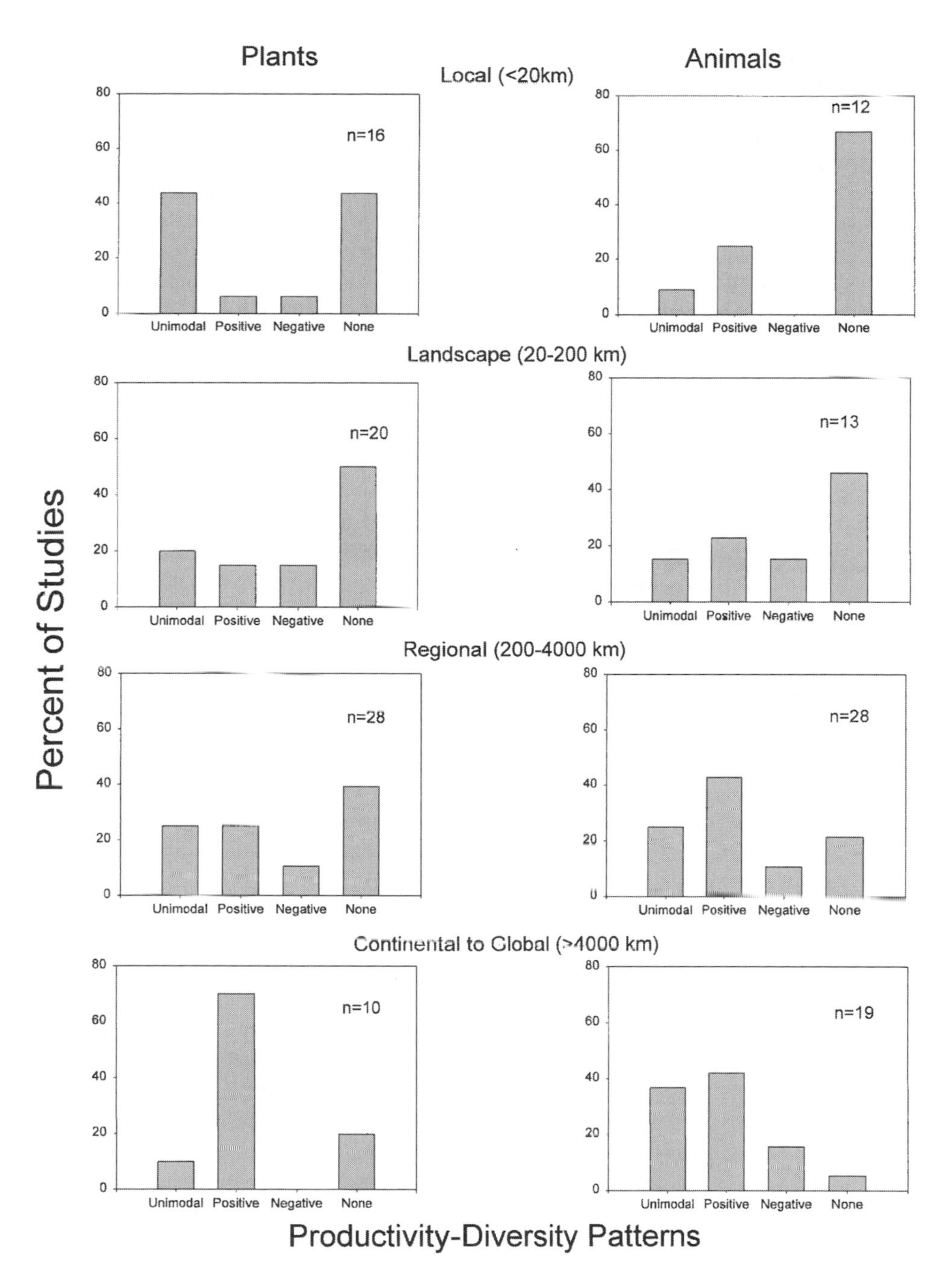

Figure 3 Percentages of published studies exhibiting particular relationships (positive linear, negative linear, unimodal, or no relationship) between species richness and productivity (or its surrogates) at each of four scales of geographic organization: local (<20 km), landscape (20–200 km), regional (200–4000 km), and continental to global (>4000 km). Patterns are illustrated separately for plants and animals. Sample sizes refer to the number of analyzed data sets in each classification.

of unimodality, relative to positive and negative relationships, declined for studies whose extents are at the level of the landscape (20–200 km) or region (200–4000 km). Again, studies showing no relationship were frequent at local to regional scales. Patterns at continental to global scales were the same as for the biotic classification; positive relationships predominated.

Studies of animals commonly exhibited no significant relationship between productivity and diversity at local to landscape scales (67% and 46%, respectively). However, when patterns occur, positive relationships between diversity and productivity were most prevalent at all geographic scales.

Our third method of classification focused on studies of vertebrates, examining productivity-diversity patterns for taxonomic groupings independent of scale. Most striking was the dominance of positive relationships for studies of birds and herpetofauna (Figure 4). In contrast, patterns were not as distinct for fish or mammals. A hump-shaped relationship was the most common pattern for fish. However, the proportions of unimodal and positive relationships were similar for fish and mammals. Most rodent diversity studies produced unimodal relationships between productivity and diversity.

When we divided studies into those concerning terrestrial and aquatic systems, striking differences became apparent. Positive relationships were more numerous in studies of terrestrial vertebrates, whereas unimodal relationships were more common in studies of aquatic vertebrates (Figure 5). Positive relationships predominated in studies of terrestrial invertebrates compared to a high percentage of unimodal relationships in studies of aquatic invertebrates. For both habitats, studies producing no relationships were numerous as well. Studies of plants in aquatic and terrestrial systems generally documented no relationship between diversity and productivity. For those studies that did show a significant relationship, a higher percentage of unimodality exists in aquatic systems.

Clearly, considerable variation characterizes the relationship between productivity and diversity, even after controlling for aspects of ecological, geographic, or taxonomic scale. Part of this variability may be a consequence of the way in which productivity was assessed for a particular site (available energy versus assimilated energy) or the power of statistical tests used to assess relationships. To assess the degree to which these factors may have affected the pattern or distribution of relationships (i.e., unimodal, positive linear, negative linear, no relationship), we conducted a hierarchical G-test (183). In general, contrasts were orthogonal and based on a priori considerations of energy, nested within habitat, nested within taxon (Figure 6). A final comparison of the pattern for all studies versus only studies with sample sizes greater than 10 was conducted for heuristic purposes (shaded portion of dendrogram in Figure 6). With one exception (aquatic animals based on all studies), the distribution of relationships was indistinguishable in contrasts between studies involving assimilated versus available energy. In addition, no significant differences in the distribution of relationships were detected for studies based on any other contrasts with respect to habitat, taxon, or data. These

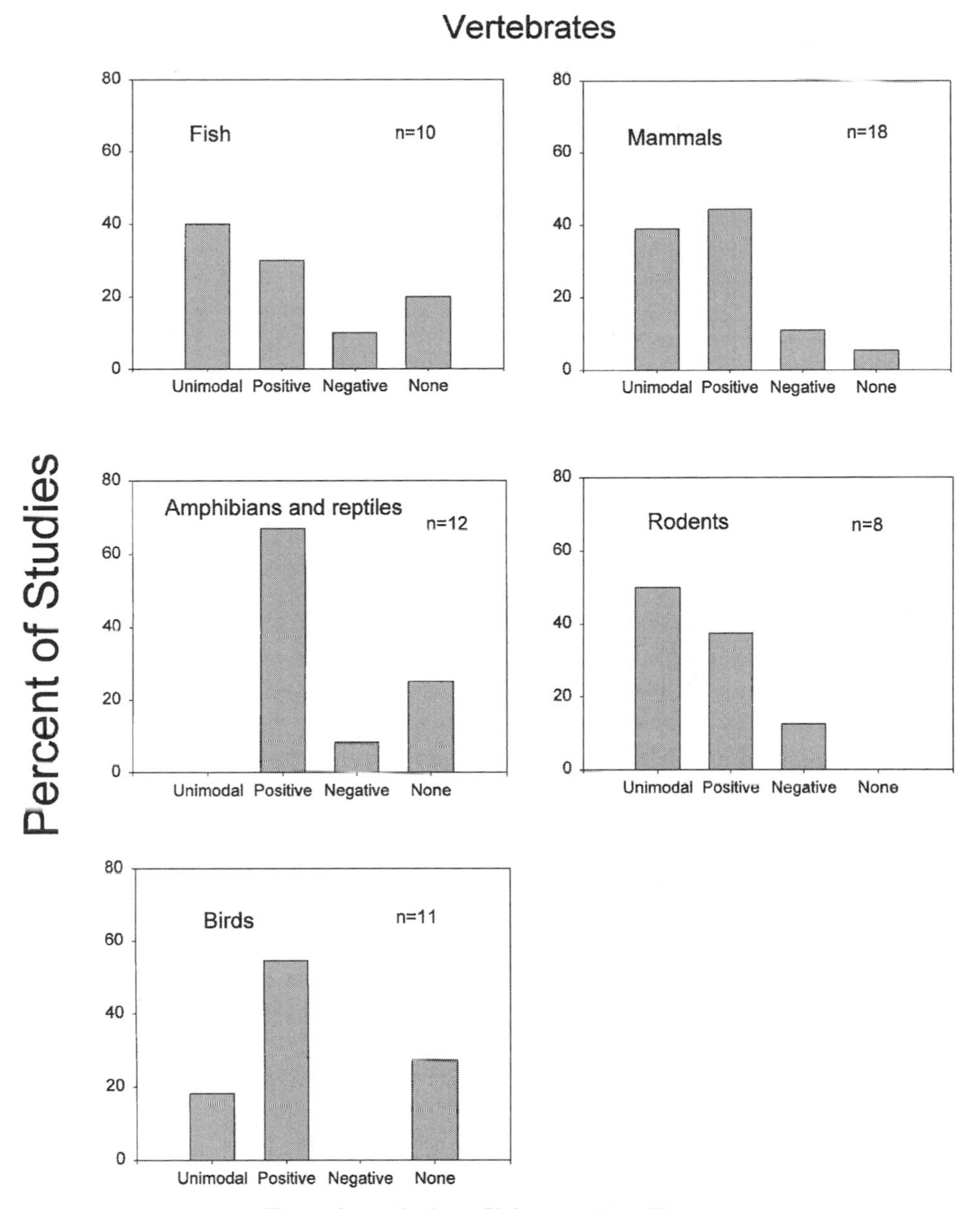

Figure 4 Percentages of published studies exhibiting particular relationships (positive linear, negative linear, unimodal, or no relationship) between species richness and productivity (or its surrogates) for each of five groups of vertebrates: fish, amphibians and reptiles, birds, mammal, and rodents. Sample sizes refer to the number of analyzed data sets in each classification.

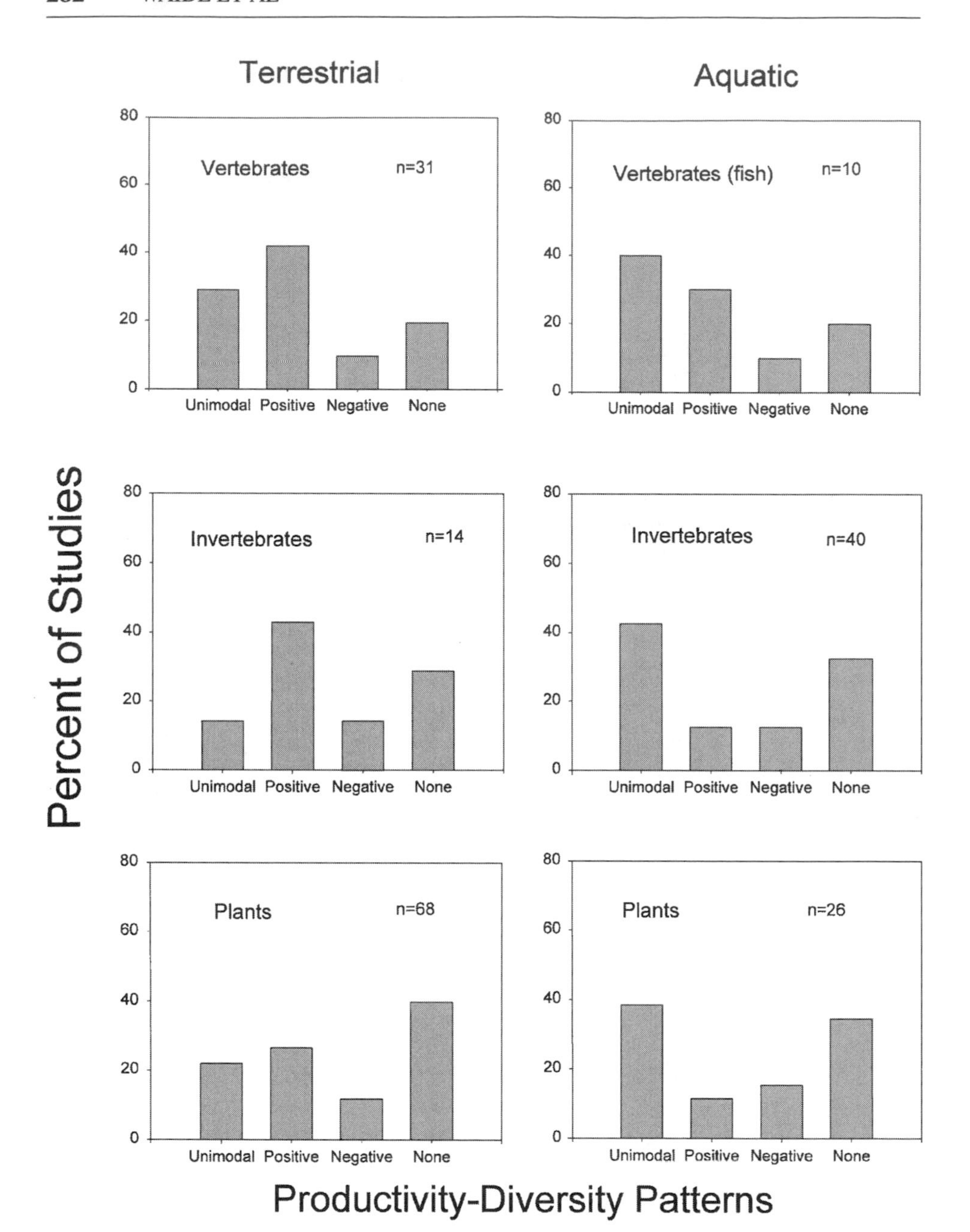

Figure 5 Percentages of published studies exhibiting particular relationships (positive linear, negative linear, unimodal, or no relationship) between species richness and productivity (or its surrogates) for each of three groups (vertebrates, invertebrates, and plants) in terrestrial and aquatic environments separately. Sample sizes refer to the number of analyzed data sets in each classification.

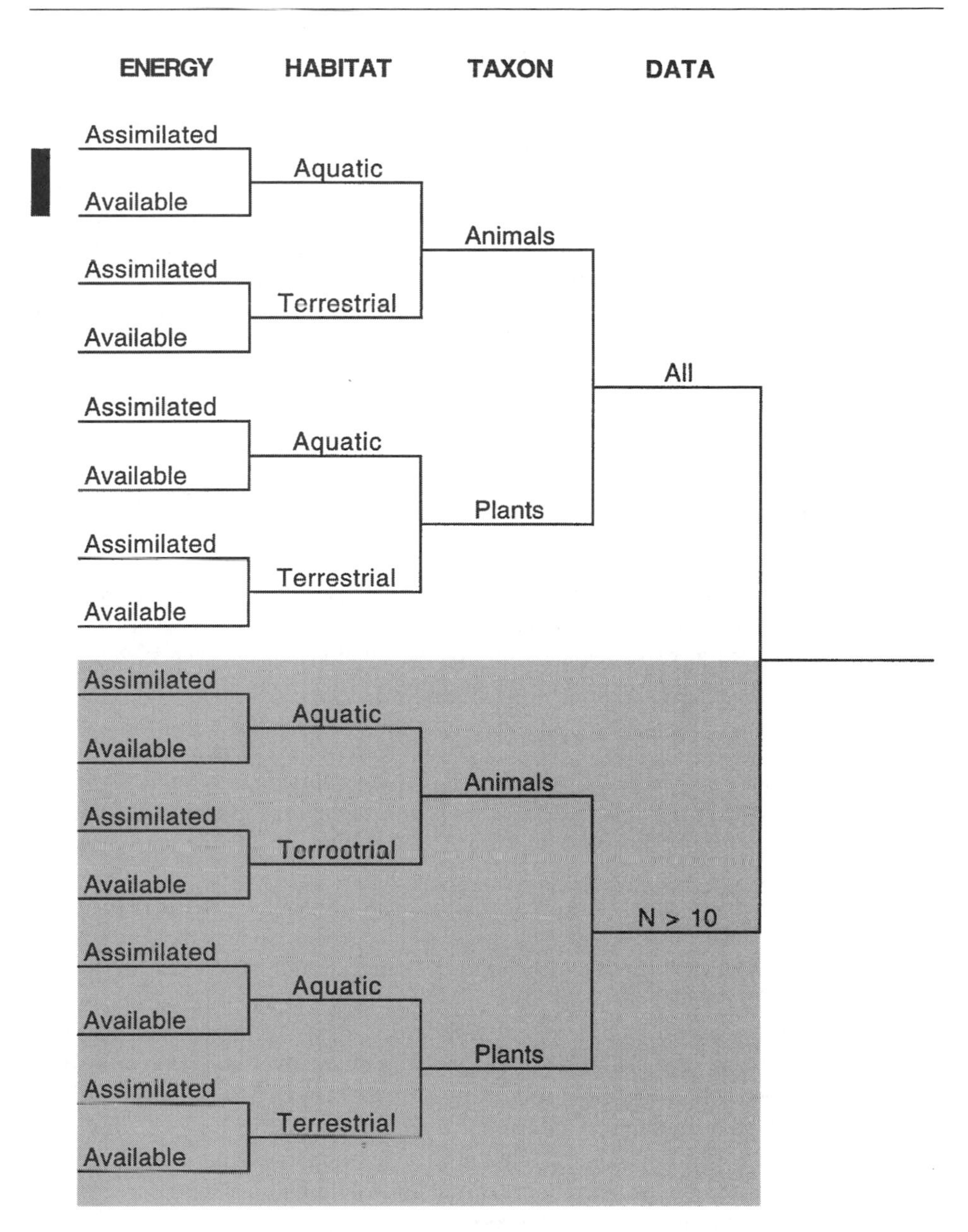

Figure 6 Dendrogram illustrating the results of a hierarchical G-test assessing differences in the distribution of relationships (positive linear, negative linear, unimodal, and no relationship) between studies classified in a nested fashion with respect to energy (available versus assimilated), habitat (aquatic versus terrestrial), taxon (animals versus plants), and data (all versus $N > 10$). Analyses conducted for heuristic purposes are shaded in gray. A statistically significant contrast ($P < 0.05$) is indicated by a black vertical bar.

results, especially when combined with those based on classifications of studies based on ecological and geographic scale, suggest that no single relationship has hegemony, and that at best, the data are insufficient to corroborate the priority of any one relationship between diversity and productivity.

Regardless of the manner of categorization, no single relationship described more than two thirds of the studies. This is surprising, given that the relationship between species richness and productivity is often characterized in the literature as unimodal (e.g., 16, 87, 167, 191). This suggests that no mechanism has a dominant role in molding patterns, that multiple mechanisms may operate simultaneously, or that confounding factors or methodological limitations (e.g., sample size or extent) conspire to producc the apparent diversity of relationships between productivity and diversity. We explore some of the factors that may be involved in producing a variety of patterns at different scales.

Conceptual Issues

Determining the manner in which productivity-diversity patterns change with scale is an important first step toward more fully understanding the applicability, limitations, and predictive power of ecological theory (111). We have benefited from several decades of experimental study of the relationship between system productivity and species richness, and a healthy body of theory complements this empirical data base. What has not been made clear is the applicability of theory to patterns at different spatial scales. The ambiguity of the proper scale of application is most obvious at the within-to-across-community scale, or at the local-to-landscape-to-regional scale. Consider as an example the graphical theories of community structure proposed by Tilman (186). When combined with resource heterogeneity, Tilman's model predicts a unimodal species diversity response to increasing nutrient supply (i.e., potential primary production). What is not clear from the model is the scale at which the pattern manifests. In fact, the theory itself can predict any type of pattern depending on range of productivity, location along a productivity gradient, heterogeneity of resources, and manner in which resource heterogeneity covaries with resource supply (2, 3, 186). Ambiguities of the scale of operation and application also may apply to alternative theories proposed to explain productivity-diversity relationships, such as the keystone predator model which also predicts hump-shaped relationships (105; for reviews of additional theories see 3, 167, 191). Determining how patterns change with scale in natural systems provides the first step in understanding the limitations and proper scales of application for theoretical frameworks.

The scale of a study may be an important factor to consider when predicting the relationship between diversity and productivity. Both geographic- and biotic-based classifications need to be considered. Unimodal patterns may emerge at relatively small spatial scales. Almost half of the studies of plant communities that reported distances between study sites of less than 20 km (local scale) produced unimodal patterns. When studies remain within a community, patterns are almost equally divided between unimodal and positive categories for plants. This clearly changes when studies cross communities: Unimodal patterns become much more pronounced. At geographic scales greater than 20 km (landscape-to-regional scales), a larger proportion of positive patterns occurs for plants.

The range of productivities encompassed within a study may explain the decoupling of geographic and biotic classifications of scale for plants. Although we expect the two to covary (i.e., studies that are spread over large geographic ranges most likely have larger ranges of productivity), the relationship need not always hold. Investigations at small spatial scales may traverse large gradients of potential productivity, as well as multiple community boundaries, (e.g., elevational gradients from woodland to tundra). Conversely, single community types may span huge distances and several geographic scales, yet may exhibit little variation in productivity or species richness. Proponents of the unimodal pattern commonly argue that patterns without humps are a result of insufficient ranges of productivity in the study (e.g., 167, 168). Restricting studies to a single community type, from the outset, constrains them to a limited range of species compositions. If unimodal relationships occur primarily across communities, then within-community studies may be sampling only portions of the whole productivity-diversity curve. There is evidence that unimodal relationships emerge only when data from different communities along the productivity gradient are accumulated (e.g., 124, 121; KL Gross et al, in litt.). Thus, future research should consider explicitly the range of productivity sampled, which may be a major factor driving the change in plant community type.

Despite the predominance of unimodal patterns at the across-community scale, they can occur within communities (about 24% of plant and animal studies), but at this scale nonsignificant relationships are also numerous (42%). In addition to smaller productivity ranges at this scale, a number of ecological processes occur at smaller spatial and within-community scales, which may result in nonsignificant relationships. For example, dispersal between patches of high species diversity and sink patches of low diversity may mask productivity-diversity patterns. Such mass or rescue effects (33, 179, 180) likely occur at smaller spatial scales in which patches are in close proximity and immigration rates are high.

Studies of animals showed no dramatic changes in the frequency distribution of relationships across biotic or geographic scales. Positive relationships always outnumbered unimodal relationships, although the differences were small at the within- and among-community scales. Differences were more pronounced across geographic scales. Almost without exception, studies of animal diversity focused on subsets of the animal community (specific taxa such as rodents or amphibians). Most theoretical explorations of productivity-diversity phenomena deal exclusively with the species richness of whole trophic levels or guilds (e.g., 105, 186). Although models can be adapted to deal with more restricted taxonomic groups, the predictions may be very different. This is especially so because most studies of animals not only deal with subsets of trophic levels, but also consider organisms that have different feeding ecologies (e.g., studies of aquatic macroinvertebrates can include primary and secondary consumers as well as detritivores).

Along gradients of productivity, taxonomic turnover may occur such that focal taxa within a trophic level drop out of the system, while other taxonomic groups (with the same feeding ecology) replace them. Although overall species diversity of the trophic level may show one pattern along the gradient, the focal group may exhibit a completely different one. Hence, the ability to assess the applicability of ecological theory and the influence of scale on productivity-diversity patterns for animals is limited.

The patterns and explanations presented thus far have dealt with studies at regional and smaller scales (<4000 km). Theory at this spatial scale deals primarily with communities assembled from presumably co-evolved regional species pools. Although our cut-off point of 4000 km is arbitrary, we hoped to distinguish between these types of studies and those whose communities may derive constituent species from different regional pools. This most likely occurs at the scale of whole continents or across continents (i.e., global scales). The results of our literature review indicate that species richness is primarily a positive function of productivity at this larger scale, for both animals and plants. Unimodal patterns were abundant for animals, though due perhaps to previously mentioned factors. These studies often include sites along gradients of latitude and can include species pools of different ages and evolutionary histories. Distinguishing the ecological effects of available energy from the evolutionary effects is difficult.

Despite the long history of interest in factors governing large-scale patterns of diversity, consensus remains elusive. Many potential problems accompany any literature review that gathers data from a wide variety of studies using disparate approaches, methods, and foci. Many of the caveats and shortcomings of our review provide guidance for future improvements in assembling productivity-diversity patterns. First, most of the studies we surveyed use a correlate of productivity, often an indicator of assimilated energy (such as standing crop biomass) or available energy (such as rainfall, latitude, evapotranspiration, or soil nutrients). In general, we expect these variables to be indicators of system productivity; nonetheless, correlations may be poor for some systems or at certain times of the year. Aboveground biomass is one of the most popular correlates in plant studies, but simple models of trophic regulation can predict complete decoupling of trophic-level biomass from productivity depending on the trophic structure of the system and the feeding efficiency of consumers (141). The studies included in our review use a great variety of quadrat and plot sizes, and rarely are area effects explicitly addressed or controlled.

Many different relationships between species diversity and productivity can be generated at a single biotic or geographic scale. Yet, the relative percentages of different patterns change with scale. Unimodal patterns have been described as textbook examples of productivity-diversity relationships (16). Our review is noteworthy for the lack of studies evincing a significant hump (despite our generous criteria for detecting one). This is especially true for animals, in which positive patterns dominate at almost all scales. Our review in no way discounts the models and mechanisms that predict hump-shaped relationships, but it does attest to the potential importance of scale when applying such models and predictions. Exciting future directions include investigating why patterns change with scale; why in some systems unimodal patterns are generated at the within-community and local scales, whereas in others unimodal patterns only emerge when crossing community boundaries or large geographical distances.

FUTURE STUDIES

Two important issues facing the scientific community are the maintenance of global biodiversity and the continuance of the ecosystem services necessary to support human life. It is clear from numerous studies that these issues are inextricably entwined (41). Modeling and empirical studies demonstrate that loss of biodiversity can influence key ecosystem characteristics such as primary productivity, predictability, and resistance to invasion by exotics (41, 116). Theory and empirical studies indicate that changes in primary productivity are related to species richness at some scales but not at others. The goal of future research must be to provide mechanistic explanations for observed patterns in the relationship between primary productivity and species richness through well-designed and carefully interpreted experiments (85) that explicitly consider spatial scale as well as local and regional mechanisms.

A key strategy for improving our understanding of the interaction of biodiversity and productivity (or other ecosystem processes) considers the integration of two common experimental approaches: the manipulation of productivity and the alteration of the number of species or functional groups. A synthesis of ideas that have developed around these two approaches is a prerequisite for the advancement of a general theory that will direct the next generation of hypotheses and experiments. Conceptual models being developed by JB Grace (68a) and M Shackak (personal communication) foreshadow this synthesis. These emerging models incorporate disturbance, plant biomass (productivity), resource heterogeneity, colonization, and the available species pool as primary factors controlling species density. Consequently, they emphasize the importance of multivariate approaches to understanding patterns of species density (Figure 7).

Central to understanding the role that humans play in the present observed high extinction rate is the relationship between anthropogenic disturbance and the natural disturbance regime (219). The first attempts to explain the control of species richness had an explicit appreciation for the importance of human activities (71, 72), which led to an integration of disturbance, environmental stress, and elements of productivity in an index of factors controlling species richness (6; 68a). Future studies need to refocus on the similarities and differences between natural and anthropogenic disturbance. Incorporation of the unique nature of human

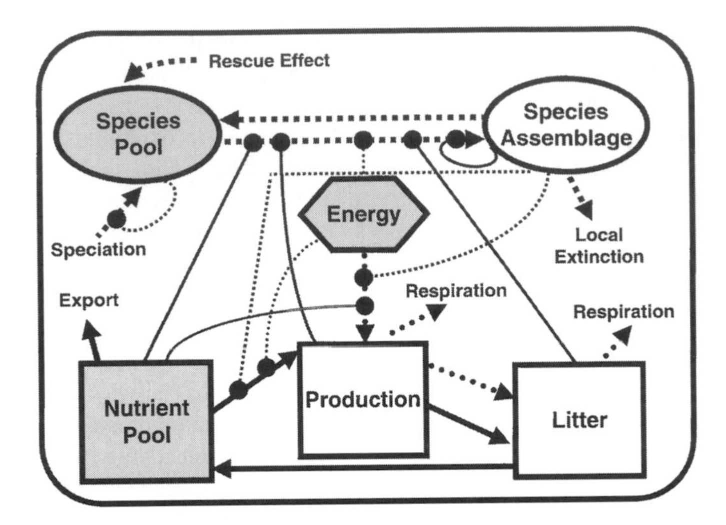

Figure 7 Conceptual model of a local ecosystem: the biota interacting with energy flow and nutrient cycling (unshaded shapes). For simplicity's sake, the model only considers the plant community. Shaded shapes represent larger-scale species pools, nutrient pools, or energy inputs. Broad arrows represent the various system currencies of species (dashed), energy (dots) or nutrients (solid); narrow lines terminating in a circle represent regulators or filters which modify the flow of species, nutrients or energy. Hence, the composition of a local assemblage is derived from a regional species pool via the action of five filters: available energy, nutrient availability, and the species already in the community, as well as structural characteristics of the production and litter. Similarly, the production of the ecosystem is affected by available energy as processed by species assemblages and constrained by nutrient availability. Because energy flow and species flow are affected by some of the same regulators, and in fact reciprocally affect each other, the relationship between them may be complex.

activities into models of the relationship between biodiversity and ecosystem processes is necessary to merge the fields of evolutionary ecology and conservation biology.

The time is appropriate for the study of the relationship between species richness and primary productivity to change focus from discerning patterns to developing mechanistic explanations, which can be tested through manipulative or observational experiments. The available evidence shows that multiple patterns exist and change with scale. The implication is also clear that multiple causal factors exist for scales, habitats, and taxa. There is reasonably strong evidence to demonstrate that productivity influences diversity at some scales, whereas functional or species diversity seems to influence productivity at other scales. Clever experiments and observations based on conceptual models of system dynamics (Figure 7) will be needed to disentangle the web of cause and effect. With this in mind, we offer in conclusion a few general ideas concerning the characteristics of future research endeavors.

- Investigators must be careful to match the scale upon which theory operates to the scale of observation. In many cases this will require the collection of new data at the appropriate scale to test theory.
- Some standardization in operational definitions is necessary for meaningful comparison. In particular, the spatial and temporal framework for the measurement of species density and productivity must be carefully controlled. Theory that is based on net primary productivity cannot be evaluated using partial measures of NPP. Similarly, theory that is based on guilds or communities cannot be evaluated using subsets of these communities.
- Multivariate approaches are needed to separate the effects of co-varying causal factors. Investigators must recognize that different species may respond to different variables along the same geographic gradient and that changes in total species richness are the sum of these species-level responses.
- Experiments must include several trophic levels and multiple ecological scales. Without this kind of experimental approach, results will be difficult to place in context.
- Theory and experimentation need to be extended to high-diversity systems. Microcosms provide a useful approach for addressing basic questions, but issues relating to the loss of taxa from species-rich systems urgently require attention.
- More sophisticated manipulations of productivity at multiple scales will be required to determine the generality of the pattern between productivity and species richness. In particular, manipulations of limiting resources that increase heterogeneity of resource availability would provide an interesting contrast to standard fertilization experiments.

ACKNOWLEDGMENTS

This work resulted from a workshop conducted at the National Center for Ecological Analysis and Synthesis (NCEAS), a Center funded by the National Science Foundation (Grant #DEB-94-21535), the University of California at Santa Barbara, and the State of California. MRW was supported in part as a Sabbatical Fellow on the same grant and received additional support from a Developmental Leave from the Office of the Provost, Texas Tech University. We appreciate the assistance provided by the staff at NCEAS, in particular Jim Reichman, Matt Jones, Mark Schildhauer, and Marilyn Snowball. The Long-Term Ecological Research Network Office of the University of New Mexico provided support for RBW through Cooperative Agreement No. DEB-9634135, as did grant DEB-9705814 from the

National Science Foundation to the Institute for Tropical Ecosystems Studies, University of Puerto Rico, and the International Institute of Tropical Forestry as part of the Long-Term Ecological Research Program in the Luquillo Experimental Forest.

The following participants in the workshop contributed to the discussion leading to the manuscript: Linda Blum, Scott Collins, Stephen Cox, Katherine Gross, Jeff Herrick, Michael Kaspari, Clarence Lehman, John Moore, Glenn Motzkin, Craig Osenberg, Michael Rosenzweig, Samuel Scheiner, Lee Turner, Maria Vernet, and Bruce Wallace. The manuscript was substantially improved as a result of comments by Steven Chown, Deborah Clark, Stephen Cox, James Grace, and Moshe Shackak. Louise Williams prepared the tables, and Leida Rohena and Saioa de Urquiza helped with the bibliography.

LITERATURE CITED

1. Aarssen LW. 1997. High productivity in grassland ecosystems: effected by species diversity or productive species? *Oikos* 80: 183–84
2. Abrams PA. 1988. Resource productivity-consumer species diversity: simple models of competition in spatially heterogeneous environments. *Ecology* 69:1418–33
3. Abrams PA. 1995. Monotonic and unimodal diversity-productivity gradients: What does competition theory predict? *Ecology* 76: 2019–27
4. Adams DA. 1963. Factors influencing vascular plant zonation in North Carolina salt marshes. *Ecology* 44:445–56
5. Agard JBR, Griffith JK, Hubbard RH. 1996. The relation between productivity, disturbance and the biodiversity of Caribbean phytoplankton: applicability of Huston's dynamic equilibrium model. *J. Exp. Mar. Biol. Ecol.* 202:1–17
6. Al-Mufti MM, Sydes CL, Furness SB, Grime JP, Band SR. 1977. A quantitative analysis of shoot phenology and dominance in herbaceous vegetation. *J. Ecol.* 65:759–91
7. Arrhenius O. 1921. Species and area. *J. Ecol.* 9:95–99
8. Auclair AN, Goff FG. 1971. Diversity relations of upland forests in the western Great Lakes area. *Am. Nat.* 105:499–528
9. Bakker JP. 1985. The impact of grazing on plant communities, plant populations and soil conditions on salt marshes. *Vegetatio* 62:391–98
10. Bamberg SA, Vollmer AT, Klienkopf GE, Ackerman TL. 1976. A comparison of seasonal primary production of Mojave Desert shrubs during wet and dry years. *Am. Midl. Natur.* 95:398–405
11. Barbour CD, Brown JH. 1974. Fish species diversity in lakes. *Am. Nat.* 108:473–89
12. Batzli GO, Collier BD, MacLean SF, Pitelka FA, White RG. 1980. The herbivore-based trophic system. In *An Arctic Ecosystem: The Coastal Tundra at Barrow, Alaska*, ed. J Brown, FL Bunnell, PC Miller, LL Tieszen, pp. 335–410. Stroudsberg, PA: Dowden, Hutchinson & Ross
13. Batzli GO, Henttonen H. 1990. Demography and resource use by microtine rodents near Toolik Lake, Alaska. *Arct. Alp. Res.* 22:51–64
14. Bazely DR, Jefferies RL. 1986. Changes in the composition and standing crop of salt-marsh communities in response to the removal of a grazer. *J. Ecol.* 74:693–706
15. Beatley JC. 1969. Biomass of desert winter annual populations in southern Nevada. *Oikos* 20:261–273

16. Begon M, Harper JL, Townsend CR. 1990. *Ecology: Individuals, Populations and Communities*. Boston: Blackwell Sci. 2nd ed.
17. Bertness MD. 1985. Fiddler crab regulation of *Spartina alterniflora* production on a New England salt marsh. *Ecology* 66:1042–55
18. Bertness MD. 1991. Interspecific interactions among high marsh perennials in a New England salt marsh. *Ecology* 72:125–37
19. Bertness MD. 1991. Zonation of *Spartina patens* and *Spartina alterniflora* in a New England salt marsh. *Ecology* 72:138–48
20. Bertness MD, Ellison AM. 1987. Determinants of pattern in a New England salt marsh community. *Ecol. Monogr.* 57:129–47
21. Billings WD. 1992. Phytogeographic and evolutionary potential of the arctic flora and vegetation in a changing climate. In *Arctic Ecosystems in a Changing Climate*, ed. FS Chapin III, RL Jefferies, JF Reynolds, GR Shaver, J Svoboda, pp. 91–109. New York: Academic
22. Billings WD, Mooney HA. 1968. The ecology of arctic and alpine plants. *Biol. Rev.* 43:481–529
23. Bliss LC. 1962. Adaptations of arctic and alpine plants to environmental conditions. *Arctic* 15:117–44
24. Bliss LC. 1977. General summary, Truelove Lowland ecosystem. In *Truelove Lowland, Devon Island, Canada: A High Arctic Ecosystem*, ed. LC Bliss, pp. 657–76. Edmonton: Univ. Alberta Press
25. Bliss LC. 1988. Arctic tundra and polar desert biome. In *North American Terrestrial Vegetation*, ed. MG Barbour, WD Billings, pp. 1–32. New York: Cambridge Univ. Press
26. Bliss LC, Bliss DI, Henry GHR, Svoboda J. 1994. Patterns of plant distribution within two polar desert landscapes. *Arct. Alp. Res.* 26:46–55
27. Bliss LC, Matveyeva NV. 1992. Circumpolar arctic vegetation. In *Arctic Ecosystems in a Changing Climate*, ed. FS Chapin III, RL Jefferies, JF Reynolds, GR Shaver, J Svoboda, pp. 59–90. New York: Academic
28. Bliss LC, Svoboda J. 1984. Plant communities and plant production in the western Queen Elizabeth Islands. *Holarctic Ecology* 7:325–44
29. Bliss, LC, Svoboda J, Bliss DI. 1984. Polar deserts, their plant cover and plant production in the Canadian High Arctic. *Holarctic Ecology* 7:305–24
30. Brenner FJ. 1993. Seasonal changes in plankton communities in three surface lakes in western Pennsylvania. *J. Penn. Acad. Sci.* 67:59–64
31. Brewer JS, Grace JB. 1990. Plant community structure in an oligohaline tidal marsh. *Vegetatio* 90:93–107
32. Brown JH. 1973. Species diversity of seed-eating desert rodents in sand dune habitats. *Ecology* 54:775–87
33. Brown JH, Kodric-Brown A. 1977. Turnover rates in insular biogeography: effect of immigration on extinction. *Ecology* 58:445–49
34. Brown S, Lugo AE. 1982. The storage and production of organic matter in tropical forests and their role in the global carbon cycle. *Biotropica* 14:161–87
35. Browne RA. 1981. Lakes as islands: biogeographic distribution, turnover rates, and species composition in the lakes of central New York. *J. Biogeog.* 8:75–83
36. Carleton TJ, Maycock PF. 1980. Vegetation of the boreal forests south of James Bay: non-centered component analysis of the vascular flora. *Ecology* 61:1199–212
37. Chabreck RH. 1972. *Vegetation, water and soil characteristics of the Louisiana coastal region. Bull. No. 664*, La. State Univ. Agric. Exp. Stn. 75 pp.
38. Chapin FS III. 1974. Morphological and physiological mechanisms of temperature compensation in phosphate absorption along a latitudinal gradient. *Ecology* 55:1180–98

39. Chapin FS III, Everett KR, Fetcher N, Kielland K, Linkins AE. 1988. Productivity and nutrient cycling of Alaskan tundra: enhancement by flowing soil water. *Ecology* 69:693–702
40. Chapin FS III, Giblin AE, Laundre JA, Nadelhoffer KJ, Shaver GR. 1995. Responses of arctic tundra to experimental and observed changes in climate. *Ecology* 76: 694–711
41. Chapin FS III, Sala OE, Burke IC, Grime JP, Hooper DU et al. 1998. Ecosystem consequences of changing biodiversity. *BioScience* 48:45–52
42. Chapin FS III, Shaver GR. 1985. Arctic. In *Physiological Ecology of North American Plant Communities*, ed. BF Chabot, HA Mooney, pp. 16–40. London: Chapman & Hall
43. Chew RM, Chew AE. 1965. The primary productivity of a desert shrub (*Larrea tridentata*) community. *Ecol. Monogr.* 35:355–75
44. Coleman BD, Mares MA, Willig MR, Hsieh Y. 1982. Randomness, area and species richness. *Ecology* 63:1121–33
45. Connell JH, Orias E. 1964. The ecological regulation of species diversity. *Am. Nat.* 98:399–414
46. Crosswhite FS, Crosswhite CD. 1982. The Sonoran Desert. In *Reference Handbook of Deserts of the World*, ed. GL Bender, pp. 163–319. Westwood, CT: Greenwood
47. Day RT, Carleton T, Keddy PA, McNeill J. 1988. Fertility and disturbance gradients: a summary model for riverine marsh vegetation. *Ecology* 69:1044–54
48. DiTommaso A, Aarssen LW. 1989. Resource manipulations in natural vegetation: a review. *Vegetatio* 84:9–29
49. Doak DF, Bigger D, Harding EK, Marvier MA, O'Malley ME, Thomson D. 1998. The statistical inevitability of stability-diversity relationships in community ecology. *Am. Nat.* 151:264–76
50. Dodson SI. 1987. Animal assemblages in temporary desert rock pools: aspects of the ecology of *Dasyhelea sublettei* (Diptera: Ceratopogonidae). *J. North Am. Benthol. Soc.* 6:65–71
51. Dodson SI. 1992. Predicting crustacean zooplankton species richness. *Limnol. Oceanogr.* 37:848–56
51a. Dodson SI, Arnott SE, Cottingham KL. The relationship in lake communities between primary production and species richness. *Ecology* (in press).
52. Ehrlich PR, Ehrlich AH. 1981. *Extinction: The Causes and Consequences of the Disappearance of Species*. New York: Random House
53. Elseth GD, Baumgardner KD. 1981. *Population Biology.* New York: Van Nostrand
54. Elton C. 1958. *The Ecology of Invasions by Animals and Plants*. London: Methuen
55. Ettershank G, Bryant M, Ettershank J, Whitford WG. 1978. Effects of nitrogen fertilization on primary production in a Chihuahuan Desert ecosystem. *J. Arid Environ.* 1:135–39
56. Floret C, Pontanier R, Rambal S. 1982. Measurement and modeling of primary production and water use in a south Tunisian steppe. *J. Arid Environ.* 5:77–90
57. Flynn KM, McKee KL, Mendelsshohn IA. 1995. Recovery of freshwater marsh vegetation after a saltwater intrusion event. *Oecologia* 103:63–72
58. Ford MF. 1996. *Impacts of fire and vertebrate herbivores on plant community characteristics and soil processes in a coastal marsh of eastern Louisiana*, USA. PhD dissertation, La. State Univ., Baton Rouge. 119 pp.
59. Ford MF, Grace JB. 1998. Effects of herbivores on vertical soil accretion, shallow subsidence, and soil elevation changes in coastal Louisiana. *J. Ecol.* 86:974–82
60. Fox JF. 1985. Plant diversity in relation to plant production and disturbance by voles in Alaskan tundra communities. *Arct. Alp. Res.* 17:199–204
61. Ganapati SV. 1940. The ecology of a temple tank containing a permanent bloom of

Microcystis aeruginosa (Kutz) Henfr. *J. Bombay Nat. Hist. Soc.* 17:65–77
62. Garcia LV, Clemente L, Maranon T, Moreno A. 1993. Above-ground biomass and species richness in a Mediterranean salt marsh. *J. Veg. Sci.* 4:417–24
63. Gentry AH. 1988. Changes in plant community diversity and floristic composition on environmental and geographical gradients. *Ann. Mo. Bot. Gard.* 75:1–34
64. Gold WG, Bliss LC. 1995. Water limitations and plant community development in a polar desert. *Ecology* 76:1558–68
65. Goldberg DE, Miller TE. 1990. Effects of different resource additions on species diversity in an annual plant community. *Ecology* 71:213–25
66. Gough L. 1996. *Plant species diversity and community structure in a Louisiana coastal marsh.* PhD dissertation, La. State Univ., Baton Rouge. 192 pp.
67. Gough L, Grace JB. 1998. Herbivore effects on plant species richness at varying productivity levels. *Ecology* 79:1586–94
68. Gough L, Grace JB, Taylor KL. 1994. The relationship between species richness and community biomass: the importance of environmental variables. *Oikos* 70:271–79
68a. Grace JB. 1999. The factors controlling species density in herbaceous plant communities: an assessment. *Perspectives in Plant Ecology, Evolution and Systematics* 2:1–28
69. Grace JB, Pugesek BH. 1997. A structural equation model of plant species richness and its application to a coastal wetland. *Am. Nat.* 149:436–60
70. Grillas T, Bonis A, van Wijck C. 1993. The effect of salinity on the dominance-diversity relations of experimental coastal macrophyte communities. *J. Veg. Sci.* 4:453–60
71. Grime JP. 1973. Competitive exclusion in herbaceous vegetation. *Nature* 242:344–47
72. Grime JP. 1973. Control of species density on herbaceous vegetation. *J. Environ. Manage.* 1:151–67
73. Guo Q, Berry WL. 1998. Species richness and biomass: dissection of the hump-shaped relationships. *Ecology* 79:2555–59
74. Gutierrez JR, Da Silva OA, Pagani MI, Weems D, Whitford WG. 1988. Effects of different patterns of supplemental water and nitrogen fertilization on productivity and composition of Chihuahuan Desert annual plants. *Am. Midl. Nat.* 119:36–43
75. Hebert PDN, Hann BJ. 1986. Patterns in the composition of arctic tundra pool microcrustacean communities. *Can. J. Fish. Aquatic Sci.* 43:1416–25
76. Hector A. 1998. The effect of diversity on productivity-detecting the role of species complementarity. *Oikos* 82:597–99
77. Henry GHR, Freedman B, Svoboda J. 1986. Effects of fertilization on three tundra plant communities of a polar desert oasis. *Can. J. Bot.* 64:2502–7
78. Hik DS, Jefferies RL. 1990. Increases in the net above-ground primary production of a salt-marsh forage grass: a test of the predictions of the herbivore-optimization model. *J. Ecol.* 78:180–95
79. Hooper DU. 1998. The role of complementarity and competition in ecosystem responses to variation in plant diversity. *Ecology* 79:704–19
80. Hooper DU, Vitousek PM. 1997. The effects of plant composition and diversity on ecosystem processes. *Science* 277:1302–5
81. Hugueny B. 1989. West African rivers as biogeographic islands: species richness of fish communities. *Oecologia* 79:236–43
82. Huston MA. 1979. A general hypothesis of species diversity. *Am. Nat.* 113:81–101
83. Huston MA. 1980. Soil nutrients and tree species richness in Costa Rican forests. *J. Biogeog.* 7:147–57
84. Huston MA. 1994. *Biological Diversity: The Coexistence of Species in Changing*

Landscapes. Cambridge, UK: Cambridge Univ. Press

85. Huston MA. 1997. Hidden treatments in ecological experiments: re-evaluating the ecosystem function of biodiversity. *Oecologia* 110:449–60
86. Huston MA. 1998. Reconciling scales and processes: the key to a general theory of species diversity. *Bull. Ecol. Soc. Am. (Supp.)* 79
87. Huston MA, DeAngelis DL. 1994. Competition and coexistence-the effects of resource transport and supply. *Am. Nat.* 144:954–77
88. Inouye RS, Huntly NJ, Stilwell M, Tester JR, Tilman D, et al. 1987. Old-field succession on a Minnesota sand plain. *Ecology* 68:12–26
89. Jacobson HA, Jacobson GL Jr. 1989. Variability of vegetation in tidal marshes of Maine, U.S.A. *Can. J. Bot.* 67:230–38
90. Jefferies RL, Klein DR, Shaver GR. 1994. Vertebrate herbivores and northern plant communities: reciprocal influences and responses. *Oikos* 71:193–206
91. Johnson KH, Vogt KA, Clark HJ, Schmitz OJ, Vogt DJ. 1996. Biodiversity and the productivity and stability of ecosystems. *Trends Evol. Ecol.* 11:372–77
92. Jonasson S. 1981. Plant communities and species distributions of low alpine *Betula nana* heaths in northernmost Sweden. *Vegetatio* 44:51–64
93. Jonasson, S. 1982. Organic matter and phytomass on three north Swedish tundra sites, and some connections with adjacent tundra areas. *Holarctic Ecology* 5:367–75
94. Jonasson S. 1992. Plant responses to fertilization and species removal in tundra related to community structure and clonality. *Oikos* 63:420–29
95. Kajimoto T, Matsuura Y, Sofronov MA, Volokitina AV, Mori S, et al. 1997. Above- and below-ground biomass and annual production rates of a *Larix gmelinii* stand near tundra in central Siberia. In *Proc. Symp. Jt. Sib. Permafrost Stud. Between Jpn. Russia 1996*, 5th, Tsukuba, Ibaraki, pp. 119–128. Japan: Natl. Inst. Environ. Stud.
96. Kazantseva TI. 1980. Productivity and dynamics of above-ground biomass of desert plants. In *Problemy Osvoeiya Pustyn*, No. 2, pp. 76–84. Komarov Bot. Inst. Acad. Sci. USSR
97. Keddy PA. 1990. Competitive hierarchies and centrifugal organization in plant communities. In *Perspectives on Plant Competition*, ed. JB Grace, D Tilman, pp. 266–90. New York: Academic
98. Kolasa J, Pickett STA (eds.). 1989. *Ecological Heterogeneity*. New York: Springer-Verlag
99. Komarkova V, Webber PJ. 1980. Two low arctic vegetation maps near Atkasook, Alaska. *Arct. Alp. Res.* 12:447–72
100. La Roi GH, Stringer MHL. 1976. Ecological studies in the boreal spruce-fir forests of the North American taiga, II. Analysis of the bryophyte flora. *Can. J. Bot.* 54:619–43
101. Latham PJ, Kitchens WM, Pearlstine LG. 1994. Species associations changes across a gradient of freshwater, oligohaline, and mesohaline tidal marshes along the Savannah River. *Wetlands* 14:174–83
102. Lawton JH. 1994. What do species do in ecosystems? *Oikos* 71:367–74
103. Lawton JH, Naeem S, Thompson LJ, Hector A, Crawley MJ. 1998. Biodiversity and ecosystem function: getting the Ecotron experiment in its correct context. *Funct. Ecol.* 12:848–52
104. Le Houerou HN. 1984. Rain use efficiency: a unifying concept in arid-land ecology. *J. Arid Environ.* 7:213–47
105. Leibold MA. 1996. A graphical model of keystone predators in food webs: trophic regulation of abundance, incidence, and diversity patterns in communities. *Am. Nat.* 147:784–812
106. Leigh EG Jr. 1965. On the relationship between productivity, biomass, diversity and stability of a community. *Proc. Natl. Acad. Sci. USA* 53:777–83

107. Loreau M. 1998. Separating sampling and other effects in biodiversity experiments. *Oikos* 82:600–2
108. Ludwig, JA. 1986. Primary production variability in desert ecosystems. In *Pattern and Process in Desert Ecosystems*, ed. WG Whitford, pp. 5–17. Albuquerque: Univ. New Mex. Press
109. Ludwig JA, Whitford WG, Cornelius JM. 1989. Effects of water, nitrogen and sulfur amendments on cover, density and size of Chihuahuan Desert ephemerals. *J. Arid Environ.* 16:35–42
110. Lutz HL. 1953. *The effects of forest fires on the vegetation of interior Alaska. USDA For. Serv., Alaska For. Res. Cent. Stn. Pap. No. 1*. Juneau, Alaska
111. Lyons SK, Willig MR. 1999. A hemispheric assessment of scale-dependence in latitudinal gradients of species richness. *Ecology.* In press
112. MacArthur RH. 1955. Fluctuations of animal populations and a measure of community stability. *Ecology* 36:533–36
113. MacArthur RH, Pianka ER. 1966. On the optimal use of a patchy environment. *Am. Nat.* 100:603–9
114. Marrs RH, Gough L, Grace JB. 1996. On the relationship between plant species diversity and biomass: a comment on a paper by Gough, Grace & Taylor. *Oikos* 75:323–26
115. Matveyeva NV, Parinkina OM, Chernov YI. 1975. Maria Pronchitsheva Bay, USSR. In *Structure and Function of Tundra Ecosystems, Ecol. Bull. 20*, ed. T Rosswall, OW Heal, pp. 61–72. Stockholm
116. McGrady-Steed J, Harris PM, Morin PJ. 1997. Biodiversity regulates ecosystem predictability. *Nature* 390:162–65
117. McKee KL, Mendelssohn IA. 1989. Response of a freshwater marsh plant community to increased salinity and increased water level. *Aq. Bot.* 34:301–16
118. McKendrick JD, Batzli GO, Everett KR, Swanson FC. 1980. Some effects of mammalian herbivores and fertilization on tundra soils and vegetation. *Arct. Alp. Res.* 12:565–78
119. McKenzie D. 1982. The northern Great Basin region. In *Reference Handbook on the Deserts of North America*, ed. GL Bender, pp. 67–102. Westport, CT: Greenwood
120. Medellin-Leal D. 1982. The Chihuahuan Desert. In *Reference Handbook on the Deserts of North America*, ed. GL Bender, pp. 321–381. Westport, CT: Greenwood
121. Medina E, Klinge H. 1983. Productivity of tropical forests and tropical woodlands. In *Physiological Plant Ecology IV, Ecosystem Processes: Mineral Cycling, Productivity and Man's Influence*, ed. OL Lange, PS Nobel, CB Osmond, H Ziegler, 9:281–303. New York: Springer-Verlag
122. Miller PC. 1982. Experimental and vegetational variation across a snow accumulation area in montane tundra in central Alaska. *Holarctic Ecol.* 5:85–98
123. Mitsch WJ, Gosselink JG. 1986. *Wetlands*. New York: Van Nostrand Reinhold
124. Moore DRJ, Keddy PA. 1989. The relationship between species richness and standing crop in wetlands: the importance of scale. *Vegetatio* 79:99–106
125. Moore JC, deRuiter PC, Hunt HW. 1993. Influence of productivity on the stability of real and model ecosystems. *Science* 261:906–8
126. Morris JT. 1991. Effects of nitrogen loading on wetland ecosystems with particular reference to atmospheric deposition. *Annu. Rev. Ecol. Syst.* 22:257–79
127. Muc M, Freedman B, Svoboda J. 1994. Aboveground standing crop in plant communities of a polar desert oasis, Alexandra Fiord, Ellesmere Island. In *Ecology of a Polar Oasis: Alexandra Fiord, Ellesmere Island, Canada*, ed. J Svoboda, B Freedman, pp. 65–74. Toronto: Captus Univ. Publ.
128. Muc M, Freedman B, Svoboda J. 1994. Vascular plant communities of a polar

desert oasis, Alexandra Fiord, Ellesmere Island. In *Ecology of a Polar Oasis: Alexandra Fiord, Ellesmere Island, Canada*, ed. J Svoboda, B Freedman, pp. 53–63. Toronto: Captus Univ. Publ.

129. Mun HT, Whitford WG. 1989. Effects of nitrogen amendment on annual plants in the Chihuahuan Desert. *Plant Soil* 120:225–31
130. Murphy PG. 1977. Rates of primary productivity in tropical grassland, savanna and forest. *Geo-Eco-Trop* 1:95–102
131. Murphy PG, Lugo AE. 1986. Ecology of tropical dry forest. *Annu. Rev. Ecol. Syst.* 17:67–88
132. Naeem S, Håkansson K, Lawton JH, Crawley MJ, Thompson LJ. 1996. Biodiversity and plant productivity in a model assemblage of plant species. *Oikos* 76:259–64
133. Naeem S, Lawlor SP, Thompson LJ, Woodfin RM. 1994. Declining biodiversity can alter the performance of ecosystems. *Nature* 368:734–36
134. Naeem S, Thompson LJ, Lawler SP, Lawton JH, Woodfin RM. 1995. Empirical evidence that declining biodiversity may alter the performance of terrestrial ecosystems. *Philos. Trans. R. Soc. London B* 347:249–62
135. Neill C. 1990. Effects of nutrients and water levels on emergent macrophyte biomass in a prairie marsh. *Can. J. Bot.* 68:1007–14
136. Neill C. 1990. Effects of nutrients and water levels on species composition in prairie whitetop (*Scolochloa festucacea*) marshes. *Can. J. Bot.* 68:1015–20
137. Nilsson SG, Ericson L. 1992. Conservation of plant and animal populations in theory and practice. In *Ecological Principles of Nature Conservation*, ed. L Hansson, pp. 71–112. London: Elsevier Appl. Sci.
138. Nyman JA, Chabreck RH, Kinler NW. 1993. Some effects of herbivory and 30 years of weir management on emergent vegetation in brackish marsh. *Wetlands* 13:165–75
139. Odum WE. 1988. Comparative ecology of tidal freshwater and salt marshes. *Annu. Rev. Ecol. Syst.* 19:147–76
140. Oechel WC, Van Cleve K. 1986. The role of bryophytes in nutrient cycling in the taiga. In *Forest Ecosystems in the Alaskan Taiga: A Synthesis of Structure and Function*, ed. K Van Cleve, FS Chapin, PW Flanigan, LA Viereck, CT Dyrness, pp. 121–137. New York: Springer-Verlag
141. Oksanen L, Arruda J, Fretwell SD, Niemela P. 1981. Exploitation ecosystems in gradients of primary productivity. *Am. Nat.* 131:424–44
142. Oksanen T, Oksanen L, Gyllenberg M. 1992. Exploitation ecosystems in heterogeneous habitat complexes II: impact of small scale heterogeneity on predator prey dynamics. *Evol. Ecol.* 6:383–98
143. Olson JS. 1975. Productivity of forest ecosystems. *Proc. Symp. Productivity World Ecosystems*, pp. 33–43. Washington, DC: Natl. Acad. Sci.
144. Orshan G, Diskin S. 1968. Seasonal changes in productivity under desert conditions. In *Functioning of Terrestrial Ecosystems at the Primary Production Level. Proc. Copenhagen Symp.*, ed. FE Echardt, pp. 191–201. Paris: UNESCO
145. Parmenter RR, Brantley SL, Brown JH, Crawford CS, Lightfoot DC, Yates TL. 1995. Diversity of animal communities on southwestern rangelands: species patterns, habitat relationships, and land management. In *Biodiversity on Rangelands. Natural Resources and Environmental Issues*. Vol. IV, ed. NE West, pp. 50–71. Logan: Utah State Univ. Press
146. Parmenter RR, Van Devender T. 1995. The diversity, spatial variability and functional roles of vertebrates in the desert grassland. In *The Desert Grassland*, ed. M McClaran T Van Devender, pp. 196–229. Tucson: Univ. Ariz. Press
147. Pastor J, Downing A, Erickson HE.

1996. Species-area curves and diversity-productivity relationships in beaver meadows of Voyageurs National Park, Minnesota, USA. *Oikos* 77:399–406
148. Patalas K. 1990. Diversity of zooplankton communities in Canadian lakes as a function of climate. *Verh. Int. Verein. Limnol.* 24:360–68
149. Patten DT. 1978. Productivity and production efficiency of an upper Sonoran Desert ephemeral community. *Am. J. Bot.* 65:891–95
150. Pearson LC. 1965. Primary productivity in grazed and ungrazed desert communities of eastern Idaho. *Ecology* 46:278–85
151. Pennings SC, Callaway RM. 1992. Salt marsh plant zonation: the relative importance of competition and physical factors. *Ecology* 73:681–90
152. Phillips OL, Gentry AH, Hall P, Sawyer SA, Vasquez R. 1994. Dynamics and species richness of tropical rain forests. *Proc. Natl. Acad. Sci. USA* 91:2805–9
153. Pianka ER. 1966. Latitudinal gradients in species diversity: a review of concepts. *Am. Nat.* 100:33–46
154. Pickett STA, Kolasa J, Jones CG. 1994. *Ecological Understanding: The Nature of Theory and the Theory of Nature*. San Diego, CA: Academic
155. Preston FW. 1962. The canonical distribution of commonness and rarity: part I. *Ecology* 43:185–215
156. Preston FW. 1962. The canonical distribution of commonness and rarity: part II. *Ecology* 43:410–32
157. Proctor J. 1984. Tropical forest litterfall II: the data set. In *Tropical Rain Forest*, ed. AC Chadwick, SL Sutton, pp. 83–113. Leeds: Leeds Philos. Lit. Soc.
158. Ruess RW, Hendrick RL, Bryant JP. 1998. Regulation of fine root dynamics by mammalian browsers in early successional taiga forests of interior Alaska. *Ecology* 79:2706–20
159. Ruess RW, Van Cleve K, Yarie J, Viereck LA. 1996. Comparative estimates of fine root production in successional taiga forests on the Alaskan interior. *Can. J. For. Res.* 26:1326–36
160. Riebesell JF. 1974. Paradox of enrichment in competitive systems. *Ecology* 55:183–87
161. Robinson CH, Wookey PA, Lee JA, Callaghan TV, Press MC. 1998. Plant community responses to simulated environmental change at a high arctic polar semi-desert. *Ecology* 79:856–66
162. Rosenzweig ML. 1971. Paradox of enrichment: destabilization of exploitation ecosystems in ecological time. *Science* 171:385–87
163. Rosenzweig ML. 1972. Stability of enriched aquatic ecosystems. *Science* 171:564–65
164. Rosenzweig ML. 1977. Aspects of ecological exploitation. *Q. Rev. Biol.* 52:371–80
165. Rosenzweig ML. 1977. Coexistence and diversity in heteromyid rodents. In *Evolutionary Ecology*, ed. B Stonehouse, C Perrins, pp. 89–99. London: Macmillan
166. Rosenzweig ML. 1992. Species diversity gradients: We know more or less than we thought. *J. Mammal.* 73:715–30
167. Rosenzweig ML. 1995. *Species Diversity in Space and Time*. Cambridge, UK: Cambridge Univ. Press
168. Rosenzweig ML, Abramsky Z. 1993. How are diversity and productivity related? In *Species Diversity in Ecological Communities: Historical and Geographical Perspectives*, ed. RE Ricklef, D Schluter, pp. 52–65. Chicago: Univ. Chicago Press
169. Rowlands P, Johnson H, Ritter E, and Endo A. 1982. The Mojave Desert. In *Reference Handbook on the Deserts of North America*, ed. GL Bender, pp. 103–162. Westport, CT: Greenwood
170. Rundel PW, Gibson AC. 1996. *Ecological Communities and Processes in a Mojave Desert Ecosystem, Rock Valley, Nevada.* New York: Cambridge Univ. Press

171. Schulze E-D, Mooney HA. 1994. Ecosystem function of biodiversity: a summary. In *Biodiversity and Ecosystem Function*, ed. E-D Schulze, HA Mooney, pp. 497–510. Berlin: Springer-Verlag
172. Shafi MI, Yarranton GA. 1973. Diversity, floristic richness, and species evenness during a secondary (post-fire) succession. *Ecology* 54:897–902
173. Sharma BM. 1982. Plant biomass in the semi-arid zone of India. *J. Arid Environ.* 5:29–33
174. Shaver GR, Cades DH, Giblin AE, Johnson LC, Laundre JA, et al. 1998. Biomass accumulation and CO_2 flux in three Alaskan wet sedge tundras: responses to nutrients, temperature, and light. *Ecol. Monogr.* 68:75–97
175. Shaver GR, Chapin FS III. 1991. Production: biomass relationships and element cycling in contrasting arctic vegetation types. *Ecol. Monogr.* 61:1–31
176. Shaver GR, Giblin AE, Laundre JA, Nadelhoffer KJ. 1996. Changes in live plant biomass, primary production, and species composition along a riverside toposequence in arctic Alaska, U.S.A. *Arct. Alp. Res.* 28:361–77
177. Shaver GR, Jonassen S. 1999. Productivity of arctic ecosystems. In *Terrestrial Global Productivity*, ed. H Mooney, J Roy, B Saugier, pp. 000–00. New York: Academic. In press
178. Shipley B, Gaudet C, Keddy PA, Moore DRJ. 1991. A model of species density in shoreline vegetation. *Ecology* 72:1658–67
179. Shmida A, Ellner SP. 1984. Coexistence of plant species with similar niches. *Vegetatio* 58:29–55
180. Shmida A, Wilson MV. 1985. Biological determinants of species diversity. *J. Biogeog.* 12:1–20
181. Silander JA, Antonovics J. 1982. Analysis of interspecific interactions in a coastal plant community—a perturbation approach. *Nature* 298:557–60
182. Smith SD, Monson RK, Anderson JE. 1997. *Physiological Ecology of North American Desert Plants.* New York: Springer-Verlag
183. Sokal RR, Rohlf FJ. 1995. *Biometry.* New York: Freeman. 3rd ed.
184. Swift MJ, Anderson JM. 1993. Biodiversity and ecosystem function in agricultural systems. In *Biodiversity and Ecosystem Function*, ed. ED Schulze, HA Mooney, pp. 15–41, Berlin: Springer-Verlag
185. Symstad A, Tilman D, Wilson J, Knops JMH. 1998. Species loss and ecosystem functioning: effects of species identity and community composition. *Oikos* 84:389–87
186. Tilman D. 1982. *Resource competition and community structure.* Princeton, NJ: Princeton Univ. Press
187. Tilman D. 1987. Secondary succession and the pattern of plant dominance along experimental nitrogen gradients. *Ecol. Monogr.* 57:189–214
188. Tilman D, Downing JA. 1994. Biodiversity and stability on grasslands. *Nature* 367:363–65
189. Tilman D, Lehman CL, Bristow CE. 1998. Diversity-stability relationships: statistical inevitability or ecological consequence? *Am. Nat.* 151:277–82
190. Tilman D, Lehman CL, Thomson KT. 1997. Plant diversity and ecosystem productivity: theoretical considerations. *Proc. Natl. Acad. Sci. USA* 94:1857–61
191. Tilman D, Pacala S. 1993. The maintenance of species richness in plant communities. In *Species Diversity in Ecological Communities: Historical and Geographical Perspectives*, eds. RE Ricklefs, D Schluter, pp. 13–25. Chicago: Univ. Chicago Press
192. Tilman D, Wedin D, Knops J. 1996. Productivity and sustainability influenced by biodiversity in grassland ecosystems. *Nature* 379:718–20
193. Valentine JW. 1976. Genetic strategies of

adaptation. In *Molecular Evolution*, ed. FJ Ayala, pp. 78–94. Sunderland, MA: Sinauer

194. Van Cleve K, Barney R, Viereck LA. 1981. Evidence of temperature control of production and nutrient cycling in two interior Alaska black spruce ecosystems. *Can. J. For. Res.* 11:258–73
195. Van Cleve K, Chapin FS III, Dyrness CT, Viereck LA. 1991. Element cycling in taiga forests: state factor control–a framework for experimental studies of ecosystem processes. *Bioscience* 41:78–88
196. Van Cleve K, Dyrness CT, Viereck LA, Fox J, Chapin FS III, et al. 1983. Taiga ecosystems in interior Alaska. *Bioscience* 33:39–44
197. Van Cleve K, Oliver LK, Schlentner R, Viereck LA, Dyrness CT. 1983. Productivity and nutrient cycling in taiga forest ecosystems. *Can. J. For. Res.* 13:747–66
198. Van Cleve K, Viereck LA. 1983. A comparison of successional sequences following fire on permafrost-dominated and permafrost-free sites in interior Alaska. *Permafrost: 4th Int. Conf. Proc.*:1286–90
199. Van Cleve K, Viereck LA, Schlentner R. 1971. Accumulation of nitrogen in alder (*Alnus*) ecosystems near Fairbanks, Alaska. *Arct. Alp. Res.* 3:101–114
200. Van Cleve K, Yarie J. 1986. Interaction of temperature, moisture, and soil chemistry in controlling nutrient cycling and ecosystem development in the taiga of Alaska. In *Forest Ecosystems in the Alaskan Taiga: a Synthesis of Structure and Function*, ed. K Van Cleve, FS Chapin, PW Flanigan, LA Viereck, CT Dyrness, pp. 160–189. New York: Springer-Verlag
201. Van Cleve K, Yarie J, Adams PC, Erickson R, Dyrness CT. 1993. Nitrogen mineralization and nitrification in successional ecosystems on the Tanana River floodplain, interior Alaska. *Can. J. For. Res.* 23:970–78
202. van de Koppel J, Huisman J, van der Wal R, Olff H. 1996. Patterns of herbivory along a productivity gradient: an empirical and theoretical investigation. *Ecology* 77:736–45
203. Vandermeer JH. 1988. *The Ecology of Intercropping*. New York: Cambridge Univ. Press
204. Vermeer JG, Berendse F. 1983. The relationship between nutrient availability, shoot biomass and species richness in grassland and wetland communities. *Vegetatio* 53:121–26
205. Viereck LA, Dyrness CT, Van Cleve K, Foote B. 1983. Vegetation, soils, and forest productivity in selected forest types in interior Alaska. *Can. J. For. Res.* 13:703–20
206. Viereck LA, Van Cleve K, Adams PC, Schlentner RE. 1993. Climate of the Tanana River floodplain near Fairbanks, Alaska. *Can. J. For. Res.* 23:899–913
207. Viereck LA, Van Cleve K, Dyrness CT. 1986. Forest ecosystem distribution in the taiga environment. In *Forest Ecosystems in the Alaskan Taiga: a Synthesis of Structure and Function*, ed. K Van Cleve, FS Chapin, PW Flanigan, LA Viereck, CT Dyrness, pp. 22–43. New York: Springer-Verlag
208. Vollenweider RA. 1974. *A Manual on Methods for Measuring Primary Production in Aquatic Environments. IBP Handbook No. 12*. Oxford, UK: Blackwell. 2nd ed.
209. Walker BH. 1992. Biodiversity and ecological redundancy. *Conserv. Biol.* 6:18–23
210. Walker MD. 1995. Patterns and causes of arctic plant community diversity. In *Arctic and Alpine Biodiversity*, ed. FS Chapin III, C Korner, pp. 1–20. New York: Springer-Verlag
211. Wardle DA, Bonner KI, Nicholson KS. 1997. Biodiversity and plant litter: experimental evidence which does not support the view that enhanced species richness improves ecosystem function. *Oikos* 79:247–58

212. Webb WL, Lauenroth WK, Szarek SR, Kinerson RS. 1983. Primary production and abiotic controls in forests, grasslands, and desert ecosystems in the United States. *Ecology* 64:134–151
213. Webber PJ. 1978. Spatial and temporal variation of the vegetation and its production, Barrow, Alaska. In *Vegetation and Production Ecology of an Alaskan Arctic Tundra*, ed. LL Tieszen, pp. 37–112. New York: Springer-Verlag
214. Wheeler BD, Giller KE. 1982. Species richness of herbaceous fen vegetation in Broadland, Norfolk in relation to the quantity of aboveground material. *J. Ecol.* 70:179–200
215. Wheeler BD, Shaw SC. 1991. Aboveground crop mass and species richness of the principal types of herbaceous rich fen vegetation of lowland England and Wales. *J. Ecol.* 79:285–301
216. Whitford WG. 1986. *Pattern and Process in Desert Ecosystems*. Albuquerque: Univ. New Mex. Press
217. Whittaker RH, Niering WA. 1975. Vegetation of the Santa Catalina Mountains, Arizona. V. Biomass, production and diversity along the elevational gradient. *Ecology* 56:771–90
218. Wiegert RG, Chalmers AG, Randerson PF. 1983. Productivity gradients in salt marshes: the response of Spartina alterniflora to experimentally manipulated soil water movement. *Oikos* 41:1–6
219. Willig MR, Walker LR. 1999. Disturbance in terrestrial ecosystems: salient themes, synthesis, and future directions. In *Ecology of Disturbed Ground*, ed. LR Walker, Amsterdam: Elsevier. In press
220. Wilson SD, Keddy PA. 1988. Species richness, survivorship, and biomass accumulation along an environmental gradient. *Oikos* 53:375–80
221. Wisheu IC, Keddy PA. 1989. Species richness-standing crop relationships along four lakeshore gradients: constraints on the general model. *Can. J. Bot.* 67:1609–17
222. Wollkind DJ. 1976. Exploitation in three trophic levels: an extension allowing intraspecies carnivore interaction. *Am. Nat.* 110:431–47
223. Wright DH, Currie DJ, Maurer BA. 1993. Energy supply and patterns of species richness on local and regional scales. In *Species Diversity in Ecological Communities: Historical and Geographical Perspectives*, ed. R Ricklefs, D Schluter, pp. 66–74. Chicago: Univ. Chicago Press
224. Yarie J, Van Cleve K, Schlentner R. 1990. Interaction between moisture, nutrients and growth of white spruce in interior Alaska. *For. Ecol. Man.* 30:73–89
225. Zedler JB, Williams P, Winfield T. 1980. Salt marsh productivity with natural and altered tidal circulation. *Oecologia* 44:236–40

Annu. Rev. Ecol. Syst. 1999. 30:301–26

ANALYSIS OF SELECTION ON ENZYME POLYMORPHISMS

Walter F. Eanes
Department of Ecology and Evolution, State University of New York, Stony Brook, New York 11794; e-mail: walter@life.bio.sunysb.edu

Key Words amino acid polymorphism, *Drosophila*, allozymes, metabolism, molecular evolution

■ **Abstract** Allozyme polymorphisms have been the focus of studies of selection at single enzyme loci, and most involve the enzymes of central metabolism. DNA sequencing of enzyme loci has shown numerous examples of multiple amino acid polymorphisms segregating within electromorphs. The amino acid heterogeneity underlying many allozyme polymorphisms should confound analysis of functional differences and selection. Metabolic control theory proposed that pathways will be insensitive to functional changes in allozymes; however, there is evidence that many polymorphisms modulate fluxes. Studies of model systems have provided detailed evidence for selection acting on enzyme polymorphisms in metabolic genes. There is also evidence that regulatory changes are superimposed on structural changes. Codon bias implies that the functional differences encountered in allozyme studies should be detectable by natural selection, however, amino acid polymorphisms may also represent weakly deleterious mutations. Future studies should connect structural changes with effects on function and stability, and they should emphasize the multilocus nature of responses to selection.

INTRODUCTION

The issues involved in studying selection on single loci, and in particular enzyme polymorphisms, have their beginnings in the early allozyme era. Lewontin (110) devoted much of his 1974 critique to the problems associated with measuring fitness differences and the observations concerning functional differences between allozymes. Although the neutralist-selection debate has faded, it spawned interest in the epistemological problems of studying natural selection on single gene polymorphisms (26, 41). This led to many different concerns, including experimental criteria for measuring kinetic parameters (65), genetic dominance and fluxes (88), and the use of DNA sequence variation to study historical selection (79, 100). Today the same questions can be asked of any amino acid polymorphism, whether involving metabolic enzymes or quantitative trait loci.

0066-4162/99/1120-0301$08.00

Gillespie (61), Watt (159), and Mitton (118) have discussed the evidence for functional adaptation and allozyme polymorphism, and Kreitman & Akashi (100) reviewed the use of sequence variation to evaluate the historical imprint of selection on genes. This review addresses issues concerning amino acid polymorphisms and adaptive evolution. It has a strong allozyme orientation because it is only recently that we have began to acquire the technical power and molecular tools to look at other classes of genes at the same level of resolution. I address recent work, although this is relative; to me a paper published in 1990 is recent history, while to a new graduate student, such a paper often constitutes part of the fossil record.

This review does not discuss the evidence for selection leading to differences between species at the amino acid level. These forces might be different from the forces affecting intraspecific polymorphism. It does not address the recent work on genes assumed a priori to be under strong selection, such as self-incompatibilty and gamete recognition genes, the MHC system, and loci involved in pesticide or microbe resistance. I discuss the current observations about the amino acid variation underlying allozymes, the issues concerning metabolic control theory and selection on protein function, observations on null alleles, and four case studies mapping biochemical phenotype to higher phenotypes. Finally, I mention several areas that hold promise for better understanding selection on enzyme polymorphisms.

The study of allozyme polymorphism is not a study of a random sample of genes. Nearly all are soluble enzymes detected after electrophoresis by methods that use the cofactors NAD or NAPD either directly or in conjunction with enzymes that are coupled to them. As a consequence, there now exists much information about polymorphisms in enzymes of the glycolytic pathway, Krebs cycle, and their branches. Figure 1 is an abridged scheme of the metabolic pathways and the enzymes that have been studied in population studies of *Drosophila melanogaster*, the species for which we have the most detailed information about allozyme polymorphisms. A number of these genes have been cloned and sequenced, and population sampling of sequencing has been carried out.

DNA SEQUENCING

Defining the Mutational Landscape for Allozyme Variation

The molecular nature of allozyme alleles is critical to understanding selection acting on them. If alleles are mixtures of many segregating and recombining amino acid polymorphisms that only share isoelectric points or net charge, then their study in the absence of sequence input will be problematic. One would like to treat individual electromorphs as mutationally unique, but this may not be realistic, and the charge-state model predicts that loci with many electromorphs will possess intraclass mutational heterogeneity (8).

In the pre-PCR era, Kreitman's 1983 (99) landmark paper on the *Adh* gene in *D. melanogaster* was the first to examine an allozyme polymorphism at the

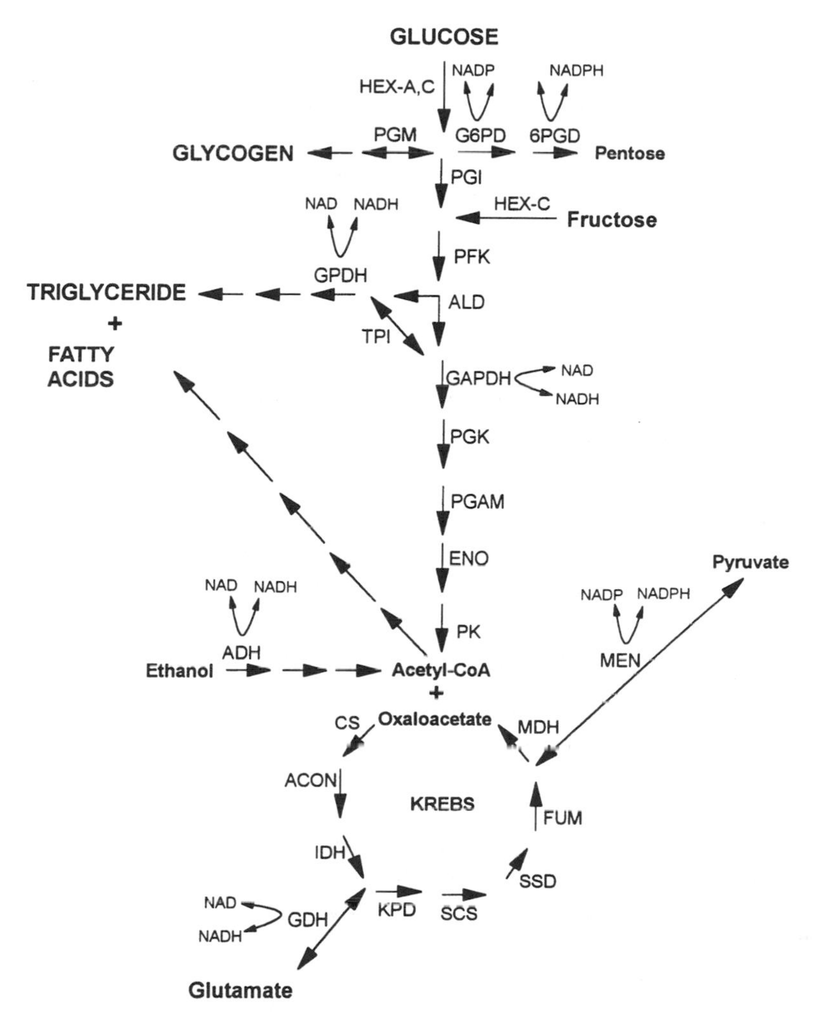

Figure 1 Abbreviated scheme of the glycolytic pathway, Krebs cycle, and their branches to show the positions of enzymes discussed in the text. Enzymes discussed are phosphoglucomutase (PGM), glucose-6-phosphate dehydrogenase (G6PD), 6-phosphogluconate dehydrogenase (6PGD), phosphoglucose isomerase (PGI), hexokinase (HEX-C), glycerol-3-phosphate dehydrogenase (GPDH), triosephosphate isomerase (TPI), phosphoglycerate kinase (PGK) , enolase (ENO), alcohol dehydrogenase (ADH), malic enzyme (MEN), glutamate dehydrogenase (GDH).

DNA level; the studies on *Est6* in *D. melanogaster* (28) and *gnd* (the gene for 6PGD) in *E. coli* (140) suggested more complex patterns of amino acid variation than the "one amino acid change–one allozyme" relationship. The introduction of PCR-facilitated sequencing has rapidly increased our appreciation of the molecular nature of allozymes. Table 1 lists what is currently known for a large number of published and unpublished cases in *Drosophila melanogaster*.

Several generalities can be recovered from Table 1. Simple allozyme polymorphisms, for example, those showing two common electromorphs with rare variants (ADH, G6PD, TPI, GPDH, 6PGD), are often defined by a single amino acid change. High resolution electrophoresis of ADH (98), G6PD (39), and GPDH (29) had failed to detect any further variation within major allozymes. In contrast, where multiple allozymes are common (EST6, PGM) and hidden variation indicated earlier, amino acid variation within major allozyme classes is pervasive (28, 71; B Verrelli, WF Eanes, unpublished). Phosphoglucomutase (PGM) is highly polymorphic metabolic enzyme, common to many allozyme studies across many taxa. In *D. melanogaster*, studies of PGM show a large number of amino acid changes associated with the three commonly recognized allozyme alleles. This possibly explains the failure to find functional differences between classes (53) or clines at this locus (125). In general, *D. simulans* shows less amino acid polymorphism (47, 112; WF Eanes, unpublished information; JH McDonald, M Kreitman, unpublished information), although single hidden amino acid polymorphisms are seen for TPI (72), and GLD (66), while, as in *D. melanogaster*, EST6 is riddled with amino acid polymorphisms (91). In *D. pseudoobscura*, the electrophoretically hypervariable *Est5* and *Xdh* loci (see discussion in 8) show as many as 27 amino acid polymorphisms (138, 154), while the monomorphic *Adh* locus shows no hidden amino acid variation (142).

Outside of *Drosophila* there are other examples. In the crickets *Gryllus veletis* and *G. pennsylvanicus*, the PGI locus shows several electromorphs that are common to each species. Katz & Harrison (92) sequenced about 70% of the *Pgi* gene in samples that included four of the electrophoretic classes in both species; they found that multiple amino acid polymorphisms were generally segregating within allozyme alleles in both species. In most cases electromorphs differed from each other by more than single amino acid change. In the killifish *Fundulus heteroclitus*, the two LDH-B electromorphs differ by two amino acid changes (135). In samples from Florida to Nova Scotia, three additional amino acid polymorphisms and numerous recombinants of the two major amino acid polymorphisms have been recovered (133). Such cases are more limited in plants, but *Adh* has been sequenced in Pearl millet (54) and *Arabidopsis* and *Arabis* (120), and there is limited sequence of PGI in *Clarkia* (153), but again multiple amino acid polymorphisms are the rule between electromorphs.

These studies emphasize the importance, no matter how challenging the task, of defining amino acid variation at a target gene. It is clear from numerous cases in Table 1 that the functional study of allozymes in the absence of information on amino acid polymorphism could be misleading. In the future, most studies of

TABLE 1 Relationship between allozyme variation and amino acid polymorphism in *Drosophila melanogaster*

Enzyme	Allozyme[a]	*n*	Amino acid changes[b]	Cline	Function[c]	Reference
ADH	2	15	T/K* (A/V, A/S)	Yes	Increase in high activity allele	99, 106
ENO	1	12	None	—	—	[d]
GDH	2	10	V/F, N/D*	Yes?	Unknown	[d]
GAPDH	1	10	None	—	—	[e]
G6PD	3	50	P/L*, G/C* (T/N)	Yes	Increase in low activity allele	47
GPDH	2		K/N*	Yes	Increase in high activity allele	152
HEX-C	2	45	E/K*, K/R (S/F, A/E)	Yes	Unknown	[d]
MEN	1	29	G/A	—	Unknown	[d]
6PGD	2	13	Q/K*	Yes	Increase in low activity allele	10
PGI	1	11	V/I (Y/H, V/A)	—	Unknown	[f]
PGK	1	12	None	—	—	[d]
PGM	3	27	12 a.a. changes	Yes	Unknown	[d]
TPI	2	25	E/K*	Yes	Unknown	72
EST6	2	16	11 amino acid polymorphisms	Yes	Yes	2, 71
SOD	2	41	K/D*	Yes?	Yes	81
GLD	2	55	A/G, L/V, L/V, R/Q*, A/T	—	Unknown	67

[a]Number of common electromorphs (1 = monomorphic).

[b]Amino acid polymorphisms observed. Singetons are given in parentheses, and the amino acid change responsible for allozyme polymorphism is marked with asterisk.

[c]Functional differences exist between allozymes and/or relationship to latitudinal cline.

[d]Unpublished, WF Eanes.

[e]M Kreitman, personal communication.

[f]Unpublished, JH McDonald & M Kreitman.

selection on single genes will be initiated from descriptions at the DNA level (such as the *per* locus, 141), so this will become a moot point. Finally, it is clear that these metabolic enzymes are not unique. Moriyama & Powell (121) list population-level sequence studies in *Drosophila* for a number of genes not associated with allozyme phenotypes, and most show amino acid polymorphisms.

Adding the Historical Dimension

The frequencies of mutations we observe in samples reflect a history of selection and demographic change. It is well known that frequency-based tests of neutrality are statistically weak to many forms of selection, and even random selection models generate frequency spectra similar to the neutral infinite allele model (61). It has been proposed that if a genealogical history could be attached to the same sample of alleles, then it might be possible to reject certain models of stochastic processes as incompatible with the observed genealogical configuration (48). DNA sequencing provided the potential to construct gene-based genealogies, or gene trees, and the *Adh* data (99) provided the first insight into a gene tree. The low sequence diversity associated with the ADH "fast" allele, along with several fixed or nearly fixed differences between fast and slow alleles, rejected the hypothesis that the polymorphism was new and that the fast allele had been historically common, despite its high frequency in contemporary populations.

Coalescence theory, which emerged in the 1980s, can be used as framework to study how selection and demographic history can distort gene trees and the associated silent nucleotide diversity (82, 89). The coalescence approach is a retrospective theory that is sample-based (see reviews 35, 78). It recognizes that genealogies can be distorted by selection on mutations within, or linked to, the sampled alleles. An important feature is the ability to use computer simulation to sample the universe of neutral genealogies generated by simple parameters extracted from the sample of sequences (78–80). Thus, it is possible to ask if the sequence diversity associated with a particular lineage, such as one marked by a specific mutation, possesses too little or too much variation for its sample frequency. This rationale was first used to investigate the hypothesis that the *D. melanogaster* superoxide dismutase *Sod*^{S} allele was under recent directional selection (81, 83). By simulation it was shown that the high frequency of the *Sod*^{S}-bearing lineage was inconsistent with the low sequence diversity of all sequences bearing this mutation. Consequently, recent directional selection was inferred. Likewise, the G6PD allozyme polymorphism in *D. melanogaster* shows the imprint of selection north of the Sahara, where the *A* allele is very high in frequency but shows very low sequence diversity (47). A similar case can be made for the amino acid polymorphism at the *Tpi* locus in *D. simulans*, where a single unique lineage, defined by a single amino acid mutation, possesses six fixed nucleotide differences, or 32% of the polymorphisms in the sample (72). One may not escape the large stochastic variance associated with such processes, and such tests are decidedly one-sided since they will be weakened by recombination, which will tend to erase genealogical structure. Nevertheless, simulations permit specific tests on a case-by-case basis and are relatively easy to program (10, 47, 72, 78–80).

A realization emerging from the sequence studies in *Drosophila* is the complexity of historical selection and the absence of genetic equilibrium at many loci. While persistent balanced polymorphisms are indicated in several cases (47, 101, 152), the ephemeral nature of selection is appearing. Many amino acid polymorphisms may reach high frequencies quickly and then disappear (10, 47, 72, 81). The idea should be entertained that fitness relationships, sufficient for a balanced polymorphism, may emerge repeatedly at loci, but that the environmental conditions determining those relationships may be short lived (less than 4N generations). Finally, superimposed on site-specific selection are the complicating effects of past population structure and genetic hitchhiking (75, 100).

PATHWAY THEORY

The study of natural selection on metabolic genes involves the theory of pathway fluxes, which is important in two respects. The first is the efficacy of functional variation at single genes to translate, via flux variation, into quantitative trait differences. These in turn must translate into fitness differences beyond the threshold where stochastic effects obscure the selective differential (70). This question first emerged in a landmark paper by Kacser & Burns in 1981, addressing the cause of genetic dominance (88). The second arena where pathway theory becomes important involves the architecture of metabolic pathways, or how metabolic fluxes might become partitioned under natural selection (93–95, 105).

Genetic Dominance

Kacser & Burns (88) proposed a simple mathematical theory, grounded in principles of metabolic control theory (87), to explain the fundamental observation that mutations having large homozygous phenotypic effects are generally recessive to the wildtype allele. This model states that the control of overall flux is a systemic feature, and activity variation at each step has little effect on overall flux unless function is markedly reduced. This theory predicts that the activity variation com monly assumed for most enzyme polymorphisms should have little effect on flux because we would expect that enzyme activity has evolved upward to a ceiling or neutral limit (70). If the effect of amino acid polymorphisms does not sufficiently reduce it below the ceiling (i.e. associated selection coefficients are less than the reciprocal of the population size), then such polymorphisms will constitute only molecular noise.

In this regard, chemostat studies in *E. coli* on the G6PD, PGI, and 6PGD allozyme polymorphisms showed that fitnesses effects were immeasurable under nonstressful conditions (34, 37, 38, 68), but in novel physical or genetic environments significant effects of variants could be unveiled (see 69). Hartl & Dykhuizen

(68) proposed that these polymorphisms may be neutral much of the time but that there is latent potential for selection under marginal conditions.

Empirical Observations on Allozymes and Flux

The relevance of control theory to adaptive responses and enzyme polymorphism has been challenged on the bases that it does not sufficiently entertain the complexities of branched pathways or deal with transient non–steady-state systems (19, 33, 158, 159). Ultimately, this is an empirical question. The testing of pathway theories requires biological models in which flux can be empirically measured, and these are not common, especially in metazoans.

There are two interconnected questions here. The first asks whether many pathway steps possess significant flux control. The second asks if allozyme variants possess functional differences great enough to affect such fluxes, even when control is high. It should be mentioned that control depends on the context in which it is defined and measured (36, 52). Thus, the "control coefficient" may be expressed with respect to an immediate flux or a more distal flux or phenotype, and these are conditioned on the physical and genetic background.

Studies using 1-^{14}C- and 6-^{14}C-labeled glucose have shown significantly higher flux associated with the dilocus high-activity allozyme genotypes at G6PD and 6PGD (21) in larvae of *D. melanogaster*. Further studies exploiting the ability of *G6pd* mutations to suppress lethal mutations at the *6Pgd* locus (40, 43, 44), and using radio-labeled glucose (102), found in vivo flux differences of about 20%, consistent with the predictions of in vitro activity measurements for G6PD. Clearly G6PD possesses a high control coefficient for flux through the pentose shunt in larvae.

Flux control through ADH in *D. melanogaster* is complex. ADH activities are twofold higher in adults than in larvae, and studies using ^{14}C-labeled ethanol found no flux differences between *Adh* genotypes in adults (117). However, studies in larvae clearly show ADH as a controlling step in the ethanol-to-lipid pathway (52, 73). The high activity ADH genotype shows higher levels of ethanol flux into lipid as well as greater flux into branching pathways off the Krebs cycle (50, 57, 74). It is also suggested that there is a loss of flux control at high ADH activities (55).

In contrast, *sn*-glycerol-3-phosphate dehydrogenase (GPDH) in *D. melanogaster* appears to possess a low flux control coefficient with respect to flight power (27). Homozygous *Gpdh* null genotypes are flightless, but flies heterozygous for full activity and null alleles (possessing 50% reductions in GPDH activity) showed wing-beat frequency phenotypes statistically indistinguishable from those of wild-type flies. GPDH has a well-characterized allozyme polymorphism (Table 1) that is clinal with latitude (126), and Barnes & Laurie-Alhberg (9) suggested that *Gpdh* allozymes possessed small temperature-dependent effects on wing-beat frequency. GPDH has tissue- and stage-specific isozymic forms that could be under selection for triglyceride synthesis in the larval or adult fat body. The high frequency of null alleles in natural population for this gene is all the more puzzling, and it is discussed below.

Significant effects, measured directly as flux differences or immediate phenotypic effects directly connected to the nature of pathway function, have been seen for PGI in the sea anemone *Metridium senile* (164), LAP and osmotic effects in the mussel *Mytilus edulis* (76, 77), the GPT polymorphism in the marine copepod *Tigriopus californicus* (18), and SOD in *D. melanogaster* (129, 149).

In summary, there appears to be support for the predictions of metabolic control theory as an explanation for genetic dominance (88). However, there are clearly many cases in which large dominance does not appear. The questions are, where do the exceptions arise, and what features of pathway architecture are associated with those exceptions? This is important in understanding how pathway architecture directs acquisition of adaptive variation. Nevertheless, even if genetic dominance in flux is considerable, activity differences may still be perceived by selection, unless the pathway or step has evolved to the hypothetical "neutral limit" (70). As discussed below, codon usage bias, formerly assumed to be associated with very subtle effects on enzyme levels and selection coefficients of nearly neutral effect, suggests that flux variations will be perceived by selection even if there are dominance effects for flux.

The Issue of Null Alleles

Null activity alleles are frequently observed in allozyme surveys, and their frequency is relevant to the issue of natural selection acting on enzyme polymorphism. Systematic screens have been devised in *D. melanogaster* to recover null alleles at many allozyme loci (104, 155), and the average frequency per locus was determined to be surprisingly high, about 3.25×10^{-3}. A study in ponderosa (*Pinus ponderosa*) and red pine (*P. resinosa*) also observed a similar overall null frequency (4). The high frequency in *D. melanogaster* implied that the fitness loss associated with heterozygous null genotypes, presumed to possess a 50% activity loss, was extremely small. The central problem is if activity losses of this magnitude cause such small fitnesses differences, how could allozyme alleles cause the large fitness effects reported by some investigators?

It appears that these studies may overestimate the frequency of null alleles. Alleles judged as null in activity often show partial activity upon further study (17) or appear to be functionally active in vivo (43). For example, a null *6Pgd* allele was also observed in many populations at frequencies of about 1% (WF Eanes, J Hey, unpublished observation). This allozyme was unstable at pH 8.9, at which it normally would be screened, but fully active at the more physiological pH 7.0. Finally, the frequency of null alleles at the *sn-Gpdh* locus is about 1% to 2%, yet this enzyme is essential to the glycerol-3-phosphate shuttle, required for dipteran flight (123). Detailed studies have now concluded that few of the null alleles from populations are true nulls (17, 60), and heterozygotes with normal activity alleles even showed significant dominance for activity. These studies question the earlier observation that alleles with large loss of function are common in natural populations.

CASE STUDIES

Although studies of large-scale patterns of sequence polymorphism and divergence permit general statements about the average relative importance of selection and drift acting on nucleotide variation (see review 100), they provide little insight into the mechanisms of selection as it acts on the biology of organisms in natural environments. This can only be accomplished by detailed dissection of the biochemistry and the proximal and distal effects of functional variation, framed in an environmental and geographic context. A handful of studies have pressed these issues in depth. I outline four such cases here.

ADH in *Drosophila melanogaster*

The most recent review of *Adh* in *Drosophila* is by Chambers (24), so discussion here focuses on several recent developments. There are clear functional differences between the two major allozymes, which are electrophoretically detected by a single charge-changing amino acid polymorphism of Lys and Thr (99). The functional differences between fast and slow allozymes result from catalytic changes affecting the k_{cat} of the reaction, and from twofold differences in ADH protein concentration. Transcript levels are the same in both genotypes (107), and a significant contributor to genotypic differences in protein level may be a small insertion-deletion polymorphism in the 5′ noncoding region (25, 106, 108).

The latitudinal cline in ADH allozyme frequencies, which is reciprocal in northern and southern hemispheres, has been used as evidence for temperature-derived selection on *Adh* (126). However, clines may also result from historical population admixture (see Ldh case below). Since the *Adh* polymorphism also involves protein level variation, and there is extensive linkage disequilibrium across the *Adh* region, it is possible that the primary target of selection is not the Thr/Lys polymorphism, but a linked regulatory site. Berry & Kreitman (13) used four-cutter restriction site variation to examine the structure of the cline with respect to the entire *Adh* gene region. While the Lys/Thr site is strongly clinal (varying in frequency about 35% from Florida to Maine), it is in nearly absolute linkage phase disequilibrium with the small indel polymorphism. Other variable sites showed weaker associations with latitude. The role that this insertion-deletion polymorphism, which is removed in the mature mRNA, would play in conjunction with the amino acid polymorphism to increase protein levels remains to be resolved. Also, in the sliding window analyses of the *Adh-Adh-dup* region (101) the indel site doesn't show an excess peak of silent polymorphism as seen around the Lys/Thr site, which is proposed to be indicative of balancing selection. This type of study is very significant. Latitudinal clines in allozyme polymorphisms are very common (see Table 1), but because it is not known how often linked and unlinked neutral sites will also be clinal, no population-specific null expectation exists (see for example 90, 113).

There has been much interest in the environmental and physiological factors that maintain the ADH polymorphism and its patterns of geographic variation. ADH studies have emphasized ethanol tolerance and detoxification (55), but strong connections have been made to lipid synthesis. In larvae reared in ethanol-supplemented media, both ADH and GPDH activity are induced severalfold (114). This is consistent with the concordant need to provide, via the GPDH branch, the triglyceride backbone for increased lipid synthesis and to balance the NAD/NADH pool (22, 50). Furthermore, there is a covarying latitudinal allozyme cline at *Gpdh*, and the high activity allele increases with latitude (126). Studies using nuclear magnetic resonance (NMR) to follow the relative fluxes of ^{13}C-labeled ethanol into lipid and other metabolites of the Krebs cycle, glycolytic pathway, and their branches indicate that in both larvae and adults, ethanol is utilized extensively, almost preferentially, in lipid synthesis (50–52). As pointed out earlier, ADH possesses a high flux control coefficient of ethanol to lipid (52) in larvae, but not adults (51). Reductions of activity associated with the allozyme alleles also show predicted drops in lipid synthesis from ethanol in larvae (57, 74). Given the higher flux control in larvae, it is surprising that ADH activity is more important in ethanol tolerance in adults than in larvae (144). The same pathway is also used to detoxify and utilize acetic acid, another major carbon source in fermenting material (23).

From these studies it is unclear if ethanol detoxification and tolerance or lipid accumulation per se is the trait under natural selection, although these traits may be interrelated. It has never been clear that populations in more temperate climates need to be more ethanol-tolerant. It is possible that the associated clines are driven by covarying pressure to increase lipid synthesis and storage in more temperate climates, and that this favors increased utilization of ethanol and acetic acid. Their use as carbon sources is preconditioned on increased tolerance. Latitudinal clines in tolerance to ethanol (32) and acetic acid (23) are known, but in adult females, triglyceride content, which is highly correlated with lipid content, also increases 38% from Florida to Vermont (WF Eanes, unpublished observation). There is at this point little debate that the *Adh* gene in *D. melanogaster* is under adaptive selection. The remaining issues involve determining the agent of selection and the interactions between *Adh* and other metabolic genes in response to selection.

G6PD in *Drosophila melanogaster*

The allozyme polymorphism at the glucose-6-phosphate dehydrogenase locus in *D. melanogaster* (*Zw* or *G6pd* are both used in the literature) is the consequence of a polymorphism for quaternary structure (150); the G6PD electrophoretically fast, *A* allele is a dimer, while the *B* variant is a tetramer. After the *Adh* polymorphism, this is globally the most widely varying allozyme polymorphism, with African populations almost exclusively *B*, while European populations are largely *A*, but harbor an endemic third electromorph (the *AF1* allozyme) at 5% to 30% frequency (39, 43, 147). Like ADH and GPDH, the polymorphism shows reciprocal latitudinal clines in North America and Australia (124). The *A* and *B* allozymes differ by a single leucine for proline in residue 382 (46).

The *A* allozyme is less active and thermally less stable (150, 163). Kinetic analysis (44) using highly purified preparations of both variants indicates that the *A* variant possesses a higher K_M for glucose-6-phosphate, and that reported V_{max} differences are due to in vitro instability of the *A* variant. The genotypes possess equal amounts of protein and equal k_{cat}'s. In aggregate the in vitro kinetic parameters predict a difference between genotypes of about 40% in catalytic efficiency.

In vivo activity differences between *G6pd* alleles were examined using a lethal suppression assay that exploits the fact that null alleles at *6Pgd* are lethal, while flies that are simultaneously null at *G6pd* and *6Pgd* are fully viable. The lethality is presumedly due to the accumulation of 6-phosphogluconate (a potent inhibitor of PGI), and the null alleles of *G6pd* block this accumulation. Using a leaky allele of *6Pgd*, differences between *G6pd* genotypes of about a 20% to 40% in in vivo activity were shown (40, 46). The same suppression assay when applied to a set of 11 rare *G6pd* variants recovered from natural populations (43) showed two functional classes of alleles. DNA sequencing established that these two classes reflect the originating allele (*A* or *B*) from which each rare variant was derived (45) and is due to the Pro/Leu amino acid difference. As mentioned above, direct flux studies using radio-labeled glucose have further established that the *G6pd* locus has a high control coefficient with respect to pentose shunt flux (102).

Restriction map variation studies of the *G6pd* gene region indicate substantial linkage disequilibrium across the region (42, 119), and sequencing showed the Pro to Leu change (46) within a domain proposed to be an NADP binding site (45) and identified as part of the dimer-dimer interface (139). Interestingly, another amino acid mutation, *AS2*, shows intermediate activity between *A* and *B* genotypes and lies in the same dimer-dimer domain only two residues from the Pro/Leu change (45). The pattern of silent site polymorphism supports the idea that this is a relatively old polymorphism, much like that seen for *Adh* (47).

After the *Adh* polymorphism, this is the best understood enzyme polymorphism in *Drosophila*. Unlike *Adh*, it is an enzyme whose function and pathway are well known. It is important to establish the cause of selection which, given the covarying polymorphism at *6Pgd* (15, 21, 124), appears to be a polymorphism for decreasing pentose shunt function with increasing latitude. Increased pentose shunt flux is intimately coupled to increased lipid synthesis on strict glucose diets (56). If selection for increased lipid storage with latitude is generating the *Gpdh* and *Adh* clines, then one would expect to see the opposite of what is observed for the clines in pentose shunt polymorphisms. Subsequent studies must address the role of malic enzyme and isocitrate dehydrogenase, additional fluxes off the Krebs cycle that may be the primary source of NADPH for lipid synthesis. However, NMR studies on ADH suggest that pentose shunt flux in larvae may be reduced in high-activity ADH genotypes, in which there is increased flux of ethanol into lipid (50), and the induction of lipid synthesis by ethanol in adults also appears to not be associated with enhanced pentose shunt flux (51). Studies to specifically address this are needed.

LDH in *Fundulus heteroclitus*

The lactate dehydrogenase B-locus (LDH) allozyme polymorphism in the killifish *Fundulus heteroclitus* has been well studied (see reviews 134, 135), and is discussed briefly here. This allozyme polymorphism, like others in *F. heteroclitus*, is associated with a steep latitudinal cline in allele frequencies along the Atlantic seaboard that appears to result from both natural selection and historical subdivision (12). As mentioned above, there are two common amino acid polymorphisms co-segregating and correlated in populations of *Fundulus heteroclitus* (135). One substitution is responsible for the allozyme mobility and the other for the thermal stability differences. Studies have shown thermally different kinetic differences between allozymes (132), and northern *F. heteroclitus* possesses twice as much LDH-B protein in the liver, resulting from higher levels of transcript (30). This was consistent with expectation that northern fish would need twice as much enzyme to thermally compensate for the 12°C difference in temperature regimes. A1 kb region immediately 5′ to the *Ldh* coding region has been sequenced in 23 lines spanning populations from Maine to Florida (145), and contains numerous small indel polymorphisms, as well as microsatellite-rich regions that show strong associations with geographic regions. The Maine sequences were much less variable than southern samples, where pairwise differences between sequences equaled the level of divergence between *F. heteroclitus* and its close relative *F. grandis*. The microsatellite regions may be enhancer sequences and when coupled to a luciferase reporter construct, the northern-derived sequences show higher transcription levels in human embryonic kidney cells (145). Adaptation in this gene appears to be a multisite phenomenon, with both catalytic and regulatory changes, much as with *Adh*. Nevertheless, the interpretation is complex because of the historical legacy of north-south population structure in *F. heteroclitus* (12). It would be interesting to see a large-scale geographic study of linkage phases for these 5′ and catalytic sites. The story for the *Ldh-B* locus is one of the best examples of selection on a metabolic gene.

PGI in *Colias* Butterflies

Watt's study in *Colias* butterflies of several allozyme polymorphisms branching off the head of glycolysis makes one of the most compelling cases for strong selection acting on metabolic genes. In particular, the polymorphisms at both PGI and PGM in *C. philidice eriphyle* and *C. eurytheme* are extraordinary in their contribution to male mating fitness (160). Earlier kinetic studies described dominant and even overdominant enzyme kinetics, and thermal stability differences, between the most common PGI electromorphs (156, 157), and these were supported by field experiments coupling other fitness-related traits to functional predictions (157, 161). In a study of the PGM and G6PD allozyme polymorphisms, substantially higher heterozygosities were observed in mating versus flying males (20). These results were again repeated in *C. meadii* for both PGI and PGM, and again overdominance was observed in the PGI kinetics (162). These are not subtle effects; G6PD heterozygosities in mating males were 20% to 40% higher. This is puzzling since, unlike PGM and PGI, G6PD is found largely in the fat body and can only be distantly connected to immediate metabolic flux in the flight muscle. A pervasive theme in this work is the tradeoff in enhanced catalytic efficiency with decreased thermal stability among genotypes.

These studies in *Colias* emphasize the potential for strong selection on polymorphisms in the central enzymes of metabolism (158). Examination of these loci at the sequence level would be interesting. High resolution electrophoretic approaches have been used to decipher different alleles (20, 162); however, given the emerging picture in *D. melanogaster* for highly polymorphic enzymes, it is likely that these allozymes will be heterogeneous at the amino acid level. It will be interesting to examine the age of these alleles given the apparently very high selection coefficients and reports of shared polymorphism between *Colias* species.

FUTURE DIRECTIONS

We are beyond the narrow dogma of the selectionist-neutralist debate (75, 100). Molecular evolution is a mixture of both neutral change and adaptive change; the rules need be discovered, and distinctions between intra- and interspecific adaptation addressed. The case studies outlined here demonstrate the stepwise approaches to investigating these questions, which have been enlightening both in their simplicity and potential complexity.

Structure-Function Relationships and Enzyme Polymorphism

Crystallographic protein models exist for most of the enzymes depicted in Figure 1. The identification of the amino acid changes associated with functional and stability differences can be used with knowledge of three-dimensional protein structures to provide insight into the mechanistic causes of functional variation. Three examples illustrate this point already. In *D. melanogaster*, the TPI amino acid polymorphism, segregating and clinal in natural populations, is a charge-reversing substitution in an otherwise highly conserved Lys residue (72). This amino acid is in a highly conserved ten-residue domain, designated in mechanistic studies as a "hinged lid" over the active site (86), and could interfere with electrostatic guidance (58). The Leu/Pro polymorphism in *D. melanogaster* G6PD is also in an important structural dimer-dimer domain (45) predicted from the 3-D structure of G6PD from the cyanobacterium *Leuconostoc mesenteroides* G6PD (139). Human G6PD deficiency mutations, which are selectively favored in malarial regions, are also predominantly found in the dimer-dimer interface. Given that many mutations throughout the molecule should disrupt G6PD activity, the bias for this domain suggests that these polymorphisms affect a specific property, such as a role in NADP/NADPH binding, or specific cell-type instabilities (14). Finally, as mentioned earlier, in *Fundulus* LDH the Ser to Ala polymorphism at residue 185 is found in the interface between subunits in the tetramer and is responsible for the thermal stability differences (135). Clearly we gain more insight into the functional effects of amino acid polymorphisms when the functional anatomy of proteins is known.

A common question in interspecific protein evolution is the issue of tradeoffs between function and stability (see review 148). Comparisons among species have documented stability-activity tradeoffs (31, 62, 63), and have shown that relatively few amino acid changes can have dramatic effects on stability. The cost of increased stability, appears to be a loss of molecular flexibility and catalytic function. In colder environments there must be temperature compensation, often accomplished through changes in substrate binding affinities (78), but the rigidity implicit in increased thermal stability decreases catalytic function. At the intraspecific level the same selection may be operating. Frequencies of allozymes with different stabilities often covary with environmental temperature (64, 130, 135, 137, 165). Why this might be a factor in some enzymes and not in others is an important question.

From studies of protein structure in normal and hyperthermophilic bacteria (84, 85, 116), and from protein engineering (111), rules are being established about the types of amino acid changes associated with gain and loss of stability. An understanding of the types of structural changes involved in enhancing stability will allow detailed exploration of reported cases of "interallelic complementation," the enhanced stability or activity of some allelic heteromers (11, 60, 156, 165). To establish a structure-function explanation of cases of apparent activity-stability dominance is important, since nonadditivity in activity leads to marginal overdominance in fitness in models of random environmental fluctuation (61).

Supergenes Revisited

In studies of enzyme polymorphism and selection, genetic linkage effects have always been a concern (41, 110). Patterns of geographic differentiation and even the maintenance of polymorphism could be the consequence of hitchhiking with other loci under selection, and experimentally measured phenotypic differences could also arise from linked genes. For phenotypes like fitness, where the genetic variance is determined by much of the genome, the effects of linked loci cannot be immediately dismissed, even if the limits of disequilibrium are narrow. However, for narrowly defined phenotypes tightly coupled to gene function, linkage effects are unlikely to be an issue. At the single gene level of resolution, it is also important to know the extent to which allozyme, or amino acid polymorphisms per se, are simply surrogate markers for larger segregating functional units. The seriousness of these concerns depends on the extent of linkage disequilibrium spanning most genes, and this was unknown in any species until studies of restriction map variation in *Drosophila* began to define its limits. The results differ from gene to gene, and the spans showing significant disequilibrium vary from hundreds of bases (72, 96, 143) to potentially tens of kilobases (5, 10, 59, 103). While these limits question the efficacy of positive hitchhiking to act across hundreds of genes, the observation of strong linkage disequilibrium within genes makes identifying the targets of selection more difficult (13), and even raises the proposition that selection is operating on a functional unit. The *Adh* activity polymorphism clearly fits a multisite model (13, 106, 108), and the functional polymorphism for *Ldh* in *Fundulus* appears to implicate both protein structural variation and regulation (145). There are exceptions. *G6pd* shows no systematic difference in protein levels between *A* and *B* alleles in *D. melanogaster*, but linkage disequilibrium is very strong spanning ten kilobases (42, 119). An important question is how frequently such polymorphic functional units build up under selection by successive independent mutational changes of both a regulatory and structural nature.

Codon Bias and Deleterious Mutation

In finite populations, fitness differentials need only be greater than the reciprocal of the population size to become biologically significant (128). Do the functional differences that we routinely observe between protein variants cause fitness differentials that are below this threshold and therefore neutral? Codon bias is relevant to this question (1–3). Since bias is strongest in abundant proteins, it is generally accepted that it results from selection favoring translational efficiency or accuracy associated with certain codons (2). The reduction in protein activity associated with the substitution of a single unfavored codon is probably experimentally immeasurable, and certainly much smaller than the activity differences seen in protein polymorphisms. Nevertheless, the small functional effects involved in codon usage have significant consequences in abundant proteins, and can keep unfavored codon fixations low (146). Most of the metabolic genes in Figure 1 possess high codon bias [among the top 5% of all genes in *Drosophila* (97)], suggesting they are under strong selection for maintaining optimal enzyme levels. Given this observation, it seems unlikely that natural selection could not act on the functional differences typically associated with many protein polymorphisms.

While codon bias argues that enzyme polymorphisms are unlikely to be hidden from selection, it does not argue they are adaptive. It is possible that most spontaneous amino acid mutations have large functional effects that are deleterious. The polymorphisms we observe may be those few reaching appreciable frequencies under genetic drift. Statistical analyses of sequence variation and other comparative arguments (6) suggest that many amino acid polymorphisms, especially those in mitochondrial genes, may be slightly deleterious (7, 122, 136, 140). However, there is also experimental evidence that many random amino acid changes, generated de novo in the lab, have very small effects on structural stability and function (111). It is surprising to not see more amino acid polymorphisms with small functional effects segregating in populations, although this could reflect a bias against reporting negative results. More work is needed to define the spectrum of functional variation associated with spontaneous amino acid mutation, and in vitro mutagenesis should continue to add to this understanding.

Getting the Big Picture–Addressing Pathways

Positive natural selection could act on metabolic enzymes in two ways. In different environments, life-history selection may favor differential accumulation of carbon into alternate pools. Alleles with significant impact on flux that facilitate partitioning will accumulate at certain loci. This rationale may apply to the cases in *D. melanogaster* discussed above, and the inhibition of PGI alleles by 6-phosphogluconate has been argued as a way of generating genetic variation in partitioning flux between glycolysis and the pentose shunt in several studies (135, 164, 165). Alternatively, selection simply favor high flux [the basic Hartl-Dykhuizen-Dean model (70)], but in varying thermal environments this may lead to the co-occurrence of alleles that are constrained by the thermodynamic tradeoff between stability and catalysis. This may be characteristic of high flux enzymes in the main glycolytic corridor (151), like PGI and PGM in *Colias* butterflies, or intertidal species that experience large thermal fluctuations. It is important to carry out comparative studies among species to determine the enzyme-specific nature of these arguments.

Finally, a theme emerging from case studies is the importance of studying selection on metabolic traits as a multilocus response (20, 50, 131, 135). Metabolic control theory also emphasizes that flux response requires multisite control (49). In *D. melanogaster* this is driven home by the three cases (G6PD, ADH, GPDH in Figure1, Table 1) where the covarying clinal polymorphisms appear to be old (47, 101, 152) and involve branchpoints associated with lipid synthesis. Add to this the geographically covarying allozyme polymorphism for 6PGD, which parallels G6PD in activity (15, 21, 124), as another point under parallel selection. There are also other side branches where NMR studies indicate ancillary fluxes (50, 51). Metabolic control theory emphasizes that unsaturated branches possess little flux control, while minor leaks, near saturation, exert control (93, 95, 105). The enzymes at the bottom of glycolysis that are likely to see high fluxes (G3PD, PGK, ENO) do not possess amino acid polymorphisms in *D. melanogaster*, and have among the highest known levels of codon bias, which suggests high rates of translation. Multi-pathway fluxes involved in ethanol utilization in *Drosophila* show that increased flux into lipid also involve enhanced fluxes to other pathways such as those to glutamate, glutamine, proline, and lactate, depending on whether it is the adult or larval stage (51). These pathways are interconnected through the NAD/NADH and NADP/NADPH pools. NMR studies indicate higher fluxes from ethanol to glutamate and pyruvate associated the higher activity ADH allele. Should this increase in flux be associated with activity variation at the GDH and MEN branchpoints? GDH is known to possess an allozyme polymorphism and appears to be clinal (147). Although in *D. melanogaster* allozyme polymorphisms do not exist in malic enzyme (MEN), sequencing reveals an amino acid polymorphism segregating in natural populations (Table 1), and it can be asked if this amino acid change affects function, and if it covaries with the other clinal polymorphisms. NMR studies also show that gluconeogenesis becomes very prevalent under ethanol diets, and the cline and polymorphism at TPI in *D. melanogaster* (72, 127) needs to be addressed with this in mind. These studies emphasize the inter-locus complexity that may have built up under selection for ethanol utilization in *Drosophila melanogaster*.

Future investigations should address amino acid polymorphisms as multilocus responses to selection on quantitative variation in metabolism and physiology. Much more information is needed on geographic variation in metabolic and life-history traits. From such patterns, predictions about the multilocus responses can be made. The use of NMR as a noninvasive method is already well established in the study of heritable metabolic disorders in humans, and this method has tremendous potential in the study of metabolic adaptation (16). Finally, DNA sequencing allows us to study genetic variation at many more loci than were screened with allozyme studies, and at much greater depth.

CONCLUSIONS

The failure of allozyme studies to resolve the neutralist-selectionist debate (109) has unfortunately detracted from how much has actually been learned about many enzyme polymorphisms (159), or more specifically, about polymorphisms in the enzymes of central metabolism. DNA sequencing has provided a detailed picture of the amino acid polymorphisms underlying allozymes, and introduced a retrospective window into patterns of selection. Studies of specific polymorphisms in *Drosophila melanogaster*, *Fundulus heteroclitus*, and *Colias* butterflies have repeatedly shown that polymorphisms are associated with functional variation measured both in vitro and in vivo. Knowledge of the three-dimensional structure of many enzymes of the major metabolic pathways will allow the determination of the structure-function relationships associated with many amino acid polymorphisms. Finally, it is apparent that metabolic responses to selection are multilocus in nature, but that all enzymes in contributing pathway may not be equal; determining the rule for these inequalities should be a major goal in studies of enzyme polymorphism.

Visit the Annual Reviews home page at http://www.AnnualReviews.org

LITERATURE CITED

1. Akashi H. 1994. Synonymous codon usage in *Drosophila melanogaster*—natural selection and translational accuracy. *Genetics* 136:927–35
2. Akashi H. 1995. Inferring weak selection from patterns of polymorphism and divergence at "silent" sites in *Drosophila* DNA. *Genetics* 139:1067–76
3. Akashi H. 1996. Molecular evolution between *Drosophila melanogaster* and *D. sim-*

ulans: reduced codon bias, faster rates of amino acid substitution, and larger proteins in *D. melanogaster*. *Genetics* 144:1297–1307
4. Allendorf FW, Knudsen KL, Blake GM. 1982. Frequencies of null alleles at enzyme loci in natural populations of ponderosa and red pine. *Genetics* 100:497–504
5. Aquadro CF. Jennings RM, Bland MM, Laurie CC, Langley CH. 1992. Patterns of naturally occurring restriction map variation, dopa decarboxylase activity variation and linkage disequilibrium in the DdC gene region of *Drosophila melanogaster*. *Genetics* 132:443–52
6. Aquadro CF, Lado KM, Noon WA.1988. The rosy region of *Drosophila melanogaster* and *Drosophila simulans*. I. Contrasting levels of naturally occurring DNA restriction map variation and divergence. *Genetics* 119:875–88
7. Ballard JWO, Kreitman M. 1994. Unraveling selection in the mitochondrial genome of *Drosophila*. *Genetics* 138:757–72
8. Barbadilla A, King LM, Lewontin RC. 1996. What does electrophoretic variation tell us about protein variation? *Mol. Biol. Evol.* 13:427–32
9. Barnes PT, Laurie-Ahlberg C. 1986. Genetic variability of flight metabolism in *Drosophila melanogaster*. III. Effects of GPDH allozymes and environmental temperature on power output. *Genetics* 112: 267–98
10. Begun DJ, Aquadro CF. 1994. Evolutionary inferences from DNA variation at the 6 phosphogluconate dehydrogenase locus in natural populations of *Drosophila*–selection and geographic differentiation. *Genetics* 136:155–71
11. Berger E. 1976. Heterosis and the maintenance of enzyme polymorphism. *Am. Nat.* 110:823–39
12. Bernardi G, Sordino P, Powers DA. 1993. Concordant mitochondrial and nuclear DNA phylogenies for populations of the teleost fish *Fundulus heteroclitus*. *Proc. Natl. Acad. Sci. USA* 90:9271–74
13. Berry A, Kreitman M. 1993. Molecular analysis of an allozyme cline: alcohol dehydrogenase in *Drosophila melanogaster* on the east coast of North America. *Genetics* 134:869–93
14. Beutler E. 1994. G6PD deficiency. *Blood* 84:3613–36
15. Bijlsma R, van der Meulen-Bruijns C. 1979. Polymorphism at the G6PD and 6PGD loci in *Drosophila melanogaster*. III. Developmental and biochemical aspects. *Biochem. Genet.* 17:1131–44
16. Brindle KM, Fulton SM, Gillham H, Williams SP. 1997. Studies of metabolic control using NMR and molecular genetics. *J. Mol. Recognit.* 10:182–87
17. Burkhart BD, Montgomery E, Langley CH, Voelker RA. 1984. Characterization of allozyme null and low activity alleles from two natural populations of *Drosophila melanogaster*. *Genetics* 107:295–306
18. Burton RS, Feldman MW. 1983. Physiological effects of an allozyme polymorphism: glutamate-pyruvate transaminase and response to hyperosmotic stress in the copepod *Tigriopus californicus*. *Biochem. Genet.* 21:239–51
19. Burton RS, Place AR. 1986. Evolution of selective neutrality: further considerations. *Genetics* 114:1033–36
20. Carter PA, Watt WB. 1988. Adaptation at specific loci. V. Metabolically adjacent loci may have very distinct experiences of selective pressures. *Genetics* 119:913–24
21. Cavener DR, Clegg MT. 1981. Evidence for biochemical and physiological differences between genotypes in *Drosophila melanogaster*. *Proc. Natl. Acad. Sci. USA* 78:4444–47
22. Cavener DR, Clegg MT. 1981. Multigenic response to ethanol in *Drosophila melanogaster*. *Evolution* 35:1–10
23. Chakir M, Capy P, Genermont J, Pla E, David JR. 1996. Adaptation to fermenting resources in *Drosophila melanogaster*:

ethanol and acetic acid tolerances share a common genetic basis. *Evolution* 50:767–76

24. Chambers GK. 1988. The Drosophila alcohol dehydrogenase gene-enzyme system. *Adv. Genet.* 25:39–107
25. Choudhary M, Laurie CC. 1991. Use of in vitro mutagensis to analyze the molecular basis of the difference in Adh expression associated with the allozyme polymorphism in *Drosophila melanogaster*. *Genetics* 129:481–88
26. Clark B. 1975. The contribution of ecological genetics to evolutionary theory; detecting the direct effects of natural selection on particular polymorphic loci. *Genetics* 79:101–13
27. Connors EM, Curtsinger JW. 1986. Relationship between a-glycerophosphate dehydrogenase activity and metabolic rate during flight in *Drosophila melanogaster*. *Biochem. Genet.* 24:245–57
28. Cooke PH, Oakeshott JG. 1989. Amino acid polymorphisms for esterase-6 in *Drosophila melanogaster*. *Proc. Natl. Acad. Sci. USA* 86:1426–30
29. Coyne JA, Eanes WF, Ramshaw JA, Koehn RK. 1982. Electrophoretic heterogeneity of α-glycerophosphate dehydrogenase among many species of *Drosophila*. *Syst. Zool.* 28:164–75
30. Crawford DL, Powers DA. 1989. Molecular basis of evolutionary adaptation at the lactate dehydrogenase locus in the fish *Fundulus heteroclitus*. *Proc. Natl. Acad. Sci. USA* 86:9365–69
31. Dahlhoff EP, Somero GN. 1993. Kinetic and structural adaptations of cytosolic malate dehydrogenases of eastern Pacific abalones (genus *Haliotis*) from different thermal habitats; biochemical correlates of geographic patterning. *J. Exp. Biol.* 185:137–50
32. David JR, Bocquet C. 1975. Similarities and differences in latitudinal adaptation of two *Drosophila* sibling species. *Nature* 257:588–90
33. Dean AM. 1994. Fitness, flux and phantoms in temporally variable environments. *Genetics* 136:1481–95
34. Dean AM, Dykhuizen DE, Hartl DL. 1986. Fitness as a function of β-galatosidase activity in *Escherichia coli*. *Genet. Res.* 48:1–8
35. Donnelly P, Tavare S. 1995. Coalescents and genealogical structure under neutrality. *Annu. Rev. Genet.* 29:401–21
36. Dykhuizen DE, Dean AM. 1990. Enzyme activity and fitness: evolution in solution. *Trends Ecol. Evol.* 5:257–62
37. Dykhuizen DE, Hartl DL. 1980. Selective neutrality of 6PGD allozymes in *E. coli* and the effects of genetic background. *Genetics* 96:801–17
38. Dykhuizen DE, Hartl DL. 1983. Functional effects of PGI allozymes in *E. coli*. *Genetics* 105:1–18
39. Eanes WF. 1983. Genetic localization and sequential electrophoresis of G6pd in *Drosophila melanogaster*. *Biochem. Genet.* 21:703–11
40. Eanes WF. 1984. Viability interactions, in vivo activity and the G6PD polymorphism in *Drosophila melanogaster*. *Genetics* 106:95–107
41. Eanes WF. 1987. Allozymes and fitness: evolution of a problem. *Trends Ecol. Evol.* 2:44–48
42. Eanes WF, Ajioka JW, Hey J, Wesley C. 1989. Restriction map variation associated with the G6PD polymorphism in *Drosophila melanogaster*. *Mol. Biol. Evol.* 6:384–97
43. Eanes WF, Hey J. 1986. In vivo function of rare *G6pd* variants from natural populations of *Drosophila melanogaster*. *Genetics* 113:679–93
44. Eanes WF, Katona L, Longtine M. 1990. Comparison of in vitro and in vivo activities associated with the G6PD allozyme polymorphism in *Drosophila melanogaster*. *Genetics* 125:845–53
45. Eanes WF, Kirchner M, Taub DR, Yoon J, Chen J. 1996. Amino acid polymorphism

and rare variants of *G6PD* from natural populations of *Drosophila melanogaster*. *Genetics* 143:401–6

46. Eanes WF, Kirchner M, Yoon J. 1993. Evidence for adaptive evolution of the G6PD gene in *Drosophila melanogaster* and *D. simulans* lineages. *Proc. Natl. Acad. Sci. USA* 90:7475–79
47. Eanes WF, Kirchner M, Yoon J, Biermann CH, Wang IN, et al. 1996. Historical selection, amino acid polymorphism and lineage-specific divergence at the *G6pd* locus in *Drosophila melanogaster* and *D. simulans*. *Genetics* 144:1027–41
48. Ewens WJ. 1977. Population genetics theory in relation to the neutralist-selectionist controversy. In *Advances in Human Genetics*, ed. H Harris, K Hirschhorn, 8:67–132. New York: Plenum
49. Fell DA, Thomas S. 1995. Physiological control of metabolic flux: the requirement for multisite modulation. *Biochem. J.* 311:35–39
50. Freriksen A, deRuiter BLA, Scharloo W, Heinstra PWH. 1994. *Drosophila* alcohol dehydrogenase polymorphism and carbon-13 fluxes: opportunities for epistasis and natural selection. *Genetics* 137:1071–78
51. Freriksen A, Seykens D, Heinstra PWH. 1994. Differences between larval and adult *Drosophila* in metabolic degradation of ethanol. *Evolution* 48:504–08
52. Freriksen A, Seykens D, Scharloo W, Heinstra PWH. 1991. Alcohol dehydrogenase controls the flux from ethanol into lipids in *Drosophila* larvae. A 13C NMR Study. *J. Biol. Chem.* 266:21399–403
53. Fucci L, Gaudio L, Rao R, Spano A, Carfagna M. 1979. Properties of the two common electrophoretic variants of phosphoglucomutase in *Drosophila melanogaster*. *Biochem. Genet.* 17:825–35
54. Gaut BS, Clegg MT. 1993. Nucleotide polymorphism in the *Adh1* locus of pearl millet (*Pennisetum glaucum*) (Poacae). *Genetics* 135:1091–97
55. Geer BW, Heinstra PW, Mckechnie SW. 1993. The biological basis of ethanol tolerance in *Drosophila*. *Comp. Biochem. Physiol.* [B] 105B:203–29
56. Geer BW, Lindel DL, Lindel DM. 1979. Relationship of the oxidative pentose shunt pathway to lipid synthesis in *Drosophila melanogaster*. *Biochem. Genet.* 17:881–95
57. Geer BW, Mckechnie SW, Bentley MM, Oakeshott JG, Quinn EM, Langevin ML. 1988. Induction of alcohol dehydrogenase by ethanol in *Drosophila melanogaster*. *J. Nutr.* 118:398–407
58. Getzoff ED, Cabelli DE, Fisher CL, Parge HE, Viezzoli MS, et al. 1996. Faster superoxide dismutase mutants designed by enhancing electrostatic guidance. *Nature* 358:347–51
59. Gibson G, Hogness DS. 1996. Effect of polymorphism in the Drosophila regulatory gene Ultrabithorax on homeotic stability. *Science* 271:200–3
60. Gibson JB, Cao A, Symonds J, Reed D. 1991. Low activity sn-glycerol-3-phosphate dehydrogenase variants in natural populations of *Drosophila melanogaster*. *Heredity* 66:75–82
61. Gillespie JH. 1991. *The Causes of Molecular Evolution*. New York, Oxford: Oxford Univ. Press. 336 pp.
62. Graves JE, Rosenblatt RH, Somero GN. 1984. Kinetic and electrophoretic differentiation of lactate dehydrogenases of teleost species-pairs from the Atlantic and Pacific Coasts of Panama. *Evolution* 37:30–37
63. Graves JE, Somero GN. 1982. Electrophoretic and functional enzyme evolution in four species of eastern Pacific barracudas from different thermal environments. *Evolution* 36:91–106
64. Hall JG. 1985. Temperature related differentiation of glucose phosphate isomerase alloenzymes isolated from the blue mussel, Mytilus edulis. *Biochem. Genet.* 23:705–28
65. Hall JG, Koehn RK. 1983. The evolution of enzyme catalytic efficiency and adaptive

inference from steady-state data. *Evol. Biol.* 16:53–69
66. Hamblin MT, Aquadro CF. 1996. High nucleotide sequence variation in a region of low recombination in *Drosophila simulans* is consistent with the background selection model. *Mol. Biol. Evol.* 13:1133–40
67. Hamblin MT, Aquadro CF. 1997. Contrasting patterns of nucleotide sequence variation at the glucose dehydrogenase (Gld) locus in different populations of *Drosophila melanogaster*. *Genetics* 145:1053–62
68. Hartl DL, Dykhuizen DE. 1881. Potential for selection among nearly neutral allozymes of 6-phosphogluconate dehydrogenase in *Escherichia coli*. *Proc. Natl. Acad. Sci. USA* 78:6344–48
69. Hartl DL, Dykhuizen DE. 1985. The neutral theory and the molecular basis of preadaptation. In *Population Genetics and Molecular Evolution*, ed. T Ohta, K Aoki, pp. 107–24. Tokyo/Berlin: Jpn. Sci. Soc. Press, Springer-Verlag
70. Hartl DL, Dykhuizen DE, Dean AM. 1985. Limits of adaptation: the evolution of selective neutrality. *Genetics* 111:655–74
71. Hasson E, Eanes WF. 1996. Contrasting histories of three gene regions associated with *In(3L)Payne* of *Drosophila melanogaster*. *Genetics* 144:1565–75
72. Hasson E, Wang IN, Zeng LW, Kreitman M, Eanes WF. 1998. Nucleotide variation in the triosephosphate isomerase (*Tpi*) locus of *Drosophila melanogaster* and *Drosophila simulans*. *Mol. Biol. Evol.* 15:756–69
73. Heinstra PWH, Geer BW. 1991. Metabolic control analysis and enzyme variation: nutritional manipulation of the flux from ethanol to lipids in *Drosophila*. *Mol. Biol. Evol.* 8:703–8
74. Heinstra PW, Scharloo W, Thorig GEW. 1987. Physiological significance of the alcohol dehydrogenase polymorphism in larvae of Drosophila. *Genetics* 117:75–84
75. Hey J. 1999. The neutralist, the fly and the selectionist. *Trends Ecol. Evol.* 14:35–38
76. Hilbish TJ, Deaton LE, Koehn RK. 1982. Effect of an allozyme polymorphism on regulation of cell volume. *Nature* 298:688–89
77. Hilbish TJ, Koehn RK. 1985. Dominance in physiological phenotypes and fitness at an enzyme locus. *Science* 229:52–54
78. Hochachka PW, Somero GN. 1984. *Biochemical Adaptation*. New Jersey: Princeton Univ. Press.
79. Hudson RR. 1990. Gene genealogies and the coalescent process. In *Oxford Surveys in Evolutionary Biology*, ed. DJ Futuyma, J Antonovics, pp. 1–44. New York/Oxford: Oxford Univ. Press
80. Hudson RR. 1993. The how and why of generating gene genealogies. In *Mechanisms of Molecular Evolution*, ed. N Takahata, AG Clark, pp. 23–36 Tokyo/Sunderland, MA: Japan Sci. Soc. Press/Sinauer
81. Hudson RR, Bailey K, Skarecky D, Kwiatowski J, Ayala FJ. 1994. Evidence for positive selection in the superoxide dismutase (Sod) region of *Drosophila melanogaster*. *Genetics* 136:1329–40
82. Hudson RR, Kaplan NL. 1988. The coalescent process in models with selection and recombination. *Genetics* 120:831–40
83. Hudson RR, Saez AG, Ayala FJ. 1997. DNA variation at the *Sod* locus of *Drosophila melanogaster*: An unfolding story of natural selection. *Proc. Natl. Acad. Sci. USA* 94:7725–29
84. Jaenicke R. 1991. Protein stability and molecular adaptation to extreme conditions. *Eur. J. Biochem.* 202:715–28
85. Jaenicke R. 1996. Glyceraldehyde-3-phosphate dehydrogenase from *Thermotoga maritima*: strategies of protein stabilization. *FEMS Microbiol. Rev.* 18:215–24
86. Joeseph DG, Petsko A, Karplus M. 1990. Anatomy of a conformational change: hinged "lid" motion of the triose phosphate isomerase loop. *Science* 249:1425–28
87. Kacser H, Burns JA. 1973. The control of flux. *Symp. Soc. Exp. Biol.* 27:65–104

88. Kacser H, Burns JA. 1981. The molecular basis of dominance. *Genetics* 97:639–66
89. Kaplan N, Hudson RR, Iizuka M. 1991. The coalescent process in models with selection, recombination and geographic subdivision. *Genet. Res.* 57:83–91
90. Karl SA, Avise JC. 1992. Balancing selection at allozyme loci in oysters: implications from nuclear RFLPs. *Science* 256:100–2
91. Karotam J, Boyce TM, Oakeshott JG. 1995. Nucleotide variation at the hypervariable esterase 6 isozyme locus of *Drosophila simulans*. *Mol. Biol. Evol.* 12:113–22
92. Katz LA, Harrison RG. 1997. Balancing selection on electrophoretic variation of phosphoglucose isomerase in two species of field cricket: *Gryllus veletis* and *G. pennsylvanicus*. *Genetics* 147:609–21
93. Keightley PD. 1989. Models of quantitative variation of flux in metabolic pathways. *Genetics* 121:869–76
94. Keightley PD. 1996. A metabolic basis for dominance and recessivity. *Genetics* 143:621–25
95. Keightley PD, Kacser H. 1987. Dominance, pleiotropy and metabolic structure. *Genetics* 117:319–29
96. Kirby DA, Stephan W. 1996. Multilocus selection and the structure of variation at the white gene of *Drosophila melanogaster*. *Genetics* 144:635–45
97. Kliman RM, Hey J. 1993. Reduced natural selection associated with low recombination in *Drosophila melanogaster*. *Mol. Biol. Evol.* 10:1239–58
98. Kreitman M. 1980. Assessment of variability within electromorphs of alcohol dehydrogenase in *Drosophila melanogaster*. *Genetics* 95:467–75
99. Kreitman M. 1983. Nucleotide polymorphism at the alcohol dehydrogenase locus of *Drosophila melanogaster*. *Nature* 304:412–17
100. Kreitman M, Akashi H. 1995. Molecular evidence for natural selection. *Annu. Rev. Ecol. Syst.* 26:403–22
101. Kreitman MK, Hudson RR. 1991. Inferring the histories of *Adh* and *Adh-dup* in *Drosophila melanogaster* from patterns of polymorphism and divergence. *Genetics* 127:565–82
102. Labate J, Eanes WF. 1992. Direct measurement of in vivo flux differences between electrophoretic variants of G6PD from *Drosophila melanogaster*. *Genetics* 132:783–87
103. Lai CG, Lyman RF, Long AD, Langley CH, Mackay TFC. 1994. Naturally occurring variation in bristle number and DNA polymorphisms at the scabrous locus of *Drosophila melanogaster*. *Science* 266:1697–1702
104. Langley CH, Voelker RA, Leigh-Brown AJ, Ohnishi S, Dickson B, Montgomery E. 1981. Null allele frequencies at allozyme loci in natural populations of *Drosophila melanogaster*. *Genetics* 99:151–56
105. LaPorte DC, Walsh K, Koshland DE. 1984. The branch point effect. Ultrasensitivity and subsensitivity to metabolic control. *J. Biol.Chem.* 259:14068–75
106. Laurie CC, Bridgham JT, Choudhary M. 1991. Associations between DNA sequence variation and variation in expression of the *Adh* gene in natural populations of *Drosophila melanogaster*. *Genetics* 129:489–99
107. Laurie CC, Stam LF. 1988. Quantitative analysis of RNA produced by Slow and Fast alleles of *Drosophila melanogaster Adh*. *Proc. Natl. Acad. USA* 85:5161–65
108. Laurie CC, Stam LF. 1994. The effect of an intronic polymorphism on alcohol dehydrogenase expression in *Drosophila melanogaster*. *Genetics* 138:379–85
109. Lewontin RC. 1991. Electrophoresis in the development of evolutionary genetics: milestone or millstone? *Genetics* 128: 657–62
110. Lewontin RC. 1974. *The Genetic Basis*

of Evolutionary Change. New York: Columbia Univ. Press

111. Matthews BW. 1993. Structural and genetic analysis of protein stability. *Annu. Rev. Biochem.* 62:139–60
112. McDonald JH, Kreitman MK. 1991. Adaptive evolution at the *Adh* locus in *Drosophila*. *Nature* 351:652–54
113. McDonald JH, Verrelli BC, Geyer LB. 1996. Lack of geographic variation in anonymous nuclear polymorphisms in the American oyster, *Crassostrea virginica*. *Mol. Biol. Evol.* 13:1114–18
114. McKechnie SW, Geer BW. 1984. Regulation of alcohol dehydrogenase in *Drosophila melanogaster* by dietary alcohol and carbohydrate. *Insect Biochem.* 14:231–42
115. Deleted in proof
116. Menendez-Arias L, Argos P. 1989. Engineering protein stability. Sequence statistics point to residue substitutions in alpha-helices. *J. Mol. Biol.* 206:397–406
117. Middleton RJ, Kacser H. 1983. Enzyme variation, metabolic flux and fitness: alcohol dehydrogenase in *Drosophila melanogaster*. *Genetics* 105:633–50
118. Mitton JB. 1997. *Selection in Natural Populations*. New York/Oxford: Oxford Univ. Press. 256 pp.
119. Miyashita NT. 1990. Molecular and phenotypic variation of the *Zw* locus region in *Drosophila melanogaster*. *Genetics* 125:407–19
120. Miyashita NT, Innan H, Terauchi R. 1996. Intra- and interspecific variation of the alcohol dehydrogenase locus region in wild plants *Arabis gemmifera* and *Arabidopsis thaliana*. *Mol. Biol. Evol.* 13:433–36
121. Moriyama EN, Powell JR. 1996. Intraspecific nuclear DNA variation in Drosophila. *Mol. Biol. Evol.* 13:261–77
122. Nachman MW, Brown WM, Stoneking M, Aquadro CF. 1996. Nonneutral mitochondrial DNA variation in humans and chimpanzees. *Genetics* 142:953–63
123. O'Brien SJ, MacIntyre RJ. 1972. The α-glycerophosphate cycle in *Drosophila melanogaster*. II. Genetic aspects. *Genetics* 71:127–38
124. Oakeshott JG, Chambers GK, Gibson JB, Eanes WF, Willcocks DA. 1983. Geographic variation in *G6pd* and *Pgd* allele frequencies in *Drosophila melanogaster*. *Heredity* 50:67–72
125. Oakeshott JG, Chambers GK, Gibson JB, Willcocks DA. 1981. Latitudinal relationships of esterase-6 and phosphoglucomutase gene frequencies in *Drosophila melanogaster*. *Heredity* 47:385–96
126. Oakeshott JG, Gibson JB, Anderson PR, Knibb WR, Chambers GK. 1982. Alcohol dehydrogenase and glycerol-3-phosphate dehydrogenase clines in *Drosophila melanogaster* on different continents. *Evolution* 36:86–96
127. Oakeshott JG, Mckechnie SW, Chambers GK. 1984. Population Genetics of the metabolically related *Adh, Gpdh*, and *Tpi* polymorphisms in *Drosophila melanogaster*. I. Geographic variation in *Gpdh* and *Tpi* allele frequencies in different continents. *Genetica* 63:21–29
128. Ohta T. 1998. The nearly neutral theory of molecular evolution. *Annu. Rev. Ecol. Syst.* 23:263–86
129. Peng TX, Moya A, Ayala FJ. 1986. Irradiation resistance conferred by superoxide dismutase: possible adaptive role of a natural polymorphism. *Proc. Natl. Acad. Sci. USA* 83:684–87
130. Phillip DP, Childers WF, Whitt GS. 1985. Correlations of allele frequencies with physical and environmental variables for populations of largemouth bass, *Micropterus salmoides*. *J. Fish. Biol.* 27: 347–65
131. Pierce VA, Crawford DL. 1997. Phylogenetic analysis of glycolytic enzyme expression. *Science* 276:256–59
132. Place AR, Powers DA. 1979. Genetic variation and relative catalytic efficiencies: lactate dehydrogenase-B allozymes of *Fundulus heteroclitus*. *Proc. Natl. Acad. Sci. USA* 76:2354–58

133. Powell MA, Crawford DL, Lauerman T, Powers DA. 1992. Analysis of cryptic alleles of *Fundulus heteroclitus* lactate dehydrogenase by a novel allele-specific polymerase chain reaction. *Mol. Mar. Biol. Biotechol.* 1:391–96
134. Powers DA, Lauerman T, Crawford D, DiMichele L. 1991. Genetic mechanisms for adaptation to a changing environments. *Annu. Rev. Genet.* 25:629–59
135. Powers DA, Smith M, Gonzalez-Villasenor I, DiMichele L, Crawford D,. et al. 1993. A multidisciplinary approach to the selectionist/neutralist controversy using the model teleost, *Fundulus heteroclitus*. In *Oxford Surveys in Evolutionary Biology*, ed. DJ Futuyma, J Antonovics, pp. 43–107. New York/Oxford: Oxford Univ. Press
136. Rand DM, Kann LM. 1996. Excess amino acid polymorphism in mitochondrial DNA: contrasts among genes from Drosophila, mice, and humans. *Mol. Biol. Evol.* 13:735–48
137. Riddoch BJ. 1993. The adaptive significance of electrophoretic mobility in phosphoglucose isomerase (PGI). *Biol. J. Linn. Soc.* 50:1–17
138. Riley MA, Kaplan SR, Veuille M. 1992. Nucleotide polymorphism at the xanthine dehydrogenase locus in *Drosophila pseudoobscura*. *Mol. Biol. Evol.* 9:56–69
139. Rowlan P, Basak AK, Gover S, Levy HR, Adams MJ. 1994. The three-dimensional structure of glucose 6-phosphate dehydrogenase from *Leuconostoc mesenteroides* refined at 2.0A resolution. *Structure* 2:1073–87
140. Sawyer SA, Dykhuizen DE, Hartl DL. 1987. Confidence interval for the number of selectively neutral amino acid polymorphisms. *Proc. Natl. Acad. Sci. USA* 84:6225–28
141. Sawyer LA, Hennessy JM, Peixoto AA, Rosato E, Parkinson H, et al. 1997. Natural variation in a Drosophila clock gene and temperature compensation. *Science* 278:2117–20
142. Schaeffer SW, Miller EL. 1992. Molecular population genetics of an electrophoretically monomorphic gene in *Drosophila pseudoobscura*: the alcohol dehydrogenase region. *Genetics* 132: 163–78
143. Schaeffer SW, Miller EL. 1993. Estimates of linkage disequilibrium and recombination parameter determined from segregating nucleotide sites in the alcohol dehydrogenase region of *Drosophila pseudoobscura*. *Genetics* 135:541–52
144. Schmitt LH, Mckechnie SW, McKechnie JA. 1986. Associations between alcohol tolerance and the quantity of alcohol dehydrogenase in *Drosophila melanogaster* isolated from a winery population. *Aust. J. Biol. Sci.* 39:59–67
145. Schulte PM, GomezChiarri M, Powers DA. 1997. Structural and functional differences in the promoter and 5′ flanking region of Ldh-B within and between populations of the teleost *Fundulus heteroclitus*. *Genetics* 145:759–69
146. Shields DC, Sharp PM, Higgins DG, Wright F. 1988. "Silent" sites in *Drosophila* genes are not neutral: evidence of selection among synonymous codons. *Mol. Biol. Evol.* 5:704–16
147. Singh RS, Rhomberg LR. 1987. A comprehensive study of genic variation in natural populations of *Drosophila melanogaster*. II. Estimates of heterozygosity and patterns of geographic differentiation. *Genetics* 117:255–72
148. Somero GN. 1995. Proteins and temperature. *Annu. Rev. Physiol.* 57:43–86
149. Staveley BE, Phillips JP, Hilliker AJ. 1990. Phenotypic consequences of copper-zinc superoxide dismutase overexpression in *Drosophila melanogaster*. *Genome* 31:867–72
150. Steele MC, Young CW, Childs B. 1968. Glucose-6-phosphate dehydrogenase in *Drosophila melanogaster*: starch gel

electrophoretic variation due to molecular instability. *Biochem. Genet.* 2:159–75
151. Suarez RK, Staples JF, Lighton JRB, West TG. 1997. Relationships between enzymatic flux capacities and metabolic flux rates: nonequilibrium reactions in muscle glycolysis. *Proc. Natl. Acad. Sci. USA* 94:7065–69
152. Takano TS, Kusakabe S, Mukai T,. 1993. DNA polymorphism and the origin of protein polymorphism at the *Gpdh* locus of *Drosophila melanogaster*. In *Mechanisms of Molecular Evolution*, ed. N Takahata, AG Clark, pp. 179–190. Tokyo/Sunderland, MA: Jpn. Sci. Soc./ Sinauer
153. Thomas BR, Frod VS, Pichersky E, Gottlieb LD. 1993. Molecular characterization of duplicate cytosolic phosphoglucose isomerase genes in *Clarkia* and comparison to the single gene in *Arabidopsis*. *Genetics* 135:895–905
154. Veuille M, King LM. 1995. Molecular basis of polymorphism at the esterase-5B locus in *Drosophila pseudoobscura*. *Genetics* 141:255–62
155. Voelker RA, Langley CH, Leigh-Brown AJ, Ohnishi S, Dickson B, et al. 1980. Enzyme null alleles in natural populations of *Drosophila melanogaster*: frequencies in a North Carolina population. *Proc. Natl. Acad. Sci. USA* 77:1091–95
156. Watt WB. 1977. Adaptation at specific loci. I. Natural selection on phosphoglucose isomerase of *Colias* butterflies: biochemical and population aspects. *Genetics* 87:177–94
157. Watt WB. 1983. Adaptation at specific loci. II. Demographic and biochemical elements in the maintenance of the *Colias* PGI polymorphism. *Genetics* 103:691–724
158. Watt WB. 1985. Bioenergetics and evolutionary genetics: opportunities for new synthesis. *Am. Nat.* 125:118–43
159. Watt WB. 1994. Allozymes in evolutionary genetics—self-imposed burden or extraordinary tool? *Genetics* 136:11–16
160. Watt WB, Carter PA, Blower SM. 1985. Adaptation at specific loci. IV. Differential mating success among glycolytic allozyme genotypes of *butterflies*. *Genetics* 109:157–75
161. Watt WB, Cassin RC, Swan MS. 1983. Adaptation at specific loci. III. Field behavior and survivorship differences among *Colias* PGI genotypes are predictable from in vitro biochcmistry. *Genetics* 103:725–39
162. Watt WB, Donohue K, Carter PA. 1996. Adaptation at specific loci. VI. Divergence vs. parallelism of polymorphic allozymes in molecular function and fitness-component effects among *Colias* species (Lepidoptera, Pieridae). *Mol. Biol. Evol.* 13:699–709
163. Williamson JH, Bentley MM. 1983. Comparative properties of three forms of glucose-6-phosphate dehydrogenase in *Drosophila melanogaster*. *Biochem. Genet.* 21:1153–66
164. Zamer WE, Hoffman RJ. 1989. Allozymes of glucose-6-phosphate isomerase differentially modulate pentose-shunt metabolism in the sea anemone *Metridium senile*. *Proc . Natl. Acad. Sci. USA* 86:2737–41
165. Zera AJ. 1987. Temperature-dependent kinetic variation among phosphoglucose isomerase allozymes from the wing-polymorphic water strider, *Limnoporus canaliculatus*. *Mol. Biol. Evol.* 4:266–85

Annu. Rev. Ecol. Syst. 1999. 30:327–62

Polymorphism in Systematics and Comparative Biology

John J. Wiens
Section of Amphibians and Reptiles, Carnegie Museum of Natural History, Pittsburgh, Pennsylvania 15213-4080; e-mail: wiensj@clpgh.org.

Key Words comparative methods, phylogenetic analysis, phylogeny, intraspecific variation, species-limits

■ **Abstract** Polymorphism, or variation within species, is common in all kinds of data and is the major focus of research on microevolution. However, polymorphism is often ignored by those who study macroevolution: systematists and comparative evolutionary biologists. Polymorphism may have a profound impact on phylogeny reconstruction, species-delimitation, and studies of character evolution. A variety of methods are used to deal with polymorphism in phylogeny reconstruction, and many of these methods have been extremely controversial for more than 20 years. Recent research has attempted to address the accuracy of these methods (their ability to estimate the true phylogeny) and to resolve these issues, using computer simulation, congruence, and statistical analyses. These studies suggest three things: that (*a*) the exclusion of polymorphic characters (as is commonly done in morphological phylogenetics) is unjustified and may greatly decrease accuracy relative to analyses that include these characters; (*b*) methods that incorporate frequency information on polymorphic characters tend to perform best, and (*c*) distance and likelihood methods designed for polymorphic data may often outperform parsimony methods. Although rarely discussed, polymorphism may also have a major impact on comparative studies of character evolution, such as the reconstruction of ancestral character states. Finally, polymorphism is an important issue in the delimitation of species, although this area has been somewhat neglected methodologically. The integration of within-species variation and microevolutionary processes into studies of systematics and comparative evolutionary biology is another example of the benefits of exchange of ideas between the fields of population genetics and systematics.

INTRODUCTION

One of the most important trends in systematics and evolutionary biology in recent years has been an increasing appreciation for the interconnectedness of these fields. For example, phylogenies are used increasingly by evolutionary biologists studying ecology and behavior (e.g. 9, 60, 82), and systematists using DNA

0066-4162/99/1120-0327$08.00

and RNA sequence data are beginning to incorporate more and more details of molecular evolutionary processes into their phylogenetic analyses (e.g. 130). One of the areas in which a phylogenetic approach has had an important impact is the study of within-species variation, particularly in the fields of phylogeography and molecular population genetics (e.g. 4, 45, 51, 62, 126). However, many unresolved questions remain as to what the study of within-species variation and microevolutionary processes might have to offer between-species systematic and comparative evolutionary studies (e.g. 58).

Heritable variation within species is the basic material of evolutionary change and the major subject of research on microevolutionary processes. Intraspecific variation is abundant in all kinds of phenotypic and genotypic traits, including morphology, behavior, allozymes, and DNA sequences. This variation is not really surprising because if characters vary between species, they must also vary within species, at least at some point in their evolution. In many cases, especially among closely related species, this instraspecific variation may persist and may be abundant. For example, among the nine species of the lizard genus *Urosaurus*, 23 of 24 qualitative morphological characters that vary between species were found to vary within one or more species as well (136).

I define polymorphism as variation within species that is (at least partly) independent of ontogeny and sex. I assume that this variation is genetically based and heritable, and for the purposes of this paper I deal primarily with variation in discrete or qualitative characters, rather than continuous variation in quantitative traits.

Despite the prevalence of intraspecific variation, phylogenetic biologists have a long and continuing tradition of ignoring polymorphism. For example, morphological systematists often exclude characters that show any or "too much" variation within species (109a). Both molecular and morphological systematists often "avoid" or minimize polymorphism by sampling only a single individual per species. When polymorphism is dealt with explicitly, as in phylogenetic analyses of allozyme data and some studies of morphology, the appropriateness of different methods for phylogenetic analysis of these data is controversial and has been the subject of heated debate for over 20 years (e.g. 11, 12, 20, 33–35, 39, 40, 43, 75, 90–93, 96, 97, 116, 129, 130, 137–139, 142, 143). The controversy over the efficacy of different methods for analyzing polymorphism is not merely academic because different methods may give very different estimates of phylogeny from the same data (Figure 1; 137).

Different phylogenetic hypotheses may have very different implications for comparative evolutionary studies. But even if the tree is stable, different methods of treating within-species variation in ancestral state reconstructions may lead to radically different hypotheses about how traits evolve (see below). Descriptions of comparative methods designed for discrete traits (e.g. 76, 104, 118) rarely mention that these traits may vary within species or what the potential impact may be of this variation on the methods or results.

Species-level systematics, or alpha taxonomy, also involves analyzing polymorphic characters. Analytically, the main task of species-level systematics is to

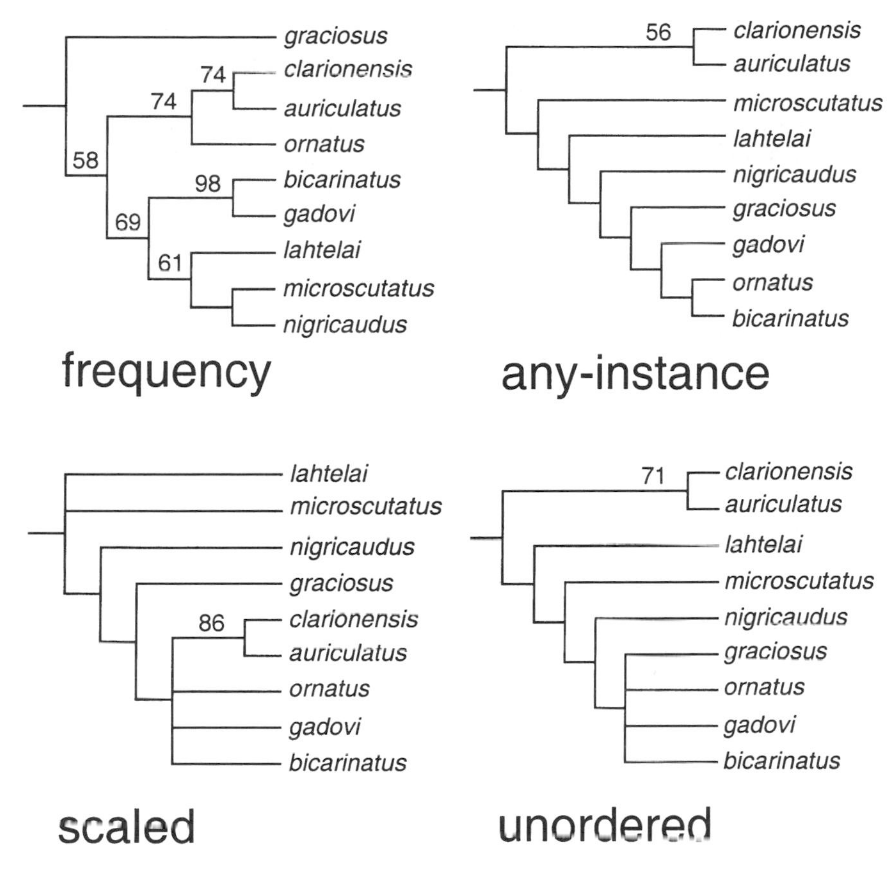

Figure 1 Different methods for coding polymorphic characters for phylogenetic analysis lead to very different hypotheses of evolutionary relationships. Results are based on morphological data for the lizard genus *Urosaurus* (136). Numbers at nodes indicate bootstrap values (42; bootstrap values <50% not shown). Each data set was analyzed with 1000 pseudoreplicates with the branch-and-bound search option.

distinguish between intraspecific and interspecific character variation. The delimitation, diagnosis, and description of species is at least as important an endeavor of systematics as phylogeny reconstruction. Yet, in contrast to phylogeny reconstruction, there has been relatively little methodological improvement in this area, especially as practiced by morphological systematists, who have described and will continue to describe most of the world's species. Alpha taxonomy is a branch of systematics that would benefit tremendously from a more explicit treatment of polymorphism.

In this paper, I review the implications of within-species variation for studies of systematics and comparative biology. I first provide an overview of common methodologies for dealing with polymorphism in phylogeny reconstruction and of some of the controversies surrounding these methods. I then describe recent studies designed to test the accuracy of these methods and resolve these controversies. I also discuss the impact, although considerably less studied, of within-species variation on comparative studies of the evolution of discrete or qualitative characters. Finally, I review the problem of delimiting species and the operational criteria and methodologies used for delimiting species and distinguishing within and between species variation.

PHYLOGENY RECONSTRUCTION

General Approaches

Polymorphism is important in reconstructing the phylogeny among species for two reasons. First, it is common in data of all types, including morphology, molecules, and behavior. Second, when polymorphism is present, it may have a significant impact on phylogenetic analyses. In particular, various methods for dealing with polymorphism may lead to very different estimates of phylogeny, even when relationships are strongly supported by one or more methods (Figure 1). The abundance and impact of polymorphism are especially clear for closely related species, but different methods for analyzing polymorphic data may affect higher-level relationships as well (e.g. relationships among genera; 138, 139). Yet, surprisingly, the issue of polymorphism is frequently ignored by systematists, particularly those working with morphological and DNA sequence data.

Systematists deal with polymorphism, or avoid dealing with polymorphism, in a number of different ways. These general approaches loosely reflect different types of data. Morphologists often exclude characters in which polymorphism is observed, and in fact this is the most common reason given for excluding characters (109a). This practice may be far more common than is apparent from the literature because morphologists rarely provide criteria for excluding or including characters (109a). The next most common exclusion criterion, excluding characters that show continuous variation, also reflects the desire to avoid characters that vary within and overlap between species. The justification for excluding polymorphic characters is rarely made clear by empirical systematists. Yet, there is a persistent idea in the systematics literature, dating back to Darwin (22), that the more variation characters show within species, the less reliable they will be for inferring the phylogeny among species (32, 86, 123). There have been few empirical tests of this idea.

Systematists working with sequence and restriction-site data typically deal with intraspecific variation by treating each individual organism (or each unique genotype or haplotype) as a separate terminal unit in phylogenetic analyses. Thus, variation within species is effectively treated in the same way as variation between species (134). However, some authors have recently suggested modifications to this general approach, specifically tailored to the problem of analyzing variation within species (e.g. 18, 19, 127, 132). Of course, one variant of this approach is to sample only a single individual from each species. This sampling regime, although obviously controversial (2, 127, 138, 142, 143), is often employed by both molecular and morphological systematists.

A third general approach involves treating each species (or population) as a terminal unit in the phylogenetic analysis. This approach incorporates intraspecific variation by different methods of coding in a parsimony or discrete character framework or by conversion of trait frequencies to genetic distances (or direct analysis of frequencies using continuous maximum likelihood; 38). This general approach is most frequently applied to allozyme data but is sometimes used for morphological data as well (12, 14, 108). A plethora of methods for dealing directly with polymorphism have been proposed and used, including at least eight parsimony coding methods (described below), two maximum likelihood methods (38, 100), and no less than 36 genetic distance methods (e.g. 114, 115, 130, 148), where each distance method is a combination of tree-building algorithm and genetic distance measure.

These parsimony, likelihood, and distance methods, designed explicitly for polymorphic data, have been the subject of considerable controversy, dating back more than 20 years (20, 39, 90, 91, 96, 97, 129, 130). Two questions have been particularly prominent. First, are frequency data appropriate for phylogenetic analysis? Many authors have argued that the frequencies of traits or alleles within species are not useful for reconstructing phylogenies among species, largely because they are thought to be too variable in space and time within species (e.g. 20, 96, 97) and are not heritable, organismal traits (e.g. 97, 122). Proponents of frequency methods have argued that frequency methods utilize valuable information ignored by other methods (e.g. a trait occurring at a frequency of 1% is different from one occurring at a frequency of 99%), even if frequencies are not stable over a macroevolutionary time scale (129, 130, 137). These authors have also argued that frequency methods downweight rare traits, and therefore they will be less subject to problems of sampling error than methods that merely treat traits as present or absent (i.e. a trait that is rare but present in several related species will be detected only sporadically with finite sample sizes, creating homoplasy, but this homoplasy will have little impact if frequency methods are used).

The second question is whether polymorphic data should be analyzed using parsimony or distance methods (e.g. 33–35, 39, 40, 43, 90, 97, 130). Most of the debate surrounding this topic has not directly involved the accuracy of the methods, but rather issues such as the meaning of branch lengths and negative distances (e.g. 33–35, 39, 40, 43).

The maximum likelihood method most widely applicable to polymorphic data (continuous maximum likelihood or CONTML; 38) has been largely ignored by empirical systematists (but see 120), presumably because it assumes a clearly unrealistic model of evolution (e.g. 71, 129). Namely, it assumes no mutations and no fixations or losses of polymorphic traits (38). However, the sensitivity of the method to violations of these assumptions has not been thoroughly explored until recently.

Methods for Coding Polymorphic Data

In this section, I briefly review some of the methods commonly used for coding polymorphic data for parsimony analysis (see also Figure 2). The terminology for these methods follows Campbell & Frost (12) and Wiens (137).

Any Instance Using this coding method, a derived trait is coded as present regardless of the frequency at which it occurs within a species (e.g. 1 to 100%). However, this method is problematic in that it can hide potentially informative reversals (12) such as the reappearance of the primitive trait as a polymorphism (e.g. a transition from 100% to 50% for the derived state). Mutation coding (96, 97) is similar to any-instance coding but potentially allows for characters with multiple derived states to be analyzed. However, its application is "frequently impossible for most loci" (96, p. 32).

Majority Using majority or "modal" coding, a species is coded as having the most common state of the polymorphic character. Potential disadvantages of this method are that it ignores the gain and loss of traits at frequencies less than 50% and that it gives a large weight to small changes in frequency close to 50% (e.g. a change from 49% to 51% has the same weight as a change from 0 to 100%).

Missing When a species that is polymorphic for a given character is coded as missing, the state is treated as unknown in the phylogenetic analysis. Any state is considered a possible assignment to the species, even if the state was not one of the ones observed to be present in the variable species (at least using PAUP). Disadvantages of the missing method are that polymorphic data cells are uninformative in tree reconstruction, and polymorphic states can be treated neither as synapomorphies nor as homoplasies.

Polymorphic Under polymorphic coding, a variable species is coded as having both states (using PAUP or MacClade). When the data are analyzed, the variable species is treated as if either state is present, but the variable cell is largely uninformative in building the tree (although some placements of the variable taxon may be considered more parsimonious than others), and the most parsimonious

→

Figure 2 Different methods for coding polymorphic characters, illustrated with a hypothetical example. Five individuals are sampled from each of four species, and the circular shape represents the primitive condition and the square shape is derived. The step matrix shows the different costs (in number of steps) for transitions between each of the states; the costs are based on the Manhattan distance between the frequencies of each species for this character. Modified from Wiens (140).

	Species A	Species B	Species C	Species D
Shape of sampled individuals	○○ ○○○	○○ ○□○	□○ □○□	□□ □□□
Proportion square	0%	20%	60%	100%
Any-instance	0	1	1	1
Majority	0	0	1	1
Missing	0	?	?	1
Polymorphic	0	(0,1)	(0,1)	1
Scaled	0	1	1	2
Unordered	0	1	1	2
Unscaled	0	1	1	2
Frequency-bin	a	f	p	y
Frequency step matrix	A	B	C	D

Frequency bins

Character state	Frequency range (%)
a	0–3
b	4–7
c	8–11
d	12–15
e	16–19
f	20–23
g	24–27
h	28–31
i	32–35
j	36–39
k	40–43
l	44–47
m	48–51
n	52–55
o	56–59
p	60–63
q	64–67
r	68–71
s	72–75
t	76–79
u	80–83
v	84–87
w	88–91
x	92–95
y	96–100

Frequency step matrix

	A	B	C	D
A	0	20	60	100
B	20	0	40	80
C	60	40	0	40
D	100	80	40	0

state assignment to the variable taxon is assigned a posteriori. As with the missing method, polymorphic states are not treated as synapomorphies, a serious disadvantage of both methods.

Scaled Using scaled coding, a species is coded as absent ("0"), polymorphic ("1"), or fixed ("2") for the derived trait. The states are ordered under the assumption that traits pass through a polymorphic stage between absence and fixed presence. If no polymorphic state is observed, it is assumed that the polymorphic stage was present but unobserved (i.e. it costs two steps to go from 0 to 2). The scaled method is equivalent to the step matrix method of Mabee & Humphries (75), but the use of a step matrix allows complex ordering of polymorphic multistate characters when there is no clear relationship among the states (as in the case of different combinations of alleles at an allozyme locus). The scaled method is advantageous in that it allows polymorphisms to act as synapomorphies (unlike the missing and polymorphic methods) and it does not mask reversals (unlike any-instance coding) or the gain and loss of rare traits (as does majority coding). However, it is potentially disadvantageous in that it utilizes no frequency information, and a change from 1% to 100% has the same weight as a change from 99% to 100%.

Unscaled The unscaled method is identical to the scaled method, except that for characters in which no polymorphism is observed it is assumed that the character did not pass through a polymorphic stage between absence and fixation. Therefore, a change from fixed absence to fixed presence has a cost of one step under unscaled coding, but a cost of two steps under scaled coding.

Unordered Unordered coding is identical to scaled and unscaled coding except that all the states are unordered, and there is an equal cost to any transition between any of the character states. As noted by Campbell & Frost (12) and Mabee & Humphries (75), the unordered method is disadvantageous in that it loses any information about the shared presence of traits (i.e. a change from trait absence to fixed presence is no more costly than a change from polymorphic presence to fixed presence).

Confidence Coding The method of Domning (27), which I dub confidence coding, is similar to the majority method but statistically incorporates sample size. For a given species, the 95% confidence interval for the frequency of the commonest trait is found, and if the lower confidence limit is >0.5, the species is coded as having the majority condition. If not (or if two traits are present at equal frequencies), the taxon is coded "whichever way was more congruent with other characters (i.e. whichever way did not imply a reversal)." The "congruence with other characters" is determined from a preliminary tree.

Frequency Parsimony Methods Frequency methods are a class of methods that use precise information on the frequency of traits within a given species, and weight changes between states based on differences in frequencies. Genetic distance methods and continuous maximum likelihood are frequency-based approaches, and there are at least three frequency parsimony methods that differ in their precision and versatility.

The most precise method is the FREQPARS program (129), which uses trait frequencies directly. However, FREQPARS has a weak tree-searching algorithm and is unlikely to find the shortest tree unless the data set has only a few taxa.

Wiens (137) used a method (suggested by D Hillis) that approximates the FREQPARS approach while still allowing for thorough tree searching, and this method was described in detail by Berlocher & Swofford (7). The method is implemented by giving each taxon that has a unique set of frequencies a different character state (Figure 2). The cost of a transition between each pair of character states is calculated by finding the Manhattan distance (129) between the frequencies; these transition costs are then entered into a step matrix (Figure 2). The step matrix allows for extremely precise frequency information to be used in character weighting. The main disadvantage of this approach is that step matrices may slow down tree searches prohibitively for large numbers of taxa.

The least precise of the three methods is the frequency-bins method (136, 137; modified from 111). This method is practical for large numbers of taxa (>100; 141) but is designed for binary characters only. With this method, each taxon is assigned one of an array of character-state bins, where each bin corresponds to a small range of frequencies of the putative derived trait (e.g. character state a = frequency of derived trait from 0–3%, b = 4–7%; Figure 2). The bins are then ordered, which forces a large number of steps between large changes in frequency and a small number of steps between small changes in frequency. The choice of bin-size relates to the maximum number of states allowed by the phylogenetic software program; most authors have used 25 bins (Figure 2; 137).

Testing Methods for Phylogenetic Analysis of Polymorphic Data

Recent work has tried to resolve some of the controversies surrounding different approaches for dealing with polymorphic data. In particular, these studies have attempted to address the accuracy of excluding versus including polymorphic characters, sampling single versus multiple individuals per species, and the relative performance of various parsimony, distance, and likelihood methods designed for analyzing polymorphic data. These studies have employed computer simulations (142, 143), congruence analyses of real data (morphology and allozymes; 138, 139), and statistical analyses of empirical data sets (morphology and allozymes; 137). Computer simulation studies involve constructing a known phylogeny, evolving characters on this tree according to some model of evolution, and testing the ability of different methods to estimate this tree given the same data (65). Congruence analyses require finding relationships that are agreed on by multiple data sets, assuming that these well-supported, congruent clades are effectively "known," and comparing the frequency with which different methods estimate these clades with the same finite data (95). Congruence studies provide a useful "reality check" on simulation studies, which make many simplifying assumptions about evolutionary processes. Comparing methods according to statistical measures that may relate to accuracy (such as bootstrapping) may be a relatively weak criterion for assessing performance (65). Nevertheless, certain methods do make assumptions that are amenable to statistical testing (e.g. whether or not polymorphic characters or frequency data contain significant nonrandom phylogenetic information; 137).

The common practice of excluding polymorphic characters implicitly assumes one or more of the following: (*a*) polymorphic characters are more homoplastic than fixed characters (characters that are invariant within species), (*b*) polymorphic characters do not contain useful phylogenetic information, and (*c*) inclusion of polymorphic characters will decrease phylogenetic accuracy (relative to excluding them and analyzing only "fixed" characters). Recent studies of empirical data suggest that polymorphic characters are more homoplastic than fixed characters (12, 137). Furthermore, there is a significant positive relationship between levels of homoplasy and intraspecific variability in morphological characters in phrynosomatid lizards (137). These two observations support the long-standing idea that more variable characters may be less useful in phylogeny reconstruction (22, 86, 123) and might be interpreted as supporting their exclusion. Yet, although they are more homoplastic than fixed characters, polymorphic characters nevertheless do contain significant phylogenetic information, as shown by the congruence between trees based on fixed and polymorphic characters (12) and randomization tests of homoplasy levels (137). Furthermore, computer simulations and congruence studies of morphology support the idea that, given a sample of fixed and nonfixed characters of realistic (i.e. limited) size, exclusion of all the polymorphic characters significantly decreases accuracy (Figure 3; 138, 142). In many cases,

Figure 3 Sample of results from simulation and congruence analyses showing the relative performance of methods for analyzing intraspecific variation (with 8 taxa and 25 characters). Data for congruence analyses are from *Sceloporus* (138), and simulations are with branch lengths varied randomly among lineages (from 0.2 to 2.0) and two alleles per locus (142, 143). Each bar represents the average accuracy from 100 replicated matrices, where accuracy is the number of nodes in common between the true and estimated phylogenies. A. Results with $n = 10$ (individuals per species) in simulations and around 10 for many species and characters in the *Sceloporus* data. B. Results with $n = 1$. The parsimony methods give identical results with $n = 1$ for the congruence analyses because heterozygotes are not detectable as such in the morphological data (so there is no polymorphism), whereas heterozygotes can be detected in the simulations. Modified from Wiens (140). CONTML, continuous maximum likelihood (38); NJ, neighbor-joining (117); FM, Fitch-Margoliash (46); Nei, Nei's (98) standard distance; CSE, Cavalli-Sforza & Edwards (13) modified chord distance.

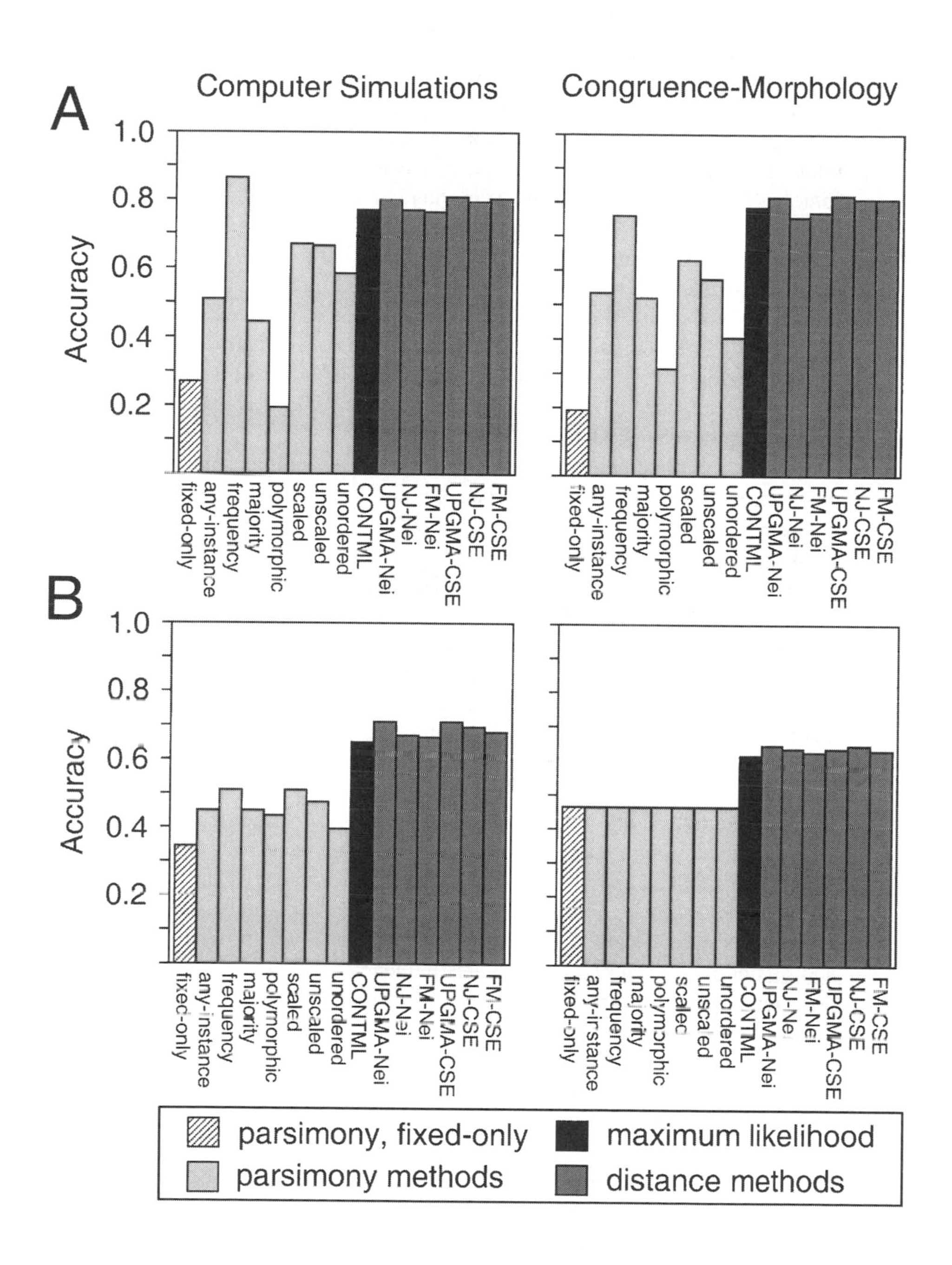
A
Computer Simulations
Congruence-Morphology
Accuracy
1.0
0.8
0.6
0.4
0.2
fixed-only
any-instance
frequency
majority
polymorphic
scaled
unscaled
unordered
CONTML
UPGMA-Nei
NJ-Nei
FM-Nei
UPGMA-CSE
NJ-CSE
FM-CSE
B
parsimony, fixed-only
parsimony methods
maximum likelihood
distance methods

twice as many known clades are recovered when polymorphic characters are included rather than excluded (138). Thus, in this trade-off between having more characters (fixed + polymorphic) and having a smaller number of characters with less homoplasy (fixed-only), it is clearly better to have more characters.

Given the observations that polymorphic characters contain useful phylogenetic information in general but that homoplasy may increase with increasing variability, some authors have proposed downweighting characters based on their level of homoplasy (e.g. using successive approximations; 12) or their degree of intraspecific variability (using a priori weighting; 32, 137). Similarly, many empirical systematists seem to delete the most polymorphic characters from their data sets, excluding characters because of "too much" intraspecific variability as opposed to any variability at all (109a). Simulation and congruence analyses suggest that, while these approaches may improve accuracy in some cases relative to some methods, they rarely improve accuracy relative to the unweighted frequency method including all polymorphic characters (138, 142).

Simulations, congruence studies, and statistical resampling studies also suggest that sample size (individuals per species) may be very important for achieving accurate results, particularly when levels of polymorphism are high (Figure 3; 2, 127, 138, 142, 143). These results argue against the sampling of a single individual per species as a general practice. For example, using congruence analyses of morphological data for spiny lizards (*Sceloporus*), Wiens (138) found that the accuracy of the "best" parsimony method is effectively cut in half by sampling only a single individual per species under some conditions (Figure 3).

An important result of recent congruence and simulation analyses is that the methods that generally perform best are those that make direct use of frequency information, whether they be parsimony, distance, or likelihood. That is, the frequency parsimony method, the genetic distance methods, and continuous maximum likelihood tend to recover more of the well-supported or known clades than do any of the nonfrequency parsimony methods. This same result is obtained for a variety of simulated branch lengths, sample sizes, numbers of taxa, and numbers of characters, and in congruence studies of both morphological (Figure 3) and allozyme (Figure 4) data. Furthermore, statistical analyses of two morphological and five allozyme data sets show that frequency-coded polymorphic characters do contain significant, nonrandom phylogenetic information (137), and that frequency methods perform best among the parsimony coding methods (or are tied for best) for a number of statistical performance criteria. These results contradict the idea that frequency data are too unstable to be used in phylogenetic analysis and that they are misleading.

Recent simulation and congruence studies also show that distance and likelihood methods may outperform all parsimony methods (both frequency and nonfrequency) in many cases. One such situation is equivalent to the "Felsenstein Zone" effect described for fixed characters (e.g. 37, 68, 69), which occurs when there are two unrelated terminal lineages with long branches separated by

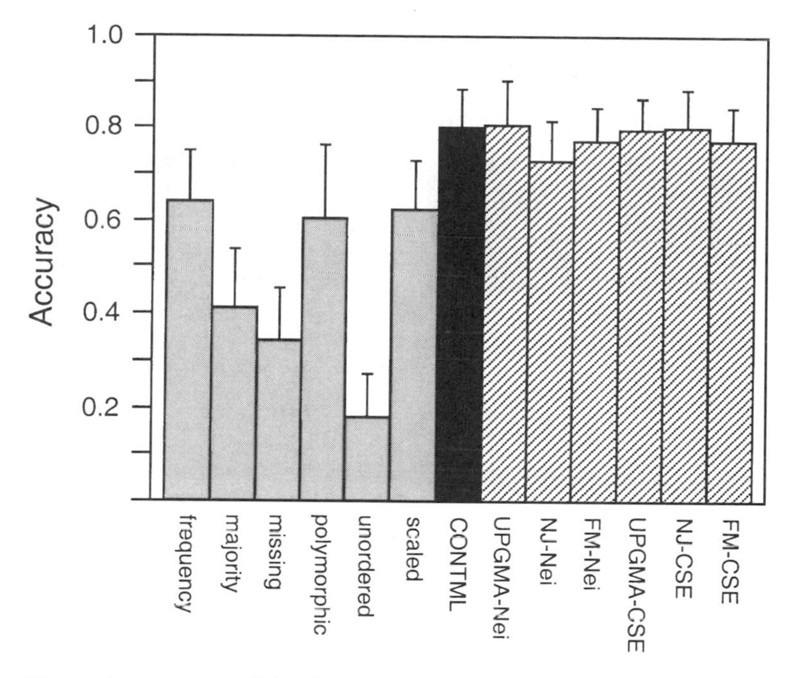

Figure 4 Accuracy of 13 phylogenetic methods averaged across eight allozyme data sets. The line above each bar indicates the standard error of each mean. Accuracy is the proportion of well-supported clades that are correctly resolved by each method. Modified from Wiens (139). See Figure 3 for abbreviations.

a short internal branch. In simulations of polymorphic data using a model in which allele frequencies evolve along branches by random genetic drift (72, 143), this Felsenstein Zone effect might occur when there are two unrelated species with small population sizes (long branches, with high probability of fixation, loss, and/or large changes in trait frequency) that are separately derived from an ancestor with a very large population size (short branch). This might correspond to a peripheral isolate model of speciation in the long-branch species. Under these conditions, parsimony and UPGMA tend to place the taxa with long branches together as sister taxa (incorrectly), even if a large number of characters are sampled (Figure 5). In contrast, continuous maximum likelihood and the additive distance methods (neighbor-joining and Fitch-Margoliash) will tend to estimate the correct tree, especially when given a large sample of characters (72, 143). The Felsenstein Zone effect for polymorphic data is interesting for several reaons: (*a*) it is very similar to the effect described for fixed characters, even though the simulated models of evolution are extremely different (i.e. the fixed character model has change as mutation only, whereas there is no mutation in the pure drift model), (*b*) increased taxon sampling to subdivide the long branches is not a potential solution

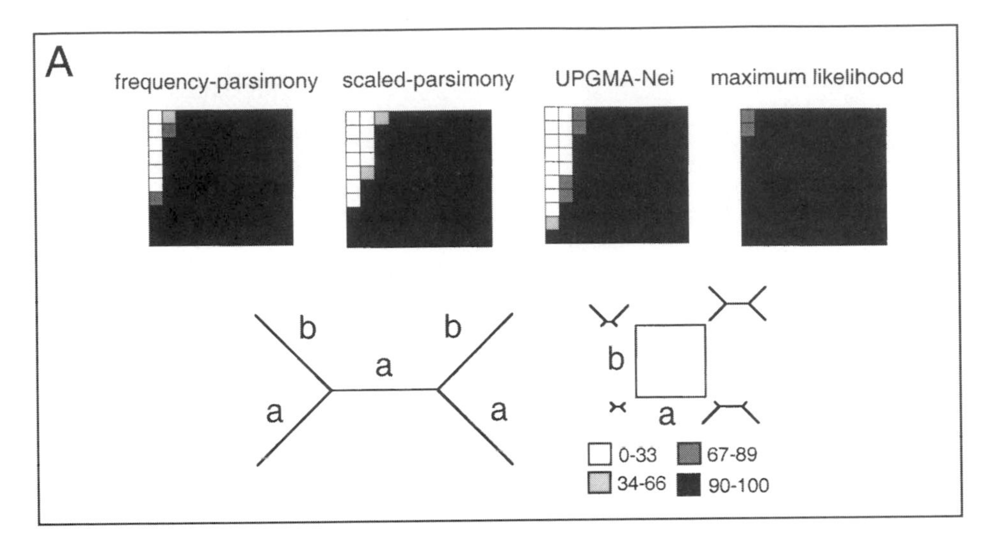
A
frequency-parsimony
scaled-parsimony
UPGMA-Nei
maximum likelihood
b
b
a
a
a
b
a
0-33
67-89
34-66
90-100

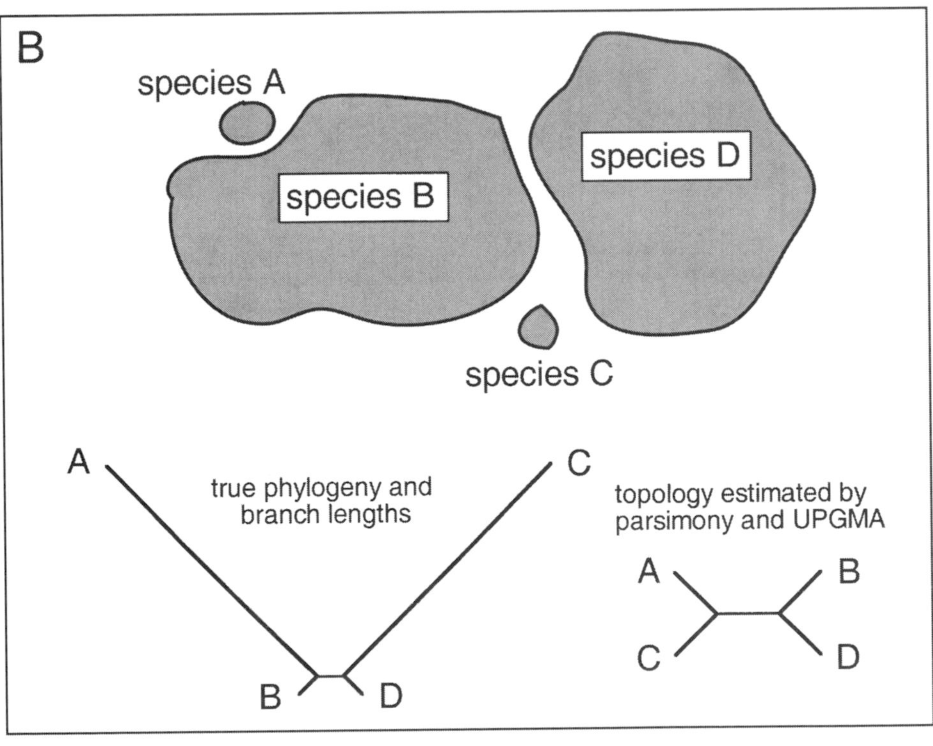
B
species A
species B
species D
species C
A
C
true phylogeny and
branch lengths
topology estimated by
parsimony and UPGMA
A
B
C
D
B
D

to the problem, and (*c*) the effect for polymorphic data has a simple biological explanation and may therefore occur commonly (143).

Of course, the Felsenstein Zone scenario described above may represent a very special case. Furthermore, the possibility of long branch repulsion (i.e. the failure of maximum likelihood to place together two long branches that are actually sister taxa; 70, 121, 150) has not been explored for polymorphic data. Yet, distance and likelihood methods outperform parsimony methods under many other conditions apart from the Felsenstein Zone. In simulations, distance and likelihood methods generally performed as well as or better than any of the parsimony methods under a variety of branch lengths, sample sizes, and numbers of characters and taxa, and the nonparsimony methods consistently outperformed parsimony when sample sizes were very small ($n = 1$ or 2; Figure 3; 142). In congruence analyses of morphology, at least some distance methods consistently performed better than any parsimony methods (138), and in congruence analyses of allozyme data sets (139), each of the distance and likelihood methods outperformed (on average) all the parsimony methods (Figure 4). The fact that continuous maximum likelihood performs as well as it does on real and simulated data sets (especially the allozyme data) is particularly interesting, given that the assumptions of this method were almost certainly not met in these data sets (e.g. the method assumes no mutation and no fixation or loss of traits). These results strongly suggest that continuous maximum likelihood will perform well even when its assumptions are violated.

In summary, the results of simulation, congruence, and statistical analyses suggest that (*a*) polymorphic characters should not be excluded, (*b*) methods that use frequency data may perform best, and there is no evidence that frequency data are misleading, and (*c*) distance and likelihood methods may be

Figure 5 The effects of branch lengths and the Felsenstein Zone effect for polymorphic data. A. The effects of branch lengths on the accuracy of four phylogenetic methods, where darker shading represents higher accuracy. The data consist of 500 loci (characters), with two alleles per locus and complete sampling of individuals within each species. Modified from Wiens & Servedio (143). B. Hypothetical example illustrating the Felsenstein Zone effect for polymorphic data. The shaded areas represent the geographic distributions (and relative population sizes) of four species. Species A and C have small geographic distributions, small population sizes, and long branch lengths under a genetic drift model. Species B and D (and the ancestors of all four species) have large population sizes and short branch lengths. Traits will tend to remain polymorphic in B and D but become fixed or lost in A and C. Parsimony methods and UPGMA will tend to put A and C together based on shared fixations, losses, and changes in trait frequency. In contrast, continuous maximum likelihood and the additive distance methods (neighbor- joining and Fitch-Margoliash) can give accurate results under these conditions, given enough characters. Modified from Wiens (140). Nei, Nei's (98) standard distance.

superior to parsimony methods for analyzing polymorphic data under many conditions.

Objections to Frequency Methods

The results of these recent studies provide some resolution to the controversies surrounding the analysis of polymorphic data. These studies show that frequency methods use more informative variation than do any of the other methods (137), are less subject to errors caused by limited sample sizes (129, 137) and unequal branch lengths (142, 143) than are nonfrequency parsimony methods, and generally give more accurate estimates of phylogeny in simulation and congruence analyses than do other methods (138, 139, 142, 143). A number of recent empirical studies have used frequency methods to include and code polymorphic characters, including studies of allozymes (e.g. 10, 31, 88), behavior (e.g. 57), and morphology (e.g. 14, 16, 55, 56, 59, 67, 87, 108, 113, 136, 141). Nevertheless, the use of frequency information in phylogenetic analysis remains controversial (e.g. 94, 97, 122).

The most common objection to the use of frequency data in phylogenetic analysis appears to be the idea that frequencies are too variable over space and time within species to be used in reconstructing relationships between species. Several authors have cited the study of Crother (20) as evidence that frequency data are unstable and therefore unusable (e.g. 12, 75, 94, 97). This example does not withstand closer scrutiny. Crother analyzed allele frequency data from four populations of *Microtus ochrogaster* and found that phylogeny estimates for these populations based on the same locus differed from year to year (using data from 50). Crother (20) concluded from this example that frequencies vary too much over time and space to be phylogenetically informative for reconstructing relationships among species. However, it should be noted that the "populations" were not natural populations from different localities but were individuals drawn from the same locality confined in four enclosures (50). Thus, there was no true phylogenetic history to be estimated for these populations, and the absence of stable phylogenetic signal in the frequency data is hardly surprising. The fact that there are different estimates of phylogeny from year to year does demonstrate that frequency methods may resolve clades that have little or no support (137). Yet, the weak support for these phylogenies is obvious from low bootstrap values ($<50\%$) and g_1 analysis (i.e. the data contain no significant phylogenetic structure; JJ Wiens, unpublished data). Extrapolating this rather artificial example to the interspecific case and generalizing the results to all applications of frequency data clearly is unjustified, especially in the face of growing evidence that frequency-coded polymorphic data do contain significant, nonrandom phylogenetic structure at the between-species level (87, 108, 137–139).

How can frequencies be highly variable within species but still informative between species? Population genetics theory (e.g. 73) suggests that traits with frequencies that are highly variable over time within species are more likely to become fixed or lost over longer evolutionary time scales. Thus, a model in which frequencies change rapidly without fixation or loss for thousands or millions of years seems unrealistic for the vast majority of characters (Figure 6). When fixations and losses do occur, theory predicts that traits at high frequencies are more likely to be fixed than lost, and vice versa. This relationship is mirrored in the weighting scheme of frequency methods. In contrast, nonfrequency methods (except majority) assume that it is just as easy to go to fixation from a frequency of 1% as it is from a frequency of 99%. Thus, in simulations, frequency methods may be superior estimators of phylogeny even when the frequencies of nonfixed traits are nearly randomized between splitting events (142). Furthermore, results of statistical and congruence analyses (137–139) imply that trait frequencies are conserved enough to contain at least some historical information.

Another objection to the use of frequency-based methods is that frequencies are not heritable and/or organismal traits (e.g. 97, 122; note that "heritable" refers to transmission from ancester to descendant, and not to the quantitative genetic meaning of the term). Although the idea that frequencies are never heritable is not strictly accurate (because frequencies *are* heritable if populations are at Hardy-Weinberg equilibrium), it is likely that nonfixed frequencies between 0 and 100% are rarely passed from an ancestor species to a descendant species without at least some change. But, from a practical perspective, it is clear that these changes in trait frequencies do not prevent frequency methods from accurately estimating phylogenies (e.g. 138, 139, 142, 143). Obviously, if frequencies never changed there would be no variation with which to reconstruct trees. The objection to nonorganismal traits appears to be questionable as well. It is true that frequencies are features of populations and species, and not of individual organisms. However, this is true for polymorphic and intraspecifically-variable quantitative data (133), not just frequency data. The fact that Kluge has argued against inclusion of frequency data is ironic because the exclusion of potentially informative data is contrary to the maxim of total evidence (74).

An objection to frequency methods somtimes raised in specific cases is that sample sizes (individuals per species) may be insufficient (e.g. 12), with the implicit assumption that frequency-based methods will be less accurate with small sample sizes than qualitative coding methods. In fact, simulations suggest that as sample sizes decrease, the performance of all methods becomes increasingly similar (i.e. if there were no polymorphism, all the polymorphism coding methods would be identical). But even with small sample sizes (e.g. $n = 1$ or 2 individuals per species), there are still noticeable differences in accuracy among methods, with frequency methods generally outperforming other coding methods (Figure 3B; 142, 143). An objection to frequency methods based on finite sample sizes is surprising because the putative robustness of frequency methods to finite sample sizes has traditionally been the major argument to justify their use (129, 137).

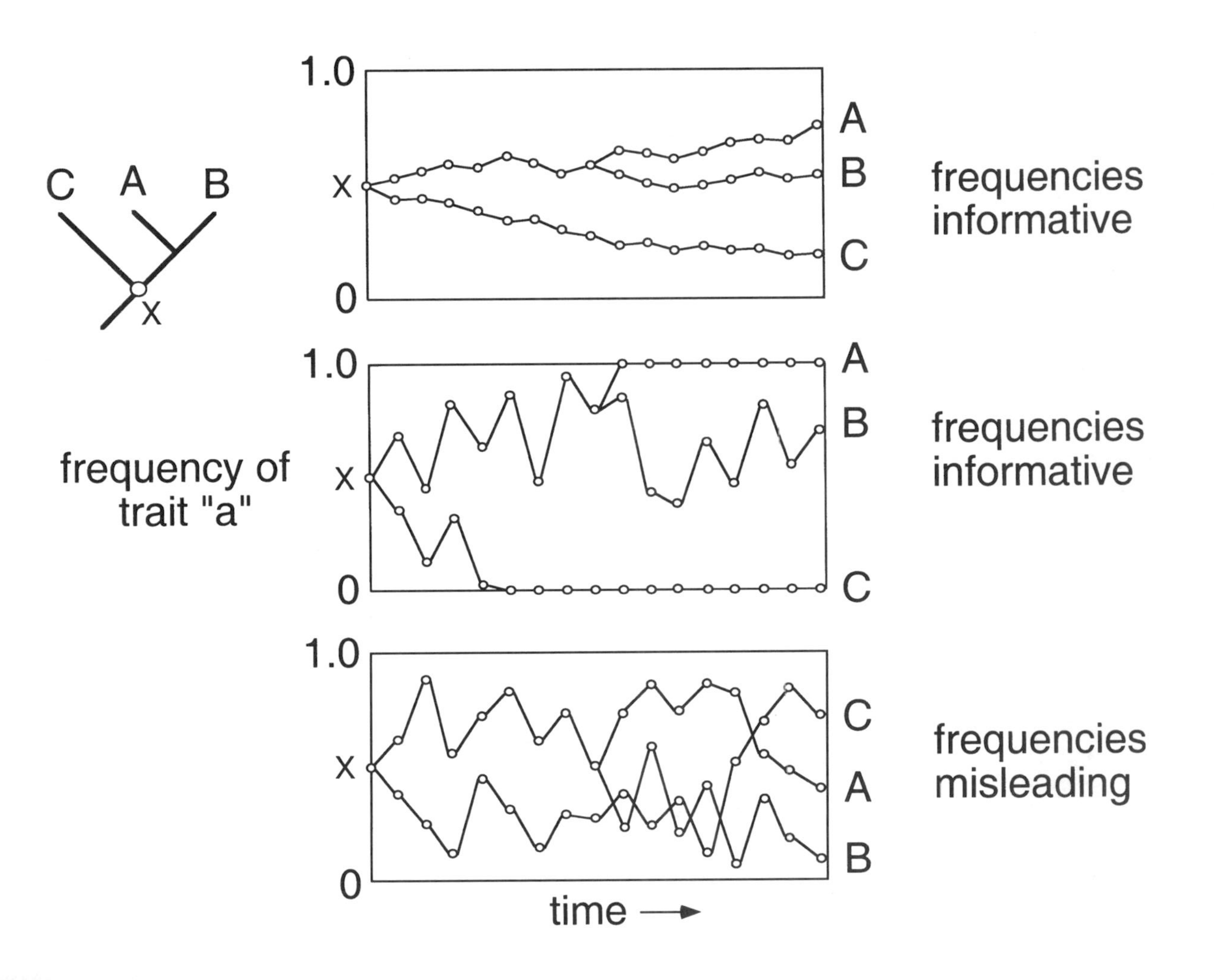

C A B
x
frequency of trait "a"
1.0
x
0
A
B
C
frequencies informative
A
B
C
frequencies informative
C
A
B
frequencies misleading
time

Areas for Future Research

Morphology Further research is needed on the phylogenetic analysis of polymorphic data for a variety of reasons. Recent studies on polymorphic morphological characters suggest that two common practices of morphological systematists, excluding polymorphic characters and ignoring frequency data, may lead to relatively poor estimates of phylogeny. But the congruence results, and those showing the information content of polymorphic characters, are so far based on morphological data from only one clade of lizards (the Phrynosomatidae). On the positive side, these conclusions are also upheld by simulations. Furthermore, recent studies in other groups of vertebrates show significant phylogenetic information in frequency-coded data, or at least frequency-based trees that are congruent with previous taxonomy or other data sets (e.g. 14, 16, 56, 67, 87, 108). Nevertheless, the generality of these conclusions should be tested in morphological data sets from other groups of organisms, especially plants and invertebrates. These conclusions will be difficult to test until more morphological systematists publish data on polymorphic characters and trait frequencies.

Previous work on intraspecific variation in morphology has focused largely on discrete, or at least qualitatively described traits. The analysis of quantitative characters with ranges of trait values overlap between species (such as meristic and morphometric variables) is also in need of study. Overlapping quantitative data, like qualitative polymorphic data, are also frequently excluded from morphological phylogenetic studies (109a), and yet numerous methods for their analysis have been proposed and debated (1, 36, 44, 52, 107, 128, 133). Using data from plants, Thiele (133) has shown that overlapping quantitative data do contain significant nonrandom covariation and produce trees that are significantly congruent with trees based on qualitative data, at least when using his gap-weighting method (which is very similar to the frequency-bins method; 137). Like the results from qualitative polymorphic characters, Thiele's results from quantitative characters support the inclusion of characters despite within-species variation, and the use of methods that treat continuously valued data (e.g. frequencies, means of quantitative traits) as continuous (e.g. frequency methods, gap weighting). However, more studies are needed, in plants and other organisms, to test the generality of these conclusions

Figure 6 Hypothetical example showing the changes in a trait "a" over time among three species (A, B, C). If frequencies change only slightly over time, trait frequencies should track the phylogeny and frequency methods should be effective (*top*). If frequencies change rapidly over time and traits go to fixation or loss then frequencies can still be informative (*middle*), because fixations and losses can be synapomorphies and will prevent further oscillations in trait frequencies. Frequencies are most likely to be misleading when frequencies change rapidly over time without becoming lost or fixed (*bottom*), but this seems unlikely without some unusual mechanism to simultaneously drive change and prevent fixation and loss (e.g. frequency dependent selection). The starting frequency at time "X" is 0.5 in all three cases.

and to compare the performance of different methods for analyzing quantitative morphological data.

Allozymes There are also many unanswered questions in the phylogenetic analysis of polymorphic allozyme data. Using congruence analyses of eight allozyme data sets, Wiens (139) found that the relative performance of different methods varies greatly from data set to data set. For example, for some data sets, the polymorphic coding method is the best of all the parsimony, distance, and likelihood methods (i.e. recovers more "known" clades than any other method), whereas for other data sets it is one of the worst. A comparable situation exists for UPGMA with Nei's distance, which on most data sets is the most accurate method (or is tied for most accurate) but on other data sets performs relatively poorly. Although the failure of these methods can be easy to explain (e.g. UPGMA is sensitive to unequal branch lengths), their strong performance on certain data sets is perplexing. This variability is particularly vexing because it makes it difficult to choose a single method that will be "the best" for every data set, or to understand which method may be preferred in a particular case. Simulation studies, with data explicitly designed to mimic allozymes, may be necessary to better understand why certain methods behave so well on some data sets but not others.

Furthermore, for both allozyme and morphological data, it is unclear whether the Felsenstein Zone effect described in simulations under a genetic drift model applies to many real data sets. Perhaps just as importantly, it is unclear whether the methods that appear to be robust to this problem in simulations (e.g. continuous maximum likelihood) will also be robust in real data sets. Simulations that utilize more complex models than those employed by Kim & Burgman (72) and Wiens & Servedio (143) may be particularly useful for addressing this question.

DNA Data Sequence data and restriction-site data are becoming widely used for inferring relationships among closely related species. This is a level where polymorphism may often have a significant impact on phylogenetic studies, but the simulation and congruence studies mentioned above may not be applicable to DNA data. These studies focused on trait frequencies at multiple, unlinked loci, whereas DNA data typically consist of linked characters at a single locus (i.e. a single nuclear gene or one or more mitochondrial or chloroplast genes). Theoretical work on the impact of within-species variation on interspecific phylogenetic inference using DNA sequence data has dealt primarily with the problem of incomplete lineage sorting of ancestral polymorphisms (e.g. 78, 99, 105, 131, 149). In this situation, the phylogeny of the gene(s) may not be congruent with the phylogeny of the species (especially when population sizes are large and/or divergence times are recent), and theoreticians have explored the effects of sampling multiple individuals and loci as possible solutions. There seems to be general agreement that sampling enough unlinked loci will resolve the problem. When only one locus is available (e.g. data from the mitochondrial or chloroplast genome), sampling multiple individuals from each species may also be helpful (131).

But how does one infer species phylogeny when multiple individuals are sampled? Surprisingly, this question has hardly been explored. In general, empirical systematists treat each haplotype (unique genotype) as a separate terminal taxon in the phylogenetic analysis, and these individuals may or may not cluster with their putatively conspecific haplotypes. The within-species phylogeny is inferred simultaneously with the among-species phylogeny, and no distinction is made between the two. However, two modifications to the haplotype-as-terminal-taxon approach have been suggested.

Using principles from population genetics and coalescent theory, Templeton et al (132) have recently developed a method (dubbed TCS) specifically tailored for within-species phylogeny reconstruction, and they have suggested its application, in combination with more traditional methods, to better infer between-species phylogeny. The algorithm was designed to overcome two major problems of within-species phylogenetics: (*a*) the scarcity of informative characters and (*b*) the problem of rooting the relatively similar within-species haplotypes with relatively divergent haplotype(s) from a different species. Crandall & Fitzpatrick (18) have combined the TCS method with more traditional among-species methods (see also 6). Using this combined approach, the most likely connections between intraspecific haplotypes are inferred using the TCS method, and these relationships are then constrained in a global parsimony or maximum likelihood search that includes all the haplotypes from all the species [although Hedin (61) found that unconstrained parsimony searches recover the same clades that are connected by the TCS method]. The combined approach seems very promising for dealing with polymorphism in DNA data, but whether it actually improves the accuracy of estimated interspecific trees has yet to be shown. Using data from a known bacteriophage phylogeny, Crandall (17) has shown that the TCS method by itself may outperform parsimony in some cases. Simulation and congruence studies are needed to further test the accuracy of the TCS and combined approaches at the interspecific level.

Smouse et al (127) proposed a method for estimating species phylogenies from restriction-site data when multiple individuals are sampled from each species. Their method involves estimating multiple phylogenies for each data set, each derived using a single individual to represent each species (i.e. the first phylogeny based on the first individual sampled from each species, the second phylogeny based on the second individual, etc). The species phylogeny is considered to be the "average" topology from among these trees, in accord with the idea of a species phylogeny as the "central tendency" of a diverse cloud of gene histories (78). The general approach of Smouse et al (127) seems readily applicable to DNA sequence data as well, but its accuracy relative to other methods has not been tested.

Nucleic acid sequence and restriction site data are not the only DNA data used in phylogenetic analysis, and recent years have seen increasing use of data from microsatellites and other hypervariable loci to estimate relationships among populations and closely-related species (e.g. 8, 89). Because of their rapid rate of evolution, the appropriateness of these markers for any but the most closely related taxa is questionable (66). Microsatellites are among the most slowly evolving of these markers and may be the most appropriate for interspecific phylogeny reconstruction. Microsatellite data are similar in many ways to allozyme data (i.e., multiple unlinked loci with high levels of polymorphism and many alleles per locus). These data have been analyzed using both an *individuals as terminal taxa* approach (using, for example, the proportion of alleles shared between individuals as a measure of distance), as commonly applied to DNA sequence data, and by analyzing the allele frequencies of populations using genetic distance methods (8). Genetic distances designed for allozymes have been used on microsatellite data, but a number of allele frequency-based distances specifically designed for phylogenetic analysis of microsatellite data have recently been developed, and the accuracy of all of these distance methods have now been tested extensively using simulated microsatellite data (e.g. 52ab, 109a, 131a).

COMPARATIVE EVOLUTIONARY STUDIES

Recent years have seen burgeoning use of phylogenies in studying patterns and processes of character evolution (9, 28, 41), and there is growing interest in testing and refining the methodologies used in comparative evolutionary studies, for both continuous (e.g. 26, 83, 84) and discrete data (e.g. 21, 77, 103, 118, 119). However, there has been little discussion of the impact of intraspecific variation in characters that are the focus of comparative studies (but see 29 and 85). Whereas studies of continuous traits typically use mean values for species, intraspecific variation in discrete traits is rarely mentioned in comparative studies. Yet, different ways of treating this variation may have a profound impact on evolutionary reconstructions and inferences.

For example, Figure 7 shows the effects of different ways of coding a single character—presence of colored female belly patches in spiny lizards (*Sceloporus*)—on evolutionary inferences. Depending on how polymorphism in this one character is coded, the trait may exhibit: (*a*) a preponderance of losses relative to gains (6 to 1), (*b*) a preponderance of gains (10 to 0), (*c*) a high degree of homoplasy (10 changes among 18 species), or (*d*) no homoplasy at all (1 change). Furthermore, the trait may be inferred to have evolved in the common ancestor of the entire clade (any-instance coding) or within a single, relatively derived species (missing coding). Clearly, the choice of methodology for dealing with polymorphism in comparative studies can be extremely important.

Given that many different methodologies are available for coding polymorphism (Figure 2), which one might be the best to apply in comparative studies?

→

Figure 7 Different methods of coding polymorphism can produce radically different hypotheses of character evolution for the same data and tree. The presence or absence of female belly patches is mapped among 18 closely related species of spiny lizard (*Sceloporus*) using MacClade (79). Data and tree from Wiens & Reeder (141). *Sceloporus tanneri* is excluded from this example because the state for this character is unknown.

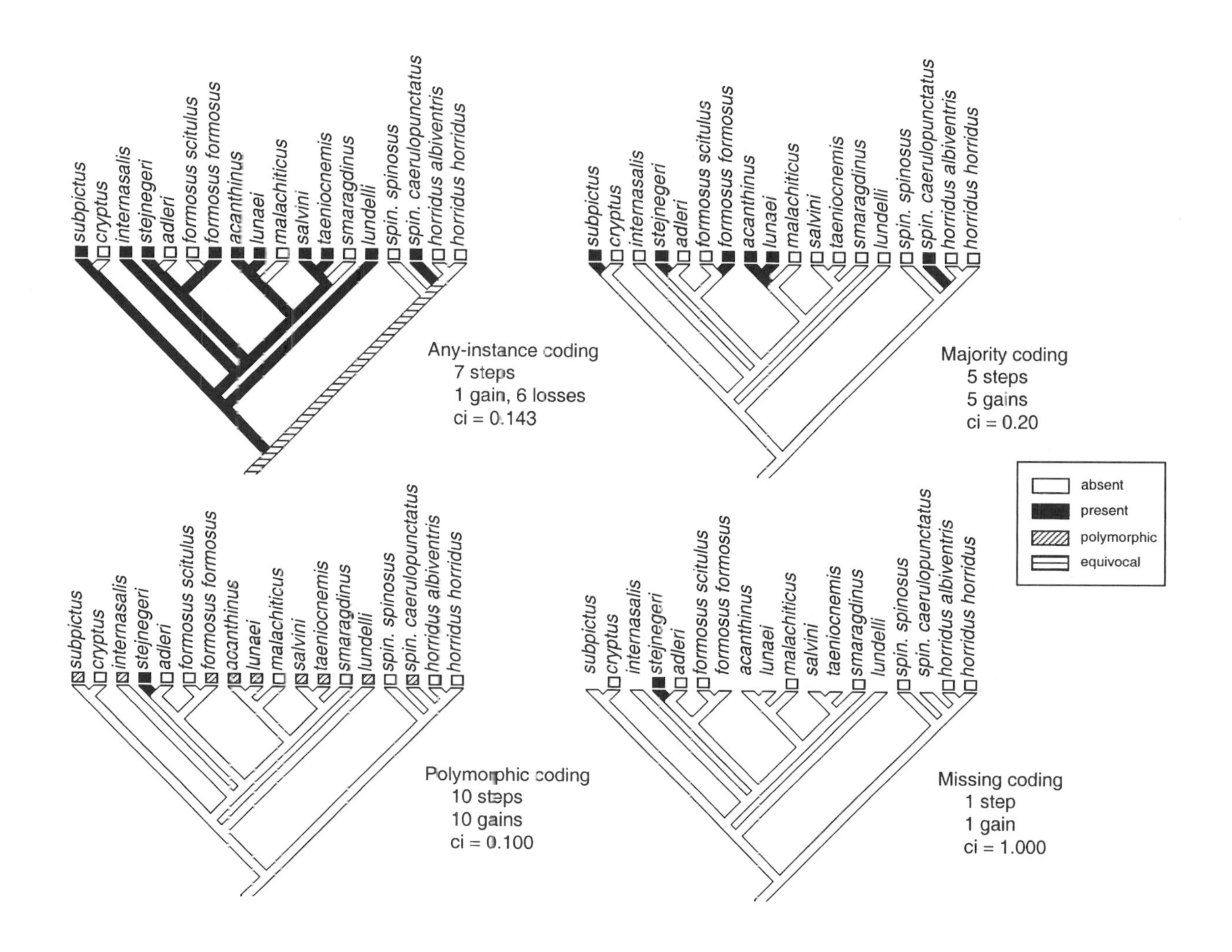
subpictus
cryptus
internasalis
stejnegeri
adleri
formosus scitulus
formosus formosus
acanthinus
lunaei
malachiticus
salvini
taeniocnemis
smaragdinus
lundelli
spin. spinosus
spin. caerulopunctatus
horridus albiventris
horridus horridus
Any-instance coding
7 steps
1 gain, 6 losses
ci = 0.143
Majority coding
5 steps
5 gains
ci = 0.20
absent
present
polymorphic
equivocal
Polymorphic coding
10 steps
10 gains
ci = 0.100
Missing coding
1 step
1 gain
ci = 1.000

The answer may depend on the specific question being asked. Many comparative studies are concerned with the timing, number, and pattern of gains and losses of qualitative traits, which require reconstructing ancestral states and (presumably) coding polymorphism in some way, regardless of whether the reconstruction method is parsimony or likelihood.

Given its strong performance in phylogeny reconstruction, frequency coding may seem an obvious choice for ancestral state reconstructions as well. However, frequency methods work by differentially weighting changes based on trait frequencies, and differential weighting is largely meaningless for reconstructing the ancestral state of a linearly ordered trait (80). Therefore, I briefly review and evaluate some of the candidate methods.

The any-instance method, which codes a derived trait as present regardless of its frequency, is problematic (12) in that it can potentially hide reversals to the primitive state (for example, if the primitive trait was regained and present at a frequency of 99%). This severely limits the effectiveness of the method for tracking the gains and losses of polymorphic traits.

The polymorphic coding method treats a variable species as having either of the two traits, but not both (during the reconstruction, taxa are treated as having whichever state is most parsimonious). Thus, ancestral nodes are never reconstructed as being polymorphic. Instead, each instance of polymorphism within a species is treated as an independent evolutionary event (as implemented by MacClade; 79), rather than allowing the possibility that polymorphisms are inherited. This approach therefore maximizes homoplasy rather than homology. This problem also applies to coding individuals as terminals when reconstructing ancestral states, or when variable species are broken up into monomorphic units (as recommended in 29, 103).

Similarly, the missing method (coding polymorphic species as unknown) also does not allow polymorphic ancestors. But in contrast to the polymorphic method, the missing method never treats polymorphisms as homoplasies, even when they seem clearly to be the result of reversal or parallelism (as in *Sceloporus spinosus caerulopunctatus* in Figure 7). Instead, polymorphic taxa coded as missing are treated as having whatever state is most parsimonious given the reconstruction of the trait based on other taxa. This property seriously compromises the ability of this method to estimate levels of homoplasy and the pattern of gains and losses in a character of interest.

The majority method (coding the commonest state as the only state present) does allow polymorphic ancestors but hides gains and losses of rare traits. In some ways, this insensitivity to rare traits may be an advantage, because the apparent gain and loss of rare traits may be due only to sampling error (129). Yet, the majority method is disadvantageous, in part because small changes in frequency close to 50% might be due to sampling error as well.

The scaled method avoids the problems of the preceding methods, although it does not downweight rare traits. The unordered method is similar to the scaled method but is problematic in that it implies no special relationship among states. Finally, the frequency method is similar to the scaled method for the purposes of ancestral state reconstruction, because, although it downweights rare traits, this differential weighting is meaningless in this context (80). The scaled or frequency method is recommended for reconstructing ancestral states in polymorphic characters. However, it should be noted that these methods are potentially sensitive to sampling error in this context, and very large sample sizes may be needed to confidently distinguish traits that are truly absent from those that are present as polymorphisms at low frequencies (112, 129, 144).

Instead of focusing specifically on the reconstruction of ancestral states, comparative evolutionary studies may also address questions about correlations between pairs of characters (e.g. 41, 76), or about differences in rates of change between characters and/or lineages (e.g. 47, 102). For these questions, the weighting of changes is important, and frequency methods may therefore be advantageous. For example, most nonfrequency methods (all but majority coding) would treat a change from 0 to 1% the same as a change from 0 to 99%, yet clearly more change has occurred in the latter case than in the former. Frequency methods are therefore recommended for studies of this kind. Treating polymorphic discrete traits as continuous frequencies may also facilitate the use of the many continuous data comparative methods (e.g. independent contrasts; 41), and the combination of discrete (polymorphic) and continuous traits. However, the performance of these methods using frequency data needs to be tested. There is also the need to modify comparative methods for discrete characters (e.g. 76, 104, 118) to accommodate polymorphism and frequency data.

POLYMORPHISM AND SPECIES DELIMITATION

Although it is tempting to equate systematics with phylogenetic analysis, the delimitation and description of species-level diversity is a major endeavor of systematics that is at least as important as reconstructing phylogenies. Species-delimitation is linked to the issue of polymorphism in that, for most alpha-level systematists, the main analytical task is distinguishing fixed (or nearly fixed) diagnostic features from those that are polymorphic. Curiously, while a voluminous literature has accumulated and continues to accumulate on the philosophical question of species concepts (what species are), the practical, methodological aspects of how we differentiate one species from another have received relatively little attention in the systematics literature (but see 23, 124), although species concepts and criteria for species recognition are often confounded (49). Recent authors have distinguished between character-based and tree-based approaches for delimiting species (e.g. 5), and this division is followed here. Character-based approaches delimit species based on character state distributions among geographic samples, whereas tree-based approaches generally use a phylogeny of haplotypes, individuals, or populations to infer species limits.

In this review of approaches to species-delimitation, I purposely avoid discussing the pros and cons of particular species concepts. However, my personal bias favors the evolutionary species concept (48, 123, 146), and I therefore follow the view that species are real entities that exist in nature, regardless of whether or not there is character or tree-based evidence that allows us to detect them (49).

Character-Based Species Delimitation

Newly discovered species are generally delimited, diagnosed, and described using a character-based approach, and this approach is widely used to test the validity of described taxa. For example, when a potentially new species is discovered, it is compared to similar species already described, and "diagnostic" (generally meaning intraspecifically invariant, or non-overlapping) character states are sought to distinguish them. This describes the basic task of most practicing alpha-level systematists for the past few hundred years. There has been surprisingly little methodological advancement in this area, especially relative to the burgeoning methods of phylogenetic analysis.

Davis & Nixon (23) described an explicit character-based methodology for delimiting species, given a set of populations with unknown species boundaries, which they called population aggregation analysis (PAA). PAA involves systematically comparing character distributions among populations, aggregating sets of populations that differ only in polymorphic traits, and considering sets of populations that differ from others by at least one fixed difference (or which share no states for a given character) to be different species. As pointed out by the authors, PAA is problematic in that: (*a*) unless many characters are sampled, the number of species present may be underestimated and (*b*) unless sample sizes are large, the number of species may be overestimated (by considering traits to be fixed that are actually polymorphic). PAA has never been modified to account for these problems, or at least to detect when the data are inadequate to make a decisive resolution. However, Wiens & Servedio (144) recently proposed a statistical test to determine whether or not sufficient characters and individuals have been sampled to argue that one or more seemingly fixed characters are truly diagnostic for a given species (i.e. the state of the other species is either absent or below a given frequency).

The methodology of PAA raises an important question: Why should fixed traits be the only ones that can delimit species? Why not differences in trait frequencies? One potential reason is that fixed differences may indicate an absence of gene flow, in a way that differences in trait frequencies may not. For example, suppose we compare two putative species and find that for a given character (e.g. flower color), all the individuals of each putative species differ from the other in this character (e.g. red versus white). The most likely explanation for this consistent difference would be that no individuals of one of the putative species are breeding with the other. Conversely, if all potentially diagnostic traits are polymorphic in one or both species, the explanation may be that there is gene flow between the species, or that the species have split too recently for any differences to become fixed.

Of course, in empirical studies, "fixed" differences are inferred from a finite sample of individuals, and very large sample sizes are necessary to be confident that fixed traits are not actually polymorphic at a low frequency. For example, if an investigator seeks a 95% probability of being able to detect a polymorphism in a putatively fixed trait occurring at a frequency of 1% in a given species, about 150 individuals would need to be sampled from that species (based on equations in 129). This estimated sample size is based on simplified assumptions about population subdivision, but more realistic conditions would likely require more sampling rather than less (112). Given the difference between the large sample sizes needed and those typically available in empirical studies, it seems unlikely that a systematist could ever say with any confidence that a diagnostic character state was actually fixed. In fact, to be truly certain that a trait was not polymorphic and present at a very low frequency would require sampling the entire population.

An alternative approach is to use frequency differences in character-based species delimitation. For example, it seems reasonable to consider a 95% frequency of red flower color in one putative species and 95% white in the other to be evidence that these taxa are distinct, although this character would not be considered informative using PAA. Surely, such large differences in frequencies must indicate that gene flow between these putative species is rare if not absent. By considering differences in frequencies, characters that would be considered uninformative for species delimitation by a strict "fixed-only" criterion could be incorporated, thus increasing the power of any test of species boundaries. Unfortunately, it is not clear what constitutes a sufficient difference in trait frequencies to be adequate for recognizing putative species as distinct, given that some discontinuity in gene flow is to be expected among conspecfic populations.

Some authors have used measures of genetic distance between populations to make species-level decisions, an approach that does incorporate frequency differences between putative species. Perhaps the simplest way this can be done is to find a standard level of distance between "good" species and then apply this value to cases that are less clear (e.g. 63, 64). However, this approach has met with considerable resistance (e.g. 48, 135), partly on the grounds that it relies on an "arbitrary measure of similarity."

A more sophisticated usage of genetic distance data (obtained from multiple populations of two or more putative species) in species delimitation involves various techniques that address the relationship between genetic and geographic distance (see review in 25). The general idea is that for conspecific populations, genetic distance should increase with geographic distance (98), but that this relationship should not hold for heterospecific populations. de Queiroz & Good (25) reviewed a number of techniques that could be used to apply the expected relationships between geographic and genetic distances to species delimitation, including the Mantel test (81) and spatial autocorrelation (15).

Porter (110) has suggested using estimates of gene flow between populations derived from population genetics (e.g. 125, 147, 148) to help determine species boundaries (i.e. certain values of F_{GT} [110] indicate that gene flow is absent or neglible between groups of populations). This approach appears to be promising for determining how much of the similarity between putative species is due to ongoing gene flow. However, all three of these frequency-based approaches are designed primarily for allozyme data, and it is not clear how successfully they could be applied or adapted to morphological data (the data used to describe and delimit most species).

Finally, Doyle (30) proposed a nonfrequency, character-based approach for species delimitation from DNA sequence data from nuclear genes, based on the sharing of alleles in heterozygous individuals. Conspecific individuals share a set of alleles not found in other species, and the combinations of alleles that define a species are seen in the heterozygous individuals. This approach seems likely to be highly sensitive to the failure to detect heterozygous individuals with finite sample sizes.

Tree-Based Species Delimitation

Much of the recent literature on "the species problem" has focused on the implications of intraspecific phylogenies (particularly of individual organisms or haplotypes) for delimiting species, and on the conceptual nature of species (3, 5, 24, 54, 101). Most authors seem to agree that when all the individuals sampled of a putative species appear as each others' closest relatives on a gene tree or trees (relative to other putative species), this is support for the presence of a distinct species. However, when the individuals of a species-level taxon fail to cluster together, the results are more difficult to interpret. Possible explanations include: (*a*) interbreeding between the putative species and other taxa (i.e. possibly suggesting the putative species is invalid); (*b*) incomplete lineage sorting of ancestral polymorphisms [i.e. possibly meaning the putative species is valid but is very recently diverged (e.g. 99)]; (*c*) the presence of multiple, unrelated species hidden by previous taxonomy (e.g. 145); and (*d*) insufficient data, such that the estimated phylogeny fails to match the gene tree.

Another issue in the tree-based approach is that it may be difficult to delimit species without reference to some extrinsic character data (e.g. 30). For example, given only a phylogeny of haplotypes, how do we determine which lineages are species and which are merely clades within species? One approach is to assume that within-species phylogenies will not be concordant between genes (because of gene flow and lineage sorting), sample multiple unlinked genes, and consider species boundaries to be the points that are congruent between gene trees (3, 5). However, the theory behind this approach has not been well explored [for example, how many genes need to be sampled before we can say that relationships are truly congruent or discordant? (3)]. A similar approach assumes that haplotype phylogenies between species will be concordant with geography, but that haplotype phylogenies within species will show discordance with geography (i.e. individuals from the same locality or population will not cluster together, suggesting gene flow between populations). Discordance between geography and gene phylogeny forms the basis for certain measures of gene flow (e.g. 126), but these methods have not been widely applied (if at all) to making species-level decisions.

Finally, it should be noted that tree-based species-delimitation is not restricted to gene trees from DNA sequence data; many authors have applied a tree-based approach to allozyme data to help infer species boundaries; using populations as terminal units and testing whether or not putative conspecifics cluster together (e.g. 53). Morphological data can be used this way as well (67), and the same morphological and/or allozyme data can be analyzed from both a tree and character-based perspective. The congruence and incongruence of population-level trees from diverse types of data (e.g. morphology, allozymes, and DNA sequences) with geography and with groups recognized by tree and character-based perspectives may be particularly revealing about species limits (3, 106, 124).

CONCLUDING REMARKS

The question of how we analyze intraspecific variation in between-species evolutionary studies lies at the intersection of the fields of population genetics, comparative biology, and systematics. In this review, I have discussed three fundamental areas of phylogenetic biology—phylogeny reconstruction, species-delimitation, and comparative studies of character evolution—where polymorphism can have a major impact. I have argued that treating polymorphic traits directly as frequencies may improve analyses in all three areas, although the methodological treatment of polymorphism in the latter two areas is very poorly explored. The frequency of a given trait within a species or population is the most basic parameter of population genetics, but one that is ignored by many systematic and comparative biologists. Future progress in this area may come not only from applying frequency information to additional questions, but also by incorporating additional information on within-species evolutionary processes into systematic and comparative analyses (e.g. 58, 110, 132).

ACKNOWLEDGMENTS

I thank Paul Chippindale, Keith Crandall, Kevin de Queiroz, Zhexi Luo, David Posada, Steve Poe, Maria Servedio, H Bradley Shafer, Andrew Simons, and Jack Sites, Jr. for helpful comments on the manuscript.

LITERATURE CITED

1. Archie JW. 1985. Methods for coding variable morphological features for numerical taxonomic analysis. *Syst. Zool.* 34:326–345
2. Archie JW, Simon C, Martin A. 1989. Small sample size does decrease the stability of dendrograms calculated from allozyme-frequency data. *Evolution* 43:678–83
3. Avise JC, Ball RM. 1990. Principles of genealogical concordance in species concepts and biological taxonomy. *Oxford Surv. Evol. Biol.* 7:45–67

4. Avise JC, Arnold J, Ball RM, Bermingham E, Lamb T, et al. 1987. Intraspecific phylogeography: the mitochondrial DNA bridge between population genetics and systematics. *Annu. Rev. Ecol. Syst.* 18:489–522
5. Baum DA, Donoghue MJ. 1995. Choosing among alternative "phylogenetic" species concepts. *Syst. Bot.* 20:560–73
6. Benabib M, Kjer KM, Sites JW Jr. 1997. Mitochondrial DNA sequence–based phylogeny and the evolution of viviparity in the *Sceloporus scalaris* group (Reptilia: Squamata). *Evolution* 51:1262–75
7. Berlocher SH, Swofford DL. 1997. Searching for phylogenetic trees under the frequency parsimony criterion: an approximation using generalized parsimony. *Syst. Biol.* 46:211–15
8. Bowcock AM, Ruiz-Linares A, Tomfohrde J, Minch E, Kidd JR, Cavalli-Sforza LL. 1994. High resolution of human evolutionary trees with polymorphic microsatellites. *Nature* 368:455–57
9. Brooks DR, McLennan DA. 1991. *Phylogeny, Ecology, and Behavior: A Research Program in Comparative Biology.* Chicago, IL: Univ. Chicago Press
10. Brumfield RT, Capparella AP. 1996. Historical diversification of birds in northwestern South America: a molecular perspective on the role of vicariant events. *Evolution* 50:1607–24
11. Buth DG. 1984. The application of electrophoretic data in systematic studies. *Annu. Rev. Ecol. Syst.* 15:501–22
12. Campbell JA, Frost DR. 1993. Anguid lizards of the genus *Abronia:* revisionary notes, description of four new species, a phylogenetic analysis, and key. *Bull. Am. Mus. Nat. Hist.* 216:1–121
13. Cavalli-Sforza LL, Edwards AWF. 1967. Phylogenetic analysis: models and estimation procedures. *Am. J. Hum. Genet.* 19: 233–57
14. Chu PC. 1998. A phylogeny of the gulls (Aves: Larinae) inferred from osteological and integumentary characters. *Cladistics* 14:1–43
15. Cliff AD, Ord JK. 1973. *Spatial Autocorrelation.* London: Pion
16. Coloma LA. 1997. *Morphology, systematics and phylogenetic relationships among frogs of the genus* Atelopus *(Anura: Bufonidae).* PhD thesis, Univ. Kansas. 287 pp
17. Crandall KA. 1994. Intraspecific cladogram estimation: accuracy at higher levels of divergence. *Syst. Biol.* 43:222–35
18. Crandall KA, Fitzpatrick JF. 1996. Crayfish molecular systematics: using a combination of procedures to estimate phylogeny. *Syst. Biol.* 45:1–26
19. Crandall KA, Templeton AR, Sing CF. 1994. Intraspecific phylogenetics: problems and solutions. In *Models in Phylogeny Reconstruction,* ed. RW Scotland, DJ Siebert, DM Williams, pp. 273–97. Oxford, UK: Clarendon
20. Crother BI. 1990. Is "some better than none" or do allele frequencies contain phylogenetically useful information? *Cladistics* 6:277–81
21. Cunningham CW, Omland KE, Oakley TH. 1998. Reconstructing ancestral character states: a critical reappraisal. *Trends Ecol. Evol.* 13:361–66
22. Darwin C. 1859. *On the Origin of Species.* Cambridge: Harvard Univ. Press
23. Davis JI, Nixon KC. 1992. Populations, genetic variation, and the delimitation of phylogenetic species. *Syst. Biol.* 41:421–35
24. de Queiroz K, Donoghue MJ. 1988. Phylogenetic systematics and the species problem. *Cladistics* 4:317–38
25. de Queiroz K, Good DA. 1997. Phenetic clustering in biology: a critique. *Q. Rev. Biol.* 72:3–30
26. Diaz–Uriarte R, Garland T. 1996. Testing hypotheses of correlated evolution using phylogenetically independent contrasts: sensitivity to deviations from Brownian motion. *Syst. Biol.* 45:27–47
27. Domning DP. 1994. A phylogenetic anal-

ysis of the Sirenia. *Proc. San Diego Soc. Nat. Hist.* 29:177–89

28. Donoghue MJ. 1989. Phylogenies and the analysis of evolutionary sequences, with examples from seed plants. *Evolution* 43:1137–56
29. Donoghue MJ, Ackerly DD. 1996. Phylogenetic uncertainties and sensitivity analysis in comparative biology. *Philos. Trans. R. Soc. Lond. B* 351:1241–49
30. Doyle JJ. 1995. The irrelevance of allele tree topology for species delimitation, and a non-topological alternative. *Syst. Bot.* 20:574–588
31. Echelle AA, Echelle AS. 1998. Evolutionary relationships of pupfishes in the *Cyprinodon eximius* complex (Atherinomorpha: Cyprinodontiformes). *Copeia* 1998:852–65
32. Farris JS. 1966. Estimation of conservatism of characters by constancy within biological populations. *Evolution* 20:587–91
33. Farris JS. 1981. Distance data in phylogenetic analysis. In *Advances in Cladistics, Volume 1. Proceedings of the First Meeting of the Willi Hennig Society,* ed. VA Funk, DR Brooks, pp. 3–23. New York. New York Bot. Gard.
34. Farris JS. 1985. Distance data revisited. *Cladistics* 1:67–85
35. Farris JS. 1986. Distances and cladistics. *Cladistics* 2:144–57
36. Farris JS. 1990. Phenetics in camouflage. *Cladistics* 6:91–100
37. Felsenstein J. 1978. Cases in which parsimony or compatibility methods will be positively misleading. *Syst. Zool.* 27:401–10
38. Felsenstein J. 1981. Evolutionary trees from gene frequencies and quantitative characters: finding maximum likelihood estimates. *Evolution* 35:1229–42
39. Felsenstein J. 1984. Distance methods for inferring phylogenies: a justification. *Evolution* 38:16–24
40. Felsenstein J. 1985. Phylogenies from gene frequencies: a statistical problem. *Syst. Zool.* 34:300–11
41. Felsenstein J. 1985. Phylogenies and the comparative method. *Am. Nat.* 125:1–15
42. Felsenstein J. 1985. Confidence limits on phylogenies: an approach using the bootstrap. *Evolution* 39:783–91
43. Felsenstein J. 1986. Distance methods: a reply to Farris. *Cladistics* 2:130–43
44. Felsenstein J. 1988. Phylogenies and quantitative characters. *Annu. Rev. Ecol. Syst.* 19:445–71
45. Felsenstein J. 1992. Estimating effective population sizes from samples of sequences: inefficiency of pairwise and segregation sites as compared to phylogenetic estimates. *Genet. Res. Camb.* 56:139–57
46. Fitch WM, Margoliash E. 1967. Construction of phylogenetic trees. *Science* 155:279–84
47. Foster SA, Cresko WA, Johnson KP, Tlusty MU, Willmott HE. 1996. Patterns of homoplasy in behavioral evolution. In *Homoplasy. The Recurrence of Similarity in Evolution,* ed. MJ Sanderson, L Hufford, pp. 245–69. San Diego, CA: Academic
48. Frost DR, Hillis DM. 1990. Species in concept and practice: herpetological applications. *Herpetologica* 46:87–104
49. Frost DR, Kluge AG. 1994. A consideration of epistemology in systematic biology, with special reference to species. *Cladistics* 10:259–94
50. Gaines MS, McLenaghan LR, Rose RK. 1978. Temporal patterns of allozyme variation in fluctuating populations of *Microtus ochrogaster*. *Evolution* 32:723–39
51. Golding B, Felsenstein J. 1990. A maximum likelihood approach to the detection of selection from a phylogeny. *J. Mol. Evol.* 31:511–23
52. Goldman N. 1988. Methods for discrete coding of variable morphological features for numerical analysis. *Cladistics* 4:59–71

52a. Goldstein DB, Linares AR, Cavalli-Sforza LL, Feldman MW. 1995. An evaluation of genetic distances for use with microsatellite data. *Genetics* 139:463–71

52b. Golstein DB, Pollock DD. 1997. Launching microsatellites: a review of mutation processes and methods of phylogenetic inference. *J. Hered.* 88:335–42

53. Good DA. 1989. Hybridization and cryptic species in *Dicamptodon* (Caudata: Dicamptodontidae). *Evolution* 43:728–44
54. Graybeal A. 1995. Naming species. *Syst. Biol.* 44:237–50
55. Grismer LL. 1999. Phylogeny, taxonomy, and biogeography of *Cnemidophorus hyperythrus* and *C. ceralbensis* (Squamata: Teiidae) in Baja California, Mexico. *Herpetologica* 55: In press
56. Gutberlet RL Jr. 1998. The phylogenetic position of the Mexican black-tailed pitviper (Squamata: Viperidae: Crotalinae). *Herpetologica* 54:184–206
57. Halloy M, Etheridge RE, Burghardt GM. 1998. To bury in sand: phylogenetic relationships among lizard species of the *boulengeri* group, *Liolaemus* (Reptilia: Squamata: Tropiduridae), based on behavioral characters. *Herpetol. Monogr.* 12:1–37
58. Hansen TF, Martins EP. 1996. Translating between microevolutionary process and macroevolutionary patterns: the correlation structure of interspecific data. *Evolution* 50:1404–17
59. Harvey M, Gutberlet RL Jr. 1999. A phylogenetic analysis of the Tropidurini (Squamata: Tropiduridae) using new characters of squamation and epidermal microstructure. *Zool. J. Linn. Soc.* In press
60. Harvey PH, Pagel MD. 1991. *The Comparative Method in Evolutionary Biology*. Oxford, UK: Oxford Univ. Press
61. Hedin MC. 1997. Speciational history in a diverse clade of habitat specialized spiders (Araneae: Nesticidae: *Nesticus*): inferences from geographic based sampling. *Evolution* 51:1929–45
62. Hey J. 1993. Using phylogenetic trees to study speciation and extinction. *Evolution* 46:627–40
63. Highton R. 1989. Biochemical evolution in the slimy salamanders of the *Plethodon glutinosus* complex in the eastern United States. Part I. Geographic protein variation. *Univ. Ill. Biol. Monogr.* 57:1–78
64. Highton R. 1998. Is *Ensatina escholtzii* a ring-species? *Herpetologica* 54:254–78
65. Hillis DM. 1995. Approaches for assessing phylogenetic accuracy. *Syst. Biol.* 44:3–16
66. Hillis DM, Mable BK, Moritz C. 1996. Applications of molecular systematics: the state of the field and a look to the future. In *Molecular Systematics*, ed. DM Hillis, C Moritz, B Mable, pp. 515–43. Sunderland, MA: Sinauer. 2nd ed.
67. Hollingsworth BD. 1998. The systematics of chuckwallas (*Sauromalus*) with a phylogenetic analysis of other iguanid lizards. *Herpetol. Monogr.* 12:38–191
68. Huelsenbeck JP, Hillis DM. 1993. Success of phylogenetic methods in the four-taxon case. *Syst. Biol.* 42:247–64
69. Huelsenbeck JP. 1995. The performance of phylogenetic methods in simulation. *Syst. Biol.* 44:17–48.
70. Huelsenbeck JP. 1998. Systematic bias in phylogenetic analysis: is the Strepsiptera problem solved? *Syst. Biol.* 47:519–37
71. Jones TR, Kluge AG, Wolf AJ. 1993. When theories and methodologies clash: a phylogenetic reanalysis of the North American ambystomatid salamanders (Caudata: Ambystomatidae). *Syst. Biol.* 42:92–102
72. Kim J, Burgman MA. 1988. Accuracy of phylogenetic-estimation methods under unequal evolutionary rates. *Evolution* 42:596–602
73. Kimura M. 1955. Random genetic drift in multi-allelic locus. *Evolution* 9:419–35
74. Kluge AG. 1989. A concern for evidence and a phylogenetic hypothesis among

Epicrates (Boidae, Serpentes). *Syst. Zool.* 38:7–25

75. Mabee PM, Humphries J. 1993. Coding polymorphic data: examples from allozymes and ontogeny. *Syst. Biol.* 42:166–81
76. Maddison WP. 1990. A method for testing the correlated evolution of two binary characters: are gains or losses concentrated on certain branches of a phylogenetic tree? *Evolution* 44:539–57
77. Maddison WP. 1995. Calculating the probability distributions of ancestral states reconstructed by parsimony on phylogenetic trees. *Syst. Biol.* 44:474–81
78. Maddison WP. 1997. Gene trees in species trees. *Syst. Biol.* 46:523–36
79. Maddison WP, Maddison DR. 1992. MacClade Ver. 3.0. *Analysis of Phylogeny and Character Evolution.* Sunderland, MA: Sinauer
80. Maddison WP, Slatkin M. 1990. Parsimony reconstructions of ancestral states do not depend on the relative distances between linearly-ordered states. *Syst. Zool.* 39:175–78
81. Mantel N. 1967. The detection of disease clustering and a generalized regression approach. *Cancer Res.* 27:209–20
82. Martins EP (ed.). 1996. *Phylogenies and the Comparative Method in Animal Behavior.* Oxford, UK: Oxford Univ. Press
83. Martins EP. 1996. Phylogenies, spatial autoregression, and the comparative method: a computer simulation test. *Evolution* 1750 65
84. Martins EP, Garland T Jr. 1991. Phylogenetic analyses of the correlated evolution of continuous characters: a simulation study. *Evolution* 45:534–57
85. Martins EP, Hansen TF. 1997. Phylogenies and the comparative method: a general approach to incorporating phylogenetic information into the analysis of interspecific data. *Am. Nat.* 149:646–67
86. Mayr E. 1969. *Principles of Systematic Zoology.* New York: McGraw-Hill
87. McGuire JA. 1996. Phylogenetic systematics of crotaphytid lizards (Reptilia: Iguania: Crotaphytidae). *Bull. Carnegie Mus. Nat. Hist.* 32:1–143
88. Mendoza-Quijano F, Flores-Villela O, Sites JW Jr. 1998. Genetic variation, species status, and phylogenetic relationships in rose-bellied lizards (*variabilis* group) of the genus *Sceloporus* (Squamata: Phrynosomatidae). *Copeia* 1998:354–66
89. Meyer E, Wiegand P, Rand SP, Kuhlmann D, Brack M, Brinkman B. 1995. Microsatellite polymorphisms reveal phylogenetic relationships in primates. *J. Mol. Evol.* 41:10–14
90. Mickevich MF. 1978. Taxonomic congruence. *Syst. Zool.* 27:143–58
91. Mickevich MF, Johnson MS. 1976. Congruence between morphological and allozyme data in evolutionary inference and character evolution. *Syst. Zool.* 25:260–70
92. Mickevich MF, Mitter C. 1981. Treating polymorphic characters in systematics: a phylogenetic treatment of electrophoretic data. In *Advances in Cladistics.* Volume 1. *Proceeding of the first meeting of the Willi Hennig Society,* ed. V. A. Funk and D. R. Brooks, pp. 45–58. New York: New York Bot. Gard.
93. Mickevich MF, Mitter C. 1983. Evolutionary patterns in allozyme data: a systematic approach. In *Advances in Cladistics.* Vol. 2. *Proceeding of the Second Meeting of the Willi Hennig Society,* ed. NI Platnick, VA Funk, pp. 169–76. New York: Columbia Univ. Press
94. Mink DG, Sites JW Jr. 1996. Species-limits, phylogenetic relationships, and origins of viviparity in the *scalaris* complex of the lizard genus *Sceloporus* (Phrynosomatidae: Sauria). *Herpetologica* 52:551–71
95. Miyamoto MM, Fitch WM. 1995. Testing species phylogenies and phylogenetic methods with congruence. *Syst. Biol.* 44:64–76

96. Murphy RW. 1993. The phylogenetic analysis of allozyme data: invalidity of coding alleles by presence/absence and recommended procedures. *Biochem. Syst. Ecol.* 21:25–38
97. Murphy RW, Doyle KD. 1998. Phylophenetics: frequencies and polymorphic characters in genealogical estimation. *Syst. Biol.* 47:737–61
98. Nei M. 1972. Genetic distance between populations. *Am. Nat.* 106:238–92
99. Neigel JE, Avise JC. 1986. Phylogenetic relationships of mitochondrial DNA under various demographic models of speciation. In *Evolutionary Processes and Theory,* ed. E Nevo, S Karlin, pp. 515–34. New York: Academic
100. Nielsen R, Mountain JL, Huelsenbeck JP, Slatkin M. 1998. Maximum likelihood estimation of population divergence times and population phylogeny in models without mutation. *Evolution* 52:669–77
101. Olmstead RG. 1995. Species concepts and plesiomorphic species. *Syst. Bot.* 20: 623–30
102. Omland KE. 1997a. Correlated rates of molecular and morphological evolution. *Evolution* 5:1381–93
103. Omland KE. 1997b. Examining two standard assumptions of ancestral reconstructions: repeated loss of dichromatism in dabbling ducks (Anatini). *Evolution* 5:1636–46
104. Pagel MD. 1994. Detecting correlated evolution on phylogenies: a general method for the comparative analysis of discrete characters. *Proc. R. Soc. Lond. B Biol. Sci.* 255:37–45
105. Pamilo P, Nei M. 1988. Relationships between gene trees and species trees. *Mol. Biol. Evol.* 5:568–83
106. Patton JL, Smith MF. 1994. Paraphyly, polyphyly, and the nature of species boundaries in pocket gophers (genus *Thomomys*). *Syst. Biol.* 43:11–26
107. Pimentel RA, Riggins, R. 1987. The nature of cladistic data. *Cladistics* 3:201–9
108. Poe S. 1998. Skull characters and the cladistic relationships of the Hispaniolan dwarf tig *Anolis*. *Herpetol. Mon.* 12:192–236
109. Poe S, Wiens JJ. 2000. Character selection and the methodology of morphological phylogenetics. In *Phylogenetic Analysis of Morphological Data,* ed. JJ Wiens, Washington, DC: Smithsonian Press. In press
109a. Pollock DD, Bergman A, Feldman MW, Goldstein DB. 1998. Microsatellite behavior with range constraints: parameter estimation and improved distances for use in phylogenetic reconstruction. *Theor. Popul. Biol.* 53:256–71
110. Porter AH. 1990. Testing nominal species boundaries using gene flow statistics: the taxonomy of two hybridizing admiral butterflies (*Limenitis*: Nymphalidae). *Syst. Zool.* 39:148–61
111. Prober S, Bell JC, Moran G. 1990. A phylogenetic and allozyme approach to understanding rarity in the "green ash" eucalypts (Myrtaceae). *Plant Syst. Evol.* 172:99–118
112. Rannala B. 1995. Polymorphic characters and phylogenetic analysis: a statistical perspective. *Syst. Biol.* 44:421–29
113. Reeder TW, Wiens JJ. 1996. Evolution of the lizard family Phrynosomatidae as inferred from diverse types of data. *Herpetol. Mon.* 10:43–84
114. Reynolds J, Weir BS, Cockerham CC. 1983. Estimation of coancestry coefficient: basis for a short–term genetic distance. *Genetics* 105:767–79
115. Rogers JS. 1986. Deriving phylogenetic trees from allele frequencies: a comparison of nine genetic distances. *Syst. Zool.* 35:297–310
116. Rohlf FJ, Wooten MC. 1988. Evaluation of the restricted maximum-likelihood method for estimating phylogenetic trees using simulated allele-frequency data. *Evolution* 42:581–95

117. Saitou N, Nei M. 1987. The neighbor-joining method: a new method for reconstructing phylogenetic trees. *Mol. Biol. Evol.* 4:406–25
118. Schluter D, Price T, Mooers AØ, Ludwig D. 1997. Likelihood of ancestor states in adaptive radiation. *Evolution* 51:1699–1711
119. Schultz TR, Crocroft RB, Churchill GA. 1996. The reconstruction of ancestral character states. *Evolution* 50:504–11
120. Shaffer HB, Clark JM, Kraus F. 1991. When molecules and morphology clash: a phylogenetic analysis of the North American ambystomatid salamanders (Caudata: Ambystomatidae). *Syst. Zool.* 40:284–303
121. Siddall ME. 1998. Success of parsimony in the four–taxon case: long–branch repulsion by likelihood in the Farris Zone. *Cladistics* 14:209–220
122. Siddall ME, Kluge AG. 1997. Probabilism and phylogenetic inference. *Cladistics* 13:313–36
123. Simpson GG. 1961. *Principles of Animal Taxonomy.* New York: Columbia Univ. Press
124. Sites JW Jr, Crandall KA. 1997. Testing species boundaries in biodiversity studies. *Conserv. Biol.* 11:1289–197
125. Slatkin M, Barton NH. 1989. A comparison of three indirect methods for estimating average levels of gene flow. *Evolution* 43:1349–68
126. Slatkin M, Maddison WP. 1989. A cladistic measure of gene flow inferred from the phylogenies of alleles. *Genetics* 123:603–13
127. Smouse PE, Dowling TE, Tworek JA, Hoeh WR, Brown WM. 1991. Effects of intraspecific variation on phylogenetic inference: a likelihood analysis of mtDNA restriction site data in cyprinid fishes. *Syst. Zool.* 40:393–409
128. Strait D, Moniz M, Strait P. 1996. Finite mixture coding: a new approach to coding continuous characters. *Syst. Biol.* 45:67–78
129. Swofford DL, Berlocher SH. 1987. Inferring evolutionary trees from gene frequency data under the principle of maximum parsimony. *Syst. Zool.* 36:293–325
130. Swofford DL, Olsen GJ, Waddell PJ, Hillis DM. 1996. Phylogeny reconstruction. In *Molecular Systematics,* ed. DM Hillis, C Moritz, B Mable, pp. 407–514. Sunderland, MA: Sinauer. 2nd ed.
131. Takahata N. 1989. Gene genealogy in three related populations: consistency probability between gene and population trees. *Genetics* 122:957–66
131a. Takezaki N, Nei M. 1996. Genetic distances and reconstruction of phylogenetic trees from microsatellite data. *Genetics* 144:389–99
132. Templeton AR, Crandall KA, Sing CF. 1992. A cladistic analysis of phenotypic associations with haplotypes inferred from restriction endonuclease mapping and DNA sequence data. III. Cladogram estimation. *Genetics* 132:619–33
133. Thiele K. 1993. The holy grail of the perfect character: the cladistic treatment of morphometric data. *Cladistics* 9:275–304
134. Vrana P, Wheeler W. 1992. Individual organisms as terminal entities: laying the species problem to rest. *Cladistics* 8:67–72
135. Wake DB, Schneider CJ. 1998. Taxonomy of the plethodontid salamander genus *Ensatina. Herpetologica* 54:279–98
136. Wiens JJ. 1993. Phylogenetic systematics of the tree lizards (genus *Urosaurus*). *Herpetologica* 44:399–420
137. Wiens JJ. 1995. Polymorphic characters in phylogenetic systematics. *Syst. Biol.* 44:482–500
138. Wiens JJ. 1998. Testing phylogenetic methods with tree congruence: phylogenetic analysis of polymorphic morphological characters in phrynosomatid lizards. *Syst. Biol.* 47:411–28

139. Wiens JJ. Reconstructing phylogenies from allozyme data: comparing method performance with congruence. *Biol. J. Linn. Soc.* Submitted
140. Wiens JJ. 2000. Coding morphological variation for phylogenetic analysis: analyzing polymorphism and interspecific variation in higher taxa. In *Phylogenetic Analysis of Morphological Data,* ed. JJ Wiens. Washington, DC: Smithsonian. In press
141. Wiens JJ, Reeder TW. 1997. Phylogeny of the spiny lizards (*Sceloporus*) based on molecular and morphological evidence. *Herpetol. Mon.* 11:1–101
142. Wiens JJ, Servedio MR. 1997. Accuracy of phylogenetic analysis including and excluding polymorphic characters. *Syst. Biol.* 46:332–45
143. Wiens JJ, Servedio MR. 1998. Phylogenetic analysis and intraspecific variation: performance of parsimony, distance, and likelihood methods. *Syst. Biol.* 47:228–53
144. Wiens JJ, Servedio MR. Species delimitation in systematics: inferring "fixed" diagnostic differences between species. *Proc. R. Soc. Lond. B Biol. Sci.* Submitted
145. Wiens JJ, Reeder TW, Nieto A. 1999. Molecular phylogenetics and evolution of sexual dichromatism among populations of the Yarrow's Spiny Lizard (*Sceloporus jarrovii*). *Evolution.* In press
146. Wiley EO. 1978. The evolutionary species concept reconsidered. *Syst. Zool.* 27:17–26
147. Wright S. 1931. Evolution in Mendelian populations. *Genetics* 16:97–159
148. Wright S. 1978. *Evolution and the Genetics of Populations.* Vol. IV. *Variation within and among Natural Populations.* Chicago: Univ. Chicago Press
149. Wu C–I. 1991. Inferences of species phylogeny in relation to segregation of ancient polymorphisms. *Genetics* 127:429–35
150. Yang Z. 1996. Phylogenetic analysis using parsimony and likelihood methods. *J. Mol. Evol.* 42:294–307

Annu. Rev. Ecol. Syst. 1999. 30:363–95

PHYSICAL-BIOLOGICAL COUPLING IN STREAMS: The Pervasive Effects of Flow on Benthic Organisms

David D. Hart and Christopher M. Finelli
Patrick Center for Environmental Research, Academy of Natural Sciences, 1900 Benjamin Franklin Parkway, Philadelphia, Pennsylvania 19103; e-mail: hart@acnatsci.org, finelli@acnatsci.org

Key Words algae, boundary layer, hydrodynamics, hydraulics, invertebrates

■ **Abstract** Flowing water has profound effects on a diverse array of ecological processes and patterns in streams and rivers. We propose a conceptual framework for investigating the multiple causal pathways by which flow influences benthic biota and focus particular attention on the local scales at which these organisms respond to flow. Flow (especially characteristics linked to the velocity field) can strongly affect habitat characteristics, dispersal, resource acquisition, competition, and predation; creative experiments will be needed to disentangle these complex interactions. Benthic organisms usually reside within the roughness layer, where the unique arrangement of sediment particles produces strongly sheared and highly three-dimensional flow patterns. Thus, accurate characterization of the local flow environments experienced by benthic organisms often requires the use of flow measurement technology with high spatial and temporal resolution. Because flow exhibits variation across a broad range of scales, it is also necessary to examine how organism-flow relationships at one scale are linked to those at others. Interdisciplinary approaches are needed in the study of physical-biological coupling; increased collaboration between ecologists and experts in fluid mechanics and hydraulic engineering is particularly desirable. A greater understanding of physical-biological coupling will not only yield deeper insights into the ecological organization of streams and rivers, it will also improve our ability to predict how flow alterations caused by various human activities affect these vital ecosystems.

INTRODUCTION

Understanding linkages between organisms and their abiotic environment is a critical step in developing predictive models regarding the structure and function of ecosystems. For example, the physical world profoundly shapes a wide array of fundamental ecological processes, including dispersal, resource acquisition, and species interactions. A more complete knowledge of the connections between

0066-4162/99/1120-0363$08.00

biological processes and physical factors is also needed to address such diverse problems as the decline of estuarine fisheries (*23*), the role of forests in global climate change (63), and the spread of exotic organisms such as zebra mussels (187). In fact, studies of physical-biological coupling have undergone a kind of renaissance in ecology, with new models, measurements, and experiments being used to examine a diverse array of phenomena and systems (*1*, *45*, 47, 56, 57, *59*, 99, 160, *169*, 198, 204) [italic indicates citations from marine systems].

In stream ecosystems, the physical world is governed by water in motion. Indeed, the central tenet of our review is that flowing water is often the dominant forcing function (or "master variable" sensu 162) to which other stream processes and patterns can be traced. For example, flow has shaped (both literally and figuratively) almost every feature of these systems, including their channel morphology and disturbance regimes, the distributions of organisms in space and time, as well as rates of energy transfer and material cycling (2, 80). Because of these pervasive effects, we suggest that flow deserves a high priority in the research agendas of stream ecologists. In particular, stream ecology would benefit from a more unified conceptual framework as well as from greater consensus about the utility of particular physical measurements and models relating ecological processes and patterns to flow conditions. Such advances will not only yield deeper insights into the structure and function of stream ecosystems, they will also aid in the development of improved methods for protecting and managing these vital systems. For instance, one of the largest impacts of human activities on streams and rivers stems from the modification of their flow regimes (132, 154, 158). Moreover, the growing demand for fresh water will only intensify the pressures on these systems. Thus, an improved understanding of physical-biological coupling in streams will also enhance our ability to solve pressing environmental problems.

Our review is organized into four parts. First, we examine the direct and indirect mechanisms by which flow can affect bottom-dwelling (or benthic) organisms; we provide a selective review of studies (including some from marine systems, *which are marked with italicized reference numbers*) that illuminate our understanding of these effects. This literature review emphasizes experimental studies that have focused primarily on invertebrates and benthic algae. Second, we consider the sources and scales of spatial and temporal flow variation in streams. Recognizing how flow varies with scale is crucial for understanding the ecological consequences of flow. Third, we evaluate alternative methods for measuring the flow characteristics experienced by benthic organisms, as well as for studying flow effects experimentally. Our understanding of organism-flow interactions will be greatly enhanced by the development and use of more accurate methods for quantifying benthic flow characteristics. Finally, we briefly place these ideas about physical-biological coupling in a broader context by considering their relevance for environmental problem-solving.

A familiarity with fluid mechanic principles is essential for studying organism-flow interactions. Owing to space limitations, we have not attempted to provide an overview of the relevant fluid mechanics and instead direct the reader to several ecologically oriented introductions to this field (41, *45*, 47, 80, 136, 198). In particular, we do not reproduce common equations and formulae because space limitations preclude an adequate discussion of the assumptions needed for their proper application. Nonetheless, we offer a few introductory comments about our use of flow terminology. In general, the flow characteristics of greatest relevance to benthic organisms are linked to time-averaged or time-varying components of the velocity time series. Thus, we focus primarily on flow characteristics related to the velocity of water past a point (measured in units of length per time–e.g., (m/s)) rather than the volumetric flow rate or discharge (measured in units of volume per time–e.g., (m^3/s)) (198). Indeed, many of the flow forces and processes affecting benthic organisms (e.g., drag, lift, diffusivity, and mass transfer) vary as a function of velocity (47, 198). Velocity varies across a broad range of space and time scales, so it is also important to define which scales are relevant to particular ecological questions. Our review emphasizes flow mechanisms operating at organismal scales, due to the role of individuals as the fundamental building-blocks of populations, communities, and ecosystems. In particular, the flow conditions experienced by benthic organisms differ from those experienced farther above the stream bed due to the presence of a velocity gradient, which is created by friction between the moving water and the stationary bed (136). Unfortunately, the complex topography of many stream beds often makes it impossible to predict near-bed velocities using simple formulae such as the log-linear relationship between velocity and height above the bed (136). Thus, when the objectives of ecological studies require the accurate estimation of the flow characteristics experienced by benthic organisms, it will often be necessary to make measurements immediately adjacent to the bed (2, 89).

FLOW EFFECTS ON ECOLOGICAL PROCESSES

The mechanisms (sensu 55) by which flow affects benthic organisms can operate via either direct or indirect paths. By direct, we mean that various hydrodynamic forces or mass transfer processes act on the organisms in question and alter their "performance." In terms of causal pathways, there are no intervening variables between flow and the organisms' response. For example, organisms can be eroded from or deposited on specific regions of the stream bed by flowing water, thus altering local population size. Indirect effects of flow, on the other hand, occur by altering some intermediate abiotic or biotic variable, which in turn affects the study organisms. For example, flow can determine the distribution of sediment particle sizes available in a stream reach, which in turn may affect biota that require specific sediment particles for shelter. This distinction between direct and indirect effects is useful if we wish to identify flow mechanisms, predict how organisms are likely to respond to altered flow fields, and interpret the degree to which the evolutionary history of organisms has entailed adaptation to flow per se.

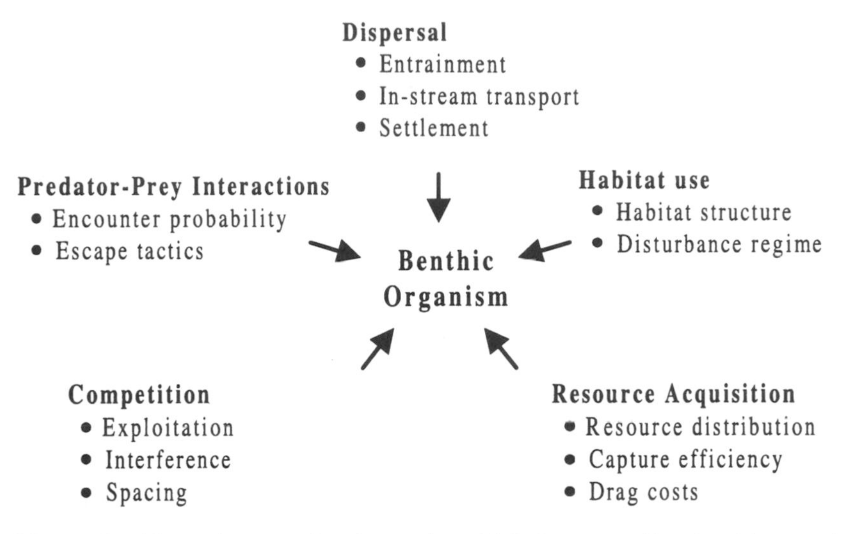

Figure 1 Alternative causal pathways by which flow can affect benthic organisms. Key components of each ecological process can be modified by flow, thereby affecting the performance, distribution, and abundance of organisms. Potential interactions among pathways not shown.

Flow can affect benthic organisms via multiple causal pathways. We illustrate these diverse pathways in a simple diagram (Figure 1) and focus first on dispersal, which can be viewed as an initializing process that delivers organisms to a particular benthic habitat. This ordering reflects the growing belief that local populations of many benthic organisms (e.g. those inhabiting a particular sediment patch) are open, in the sense that they are affected by immigration and emigration processes occurring over much larger spatial scales such as the channel reach or catchment (54, 111, 140, 169). Thus, we begin our discussion by examining how flow influences the dispersal of benthic organisms into and out of local areas. Next, we consider how flow affects various components of the abiotic environment that determine patterns of habitat use. In addition to the effects of flow on these abiotic conditions, it can also modify the acquisition of limiting resources that determine rates of growth and reproduction as well as competitive interactions. Finally, flow can impact benthic organisms by mediating the effects of predators or pathogens. For most benthic organisms, flow characteristics such as average velocity affect patterns of distribution and abundance via multiple (and sometimes countervailing) mechanisms that can operate at different spatial and temporal scales. As a result, ecologists will need considerable ingenuity to disentangle the multifarious causal pathways that link the biology and physics of stream ecosystems.

Dispersal

Stream organisms are often carried downstream via water currents (128). Indeed, these suspended or "drifting" organisms can provide a significant supply of immigrants to newly available substrates (192). Such drifting organisms are rarely adapted to planktonic life, which suggests that they are either accidentally dislodged from the bed or they actively entered the water column in search of more favorable benthic habitats elsewhere in the stream (2). In either case, an explicit focus on the hydrodynamic mechanisms governing water column entry, instream transport, and settlement is critical for understanding how such transport affects the dispersal and dynamics of benthic populations.

Water Column Entry The first step required for flow-mediated transport is entry to the water column. The simplest way in which benthic organisms are removed from the bed and transported in the water column is via the same passive entrainment mechanisms that move sediment particles (80). Specifically, wherever water flows around a solid body, one or more forces (drag, lift, and the acceleration reaction) act on the object (*45*). The magnitude of such forces generally increases nonlinearly with velocity and body size (47, 198).

Although hydraulic engineers and fluvial geomorphologists have quantified the drag forces or critical erosion velocities required to move sediment particles of different size (80), such information is much less common for benthic stream organisms. One study of benthic meiofauna demonstrated that their critical entrainment velocities were lower than those of the sandy sediments they inhabited (139). Another study found that the shear stress required to dislodge benthic algae varied by more than 25 fold, depending on the taxonomic composition and physiognomy of the assemblage (15). More extensive studies in marine environments have emphasized the importance of peak, as opposed to average, forces in causing the dislodgement of benthic organisms (*44, 46, 61, 143*). These results reinforce the conclusions of work on sediment transport in turbulent flows, which have found that the majority of transport events are associated with infrequent, high velocity turbulent motions near the bed (133, 168). Such turbulent flow structures are also likely to play an important role in the entrainment of stream benthic organisms.

In contrast to the passive entrainment of sediment particles, many benthic organisms are able to lower their dislodgement risk by various morphological and behavioral traits (96, 99, 198). For example, the streamlined bodies of some stream invertebrates and fish (198) and flexible stems of stream macrophytes (99) reduce the drag forces experienced by these organisms. In addition, microorganisms may be protected from dislodgement by their small size, which allows them to inhabit zones of greatly reduced velocity that often exist within a few hundred micrometers of the surfaces to which they attach (175). Other benthic organisms have specialized structures that reduce the probability of dislodgement, including mucilage (13), hooks (39), and suckers (77). Finally, stream invertebrates and fish exhibit a wide range of behaviors that lessen the flow forces they experience. For example, when exposed to high flows, both crayfish (118) and fish (201) modify their posture to avoid dislodgement. Other animals such as flatworms reduce the chance of dislodgement by their behavioral avoidance of microhabitats with high flows (84).

In spite of various adaptations to reduce dislodgement, benthic organisms may be unable to avoid entrainment during bed-moving floods (14, 37, 161, 193). Such extreme disturbances can lead to catastrophic drift (2), in which a large fraction of the biota may be transported downstream. On the other hand, the presence of benthic organisms on the stream bed can sometimes reduce the probability of sediment entrainment. For example, the silken shelters created by large aggregations of hydropsychid caddis larvae can dramatically lower the probability of sediment entrainment compared to sediment where such larvae are absent (179).

Rather than avoid flow dislodgement, some benthic organisms seem to actively exploit the flow as an opportunity for dispersal. For example, the rate at which larval black flies enter the water column decreases with increasing water velocity (75), which clearly conflicts with a passive entrainment model. Similarly, grazing mayflies drifted at higher rates when algal abundances were low, even though flow remained constant (101). In both studies, the investigators suggested that water column entry represented an active foraging strategy by which individuals sought to locate better feeding areas. Other species use drift as a predator-avoidance behavior (116, 147). Stream algae also appear to use fluid-mediated dispersal to avoid unfavorable habitats. For instance, Bothwell et al (17) demonstrated that benthic diatoms selectively emigrated from experimentally darkened flumes by altering their buoyancy or form resistance, thereby increasing their entrainment into the flow.

Instream Transport Once an organism enters the water column, several factors can influence its travel distance and transit time. Recent theoretical developments provide a valuable framework for examining the processes governing instream transport. Specifically, advection-diffusion models demonstrate that transport distances and times are controlled by the organism's settling velocity as well as the degree of turbulent mixing (*48, 62, 81*, 120). These investigations have identified the Rouse number (i.e., a dimensionless ratio of settling velocity to shear velocity) as a critical parameter determining the distribution of organisms in the water column and their potential delivery to the bed (*81*, 120). Such advection-diffusion models can also provide a basis for constructing null hypotheses regarding the expected settlement patterns of benthic organisms if they behaved as passive particles (*177*). Furthermore, differences between observed and expected patterns can yield valuable insights into the adaptations used by organisms to modify their settlement rates.

In contrast to these mechanistic transport models, studies of stream drift transport have largely been empirically based. For example, two early field studies observed a characteristic negative exponential relationship between the fraction of organisms remaining in the drift and the distance downstream from their point of entry to the water column (64, 119). This rate of exponential decay was, in turn, inversely related to average velocity. These earlier models have been extended to investigate variation in drift distances due to channel geomorphology (107) and invertebrate behavior (3). For example, studies comparing the drift of live and dead invertebrates have demonstrated that swimming behavior and related adjustments of posture while in transport can either decrease (3, 32, 139) or increase (3, 28) the distance traveled relative to a passive particle. Thus, organisms may avoid settling in unfavorable habitats by decreasing their settling velocity, whereas they may avoid drift-feeding fish by increasing their settling velocity to rapidly exit the water column.

Settlement The fate of dispersing organisms depends on the location and timing of settlement, which can be loosely defined as a process that includes contact with and reversible attachment to the bed. Although it can be difficult to distinguish between the mechanisms that affect water column transport vs. settlement per se, the latter process necessarily involves direct contact with the substratum. Thus, settlement is much more likely to be affected by local conditions and steep gradients in the interfacial world between the water column and the bed.

Settlement processes have received considerable attention in marine systems (recent reviews in *1, 26*), whereas less is known about these processes in streams (73, 140). Flow affects settlement by altering the hydrodynamic forces that deliver organisms to the bed as well as by providing or mediating various cues related to bed suitability (*1*). Near-bed flow characteristics can strongly affect the location of settlement, particularly for small organisms that are also weak swimmers (*26*, 73). For example, areas of flow separation and reattachment may enhance deposition of organisms onto the bed (*129*, 183). In marine systems, flow serves as a direct or indirect settlement cue. For instance, the ability of settling organisms to detect waterborne chemical cues that provide information about bed suitability (e.g., the presence of food resources or conspecifics) depends in part upon rates of turbulent mixing (*1, 146*). Although near-bed flow characteristics in marine systems and streams are not always comparable, stream researchers could nonetheless benefit from more careful attention to the approaches and results of their marine colleagues (recent reviews by *1, 26, 146*) as they begin examining the potential importance of such cues.

Valuable insights about the effect of flow on settlement in streams have come from studies of benthic algal immigration (185). Several investigations have demonstrated an inverse relationship between the immigration rate of algal cells and velocity (150, 183). Further mechanistic studies are needed to determine whether the reduced immigration observed at high velocities is caused by lower rates of cell delivery to the bed or higher post-contact rates of cell removal (i.e., entrainment).

If the dispersal of benthic organisms is strongly affected by flow, then benthic distributions may reflect hydrodynamic processes rather than post-settlement habitat preferences. Although these "supply-side" processes can be quite important in marine systems (*26, 169, 177*), stream ecologists have generally assumed that the high mobility of benthic invertebrates should override the effects of initial settlement in controlling benthic distributions (but see 54, 140). In one of the few explicit tests of this assumption, Fonseca examined the mechanisms and consequences of flow-mediated dispersal in larval black flies. She found that larvae had difficulty attaching to the bed in all but the slowest flows, which led to a reduced settlement probability in faster flows (i.e., those preferred for feeding) (74). Moreover, related field experiments in which she manipulated larval settlement rates demonstrated that these dispersal constraints caused larval abundances to be lower than expected in their preferred feeding sites (73).

Overall, there are many parallels between the flow processes governing sediment transport and organismal dispersal (168). These similarities strongly suggest that ecologists would benefit from closer interaction with experts in hydraulic engineering and fluvial geomorphology as they seek to understand the ways that flow affects the entrainment, transport, and settlement of stream benthic organisms.

Habitat Use

Upon settlement, a benthic organism will encounter a suite of local abiotic conditions. In turn, these habitat characteristics are often affected by flow, which can lead to flow-dependent patterns of habitat use. Of course, many local habitat features ultimately derive from broad-scale variations in climate and geology that control the hydrology and water quality of streams within a catchment or region (10). Moreover, regional differences in hydrological disturbance regimes can act as a powerful filter that limits the pool of species (and species traits) available to colonize a particular catchment (152). Thus, disturbances associated with floods and droughts are often regarded as a primary determinant of broad-scale variations in the structure and function of stream ecosystems (2, 10, 155, 193, 207).

At more local scales, flow controls the erosional and depositional processes that determine bed form and composition (80, 110). Consequently, flow influences many habitat features of potential importance to benthic organisms, including channel sinuosity, pool-riffle sequences, and the abundances and arrangements of different sizes of sediment particles. The strong covariation between bed characteristics and flow can make it difficult to identify the causal pathways underlying relationships between benthic organisms and habitat structure. Some ecological patterns are probably linked directly to habitat characteristics per se. For example, certain benthic organisms may prefer particular sediment particle sizes for attachment, shelter, or locomotion (40, 96, 122), regardless of local flow conditions.

On the other hand, flow mechanisms probably play an important role in many correlations between benthic organisms and various habitat features (*177*). For instance, several passive suspension feeders are more abundant on large substrates, in part because these are commonly characterized by higher velocities that are preferred for foraging (38). Similarly, some detritivores are more common in sediments containing high concentrations of fine particulate organic matter (122), which in turn are more likely to be retained in microhabitats with slow flows. Still other organism-sediment associations may stem from the residual effects of sediment-specific differences in the intensity of past disturbances (76).

Some of the most dramatic effects of flow on physical habitat occur as the result of floods and droughts. In particular, the channel scouring that occurs during large floods often results in extensive displacement and mortality of benthic organisms (8, 9, 79, 172, 174). Relatively little study has focused on the precise mechanisms by which floods affect benthic organisms, however (but see 14). Some benthic organisms are presumably crushed by bed load, whereas others may be abraded by suspended sediment (11, 76). Mortality of organisms that are dislodged from the bed during a flood can occur either during transport or after deposition in unfavorable environments such as deep pools or receding flood plains.

The abundances of benthic organisms sometimes recover relatively quickly after floods (37, 184), which implies the existence of spatial refuges where the negative effects of disturbance are ameliorated. At a microscopic scale, some diatoms can persist within protected crevices on the surface of individual sand grains, whereas they are scoured from the more exposed surfaces of those sediment particles (7, 121). At a larger scale, Lancaster & Hildrew (106) have identified portions of the streambed known as hydrodynamic dead zones (167), in which bed shear stress undergoes little or no increase during a flood. They suggest that these areas of minimal flow change can serve as refugia for organisms that would otherwise be dislodged or harmed by the high shear stress that occurs elsewhere in the channel. Indeed, their studies have demonstrated that some benthic invertebrate taxa are relatively more abundant in flow refugia after a storm than before it (see also 76). It is not always clear whether this increase reflects an active immigration of individuals to flow refugia, a passive deposition of individuals within flow refugia, or a reduction in the abundance of individuals from areas outside those refugia. One recent study has demonstrated active use of flow refuges during floods (90). Specifically, larval black flies moved to more sheltered sites on boulders and artificial substrata within minutes after near-bed velocities began increasing during either natural or experimentally created floods. After flood waters receded, the larvae returned to more exposed sites on those same substrata.

Another potential refuge for stream organisms during floods is the hyporheic zone. Palmer et al (142) conducted one of the few explicit tests of this hypothesis by focusing on the vertical distribution of benthic meiofauna before and after floods. Although these invertebrates are very susceptible to entrainment by flows and thus would benefit from moving into the hyporheos, they showed no tendency to move deeper into stream bed sediments to avoid high flows. Further studies are needed to determine whether other benthic organisms make use of such hyporheic refuges.

Droughts, which are characterized by extremely low flows, impose a very different set of stresses on benthic organisms (19, 206). In comparison to floods, much less study has focused on the ecological responses to such low flows. As flows decline, some benthic organisms may experience greater resource limitation or physiological stress due to reduced rates of mass transfer (see below). Habitat availability also declines during droughts, which can lead to increased intensities of either competition or predation. Experimental studies are needed to disentangle the flow-related mechanisms by which droughts affect benthic populations and communities.

Resource Acquisition

To grow and reproduce, all organisms must obtain resources such as dissolved nutrients or particulate foods from their environment. Flow can enhance or hinder the rate and efficiency of resource acquisition via its effects on the distribution of resources as well as the ability of organisms to locate and gather those resources. For example, the water column distribution of various dissolved substances used by benthic organisms is affected by the magnitude of turbulent mixing, whereas the vertical concentration profile of suspended particles is determined by both turbulence and particle settling velocities (see Dispersal, above). In general, low settling velocities of suspended material and high levels of turbulent mixing tend to homogenize the concentration of these resources in the water column (120, *130*). Therefore, in the absence of local sources or sinks, spatial heterogeneity in flow characteristics will be the primary determinant of any variation in the flux of these limiting resources to organisms. Further studies are needed, however, to validate this assumption of resource homogeneity. For example, tributaries carrying high concentrations of seston can create locally enriched resource plumes that affect the distribution of suspension feeders (67). Similarly, the discharge of nutrient-rich groundwater can produce localized regions of high algal growth (195). Recent studies have even suggested that dissolved organic matter can be converted to particulate organic matter by turbulence-induced flocculation, which can potentially modify the quantity and quality of resources available to passive suspension feeders (34).

In addition to altering the distribution of suspended or dissolved materials, flow can influence the ability of organisms to locate or obtain these resources. The best examples of this effect in streams come from studies of nutrient uptake, growth, and photosynthesis in benthic algae (16, 185 and references therein). For example, the nutrient uptake rates of benthic algae and aquatic plants are sometimes limited by the rates of molecular diffusion of dissolved materials across the laminar (viscous) sublayer (198), which typically surrounds organisms. Because the thickness of this layer is inversely proportional to velocity (47, 198), increases in flow can enhance the rate at which limiting nutrients are exchanged with the water column (*82*). This stimulatory effect of velocity is reduced for thinner mats (185) and nutrient-replete cells (16).

Several authors have tried to establish a mechanistic relationship between water flow and uptake rates of limiting nutrients through the analysis of dimensionless parameters such as the Sherwood number (*145*) and the Stanton number (*189*). These parameters describe the enhancement of diffusion across a surface in terms of flow velocity, surface roughness, and nutrient-specific diffusion coefficients. Although diffusion limitation is often associated with photosynthetic processes, it can also pose a challenge for animals. For example, in microhabitats with low velocities, many benthic invertebrates actively circulate water past respiratory organs to decrease the thickness of the laminar sublayer, thereby increasing the rate of gas exchange (66, 205).

In addition to their role as nutrients, dissolved chemical compounds can also serve as olfactory cues. In fact, olfaction is the most common sensory system used in communication and directed search (58), and many stream inhabitants use olfaction to avoid predators, find prey, or select mates (see 52 for a recent review). Because odors are usually transported by turbulent water flow, there is a strong link between odor-mediated searching behavior and ambient flow conditions. First, the temporal and spatial distribution of odorant within odor plumes is determined by turbulent mixing and dilution processes (*71, 123, 203*). Second, many organisms use the direction and strength of flow as ancillary rheotactic cues to navigate successfully in a turbulent odor plume (*72, 212*). The interaction between these two flow effects often produces a unimodal relationship between search success and velocity (*72, 203*). For example, when flow is slowest, searching success is low because there is insufficient mechanical stimulation for rheotaxis, even though odor dilution is minimal. At higher velocities, search success increases as the strength of potential rheotactic cues increases. In very fast water, foraging efficiency is again reduced because rates of dilution are so high that the odor is rapidly diluted below detectable levels.

Such unimodal relationships between measures of organismal performance and velocity may be widespread in benthic environments. For example, flow can have countervailing effects on the accrual of benthic algae (12). At the lowest velocities, nutrient uptake by algae is limited by the thickness of the viscous sublayer and by low levels of turbulent flux. In contrast, the greater drag associated with high velocities causes increased algal sloughing. Maximum levels of algal biomass often occur at intermediate velocities as a result of this type of subsidy-stress relationship (12).

Flow has similar effects on the collection of particulate resources by passive suspension feeders (115, 124, 173, 199, 204). For example, ingestion rates of these consumers are often limited by low flux rates of seston when velocities are slow, whereas high drag may impair the performance of the feeding structures in fast flows. These offsetting mechanisms commonly produce a unimodal relationship between ingestion rate and velocity (115, 124, 173, 204). Further support for these conclusions comes from theoretical (31, 114, 170) and experimental (21, 22, 104) studies of the mechanisms by which organisms remove food particles from suspension. Moreover, recent studies have demonstrated flow-dependent phenotypic plasticity in the morphology of feeding appendages, which suggests that the design of these structures represents a balance between maximizing particle encounter rates and minimizing drag costs (210; see also 93).

Competition

When flow controls the supply rate of limiting resources such as nutrients or suspended particles, it can also potentially mediate the intensity and outcome of competitive interactions. One striking feature of such interactions is their unidirectional nature. For example, upstream organisms can reduce the availability of resources that might otherwise be used by individuals located farther downstream, but not vice versa. The mechanisms by which upstream consumers alter resource availability can involve either exploitation or interference (sensu 86). In exploitation competition, resources are directly consumed by upstream individuals, leading to a progressive reduction in resource availability as a water mass passes over an array of organisms. Such exploitation competition has been modeled in marine bivalves by using advection-diffusion equations to describe how the concentration of seston declines in response to resource uptake (*27*, 204). High levels of resource depletion are most likely to occur where consumer densities and per capita consumption rates are both high, and where bulk mixing due to turbulence is low. There is also evidence that such resource depletion occurs in some streams (108, 125, 208), although few investigations have focused on the flow mechanisms that govern this process.

A second mechanism by which consumers modify resource availability to organisms located farther downstream involves alterations of flow characteristics rather than resource consumption per se. Specifically, flow patterns are modified by the shapes and activities of benthic organisms, which can in turn impact downstream patterns of resource flux (*60*). In the case of suspension-feeding black flies, Clark & Hart (35) demonstrated that flow disruptions caused by upstream larvae reduced the local mean velocity and increased the relative turbulence intensity experienced by downstream neighbors, which in turn can lower their ingestion rates (33). Similarly, Hemphill (94) and Englund (65) reported that velocities were reduced by as much as 50% several body lengths downstream from net-spinning caddis flies.

Whether via exploitation or interference, the reduction in resource availability caused by upstream organisms can alter interactions among consumers. For example, Hart (87) observed that black fly larvae often behaved aggressively toward nearby upstream neighbors in an effort to displace them. Moreover, such aggressive behavior declined in response to an experimental increase in seston concentration, thereby suggesting that food concentration and velocity can be viewed as partially substitutable resources (sensu 190) for these passive suspension feeders. Similarly, the aggressive interactions typical of many salmonids stem from competition for preferred feeding sites that are governed by water-borne delivery of invertebrate prey (69).

When flow mediates competitive interactions, it can also affect the spatial distribution of consumers. For example, several authors have shown that passive suspension feeders avoid sites located immediately downstream from other individuals (33, 65, 94). Likewise, Matczak & Mackay (117) demonstrated that nearest-neighbor distances in territorial net-spinning caddisflies declined in response to experimental increases in either velocity or food concentration. Because flow also mediates density-dependent emigration from foraging sites, a heterogeneous flow environment can in turn give rise to spatially patchy distributions of consumers (75). It is not yet clear how such flow-mediated competitive interactions contribute to patterns of resource partitioning (114).

Physical disturbances caused by extremely high or low flows can also alter the outcome of interspecific competition. For example, Hemphill & Cooper (95) demonstrated that winter floods modified the outcome of competition for space between net-spinning caddisflies and black flies. Specifically, scouring during floods created open space that was rapidly colonized by competitively subordinate black flies. In the absence of disturbance, such space was usually monopolized by larval caddisflies. Zhang et al (211) also found that black flies benefited from hydraulic disturbances that occur below dammed sites on rivers, apparently due to the reduced impact of less disturbance-tolerant predators and competitors.

Predator-Prey Interactions

Flow can affect the outcome of predator-prey interactions by altering either predator-prey encounter rates or the predator's ability to successfully capture prey following an encounter. For example, Hansen et al (84) and Hart & Merz (92) performed lab and field experiments, respectively, examining how predator-prey interactions between flatworms and black fly larvae varied as a function of velocity. Flatworms were unable to tolerate the high velocities preferred by black fly larvae, thereby providing larvae with a flow-mediated refuge from these predators. Moreover, even where velocities were slow enough to allow encounters between flatworms and larvae, the probability of successful capture declined markedly with increasing flow owing to difficulties in handling prey at higher velocities. Similar reductions in predator impact with increasing velocities have been observed in stoneflies (115, 148). Hart & Merz (92) suggested that such flow-mediated refuges may be common in many benthic communities because of the tendency for prey to be smaller than their predators, which would expose them to lower drag forces and thereby reduce their risk of dislodgement in high flows.

Interactions between benthic algae and grazers can also be mediated by flow. DeNicola & McIntire (43) conducted studies in a laboratory flume demonstrating that high flows prevented snails from grazing on algae occurring in exposed microhabitats. Poff & Ward (156, 157) found that mobile caddisfly grazers moved more slowly as flow rate increased, and that their negative effect on algal abundance was greater in slow flows. Similarly, Hart (88) showed that grazing crayfish were able to eliminate filamentous green algae from microhabitats characterized by velocities <20 cm/s, whereas the algae flourished at velocities >50 cm/s. Indeed, this latter velocity corresponded closely to the flow threshold at which the crayfish had difficulty maintaining their hold on the stream bed (118).

In summary, the potent and diverse flow effects documented in the preceding sections strongly support the view that flow is the fundamental abiotic factor controlling ecological processes and patterns in streams. Developing a more comprehensive understanding of these important effects depends not only on greater attention to the mechanisms by which organisms respond to flow, but also the range of flow conditions that benthic organisms are likely to experience. Accordingly, we first turn our attention to the nature of flow heterogeneity in benthic environments, and subsequently examine methods for quantifying flow characteristics in these settings.

SOURCES AND SCALES OF FLOW VARIATION

Flow characteristics in a particular stream vary over a broad range of space and time scales. Velocity exhibits spatial variation from scales as short as the Kolmogorov scale ($\sim 10^{-4}$ m) at which turbulence is completely dissipated to heat, to scales as long as those describing channel forms such as the meander wavelength ($\sim 10^{2}$ m). Temporal variation in velocity occurs at scales as short as those associated with the smallest turbulent eddies ($\sim 10^{-2}$ s) to scales as large as the recurrence intervals of bankfull floods ($\sim 10^{7}$ s). One of the central challenges in the study of organism-flow interactions is to determine which of these space and time scales, which span more than six orders of magnitude, are most important for understanding particular ecological processes and patterns. Given our focus on the flow environnments experienced by individual benthic organisms, we are particularly interested in the range of flow conditions an organism would encounter over the array of microhabitats it can occupy during its lifespan.

In the last two decades, several different approaches have been followed in applying principles from fluid mechanics and hydraulic engineering to predict benthic flow characteristics and examine their effects on organisms, including: (*a*) the application of boundary layer theory (41, *45*, 136, 175, 198); (*b*) the classification of near-bed flow fields depending on velocity and depth, as well as the size and spacing of roughness elements (42, 209); and (*c*) prediction of benthic flow characteristics from coarse-scale hydraulic engineering models (180). Although these approaches are often useful for predicting flow characteristics in simplified settings (e.g., in pipes or on flate plates) or as the spatial average for an entire channel reach, the physical models on which they are based were not designed to predict the local flow environments actually experienced by benthic organisms that inhabit the topographically variable surface of a natural stream bed.

Some of the challenges involved in using physical models to predict flow patterns in benthic environments can be illustrated by examining the vertical gradient in flow characteristics that exists in many streams. Recently, Nikora et al (134) developed a simplified hydraulic model for spatially averaged open channel flow over a rough bed that subdivides the flow into several vertical layers (Figure 2) (see also 163). The model's focus on hydraulically rough flow is particularly relevant because such flows are the norm for most natural streams (29, 80), with turbulent eddies extending to the substratum surface where they disrupt the formation of a viscous sublayer. When the depth of the flow is much greater than the height of the roughness elements, an outer and logarithmic layer will exist. In the logarithmic layer, average velocity exhibits a log-linear relationship with height above the bed. Flow characteristics in this layer can be readily predicted according to the "law of the wall" (136), but this is not the layer inhabited by most

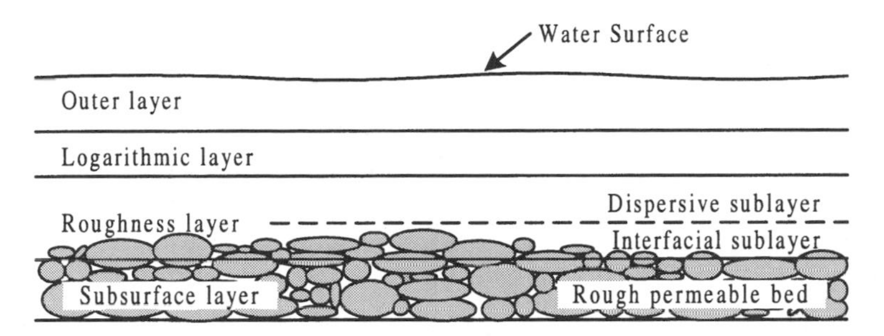

Figure 2 Subdivision of hydraulically rough open-channel flow into horizontal layers (modified from 134). Importantly, flow velocities within the "roughness layer" are unpredictable based solely on knowledge of overlying flow (e.g., logarithmic layer). This figure is not drawn to scale.

benthic organisms. Rather, stream benthic organisms usually live in the roughness layer, which includes a zone that extends above the roughness crests where three-dimensional flows are present (the dispersive sublayer) as well as a zone among the roughness elements per se (the interfacial sublayer). Unfortunately, no models are currently capable of predicting flow characteristics at any specific location within this roughness layer (i.e., in a particular microhabitat where an organism might reside), because local flow patterns are highly three-dimensional and uniquely dependent on the exact shape, size, and arrangement of these roughness elements.

An alternative means of describing near-bed flow patterns was proposed by Davis & Barmuta (42; see also 209), who built on earlier work (126) to develop a classification system based on the height and spacing of roughness elements. Irregularity in the arrangement of these elements on natural stream beds, however, makes the classification of near bed flow into well-defined categories more challenging. Moreover, a large range of flow microhabitats are likely to occur within each flow category (e.g., wake-interference flow), so these classification systems may lack adequate resolution for describing the flow field experienced within a particular microhabitat.

Of course, spatial flow variation also occurs on scales larger and smaller than those associated with sediment particles in the roughness sublayer. For example, an alternating sequence of riffles and pools creates heterogeneity in both bed slope and roughness on the scale of tens to hundreds of meters. On the other hand, the pitted and grooved surfaces of individual substrata produce heterogeneity on a millimeter scale. Collectively, these complex topographic features cause marked heterogeneity in benthic flow characteristics, which is likely to produce an equally heterogeneous array of ecological processes and patterns.

Temporal flow variation is also a conspicuous feature of stream benthic environments. Considerable attention has focused on the ecological consequences of extreme flow variations associated with floods and droughts that operate on relatively long time scales (36, 97, 153). Paradoxically, ecologists have focused much

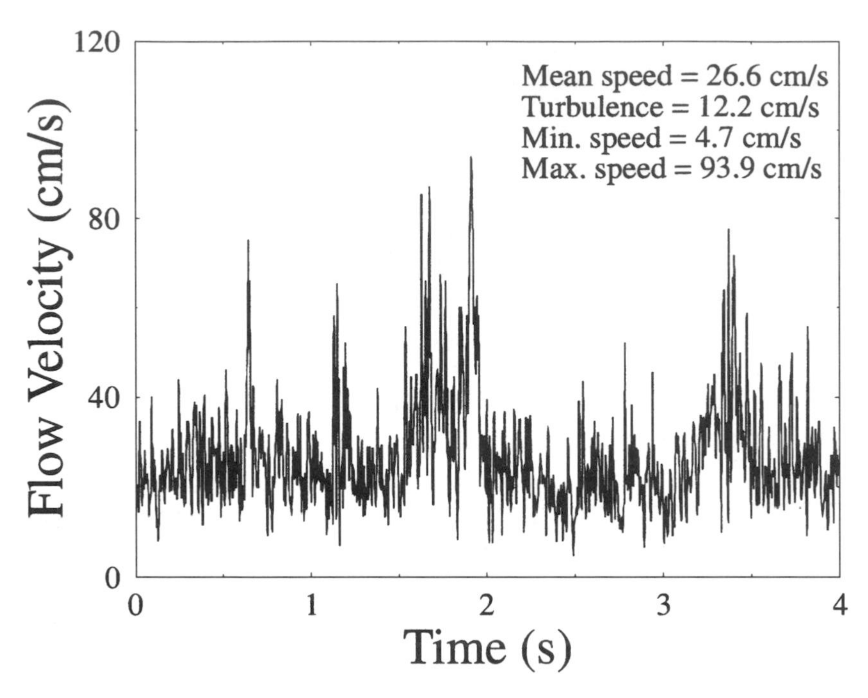

Figure 3 Four-second velocity time series collected using a hot-film velocimeter. Data were collected at 256 Hz from a 2-mm height above the surface of a natural stone inhabited by black fly larvae in Taylor Run (Chester County, PA). Turbulence is measured as the standard deviation of the time series, or the root mean square (RMS) turbulence intensity.

less on the effects of short-term flow variation associated with turbulence (Figure 3), even though it is a more prevalent phenomenon (80). Indeed, except for the smallest benthic organisms living in the slowest flows, turbulence is the rule rather than the exception in streams and rivers (29). Turbulent flows are highly irregular and unpredictable in nature, and they are characterized by large, rotating eddies that produce intense mixing (188). Benthic organisms will often be subjected to shear-induced turbulence and may also experience quasi-periodic oscillations in the flow due to vortex shedding from upstream objects. Turbulent flows can be characterized by several different quantities in three dimensions, including turbulent stresses, kinetic energy flux, and various scales (e.g., Kolmogorov, Taylor, integral) of eddy lengths (20, 166, 176, 188).

It is sometimes incorrectly assumed that turbulent fluctuations are relatively unimportant in near-bed flows due to the general reduction in velocity and corresponding inertial forces immediately above the stream bed. As demonstrated by Figure 3, time series of velocity recorded just 2 mm above the stream bed can reflect a high degree of turbulence, with velocities varying by more than an order of magnitude (89). Some characteristics of turbulence can be predicted in simplified settings, such as the development of a turbulent boundary layer along a flate plate (136). In the roughness layer where most benthic organisms live, the generation of turbulence involves a more complex array of processes. In particular, high levels of turbulence can arise when flow separation around upstream roughness elements produces vortices that are advected downstream (42, 133, 209). Indeed, such flow disturbances are caused by benthic biota as well as bed features per se (135, 171). Analysis of vertical and horizontal components of turbulent flow fluctuations over coarse gravel beds has also demonstrated bursting phenomena, which include infrequent sweeps of high velocity water toward the bed and associated ejections of low velocity water away from the bed (168). These bursting motions are known to play important roles in sediment entrainment (136) and may be equally important in governing the transport of benthic organisms to and from the bed (45).

Although traditional models from fluid mechanics and hydraulic engineering are useful for describing flow variation in geometrically simple settings or at larger scales, their utility for characterizing flow patterns within the roughness layer is much more limited. Thus, when these models are used to make predictions about benthic flow characteristics, they are likely to produce two kinds of errors that pose problems for the study of organism-flow interactions. First, the flow characteristics predicted at any particular location on the bed will usually differ from the true conditions because the models are unable to account for the complex 3-D flow patterns in the roughness layer. Moreover, the sign and magnitude of difference between predicted and actual flow characteristics is likely to vary dramatically depending on the local setting to which the model is applied. Ultimately, such inaccuracies can greatly distort our estimation of the intensity and importance of organism-flow relationships (sensu 202).

A second, and perhaps more serious, consequence of the misapplication of traditional engineering models is that they will usually underestimate the spatial and temporal heterogeneity of flow fields available to benthic organisms. Of course, the tractability of such models is enhanced when they assume a more homogeneous flow setting, but what is the price of such simplification? A growing body of theory suggests that environmental heterogeneity per se is a critical factor governing ecological processes and patterns (85, 98, 144, 191). Indeed, the mosaic of microhabitats created by flow heterogeneity is likely to be an essential property of stream ecosystems, given the role of flow in controlling the availability of refuges (92, 105), affording opportunities for niche diversification (114), mediating dispersal (140), and constraining food web dynamics (162). Thus, although spatial and temporal heterogeneity in flow creates numerous challenges for ecological research, such heterogeneity is too important to be ignored.

Two more recent approaches to flow modeling may enhance our ability to characterize heterogeneous flow environments and examine their effects on benthic organisms. In particular, advances in computational methods have made it possible to develop more realistic models of flow patterns. The first approach toward a more realistic description of flow involves the use of computational fluid dynamic models. This class of computer models can provide detailed predictions of flow fields in one, two, or (occasionally) three dimensions via numerical solution of the Navier-Stokes equations (the fundamental equations governing mass and momentum transfer in a moving fluid) (4). The power of these models stems from their versatility. For example, predictions of flow patterns can be obtained for flows around relatively simple objects such as cylinders (4, 91), or very complex structures such as a moth wing in flight (113). Thus, such models offer a potentially valuable bridge between the direct, but usually labor-intensive measurement of the benthic flow characteristics at a particular locale, and the ultimate goal of predicting near-bed flows for much larger portions of the stream bed (30). Moreover, such models are likely to provide valuable insights regarding scale-dependent patterns of physical-biological coupling.

A second approach involves the application of advection-diffusion (= dispersion) theory to study the turbulent transport of suspended particles and benthic organisms. For example, a number of research groups have developed hydrodynamically explicit models for predicting the distribution of organisms in the water column under various flow conditions (*48, 59, 62, 81*, 120, *194*). Although these models still require simplifying assumptions, they also increase the level of realism through incorporation of more complex topographic relief (*194*), organismal behavior (*62, 194*), and tidal forcing (*62, 81, 194*).

QUANTIFICATION AND MANIPULATION OF BENTHIC FLOW FIELDS

The direct measurement of benthic flows can play at least two important roles in advancing ecological research in streams. First, there is a critical need for the validation of physical flow models. As indicated above, we need to quantify the accuracy of these models by determining how predicted near-bed flows differ from those actually experienced by benthic organisms. Second, when accurate flow models are unavailable, direct measurements provide the only means of quantifying the benthic flow environments inhabited by stream organisms. Although a diverse array of instruments and techniques are available for quantifying flow, the investigator must carefully consider various trade-offs regarding spatial and temporal resolution, cost, and difficulty of use (Table 1). For example, the existence of complex velocity gradients within the roughness layer places a premium on the use of instruments with fine spatial resolution (29). This challenge was highlighted by Hart et al (89), who used hot-film anemometry to measure velocities in the vicinity of larval black flies inhabiting natural stone surfaces. They found no statistical relationship between velocities measured 2-mm above the stone (i.e., the height at which larvae hold their filtering appendages), and those measured at 10-mm height (see also 12). Moreover, larval abundance was strongly related to the 2-mm height velocities, but unrelated to those at 10-mm height. These results suggest that flow measurements made with coarse spatial resolution may fail to detect significant organism-flow relationships. The results also emphasize how a knowledge of the study organism's natural history can help guide decisions about flowmeter resolution. Considerations about the temporal resolution of flow measurements are particularly important for assessing the effects of turbulence on benthic organisms. Although turbulence affects many critical processes, including dislodgement (133), settlement (120), resource acquisition (16), and competitive interactions (35), the flowmeters used by stream ecologists often lack adequate temporal resolution for characterizing turbulent flows. Fortunately, flowmeters with sufficiently high spatial resolution for making measurements at scales relevant to benthic organisms also tend to have high sampling frequencies for quantifying turbulence, although there are exceptions to this pattern.

Rather than espouse any one device or technique, we urge investigators to consider carefully the operating principles and limitations of different devices before making a selection (Table 1). As mentioned above, the size of the study organism may dictate the necessary spatial resolution of the flowmeter (hot-film or hot-bead anemometry for very small invertebrates, propeller or electromagnetic flowmeters for large aquatic macrophytes). Moreover, before any new technique is adopted for measuring benthic flows, it should be thoroughly calibrated against a reliable standard in flow fields similar to those where it will be deployed (e.g., strongly sheared, near-wall conditions). For example, acoustic Doppler velocimeters (ADV) have recently been adopted for use by benthic stream ecologists (18, 135). Although the performance of ADV has been carefully validated for use in open water (5), its spatial resolution may be too large to provide accurate flow measurements in the first few mm above the bed where many benthic organisms live (70). It is also important to be clear about the flow parameter being measured. For example, dissolution techniques measure the rate of mass transfer at a given point but may not provide reliable information on average velocity due to the confounding effects of turbulence. Similarly, some methods only measure speed (a scalar) rather than velocity (a vector defined by both speed and direction), although the difference between these two measures in strongly unidirectional flows may be small, given suitable instrument alignment. In sum, we believe that greater attention to the suitability of alternative methods for quantifying benthic flows will markedly enhance our understanding of physical-biological coupling in streams.

Experimental studies are needed to understand the mechanisms underlying organism-flow relationships. It is important to be aware of the many trade-offs inherent in conducting such experiments in the laboratory versus the field (49, 165). The ability to control, manipulate, and quantify flow will almost always be easier when studies are conducted in laboratory flumes (*45*, 137). For example, the assumptions of uniform and fully developed flow can often be met in the laboratory (136, 137). Moreover, by manipulating substratum geometry, it is possible to create a variety of flow microhabitats (103, *129*). It is even possible to decouple average velocity and turbulence characteristics in the laboratory via the use of upstream grids, weirs, and varied levels of bed roughness (109, 136, 137). For instance, Weissburg & Zimmer-Faust (*203*) were able to distinguish the roles of

TABLE 1 Flow measurement devices/methods.

Device/method[a]	Max. spatial resolution[b]	Max. frequency response[c]	Cost[d]	Lab/Field	Pros	Cons	Suppliers[e]	References[f]
Flow visualization/ Particle image velocimetry	variable	30 Hz. typical NTSC video	$-$$$$$	both	Non-invasive, can be used in boundary layers. 3-D. Frequency response can be increased with high speed imaging systems.	Boundary layer and high resolution applications may require costly cameras or lights and may be restricted to lab.	*24*, Dantec, TSI	*25*, 75, 178
Volumetric flow measurements	variable	0.1 Hz	$	both	Easy method for estimating average velocity through a known cross sectional area.	Variable precision. No turbulence or boundary layer information. 1-D.	NA	124, 186
Dissolution methods	variable	<0.01 Hz	$	both	Easy method for measuring average flow or mass transfer rates at a point.	Slow response time. Usually coarse spatial resolution; invasive. Measurement of velocity easily confounded by turbulence. 1-D.	131	*53*, *100*, 151
Fliesswasser-stammtisch (FST) Hemisperes	30 cm^3	<0.01 Hz	$-$$	both	Direct integrated estimate of shear stress on platform near the bed.	Highly invasive, sensitive to bed roughness, no boundary layer or turbulence information. 1-D.	182	50, 78, 106
Drag deflection, strain	1 cm^3	1 Hz	$-$$	both	Robust method of estimating mean or maximum velocity.	Unsuitable for most boundary layer conditions, invasive. 1-D.	*6*	65
Bentzel tubes	0.5 cm^2	0.1 Hz	$	both	Simple measurement of mean velocity.	Not commercially available or commonly used in ecological research, invasive. 1-D.	68	94
Propeller Bucket wheel	0.25 cm^3	1 Hz	$-$$$	both	Easy, sturdy method of measuring mean velocity in free stream.	Unsuitable for most boundary layer conditions, invasive. 1-D.	Nixon, Pygmy, Schildknecht	88, 92, 156

Thermistors	approx. 0.002 cm^3	100 Hz	$	both	Sturdy method for measuring fine scale flows in boundary layer and free stream.	Electronic and thermal drift, build-it-yourself technology, limited range of sensitivity, invasive. 1-D.	102	51, 200
Pitot tubes	0.05 cm^3	10 Hz	$-$$$	lab	Can be fine scale, operates by well understood physical principle.	Not commonly used in ecological research, invasive, require steady, non-accelerating flows. 1-D. Insensitive in slow flow.	Midwest instruments, Dwyer instruments, 197	*46*, 164
Electromagnetic meters	0.5 cm^3	10 Hz	$$-$$$	both	Sturdy, 2-D, directional, no calibrating.	Invasive. Limited boundary layer applications. Poorly defined sample volume.	Marsh-McBirney, InterOcean Systems	*71*, *72*, 139, *212*
Hot film	approx. 0.0004 cm^3	10,000 Hz	$$$$	both	Suitable for many boundary layer measurements, high spatial and temporal resolution. 1 to 3-D.	Electronic and thermal drift, frequent calibration, fragile sensors, invasive, non-directional.	Dantec, TSI	89, 171
Hot film shear probes	0.1 cm^2	1000 Hz	$$$$	both	Suitable for many boundary layer measurements, direct measurement of friction velocity and skin friction.	Electronic and thermal drift, frequent calibration, non-directional, not-suited for oscillatory flow. 1-D.	Dantec, TSI	*61*, *83*, *143*
Acoustic Doppler	approx. 0.25 cm^3	25 Hz	$$$$	both	Sturdy, 3-D, non-invasive, no calibrating, direct measurement of Reynolds stresses, directional.	Strong backscatter from solid bed, may be unsuitable for some near-bed measurements.	SonTek, NorTek, Med-systems Inc.	18, 70, 135
Laser Doppler	approx. 0.001 cm^3	1000 Hz	$$$$	lab	3-D, non-invasive, excellent for boundary layer measurements.	Requires high powered lasers and optics. Unsuitable for field use due to beam alignment and other problems.	Dantec, TSI, Medtronics	181, 209

[a]Depending on flow field and deployment method, some techniques measure flow speed rather than velocity.

[b]Maximum spatial resolution is defined by the volume occupied by measurement device or area of probe exposed to flow.

[c]Maximum frequency responses are for typical deployment.

[d]Costs are estimated in US Dollars: $ = < $500, $$ = $500 to $2,500, $$$ = $2,500 to $5,000, $$$$ = $5,000 to $15,000, $$$$$ = >$15,000.

[e]Suppliers are for commonly used systems in aquatic research. Where citations are given as a supplier, the referenced paper gives details on construction or principles of operation.

[f]References are for field and lab deployments in ecological research.

mean velocity and turbulence in governing an odor-mediated predator-prey interaction by varying velocity and bed roughness in a laboratory flume.

Although field experiments generally offer greater realism than the laboratory, it can be much more difficult to quantify the benthic flow characteristics produced in such experiments. This is particularly true when field experiments are conducted within the roughness layer. One compromise is to deploy appropriately shaped substrata that project well above the surrounding bed or that are raised off the bed entirely. This technique not only reduces the problems stemming from vortex shedding by upstream roughness elements, it can also take advantage of the relatively predictable flow behavior over simple geometrical shapes (91). Indeed, this approach may offer one of the best opportunities for predicting near-bed flow characteristics from flow measurements made at coarser scales.

Even when flow is not the primary focus of study, the experimentalist may need to be concerned with flow effects. For example, field experiments examining species interactions commonly use barriers (metal cages or mesh fences) to prevent the emigration and immigration of organisms. Unfortunately, these barriers will often change the local flow environment as well (149, *196*). Particularly where different kinds of cages are used for different treatments (cages with open sides vs. those with closed sides), it can be difficult to determine whether observed treatment effects are due to species interactions or to the confounding effects of different flows. Thus, investigators need to be aware of flow modifications (even inadvertent ones) to attribute cause and effect correctly in a variety of experimental studies.

ORGANISM-FLOW INTERACTIONS AND ENVIRONMENTAL PROBLEM-SOLVING

Many human activities modify the natural flow regime in streams and rivers (154, 158). For example, flow patterns have been altered by dams, channelization, and urbanization. Given the pervasive effects of flow on the structure and function of stream ecosystems, such activities may have undesirable consequences. Thus, it is imperative that we examine the ways in which human activities alter natural flow patterns, and the consequences of these alterations for the health of stream ecosystems. Two brief examples serve to illustrate the range of environmental problems for which a better understanding of organism-flow interactions is required.

Dams provide many social benefits (e.g., hydropower, flood control, and recreation), but they can also damage streams and their biota (154). In the past, the magnitude and timing of flow releases from hydropower dams has often been dictated primarily by the economics of power generation, with much less concern for the instream flows needed to provide critical ecosystem goods and services (159). Unfortunately, the flow modifications caused by dams can negatively affect sediment transport, resource availability, and species interactions (112, 162). If we had a more complete understanding of the ways that spatial and temporal flow variations affected the structure and function of stream ecosystems, it might be possible to manage water releases from dams in a manner that would achieve a better balance between economic and environmental benefits.

An improved understanding of flow effects would also enhance efforts to restore streams and rivers (138). Many streams have been degraded as the result of channelization, removal of riparian vegetation, water abstraction, pollutant discharges, and stormwater runoff. Restoration efforts often focus on improving physical habitat within the channel, replanting streamside forests, and building stormwater retention structures. Because one of the goals of these restoration practices is to create flow conditions that improve the health of streams, it is clear that such efforts would benefit from a more complete knowledge of physical-biological coupling (141).

SUMMARY AND FUTURE DIRECTIONS

Flow affects the biota of streams in so many different and powerful ways that it should be viewed as the primary environmental factor determining the essential character of these important ecosystems. Although some areas of stream research have already adopted this view (127, 155), we hope that this explicit statement will encourage more ecologists to identify the critical information about flow effects needed to develop improved models of stream ecosystem behavior. We see these information needs as threefold. First, research must focus on the direct and indirect mechanisms by which flow affects organisms. Indeed, creative experimental designs will be needed to disentangle the multiple causal pathways by which such flow effects are manifested. Second, we must rigorously measure the natural flow fields actually experienced by stream organisms. In spite of the large technical challenges involved in quantifying benthic flow characteristics in streams, we are convinced that this approach will yield substantial rewards. Third, research on physical-biological coupling in streams needs to become more scale-explicit. Our review has emphasized the understudied, but critically important, organismal scales at which many flow effects operate. Yet flow varies across a broad range of interconnected space and time scales. Thus, more attention needs to focus on how physical-biological coupling at coarse scales is linked to fine-scale coupling (160). Given the interdisciplinary nature of this entire field, we believe that collaborations between ecologists and experts in fluid mechanics (e.g., hydraulic engineers, fluvial geomorphologists) are likely to be particularly fruitful.

Finally, an improved understanding of flow effects should produce important societal benefits. Streams and rivers are among the most intensively modified ecosystems on our planet, and many human activities alter natural flow regimes. By developing a better understanding of organism-flow interactions, as well as predictive models of such physical-biological coupling, ecologists can help improve our ability to protect and manage these valuable systems.

ACKNOWLEDGMENTS

Our work on organism-flow interactions has been generously supported by grants from the National Science Foundation to DDH (DEB-8918608, DEB-9224214, DEB-9627463), as well as a postdoctoral fellowship to CMF from the Academy of Natural Sciences. During the preparation of this manuscript, DDH was also supported by a Fulbright Senior Scholar Award and a Rosenbaum Fellowship. We received valuable comments on our manuscript from Angie Bednarek, Barry Biggs, Shane Brooks, Marie Bundy, Brian Clark, Dina Fonseca, Ian McEwan, Bjorn Malmqvist, Vladimir Nikora, and LeRoy Poff. DDH also thanks Barry Biggs and the National Institute of Water and Atmospheric Research for facilitating the completion of this paper.

LITERATURE CITED

1. Abelson A, Denny M. 1997. Settlement of marine organisms in flow. *Annu. Rev. Ecol. Syst.* 28:317–39
2. Allan JD. 1995. *Stream Ecology: Structure and Function of Running Waters.* London: Chapman & Hall. 388 pp
3. Allan JD, Feifarek BP. 1989. Distances traveled by drifting mayfly nymphs: factors influencing return to the substrate. *J. N. Am. Benthol. Soc.* 8:322–30
4. Anderson DA, Tannehill JC, Pletcher RH. 1984. *Computational Fluid Mechanics and Heat Transfer.* Washington, DC: Hemisphere
5. Anderson S, Lohrmann A. 1995. Open water test of the SonTek acoustic Doppler velocimeter. *Proc. IEEE Fifth Work. Conf. on Curr. Measurements*, pp. 188–92. St. Petersburg, FL
6. Bell EC, Denny MW. 1994. Quantifying "wave exposure": a simple device for recording maximum velocity and results of its use at several field sites. *J. Exp. Mar. Biol. Ecol.* 181:9–29
7. Bergey EA. 1999. Crevices as refugia for stream diatoms: effect of crevice size on abraded substrates. *Limnol. Oceanogr.* 44: In press
8. Biggs BJF. 1995. The contribution of disturbance, catchment geology and land use to the habitat templet of periphyton in stream ecosystems. *Freshw. Biol.* 33:419–38
9. Biggs BJF. 1996. Hydraulic habitat of plants in streams. *Regulated Rivers Res. Manag.* 12:131–44
10. Biggs BJF. 1996. Patterns in benthic algae of streams. *In Algal Ecology: Freshwater Benthic Ecosystems,* ed. RJ Stevenson, ML Bothwell, RL Lowe, pp. 31–56. New York: Academic
11. Biggs BJF, Duncan MJ, Francoeur SN, Meyer WD. 1997. Physical characterization of microform bed cluster refugia in 12 headwater streams, New Zealand. *New Zealand J. Mar. Freshw. Res.* 31:413–22
12. Biggs BJF, Goring DG, Nikora VI. 1998. Subsidy and stress responses of stream periphyton to gradients in water velocity as a function of community growth form. *J. Phycol.* 34:598–607
13. Biggs BJF, Hickey CW. 1994. Periphyton responses to a hydraulic gradient in a regulated river in New Zealand. *Freshw. Biol.* 32:49–59
14. Biggs BJF, Smith RA, Duncan MJ. 1999. Velocity and sediment disturbance of periphyton in headwater streams: biomass and metabolism. *J. N. Am. Benthol. Soc.* In press

15. Biggs BJF, Thomsen H. 1995. Disturbance of stream periphyton by perturbations in shear stress: time to structural failure and differences in community resistance. *J. Phycol.* 31:233–41
16. Borchardt MA. 1996. Nutrients. In *Algal Ecology: Freshwater Benthic Ecosystems*, ed. RJ Stevenson, ML Bothwell, RL Lowe, pp. 184–228. New York: Academic
17. Bothwell ML, Suzuki KE, Bolin MK, Hardy FJ. 1989. Evidence of dark avoidance by phototrophic periphytic diatoms in lotic systems. *J. Phycol.* 25:85–94
18. Bouckaert FW, Davis JA. 1998. Microflow regimes and the distribution of macroinvertebrates around stream boulders. *Freshw. Biol.* 40:77–86
19. Boulton AJ, Lake PS. 1990. The ecology of two intermittent streams in Victoria, Australia. I. Multivariate analyses of physicochemical features. *Freshw. Biol.* 24:123–41
20. Bradshaw P. 1971. *An Introduction to Turbulence and its Measurement.* New York: Pergamon. 218 pp.
21. Braimah SA. 1987. The influence of water velocity on particle capture by the labral fans of larvae of *Simulium bivittatum* Malloch (Diptera: Simuliidae). *Can. J. Zool.* 65:2395–99
22. Braimah SA. 1987. Pattern of flow around filter-feeding structures of immature *Simulium bivittatum* Malloch (Diptera: Simuliidae). *Can. J. Zool.* 65:514–21
23. Breitburg DL. 1992. Episodic hypoxia in the Chesapeake Bay: interacting effects of recruitment, behavior, and physical disturbance. *Ecol. Monogr.* 62:525–46
24. Breithaupt T, Ayers J. 1996. Visualization and quantitative analysis of biological flow fields using suspended particles. In *Zooplankton: Sensory Ecology and Physiology*, ed. PH Lenz, DK Hartline, JE Purcell, DL Macmillan, pp. 117–29. Amsterdam: Gordon & Breach
25. Breithaupt T, Ayers J. 1998. Visualization and quantification of biological flow fields through video-based digital motion–analysis techniques. *Mar. Fresh. Behav. Physiol.* 31:55–61
26. Butman CA. 1987. Larval settlement of soft-sediment invertebrates: the spatial scales of pattern explained by active habitat selection and the emerging role of hydrodynamic processes. *Oceanogr. Mar. Biol. Annu. Rev.* 25:113–65
27. Butman CA, Frechette M, Geyer WR, Starczak VR. 1994. Flume experiments on food supply to the blue mussel *Mytilus edulis* as a function of boundary-layer flow. *Limnol. Oceanogr.* 39:1755–68
28. Campbell RNB. 1985. Comparison of the drift of live and dead Baetis nymphs in a weakening water current. *Hydrobiologia* 126:229–36
29. Carling PA. 1992. The nature of the fluid boundary layer and the selection of parameters for benthic ecology. *Freshw. Biol.* 28:273–84
30. Carter G, Duncan M, Biggs B. 1998. Numerical hydrodynamic modeling of mountain streams for assessing instream habitat. In *Hydrology, Water Resources and Ecology in Headwaters: Proceedings of Headwater '98*, ed. K Kovar. Merano, Italy, April 1998. 223 pp.
31. Cheer AYL, Koehl MAR. 1987. Paddles and rakes: fluid flow through bristled appendages of small organisms. *J. Theor. Biol.* 129:17–39
32. Ciborowski JJH. 1983. Downstream and lateral transport of nymphs of two mayfly species (Ephemeroptera). *Can. J. Fish. Aquat. Sci.* 40:2025–29
33. Ciborowski JJH, Craig DA. 1989. Factors influencing dispersion of larval black flies (Diptera: Simuliidae): effects of current velocity and food concentration. *Can. J. Fish. Aquat. Sci.* 46:1329–41
34. Ciborowski JJH, Craig DA, Fry KM. 1997. Dissolved organic matter as food for black fly larvae (Diptera: Simuliidae). *J. N. Am. Benthol. Soc.* 16:771–80

35. Clark BD, Hart DD. 1995. Disturbances to flow introduced by feeding black fly larvae. *Am. Zool.* 35:113A
36. Clausen B, Biggs BJF. 1997. Relationships between benthic biota and flow indices in New Zealand rivers. *Freshw. Biol.* 38:327–42
37. Cobb GG, Galloway TD, Flannagan JF. 1992. Effects of discharge and substrate stability on density and species composition of stream insects. *Can. J. Fish. Aquat. Sci.* 49:1788–95
38. Covich AP. 1988. Atyid shrimp in the headwaters of the Luquillo Mountains, Puerto Rico: filter feeding in artificial and natural streams. *Verh. Int. Verein. Limnologie* 23: 2108–13
39. Crosskey RW. 1990. *The Natural History of Blackflies*. New York: John Wiley
40. Cummins KW. 1966. A review of stream ecology with special emphasis on organism—substrate relationships. *Pymatuning Lab. of Ecol. Spec. Publ.* 4:2–51
41. Davis JA. 1986. Boundary layers, flow micro-environments and stream benthos. In *Limnology in Australia*, ed. P DeDeckker, WD Williams, pp. 293–312. Melbourne: CSRIO
42. Davis JA, Barmuta LA. 1989. An ecologically useful classification of mean and near-bed flows in streams and rivers. *Freshw. Biol.* 21:271–82
43. DeNicola DM, McIntire CD. 1991. Effects of hydraulic refuge and irradiance on grazer-periphyton interactions in laboratory streams. *J. N. Am. Benthol. Soc.* 10:251–62
44. Denny MW. 1985. Wave forces on intertidal organisms: a case study. *Limnol. Oceanogr.* 30:1171–87
45. Denny MW. 1988. *Biology and the Mechanics of the Wave-Swept Environment*. Princeton, NJ: Princeton Univ. Press. 329 pp.
46. Denny MW. 1989. A limpet shell shape that reduces drag: laboratory demonstration of a hydrodynamic mechanism and an exploration of its effectiveness in nature. *Can. J. Zool.* 67:2098–2106
47. Denny MW. 1993. *Air and Water: the Biology and Physics of Life's Media*. Princeton, NJ :Princeton Univ. Press. 341 pp.
48. Denny MW, Shibata MF. 1989. Consequences of surf-zone turbulence for settlement and external fertilization. *Am. Nat.* 134:859–89
49. Diamond J. 1986. Overview: laboratory experiments, field experiments, and natural experiments. In *Community Ecology*, ed. J Diamond, T Case, pp. 3–22. New York: Harper Row
50. Dittrich A, Schmedtje U. 1995. Indicating shear stress with FST–hemispheres-effects of stream-bottom topography and water depth. *Freshw. Biol.* 34:107–21
51. Dodds WK, Hutson RE, Eichem AC, Evans MA, Gudder DA, Fritz KM, Gray L. 1996. The relationship of floods, drying, flow and light to primary production and producer biomass in a prairie stream. *Hydrobiologia* 333:151–59
52. Dodson SI, Crowl TA, Peckarsky BL, Kats LB, Covich AP, Culp JM. 1994. Non-visual communication in freshwater benthos: an overview. *J. N. Am. Benthol. Soc.* 13:268–82
53. Doty MS. 1971. Measurement of water movement in reference to benthic algal growth. *Bot. Mar.* 14:32–5
54. Downes BJ, Keough MJ. 1998. Scaling of colonization processes in streams: parallels and lessons from marine hard substrata. *Aust. J. Ecol.* 23:8–26
55. Dunham AE, Beaupre SJ. 1998. Ecological experiments: scale, phenomenology, mechanism, and the illusion of generality. In *Experimental Ecology: Issues and Perspectives,* ed. WJ Resetarits Jr, J Bernardo, pp. 27–40. Oxford, UK: Oxford Univ. Press
56. Dunham AE, Overall KL. 1994. Population responses to environmental change: life history variation, individual-based models, and the population dynamics of

short-lived organisms. *Am. Zool.* 34:382–96
57. Dunson WA, Travis J. 1991. The role of abiotic factors in community organization. *Am. Nat.* 138:1067–91
58. Dusenbery DB. 1992. *Sensory Ecology.* New York: WH Freeman
59. Eckman JE. 1994. Modeling physical-biological coupling in the ocean: the U.S. GLOBEC program. *Deep Sea Res. II* 41:1–5
60. Eckman JE, Okamura B. 1998. A model of particle capture by bryozoans in turbulent flow: significance of colony form. *Am. Nat.* 152:861–80
61. Eckman JE, Savidge WB, Gross TF. 1990. Relationship between duration of cyprid attachment and drag forces associated with detachment of *Balanus amphitrite* cyprids. *Mar. Biol.* 107:111–18
62. Eckman JE, Werner FE, Gross TF. 1994. Modeling some effects of behavior on larval settlement in a turbulent boundary layer. *Deep Sea Res. II* 41:185–208
63. Ehleringer JR, Field CB. 1992. *Scaling Physiological Processes: Leaf to Globe.* New York: Academic
64. Elliott JM. 1971. The distances traveled by drifting invertebrates in a Lake District stream. *Oecologia* 6:191–220
65. Englund G. 1991. Asymmetric resource competition in a filter feeding stream insect (*Hydropsyche siltalai;* Trichoptera). *Freshw. Biol.* 26:425–32
66. Eriksen CH, Lamberti GA, Resh VH. 1996. Aquatic insect respiration. In *An Introduction to the Aquatic Insects of North America*, ed. RW Merritt, KW Cummins, 3:29–40. Dubuque, IA: Kendall/Hunt
67. Erman DC, Chouteau WC. 1979. Fine particulate organic carbon output from fens and its effect on benthic macroinvertebrates. *Oikos* 32:409–15
68. Everest JM. 1967. Midget Bentzel current speed tube for ecological investigations. *Limnol. Oceanogr.* 12:179–80
69. Fausch KD, White RJ. 1981. Competition between brook trout (*Salvelinus fontinalis*) and brown trout (*Salmo trutta*) for positions in a Michigan stream. *Can. J. Fish. Aquat. Sci.* 38:1220–27
70. Finelli CM, Hart DD, Fonseca DM. 1999. Evaluating the spatial resolution of an acoustic Doppler velocimeter and the consequences for measuring near-bed flows. *Limnol. Oceanogr.* 44: In press
71. Finelli CM, Pentcheff ND, Zimmer-Faust RK, Wethey DS. 1999. Odor transport in turbulent flows: constraints on animal navigation. *Limnol. Oceanogr.* 44:1056–71
72. Finelli CM, Pentcheff ND, Zimmer-Faust RK, Wethey DS. 1999. Physical constraints on ecological processes: a field test of the effects of prey odor and flow speed on foraging. *Ecology.* In press
73. Fonseca DM. 1996. *Fluid mediated dispersal: effects on the foraging behavior and distribution of stream insects.* PhD Thesis. Univ. Penn. Philadelphia. 163 pp.
74. Fonseca DM. 1999. Fluid-mediated dispersal in streams: models of settlement from the drift. *Oecologia.* In press
75. Fonseca DM, Hart DD. 1996. Density-dependent dispersal of black fly neonates is mediated by flow. *Oikos* 75:49–56
76. Francoeur SN, Biggs BJF, Lowe RL. 1998. Microform bed clusters as refugia for periphyton in a flood-prone headwater stream. *New Zealand J. Mar. Freshw. Res.* 32:363–74
77. Frutiger A. 1998. Walking on suckers-new insights into the locomotory behavior of larval net-winged midges (Diptera: Blephariceridae). *J. N. Am. Benthol. Soc.* 17:104–20
78. Frutiger A, Schib JL. 1993. Limitations of FST hemispheres in lotic benthos research. *Freshw. Biol.* 30:463–74
79. Giller PS, Sangpradub SN, Twomey H. 1991. Catastrophic flooding and macroinvertebrate community structure. *Int. Ver. Theor. Limnol. Verh.* 24:1724–29
80. Gordon ND, McMahon TA, Finlayson BL.

1992. *Stream Hydrology. An Introduction for Ecologists*. Chichester, UK: John Wiley
81. Gross TF, Werner FE, Eckman JE. 1992. Numerical modeling of larval settlement in turbulent bottom boundary layers. *J. Mar. Res.* 50:611–42
82. Gundersen JK, Jorgensen BB. 1990. Microstructure of diffusive boundary layers and the oxygen uptake of the sea floor. *Nature* 345:604–07
83. Gust G. 1982. Tools for oceanic small-scale, high frequency flows: metal-clad hot wires. *J. Geophys. Res.* 87:447–55
84. Hansen RA, Hart DD, Merz RA. 1991. Flow mediates predator-prey interactions between triclad flatworms and larval black flies. *Oikos* 60:187–96
85. Hansson L, Fahrig L, Merriam G. 1995. *Mosaic Landscapes and Ecological Processes*. New York: Chapman & Hall
86. Hart DD. 1983. The importance of competitive interactions within populations and communities. In *Stream Ecology: Application and Testing of General Ecological Theory*, ed. JR Barnes, GW Minshall, pp. 99–136. New York: Plenum
87. Hart DD. 1986. The adaptive significance of territoriality in filter-feeding larval blackflies (Diptera: Simuliidae). *Oikos* 46:88–92
88. Hart DD. 1992. Community organization in streams: the importance of species interactions, physical factors, and chance. *Oecologia* 91:220–28
89. Hart DD, Clark BD, Jasentuliyana A. 1996. Fine-scale field measurement of benthic flow environments inhabited by stream invertebrates. *Limnol. Oceanogr.* 41:297–308
90. Hart DD, Fonseca DM. 1997. Seeking shelter from the storm: behavioral responses of stream invertebrates to natural and experimental floods. *Bull. North Am. Benthol. Soc.* 14:102
91. Hart DD, Fonseca DM, Finelli CM, Farouk B. 1999. A multi-scale analysis of consumer responses to spatial resource variation in streams. *Ecology*. Submitted
92. Hart DD, Merz RA. 1998. Predator-prey interactions in a benthic stream community: a field test of flow-mediated refuges. *Oecologia* 114:263–73
93. Hart DD, Merz RA, Genovese SJ, Clark BD. 1991. Feeding postures of suspension-feeding larval black flies: the conflicting demands of drag and food acquisition. *Oecologia* 85:457–63
94. Hemphill N. 1988. Competition between two stream dwelling filter-feeders, *Hydropsyche oslari* and *Simulium virgatum*. *Oecologia* 77:73–80
95. Hemphill N, Cooper SD. 1983. The effect of physical disturbance on the relative abundances of two filter-feeding insects in a small stream. *Oecologia* 58: 378–82
96. Hynes HBN. 1970. *The Ecology of Running Waters*. Liverpool, UK: Univ. Liverpool Press
97. Jowett IG, Biggs BJF. 1997. Flood and velocity effects on periphyton and silt accumulation in two New Zealand rivers. New Zealand *J. Mar. Freshw. Res.* 25:187–99
98. Kareiva P. 1990. Population dynamics in spatially complex environments: theory and data. *Philos. Trans. R. Soc. Lond. B.* 330:175–90
99. Koehl MAR. 1996. When does morphology matter? *Annu. Rev. Ecol. Syst.* 27: 501–42
100. Koehl MAR, Alberte RS. 1988. Flow, flapping and photosynthesis of *Nereocystis luetkeana*: a functional comparison of undulate and flat blade morphologies. *Mar. Biol.* 99:435–44
101. Kohler SL. 1985. Identification of stream drift mechanisms: an experimental and observational approach. *Ecology* 66:1749–61
102. LaBarbera M, Vogel S. 1976. An inexpensive thermistor flowmeter for aquatic biology. *Limnol. Oceanogr.* 21:750–56

103. Lacoursiere JO. 1992. A laboratory study of fluid flow and microhabitat selection by larvae of the black fly *Simulium vittatum* (Diptera: Simuliidae). *Can. J. Zool.* 70:582–96
104. Lacoursiere JO, Craig DA. 1993. Fluid transmission and filtration efficiency of the labral fans of black fly larvae (Diptera: Simuliidae): hydrodynamic, morphological, and behavioural aspects. *Can. J. Zool.* 71:148–62
105. Lancaster J, Belyea LR. 1997. Nested hierarchies and scale-dependence of mechanisms of flow refugium use. *J. North Am. Benthol. Soc.* 16:221–38
106. Lancaster J, Hildrew AG. 1993. Flow refugia and microdistribution of lotic macroinvertebrates. *J. N. Am. Benthol. Soc.* 12:385–93
107. Lancaster J, Hildrew AG, Gjerlov C. 1996. Invertebrate drift and longitudinal transport processes in streams. *Can. J. Fish. Aquat. Sci.* 53:572–82
108. Lauritsen DD. 1986. Filter-feeding in *Corbicula fluminea* and its effect on seston removal. *J. N. Am. Benthol. Soc.* 5: 165–72
109. Laws EM, Livesey JL. 1978. Flow through screens. *Annu. Rev. Fluid Mech.* 10:247–66
110. Leopold LB, Wolman MG, Miller JP. 1964. *Fluvial Processes in Geomorphology.* San Francisco: WH Freeman
111. Lewin R. 1986. Supply-side ecology. *Science* 234:25–27
112. Ligon FK, Dietrich WE, Trush WJ. 1995. Downstream ecological effects of dams. *BioScience* 45:183–92
113. Liu H, Ellington CP, Kawachi K, Van den Berg C, Willmott AP. 1998. A computational fluid dynamic study of hawkmoth hovering. *J. Exp. Biol.* 201:461–77
114. Louden C, Alstad DN. 1990. Theoretical mechanics of particle capture: predictions for hydropsychid caddisfly distributional ecology. *Am. Nat.* 135:360–81
115. Malmqvist B, Sackman G. 1996. Changing risk of predation for a filter-feeding insect along a current velocity gradient. *Oecologia* 108:450–8
116. Malmqvist B, Sjöström P. 1987. Stream drift as a consequence of predator disturbance: field and laboratory experiments. *Oecologia* 74:396–403
117. Matczak TZ, Mackay RJ. 1990. Territoriality in filter-feeding caddisfly larvae: laboratory experiments. *J. N. Am. Benthol. Soc.* 9:26–34
118. Maude SH, Williams DD. 1983. Behavior of crayfish in water currents: hydrodynamics of eight species with reference to their distribution patterns in southern Ontario. *Can. J. Fish. Aquat. Sci.* 40:68–77
119. McLay CL. 1970. A theory concerning the distance traveled by animals entering the drift of a stream. *J. Fish. Res. Board. Can.* 27:359–70
120. McNair JN, Newbold JD, Hart DD. 1997. Turbulent transport of suspended particles and dispersing benthic organisms: How long to hit bottom? *J. Theor. Biol.* 188:29–52
121. Miller AR, Lowe RL, Rotenberry JT. 1987. Succession of diatom communities on sand grains. *J. Ecol.* 75:693–709
122. Minshall GW. 1984. Aquatic insect-substratum relationships. In *The Ecology of Aquatic Insects*, ed. VH Resh, DM Rosenberg, pp. 358–400. New York: Praeger Sci.
123. Moore PA, Weissburg MJ, Parrish JM, Zimmer-Faust RK, Gerhardt GA. 1994. Spatial distribution of odors in simulated benthic boundary layer flows. *J. Chem. Ecol.* 20:255–79
124. Morin A, Back C, Chalifour A, Boisvert J, Peters RH. 1988. Empirical models predicting ingestion rates of black fly larvae. *Can. J. Fish. Aquat. Sci.* 45:1711–19
125. Morin A, Back C, Chalifour A, Boisvert J, Peters RH. 1988. Effects of blackfly ingestion and assimilation on seston trans-

port in a Quebec lake outlet. *Can. J. Fish. Aquat. Sci.* 45:705–14
126. Morris HM. 1955. Flow in rough conduits. *Trans. ASCE* 120:373–98
127. Mulholland PJ. 1996. Role of nutrient cycling in streams. In *Algal Ecology: Freshwater Benthic Ecosystems*, ed. RJ Stevenson, ML Bothwell, RL Lowe, pp. 609–639. New York: Academic
128. Muller K. 1974. Stream drift as a chronobiological phenomenon in running water ecosystems. *Annu. Rev. Ecol. Syst.* 5:309–23
129. Mullineaux LS, Butman CA. 1991. Initial contact, exploration and attachment of barnacle (*Balanus amphitrite*) cyprids settling in flow. *Mar. Biol.* 110:93–103
130. Mushenheim DK. 1987. The dynamics of near-bed seston flux and suspension-feeding benthos. *J. Mar. Res.* 45:473–96
131. Muus BJ. 1968. A field method for measuring "exposure" by means of plaster balls, a preliminary account. *Sarsia* 34: 61–8
132. Naiman RJ, Magnuson JT, McKnight DM, Stanford JA. 1995. *The Freshwater Imperative: A Research Agenda.* Washington, DC: Island
133. Nelson JM, Shreve RL, McLean SR, Drake TG. 1995. Role of near-bed turbulence structure in bed load transport and bed form mechanics. *Water Resources Res.* 31:2071–86
134. Nikora V, Goring D, McEwan I, Griffiths G. 1999. Spatially-averaged open-channel flow over rough beds. *J. Hydraul. Eng.* Submitted
135. Nikora VI, Suren AM, Brown SLR, Biggs BJF. 1998. The effects of the moss *Fissidens rigidulus* (Fissidentaceae: Musci) on near-bed flow structure in an experimental cobble bed flume. *Limnol. Oceanogr.* 43:1321–31
136. Nowell ARM, Jumars PA. 1984. Flow environments of aquatic benthos. *Annu. Rev. Ecol. Syst.* 15:303–28
137. Nowell ARM, Jumars PA. 1987. Flumes: theoretical and experimental considerations for simulation of benthic environments. *Oceanogr. Mar. Biol. Annu. Rev.* 25:91–112
138. Osborne LL, Bayley PB, Higler LWG, Statzner B, Triska F, Moth Iverson T. 1993. Restoration of lowland streams: an introduction. *Freshw. Biol.* 29:187–94
139. Palmer MA. 1992. Incorporating lotic meiofauna into our understanding of faunal transport processes. *Limnol. Oceanogr.* 37:329–41
140. Palmer MA, Allan JD, Butman CA. 1996. Dispersal as a regional process affecting the local dynamics of marine and stream benthic invertebrates. *Trends Ecol. Evol.* 11:322–26
141. Palmer MA, Ambrose RF, Poff NL. 1997. Ecological theory and community restoration ecology. *Rest. Ecol.* 5:291–300
142. Palmer MA, Bely AE, Berg KE. 1992. Responses of invertebrates to lotic disturbances: a test of the hyporheic refuge hypothesis. *Oecologia* 89:182–94
143. Palmer MA, Gust G. 1985. Dispersal of meiofauna in a turbulent tidal creek. *J. Mar. Res.* 43:179–210
144. Palmer MA, Poff NL. 1997. The influence of environmental heterogeneity on patterns and processes in streams. *J. N. Am. Benthol. Soc.* 16:169–73
145. Patterson MR. 1992. A chemical engineering view of cnidarian symbioses. *Am. Zool.* 32:566–82
146. Pawlik JR. 1992. Chemical ecology of the settlement of benthic marine invertebrates. *Oceanogr. Mar. Biol. Annu. Rev.* 30:273–335
147. Peckarsky BL. 1980. Predator-prey interactions between stoneflies and mayflies: behavioral observations. *Ecology* 61:932–43
148. Peckarsky BL, Horn SC, Statzner B. 1990. Stonefly predation along a hydraulic gradient: a field test of the harsh-

benign hypothesis. *Freshw. Biol.* 24:181–91

149. Peckarsky BL, Penton MA. 1990. Effects of enclosures on stream microhabitat and invertebrate community structure. *J. N. Am. Benthol. Soc.* 9:249–61
150. Peterson C, Stevenson RJ. 1989. Substratum conditioning and diatom colonization in different current regimes. *J. Phycol.* 25:790–93
151. Petticrew EL, Kalff J. 1991. Calibration of a gypsum source for freshwater flow measurements. *Can. J. Fish. Aquat. Sci.* 48:1244–49
152. Poff NL. 1997. Landscape filters and species traits: towards mechanistic understanding and prediction in stream ecology. *J. N. Am. Benthol. Soc.* 16:391–409
153. Poff NL, Allan JD. 1995. Functional organization of stream fish assemblages in relation to hydrologic variability. *Ecology* 76:606–27
154. Poff NL, Allan JD, Bain MB, Karr JR, Prestegaard KL, et al. 1997. The natural flow regime. *BioScience* 47:769–84
155. Poff NL, Ward JV. 1989. Implications of streamflow variability and predictability for lotic community structure: a regional analysis of streamflow patterns. *Can. J. Fish. Aquat. Sci.* 46:1805–18
156. Poff NL, Ward JV. 1992. Heterogeneous currents and algal resources mediate in situ foraging activity of a mobile stream grazer. *Oikos* 65:465–78
157. Poff NL, Ward JV. 1995. Herbivory under different flow regimes: a field experiment and test of a model with a benthic stream insect. *Oikos* 71:179–88
158. Postel S. 1997. *Last Oasis: Facing Water Scarcity.* New York: Norton. 2nd ed.
159. Postel S, Carpenter S. 1997. Freshwater ecosystem services. In *Nature's Services: Societal Dependence on Natural Ecosystems*, ed. GC Daily, pp. 195–214 Covelo, CA: Island
160. Powell TM. 1989. Physical and biological scales of variability in lakes, estuaries, and coastal oceans. In *Ecological Theory*, ed. J Roughgarden, RM May, SA Levin, pp. 157–77. Princeton, NJ: Princeton Univ. Press
161. Power ME, Stewart AJ. 1987. Disturbance and recovery of an algal assemblage following flooding in an Oklahoma river. *Am. Midl. Nat.* 117:333–45
162. Power ME, Sun A, Parker M, Dietrich WE, Wootton JT. 1995. Hydraulic food-chain models: an approach to the study of food-web dynamics in large rivers. *BioScience* 45:159–67
163. Raupach MR, Antonia RA, Rajagopaian S. 1991. Rough-wall turbulent boundary layers. *Appl. Mech. Rev.* 44:1–25
164. Reiter MA, Carlson RE. 1986. Current velocity in streams and the composition of benthic algal mats. *Can. J. Fish. Aquat. Sci.* 43:1156–62
165. Resetarits WJ, Bernardo J. 1999. *Experimental Ecology: Issues and Perspectives.* Oxford, UK: Oxford Univ. Press
166. Reynolds AJ. 1974. *Turbulent Flows in Engineering.* New York: John Wiley
167. Reynolds CS, Carling PA, Beven KJ. 1991. Flow in river channels: new insights into hydraulic retention. *Arch. Hydrobiol.* 121:171–79
168. Robert A. 1993. Bed configuration and microscale processes in alluvial channels. *Prog. Phys. Geogr.* 17:123–36
169. Roughgarden J, Gaines SD, Pacala SW. 1987. Supply side ecology: the role of physical transport processes. In *Organization of Communities Past and Present*, ed. JHR Gee, PS Giller, pp. 491–518. Oxford: Blackwell
170. Rubenstein DI, Koehl MAR. 1977. The mechanisms of filter feeding: some theoretical considerations. *Am. Nat.* 111:981–94
171. Sand-Jensen K, Mebus JR. 1996. Fine-scale patterns of water velocity within macrophyte patches in streams. *Oikos* 76:169–80
172. Scarsbrook MR, Townsend CR. 1993.

Stream community structure in relation to spatial and temporal variation: a habitat templet study of two contrasting New Zealand streams. *Freshw. Biol.* 29:395–410
173. Shimeta J, Jumars PA. 1991. Physical mechanisms and rates of particle capture by suspension-feeders. *Oceanogr. Mar. Biol. Annu. Rev.* 29:191–257
174. Siegfreid CA, Knight AW. 1977. The effects of washout in a Sierra foothill stream. *Am. Midl. Nat.* 98:200–7
175. Silvester NR, Sleigh MA. 1985. The forces on microorganisms at surfaces in flowing water. *Freshw. Biol.* 15:433–48
176. Smith IR. 1975. *Turbulence in lakes and rivers. Freshw. Biol. Assoc. Pub. No. 29*
177. Snelgrove PVR, Butman CA. 1994. Animal-sediment relationships revisited: cause versus effect. *Oceanogr. Mar. Biol. Annu. Rev.* 32:111–77
178. Stamhuis EJ, Videler JJ. 1995. Quantitative flow analysis around aquatic animals using laser sheet particle image velocimetry. *J. Exp. Biol.* 198:283–94
179. Statzner B, Arens MF, Champagne JY, Morel R. 1998. Lotic organisms as engineers of hydraulic habitat conditions. *Bull. N. Am. Benthol. Soc.* 15:92
180. Statzner B, Gore JA, Resh VH. 1988. Hydraulic stream ecology: observed patterns and potential applications. *J. N. Am. Benthol. Soc.* 7:307–60
181. Statzner B, Holm TF. 1989. Morphological adaptation of shape to flow: microcurrents around lotic macroinvertebrates with known Reynolds numbers at quasi-natural flow conditions. *Oecologia* 78:145–57
182. Statzner B, Muller R. 1989. Standard hemispheres as indicators of flow characteristics in lotic benthos research. *Freshw. Biol.* 21:445–59
183. Stevenson RJ. 1983. Effects of current and conditions simulating autogenically changing microhabitats on benthic diatom immigration. *Ecology* 64:1514–24
184. Stevenson RJ. 1990. Benthic algal community dynamics in a stream during and after a spate. *J. N. Am. Benthol. Soc.* 9:277–88
185. Stevenson RJ. 1996. The stimulation and drag of current. In *Algal Ecology: Freshwater Benthic Ecosystems*, ed. RJ Stevenson, ML Bothwell, RL Lowe, pp. 321–41. New York: Academic
186. Stevenson RJ, Glover R. 1993. Effects of algal density and current on ion transport through periphyton communities. *Limnol. Oceanogr.* 38:1276–81
187. Stoeckel JA, Schneider DW, Soeken LA, Blodgett KD, Sparks RE. 1997. Larval dynamics of a riverine metapopulation: implications for zebra mussel recruitment, dispersal, and control in a large–river system. *J. N. Am. Benthol. Soc.* 16: 586–601
188. Tennekes H, Lumley JL. 1972. *A First Course in Turbulence.* Cambridge, MA: MIT Press
189. Thomas FIM, Atkinson MJ. 1997. Ammonium uptake by coral reefs: effects of water velocity and surface roughness on mass transfer. *Limnol. Oceanogr.* 42:81–88
190. Tilman D. 1982. *Resource Competition and Community Structure.* Princeton, NJ: Princeton Univ. Press
191. Townsend CR. 1989. The patch dynamics concept of stream community ecology. *J. N. Am. Benthol. Soc.* 8:36–50
192. Townsend CR, Hildrew AG. 1976. Field experiments on the drifting, colonization, and continuous redistribution of stream benthos. *J. Anim. Ecol.* 45:759–72
193. Townsend CR, Scarsbrook MR, Doledec S. 1997. Quantifying disturbance in streams: alternative measures of disturbance in relationship to macroinvertebrate species traits and species richness. *J. N. Am. Benthol. Soc.* 16:531–44
194. Tremblay MJ, Loder JW, Werner FE, Naime CE, Page FH, Sinclair MM. 1994. Drift of sea scallop larvae *Placopecten*

magellanicus on Georges Bank: a model study of the roles of mean advection, larval behavior and larval origin. *Deep Sea Res. II.* 41:7–49

195. Vallett HM, Fisher SG, Grimm NB, Camill P. 1994. Vertical hydrologic exchange and ecological stability of a desert stream ecosystem. *Ecology* 75:548–60
196. Virnstein RW. 1977. Predator caging experiments in soft sediments: caution advised. In *Estuarine Interactions*, ed. ML Wiley, pp. 261–73. New York: Academic
197. Vogel S. 1981. *Life in Moving Fluids.* Princeton, NJ: Princeton Univ. Press. 1st ed.
198. Vogel S. 1994. *Life in Moving Fluids.* Princeton, NJ: Princeton Univ. Press. 2nd ed.
199. Wallace JB, Merritt RW. 1980. Filter-feeding ecology of aquatic insects. *Annu. Rev. Entomol.* 25:103–32
200. Way CM, Burky AJ, Bingham CR, Miller AC. 1995. Substrate roughness, velocity refuges, and macroinvertebrate abundance on artificial substrates in the lower Mississippi River. *J. N. Am. Benthol. Soc.* 14:510–18
201. Webb PW. 1989. Station-holding by three species of benthic fishes. *J. Exp. Biol.* 145:303–20
202. Weldon CW, Slauson WL. 1986. The intensity of competition versus its importance: an overlooked distinction and some implications. *Q. Rev. Biol.* 61:23–44
203. Weissburg MW, Zimmer-Faust RK. 1993. Life and death in moving fluids: hydrodynamic effects on chemosensory-mediated predation. *Ecology* 74:1428–43
204. Wildish D, Kristmanson D. 1997. *Benthic Suspension Feeders and Flow.* Cambridge: Cambridge Univ. Press. 409 pp.
205. Wiley MJ, Kohler SL. 1980. Positioning changes of mayfly nymphs due to behavioral regulation of oxygen consumption. *Can. J. Zool.* 58:618–22
206. Williams DD. 1996. Environmental constraints in temporary fresh waters and their consequences for insect fauna. *J. N. Am. Benthol. Soc.* 15:634–50
207. Wootton JT, Parker MS, Power ME. 1996. Effects of disturbance on river food webs. *Science* 273:1558–61
208. Wotton RS. 1992. Feeding by blackfly larvae (Diptera: Simuliidae) forming dense aggregations at lake outlets. *Freshw. Biol.* 27:139–49
209. Young WJ. 1992. Clarification of the criteria used to identify near-bed flow regimes. *Freshw. Biol.* 28:383–91
210. Zhang Y, Malmqvist B. 1997. Phenotypic plasticity in a suspension-feeding insect, *Simulium lundstromi* (Diptera: Simuliidae), in response to current velocity. *Oikos* 78:503–10
211. Zhang Y, Malmqvist B, Englund G. 1998. Ecological processes affecting community structure of blackfly larvae in regulated and unregulated rivers: a regional study. *J. Applied. Ecol.* 35:673–86
212. Zimmer-Faust RK, Finelli CM, Pentcheff ND, Wethey DS. 1995. Odor plumes and animal navigation in turbulent water flow: a field study. *Biol. Bull.* 188:111 16

ASTROBIOLOGY: Exploring the Origins, Evolution, and Distribution of Life in the Universe

D. J. Des Marais
Ames Research Center, NASA, Moffett Field, California 94035-1000; e-mail: ddesmarais@mail.arc.nasa.gov

M. R. Walter
School of Earth Sciences, Macquarie University, North Ryde, N.S.W., Australia; e-mail: gc_walter@hope.ocs.mq.edu.au

Key Words microorganisms, Precambrian, biospheres, solar system, planets, origin of life

■ **Abstract** The search for the origins of life and its presence beyond Earth is strengthened by new technology and by evidence that life tolerates extreme conditions and that planets are widespread. Astrobiologists learn how planets develop and maintain habitable conditions. They combine biological and information sciences to decipher the origins of life. They examine how biota, particularly microorganisms, evolve, at scales from the molecular to the biosphere level, including interactions with long-term planetary changes. Astrobiologists learn how to recognize the morphological, chemical, and spectroscopic signatures of life in order to explore both extraterrestrial samples and electromagnetic spectra reflected from extrasolar planets.

INTRODUCTION

Humanity has long been fascinated with the possibility that life exists beyond Earth. Recently, the discipline of astrobiology has been created to study the origin, evolution, and distribution of life in the Universe, including the prospects for Earth-based life to move beyond its planet of origin. Astrobiology is broadly interdisciplinary because, in its quest to understand the origins and evolution of habitable planets and biospheres, its agenda integrates investigations across the biological, chemical, geological, and space sciences.

So what is novel about astrobiology? What does it offer beyond, for example, earlier programs in exobiology that addressed the origin and early evolution of life and searched for extraterrestrial life? Perhaps the major differences between astrobiology and previous efforts are a more concerted attempt to incorporate related disciplines into a more tightly integrated program, and the substantial involvement of new communication and information sciences and technologies to achieve this integration. The motivation to move forward now has been triggered by the remarkable array of key recent discoveries. For example, a resonance is developing between studies of the origin of life, molecular biology, and bioengineering. Laboratory demonstrations of self-replication and evolution of biological macromolecules (48), now an integral part of the search for new pharmaceutical products, are being employed to explore the origins of molecular replication. Computational simulations of membrane function that have provided insights into mechanisms of drug activity (76) also explore the roles played by membranes in the earliest cells (protocells). The rapidly developing field of information sciences includes, among its many applications, a mandate to define the properties that emerge from complex systems having many interacting components. Such systems might include regional electrical power grids, spacecraft, or, for that matter, ecosystems and the molecular machinery of cells. Life is an information-rich entity, and its origins are a prime candidate for study.

Recent Discoveries Create a Mandate

New tools for molecular analyses now permit exploration of the enormous diversity of microbial life, and this search has expanded the known range of environments in which microorganisms can thrive. These discoveries also expand the range of habitable conditions within which we might expect to find life beyond the Earth. Life may have started soon after Earth became habitable (82, 84, 70), which encourages the thought that it could also have arisen rapidly elsewhere. Liquid water may have persisted for perhaps hundreds of millions of years in other planets of our own solar system (Mars: 15, Europa? 77). Earth's subsurface harbors habitable environments and an extensive biota (30) based upon nonphotosynthetic sources of energy and perhaps persisting to depths where it encounters its thermal limit (33). Accordingly, Mars (12) and Europa (77) may still have subsurface habitable conditions. Because habitable environments might have developed beyond the Earth and persisted sufficiently long for life to begin, we are encouraged to search for those environments as a prelude to our search for other biospheres.

Astrobiology's mandate arises from discoveries of potential habitats beyond our own solar system. Models for solar system formation indicate that planets probably accompany billions of other stars (99). Such models have been supported by recent discoveries of extrasolar planets (65a). While these newly found planets may not be habitable, they indicate that a broad diversity exists among planetary systems. New methods for finding smaller, more habitable planets are being developed (7), and so we shall soon chart the frequency of habitable planets within that diversity of solar systems.

Astrobiology will lead inevitably to a deeper understanding of life itself. Current definitions of life focus upon its key properties (71) and tend to reflect the particular perspectives of their authors. For example, molecular and cell biologists cite the ability of cells to harvest energy and metabolize, to replicate and to evolve. Those with an ecological perspective emphasize that ecosystems are fundamental units of life. Advocates of the Gaia Hypothesis mandate that nothing less than the entire biosphere-Earth system is the most fundamental unit of life (63). Our search for life beyond Earth clearly compels a scrutiny of our definition of life, as that definition establishes the criteria by which life might be identified elsewhere. If extraterrestrial life is indeed discovered, the diversity of known life-forms will increase. For example, life as we know it utilizes only a small fraction of the possible organic compounds. It seems likely that alien life forms will have explored alternative possibilities.

Key Components of the Discipline

Operationally, astrobiology integrates key research disciplines into a program that combines technology development, remote observation (space missions), model building, and the extensive involvement of educators and the public. This agenda addresses the following three canonical questions: How does life begin and develop? Does life exist elsewhere in the Universe? What is the future of life on Earth and in space? This review emphasizes the first two questions, although the three are interdependent. For example, an improved understanding of the morphological, chemical, and isotopic traces of early life on Earth (first question) prepares us to analyze samples returned from Mars and elsewhere (second question). Studies of habitable environments and the potential for life beyond Earth (second question) help us chart our future (third question). The search for past and present habitable environments on Mars will locate the resources necessary to sustain future human exploration. The size and scope of this agenda necessarily make it an international endeavor. The requisite scientific and technical expertise spans international borders, and multigovernmental support will be essential for the long-term exploration of the full range of extreme environments on Earth and beyond.

Astrobiology recognizes that students and the public are essential ongoing participants. The search for the origins and distribution of life in the Universe has great religious and philosophical significance. The universal human imperative to explore will both strengthen and shape the effort. Our stewardship of the planetary environments we explore will reflect the ethics of our global society. Public health concerns will arise during, for example, the return of samples from Mars, mandating a plan of exploration and research that reflects public input.

Astrobiology is fundamentally important for education. It motivates students to acquire scientific and technical skills that create a broad array of economically important careers. Astrobiology strengthens linkages between science, technology, and the humanities, creating an integrated view of our world that will be beneficial for helping to define the roles that future generations will play as stewards of our global environment and its resources.

Life in the Universe: The Basic Issues

Astrobiology depends critically upon our understanding of the processes that control the origin and evolution of both habitable environments and life. The range of relevant issues are succinctly highlighted by the Drake Equation (80), which assesses the distribution of intelligent, communicative civilizations in our galaxy, as follows:

$$N_C = R_S f_P n f_L f_I f_C L, \quad 1.$$

where N_C is the number of civilizations in our galaxy with whom we might communicate, R_S is the rate of formation of stars in the galaxy, f_P is the fraction of stars that have planetary systems, n is the average number of habitable planets within a solar system, f_L is the fraction of habitable planets on which life arises, f_I is the fraction of biospheres that developed intelligent species, f_C is the fraction of intelligent species that are interested in communicating with other civilizations, and L is the average lifetime of a civilization.

If this assessment is broadened to include all biospheres in our galaxy, a "Biosphere Equation" might take the following form:

$$N_B = R_S f_P n f_L L_B, \quad 2.$$

where N_B is the number of biospheres in our galaxy, and L_B is the average lifetime of a biosphere. Clearly all of these terms are important for a comprehensive assessment of life's cosmic distribution; however, astrobiological research focuses particularly upon the latter terms of these equations (from the term "n" on to the right), as these directly address planetary habitability and life. The sections that follow summarize how our current understanding of these factors might shape the research agenda.

RESEARCH AND EXPLORATION

Formation and Distribution of Habitable Planets

Abundance of Planetary Systems

The view that planetary systems are abundant has been strengthened recently by discoveries of several protoplanetary nebular disks in star-forming regions (11). Perhaps one quarter to one half of very young stars have disks similar to the one from which our own solar system formed. The term "f_P" in the above equations might possibly be as large as 0.2 to 0.5. Extrasolar planets are now being discovered at a rapid rate (see 65a), although these are not Earth-like and habitable. This is because the currently most successful detection method identifies planet-induced oscillatory shifts in the star's spectra and thus is most sensitive to closely orbiting planets that are about Jupiter-size or larger, probably gas-rich, and thus not habitable. Although the apparent multitude of solar nebulae indicates that Earth-like planets might be widespread, the most immediate benefit of their discovery is that we can now study diverse examples of the planet-forming process. Someday we will understand in detail the context for the formation of planets, including Earth-like ones, and thus quantify the "f_P" and "n" terms in the above Drake and Biosphere Equations. Such an endeavor is a key focus of NASA's new Origins Program (7).

Controls on the Habitability of Planetary Surfaces

A rigorous search for life should include a strategy for locating, describing, and modeling habitable planetary environments and their evolution. The life-enabling commodities that most sharply distinguish Earth from other planets in our solar system are liquid water and the climates that maintain it (66). We have not yet located *any* liquid water beyond Earth, although it is anticipated. Mars and Europa are exploration targets in part because liquid water apparently once existed at or near their surfaces. A planetary habitat also must provide biologically useful sources of energy (46, 95), chemical nutrients, and a degree of environmental stability that allows life to arise and evolve at a rate sufficient to survive environmental changes that do occur.

Planetary habitability depends upon several factors. One is the amount of sunlight received by a planet; this has been used to delineate the habitable zone (43), namely, that region around a star within which planetary surfaces are habitable. Because the luminosity of main-sequence stars slowly increases, such a habitable zone will uniformly migrate away from the star. The region that sustains a habitable zone continuously during the history of the solar system is called the continuously habitable zone (40).

Planetary processes also influence the habitable zone. Planetary surface temperatures reflect the energy balance of the atmosphere, which is influenced strongly by greenhouse gases (40, 55). Water vapor achieves most of the greenhouse warming on Earth today. However, CO_2 also contributes substantial greenhouse warming, and it also has played a long-term role in stabilizing Earth's climate and surface temperature (50). Earth's interior thermal activity (volcanism and tectonics) adds CO_2 to the oceans and atmosphere, and CO_2 is removed via aqueous rock weathering, water transport, and burial in sediments as organic matter and carbonates. Climate regulation is achieved because the processes that remove CO_2 are temperature dependent. Temperature increases are offset by greater rates of removal of this greenhouse gas from the atmosphere. As temperatures decline, slower removal rates allow Earth's thermal sources of CO_2 to increase atmospheric levels of CO_2 and thus enhance greenhouse warming.

Greenhouse gases have strongly influenced the extent and the evolution of the habitable zone (55). Because the ancient sun was less luminous, Venus might have once enjoyed a habitable surface environment. However, CO_2-water–mediated climate regulation eventually failed on Venus because increased solar heating allowed water vapor to invade the Venusian stratosphere and then escape to space (54). With the loss of aqueous processes for removing CO_2, "runaway greenhouse" conditions forever trapped Venus in an uninhabitably hot environment. Mars also illustrates the role of planetary processes. Mars' surface environment is now uninhabitable (19), owing principally to low temperatures and the absence of liquid water. Because the ancient sun was less luminous, it would follow that Mars' early surface environment was even colder than it is today. However, evidence of elevated erosion rates, rivers, and standing water on early Mars indicates that its early environment was apparently wetter and at least somewhat warmer (15, 39). This is possible if, early in Mars' history, thermal processes added CO_2 more rapidly to the surface environment than they do today. As the interiors of rocky planets cooled because radioactivity-based heating declined, volcanism and other thermal processes delivered less CO_2 to the surface environment (23). The surfaces of both Venus and Mars once might have been habitable; thus, planetary processes might have sustained a habitable zone in our early solar system that was wider than it is today.

For astrobiology, a more rigorous assessment of the habitability of Mars' surface environment over time is necessary to define better the evolution of our solar system's habitable zone and continuously habitable zone. The physical and chemical composition of Mars' ancient crustal materials must be assessed (68). This characterization is a major objective of NASA's Mars Surveyor Program, which has been chartered to assess Mars' climate history, its past and present potential to support life, and its resources for future human exploration (68). All of these objectives require that the abundance and physical state of water be assessed. This program combines orbital mapping (1), surface reconnaissance, and a series of sample returns.

Extrasolar Habitable Zones and Planets

As we explore habitable zones around other stars, we must consider the range of stellar sizes as well as multiple star systems. Larger stars burn much more brightly and evolve more quickly through their main sequence phase, and therefore their habitable zones are shorter lived (55). For example, a star 0.5 times as massive as our sun (0.5 M_S) dwells on the main sequence for much longer than ten billion years; a star 1.5 times as massive as our sun (1.5 M_S) dwells there for less than two billion years. Perhaps two thirds or more of the stars within the size range 0.5 M_S to 1.5 M_S occur in binary and multiple star systems (27). It was once believed that planetary orbits would be unstable in such systems. However, stable planetary orbits can exist either at radii that are more than five times the distance between binary stars (external binary: planet orbits the center of mass of both stars), or at radii that are less than 20% of the distance between binary stars (internal binary: planet orbits only one star) (75). Perhaps 5% of external and 50% of internal binary systems, respectively, might support a habitable zone (55). Because stars in the mass range 0.5 M_S to 1.5 M_S are so abundant, circumstellar habitable zones may be widespread.

We can apply to extrasolar planets the same factors affecting climates in our solar system. The distance between a star and its habitable zone increases with stellar size and luminosity. The habitable zones of a 0.5 M_S and a 1.5 M_S star are centered at 0.3 A.U. (one astronomical unit is the distance from Earth to our sun) and 2.5 A.U., respectively (55). Current models predict that terrestrial-like planets indeed develop frequently between 0.3 A.U. to several A.U. (99).

Giant planets can significantly affect the habitable zone. Recent discoveries of Jupiter-sized planets having very small orbits (65) indicate that gas giants, which form farther out in the nebular disk, can migrate inward (61), perhaps sweeping up terrestrial-like planets in their path. We cannot yet determine the relative abundances of more stable solar systems like our own, versus those where gas giant planets have spiraled inward. Giant planets also can significantly affect the delivery of planetesimals to the inner solar system. For example, a Jupiter-like planet greatly reduces the flux of comets (98). In simulations where Jupiter and Saturn were replaced by Neptune-class bodies, the comet flux from the Kuiper belt (the region beyond the planet Pluto) increased by roughly three orders of magnitude! Large Jupiter-sized planets shield inner habitable planets from a rain of environmentally disruptive impactors. Are such shepherd planets necessary for biospheres to begin and develop? If so, then the need for a shepherd planet might lower our estimates of the total number of continuously habitable solar systems.

Astrobiologists will benefit from the efforts of astronomers to map the distribution of various types of solar systems. Soon, even Earth-sized planets will be found using the photometric method for detecting the reduction in starlight that accompanies the transit of planets in front of their stars. Interferometric astronomical techniques that greatly reduce the light from the star might capture family portraits of extrasolar planetary systems within the next decade or two (7). Evidence of atmospheric CO_2 and abundant water vapor, hence habitability (24), seems obtainable from the infrared spectra of Earth-like planets. The first detection of an extrasolar habitable planet will be a major step toward discovering the first extrasolar biosphere.

Subsurface Habitable Environments

Earth's subsurface environments (aquifers, oil fields, and hydrothermal systems) harbor an extensive biota that can be partially or totally independent of photosynthetically derived energy (30). Life might persist to depths where it encounters its thermal limit, and subsurface biomass might equal or even exceed the biomass at Earth's surface (33, 34). Abundant subsurface life is significant for astrobiology because it implicates additional planets and environments that previously had not been considered seriously as sites for life.

For example, the surface of Mars is currently uninhabitable (19); therefore strategies to search for life once focused principally upon a search for fossil evidence in ancient terrains (66). However, because life can pursue a subsurface chemosynthetic life style (12) and because subsurface martian aquifers might be sustained by geothermal heat (20) or by residual heat from impacts (73), life may have persisted after the surface environment became hostile. Have subsurface habitable environments survived until today? One approach is to examine all martian meteorites for additional evidence of hydrothermal activity (60). Inactive thermal spring sites should be explored as part of the Mars Surveyor Program. Hydrothermal mineral deposits might reveal when these springs were active, what the conditions were, and whether life existed there (95). A third approach is to search directly for subsurface aquifers. In the Mars Surveyor mission scheduled for launch in 2003, an Italian geophysical sounder may search at crustal depths of hundreds of meters to kilometers.

Liquid water might exist within planetary satellites such as Jupiter's moon Europa (77), owing to the deposition of energy by tidal stresses. Europa's sparsely cratered landscape indicates that its surface remains active. Satellites in the Jovian system fall completely outside the habitable zone defined according to the availability of solar energy (43). An orbiter has been proposed to search for evidence of a subsurface "ocean" using geophysical methods (18). Subsurface chemical processes might be revealed by, among other things, the variably colored deposits decorating the linear fracture systems that criss-cross Europa's surface.

Can subsurface Europa-like biospheres be discovered in extrasolar planets? Such a search seems extraordinarily difficult, given the great distances to these planets, the subtlety of evidence for subsurface habitability, and the extreme difficulty in imaging Europa-like planetary satellites next to planets as large as Jupiter. One perspective on this challenge is offered by comparing biospheres driven by photosynthesis with those driven only by energy sources for chemosynthesis. Detecting even extrasolar biospheres sustained by photosynthesis will be extraordinarily challenging, especially if these biospheres are not accompanied by an abundant atmospheric inventory of O_2 (7, 24). If Earth's biosphere depended only upon chemical energy derived from hydrothermal activity and weathering, global productivity levels would be less than one percent of the levels supported by photosynthesis (26). Detecting an extrasolar biosphere that lacks photosynthesis seems impossible in the forseeable future.

The Origins of Life

How Does Life Begin?

Phrasing this question in the present tense reflects the hypothesis that the origin of life is a deterministic process, that life might begin wherever and whenever the necessary environmental conditions occur (71). Thus, the origin of life might be an ongoing process in the Universe; indeed, perhaps someday it will be an ongoing process in research laboratories engaged in assembling model protocells. Historically, origin of life research has focused upon synthesizing organic monomers in reducing atmospheres and examining their assembly into structures ("proteinoids") and macromolecules (i.e. peptides and polynucleotides) whose roles embody life's most diagnostic properties, namely, its ability to replicate and to evolve (69). Following naturally from the traditional theory that life arose within a reducing organic-rich aqueous broth, the earliest cells were assumed to have been fermentative heterotrophs (69). A long-standing paradox has been that the functions performed by the nucleic acid–protein translation apparatus were assumed to be essential for life to begin; yet this apparatus seems too complex to have arisen in the prebiotic milieu. The discovery of "ribozymes" (16) offers one potential solution to this paradox because both information storage and catalysis might have been achieved by a single class of compounds early in life's history. Even so, it seems extraordinarily difficult for ribozymes to have been synthesized in the absence of cellular life.

Current research explores in more detail the potential roles played by environmental conditions in prebiotic evolution and the origin of life. Some studies have recognized that prebiotic chemical processes required sustained energy sources that were actually available in the environment. For example, Wächtershäuser (93) developed a model whereby prebiotic evolutionary biochemistry was driven by oxidation/reduction reactions involving iron and sulfur species. Such energy-rich species occur abundantly in hydrothermal systems within planetary crusts. Amphiphilic compounds (lipids) assemble spontaneously into vesicles that resemble cellular membranes and that create chemical microenvironments favorable for the development of "protometabolism" (21). Similar amphiphilic compounds have been identified in meteorites. The increase in complexity of these molecular systems imparts emergent properties (structures, specific molecular interactions) that probably led to life's origin but that require specific attention by theorists and experimentalists. This relationship between theories of complexity and the origin of life has been explored (56).

Origin of Life Within a Planetary Context

Astrobiology contributes to research on the origin of life by addressing both theoretical and experimental approaches in the context of those planetary environments, be they on Earth or elsewhere, that favored the origin of life. For example, extraterrestrial organic matter from cosmic dust and ice rained down on the early Earth (17). If this external supply was abundant only early in a solar system's history yet was essential for life to begin, it imposes a key constraint on life's distribution in the cosmos. Perhaps planetary processes also created prebiotic organic matter. A mildly reducing early atmosphere, sustained by thermal processes acting upon a more reduced crust and upper mantle, might have sustained prebiotic organic synthesis (53). Hydrothermal systems may have hosted prebiotic organic synthesis (78, 87). A cold early surface environment would have allowed prebiotically important species to survive and accumulate (3). It therefore becomes important to define the nature of early habitable planetary environments on Earth and elsewhere (45, 79). Assessing the relative importance of various organic sources should at least partially constrain the range of plausible prebiotic scenarios.

The Value of a Second Example of Life

There is a spectrum of opinion as to whether the origins of complex living systems was a directed (deterministic) process or was more random (driven by contingency). For example, perhaps chance plays a role, but only within limits set by the physical and chemical properties of life (22). Evolution looks random when the whole range of species is viewed, but trim the tree of life "of this outer diversity and you are left with a stark trunk delineated by a relatively small number of major forks...." While there is plenty of scope for evolution on another planet to have taken different pathways, "certain directions may carry such decisive selective advantages as to have a high probability of occurring elsewhere as well..." (22). In contrast, the process of evolution may be based upon an infinite number of contingent events (35). That is, each new species evolves when a local environmental opportunity exists, and others become extinct for any one of many causes. Thus, if we were able to "rewind the tape of life and replay it," we would get a different result every time. One major barrier to resolving these contrasting views is that we know only one example of a tree of life. If we had other examples, we would not have to argue in the abstract; we could directly analyze the degrees of similarity between them and from that begin to discern general principles of evolution, including the role of pure chance. This circumstance creates a powerful scientific argument for looking for life elsewhere.

Accordingly, we must broaden the agenda beyond attempting to recreate precisely the events leading up to the origin of life on Earth. To the extent that the earliest phase of our existence was driven by contingency, including the destruction of the earliest fossil record, we are unlikely ever to know the exact history of our own origins. Furthermore, laboratory studies on a diverse array of molecules might reveal principles of molecular evolution that are otherwise too subtle to discern through studies restricted to our own biomolecules. An effective search for life elsewhere also demands a broader approach. If life beyond Earth obeys similar principles but uses different molecules, we must devise more flexible strategies to detect it.

Even with the need for a diversity of studies, a focus upon understanding how the first protocellular structures were assembled remains a centrally important goal. The spontaneous self-assembly of amphiphilic molecules into vesicles may have been an important part of this process (21), as it perhaps created favorable chemical gradients as well as microenvironments. However, the processes leading to the first energy-harvesting metabolism (71, 93) and the mechanisms for self-replication also merit attention. The development of such systems leads to dramatic increases in complexity and information content, increases that were made possible by inputs of chemical energy. Theoretical and experimental models for the coupling of such energy to the emergence of molecular complexity lie at the heart of this effort. Also, laboratory-based molecular systems that are capable of self-replication and evolution are excellent testbeds for exploring prebiotic evolution.

Evolution on Molecular, Organism, Ecosystem, and Biosphere Levels

Microorganisms deserve particular attention in astrobiology because our biosphere was exclusively microbial for more than three quarters of its history (82) and because microorganisms continue to dominate our biosphere. Still, only a small percentage of all molecular biologists and paleontologists are specialists in microbial evolution. However, new methods for identifying, manipulating, and analyzing biomolecules and fossiliferous rocks allow microbial life to be explored from molecular to global scales. Our knowledge of the diversity of the microbial world is still very incomplete (74). An objective search and analysis of natural microbial ecosystems will identify the environmental and biological drivers of evolution and reveal how life diversified and adapted to extreme conditions. If the full range of Earth's habitable conditions can be precisely defined, it will help to identify more fully the range of environments to be explored beyond Earth.

The early record of our biosphere exists in two forms. One is the living record of life preserved within the structure and function of its biopolymers and metabolic pathways. For example, the sequences of monomers in highly conserved biopolymers reflect their descent from their molecular ancestors (100, 101). The second record is the remnants within rocks of fossilized cells, ecosystems, and chemical and isotopic compositions that reflect biological activity.

The Living Record of Early Life

The increasing ease of sequencing macromolecules and the rapid growth of sequence databases are creating unprecedented opportunities for evolutionary studies. However, these databases have been developed principally for their biotechnological applications; therefore, many microorganisms having great potential value for evolutionary research still await study. Furthermore, evolutionary interpretations of sequence data require specialized software and groups of investigators, both currently in short supply. Such teams could examine how genes and gene families are created (via gene duplication), rearranged within genomes, and transferred between organisms.

Studies of molecular phylogeny should be extended to include additional key biomolecules, particularly those enzyme systems that exerted profound impacts upon the environment. For example, enzymes for CO_2 assimilation arose early in evolution and very likely played key biogeochemical roles on the early Earth (32, 97). Accordingly, the phylogeny of ribulose bisphosphate carboxylase-oxygenase (RUBISCO) has been studied extensively. However, several other CO_2-assimilating enzymes merit comparable attention, for example, those associated with the reverse-TCA cycle, acetyl coenzyme-A pathway, etc. Genetic analyses might identify key "nodes" within the phylogenetic trees of these enzyme systems, for example, a radiation of enzyme lineages. It is important to identify the factors that contributed to such evolutionary events. Perhaps environmental changes played a role in many cases.

Understanding how microbial life adapts to environmental gradients and extremes is essential not only for understanding evolution, but also for defining the limits of habitable conditions to be sought beyond Earth. Environmental parameters include temperature, availability of chemical and/or light energy, water potential, solute composition (pH, nutrients, toxins, etc), and environmental variability. Adaptation to environmental gradients and variability has contributed at least in part to diversity. For example, "guilds" of thermophilic cyanobacteria apparently have diversified along thermal gradients in hot spring streams (96). Thermophilic microorganisms have stabilized their molecular machinery against the challenges of high temperatures (10, 91), but the details of these adaptations by various cellular "subsystems" are still poorly known.

Modern microbial ecosystems can help us interpret both the living and rock records of our biosphere. Many types of fossils (morphological, chemical, isotopic) are emergent properties of ecosystems. For example, stromatolites, which are macroscopic laminated forms built by trapping and binding of sediment by microbial mat (biofilm) communities, are the most ancient and widespread morphological evidence of pre-Phanerozoic life (94). Stromatolite morphology probably arises from complex interactions that occur between microbial mats and their environment (38, 94). The microenvironments within microbial mats reflect the structure and function of these ecosystems, for example, the sharing of wavelengths of light (47) and the exchange of metabolic substrates (14). Mat microenvironments can differ markedly from their surroundings and thereby shape both the adaptation and the evolution of the microorganisms as well as the nature of the microbial remains that become incorporated into the fossil record (88).

The Geologic Record of Early Life

Our biosphere has altered the global environment by influencing the chemistry of those elements that are important for life–carbon, nitrogen, sulfur, oxygen, phosphorus, and transition metals such as iron and manganese. Such interactions can be viewed as part of a system of "biogeochemical cycles." Such cycles are networks consisting of elemental reservoirs (within the biosphere, atmosphere, oceans, crust, and mantle) that are linked by geological, biological, and other environmental processes (36).

For example, the coupling of oxygenic photosynthesis with the burial in sediments of photosynthetic organic matter has increased the oxidation state of the oceans and atmosphere and also created highly reduced conditions within sedimentary rocks that have extensively affected the chemistry of other elements (31). Nonbiological processes have also influenced the course of biological evolution. For example, the decline of volcanism during Earth's history has decreased the flux of reduced chemical species that reacted with photosynthetically produced O_2 (51, 58). Therefore, atmospheric O_2 levels were destined to increase. To the extent that the development of modern aerobic bacteria, algae, and multicellular life has certain minimum O_2 requirements, the long-term net accumulation of photosynthetic O_2 via biogeochemical processes has influenced our atmosphere and biosphere profoundly (58).

Microorganisms have also affected the precipitation of minerals (5), thereby altering Earth's crust and creating a record of microbial activities in ancient sediments. For example, the precipitation of carbonate in the marine environment has apparently come under stronger biological control over the past two to three billion years (37). One important aspect of mineral formation lies in the tendency for patterns of precipitation that develop at the molecular or microscopic level to be propagated to macroscopic scales (millimeters to centimeters or larger; 44). Perhaps the macroscopic morphologies of some carbonate stromatolites originated as microscopic phenomena whose biological control is uncertain (38). To the extent that sulfidic ores form as a consequence of the accumulation of sedimentary organic matter and biogenic sulfides, processes operating at the microscale can create sulfide ore bodies that can attain kilometer-scale dimensions. Understanding such mineralogical effects is important, not only for understanding the evolution of our own biosphere, but also for locating and interpreting fossil evidence of life on Earth and elsewhere.

Interactions between the long-term evolution of the biosphere and its environment can be explored by focusing upon time intervals that witnessed profound environmental changes. For example, atmospheric O_2 levels apparently rose dramatically between 2.2 and 2.0 billion years ago (42). Can the biological consequences of this event be identified in the living molecular biological record? Major glaciations in the late Proterozoic (800 to 600 million years ago) appear to be correlated with major excursions in the diversity of eukaryotic plankton (41, 92). Can a cause-and-effect relationship be demonstrated?

Our own biosphere would share one important circumstance with other biospheres. Silicate-rich planets (Venus, Earth, and Mars) experience evolutionary trends that are most similar early in their history. For example, the inner planets of our solar system all experienced an early heavy meteoritic bombardment, similar styles of formation and chemical evolution of the early crust, a long-term decline in volcanic activity, and a long-term increase in solar luminosity (see summary in 26). To the extent that similar geological processes have affected the evolution of all habitable planets, those aspects of our own biosphere's history should assist our search for extraterrestrial life.

Recognizing Signatures of Life Beyond Earth

The Challenge

How will we recognize life elsewhere? The discovery of habitable environments beyond Earth would indeed set the stage for learning whether we are truly alone in the Universe. However, habitable environments are geologically and climatologically active places. How can we distinguish between this nonbiological activity and life? We can start with the following basic definition of life: One, life maintains metabolism, that is, a network of chemical reactions that harvest energy for the biosynthesis and maintenance of cellular components; two, life is capable of self-replication; and three, life is capable of Darwinian evolution. This definition lists a few key properties of life, more specifically, its key processes. Such a definition aids our search for life principally to the extent that we can directly observe biological processes. Accordingly, the Viking life-detection experiment was designed to establish life's presence on Mars specifically by observing processes of metabolism and growth (reproduction) (57). However, most astrobiologists agree that the Viking mission did not deliver the final verdict about the history of life on Mars. Our approach cannot be restricted only to observing biological processes directly. The discussion that follows summarizes additional approaches for broadening the search for life on Mars and beyond.

A more effective search for evidence of life addresses the following circumstances that make a Viking-like approach inconclusive and/or impossible:

1. It is possible to detect certain chemical consequences of a biosphere's activity, but it is not possible to observe organisms directly. One hypothetical example is a martian biosphere that resides several kilometers beneath the surface and is currently inaccessible but that alters the trace gas composition of the atmosphere. A second example is a biosphere on a distant extrasolar planet detectable only through its effect upon the composition of its atmosphere (59).
2. Extraterrestrial life might differ substantially from life on Earth. If living organisms existed on the martian surface but simply did not respond to the particular incubation experiments deployed by Viking, they clearly could have escaped detection (57). The nondetection of organic compounds by the Viking mass spectrometer did set upper limits on the quantity of organisms present, but scenarios can be envisioned in which even traces of Earth-like life might have escaped detection (8a). Because biological diversity in the Universe probably exceeds the diversity on Earth, our search strategy must be appropriately broadened.
3. Life has become extinct, or it was once active in a sample or landing site and has since retreated elsewhere. This circumstance compels us to become paleontologists because we must search for fossil evidence of life. A recent example of this approach has been the examination of the martian meteorite ALH84001 (67) and the claim that an array of morphological, chemical, and mineralogical features collectively indicate the presence of life on early Mars.

The Universal Properties of Life Guide Our Search

It is useful to consider those attributes that are universal for life on Earth and thus potentially form the basis for recognizing biological indicators of life ("biomarkers") elsewhere. In his book *Beginnings of Cellular Life*, Morowitz delineated the following 15 universal features of life (71):

1, 2. "*All life is cellular in nature. There is a universal type of membrane structure used in all biological systems.* A cell is the most elementary unit that can sustain life. The chemical reactions that produce the molecules of living organisms take place in cells. Every cell is surrounded by a selectively permeable membrane barrier that keeps it separate from other cells and from the environment." Because cellular forms can be preserved upon burial in sedimentary rocks and yield diagnostic microfossils, the cellular habit has created a key category of biomarker for detecting life's former presence (85).

3. "*The water content of functioning living forms varies from 50% to over 95%.* The chemistry of life is carried out in aqueous solutions or at water interfaces." Cells can survive the removal and restoration of cellular water, but water is essential to cellular function. This attribute reiterates the necessity of liquid water for ensuring life's long-term survival; therefore we should search the cosmos for evidence of liquid water.

4. "*The major atomic components in the covalently bonded portions of all functioning biological systems are carbon, hydrogen, nitrogen, oxygen, phosphorus and sulfur.*" Accordingly, enrichments of phosphorus or organic carbon and nitrogen in terrestrial geological deposits have been interpreted to be strong indicators of biological activity. Certain sulfide ore deposits might also owe their origin to biogenic organic matter.

2, 5, 6, 7, 8, 9, 11. "*There is a universal set of small organic molecules that constitutes a large portion of the total mass of all cellular systems. There exists a universal network of intermediate reactions such that the metabolic chart of any extant species is a subset of the universal chart. Most of the nonaqueous portion of functioning biological systems consists of proteins, lipids, carbohydrates and nucleic acids. The flow of energy in the biosphere is accompanied by the formation and hydrolysis of phosphate bonds. Every replicating cell has a genome made of deoxyribonucleic acid.... All growing cells have ribosomes.... The translation of information from nucleotides [to proteins] takes place through specific activating enzymes and transfer RNAs. There is a universal type of membrane structure used in all biological systems.*" In contrast to the enormous diversity of organisms, the variety of biochemical pathways, metabolic intermediates, and classes of macromolecules and membrane lipids universally employed by life is remarkably small. Perhaps for purposes of functional efficiency, life has restricted itself to a very small subset of the near-infinite array of organic compounds that might exist. Therefore, the "unusually" high relative abundance of specific organic compounds in an environmental sample might by itself be highly diagnostic evidence of a biogenic origin of those compounds.

12. "*Biological information is structural.*" If some organisms are chilled to near-absolute zero or are desiccated in a way that preserves the structural integrity of their membranes and macromolecules, they can revive upon returning to normal conditions. This observation indicates that life on other planets might survive periodic freezing and/or desiccation; therefore we should broaden our search for life to include environments that are only periodically habitable.

13. "*Those reactions that proceed at appreciable rates in living cells are catalyzed by enzymes.*" One of life's hallmarks is its ability to accelerate organic chemical reactions, which, in the absence of life, are typically quite sluggish. Accordingly, for example, life can accelerate a variety of oxidation-reduction reactions in nature in order to harvest chemical energy. Biologically mediated interactions between organic and inorganic compounds can affect the deposition of minerals (carbonates and transition metal oxides and sulfides) in ways that create mineralogical evidence of life (5). Characteristic isotopic patterns among carbon

and sulfur species reflect the enzymatic acceleration at ambient temperatures (0 to 30°C) of oxidation-reduction reactions that, in the absence of life, would proceed at significant rates only above 200°C (25). Such "biogenic" isotopic patterns have been preserved in ancient sedimentary rocks.

14. "*Sustained life is a property of an ecological system rather than a single organism or species.*" More than one species are necessary for primary production and organic degradation and nutrient regeneration. Such cycling of chemical constituents often seems required to maintain the flow-through of energy at the rate needed to sustain the ecosystem. Communities of organisms (biofilms) can create chemical products (minerals) and physical features (sedimentary structures) that are much larger than individual cells and/or that persist long after biological activity has ceased. For example, iron oxides and sulfides can indicate previous biological activity (29). Sedimentary microfabrics such as laminae and larger structures such as stromatolites can indicate the former presence of biofilms (94).

15. "*All populations of replicating biological systems give rise to altered phenotypes that are the result of mutated genotypes.* This is the empirical generalization that is a sine qua non for the process of evolution as well as the science of genetics." Progressive changes in fossil biomarkers (fossil morphology) over time can record aspects of the processes of biological evolution.

Biomarkers An effective search for extraterrestrial life must create more effective means for recognizing and interpreting the full range of biomarkers, as mentioned above. Summarizing the foregoing discussion, categories of biomarkers that can survive in geological deposits, or else be detected remotely, include the following:

1. Cellular remains
2. Textural fabrics in sediments that record structure and/or function of biological communities (e.g. stromatolites; 94)
3. Biogenic organic matter, including hydrocarbons
4. Minerals whose deposition has been affected by biological processes
5. Stable isotopic patterns that reflect biological activity
6. Atmospheric constituents whose concentrations require a biological source.

Precambrian paleontology offers a rich legacy of experience in the study of microbial biomarkers, including the effects of biological processes upon the atmosphere (42, 83, 85).

Other Biospheres in Our Solar System?

Mars

The debate surrounding the study of martian meteorite ALH84001 offers a recent example of the issues regarding the effective use of biomarkers to search for evidence of life beyond Earth. Citing an array of significant observations of ALH84001, a martian meteorite recovered from Antarctica, McKay et al (67) proposed that martian life existed, at least at the time that the carbonate formed in the meteorite. A brief summary of the ensuing debate follows.

Establishing Whether Habitable Conditions Existed The meteorite ALH84001 is an igneous rock; therefore, it formed at temperatures that were too high for life. However, it contains disc-shaped carbonates that subsequently developed along fractures and that contain the features cited as evidence for life (67). Did these carbonates form under habitable conditions—namely, with liquid water and cooler temperatures (known organisms on Earth are restricted to $<113°C$; 10)? Unfortunately, estimates of the temperatures of formation of these carbonates are poorly constrained, in part because the meteorite had a complex history (90) that included multiple heating events due to meteorite impacts. Thus, it is uncertain whether geochemical estimates of temperature reflect the original conditions of carbonate formation or some later event. Also, not enough is known about the broader geological context of the environment in which these carbonates formed. For example, because the composition of the fluids that deposited the carbonate (60) is uncertain, one can propose either low temperatures of formation assuming one fluid composition or high temperatures ($>200°C$) assuming another composition. A key objective for astrobiology is to ensure that future missions to Mars acquire samples that provide firmer constraints for early, potentially habitable conditions.

Cell-Shaped Objects Perhaps the most visually compelling life-like features in ALH84001 are the submicron cell-shaped objects that resemble microorganisms. However, the sizes of many of these objects seem too small to accommodate the biochemical machinery required for free-living microorganisms (64, 72). Some cell-shaped objects may be minerals oriented crystallographically along carbonate substrates (13). Cell-shaped objects have been reported in lunar meteorites found in Antarctica (86), even though these rocks never experienced habitable conditions on the Moon. These observations highlight the challenge to avoid terrestrial contamination in samples returned to Earth and, furthermore, to develop diagnostic criteria (internal morphology and composition of cell-shaped objects) that provide definitive interpretations of the origins of cell-shaped objects.

Organic Matter The observation of polycyclic aromatic hydrocarbons (PAH) in ALH84001 (67) raised two issues: Did these compounds come from Mars, and are they biogenic? The absence of PAH from the external fusion (melt) crust of the meteorite indicates that they were not acquired on Earth (67), whereas their close association with the carbonate globules and rims indicates that they once resided on Mars (28). However, PAH also occur in interplanetary dust particles and in chondritic meteorites (8), meteorites that were probably delivered to the martian surface in the past. It is not possible to interpret from their structures whether these PAH were derived from biogenic matter. However, other organic compounds in ALH84001 are clearly terrestrial contaminants (4, 49). The challenge for astrobiology, therefore, is to avoid organic contamination by terrestrial organic matter and also to devise criteria for determining the mechanisms for the synthesis of extraterrestrial organic matter (biological or abiotic?).

Minerals The structure and/or composition of some minerals can be influenced by biological activity; therefore, they can become indicators of life (6). Multiple populations of magnetite (Fe_3O_4) have been reported in the carbonate globules of ALH84001 (89), and these might reflect multiple mechanisms of formation. For example, some have clearly grown simultaneously with the surrounding carbonate (13, 9) and therefore could not have developed inside a microorganism. Others, upon their removal from the carbonate for analysis, revealed shapes otherwise reported only from bacteria (89). Minerals have the potential to reveal life's former presence in samples that have lost other evidence. However, the relationships between the processes of formation and the composition and shapes of potentially biogenic minerals requires much further study.

The Geological Context The ALH84001 debate illustrates the need to develop further the basis for interpreting potential biomarker features in extraterrestrial materials. It also indicates that, during our campaigns to return samples from Mars and elsewhere for astrobiological study, we must constrain as much as possible our interpretation about the geological context of those samples. The remarkable preservation of minerals and morphology at the nanometer scale in ALH84001 indicates that the martian crust has indeed retained, at least at some localities, a rich storehouse of information about the history of its earliest environments, environments that might have witnessed a second example of life in our solar system. The now-international Mars Surveyor Program is challenged to continue the exploration of that storehouse (68), through a series of missions to conduct global surveillance, rover-based exploration of the surface, and sample return.

Europa

Many of the considerations outlined for Mars exploration are relevant also to a search for evidence of life on Europa, the satellite of Jupiter. Materials (ice, etc) that might have experienced a habitable subsurface Europa "ocean" should be examined for evidence of life. A search for evidence of life within Europa might be conducted in several ways. An orbiting IR spectrometer could analyze surface ices for evidence of organic components that emerged from within the planet. A lander might examine surface ices locally and in more detail. A flyby orbiter could, as it approaches, release an impacting projectile and then, after it orbits Europa, it could sample ejected materials for return to Earth. As with Mars, we are challenged to distinguish between the formation of minerals and organic components by nonbiological versus biological processes.

Biospheres in Other Solar Systems?

Direct Observations are Challenging but Essential

Although a planet's presence can be inferred by indirect means (discussed above), the remote search for evidence of life requires direct observation of photons from a planet. Viewed from distances of several light-years, a habitable planet would appear quite faint in the visible wavelength range (10^{-10} times as bright as its star), yet extremely close (about 0.1 second of arc) to the star (59). Such a planet appears relatively brighter in the mid-infrared (MIR) range (10^{-7} to 10^{-6} times as bright); however, a MIR telescope must be located above Earth's atmosphere in order to avoid interferences from H_2O in Earth's atmosphere. Optical interferometry can image the planet by filtering out direct starlight (2, 7). Within the next two decades, space-based interferometric telescopes might obtain low-resolution MIR (6 to 18 μm) spectra of the atmospheres of habitable planets located within 20 parsecs (light-years) (7).

Spectroscopic Indicators of Life

The challenge to detect an extrasolar biosphere is therefore to identify, within MIR spectra of extrasolar planetary atmospheres, absorption features whose presence requires a biological source (59, 24). Ideally, atmospheric "biomarker" compounds would be well-known biological products, and they would create an overall atmospheric composition that is out of thermodynamic equilibrium with the planet's crustal composition (62). For example, evidence of life on Earth was "confirmed," using data from the Galileo spacecraft's near-IR mapping spectrometer, by detecting the presence of both oxidized and reduced biogenic gases in the atmosphere (O_2, CH_4, and N_2O) (81). It will not soon be possible to explore the near IR region in the faint light from extrasolar planets. However, it should be possible to search for O_2 because its abundance is closely related to that of O_3, which exhibits a strong MIR band at 9.6 μm (59). Atmospheric O_2 can accumulate solely by nonbiological processes under certain conditions. For example, a Venus-like planet with a runaway greenhouse climate, which can attain several hundred degrees Celsius, can lose H to space as atmospheric H_2O is photodissociated, thus allowing O_2 to accumulate in the atmosphere (54). However, a runaway greenhouse can be independently inferred if the planet lies inside the habitable zone, as does Venus. High temperatures can also be indicated independently by the IR spectra of CO_2 and H_2O.

If a habitable, Earth-like planet lacked a biosphere, its atmospheric O_2 levels would be very low because the small amount of O_2 from photodissociation of H_2O would be rapidly consumed by reduced volcanic chemical species (51). Thus, a planet that is both habitable at its surface and has abundant atmospheric O_2 is indeed inhabited. However, a habitable planet lacking abundant O_2 is not necessarily uninhabited. Earth's fossil record extends back at least 3.5 billion years (85); however, atmospheric O_2 levels probably became substantial only about 2.1 billion years ago (42). How might we detect an early-Earth-like biosphere around another star? Atmospheric methane from methanogenic bacteria might be one indicator, but nonbiological sources of methane might be substantial on young habitable planets (53). Anaerobic bacteria produce other reduced gases that might be more diagnostic of life (24); however, very little modeling has been done of trace gas compositions in Earth's early anoxic atmosphere. An astrobiology research program that addresses the biological, geological, and photochemical aspects of anoxic atmospheres will benefit studies of our early biosphere as well as our search for extrasolar biospheres.

ASTROBIOLOGY: The Whole Exceeds the Sum of Its Parts

Just as life is an emergent property of highly complex and coordinated molecular systems, astrobiology must emerge as a highly coordinated program of research, exploration, and education, a program sustained by biological, chemical, earth, and space scientists, engineers, and educators. The natural universe works as a unified whole, and so we must unify our efforts in order to learn its deepest and most valuable secrets. Foremost among those secrets is an understanding of the origin, evolution, and distribution of life in the universe.

LITERATURE CITED

1. Albee AL, Pallueoni FD, Arvidson RE. 1998. Mars Global Surveyor Mission: overview and status. *Science* 279:1671–72
2. Angel JRP, Woolf NJ. 1996. Searching for life on other planets. *Sci. Am.* 274:60–66
3. Bada JL, Bigham C, Miller SL. 1994. Impact melting of frozen oceans on the early Earth: implications for the origin of life. *Proc. Natl. Acad. Sci. USA* 91:1248–50
4. Bada JL, Glavin DP, McDonald GD, Becker L. 1998. A search for endogenous amino acids in Martian meteorite ALH84001. *Science* 279:362–65
5. Banfield J, Nealson K, eds. 1997. *Geomicrobiology.* Washington, DC: Mineralogical Soc. Am.
6. Bazylinski DA, Moskowitz BM. 1997. Microbial biomineralization of magnetic iron minerals: microbiology, magnetism and environmental significance. In *Geomicrobiology: Interactions Between Microbes and Minerals,* ed. JF Banfield, KH Nealson. 35:181–223, Washington, DC: Mineralogical Soc. Am.
7. Beichman CA. 1996. *A road map for the exploration of neighboring planetary systems. Jet Propulsion Lab. Pub. #96-22*
8. Bell JF. 1996. Evaluating the evidence for past life on Mars. *Science* 274:2121–22

8a. Benner SA. 1999. The missing organic molecules of Mars, alternative interpretation of experiments from the Viking 1976 lander. *Proc. Natl. Acad. Sci. USA.* In press

9. Blake D, Treiman A, Cady S, Nelson C, Krishman K. 1998. Characterization of magnetite within carbonate in ALH84001. In *Lunar and Planetary Science, XXIX, abstract #1347,* CD-ROM. Houston: Lunar & Planetary Inst.

10. Blochl E, Rachel R, Burggraf S, Hafenbradl D, Jannasch HW, et al. 1997. *Pyrolobus fumarii,* gen. and sp. nov., represents a novel group of archaea, extending the upper temperature limit for life to 113 degrees C. *Extremophiles* 1:14–21
11. Boss AP. 1998. The origin of protoplanetary disks. In *Origins, Astron. Soc. Pac. Conf. Ser.,* ed. CE Woodward, JM Shull, HA Thronsen, 148:314–26. San Francisco:
12. Boston PJ, Ivanov MV, McKay CP. 1992. On the possibility of chemosynthetic ecosystems in subsurface habitats on Mars. *Icarus* 95:300–8
13. Bradley J, McSween HY, Harvey RP. 1998. Epitaxial growth of nanophase magnetite in Martian meteorite Allan Hills 84001: implications for biogenic mineralization. *Meteor. Planet. Sci.* 33:765–73
14. Canfield DE, Des Marais DJ. 1993. Biogeochemical cycles of carbon, sulfur, and free oxygen in a microbial mat. *Geochim. Cosmochim. Acta* 57:3971–84
15. Carr MH. 1996. Water erosion on Mars and its biologic implications. *Endeavour* 20:56–60
16. Cech TR. 1989. RNA as an enzyme. *Biochem. Int.* 18:7–14
17. Chyba C, Sagan C. 1992. Endogenous production, exogenous delivery and impact-shock synthesis of organic molecules: an inventory for the origins of life. *Nature* 355:125–32
18. Chyba CF. 1998. Radar detectability of a subsurface ocean on Europa. *Icarus* 134:292–302
19. Clark B. 1998. Surviving the limits of life at the surface of Mars. *J. Geophys. Res.* 103:28,545–55
20. Clifford SM. 1993. A model for the hydrologic and climatic behavior of water on Mars. *J. Geophys. Res.* 98:10973–11016
21. Deamer DW, Oro J. 1980. Role of lipids in prebiotic structures. *BioSystems* 12:167–75
22. DeDuve C. 1995. *Vital Dust.* New York: Basic Books. 362 pp.
23. Des Marais DJ. 1985. Carbon exchange between the mantle and crust and its effect upon the atmosphere: today compared to Archean time. In *The Carbon Cycle and Atmospheric CO_2: Natural Variations Archean to Present,* ed. ET Sundquist, WS Broecker, 32:602–11. Washington, DC: Am. Geophys. Union
24. Des Marais DJ. 1996. The Blue Dot workshop: spectroscopic search for life on extrasolar planets. *NASA Conf. Publ. 10154,* Ames Res. Cent., Moffett Field, CA
25. Des Marais DJ. 1996. Stable light isotope biogeochemistry of hydrothermal systems. In *Evolution of Hydrothermal Ecosystems on Earth (and Mars?),* ed. GR Bock, JA Goode. *Ciba Found. Symp.* 202:83–98. Chichester, UK: Wiley
26. Des Marais DJ. 1997. Long-term evolution of the biogeochemical carbon cycle. In *Geomicrobiology,* ed. J Banfield, K Nealson, 35:429–45. Washington, DC: Mineralogical Soc. Am.
27. Duquennoy A, Mayor M. 1991. Multiplicity among solar-type stars in the solar neighborhood. II. Distribution of the orbital elements in an unbiased sample. *Astron. Astrophys.* 248:485–524
28. Flynn GJ, Keller LP, Miller MA, Jacobsen C, Wirrick S. 1998. Organic compounds associated with carbonate globules and rims in the ALH84001 meteorite. In *Lunar and Planetary Science, XXIX, abstract #1156,* CD-ROM. Houston: Lunar & Planetary Inst.
29. Fortin D, Ferries FG, Beveridge TJ. 1997. Surface-mediated mineral development by bacteria. In *Geomicrobiology: Interactions Between Microbes and Minerals,* ed. JF Banfield, KH Nealson, 35:161–77. Washington, DC: Mineralogical Soc. Am.
30. Fredrickson JK, Onstott TC. 1996. Microbes deep inside the Earth. *Sci. Am.* 275:68–73
31. Garrels RM, Perry EA Jr. 1974. Cycling of carbon, sulfur, and oxygen through

geologic time. In *The Sea*, ed. ED Goldberg, 5:303–36. New York: Wiley
32. Gogarten JP. 1998. Origin and early evolution of life: deciphering the molecular record. In *Origins,* ed. CE Woodward, JM Shull, HAJ Thronson. San Francisco: Astronom. Soc. Pacific Conf. Ser. 148.
33. Gold T. 1992. The deep, hot biosphere. *Proc. Natl. Acad. Sci. USA* 89:6045–49
34. Gold T. 1999. *The Deep Hot Biosphere.* New York: Copernicus
35. Gould SJ. 1996. *Full House.* New York: Three Rivers
36. Gregor CB, Garrels RM, MacKenzie FT, Maynard JB. 1988. *Chemical Cycles in the Evolution of the Earth.* New York: Wiley
37. Grotzinger JP, Kasting JF. 1993. New constraints on Precambrian ocean composition. *J. Geol.* 101:235–43
38. Grotzinger JP, Rothman DH. 1996. An abiotic model for stromatolite morphogenesis. *Nature* 383:423–25
39. Haberle RM. 1998. Early Mars climate models. *J. Geophys. Res.* 103:28467–80
40. Hart NH. 1978. The evolution of the atmosphere of the Earth. *Icarus* 33:23–39
41. Hoffman PF, Kaufman AJ, Halverson GP, Schrag DP. 1998. A neoproterozoic snowball Earth. *Science* 281:1342–46
42. Holland HD. 1992. Distribution and paleoenvironmental interpretation of Proterozoic paleosols. In *The Proterozoic Biosphere: a Multidisciplinary Study*, ed. JW Schopf, C Klein, pp. 153–55. New York/ Cambridge, UK: Cambridge Univ. Press
43. Huang AS. 1959. Occurrence of life in the Universe. *Am. Sci.* 47:397–402
44. Hyde ST, Andersson S, Blum Z, Lidin S, Larsson K, et al. 1997. *The Language of Shape.* Amsterdam: Elsevier Sci. B.V.
45. Jakosky B. 1998. *The Search for Life on Other Planets.* Cambridge, UK: Cambridge Univ. Press
46. Jakosky BM, Skock EL. 1998. The biological potential of Mars, the early Earth, and Europa. *J. Geophys. Res.* 103:19359–64
47. Jørgensen BB, Des Marais DJ. 1988. Optical properties of benthic photosynthetic communities: fiber optic studies of cyanobacterial mats. *Limnol. Oceanogr.* 33:99–113
48. Joyce GF. 1992. Directed molecular evolution. *Sci. Am.* 267:90–97
49. Jull AJT, Courtney C, Jeffrey DA, Beck JW. 1998. Isotopic evidence for a terrestrial source of organic compounds found in Martian meteorites Allan Hills 84001 and Elephant Moraine 79001. *Science* 279:366–69
50. Kasting JF. 1993. Earth's early atmosphere. *Science* 259:920–26
51. Kasting JF. 1997. Habitable zones around low mass stars and the search for extraterrestrial life. *Orig. Life* 27:291–307
52. Deleted in proof
53. Kasting JF, Brown LL. 1999. Setting the stage: the early atmosphere as a source of biogenic compounds. In *The Molecular Origins of Life: Assembling the Pieces of the Puzzle,* ed. A Brack, pp. 35–96. New York: Cambridge Univ. Press
54. Kasting JF, Pollack JB, Ackerman TP. 1984. Response of Earth's atmosphere to increases in solar flux and implications for loss of water from Venus. *Icarus* 57:335–55
55. Kasting JF, Whitmire DP, Reynolds RT. 1993. Habitable zones around main sequence stars. *Icarus* 101:108–28
56. Kauffman S. 1995. *At Home in the Universe.* New York: Oxford Univ. Press. 321 pp.
57. Klein H. 1998. The search for life on Mars: What we learned from Viking. *J. Geophys. Res.* 103:28,463–28,466
58. Knoll AH, Holland HD. 1995. Oxygen and Proterozoic evolution: an update. In *Effects of Past Global Change on Life,* ed. S Stanley, pp. 21–33. Washington, DC: Natl. Acad. Press
59. Leger A, Pirre M, Marceau FJ. 1993. Search for primitive life on a distant planet: relevance of O_2 and O_3 detections. *Astron. Astrophys.* 277:309–13

60. Leshin LA, Epstein S, Stolper EM. 1996. Hydrogen isotope geochemistry of SNC meteorites. *Geochim. Cosmochim. Acta* 60:2635–50
61. Lin DNC, Bodenheimer P, Richardson DC. 1996. Orbital migration of the planetary companion of 51 Pegasi to its present location. *Nature* 380:606–7
62. Lovelock JE. 1965. A physical basis for life detection experiments. *Nature* 207:568–70
63. Lovelock JE. 1979. *Gaia: a New Look at Life on Earth.* Oxford, UK: Oxford Univ. Press
64. Maniloff J, Nealson KH, Psenner R, Loferer M, Folk RL. 1997. Nanobacteria: size limits and evidence. *Science* 176:1773–76
65. Marcy GW, Butler RP. 1996. A planetary companion to 70 Virginis. *Astrophys. J.* 464:L147–51
65a. Marcy GW, Butler RP. 1998. Detection of extrasolar giant planets. *Annu. Rev. Astron. Astrophys.* 36:57–97
66. McKay CP, Stoker CR. 1989. The early environment and its evolution on Mars: implications for life. *Rev. Geophys.* 27: 189–214
67. McKay DS, Gibson EK Jr., Thomas-Keprta KL, Vali H, Romanek CS, et al. 1996. Search for past life on Mars: possible relic biogenic activity in Martian meterorite ALH84001. *Science* 273:924–30
68. Meyer MC, Kerridge JF. 1995. *An exobiological strategy for Mars exploration, NASA Spec. Publ. 530*
69. Miller S, Orgel LE. 1974. *The Origins of Life on the Earth.* New York: Prentice Hall
70. Mojzsis SJ, Arrhenius G, McKeegan KD, Harrison TM, Nutman AP, et al. 1996. Evidence for life on Earth before 3,800 million years ago. *Nature* 384:55–59
71. Morowitz HJ. 1992. *Beginning of Cellular Life.* New Haven, CT: Yale Univ. Press
72. Nealson K. 1997. The limits of life on Earth and searching for life on Mars. *J. Geophys. Res.* 102:23,675–86
73. Newsom HE, Brittelle GE, Hibbitts CA, Crossey LJ, Kudo AM. 1996. Impact crater lakes on Mars. *J. Geophys. Res. Planets* 101:14951–55
74. Pace NR. 1997. A molecular view of microbial diversity and the biosphere. *Science* 276:734–40
75. Pendleton YJ, Black DC. 1983. Further studies on criteria for the onset of dynamical instability in general three body systems. *Astron. J.* 88:1415–19
76. Pohorille A, New MH, Schweighofer K, Wilson A. 1999. Insights from computer simulations into the interactions of small molecules with lipid bilayers. In *Membrane Permeability: 100 Years Since Ernst Overton*, ed. D Deamer, pp. 49–76. San Diego, CA: Academic Press
77. Reynolds RT, Squyres SW, Colburn DS, McKay CP. 1983. On the habitability of Europa. *Icarus* 56:246–54
78. Russell MJ, Hall AJ. 1997. The emergence of life from iron monosulfide bubbles at a submarine hydrothermal redox and pH front. *J. Geol. Soc. London* 154:377–402
79. Sagan C. 1994. The search for extraterrestrial life. *Sci. Am.* Oct. 1994:92–99
80. Sagan C, Drake F. 1975. The search for extraterrestrial intelligence. *Sci. Am.* 232:80–89
81. Sagan C, Thompson WR, Carlson R, Gurnett D, Hord C. 1993. A search for life on Earth from the Galileo spacecraft. *Nature* 365:715–21
82. Schopf JW. 1983. *Earth's Earliest Biosphere.* Princeton, NJ: Princeton Univ. Press. 543 pp.
83. Schopf JW, Klein C, eds. 1992. *The Proterozoic Biosphere: a Multidisciplinary Study.* New York: Cambridge Univ. Press
84. Schopf JW, Packer BM. 1987. Early Archean (3.3 billion to 3.5 billion-year-old) microfossils from the Warrawoona

Group, Western Australia. *Science* 237: 70–73
85. Schopf JW, Walter MR. 1983. Archean microfossils: new evidence of ancient microbes. In *Earth's Earliest Biosphere, Its Origin and Evolution,* ed. JW Schopf, pp. 214–38. Princeton, NJ: Princeton Univ. Press
86. Sears DWG, Kral TA. 1998. Martian "microfossils" in lunar meteorites? *Meteor. Planet. Sci.* 33:791–94
87. Shock EL, Schulte MD. 1998. Organic synthesis during fluid mixing in hydrothermal systems. *J. Geophys. Res.* 103:28,513–28,527
88. Stal LJ, Caumette P. 1994. *Microbial mats: structure, development and environmental significance. Series G. Ecological Sciences.* Heidelberg: Springer Verlag
89. Thomas-Keprta KL, Bazlinski DA, Wentworth SJ, McKay DS, Golden DC, et al. 1998. *Mineral biomarkers in Martian meteorite Allan Hills 84001? Martian meteorites: Where do we stand and where are we going? In LPI, Contribution #956, 51-3* Houston, TX: Lunar Planetary Inst.
90. Treiman A. 1998. The history of Allan Hills 84001 revised: multiple shock events. *Meteor. Planet. Sci.* 33:753–64
91. Trent JD. 1996. A review of acquired thermotolerance, heat-shock proteins, and molecular chaperones in archaea. *FEMS Microbiol. Rev.* 18:249–58
92. Vidal G, Knoll AH. 1982. Radiations and extinctions of plankton in the late Precambrian and early Cambrian. *Nature* 197:57–60
93. Wächtershäuser G. 1992. Groundwork for an evolutionary biochemistry: the iron-sulphur world. *Prog. Biophys. Molec. Biol.* 58:85–201
94. Walter MR, ed. 1976. *Stromatolites.* Amsterdam: Elsevier
95. Walter MR, Des Marais DJ. 1993. Preservation of biological information in thermal spring deposits: developing a strategy for the search for fossil life on Mars. *Icarus* 101:129–43
96. Ward DM, Ferris MJ, Nold SC, Bateson MM. 1999. Microbial biodiversity within hot spring cyanobacterial mat communities: an evolutionary ecology view. *Microbial Mol. Biol. Rev.* In press
97. Watson GMF, Yu JP, Tabita R. 1999. Unusual ribulose 1,5-bisphosphate carboxylase/oxygenase of anoxic Archaea. *J. Bacteriol.* 181:1569–75
98. Wetherill GW. 1994. Provenance of the terrestrial planets. *Geochim. Cosmochim. Acta* 58:4513–20
99. Wetherill GW. 1996. The formation and habitability of extra-solar planets. *Icarus* 119:219–38
100. Woese CR. 1998. The universal ancestor. *Proc. Natl. Acad. Sci. USA* 95:6854–59
101. Woese CR, Fox GE. 1977. Phylogenetic structure of the prokaryotic domain: the primary kingdoms. *Proc. Natl. Acad. Sci. USA* 74:5088–90

Annu. Rev. Ecol. Syst. 1999. 30:421–55

EVOLUTION OF EASTERN ASIAN AND EASTERN NORTH AMERICAN DISJUNCT DISTRIBUTIONS IN FLOWERING PLANTS

Jun Wen
Department of Biology, Colorado State University, Fort Collins, Colorado 80523; e-mail: jwen@lamar.colostate.edu

Key Words disjunct distribution, eastern Asia, eastern North America, evolution, biogeography

■ **Abstract** The disjunct distributions of morphologically similar plants between eastern Asia and eastern North America have fascinated botanists and biogeographers since the Linnaean era. This biogeographic pattern is currently recognized by the disjunct distributions of some species, approximately 65 genera, and a few closely related genera in these two widely separated areas. Early workers treated many disjuncts as conspecific, but most were later recognized as intercontinental species pairs. Recent phylogenetic studies confirm affinities between many of the disjunct taxa but also indicate that the disjunct pairs of species are rarely each other's closest relatives. Instead, a pattern of further diversification of species on one or both continents is commonly found. Phylogenetic, molecular, geologic, and fossil data all support the hypothesis that the eastern Asian and eastern North American disjunct distributions are relicts of the maximum development of temperate forests in the northern hemisphere during the Tertiary. Fossil and geologic evidence supports multiple origins of this pattern in the Tertiary, with both the North Atlantic and the Bering land bridges involved. In many genera of flowering plants, current estimates of divergence times using molecular and fossil data suggest that the disjunct patterns were established during the Miocene. Morphological stasis, evidenced by the minimal morphological divergence of species after a long time of separation, must have occurred in some of the disjunct groups in the north temperate zone.

INTRODUCTION

One of the major patterns in the worldwide distribution of plants is that many taxa are represented in both eastern Asia and eastern North America (7, 19, 38, 123, 124 141). This pattern was discovered in the ginseng genus (*Panax*, Araliaceae) by a French Jesuit, Father Joseph Francis Lafitau, who found American ginseng near Montreal, Canada, in 1716 after reading a description of the morphology, habitat, and use of native ginseng from Manchuria, China (118). One of the pupils of Linnaeus, Jonas P. Halenius, first discussed this phenomenon in scientific literature

(48). In the mid-nineteenth century, Asa Gray made floristic comparisons among eastern North America, western North America, Japan, and Europe (41–45) and concluded that eastern North America showed a floristic affinity closer to Japan than to western North America. This pattern is occasionally referred to as the ASA GRAY disjunction (78). It is currently recognized not only for many plant genera peculiar to the two widely separated areas (31, 38, 60, 65, 66, 70, 73, 88, 170) but also for fungi (61, 169), arachnids (138), millipedes (26), insects (111), and freshwater fishes (117).

A close floristic relationship between the two widely isolated areas has been recognized historically with many intercontinental species pairs (13, 29, 59, 60, 69 88, 89). But recent phylogenetic analyses have shown that in most genera with three or more species the presumed intercontinental species pairs are not sister taxa (121, 131, 148, 150, 153, 155, 156). This lack of direct intercontinental sister species relationships may be caused by three possible scenarios. First, the species pair may be a relict of a formerly widespread distribution that has experienced further diversification in eastern Asia and eastern North America after its separation. Thus, the intercontinental species pairs may have been sister species originally but then became further diversified. Second, the morphologically similar species pairs may be the product of a combination of a common history and convergence in distant but similar environments. In this case, they may never have been sister species. Third, the presumed species pairs may not be as similar to each other morphologically as previously claimed. Unfortunately few morphological comparisons have been made for the disjuncts. Studies are needed to address the following questions: First, are the eastern Asian and eastern North American taxa morphologically more similar to each other than to their western North American or European counterparts, as previous workers suggested (e.g., 13, 88, 89, 94)? And second, do the patterns of overall morphological similarity correlate with phylogenetic relationships?

Some traditionally recognized disjunct taxa have recently been shown to be paraphyletic (120, 152, 155) or polyphyletic (121, 135, 136). Thus, these taxa are based on symplesiomorphies or evolutionary convergences. The extent of paraphyly and polyphyly needs to be examined, and taxonomic reevaluations need to be made based on phylogenetic evidence. Morphological stasis (i.e., low level of morphological change after the separation of taxa) has been suggested to be common among the eastern Asian and eastern North American disjuncts (e.g., 116, 122). The paraphyly of some disjunct taxa may indicate that the morphological cohesion (or stasis) may be the result of evolutionary constraints; and the morphological stasis in the polyphyletic taxa may be attributable to convergence in similar habitats.

Recent studies of taxa with disjunct distributions in eastern North America, western North America, and eastern Asia (reviewed in 173) have shown a closer relationship between eastern North America and western North America than between eastern North America and eastern Asia. These results suggest that the traditional view of the close floristic relationship between eastern North America and eastern Asia needs to be modified.

This general phytogeographic pattern can be readily explained by the widespread distribution of temperate forest elements in the northern hemisphere during the mid-Tertiary and subsequent extirpations in western North America and western Europe in response to the late Tertiary and Quaternary climatic cooling (40, 98). Both the Bering (63) and the North Atlantic (102, 103, 144) land bridges probably contributed to the floristic exchanges between eastern Asia and eastern North America and to the maximum development of the deciduous forests in the mid-Tertiary. Tiffney (143) proposed five major periods in which floristic exchanges between eastern Asia and eastern North America were most likely. They are the Pre-Tertiary, the Early Eocene, the Late Eocene-Oligocene, the Miocene, and the Late Tertiary-Quaternary periods. These proposed periods have served as valuable working hypotheses for investigations of the origin and evolution of this disjunction (83, 155). The hypothesis of multiple origins of the biogeographic pattern is supported by recent molecular evidence (16, 83, 122, 148, 149, 151, 153, 171), but further tests are needed.

Li (88) reviewed the early studies and documented the disjunction for seed plants and pteridophytes. Iwatsuki (70) investigated the relationships between the two moss floras. Graham (38) compiled a collection of papers that discussed the history and development of the disjunct pattern (39, 84, 145, 162, 178), documented the disjunction in various plant groups (15, 71, 80, 89, 106, 110), and examined the differentiation patterns among disjunct species (51). Twenty-one papers by scientists from China, Japan, Canada, and the United States that appeared in the *Annals of the Missouri Botanical Garden* (volume 70, numbers 3 and 4, 1983) discussed the evolution of the disjuncts based on geologic, paleontological, zoological, and botanical data available to the early 1980s. Tiffney (143, 144) synthesized fossil, geologic, and climatic evidence and provided an excellent discussion of the possible origin and development of floristic similarity between eastern Asia and eastern North America. Parks & Wendel (116), who pioneered the use of molecular, breeding, and fossil data to examine the origin and evolution of a vicariant species pair in *Liriodendron*, suggested that morphological stasis may be common among the eastern Asian and eastern North American disjuncts.

This contribution attempts to review the current understanding of the origin and development of eastern Asian and eastern North American disjunct distributions and examines phylogenetic relationships and patterns of differentiation of the disjunct species. Insights from the most recent phylogenetic analyses and molecular divergence data of flowering plants are emphasized.

PLANT TAXA DEMONSTRATING AN EASTERN ASIAN AND EASTERN NORTH AMERICAN DISJUNCT PATTERN

The disjunct biogeographic pattern between eastern Asia and eastern North America has been recognized by the discontinuous distribution of some species, many genera, and a few closely related genera in these two widely separated areas. The high level of resemblance between the two floras was emphasized historically, and many discontinuous species were recognized by Good (34), Gray (42, 45), Halenius (48), Hara (50), Hong (60), and Iwatsuki & Ohba (69). Hong (60) recognized eight discontinuous species of flowering plants between eastern Asia and eastern North America. They are *Brachyelytrum erectum*, *Caulophyllum thalictroides*, *Diphylleia cymosa*, *Mitchella repens*, *Mitella nuda*, *Phryma leptostachya*, *Symplocarpus foetidus*, and *Trautvetteria caroliniensis*. A variety of *Trautvetteria caroliniensis* (var. *occidentalis*) also occurs in western North America (56). Most disjunct "species" listed by Hong (60) have been divided into two species (75, 83, 154, 173). In only a few cases are the disjuncts still recognized to be differentiated at the infraspecific level. Examples are *Toxicodendron radicans* ssp. *radicans* (eastern North America) and *T. radicans* ssp. *hispidum* (eastern Asia); and *Phryma leptostachya* var. *leptostachya* (eastern North America) and *P. leptostachya* var. *asiatica* (eastern Asia). Lee et al (83) reported a strikingly high level of divergence of allozyme loci (Nei's genetic identity of 0.291) and sequences of the internal transcribed spacer (ITS) regions of nuclear ribosomal DNA (4.46%) between the two varieties of *P. leptostachya*. Such a high level of genetic divergence suggests the need for re-evaluating the taxonomic status of the two taxa of *Phryma*.

Li (88) recorded 52 plant genera disjunct between eastern Asia and eastern North America. Wu (170) included about 120 genera in his table showing the eastern Asian and eastern North American intercontinental discontinuities. Many genera in Wu's list have an eastern Asian and western North American or a broader Asian and North American disjunction. Hong (60) listed 91 genera with a discontinuous distribution between eastern Asia and North America. Approximately 65 genera of seed plants are confirmed to be distributed disjunctly in eastern Asia and eastern North America (Table 1).

The eastern Asian and eastern North American disjunction is occasionally recognized between presumably closely related genera of the same family, as exemplified by *Weigela* (11–12 spp., Caprifoliaceae) from eastern Asia and *Diervilla* (3 spp.) from eastern North America; and *Glyptostrobus* (1 sp., Taxodiaceae) from eastern Asia and *Taxodium* (2 spp.) from eastern North America. However, recent phylogenetic studies have shown that the eastern North American *Diervilla* is nested within the Asian *Weigela* (76). The eastern Asian *Glyptostrobus* is reported to be more closely related to the eastern Asian *Cryptomeria* than to the eastern North American *Taxodium* (9). Morphological similarities may be largely attributable to symplesiomorphies in *Weigela* and *Diervilla*, and to convergences in *Glyptostrobus* and *Taxodium*. The similar habitat of *Glyptostrobus* and *Taxodium* may have caused morphological convergence among species of the two genera.

A somewhat neglected aspect is that many north temperate disjunct genera include some morphologically similar species from eastern Asia and eastern North America. Thunberg (142) listed 20 species from Japan that were originally described from North America, many of which are members of more widely distributed genera. Li (89) also discussed a few species pairs embedded within large

TABLE 1 Genera of seed plants with an eastern Asian-eastern North American disjunction

Family	Genus	Species in EAs/ENAm	Notes
Anacardiaceae	*Cotinus*	2/1	One spp. to S Europe
Apocynaceae	*Trachelospermum*	19/1	
Araceae	*Symplocarpus*	2/1	
Araliaceae	*Aralia* sect. *Dimorphanthus*	24/1	
Araliaceae	*Panax*	10/2	
Berberidaceae	*Caulophyllum*	1/1–2	
Berberidaceae	*Diphylleia*	2/1	
Berberidaceae	*Jeffersonia*	1/1	
Berberidaceae	*Podophyllum*	1/1	
Bignoniaceae	*Campsis*	1/1	
Bignoniaceae	*Catalpa*	4/7	Five spp. in West Indies
Buxaceae	*Pachysandra*	2/1	
Caprifoliaceae	*Triosteum*	4/3	
Cornaceae	*Cornus* subgen. *Mesomora*	1/1	
Cornaceae	*Nyssa*	4–5/4	One sp. in Cost Rica
Diapensiaceae	*Shortia*	5/1	
Ericaceae	*Epigaea*	2/1	
Ericaceae	*Lyonia*	9/21	
Ericaceae	*Pieris*	5/2	
Fabaceae	*Apios*	6/4	
Fabaceae	*Cladrastis*	7/1	
Fabaceae	*Gleditsia*	10/2	One sp. in S America; some Asiatic spp. extending to SE Asia
Fabaceae	*Gymnocladus*	3/1	
Fabaceae	*Lespedeza*	65+/15+	S to Australia
Fabaceae	*Wisteria*	7/3	
Fumariaceae	*Adlumia*	1/1	
Grossulariaceae	*Itea*	14/1	Asiatic spp. extending to Himalayas and W. Malaysia
Hamamelidaceae	*Hamamelis*	2/2–4	
Hydrangeaceae	*Decumaria*	1/1	
Hydrangeaceae	*Hydrangea* sect. *Hydrangea*		
Illiciaceae	*Illicium*	40/2	Throughout Asia; also in Mexico and West Indies

(Continued)

TABLE 1 (*Continued*)

Family	Genus	Species in EAs/ENAm	Notes
Juglandaceae	*Carya*	2/16	To central America
Lamiaceae	*Meehania*	1/1	
Lauraceae	*Lindera*	80/2	To tropical Asia
Lauraceae	*Sassafras*	2/1	
Liliaceae	*Aletris*	13/5	
Loganiaceae	*Gelsemium*	1/2	Extending to C America and SE Asia
Magnoliaceae	*Liriodendron*	1/1	
Magnoliaceae	*Magnolia*	30+/10+	To tropical Asia and tropical Americas
Menispermaceae	*Menispermum*	1/1+1	One sp. in Mexico
Nelumbonaceae	*Nelumbo*	1/1	To Australia and C America
Oleaceae	*Osmanthus*	14/1–2	Extending to Hawaii
Orchidaceae	*Tipularia*	2/1	
Papaveraceae	*Stylophorum*	2/1	
Poaceae	*Brachyelytrum*	1/1	
Poaceae	*Diarrhena*	1/2	
Poaceae	*Zizania*	1/3	
Polygonaceae	*Antenoron*	3/1	S to the Philippines
Ranunculaceae	*Hydrastis*	1/1	
Rubiaceae	*Mitchella*	1/1	
Santalaceae	*Buckleya*	3/1	
Santalaceae	*Pyrularia*	2/2	
Saururaceae	*Saururus*	1/1	
Saxifragaceae	*Astilbe*	12/2	
Saxifragaceae	*Penthorum*	1/1	
Schisandraceae	*Schisandra*	24/1	
Solanaceae	*Leucophysalis*	4/2	
Stemonaceae	*Croomia*	2/1	
Styracaceae	*Halesia*	1/3	
Theaceae	*Gordonia*	35+/1	S to SE Asia
Theaceae	*Stewartia*	6+/2	
Verbenaceae (or Phrymaceae)	*Phryma*	1/1	
Vitaceae	*Ampelopsis*	15/4	
Vitaceae	*Parthenocissus*	9/3	

and widely distributed genera. Iwatsuki & Ohba (69) provided a list of the corresponding taxa between eastern North America and Japan. Examples of widely distributed genera with eastern Asian and eastern North American floristic connections are: *Abelia*, *Abies*, *Acer*, *Adoxa*, *Aesculus*, *Alisma*, *Alnus*, *Amelanchier*, *Anemone*, *Angelica*, *Asarum*, *Betula*, *Carpinus*, *Castanea*, *Circaea*, *Clematis*, *Clintonia*, *Cornus*, *Corylus*, *Crataegus*, *Elliottia*, *Epilobium*, *Fagus*, *Fraxinus*, *Hydrangea*, *Ilex*, *Juglans*, *Juniperus*, *Lilium*, *Lonicera*, *Malus*, *Osmorhiza*, *Ostrya*, *Pedicularis*, *Picea*, *Pinus*, *Platanus*, *Polygonatum*, *Polygonum*, *Prunus*, *Quercus*, *Rhododendron*, *Rhus*, *Ribes*, *Sagittaria*, *Sambucus*, *Sanicula*, *Saxifraga*, *Smilax*, *Sorbus*, *Staphylea*, *Styrax*, *Symphoricarpos*, *Symplocos*, *Taxus*, *Teucrium*, *Tilia*, *Torreya*, *Toxicodendron*, *Ulmus*, *Vaccinium*, *Valeriana*, and *Viburnum*.

Most disjunct genera between eastern Asia and eastern North America are north temperate elements with only a few occurring in subtropical and tropical regions (175). Examples of the latter are *Gordonia*, *Illicium*, *Magnolia*, *Osmanthus*, and *Schisandra*. Most disjuncts are woody and some are perennial herbs. *Adlumia* is the only known biennial genus. Only one species (one of the five eastern Asian species of *Meehania*) is annual (60). So far all recognized discontinuous species are herbaceous (60).

Most studies refer to the biogeographic discontinuity vaguely as an eastern Asian–eastern North American disjunction. A few workers, including Boufford (6), Hong (60), and Iwatsuki & Ohba (69), have recently discussed a major difference in the forest composition in eastern Asia and eastern North America. The eastern Asian disjuncts are mostly elements of mixed deciduous and evergreen forests composed of many taxa in families such as Araliaceae, Daphniphyllaceae, Ericaceae, Fagaceae, Illiciaceae, Lauraceae, Magnoliaceae, Schisandraceae, Symplocaceae, and Theaceae. Most of the eastern North American disjuncts occur in moist mixed deciduous forests such as in the Appalachians. Boufford (6) pointed out that the eastern North American forests most similar to those of eastern Asia occur along the Appalachicola River in the panhandle region of northern Florida. A few disjunct genera, such as *Croomia*, *Illicium*, *Osmanthus*, and *Schisandra*, are restricted to the panhandle of Florida. However, they represent only a small proportion of the disjuncts. In eastern North America, generally most disjunct taxa occur in the deciduous forests of the Appalachians.

In eastern Asia, most eastern Asian–eastern North American disjuncts occur within the Sino-Japanese Floristic Region (6). This region includes the area from western Yunnan and Sichuan provinces, through central, eastern, and most of the southern parts of China, and eastward to Korea and Japan (168). The Sino-Japanese Floristic Region is also rich in endemics, including plant families such as Bretschneideraceae, Cephalotaxaceae, Cercidiphyllaceae, Eucommiaceae, Eupteleaceae, and Tetracentraceae, as well as a large number of endemic genera, such as *Cathaya*, *Chimonanthus*, *Corylopsis*, *Cuninghamia*, *Engelhardia*, *Ginkgo*, *Glyptostrobus*, *Hosta*, *Kingdonia*, *Metasequoia*, *Ostryopsis*, *Platycarya*, *Pseudolarix*, *Pterocarya*, *Saruma*, *Sciadopitys*, *Sinowilsonia*, and *Zelkova* (168, 177).

PHYLOGENETIC RELATIONSHIPS BASED ON MOLECULAR AND MORPHOLOGICAL DATA

Phylogenetic analyses have been reported for 11 taxa with disjunct distributions in eastern Asia and eastern North America, including *Aralia* sect. *Dimorphanthus* (153), *Gleditsia* (131), *Gordonia* (120), *Hamamelis* (91, 156), *Hydrangea* sect. *Hydrangea* (136), *Magnolia* sect. *Rytidospermum* (121), *Nyssa* (150), *Panax* (152), *Symplocarpus* (154), *Triosteum* (36), and the *Weigela-Diervilla* complex (76). Brunsfeld et al (9) included the presumed disjunct genus-pair *Glyptostrobus* and *Taxodium* in the analysis of Taxodiaceae. These phylogenetic analyses have provided important insights into how the disjuncts are related, whether the intercontinental species pairs are sister species, and whether the traditionally recognized disjunct taxa (genera, sections, and series) are monophyletic.

Overview of Phylogenetic Data

Aralia sect. *Dimorphanthus* (Araliaceae)—A phylogeny of *Aralia* sect. *Dimorphanthus* was constructed using sequences of the ITS regions of nuclear ribosomal DNA (Figure 1*a*) (153). The section consists of 25 species with one species from North America and 24 from eastern and southeastern Asia. Based on morphology, Wen (147) recognized three taxonomic series within *Aralia* sect. *Dimorphanthus:Chinensis*, *Dimorphanthus*, and *Foliolosae*. Series *Dimorphanthus* is distributed disjunctly in eastern Asia and eastern North America, and species from series *Chinensis* and series *Foliolosae* are mostly from subtropical and tropical Asia. The ITS study sampled 16 species from all three series. Li (87–89) suggested a species pair relationship between the eastern North American *A. spinosa* and the eastern Asian *A. chinensis*. *Aralia chinensis* sensu Li (87–89) includes two taxonomic entities: the widely distributed Asiatic *A. elata* and *A. stipulata* from western China (147).

The ITS phylogeny (Figure 1*a*) shows that the tropical and subtropical series *Chinensis* and series *Foliolosae* are nested within the eastern Asian and eastern North American disjunct series *Dimorphanthus*, which is paraphyletic. Furthermore, the North American *A. spinosa* is cladistically basal in sect. *Dimorphanthus*. Therefore, the presumed species pair (the eastern Asian *A. elata* or *A. stipulata* and the eastern North American *A. spinosa*) are not sister species.

Diphylleia (Berberidaceae)—*Diphylleia*, which has been recently monographed (176), includes three species: *D. cymosa* from eastern North America, and *D. grayi* and *D. sinensis* from eastern Asia. A chloroplast DNA (cpDNA) phylogeny of Berberidaceae (75) suggests its paraphyly, with *Dysosma* and *Podophyllum* nested within *Diphylleia*.

Gleditsia (Fabaceae)—*Gleditsia* contains 13 species, with eight from eastern Asia, two from eastern North America, one in the Himalayas, one in western Asia, and one in South America. A phylogeny of *Gleditsia* was constructed based on *ndh*F and *rpl*16 chloroplast gene sequences (Figure 1*c*) (131). Previous taxonomic work (35) suggested an eastern Asian-eastern North American vicariad species pair (*G. aquatica* from eastern United States and *G. microphylla* from China). In addition, the eastern North American *G. triacanthos* was regarded as closely related to the eastern Asian *G. japonica*, *G. delavayi*, and the western Asian *G. caspica* (35). Molecular analysis (131) supports the monophyly of *Gleditsia* and suggests only a single Asian–North American disjunction and no intercontinental sister species relationships.

Gordonia (Theaceae)—*Gordonia* s. l. has at various times included all Asian species of *Gordonia*, *Polyspora*, and *Laplacea*, and the southeastern North American *G. lasianthus*. The *rbc*L and *mat*K data (120) put *Gordonia lasianthus* of eastern North America in a lineage with *Schima* and *Franklinia*. The Asian *Polyspora* and *Laplacea* representatives all fell in the clade with *Camellia* and *Tutcheria*. Thus, *Gordonia* s.l. is paraphyletic.

Hamamelis (Hamamelidaceae)—*Hamamelis* (the witch hazel genus) consists of four to six species with two from eastern Asia and two to four species from eastern North America. The phylogeny of *Hamamelis* was constructed with the ITS sequences (Figure 1*b*) (156) as well as the combined nrITS and the chloroplast *mat*K sequences (91). The combined analyses (91) generated the same topology as the ITS phylogeny (156) and the *mat*K sequences of *Hamamelis* showed little variation. Although several workers suggested a close relationship between the eastern Asian *H. japonica* and the North American *H. vernalis* (89, 104), the molecular phylogeny (91, 156) shows that the two Asiatic species are phylogenetically basal and the North American species form a monophyletic group, suggesting a single intercontinental disjunction. Furthermore, there is no intercontinental sister species relationship between eastern Asia and eastern North America in *Hamamelis*.

Hydrangea sect. *Hydrangea* (Hydrangeaceae)—In a phylogenetic study of Hydrangeaceae, Soltis et al (136) sampled four species of the Asian and eastern North American *Hydrangea* sect. *Hydrangea*. The *rbc*L phylogeny suggests the polyphyly of sect. *Hydrangea*.

Magnolia sect. *Rytidospermum* (Magnoliaceae)—Qiu et al (121) examined the phylogeny of *Magnolia* sect. *Rytidospermum* using cpDNA restriction site variation. The section is composed of six species of temperate deciduous trees, which are morphologically similar. The cpDNA analysis shows that *Magnolia* sect. *Rytidospermum* is polyphyletic (Figure 2*a*). The similarity in leaf morphology and wood anatomy, which define the eastern Asian–eastern North American disjunct section, is attributable to convergence. No direct sister species relationships are detected in this section.

Nyssa (Cornaceae)—Wen & Stuessy (150) analyzed *Nyssa* using morphological characters (Figure 1*e*). *Nyssa* has eight species with four from eastern Asia, three from eastern North America, and one from Costa Rica. Two phylogenetic connections are detected between Asia and the Americas. The presumed species pair relationship between *N. sylvatica* from eastern North America and *N. sinensis* from eastern Asia (29) is not supported by the phylogenetic analysis. *Nyssa*

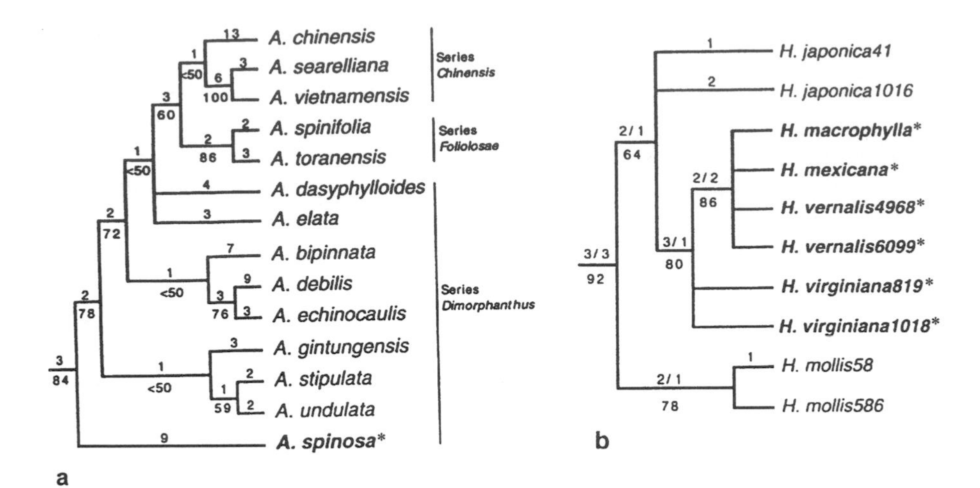

A. chinensis
A. searelliana
A. vietnamensis
Series Chinensis
A. spinifolia
A. toranensis
Series Foliolosae
A. dasyphylloides
A. elata
A. bipinnata
A. debilis
A. echinocaulis
Series Dimorphanthus
A. gintungensis
A. stipulata
A. undulata
A. spinosa*
a
H. japonica41
H. japonica1016
H. macrophylla*
H. mexicana*
H. vernalis4968*
H. vernalis6099*
H. virginiana819*
H. virginiana1018*
H. mollis58
H. mollis586
b

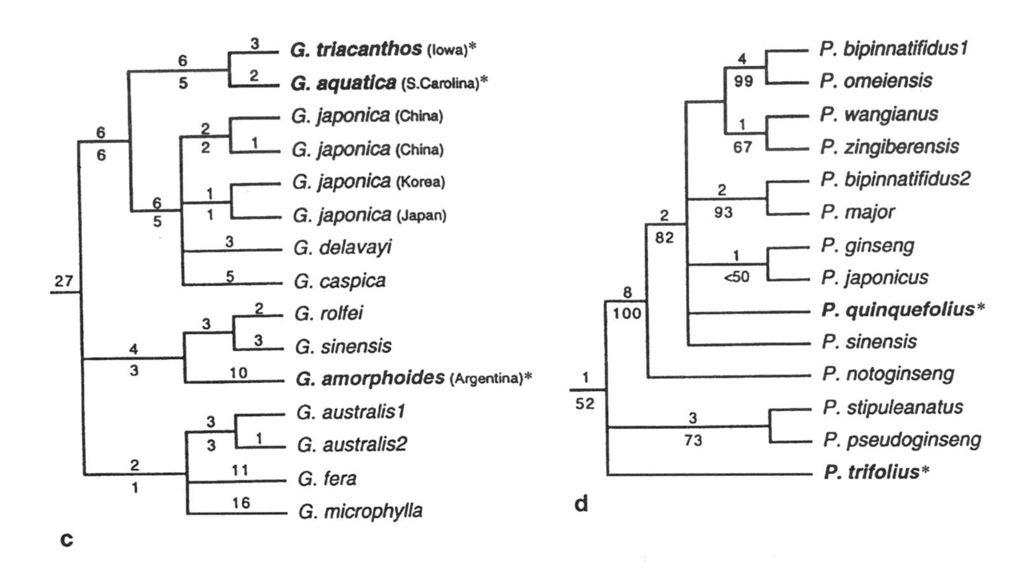

G. triacanthos (Iowa)*
G. aquatica (S.Carolina)*
G. japonica (China)
G. japonica (China)
G. japonica (Korea)
G. japonica (Japan)
G. delavayi
G. caspica
G. rolfei
G. sinensis
G. amorphoides (Argentina)*
G. australis1
G. australis2
G. fera
G. microphylla
c
P. bipinnatifidus1
P. omeiensis
P. wangianus
P. zingiberensis
P. bipinnatifidus2
P. major
P. ginseng
P. japonicus
P. quinquefolius*
P. sinensis
P. notoginseng
P. stipuleanatus
P. pseudoginseng
P. trifolius*
d

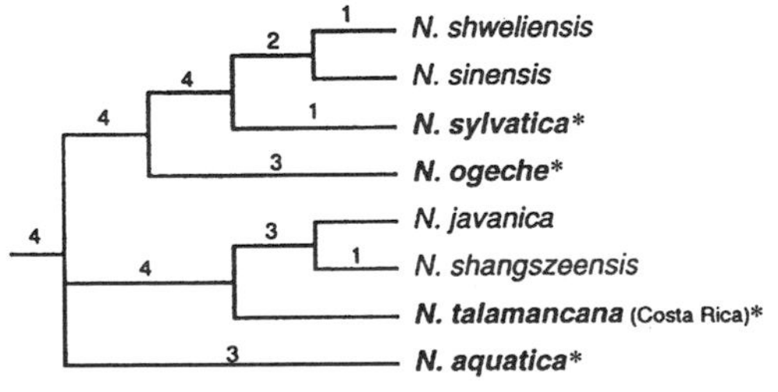

N. shweliensis
N. sinensis
N. sylvatica*
N. ogeche*
N. javanica
N. shangszeensis
N. talamancana (Costa Rica)*
N. aquatica*
e

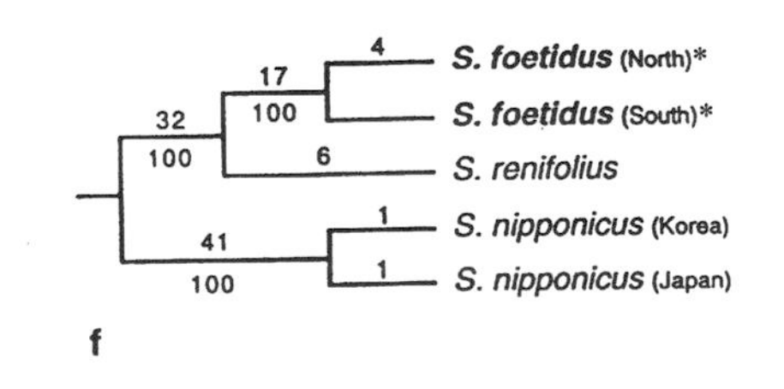

S. foetidus (North)*
S. foetidus (South)*
S. renifolius
S. nipponicus (Korea)
S. nipponicus (Japan)
f

sinensis is a sister species to *N. shweliensis* from eastern Asia. *Nyssa sylvatica* is basal to the *N. shweliensis–N. sinensis* clade. *Nyssa sylvatica* is diversified into two varieties, var. *sylvatica* and var. *ursina*.

Panax (Araliaceae)—A phylogeny of *Panax*, the ginseng genus, was constructed using nrITS sequences (Figure 1*d*) (152). *Panax* consists of approximately 12 species with about ten from eastern Asia and two from eastern North America (*P. quinquefolius* and *P. trifolius*). *Panax quinquefolius* is more closely related to the eastern Asian species in the ITS tree, while *P. trifolius* is phylogenetically isolated. Although the eastern Asian *P. ginseng* was regarded as most closely related to the eastern North American *P. quinquefolius* (62, 87, 88, 180), *P. ginseng* is sister to *P. japonicus* from Japan. No intercontinental sister species relationships are found in *Panax*.

Symplocarpus (Araceae)—Chloroplast DNA variation was surveyed with 20 restriction endonucleases for *Symplocarpus*, which has three species, two from eastern Asia and one from eastern North America (154). The cpDNA phylogeny (Figure 1*f*) reveals a sister group relationship between *S. foetidus* from eastern North America and *S. renifolius* from eastern Asia.

Triosteum (Caprifoliaceae)—A phylogenetic analysis of *Triosteum* was performed using nrITS and the *waxy* gene sequences (36). *Triosteum* has seven species with three from eastern North America and four from eastern Asia. No intercontinental sister species relationships are detected in the molecular phylogenies.

The *Weigela-Diervilla* generic pair (Caprifoliaceae)—*Weigela* consists of 11–12 species from eastern Asia, and *Diervilla* has three species from eastern North America. An ITS phylogeny of the complex (76) suggests that *Weigela* is paraphyletic with *Diervilla* nested within it. The three species of *Diervilla* are morphologically distinct from their eastern Asian *Weigela* relatives in their habit and floral morphology.

Phylogenetic Relationships

Phylogenetic analyses have been done on only a few disjunct taxa, and most are based on a single molecular marker or morphology. The robustness of the phylogenies needs to be tested. Nevertheless, it is still possible to draw some inferences, which may serve as working hypotheses for future studies.

Figure 1 Phylogenetic relationships of examples of eastern Asian–eastern North American disjuncts: (*a*) *Aralia* sect. *Dimorphanthus* (153); (*b*) *Hamamelis* (156); (*c*) *Gleditsia* (131); (*d*) *Panax* (152); (*e*) *Nyssa* (150); (*f*) *Symplocarpus* (154). The trees are redrawn from the references indicated in parentheses. Names of New World species are marked with "*". Numbers above lines indicate minimum character support in *a*, *c*, *d*, *e*, and *f*; and numbers below lines are bootstrap values in *a*, *b*, *d*, and *f*, and decay indices in *c*. In *b*, numbers above lines are the branch lengths (*left*, or the sole number) and decay indices (*right*).

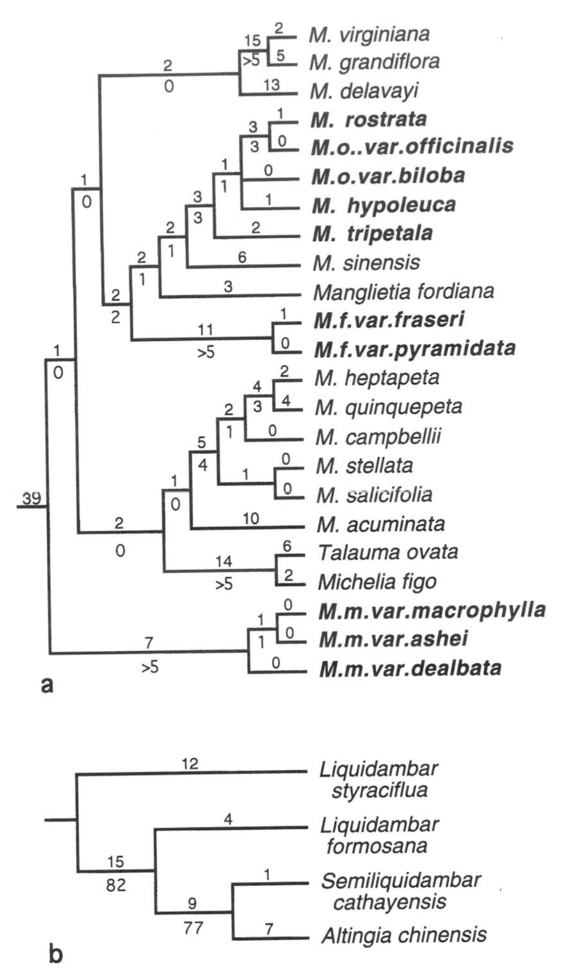

Figure 2 Examples of polyphyletic and paraphyletic disjunct taxa: (*a*) *Magnolia* sect. *Rytidospermum*; (*b*) *Liquidambar*. The bold faced taxa in (*a*) are members of sect. *Rytidospermum*.

The presumed species pairs are not sister species in most eastern Asian–eastern North American disjunct genera with three or more species. In all taxa examined (except *Symplocarpus*), there is a lack of direct intercontinental sister species relationship (Figures 1*a–e*). Species apparently have undergone further speciation either before or after the formation of the disjunct pattern in one or both continents. Thus, the traditional view, emphasizing the disjunction at the level of the same species or closely related species pairs, needs to be modified. The disjunct pattern was first thought to be the discontinuous distribution of the same species in these geographically isolated areas. For example, Halenius (48) listed nine species from both Kamchatka and North America in 1750; Thunberg (142) included 20 species in both Japan and North America; and Gray (44) reported 134 "identical species" between Japan and eastern North America. Subsequently almost all of Gray's "identical species" were shown to be morphologically distinct and were treated as disjunct congeners (e.g., 27, 50, 88). Many later workers recognized species pairs between eastern Asia and eastern North America (13, 29, 60, 69, 77, 88, 89). Phylogenetic analyses have shown that the disjunction is mostly among closely related species, but usually not sister species in those taxa with three or more species examined. Additional genera need to be analyzed before a generalization can be reached.

In disjunct taxa with only two species, the sister species relationship is, of course, supported as in *Caulophyllum* and *Jeffersonia* of Berberidaceae (75) and *Cornus* subgenus *Mesomora* of Cornaceae (109, 172). The two species of *Caulophyllum* exhibit a relatively high level of sequence divergence as measured from allozyme (Nei's genetic identity is 0.546) and ITS sequence data (1.31% divergence). The two *Caulophyllum* species thus represent relatively ancient disjuncts, which diverged perhaps in the late Miocene (75). Some disjunct taxa with two vicariad species such as *Campsis* (151), *Cornus* subgenus *Mesomora* (30), and *Liriodendron* (5, 116) are relicts, as suggested by fossil evidence. Sister species in these taxa may not have a direct progenitor-derivative relationship.

In *Aralia* sect. *Dimorphanthus*, *Gleditsia*, and *Hamamelis*, only a single intercontinental disjunction has been detected. In some species-rich disjunct genera, multiple patterns are evident. For example, two or more disjunctions are found in *Nyssa* and *Panax*.

In some instances, the perceived disjunction may be due to convergence. *Magnolia* sect. *Rytidospermum* (Figure 2*a*) (121) and *Hydrangea* sect. *Hydrangea* (136) are polyphyletic. Qiu et al (121) discussed the general habitat similarity of the taxa in eastern Asia and eastern North America. The similar habitats may have exerted similar selection pressures and led to similar morphological adaptations among the six species of this section. Wen et al (155) showed the biphyly of *Aralia* sect. *Aralia*, which has a disjunct distribution among eastern Asia, eastern North America, and western North America. Because the extent of polyphyly among the disjunct genera is not known, it is necessary to include close relatives in phylogenetic analyses of disjunct genera between eastern Asia and eastern North America.

The eastern Asian and the eastern North American disjunct *Diphylleia* and series *Dimorphanthus* of *Aralia* sect. *Dimorphanthus* are paraphyletic (75, 153). The morphological cohesion of these paraphyletic disjunct groups may be attributable to symplesiomorphies. These two disjunct groups represent grades rather than clades. Morphologically the temperate group of *Aralia* sect. *Dimorphanthus*, i.e., series *Dimorphanthus*, is cohesive, and their tropical and subtropical relatives are distinct, recognized taxonomically as series *Chinensis* and series *Foliolosae*. This pattern of paraphyly, with some tropical and subtropical species derived from the north temperate disjuncts, suggests that the evolution of the eastern Asian and eastern North American disjuncts should be examined in the context of global biogeography (82a).

A sister group relationship between the Japanese and North American species is reported for *Hamamelis* (91) and *Viburnum* sect. *Pseudotinus* (36). Gould & Donoghue (36) and Li et al (91) suggested that this pattern may have resulted from vicariance following the Beringian connection between Asia and North America. The intercontinental "SISTER GROUP" relationship does not necessarily represent direct intercontinental "SISTER SPECIES" relationship. For example, *Hamamelis japonica* from Japan has a sister group relationship with the eastern North American clade (Figure 1*b*), but it does not have a direct sister species relationship with any of the North American species.

EXAMINATION OF EASTERN ASIAN AND EASTERN NORTH AMERICAN DISJUNCTION IN A BROADER BIOGEOGRAPHIC CONTEXT

Phylogenetic studies of genera that are distributed widely in the northern hemisphere, especially in eastern Asia, eastern North America, western North America, western Asia, and southeastern Europe, might provide additional insight into the floristic migration between eastern Asia and eastern North America (143). Based on phylogenetic studies of *Aralia* sect. *Aralia* and *Calycanthus* with a disjunct distribution in eastern Asia and eastern and western North America, Wen et al (149, 153, 155) suggested a close biogeographic relationship between eastern and western North America, with eastern Asia basal to the North American clade. A close floristic relationship between eastern and western North America has been detected in the fern *Adiantum pedatum* complex (113, 114). Xiang et al (173) found this pattern in *Boykinia*, *Cornus*, *Tiarella*, and *Trautvetteria* and provided an excellent synthesis of the disjunction between eastern and western North America, and eastern Asia. The close biogeographic relationship between eastern and western North America is also seen in *Trillium* (74). This pattern has not been emphasized owing to the existence of past biogeographic barriers. The North American Mid-Continental Seaway separated eastern and western North America in the late Cretaceous and then retreated to the north and south by the Paleocene. This regression occurred concomitantly with the uplifting of the Rocky Mountains in the early Tertiary. The North American epicontinental seaway and the subsequent formation of the Rockies are considered to have created biogeographic barriers for floristic exchanges between eastern and western North America (40). The phylogenetic data, however, support a strong biogeographic relationship between eastern and western North America, as previously proposed by Wood (166, 167). Both fossil and molecular data (83, 143, 153, 173) suggest that the above pattern of disjunction evolved multiple times (polychronically). Xiang et al (173) refer to the polychronic biogeographic pattern as "pseudocongruence," following Cunningham & Collins (18). The common occurrence of this pattern suggests that the traditional emphasis on the floristic similarity between eastern Asia and eastern North America needs to be enlarged to include western North America.

A closer link between eastern North America and western Asia than between North America and eastern Asia is reported in *Liquidambar* based on allozyme divergence data (57, 58). *Liquidambar* consists of four species with one from eastern North America, one from western Asia, and two from eastern Asia. Hoey & Parks (57) have shown that *L. styraciflua* from eastern North America and *L. orientalis* from western Asia represent the most closely related intercontinental species pair. This relationship was recently supported by a phylogenetic analysis using chloroplast *mat*K sequences (90). The most likely explanation for the pattern in *Liquidambar* is the migration of species along the Tethys seaway in the Tertiary and across the North Atlantic land bridge (57).

Xiang et al (174) interpreted the phylogenetic and biogeographic history of *Aesculus*, which consists of 13–19 species in eastern Asia, eastern and western North America, and Europe, using nrITS and chloroplast *mat*K sequences, and morphology. Several differences were found between the ITS and *mat*K phylogenies. Based on evidence from phylogeny, molecular sequence divergence, and fossils, the authors suggest that *Aesculus* originated at a high latitude in eastern Asia and migrated into Europe and North America as part of the boreotropical flora. The current intercontinental disjunction in *Aesculus* is interpreted to have resulted from geologic and climatic changes during the Tertiary. Both the Bering and the North Atlantic land bridges contributed to the widespread distribution of *Aesculus* in the Tertiary. The disjunction between Eurasia (Old World) and North America (New World) must have occurred first in the genus after the breakup of the North Atlantic land bridge. The disjunction between eastern and western North America is assumed to have been before the separation of the Japanese and the European species in *Aesculus*. A strong correlation is found between the phylogenetic relationships and the geographic distribution (174).

Boufford et al (8) conducted a phylogenetic analysis of *Circaea*, a herbaceous perennial genus of seven species from the northern hemisphere. The closest relative of the eastern North American *C. lutetiana* ssp. *canadensis* is *C. lutetiana* ssp. *quadrisulcata* from eastern Asia (barely extending to eastern Europe). The western North American *C. alpina* ssp. *pacifica* is sister to the clade consisting of the Chinese *C. alpina* ssp. *micrantha* and the circumboreal *C. alpina* ssp. *alpina*.

Phylogenetic and/or biogeographic studies of additional northern hemisphere intercontinental disjunct taxa such as *Acer* (52, 165), *Crataegus* (119), Juglandaceae (95, 98), *Prunus sambucus* (S T Lee, J Wen, unpublished), *Styrax* (33), *Symplocos* (78), Ulmaceae (96), and *Viburnum* (20–23) suggest complex biogeographic relationships among taxa from five major areas of the northern hemisphere: eastern Asia, eastern North America, western North America, western Asia, and southeastern Europe (e.g., 33, 119, 165). Detailed biogeographic analyses of these genera have been hindered by the difficulties in sampling throughout their distributional ranges. Another obstacle is the lack of analytical tools for reconstructing complex biogeographic patterns (108).

Phylogenetic evidence from these more broadly distributed groups shows that eastern Asian taxa may not be most closely related to eastern North American relatives. Closer ties are found between eastern and western North America in many genera (see 173 for a review), between eastern North America and western Asia (57, 58, 90), and between eastern Asia and southeastern Europe (174). Only one clade in *Circaea* and one clade in *Corylus* shows eastern Asia closest to eastern North America. Thus, phylogenetic studies support the diffuse origin of floras in eastern Asia and eastern North America (143). Eastern Asian and/or eastern North American floristic elements may be closely related to those of western North America, Europe, western Asia, tropical Asia, tropical America, and circumpolar regions. The diffuse origin hypothesis was previously suggested based on fossil data (e.g., 78, 143) and subsequently supported by the phylogenetic evidence.

PATTERNS OF DIFFERENTIATION

Morphological Stasis

Morphological stasis is the lack of significant morphological change in a species for long periods of time (158). Evolutionary stasis has been discussed repeatedly in relation to the "punctuated equilibrium" mode for evolutionary change and remains a highly debated topic (e.g., 24, 37, 85, 86, 99, 100, 158). Morphological stasis is suggested to be common in the evolution of eastern Asian and eastern North American disjunct plants (57, 58, 116, 121, 122). Parks & Wendel (116) reported that the two morphologically similar disjunct species of *Liriodendron* (*L. chinense* and *L. tulipifera*) show a high level of allozyme and cpDNA divergence. Molecular and fossil data suggest the divergence time of the two species to be 10–16 mya (middle Miocene). The morphological similarity of the two species of *Liriodendron* after such a long period of separation is thought to be due to stasis (116). This mode of evolution is also proposed for the Asian and North American species of *Liquidambar* (57) and *Magnolia* sect. *Rytidospermum* (122). Liston et al (93) reported morphological stasis in *Datisca*, a genus disjunct between western North America and central and western Asia. Based on correlation of geographic ranges in eastern Asia and North America, Ricklefs & Latham (126) suggested stasis in ecological attributes in disjunct perennial herbaceous genera generally.

Constraints and stabilizing selection are proposed to be the two major mechanisms for evolutionary stasis (158). Stasis by constraints is the lack of significant changes through intrinsic features of the genetic and developmental architecture, and stasis via stabilizing selection is through the elimination of all phenotypes that deviate significantly from the populational mean. Parks and his collaborators (57, 116, 121, 122) have emphasized the similar habitats occupied by these disjuncts; thus they have preferred stabilizing selection as the rationale for the morphological stasis of the disjuncts between eastern Asia and eastern North America. Williamson (158) stressed that stabilizing and constraining mechanisms may act in concert. The relative contributions of these two mechanisms to morphological stasis in the disjuncts are not well understood, and the extent of morphological stasis among the disjuncts needs to be studied. The general lack of reasonably good fossil record for most disjunct groups presents an obstacle for understanding this mode of evolution. Stasis may be operating only at the morphological level in the Asian–North American disjuncts, or there may be evolutionary stasis at both morphological and molecular levels.

Wen (148) suggested that morphological stasis may be investigated by examing patterns of morphological similarity of the disjunct taxa and their derivatives in a phylogenetic framework. Morphological stasis among the eastern Asian and eastern North American disjuncts predicts both a high level of morphological similarity among the disjunct taxa and morphological divergence between the disjunct taxa and their derivatives from other geographic areas. Comparing the patterns of morphological similarity and phylogenetic relationships, Wen (148) suggested that morphological stasis may have occurred in the north temperate species in the Altingioideae (Hamamelidaceae). The Altingioideae includes the north temperate genus *Liquidambar* and the subtropical and tropical *Altingia* and *Semiliquidambar* (25). *Liquidambar* consists of four morphologically cohesive species. Its tropical and subtropical relatives are morphologically distinct and recognized as the genera *Altingia* and *Semiliquidambar*. *Liquidambar* has recently been shown to be paraphyletic (Figure 2*b*) (132), with *Altingia* and *Semiliquidambar* nested within it. The paraphyly of *Liquidambar* is compatible with predictions from the hypothesis of morphological stasis in the north temperate species in *Liquidambar*. Morphological stasis of the eastern Asian and eastern North American disjuncts is also supported by phylogenetic evidence from *Aralia* sect. *Dimorphanthus* (148) and *Osmorhiza* (J Wen, MD McBroom, P Lowry, unpublished).

Discordance Between Overall Similarity and Phylogenetic Relationships

Discordance between the overall DNA sequence similarity and phylogenetic relationship is detected in several disjunct taxa including *Aralia* sect. *Aralia* (155), *Aralia* sect. *Dimorphanthus* (153), *Magnolia* sect. *Rytidospermum* (121, 122), and *Panax* (152). In these taxa, disjunct species between eastern Asia and eastern North America have a high overall DNA sequence similarity but lack sister species relationships. Overall similarity among species may be attributed to synapomorphies (shared derived characters), symplesiomorphies (shared primitive characters), and convergence (independently evolved similar characters) (54). Symplesiomorphies and convergence contribute to overall similarity, but not to phylogenetic (i.e., cladistic) relationships. The incongruence between overall similarity and phylogenetic pattern may be due to dominance of symplesiomorphies and convergence among the disjunct taxa.

There may be a disparity between the level of morphological similarity and the phylogenetic relationships among the eastern Asian and eastern North American disjunct plants and their relatives in western North America. In genera with a broad distribution in eastern Asia, eastern North America, and western North America, many species pairs are proposed between eastern Asia and eastern North America based on morphological similarity. For example, Li (88, 89) suggested a species pair relationship between the eastern Asian *Aralia cordata* and the eastern North American *A. racemosa*, with the western North American *A. californica* thought to be more distinct. Another example is found in *Osmorhiza* of the Apiaceae. Lowry & Jones (94) treated the eastern Asian *O. aristata* in the same taxonomic section as the eastern North American *O. claytonii* and *O. longistylis*. The western North American and South American *Osmorhiza* species were placed in other sections. Phylogenetic studies of *Aralia* sect. *Aralia* (155) showed that the eastern North American *A. racemosa* is sister to the western North American *A. californica*, and the eastern Asian species are basal to the North American clade. This result is striking because of the discordance between the reported morphological similarity and the phylogenetic relationship. Without detailed morphological comparisons, it is not known if the discordance is real. Careful analyses of the patterns of morphological differentiation of the eastern Asian and eastern North American disjunct species are needed in comparison with their relatives in western North America and other geographic areas.

Cross Compatibility Among Eastern Asian and Eastern North American Disjuncts

Eastern Asian and eastern North American disjunct plant species demonstrate cross compatibility under artificial conditions (116, 122). Viable hybrids are reported for the intercontinental disjuncts in *Campsis* (125), *Catalpa* (112), *Liriodendron* (53, 115, 128), and *Magnolia* sect. *Rytidospermum* (116). Artificial hybrids between the two disjunct species of *Campsis* are designated as *C. tagliabuana* (125, 130) and are cultivated as an attractive ornamental. *Catalpa x erubescens* is a hybrid between the eastern Asian *C. ovata* and the eastern North American *C. bignonioides*, and is often cultivated in Europe and North America. The hybrid between the eastern Asian *C. ovata* and the eastern North America *C. speciosa* is known as *C. x galleana* (112). Vigorous hybrids are obtained between Japanese and eastern North American species of *Magnolia* sect. *Rytidospermum* and can be successfully backcrossed with their American parent (116). Artificial crosses between intercontinental disjuncts of the north temperate zone have been reported in *Aesculus* (146), *Betula* (72), *Castanea* (72), *Juglans* (72), *Liquidambar* (53, 127), *Pinus* (105), *Platanus* (129, 130), and *Taxus* (72).

Molecular and fossil evidence has shown the antiquity of most eastern Asian and eastern North American disjuncts. The two species of *Campsis* may have been separated for more than 24 million years (151). The divergence time for the two species of *Liriodendron* is estimated to be 10–16 mya (116). Parks & Wendel (116) point out that cross compatibility does not indicate recent divergence.

Higher Species Diversity in Eastern Asia

Of the approximately 65 disjunct genera between eastern Asia and eastern North America, most have more species in eastern Asia (Table 1). The richness of taxonomic diversity in eastern Asia is also revealed in recent comparisons of terrestrial bird and mammal fauna (107), trees (81, 82), and vascular plants in general (46, 92). Tiffney (144) suggested that the Paleocene forests of eastern North America may have been as rich as those of eastern Asia. The present low diversity is "a function of the subsequent Eocene-Oligocene and/or Pleistocene climatic fluctuations" (144, p. 248).

Eastern Asia has a more complex topography than does eastern North America. The collision of the Indian plate with Eurasia in the Tertiary caused numerous orogenies in eastern Asia, especially in China (64). Furthermore, most mountain ranges in eastern Asia have an east-west orientation (179), which prevented the southward migration of glaciers in the Quaternary (64). The mountains also provided barriers for speciation. The higher species diversity in eastern Asia may be due to both the lesser impact of the Quaternary glaciations and more active speciation. The difficulties in delimiting species in many eastern Asian plants, e.g., *Aralia*, *Astilbe*, *Betula*, *Carpinus*, *Gleditsia*, *Lespedeza*, *Lindera*, *Magnolia*, *Panax*, *Schisandra*, and *Stewartia*, may reflect their recent divergence.

Axelrod et al (4) recently proposed two additional reasons to explain the floristic richness of China in comparison with Europe and the United States. First, China has more extensive tropical rainforest, which is species-rich. Second, China has vegetational continuity between tropical, subtropical, temperate, and boreal forests. This unbroken continuity is considered important in the formation of rich plant associations. These historical and ecological explanations are also discussed in a few recent papers (e.g., 46, 47, 79, 81, 82, 92).

POSSIBLE ORIGINS AND DEVELOPMENT OF THE DISJUNCT PATTERN

The origin and development of eastern Asian and eastern North American disjunctions have been discussed based on paleobotanical, geologic, paleoclimatic, and molecular evidence. Most eastern Asian and eastern North American disjuncts are apparently remnants of widely distributed genera in the north temperate zone during the Tertiary, which disappeared in western North America owing to the orogenies and climatic changes during the late Tertiary and Quaternary, and then became extinct in western Europe during the Quaternary glaciations (40, 64, 98). The wider Tertiary distributions of these genera in the northern hemisphere were due to the Bering (63) and the North Atlantic (102, 103, 144) land bridges. The relative significance of these two routes in the development of the disjunct pattern is controversial. A brief review of the evidence from geology, fossils, phylogenetic patterns, and molecular divergence data is provided below.

Bering Land Bridge

The Bering land bridge has connected northeastern Asia and northwestern North America several times since the Mesozoic. Floristic and faunal migrations over the land bridge were primarily controlled by climate (63, 103). The latitude of the bridge in the early Tertiary was approximately 75°N, which is about 7° farther north than its present position. During warm intervals in the early and late Eocene, some tropical and subtropical species could have migrated across the bridge. Andrews (1) provided new estimates of positions for the pole using hot spots for reference points. Using Andrews' data, Taylor (140) estimated that the latitude of Beringia was approximately 69° during the Paleocene in comparison with 75° then (102). Beringia was thus farther south than previously thought. During the Miocene, the Bering land bridge was suitable for exchanges of temperate deciduous plants and remained available for floristic exchanges until about 3.5 mya. In addition, an Aleutian land bridge may have existed in the Tertiary south of the Bering land bridge. The geologic history of that area is complex, composed of terranes accreted during the late Mesozoic to early Cenozoic (49).

North Atlantic Land Bridge

Thc North Atlantic land bridge may have played a major part in the spread of thermophilic flora and the development of the floristic similarity between eastern Asia and eastern North America (143, 144). Its significance was proposed by paleozoologists (e.g., 101–103). For an excellent review of the North Atlantic land bridge, see Tiffney (144), from which this discussion is mostly derived. Floristic exchanges between eastern Asia and eastern North America via the North Atlantic land bridge require two major land links: one between Europe and eastern North America, and the other between Europe and eastern Asia. Eastern North America was connected to Europe in the early Tertiary via Greenland, with a northern connection in the high arctic region and a southern connection through southern Greenland. The latitude of the southern route was approximately 60° (140), thus farther south than the Bering land bridge. Eastern Asia and Europe were connected via the land bridge along the coasts of the Tethys Seaway in the early Tertiary or after the disappearance of the Turgai Straits in the early Oligocene.

Tiffney (143, 144) postulated that floristic migration via the North Atlantic land bridge was possible during the Paleocene and Eocene. By the Miocene, they may have permitted the filtering of some deciduous taxa through a series of "stepping stones." Although the migrations of thermophilic species across the North Atlantic in the early Tertiary are viewed as more likely than across the high-latitude Beringia (101, 102, 144), Manchester (97) pointed out the early Tertiary fossil floras studied so far from Greenland and Spitzbergen do not contain many thermophilic taxa.

Historical Explanations

Gray (45) employed a multidisciplinary approach and proposed that many plant taxa were widely distributed in the northern hemisphere during the Tertiary as a polar flora, and that subsequent glaciations had interrupted this continuous distribution, leading to the current eastern Asian and eastern North American disjunction. Chaney (11) and Axelrod (3) updated Gray's hypothesis and suggested that these floristic similarities had their origin in the range restrictions of the Arcto-Tertiary

Geoflora of the northern hemisphere because of climatic changes in the late Tertiary and Quaternary. A geoflora is "a group of plants which has maintained itself with only minor changes in composition for several epochs or periods of earth history, during which time its distribution has been profoundly altered although the area it has covered at any one time may not have varied greatly in size" (12, p. 12). Further paleobotanical and geologic discoveries challenged the geoflora concept and demonstrated that there was no evidence of a homogeneous high latitude flora similar to extant eastern Asian–eastern North American temperate floras in the early Tertiary (55, 161–164). Wolfe (163) showed that the neotropical elements in the early Tertiary had affinities with extant genera in southeastern Asia. He proposed a northern subtropical and tropical flora known as the boreotropical flora in the early Tertiary, extending from North America to Europe, and along the shores of the Tethys seaway into Asia. Differing from the Arcto-Tertiary Geoflora, but compositionally similar to the Paleotropical-Tertiary Geoflora, the boreotropical flora contained many thermophilic tropical and subtropical taxa and had a diffuse origin (143). One of the difficulties lies in defining thermophilic taxa. Many disjunct genera such as *Aralia*, *Lindera*, *Magnolia*, *Nyssa*, *Parthenocissus*, and *Schisandra* are currently distributed in the temperate and subtropical to tropical regions. These genera thus have both temperate deciduous and thermophilic elements/species. An understanding of phylogenetic relationships of these genera is needed to determine if the disjunction between eastern Asia and eastern North America is among mostly the thermophilic species or the temperate deciduous species, or if there exist more complex patterns among the temperate and thermophilic elements.

Tiffney (143, 144) reviewed the fossil, geographic, geologic, and climatic evidence and hypothesized a multiple origin of the eastern Asian and eastern North American disjunct pattern throughout the Tertiary. He (143) proposed five major periods/patterns (the pre-Tertiary, early Eocene, late Eocene-Oligocene, Miocene, and late Tertiary-Quaternary), during which many disjuncts migrated between Asia and North America via the Bering land bridge or the North Atlantic land bridge. The pre-Tertiary pattern may be dominant in some aquatic angiosperms and monocots, conifers, bryophytes, and pteridophytes. The early Eocene pattern may be responsible for many of the evergreen disjuncts such as the Magnoliaceae, Lauraceae, and Theaceae, accompanied by many herbs of the forest floor or disturbed forest sites, and some deciduous trees. The late Eocene-Oligocene pattern fits well with some taxa adapted to cooler and more seasonal sites. The Miocene pattern may explain many deciduous and some herbaceous disjuncts. The late Tertiary-Quaternary pattern may explain the disjuncts with the arctic and alpine forms. An extensive review of Tertiary floras of North America in comparison with Europe and Asia by Manchester (98) also supports the multiple patterns in the timings and routes for floristic exchanges between Asia and North America.

Iltis (68) argued for the importance of long distance dispersal in the evolution of the eastern Asian and eastern North American disjunction. This proposal has been generally rejected. However, the long distance dispersal hypothesis may explain disjunctions of certain ferns (134) and some herbaceous angiosperm taxa (88).

Estimates of Times of Divergence Based on Molecular Data

Estimates of divergence times between eastern Asian and eastern North American disjunct taxa may be obtained from fossil evidence and/or molecular clock calculations. Most disjunct genera have a poor fossil record (reviewed in 143), but a few recent studies have provided estimates of the times of divergence based on molecular data (Table 2). Congruent estimates are obtained from two molecular markers (see Table 2) for *Caulophylum*, *Gleditsia*, *Gymnocladus*, *Liriodendron*, *Magnolia*, *Penthorum*, and *Phryma*. Most estimates are consistent with available fossil evidence. However, Lee et al (83) reported a disconcordant estimate of divergence times between allozymes and ITS sequences of nuclear ribosomal DNA in *Menispermum*. The allozyme and ITS sequence data suggest the divergence times as 19.5–29 mya and 3.33 mya, respectively. This incongruence may be due to concerted evolution for the ribosomal DNA (181). The fossil record of *Gleditsia* indicates its occurrence in both Asia and North America throughout the Miocene, contrary to molecular estimates of divergence times as the early Pliocene. This conflict may be due to a slower rate of sequence evolution of *ndh*F and *rpl*16 genes in *Gleditsia* or the extinction of an earlier lineage and the reestablishment of *Gleditsia* in North America at a later stage (ca. 4–5 mya, 131). Only further fossil evidence and more accurate calibration of sequence substitution rates can resolve such conflicts.

A wide range of divergence times has been established for different disjunct genera between eastern Asia and eastern North America (Table 2). The molecular data suggest that most disjunct genera have divergence times in the range of 2–25 mya. Among the general patterns proposed by Tiffney (144), the Miocene pattern appears to predominate for the genera studied so far, as was found in *Calycanthus*, *Campsis*, *Caulophyllum*, *Gymnocladus*, *Hamamelis*, *Liriodendron*, *Penthorum*, and *Phryma*. A divergence time of only 1.7–5.5 mya is reported for *Magnolia: M. officinalis* var. *biloba* from eastern Asia and *M. tripetala* from eastern North America (122). This disjunction in these warm temperate *Magnolia* taxa may have evolved during the last warm interval of the late Tertiary (5–6 mya) via the Bering land bridge. An alternative hypothesis is long-distance dispersal, if the lower bounds of the estimate (1.7 mya) in the Pleistocene are considered (122).

Estimates of times of disjunction from molecular data require the assumption of molecular clocks. Although correlations between the amount of molecular divergence and times of origin have been reported (32, 133, 159, 160, 182), there are problems in using a molecular clock to estimate divergence times (2, 14). A further obstacle for molecular estimates is the lack of accurate calibrations for evolutionary rates of different genes in different lineages. Future studies are needed to provide calibrations of molecular clocks using fossil evidence and a phylogenetic framework. Inferences of divergence times from molecular data should therefore be viewed with caution.

TABLE 2 Estimated times of divergence among closely related disjunct species between eastern Asia and North America[a]

Species	Divergence time (mya)	Molecular divergence						Calibration
		Allozyme	cpDNA	*rbc*L	*ndh*F	*rpl*16	ITS	
Calycanthus chinensis-C. floridus	6.6		0.66					10^{-9} substitutions per site per year
Calycanthus chinensis-C. occidentalis	5.5		0.55					10^{-9} substitutions per site per year
Campsis grandiflora-C. radicans	24.4		2.44					10^{-9} substitutions per site per year
Caulophyllum robustum-C. thalictroides	6–6.5 (9–12)	0.546						Table 9.2 of Nei (109a), or 1 unit Nei's genetic distance per 15–20 my (in parentheses)
	3.33						1.31	3.9×10^{-9} substitutions per site per year
G. japonica-G. aquatical/G. triacanthos	3.9–4.7							4×10^{-10} and 6×10^{-10} substitutions per site per year for *ndh*F and *rpl*16, respectively
Gymnocladus chinensis-G. dioica	17.5–13.8				1.5	2.1		4×10^{-10} and 6×10^{-10} substitutions per site per year for *ndh*F and *rpl*16, respectively
[b]*Hamamelis japonica-H. virginiana*	8.5						0.6–0.7	3.5–4.1×10^{-10} substitutions per site per year
Liriodendron chinense-L. tulipifera	10 (16)	0.434						Equation 9.62 of Nei (109a), or 1 unit Nei's genetic distance per 15–20 my (in parentheses)

	12.4 + 1.45		1.24 + 0.145		10^{-9} substitutions per site per year
Magnolia officinalis var. *biloba-M. tripetala*	1.9 (4.1–5.5)	0.712			Equation 9.62 of Nei (109a), or 1 unit Nei's genetic distance per 15–20 my (in parentheses)
	1.7 ± 0.8		0.083		5×10^{-10} substitutions per site per year
Menispermum dauricum-M. canadense	29 (19.5–26)	0.273			Table 9.2 of Nei (109a), or 1 unit Nei's genetic distance per 15–20 my (in parentheses)
	2.35			0.93	3.9×10^{-10} per site per year
Penthorum chinense-P. sedoides	6–6.5 (9.5–12.6)	0.534			Table 9.2 of Nei (109a), or 1 unit Nei's genetic distance per 15–20 my (in parentheses)
	4.21			1.68	3.9×10^{-9} per site per year
Phryma leptostachya var. *asiatica-P. leptostachya* var. *leptostachya*	25 (18.5–24.6)	0.291			Table 9.2 of Nei (109a), or 1 unit Nei's genetic distance per 15–20 my (in parentheses)
	12.35			4.46	3.9×10^{-9} per site per year
Symplocarpus foetidus-S. renifolius	3.4	0.34			10^{-9} substitutions per site per year

[a]Sources of data are *Calycanthus* (153), *Campsis* (151), *Caulophyllum* (873), *Gleditsia* (131), *Hamamelis* (156), *Liriodendron* (116), *Magnolia* (122), *Menispermum* (83), *Penthorum* (83), *Phryma* (83), and *Symplocarpus* (154).

[b]The divergence time of *Hamamelis japonica* and *H. virginiana* was obtained based on fossil evidence. The rate of substitutions was subsequently estimated using that divergence time.

CONCLUSIONS AND FUTURE DIRECTIONS

A disjunct pattern between eastern Asia and eastern North America occurs in 65 genera of flowering plants. Recent phylogenetic analyses have shown that most intercontinental species pairs studied so far are not sister species. Further diversification of species has occurred in one or both continents. A few disjuncts have even differentiated into distinct genera.

Some of the genera, sections, and series traditionally treated as disjunct taxa are now known to be polyphyletic or paraphyletic. The general morphological similarity of the paraphyletic and polyphyletic taxa may be attributable to symplesiomorphies and convergences (148, 153). The extent of polyphyly or paraphyly among the disjunct groups is not known.

It is important to include the close relatives from tropical and subtropical regions in phylogenetic studies of eastern Asian–eastern North American disjuncts (82a, 148). We may be misled by the morphological cohesion of temperate disjuncts and the morphological divergence of their close relatives in the tropical and subtropical regions. Some of the tropical and subtropical relatives may have arisen from these north temperate disjuncts, as documented in *Aralia* sect. *Dimorphanthus* (153) and in the subfamily Altingioideae of Hamamelidaceae (132).

Examination of the eastern Asian and eastern North American disjunct pattern in a broader northern hemisphere biogeography suggests a closer biogeographic relationship between eastern North America and western North America than between eastern North America and eastern Asia (153, 155, 173). These data suggest that the traditional view of the floristic similarity between eastern Asia and eastern North America needs to be modified. Additional phylogenetic studies are needed for genera with wide distributions in the northern hemisphere to further test the close floristic relationship between eastern Asia and eastern North America.

Fossil, geologic, molecular, and phylogenetic evidence suggests complex origins of the disjunct pattern between eastern Asia and eastern North America (98, 144, 148, 153, 173). The "Asa Gray disjunction" probably originated at different times and via different pathways. Tiffney (144) proposed five major periods during which floristic exchanges between eastern Asia and eastern North America were most likely. Molecular and fossil data suggest that the disjunct pattern dates back at least to the Miocene. It is generally accepted that the disjunct pattern represents a relict distribution of taxa with a wider geographic range due to the maximum development of the temperate forests in the mid-Tertiary, especially in the Miocene. Both the Bering and the North Atlantic land bridges probably played important roles in the evolution of the Miocene deciduous forests. The deterioration of the temperate flora in western North America and in Europe during the Pliocene and the Quaternary caused the present distributional pattern (40). The disjunct pattern between eastern Asia and eastern North America is the product of vicariance, dispersal, extinction, and speciation. Examination of large genera in the north temperate zone, such as *Acer*, *Betula*, *Carpinus*, *Malus*, and *Prunus*, using an integrated phylogenetic and paleobotanical approach should provide important insights into the past floristic exchanges between eastern Asia and eastern North America.

Detailed documentation of the distribution and phylogenetic affinities of carefully studied fossil species for the disjunct groups or lineages is needed. Synthesis of data on fossil families by Tiffney (143, 144) and on genera by Manchester (98) has provided insights into times and routes of floristic exchanges between Asia and North America. Synthesis of fossil data at the species level for lineages showing the disjunction in combination with phylogenetic data will provide a more accurate measure of the patterns and times of floristic exchanges between eastern Asia and eastern North America and provide additional insights into the evolution and diversification of the disjuncts.

Comparisons of eastern Asia and eastern North America are needed at the levels of vegetation structure and community dynamics. Historically the biogeographic pattern has been vaguely referred to as an eastern Asian and eastern North American disjunction. However, a major difference is reported in species diversity (e.g., 46, 47, 81, 82) and vegetation composition (6) between the two continents. The biodiversity needs to be evaluated comparatively on multiple-scales (i.e., α-, β- and γ-diversity; 46, 157). These ecological comparisons should provide insights into the mechanisms of evolutionary stasis among the disjuncts.

The taxonomy of the Asian and North American disjunct groups is still not well studied, perhaps owing to the wide geographic range involved. As pointed out by Boufford (6), more accurate distributional data are needed, and important insights may be obtained by comparing the patterns of distribution among the disjunct taxa. Future taxonomic monographs should provide a framework for in-depth biogeographic and evolutionary studies. Collaborations among phylogeneticists, monographers, paleobotanists, and geologists are needed to gain insights into the origin and evolution of the eastern Asian–eastern North American disjunct pattern.

ACKNOWLEDGMENTS

I thank WC Black, IV, DE Boufford, MJ Donoghue, P Fritsch, HJ Hutcheson, RK Jansen, Y-D Kim, SR Manchester, AJ Miller, CR Parks, L Prince, T Sang, A Schnabel, BH Tiffney, IN Whitcher, DA Young, and Q Xiang for their helpful comments, advice, and/or discussions. This study was supported by the National Science Foundation (DEB-9726830) and the National Geographic Society.

LITERATURE CITED

1. Andrews JA. 1985. True polar wander: an analysis of Cenozoic and Mesozoic paleomagnetic poles. *J. Geophys. Res. B* 90: 7737–50
2. Avise JC. 1994. *Molecular Markers, Natural History and Evolution*. New York: Chapman & Hall
3. Axelrod DI. 1960. The evolution of flower

ing plants. In *Evolution After Darwin*, Vol. 1, ed. S. Tax, pp. 227–305. Chicago: Univ. Chicago Press
4. Axelrod DI, Al-Shehbaz I, Raven RH. 1998. History of the modern flora of China. In *Floristic Characteristics and Diversity of East Asian Plants*, ed. A-L Zhang, S-G Wu, pp. 43–55. Beijing: China Higher Educ. Press
5. Baghai NL. 1988. *Liriodendron* (Magnoliaceae) from the Miocene Clarkia flora of Idaho. *Am. J. Bot.* 75:451–64
6. Boufford DE. 1998. Eastern Asian–North American plant disjunctions; opportunities for further investigation. *Kor. J. Plant Tax.* 28:49–61
7. Boufford DE, Spongberg SA. 1983. Eastern Asian-Eastern North American phytogeographical relationships–a history from the time of Linnaeus to the twentieth century. *Ann. Missouri Bot. Gard.* 70:423–39
8. Boufford DE, Crisci JV, Tobe H, Hoch PC. 1990. A cladistic analysis of *Circaea* (Onagraceae). *Cladistics* 6:171–82
9. Brunsfeld SJ, Soltis PS, Soltis DE, Gadek PA, Quinn CJ, et al. 1994. Phylogenetic relationships among the genera of Taxodiaceae and Cupressaceae: evidence from *rbc*L sequences. *Syst. Bot.* 19:253–62
10. Deleted in proof
11. Chaney RW. 1947. Tertiary centers and migration routes. *Ecol. Monogr.* 17:139–48
12. Chaney RW. 1959. Composition and interpretation. In *Miocene Floras of the Columbia Plateau*, ed. RW Chaney, DI Axelrod. Washington, DC: Carnegie Inst. Wash. Publ. 617
13. Chen S-C. 1983. A comparison of orchid floras of temperate North America and eastern Asia. *Ann. Missouri Bot. Gard.* 70:713–23
14. Clegg MT, Gaut BS, Learn GH Jr, Morton BR. 1995. Rates and patterns of chloroplast DNA evolution. In *Tempo and Mode in Evolution: Genetics and Paleontology 50 Years after Simpson*, ed. WM Fitch, FJ Ayala, pp. 215–34. Washington, DC: Natl. Acad. Press
15. Constance L. 1972. Patterns in the distribution of Japanese-American Umbelliferae. In *Floristics and Paleofloristics of Asia and Eastern North America*, ed. A Graham, pp. 93–99. Amsterdam: Elsevier
16. Crawford DJ, Lee NS, Stuessy TF. 1992. Plant species disjunctions: perspectives from molecular data. *Aliso* 13:395–409
17. Deleted in proof
18. Cunningham CW, Collins TM. 1994. Developing model systems for molecular biogeography: vicariance and interchange in marine invertebrates. In *Molecular Ecology and Evolution, Approaches and Application*, ed. B Schierwater, B Streit, GP Wagner, R DeSalle, pp. 405–33. Switzerland: Birkhauser Verlad Base
19. Davidse G. 1983. Biogeographical relationships between temperate eastern Asia and temperate eastern North America: the twenty-ninth annual systematics symposium. *Ann. Missouri Bot. Gard.* 70:421–22
20. Donoghue MJ. 1983. The phylogenetic relationships of *Viburnum*. In *Advances in Cladistics*, Vol. 2, ed. NI Platnick, VA Funk, pp 143–66. New York: Columbia Univ. Press
21. Donoghue MJ. 1983. A preliminary analysis of phylogenetic relationships in *Viburnum* (Caprifoliaceae s.l.). *Syst. Bot.* 8:45–58
22. Donoghue MJ, Baldwin BG. 1993. Phylogenetic analysis of *Viburnum* based on ribosomal DNA sequences from the internal transcribed spacer regions. *Am. J. Bot.* 80 (6):145
23. Donoghue MJ, Sytsma KJ. 1993. Phylogenetic analysis of *Viburnum* based on chloroplast DNA restriction-site data. *Am. J. Bot.* 80 (6):146
24. Eldredge N, Gould SJ. 1972. Punctuated equilibria: an alternative to phyletic gradualism. In *Models in Paleobiology*, ed. TJM Schopf, pp. 82–115. San Francisco: Freeman Cooper

25. Endress PK. 1989. A suprageneric taxonomic classification of the Hamamelidaceae. *Taxon* 38:371–76
26. Enghoff H. 1993. Phylogenetic biogeography of a Holarctic group: the julidan millipedes. Cladistic subordinateness as an indicator of dispersal. *J. Biogeogr.* 20:525–36
27. Engler A. 1879. *Versuch einer Entwicklungsgeschichte der Pflanzenwelt, inbesondere der Florengebiete seit der Tertiärperiode, 1. Die extratropischen Gebiete der nordlichen Hemisphäre*. Leipzig: W. Engelmann
28. Eriksson T, Donoghue MJ. 1997. Phylogenetic relationships of *Sambucus* and *Adoxa* (Adoxoideae, Adoxaceae) based on nuclear ribosomal ITS sequences and preliminary morphological data. *Syst. Bot.* 22:555–73
29. Eyde RH. 1963. Morphological and paleobotanical studies of the Nyssaceae. I. The modern species and their fruits. *J. Arnold Arbor.* 44:1–59
30. Eyde RH. 1988. Comprehending *Cornus*: puzzles and progress in the systematics of the dogwoods. *Bot. Rev.* 233–351
31. Fernald ML. 1931. Specific segregations and identities in some floras of eastern North America and the Old World. *Rhodora* 33:25–63
32. Fitch WM, Ayala FJ. 1995. The superoxide dismutase molecular clock revisited. In *Tempo and Mode in Evolution: Genetics and Paleontology 50 Years after Simpson*, ed. WM Fitch, FJ Ayala, pp. 235–49. Washington, DC: Natl. Acad. Press
33. Fritsch PW. 1998. Phylogeny of *Styrax* (Styracaceae) based on morphological characters, with implications for biogeography and infrageneric classification. *Am. J. Bot.* 85 (6):130
34. Good R. 1974. *The Geography of the Flowering Plants*. London: Longman
35. Gordon D. 1966. *A revision of the genus Gleditsia (Leguminosae)*. PhD thesis. Indiana Univ., Bloomington
36. Gould KR, Donoghue MJ. 1998. Phylogeny and biogeography of *Triosteum* (Dipsacales, Lonicereae). *Am. J. Bot.* 85 (6):133
37. Gould SJ, Eldredge N. 1993. Punctuated equilibrium comes of age. *Nature* 366:223–27
38. Graham A, ed. 1972. *Floristics and Paleofloristics of Asia and Eastern North America*. Amsterdam: Elsevier
39. Graham A. 1972. Outline of the origin and historical recognition of floristic affinities between Asia and eastern North America. See Ref. 15, pp. 1–16
40. Graham A. 1993. History of the vegetation: Cretaceous (Maastrichtian)-Tertiary. In *Flora of North America North of Mexico*, Vol. 1, ed. Flora of North America Editorial Committee, pp. 57–70. New York: Oxford Univ. Press
41. Gray A. 1840. Dr. Siebold, Flora Japonica; sectio prima. Plantas ornatui vel usui inservientes; digessit Dr. J. G. Zuccarini: fasc. 1–10, fol. (A review). *Am. J. Sci.* 39: 175–76
42. Gray A. 1846. Analogy between the flora of Japan and that of the United States. *Am. J. Sci. Arts, ser. 2,* 2:135–36
43. Gray A. 1856, 1857. Statistics of the flora of the northern United States. *Am. J. Sci. Arts, ser. 2,* 22:204–32. 1856; 23:62–84, 369–403. 1857
44. Gray A. 1859. Diagnostic characters of phanerogamous plants, collected in Japan by Charles Wright, botanist of the U. S. North Pacific Exploring Expedition, with observations upon the relationship of the Japanese flora to that of North America and of other parts of the northern temperate zone. *Mem. Am. Acad. Arts* 6:377–453
45. Gray A. 1878. Forest geography and archaeology, a lecture delivered before the Harvard Univ. Nat. Hist. Soc. *Am. J. Sci. Arts,* ser. 3, 16:85–94, 183–96
46. Guo Q. 1999. Ecological comparisons between Eastern Asia and North America: historical and geographical perspectives. *J. Biogeogr.* In press

47. Guo Q, Ricklefs RE, Cody ML. 1998. Vascular plant diversity in eastern Asia and North America: historical and ecological explanations. *Bot. J. Linn. Soc.* 128:123–36
48. Halenius J. 1750. *Plantae Rariores Camschatcenses*. PhD thesis. Univ. Uppsala, Uppsala
49. Hamilton W. 1983. Cretaceous and Cenozoic history of the northern continents. *Ann. Missouri Bot. Gard.* 70:440–58
50. Hara H. 1952, 1956. Contributions to the study of variations in the Japanese plants closely related to those of Europe or North America, Parts 1 and 2. *J. Fac. Sci., Univ. Tokyo,* sect. 3, Botany, 6:29–96, 343–91
51. Hara H. 1972. Patterns of differentiation in flowering plants. See Ref. 15, pp. 55–60
52. Hasebe M, Ando T, Iwatsuki K. 1998. Intrageneric relationships of maple trees based on the chloroplast DNA restriction fragment length polymorphisms. *J. Plant Res.* 111:441–51
53. He S, Santamour FS Jr. 1983. Isoenzyme verification of American-Chinese hybrids of *Liquidambar* and *Liriodendron*. *Ann Missouri Bot. Gard.* 70:748–49
54. Hennig W. 1966. *Phylogenetic Systematics*. Translated by DD Davis, R Zangerl. Urbana: Univ. Ill. Press
55. Hickey LJ, West RM, Dawson MR, Choi DK. 1983. Arctic terrestrial biota: paleomagnetic evidence of age disparity with mid-northern latitudes during the Late Cretaceous and early Tertiary. *Science* 221:1153–56
56. Hickman JC. 1993. *The Jepson Manual, Higher Plants of California*. Berkeley: Univ. Calif. Press
57. Hoey MT, Parks CR. 1991. Isozyme divergence between eastern Asian, North American and Turkish species of *Liquidambar* (Hamamelidaceae). *Am. J. Bot.* 78:938–47
58. Hoey MT, Parks CR. 1994. Genetic divergence in *Liquidambar styraciflua*, *L. formosana*, and *L. acalycina* (Hamamelidaceae). *Syst. Bot.* 19:308–16
59. Hong D-Y. 1983. The distribution of Scrophulariaceae in the Holarctic with special reference to the floristic relationships between eastern Asia and eastern North America. *Ann. Missouri Bot. Gard.* 70:701–12
60. Hong D-Y. 1993. Eastern Asian–North American disjunctions and their biological significance. *Cathaya* 5:1–39
61. Hongo T, Yokoyama K. 1978. Mycofloristic ties of Japan to the continents. *Mem. Fac. Educ., Shiga Univ.* 28:75–80
62. Hoo G, Tseng CJ. 1978. *Angiospermae, Dicotyledoneae, Araliaceae. Flora Reipublicae Popularis Sinicae*, Vol. 54. Beijing: Science Press
63. Hopkins DM, ed. 1967. *The Bering Land Bridge*. Stanford, CA: Stanford Univ. Press
64. Hsü J. 1983. Late Cretaceous and Cenozoic vegetation in China, emphasizing their connections with North America. *Ann. Missouri Bot. Gard.* 70:490–508
65. Hu HH. 1935. A comparison of the ligneous flora of China and eastern North America. *Bull. Chinese Bot. Soc.* 1:79–97
66. Hu HH. 1936. The characteristics and affinities of Chinese flora. *Bull. Chinese Bot. Soc.* 2:67–84
67. Deleted in proof
68. Iltis HH. 1983. Biogeographical relationships between temperate eastern Asia and temperate eastern North America. In *Proc. U.S.-Japan Sci. Semin. on the Origin of the Eastern Asian and North American Floras from Arctotertiary Precursors*, pp. 49–51. New York: US-Jpn. Coop. Sci. Prog.
69. Iwatsuki K, Ohba H. 1994. The floristic relationship between East Asia and eastern North America. In *Vegetation in Eastern North America*, ed. A Miyawaki, K Iwatsuki, MM Grandtner, pp. 61–74. Tokyo: Univ. Tokyo Press
70. Iwatsuki Z. 1958. Correlations between the moss floras of Japan and of the southern Appalachians. *J. Hattori Bot. Lab.* 20:304–52
71. Iwatsuki Z. 1972. Distribution of bryo-

phytes common to Japan and the United States. See Ref. 15, pp. 107–37
72. Johnson LPV. 1939. A descriptive list of natural and artificial hybrids in North American forest-tree genera. *Can. J. Res.* 17C:411–44
73. Kato M, Iwatsuki K. 1983. Phytogeographic relationships of pteridophytes between temperate North America and Japan. *Ann. Missouri Bot. Gard.* 70:724–33
74. Kato H, Kawano S, Terauchi R, Ohara M, Utech FH. 1995. Evolutionary biology of *Trillium* and related genera (Trilliaceae) I. Restriction site mapping and variation of chloroplast DNA and its systematic implications. *Plant Species Biol.* 10:17–29
75. Kim Y-D, Jansen RK. 1998. Chloroplast DNA restriction site variation and phylogeny of the Berberidaceae. *Am. J. Bot.* 85:1766–78
76. Kim Y-D, Kim S-H. 1999. Phylogeny of *Weigela* and *Diervilla* (Caprifoliaceae) based on nuclear rDNA internal transcribed spacer sequences. *J. Plant Res.* In press
77. Koyama T, Kawano S. 1963. Critical taxa of grasses with North American and eastern Asiatic distribution. *Can. J. Bot.* 42:859–84
78. Krutzsch W. 1989. Paleogeography and historical phytogeography (paleochorology) in the Neophyticum. *Plant Syst. Evol.* 162:5–61
79. Kubitzki K, Krutzsch W. 1998. Origins of east and south east Asian plant diversity. See Ref. 4, pp. 56–70
80. Kurokawa S. 1972. Probable mode of differentiation of lichens in Japan and eastern North America. See Ref. 15, pp. 139–46
81. Latham RE, Ricklefs RE. 1993. Continental comparisons of temperate-zone tree species diversity. In *Species Diversity in Ecological Communities: Historical and Geographical Perspectives*, ed. RE Ricklefs, D Schluter, pp. 294–314. Chicago: Univ. Chicago Press
82. Latham RE, Ricklefs RE. 1993. Global patterns of tree species richness in moist forests: energy-diversity theory does not account for variation in species richness. *Oikos* 67:325–33
82a. Lavin M, Luckow M. 1993. Origins and relationships of tropical North America in the context of the boreotropics hypothesis. *Am. J. Bot.* 80:1–14
83. Lee NS, Sang T, Crawford DJ, Yeau SH, Kim S-C. 1996. Molecular divergence between disjunct taxa in eastern Asia and eastern North America. *Am. J. Bot.* 83:1373–78
84. Leopold EB, MacGinitie HD. 1972. Development and affinities of Tertiary floras in the Rocky Mountains. See Ref. 15, pp. 147–200
85. Levinton JS. 1983. Stasis in progress: the empirical basis of macroevolution. *Annu. Rev. Ecol. Syst.* 14:103–37
86. Levinton JS, Simon C. 1980. A critique of the punctuated equilibrium model and implications for the detection of speciation in the fossil record. *Syst. Zool.* 29:130–42
87. Li H-L. 1942. The Araliaceae of China. *Sargentia* 2:1–134
88. Li H-L. 1952. Floristic relationships between eastern Asia and eastern North America. *Trans. Am. Philos. Soc.* 42:371–429
89. Li H-L. 1972. Eastern Asia–Eastern North America species-pairs in wide ranging genera. See Ref. 15, pp. 65–78
90. Li J-H, Bogle AL, Klein AS. 1997. Interspecific relationships and genetic divergence of the disjunct genus *Liquidambar* (Hamamelidaceae) inferred from DNA sequences of plastid gene *mat*K. *Rhodora* 99:229–40
91. Li J-H, Donoghue MJ, Bogle AL, Klein AS. 1998. Phylogeny and biogeography of *Hamamelis* (Hamamelidaceae): an update. *Am. J. Bot.* 85 (R):141
92. Li S, Adair KT. 1994. Species pools in eastern Asia and North America. *Sida* 16:281–99
93. Liston A, Rieseberg LH, Elias TS. 1989. Morphological stasis and molecular diver-

gence in the intercontinental disjunct genus *Datisca* (Datiscaceae). *Aliso* 12:525–42
94. Lowry PPII, Jones AG. 1984. Systematics of *Osmorhiza* Raf. (Apiaceae: Apioideae). *Ann. Missouri Bot. Gard.* 71:1128–71
95. Manchester SR. 1987. The fossil history of Juglandaceae. *Missouri Bot. Gard. Monogr.* 21:1–137
96. Manchester SR. 1989. Systematics and fossil history of the Ulmaceae. In *Evolution, Systematics, and Fossil History of the Hamamelidae*, Vol. 1: *Introduction and 'Lower' Hamamelidae*, ed. PR Crane, S Blackmore, pp. 221–51. Oxford UK: Clarendon
97. Manchester SR. 1994. Fruits and seeds of the Middle Eocene Nut Beds flora, Clarno Formation, Oregon. *Palaeontogr. Am.* 58:1–205
98. Manchester SR. 1999. Biogeographical relationships of North American Tertiary floras. *Ann. Missouri Bot. Gard.* In press
99. Maynard Smith J. 1983. Current controversies in evolutionary biology. In *Dimensions of Darwinism*, ed. M Grene, pp. 273–86. Cambridge, UK: Cambridge Univ. Press
100. Maynard Smith J. 1983. The genetics of stasis and punctuation. *Annu. Rev. Genet.* 17:11–25
101. McKenna MC. 1975. Fossil mammals and early Eocene North Atlantic continuity. *Ann. Missouri Bot. Gard.* 62:335–53
102. McKenna MC. 1983. Cenozoic paleogeography of North Atlantic land bridges. In *Structure and Development of the Greenland-Scotland Ridge: New Concepts and Methods*, ed. MHP Bott, S Saxov, M Talwani, J Thiede, pp. 351–99. New York: Plenum
103. McKenna MC. 1983. Holarctic landmass rearrangement, cosmic events, and Cenozoic terrestrial organisms. *Ann. Missouri Bot. Gard.* 70:459–89
104. Meyer F. 1997. Hamamelidaceae R. Brown—witch-hazel family. In *Flora of North America North of Mexico*, vol. 3, ed. Flora of North America Edit. Com., pp. 362–67. New York: Oxford Univ. Press
105. Mirov NT. 1967. *The Genus Pinus*. New York: Ronald
106. Mizushima M. 1972. Taxonomic comparison of vascular plants found in western North America and Japan. See Ref. 15, pp. 83–92
107. Mönkkönen M, Viro P. 1997. Taxonomic diversity of the terrestrial bird and mammal fauna in temperate and boreal biomes of the northern hemisphere. *J. Biogeogr.* 24:603–12
108. Morrone JJ, Crisci JV. 1995. Historical biogeography: introduction to methods. *Annu. Rev. Ecol. Syst.* 26:373–401
109. Murrell ZE. 1993. Phylogenetic relationships in *Cornus* (Cornaceae). *Syst. Bot.* 18:469–95
109a. Nei M. 1987. *Molecular Evolutionary Genetics*. New York: Columbia Univ. Press
110. Nishida M. 1972. An outline of the distribution of Japanese ferns. See Ref. 15, pp. 101–5
111. Nordlander G, Liu Z-W, Ronquist F. 1996. Phylogeny and historical biogeography of the cynipoid wasp family Ibaliidae (Hymenoptera). *Syst. Entomol.* 21:151–66
112. Paclt J. 1952. Synopsis of the genus *Catalpa* (Bignoniaceae) III. *Candollea* 13:241–78
113. Paris CA. 1991. *Molecular systematics of the* Adiantum pedatum *complex: phylogeny and biogeography*. PhD diss. Univ. Vermont, Burlington.
114. Paris CA, Haufler CH. 1994. Geographic isolation and allopatric speciation in ferns with a disjunct distribution in eastern Asia and eastern North America. *Am. J. Bot.* 81:137
115. Parks CR, Miller NG, Wendel JF, McDougal KM. 1983. Genetic divergence within the genus *Liriodendron* (Magnoli-

aceae). *Ann. Missouri Bot. Gard.* 70:658–66

116. Parks CR, Wendel JF. 1990. Molecular divergence between Asian and North American species of *Liriodendron* (Magnoliaceae) with implications of fossil floras. *Am. J. Bot.* 77:1243–56
117. Patterson C. 1981. The development of the North American fish fauna—a problem of historical biogeography. In *Chance, Change and Challenge*, vol. 2, *The Evolving Biosphere*, ed. PL Forey, pp. 265–81. London: Br. Mus. (Nat. Hist.) and Cambridge Univ. Press
118. Persons WS. 1994. *American Ginseng: Green Gold*. Asheville, NC: Bright Mountain Books. Rev. ed.
119. Phipps JB. 1983. Biogeographic, taxonomic, and cladistic relationships between East Asiatic and North American *Crataegus*. *Ann. Missouri Bot. Gard.* 70:667–700
120. Prince LM, Parks CR. 1998. An evaluation of tribal and generic classification of Theoideae (Theaceae) based on DNA sequence data. *Am. J. Bot.* 85 (6):152–53
121. Qiu Y-L, Chase MW, Parks CR. 1995. A chloroplast DNA phylogenetic study of the eastern Asia–eastern North America disjunct section *Rytidospermum* of *Magnolia* (Magnoliaceae). *Am. J. Bot.* 82:1582–88
122. Qiu Y-L, Parks CR, Chase MW. 1995. Molecular divergence in the eastern Asia–eastern North America disjunct section *Rytidospermum* of *Magnolia* (Magnoliaceae). *Am. J. Bot.* 82:1589–98
123. Raven PH. 1972. Plant species disjunctions: a summary. *Ann. Missouri Bot. Gard.* 59:234–46
124. Raven PH, Axelrod DI. 1974. Angiosperm biogeography and past continental movements. *Ann. Missouri Bot. Gard.* 61:539–673
125. Rehder A. 1932. New species, varieties and combinations from the herbarium and the collections of the Arnold Arboretum. *J. Arnold Arbor.* 13:337–41
126. Ricklefs RE, Latham RE. 1992. Intercontinental correlation of geographical ranges suggests stasis in ecological traits of relict genera of temperate perennial herbs. *Am. Nat.* 139:1305–21
127. Santamour FS Jr. 1972. Interspecific hybridization in *Liquidambar*. *Forest. Sci.* 18:23–26
128. Santamour FS Jr. 1972. Interspecific hybrids in *Liriodendron* and their chemical verification. *For. Sci.* 18: 233–36
129. Santamour FS Jr. 1972. Interspecific hybridization in *Platanus*. *For. Sci.* 18:236–39
130. Sax K. 1933. Species hybrids in *Platanus* and *Campsis*. *J. Arnold Arbor.* 14: 274–78
131. Schnabel A, Wendel JF. 1998. Cladistic biogeography of *Gleditsia* (Leguminosae) based on *ndh*F and *rpl*16 chloroplast gene sequences. *Am. J. Bot.* 85:1753–65
132. Shi S, Chang H-T, Chen Y, Qu L-H, Wen J. 1998. Phylogeny of the Hamamelidaceae based on the ITS sequences of nuclear ribosomal DNA. *Biochem. Syst. Ecol.* 25:55–69
133. Shields GF, Wilson AC. 1987. Calibration of mitochondrial DNA evolution in geese. *J. Mol. Evol.* 24:212–17
134. Smith AR. 1993. Phytogeographic principles and their use in understanding fern relationships. *J. Biogeogr.* 20:255–64
135. Soltis DE, Kuzoff R. 1995. Discordance between nuclear and chloroplast phylogenies in the *Heuchera* group (Saxifragaceae). *Evolution* 49:727–42
136. Soltis DE, Xiang Q-Y, Hufford L. 1995. Relationships and evolution of Hydrangeaceae based on *rbc*L sequence data. *Am. J. Bot.* 82:504–14
137. Soltis DE, Soltis PS, Collier TG, Edgerton ML. 1991. Chloroplast DNA variation within and among genera of the Heuchera group (Saxifragaceae): evi-

dence for chloroplast transfer and paraphyly. *Am. J. Bot.* 78:1091–1112
138. Suzuki S, Tomishima K, Yano S, Tsurusaki N. 1977. Discontinuous distributions in relict harvestmen (Opiliones, Arachnida). *Acta Arachnol.* 27:121–38 (In Japanese with English abstract)
139. Deleted in proof
140. Taylor DW. 1990. Paleobiogeographic relationships of angiosperms from the Cretaceous and early Tertiary of the North American area. *Bot. Rev.* 56:279–417
141. Thorne RF. 1972. Major disjunctions in the geographical ranges of seed plants. *Q. Rev. Biol.* 47:365–411
142. Thunberg CP. 1784. *Flora Japonica.* Leipzig: I. G. Mulleriano
143. Tiffney BH. 1985. Perspectives on the origin of the floristic similarity between Eastern Asia and eastern North America. *J. Arnold Arbor.* 66:73–94
144. Tiffney BH. 1985. The Eocene North Atlantic land bridge: its importance in Tertiary and modern phytogeography of the northern hemisphere. *J. Arnold Arbor.* 66:243–73
145. Tuyama T. 1972. The status of the Bonin Islands flora in the Pacific. See Ref. 15, pp. 79–82
146. Upcott M. 1936. The parents and progeny of *Aesculus carnea. J. Genet.* 33:135–49
147. Wen J. 1991. *Systematics of Aralia L. (Araliaceae).* PhD diss. Ohio State Univ., Columbus
148. Wen J. 1998. Evolution of the eastern Asian and eastern North American disjunct pattern: insights from phylogenetic studies. *Korean J. Plant Taxon.* 28:63–81
149. Wen, J, Jansen RK. 1992. Phylogenetic and biogeographic studies on eastern Asian and eastern North American disjunct taxa: integrating morphological and chloroplast DNA data. *Am. J. Bot.* 79 (6):9
150. Wen J, Stuessy TF. 1993. Phylogeny and biogeography of *Nyssa* (Cornaceae). *Syst. Bot.* 18:68–79
151. Wen J, Jansen RK. 1995. Morphological and molecular comparisons of *Campsis grandiflora* and *C. radicans* (Bignoniaceae), an eastern Asian and eastern North American vicariad species pair. *Plant Syst. Evol.* 196:73–83
152. Wen J, Zimmer EA. 1996. Phylogeny and biogeography of *Panax* L. (the ginseng genus, Araliaceae): inferences from ITS sequences of nuclear ribosomal DNA. *Mol. Phylogenet. Evol.* 6:166–77
153. Wen J, Jansen RK, Zimmer EA. 1996. Phylogenetic relationships and DNA sequence divergence of eastern Asian and eastern North American disjunct plants. In *Current Topics in Molecular Evolution*, ed. M Nei, N Takahata, pp. 37–44. Hayama (Japan): publ. jointly by Penn. State Univ., USA, and Grad. Sch. for Adv. Stud., Hayama, Japan
154. Wen J, Jansen RK, Kilgore K. 1996. Evolution of the eastern Asian and eastern North American disjunct genus *Symplocarpus* (Araceae): insights from chloroplast DNA restriction site data. *Biochem. Syst. Ecol.* 24:735–47
155. Wen J, Shi S, Jansen RK, Zimmer EA. 1998. Phylogeny and biogeography of *Aralia* sect. *Aralia* (Araliaceae). *Am. J. Bot.* 85:866–75
156. Wen J, Shi S. 1999. A phylogenetic and biogeographic study of *Hamamelis* (Hamamelidaceae), an eastern Asian and eastern North American disjunct genus. *Biochem. Syst. Ecol.* 27:55–66
157. White PS. 1983. Eastern Asian-North American floristic relations: the plant community level. *Ann. Missouri Bot. Gard.* 70:734–47
158. Williamson PG. 1987. Selection or constraint?: a proposal on the mechanism for stasis. In *Rates of Evolution*, ed. KSW Campbell, MF Day, pp. 129–42. London: Allen & Unwin
159. Wilson AC, Carlson SS, White TJ. 1977. Biochemical evolution. *Annu. Rev. Biochem.* 46:573–639

160. Wilson AC, Cann RL, Carr SM, George M Jr, Gyllensten UB, et al. 1985. Mitochondrial DNA and two perspectives on evolutionary genetics. *Biol. J. Linn. Soc.* 26:375–400
161. Wolfe JA. 1969. Neogene floristic and vegetational history of the Pacific Northwest. *Madroño* 20:83–110
162. Wolfe JA. 1972. An interpretation of Alaskan Tertiary floras. See Ref. 15, pp. 201–33
163. Wolfe JA. 1975. Some aspects of plant geography of the northern hemisphere during the Late Cretaceous and Tertiary. *Ann. Missouri Bot. Gard.* 62:264–79
164. Wolfe JA. 1977. Paleogene floras from the Gulf of Alaska region. *US Geol. Surv. Prof. Pap.* 997:1–108
165. Wolfe JA. 1981. Vicariance biogeography of angiosperms in relation to paleobotanical data. In *Vicariance Biogeography*, ed. G Nelson, DE Rosen, pp. 413–27. New York: Columbia Univ. Press
166. Wood CE. 1971. Some floristic relationships between the southern Appalachians and western North America. In *The Distributional History of the Biota of the Southern Appalachians*, Part II. *Flora*, ed. PC Holt, pp. 331–404. Blacksburg: Va. Polytech. Inst. & State Univ.
167. Wood CE. 1972. Morphology and phytogeography: the classical approach to the study of disjunctions. *Ann. Missouri Bot. Gard.* 59:107–24
168. Wu C-Y. 1998. Delineation and unique features of the Sino-Japanese Floristic Region. *Bull. Univ. Mus. Univ. Tokyo* 37:1–35
169. Wu Q, Mueller GM. 1997. Biogeographic relationships between the macrofungi of temperate eastern Asia and eastern North America. *Can. J. Bot.* 75:2108–16
170. Wu Z-Y. 1983. On the significance of Pacific intercontinental discontinuity. *Ann. Missouri Bot. Gard.* 70:577–90
171. Xiang Q-Y, Soltis DE, Soltis PS. 1994. Phylogenetic relationships and genetic divergence of disjunct taxa from eastern Asia and North America inferred from molecular data: examples from Cornaceae, Hydrangeaceae, and Saxifragaceae. *Am. J. Bot.* 81 (6):138
172. Xiang Q-Y, Soltis DE, Soltis PS. 1998. Phylogenetic relationships of Cornaceae and close relatives inferred from *mat*K and *rbc*L sequences. *Am. J. Bot.* 85:285–97
173. Xiang Q-Y, Soltis DE, Soltis PS. 1998. The eastern Asian, eastern and western North American floristic disjunction: congruent phylogenetic patterns in seven diverse genera. *Mol. Phylogen. Evol.* 10:178–90
174. Xiang Q-Y, Crawford DJ, Wolfe AD, Tang Y-C, DePamphilis CW. 1998. Origin and biogeography of *Aesculus* L. (Hippocastanaceae): a molecular phylogenetic perspective. *Evolution* 52:988–97
175. Ying T-S. 1983. The floristic relationships of the temperate forest regions of China and the United States. *Ann. Missouri Bot. Gard.* 70:597–604
176. Ying T-S, Terabayashi S, Boufford DE. 1984. A monograph of *Diphylleia* (Berberidaceae). *J. Arnold Arbor.* 65:57–94
177. Ying T-S, Zhang Y-L, Boufford DE. 1993. *The Endemic Genera of Seed Plants of China*. Beijing: Science Press
178. Yurtsev BA. 1972. Phytogeography of northeastern Asia and the problem of tansberingian floristic interrelations. See Ref. 15, pp. 19–54
179. Zhang ZM, Liou JG, Coleman RG. 1984. An outline of the plate tectonics of China. *Geol. Soc. Am. Bull.* 95:295–312
180. Zhou J, Huang W-G, Wu M-Z, Yang C-R, Feng K-M, Wu Z-Y. 1975. Triterpenoids from *Panax* Linn. and their relationship with taxonomy and geographical distribution. *Acta Phytotax. Sin.* 13 (2):29–45, pls 6-7

181. Zimmer EA, Martin SL, Beverley SM, Kan YW, Wilson AC. 1980. Rapid duplication and loss of genes coding for the alpha chains of hemoglobin. *Proc. Natl. Acad. Sci. USA* 77:2158–62

182. Zuckerkandl E, Pauling L. 1965. Evolutionary divergence and convergence in proteins. In *Evolving Genes and Proteins*, ed. V Bryson, HJ Vogel, pp. 97–106. New York: Academic Press

Annu. Rev. Ecol. Syst. 1999. 30:457–77

FULL OF SOUND AND FURY: The Recent History of Ancient DNA

Robert K. Wayne*, Jennifer A. Leonard*, and Alan Cooper+

**Department of Organismic Biology, Ecology and Evolution, University of California, Los Angeles California 90095-1606; e-mail: rwayne@ucla.edu; and +Department of Biological Anthropology, University of Oxford, Oxford OX2 6QS, United Kingdom*

Key Words aDNA, PCR, molecular paleontology, archeology, non-invasive typing

■ **Abstract** The discovery that DNA survives in ancient remains and can be amplified by the polymerase chain reaction has added a direct temporal dimension to evolutionary studies. Initial reports suggested that the time period open to investigation was vast, extending back into the Cretaceous period. However, attempts to replicate of results involving DNA purported to be over a million years old have not succeeded. Theoretical studies suggest that DNA is unlikely to survive intact more than about 100,000 years. However, even over this time period, the evolutionary questions that can be addressed are far reaching and include systematics, paleoecology, the origin of diseases, and evolutionary processes at the population level.

INTRODUCTION

"Would that I could discover truth as easily as I can uncover falsehood."

Cicero, *De Natura Deorum* (44 BC)

The use of genetic data has greatly expanded evolutionary studies of extant organisms, with one major limitation: History must be reconstructed by extrapolation from current genetic patterns rather than directly observed in the fossil record. Systematists utilize character state data and phylogenetic methods to reconstruct evolutionary history, but this approach is fraught with problems caused by character reversals and parallelisms (134). Similarly, population genetics cannot deal well with the confounding effects of natural selection, genetic drift, gene flow, and population history without resorting to simplified models and assumptions (141, 151). Molecular evolutionary biologists have envied the paleontologist's unique access to historical information and longed to utilize the rich information inherent in ancient DNA sequences. Such was the promise of research on DNA preserved in the remains of extinct organisms (19, 26, 101, 103, 135). However, the intense excitement at the first reports of ancient DNA greater than a million years old was followed by confusion and disillusionment when the claims were severely criticized (3, 58, 124, 135, 136, 155). As a result, the field has matured considerably, and while it remains thematically diverse, rigorous authentication procedures provide a unifying methodological focus. In this review, we attempt to reduce the confusion surrounding ancient DNA studies by first providing a historical overview and then focusing on studies that define directions for future research.

Ancient DNA: Definition and Scope

Ancient DNA deals exclusively with deoxyribonucleic acid isolated from plant or animal remains. It is therefore a subdiscipline of molecular paleontology, which more broadly concerns any biomolecule that can be extracted from ancient tissues such as pigments or polysaccharides from plant tissues (12, 62, 85, 118) or collagen and hemoglobin from animal remains (2, 18, 32, 86, 123). The time frame is broad for ancient DNA and includes the study of extant populations through the collection of hair, feces, or seeds (60, 73, 93), an endeavor often called "noninvasive sampling" (92). More commonly, however, ancient DNA has been used to describe a wide range of research efforts involving: (*a*) museum and archeological specimens (23, 25, 41, 48, 59, 61, 68, 100, 119, 142, 147); (*b*) subfossil remains from the late Pleistocene (51, 64, 71); and (*c*) fossils greater than a million years old (14, 29, 35, 110, 126).

The questions addressed by ancient DNA research concern any issue that benefits from a direct historical perspective. These areas include (*a*) systematics (23, 29, 36, 51, 61, 64, 67); (*b*) changes in genetic diversity as a function of time and environmental change (40, 52, 143); (*c*) migration and admixture (79, 91, 120); (*d*) ecology and paleoecology (73, 111); (*e*) the origin and spread of disease (8, 121); and (*f*) tempo and mode of mutations in populations (55). Consequently, the potential impact of ancient DNA studies is considerable; hence the enthusiasm and anticipation that greeted the first reports of DNA isolation from ancient remains (26, 103, 135).

THE HISTORY OF ANCIENT DNA

> "Life's but a walking shadow ... It is a tale told by an idiot, full of sound and fury, signifying nothing."
>
> Shakespeare, *Macbeth* (1605–1606)

From the beginning, ancient DNA research was a populist science. Reports of DNA from ancient remains led to wild speculation in the press and film that life could be restored to ancient creatures. Each new discovery served to reconfirm the public impression that scientists were moving quickly toward this goal. New reports of ancient DNA, although often of limited evolutionary significance, were published in the most prestigious journals. The first reports in 1985 were based on fairly recent material in museum vertebrate collections or human mummies and utilized cloning technology (59, 100). The initial report on mummies was spectacular because the length of DNA cloned was 3.4 kilobases (100). The preservation of such long fragments boded well for ancient DNA research; potentially entire genes could be resurrected. However, cloning ancient fragments requires a substantial source of DNA and is hit-or-miss with regard to specific single copy genes (101). In fact, the cloned sequences were uninformative, highly repetitive DNA later shown to be a part of the human HLA system and a probable contaminant (27).

Several years later, the advent of the polymerase chain reaction (PCR) permitted specific sequences to be amplified from only a few template molecules (94). As a result, the requirement for a substantial source of intact DNA was greatly reduced, and specific, phylogenetically informative genes could be targeted (61). For animals, the genes chosen were from the mitochondrial genome because there are thousands of copies in each cell. In addition, nearly a decade of research on genes such as *cytochrome b* and the hypervariable control region had demonstrated their utility for evolutionary questions ranging from population genetics to higher order systematics (5). For plants, the focus was on a chloroplast gene, 1,5 *bisphosphate carboxylase oxygenase* (*RbcL*), which had been sequenced in a wide variety of plant species owing to the availability of universal primers (22). In fact, the first DNA sequences from specimens more than a million years old were claimed for remarkably well-preserved 17–20-million-year-old magnolia leaf compression fossils from clay sediments of a Miocene shallow freshwater lake (35). Visible tissue structures and even cell walls and intracellular organelles could be seen in these ancient leaves. Phylogenetic analysis of supposed ancient sequences showed them to be most closely related to living *Magnolia* species, thus adding considerable weight to the claim of authenticity. Until the publication of the magnolia paper, ancient DNA sequencing was confined to specimens only a few thousand years old such as museum skins of the extinct Tasmanian wolf, *Thylacinus cynocephalus* (142), human mummies, and bog-preserved human brain tissue (30, 100). A year later, the presence of ancient DNA sequences in the magnolia leaves was apparently supported by an independent group using bald cypress specimens from the same deposit (126). However, the *Magnolia* study was performed without PCR controls (83), and other attempts to replicate these results have failed, raising strong doubts (105, 124, 127).

The magnolia study broke the million-year barrier, and with considerable excitement researchers turned to long extinct charismatic creatures such as dinosaurs as well as more mundane insects preserved in amber. A strong case could be made for amber preserved specimens because researchers believed that amber rapidly desiccates trapped remains and inhibits bacterial activity, two prerequisites for preservation of ancient DNA (14, 109). The list of amber preserved specimens from which ancient sequences were recovered is impressive and includes stingless bees, termites, honeybees, and plants (14–16, 29, 110). Even fossil bacterial sequences were claimed from amber (13), and recovered spores were viably cultured (17), although these sequences were subsequently shown to be probable contaminants (154). Additionally, the presence in extracts of contaminating sequences from other extant insect species was worrisome (28, 29), although the practice of reporting such contaminants was not widespread. More recent studies on chemical preservation of plant and animal remains in Oligocene amber show DNA is unlikely to survive (129).

The grand culmination of studies using specimens in the million-year–plus range was the report of dinosaur mtDNA sequences from an unlikely source: partially carbonized material from a coal deposit in Idaho (150). The sequences appeared neither mammalian nor reptilian and were therefore assumed to be endogenous. Science fiction was finally fact. However, although the methods and results seemed reasonable, given past research, existing authentication standards were not met (83), and the dinosaur DNA was soon shown by several groups to be an artifact (1, 56, 58, 153, 155). Comprehensive phylogenetic analysis showed that the supposed dinosaur sequences were classified with mammals and were closest to humans. An elegant study from the laboratory of Svante Pääbo, the leading critic and practitioner of ancient DNA research, showed that the putative dinosaur DNA was probably a previously undescribed human intron (156). A similar fate befell the study of DNA from Cretaceous dinosaur eggs (146).

Pääbo and colleagues have also criticized ancient DNA research on other grounds such as the lack of suitable controls, facilities, and techniques to limit contamination (102, 103). The power of PCR was and is the problem: A single intact contaminating sequence from a recent source can potentially outcompete ancient degraded and damaged DNA in the process of amplification. Remarkably, recent contaminating DNA could be mixed into ancient sequences by "jumping PCR" (103). Chimeric molecules are formed when damaged ancient DNA templates are incompletely copied during a PCR cycle, allowing the incomplete extension product itself to act as a primer in the next round of PCR. Damaged DNA can cause many problems during PCR including erroneous insertion of adenosine residues (104) and inaccurate amplification of repetitive motifs as well as the production of chimeric molecules from endogenous and contaminating sequences (28). On phylogenetic grounds, chimeric sequences could appear quite reasonable and thus could be mistaken as authentic. The sources of contamination are numerous and unpredictable, ranging from those at the point of excavation to curatorial activities and laboratory reagents and aerosols (48).

Yet another problem is "PCR carry-over," which occurs when PCR products from one reaction contaminate those of another (78). The scale of this problem can be considerable. For example, a single 100μl PCR reaction produces enough template molecules that, if diluted in an Olympic size swimming pool, each 100μl from the pool would still contain 400 intact copies (78). The possibility, therefore, of contaminants from clothes, reagents, and the atmosphere in genetic laboratories is immense. This problem is less severe in studies of nonhumans because a laboratory can be chosen in which the target taxon and related species have never been studied. However, even animal studies are not immune from contamination; despite careful controls, sequences related to *Drosophila* have been reported in extracts of amber termites (29).

Contamination problems take on a new dimension when human remains are used as a source of ancient DNA (24, 102, 133). However, numerous researchers have claimed success with ancient human specimens hundreds to thousands of years old (97). The early reports simply documented the presence of endogenous sequences (30, 80, 99, 100, 149), whereas more recent studies have addressed questions of kinship, sex ratios, disease, and large-scale patterns of human colonization (8, 41, 44, 46, 90). Notably, studies on ancient Americans imply that they had a different genetic composition than do present-day native Americans, thus casting doubt on colonization theories based on recent samples alone (79, 90, 91, 107, 122, 132). The pattern of settlement of Polynesia also is controversial. It appears to have been settled in several waves, contradicting the "fast track" theory of settlement from Asia rather than Melanesia; however, no consensus has been reached (42, 44). Few of these studies have been independently replicated. Importantly, DNA analysis of ancient remains from areas separated by only a few hundred kilometers do not agree (30, 107). These discrepancies have been variously explained as sampling effects or evidence of local spatial or temporal differentiation (97).

The possibility of recent human DNA contamination and the likelihood of jumping PCR cast doubt on the verity of the human results (48). Some sequences or parts thereof may be authentic, but it is difficult to assess without independent replication how much is truth or falsehood. Very troubling is the absence of success of ancient human studies in the Pääbo laboratory where conditions were carefully controlled for contamination: 110 ancient Egyptian bone and tissue samples and 45 ancient remains from the New World yielded little amplifiable DNA (49, 65). More convincing are cases where some genetic information is known about the specimens in advance. For example, bones sexed by morphologic methods had the expected sex chromosome markers, strongly suggesting the DNA was endogenous (131). No simple and elegant solution has emerged for the problems of contamination (3, 4), although independent replication is essential. The minimum requirements for research with ancient human remains are to extract DNA from ancient samples in a separate facility with air filtration, UV sterilization, and protective sterile clothing and to use a carefully considered set of negative and positive controls. Even then, contaminants are common (50, 76) or sometimes surprisingly absent (41, 63, 79).

Cloning and sequencing of PCR products have revealed recombinants and allowed ancient sequences to be reconstructed by separating the contaminants from endogenous sequences (28). However, this approach assumes that the contaminating partial sequences can be identified. Importantly, recombinant sequences should be highly context dependent; thus, a separate amplification from the original extract or an independent laboratory should not obtain the same recombinant sequences. Therefore, a first proof of authenticity is replication. Sadly, no million-year-old

DNA has been replicated in this fashion, despite considerable effort (4, 145). Since 1995, no reliable reports of DNA past the million-year mark have been published. Given theoretical and empirical results, the view now is that specimens much older than 100,000 years in age are unlikely to contain endogenous DNA that can be amplified (82, 84).

The gold standard of authentication in studies of ancient DNA has been set by Pääbo and his collaborators (49, 50, 76, 77). Their general experimental design is exemplified by DNA analysis of the "Tyrolean Ice Man," a 5000-year-old human discovered frozen in the Italian Alps (50). DNA was extracted in a special facility for ancient DNA with extraction controls; results were replicated independently, and PCR products were cloned and screened for contaminants. However, the theoretical importance of the study for human evolution was negligible; their considerable effort showed only that the iceman was European rather than a South American mummy hoax. More consequential was the carefully documented DNA analysis of the Neanderthal-type specimen (76, 77). Here, PCR products were originally cloned and sequenced by the Pääbo laboratory, and sequences were examined for persistent motifs that differed from those in living humans. These sequence differences were then used to design Neanderthal-specific PCR primers that would not amplify contaminating human DNA. Using these specific primers, the resulting Neanderthal mitochondrial control region sequence was replicated in an independent laboratory and was shown to be phylogenetically divergent from recent humans. Consequently, the results imply that the Neanderthals were a separate hominid lineage that did not significantly contribute genetically to modern humans.

THE REAL ANCIENT DNA

Theoretical and empirical studies have demonstrated that DNA is best preserved in cold and dry environments protected from UV exposure and in remains less than 100,000 years old (82, 84, 101, 112). DNA has been recovered from animal remains preserved in the past 50,000 years from the Arctic permafrost by several groups (47, 65, 96, 98, 152), confirming that DNA is well preserved under these conditions. Other environments conducive to preservation are those in high altitude cold, dry caves (40, 64). DNA has also been reported from ancient human remains only a few thousand years old under a variety of conditions, including exposure to water where hydrolytic damage to DNA might be expected (30, 55). However, as discussed above, considerable uncertainty exists about the actual presence of endogenous DNA sequences. The more convincing studies are those in which DNA sequences match that expected from independent species identification of archeological material (21). For example, pig and sheep bones recovered from the *Mary Rose*, a British ship sunk in the sixteenth century, provided the expected sequences for these species as well as those of human contaminants (43).

Despite the limitations imposed by time and preservation conditions, some deposits hold great potential for the preservation of DNA. Most promising are remains preserved in the Arctic permafrost. Hundreds of thousands of bones recovered from permafrost deposits are housed in museum collections in North America and Siberia, representing large numbers of mammal and bird species. Recent research has focused on the systematics of the largest of the permafrost species, the mammoth and mastodons of Alaska and Siberia. Mammoths appear most closely related to recent elephants, as expected, although the topology of this clade is still unresolved (47, 67, 98, 152). Other species such as horses and woolly rhinos also have provided DNA sequences (9, 65). The hope is that with further sampling, population-level questions can be resolved as well. Most permafrost samples range in age from about 10,000 to 50,000 years before present (38), and hence they provide a genetic record of diversity before and after the last glaciation. Population declines associated with the last glaciation and the Late Pleistocene large mammal extinctions (37, 39) are likely to have had significant effects on genetic diversity. Ancient DNA sequences from a dated series of specimens would provide a direct record of these genetic changes and of faunal interchanges between the Old and New World across the Bering land bridge.

The most spectacular record of Late Pleistocene carnivores is without question that preserved in the Rancho La Brea tar pits at the George C. Page Museum in Los Angeles. Several thousand specimens of carnivores and, to a much lesser extent, large ungulates, are housed in the museum collection and span a time period from about 8,000 to 40,000 years before present. DNA was reported from marrow bone samples of a saber-toothed cat, *Smilodon fatalis*, after removal of embedded tar using kerosene (71). DNA segments from mitochondrial *12S rRNA* and the nuclear *FLA-I* gene were amplified and shown to be cat-like. Phylogenetic analysis of the sequences appeared to group the saber-tooth cats closely with modern cats rather than paleofelids (Nimravidae). However, the sequence data did not provide enough resolution to definitively align the saber-tooth with a specific group of living cats and was not replicated. Consequently, the possibility of large-scale analysis of tar pit fossils to address systematic or population level questions remains an interesting possibility.

DNA sequences have been amplified from a diversity of species preserved in caves at a variety of locations. An early and dramatic report concerned the isolation and amplification of mitochondrial DNA sequences from five species of extinct New Zealand moa (Dinornithiformes) recovered principally from caves (23). New Zealand has been separated from Australia for about 80 million years, and it was therefore assumed that these flightless birds shared a common ancestor with the other endemic ratite, the kiwi. However, phylogenetic analysis of 12s RNA sequences showed that the moa represents an early offshoot of the ratites, whereas the kiwi was closely related to the Australian emu and cassowary. Consequently the data indicate that ratite birds had independently colonized New Zealand twice. DNA from teeth of pocket gophers (40), cave bears (51), rabbits (54) and ground sloths (64) are other examples of important studies based on cave material.

Archeological material preserved by burial in temperate soils has provided a source of DNA that can be amplified by PCR (87). The European rabbit,

Oryctolagus cuniculus, has been best studied in this regard. Mitochondrial DNA has been amplified from 90 rabbit bones from 22 localities, ranging over the past 11,000 years (52). Additionally, an island population off the Tunisian coast that was founded over 1400 years ago was studied (53). The research showed genetic continuity over much of the recent history of the European rabbit, but during the Middle Ages, a new group of mitochondrial *cytochrome b* genotypes appeared in France that are ancestral to modern domestic rabbits. This result suggests that rabbits from this lineage were favored and may reflect the establishment of warrens in the tenth century. Finally, insight into the process of domestication was gained through the study of ancient cattle, *Bos primigenius*, and their domestic descendants *B. taurus* (6). Sequences from remains of primitive cattle as old as 12,000 years before present were directly ancestral to modern European breeds. This further supports genetic analyses of recent cattle that suggested that European cattle were domesticated from a different subspecies than were Asian breeds.

THE FUTURE OF ANCIENT DNA RESEARCH

The number of ancient DNA papers has declined in the past few years as has the number of reports appearing in high profile journals. In the first decade of ancient DNA research, 35% of articles appeared in *Nature, Science*, or the *Proceedings of the National Academy of Science of the USA*, but in the last three years this proportion has declined to less than 15%. The more recent proportion is similar to that for 300 randomly chosen papers in evolutionary biology (7.6%). The honeymoon period has passed for ancient DNA research, and the difficulties associated with a maturing field need confronting. The absence of replication of million-year-old DNA studies and the theoretical and empirical research on DNA preservation all suggest that DNA is unlikely to be preserved much beyond 100,000 years (82, 112). Therefore, only a small fraction of time can be traced through the study of ancient DNA. Nevertheless, as outlined below, we feel that several research topics can dependably be addressed by studies of ancient DNA from specimens younger than 100,000 years.

Systematics of Recently Extinct Species

The most common use of ancient DNA has been directed at the systematics of extinct species and was first demonstrated by a study of the extinct Tasmanian wolf (*Thylacinus cynocephalus*), a marsupial analog of the gray wolf that vanished in the early part of the twentieth century (142). *Cytochrome b* sequences demonstrated that the species was closely related to other Australian marsupials rather than to South American carnivorous marsupials as some had supposed. More extensive analysis has confirmed this initial result (75). Similarly, *cytochrome b* sequences amplified from museum specimens of the extinct quagga (*Equus quagga*) clearly demonstrated its relationship to the mountain zebra (*E. zebra*) rather than to the horse (*E. caballus*) (59, 61).

Cave deposits have provided important systematic insights for mammals as well. A control region sequence from a Europeam late Pleisocene cave bear (*Ursus spelaeus*) sample was shown to be from a distinct species with a sequence basal to those found in Old and New World populations of brown bears, *Ursus arctos* (51). However, cave material often may not contain DNA that can be amplified; only two of 35 samples of the ground sloth *Mylodon darwinii* from a variety of settings in North and South America provided DNA that could be amplified (64). The two successful 13,000-year-old samples were from Mylodon cave, a cold, dry cave in southern Chile. Phylogenetic analysis of mitochondrial genes encoding for *12S* and *16S rRNA* clearly showed that the ground sloth was unexpectedly sister to two-toed rather than three-toed sloths, implying that arboreality in sloths evolved twice. This type of study will require more effort in the future, as there are severe limitations on phylogenetic analysis using very short stretches of DNA. Longer sequences, composed of multiple overlapping fragments, will be necessary to improve the resolution and confidence of phylogenies.

Genetic Diversity Over Time

One of the earliest, underappreciated reports on ancient DNA concerned a study of temporal continuity in kangaroo rats (*Dipodomys panamintinus*) collected in 1911, 1917, and 1937 from three areas in central California and representing three different subspecies (143). Researchers sampled the three localities in 1988, after some 25 to 50 generations had elapsed. This is sufficient time for dominance of a mitochondrial control region matriline to change through drift or through recolonization after an extinction event. However, in each population, the genetic diversity appeared unchanged, and the distribution of control region genotypes was not significantly different from that expected in a population with high levels of genetic diversity.

This research set the stage for other efforts where temporal changes were explicitly expected. Endangered species are a model group in this regard because many have had their population sizes reduced dramatically within the past hundred years, and their declines have been periodically sampled by museum collectors. In fact, some problematic carnivore species that disappeared from the United States such as the gray wolf (*Canis lupus*) and eastern puma (*Felis concolor*) were exterminated by government officials who incidentally sent some specimens to museums. The red wolf (*Canis rufus*) is an example. Analysis of early twentieth century skins showed that mitochondrial *cytochrome b* sequences were not distinct from those in gray wolves (*C. lupus*) or coyotes (*Canis latrans*), suggesting that the red wolf was a hybrid species (147). Later, more extensive research on nuclear and other mitochondrial genes confirmed the early report (120).

By comparison with more abundant related species, endangered populations are often found to be low in genetic variability (34). However, differences in life history or low predisturbance abundance may be the reason for current lower levels of diversity. Hence a direct temporal perspective is preferred. A study of the endangered Australian northern hairy-nosed wombat (*Lasiorhinus krefftii*) demonstrated the problems with the comparative approach (140). Here, samples from the last existing wombat colony of about 50 individuals had low levels of variation relative to a related species. However, the levels of variation were similar to that of an extinct pre-bottleneck northern hairy-nosed wombat population represented by museum skins. Thus, low levels of variation may have been characteristic of northern hairy-nosed wombats before the decline. In contrast, historic feather samples from the endangered San Clemente Island loggerhead shrike (*Lanius ludovicianus mearnsi*) and greater prairie chicken (*Tympanuchus cupido*) demonstrated genetic continuity but marked reductions in genetic variation in the extant populations (10, 95).

Reconstructing the past population relationships of endangered species also is important for reintroduction programs. In the wombat study, the past geographic range was uncertain because partial skeletal remains were not definitive. Mitochondrial DNA sequencing clearly established that a far southern population once existed. Similarly, subfossil bones of the Laysan duck (*Anas laysanensis*) were identified by mitochondrial DNA analysis from lava tubes on the main Hawaiian islands where they apparently had gone extinct (25). These data justified reintroduction and suggested that many island endemics may be relics of former cosmopolitan species (130). Museum collections have become a source of DNA from populations that are difficult to sample or are extinct, so much so that most museums now have a specific loan policy and review board for destructive DNA sampling (20, 148).

Migration and Admixture

Some fossil deposits hold the potential for temporal analysis of genetic changes at the population level. For example, owl pellet deposits in caves provide a temporal sampling of rodent populations and preserve DNA well (138). Owls roosting near cave openings and depositing pellets create a rich stratigraphic collection of their prey. The high altitude Lamar Cave in Yellowstone National Park was excavated and yielded several thousand pocket gopher (*Thomomys talpoides*) mandibles from 16 dated stratigraphic levels (40). *Cytochrome b* sequences were obtained from 73 of 88 teeth representing these stratigraphic levels and showed that genetic continuity characterized the three-thousand-year history of the deposit. The population near Lamar Cave was apparently isolated from genetic exchange from populations only a few hundred kilometers distant. In contrast, phenotypic changes occurred in this population that reflected changing climates. Such phenotypic change must be due to within-population processes and does not involve the differential migration of individuals having morphologic attributes more appropriate for the altered environment. The result has important conservation implications (40, 81); it implies that in small nonvagile species, phenotypic changes may occur in response to climate change but are limited by the existing genetic variation within populations.

Ecology and Paleoecology

DNA sequences can often be retrieved from remains deposited by organisms as part of their life cycle. For vertebrates, this might include hair, feathers, scales, feces, saliva, and urine (92). For invertebrates, pupal casts, eggs, and sperm could provide a source of DNA as well as museum specimens. A variety of tissues are discarded by plants; DNA sequences have been retrieved from herbarium samples of leaves, flowers, and seeds (128). Noninvasive sampling is often the only means to do genetic analysis in endangered species and has been used to understand population relationships and diversity in the brown bear (*Ursus arctos*) from hair samples left on scrapes (137). However, feces provide a more comprehensive source of information as they are often left in predictable areas and contain a genetic record not only of the DNA of an individual but also that of ingested foods and internal microorganisms (66, 73, 114).

The spatial distribution of feces, identified by genetic techniques to an individual, provides information about population density, home range, and movement patterns. A systematic collection of coyote (*Canis latrans*) feces from a southern California parkland permitted investigators to more accurately estimate population size, sex ratio, and movement patterns that were not apparent from radiotelemetry studies (74). Fecal analysis is not limited to modern samples. Fecal DNA was elegantly exploited in a study on 11,000-year-old ground sloth dung (111). Here aspects of diet and altitudinal movement patterns could be inferred from DNA analysis of feces. Considering that coprolites of large mammals are not uncommon, they may provide a unique window into paleoecology of extinct Late Pleistocene mammals. Other studies have used hair samples from chimpanzee (*Pan troglodytes*) night nests to deduce paternity and mating patterns (93), and samples of whale meat sold in Japanese markets have identified endangered specimens banned under international convention (7, 106). Forensic applications of ancient DNA analysis are numerous (33, 45, 72).

Disease

Disease outbreaks often begin with a few cases over a limited geographic area. However, the pathogenic organisms causing the disease may have a wider distribution for some time previous to the first reported cases. DNA analysis of museum or archeological specimens can provide a more exact temporal and spatial perspective on the initiation and spread of disease. For example, analysis of ticks from deer skins in museum collections showed that Lyme disease (*Borrelia burgdorferi*) was present nearly a decade earlier than the first reported cases in the United States and Europe (88, 108). Sequences from *Mycobacterium tuberculosis, M. leprae, Yersinia pestis*, and the 1918 "Spanish" influenza have variously been reported from samples of ancient bone, and teeth, medical samples, and permafrost-preserved soft tissue (8, 11, 31, 113, 115, 121, 139), suggesting a potential field of genetic paleopathology. Unfortunately, some pathogens are common in the extant human population (e.g. *M. tuberculosis*) or have congeneric relatives common in soil or other animal hosts (e.g. *Mycobacterium bovae*), raising the possibility of contamination through handling, burial, or the use of animal products such as BSA. The possibility of identifying genetic disease in historical figures such as the Marfan syndrome in Abraham Lincoln has been discussed (89). Disease greatly impacts the dynamics of wild populations, and much can be learned from understanding the geographic origin of pathogenic organisms (116). Disease organisms that leave a DNA record in preserved tissue can potentially be detected if the analysis is done with appropriate controls and in laboratories that have had no history of research on the disease-causing organism.

Mutational Dynamics

Some populations are isolated for evolutionary time periods. Therefore, mutations that appear in these populations provide a direct estimate of mutation and fixation rates. Considerable controversy exists on the mutation rate of mitochondrial genes as there is an order of magnitude difference in mutation rates from germline studies and those based on evolutionary divergence (69). Dependence between observed rates and time is implied. As both mutation rate estimates involve extrapolation, a direct measure of mutation rates in an isolated population afforded by sampling over time would be more desirable. A few studies have hinted at this possibility. For example, a rate of control region evolution was estimated from mutations appearing in 2000 years of human remains at a single site (55) and from remains of primitive cattle (6). In both cases, the estimated rate was similar to that predicted by indirect phylogenetic methods. However, neither population was likely to be geographically isolated, and thus gene flow remains a source of new alleles. Studies of isolated populations in which the temporal record is good are needed. For example, the Great Salt Lake deposits of Utah are dominated by cysts from brine shrimp, *Artemia*, that may contain preserved DNA as well (MT Clegg, personal communication). Genetic analysis of these shrimp over thousands of years is conceivable with large population samples of each time interval.

PERSPECTIVES AND CONCLUSIONS

The rise and decline of ancient DNA research has been dependent on the development of new techniques, the limits of which subsequently were defined by critical analysis and replication. The development of new techniques and methods of analysis still promises new avenues of research (e.g. sloth dung), but the realm of questions is now circumscribed by the likelihood that DNA will not persist much beyond 100,000 years. Given this reality, efforts need to focus on the application of existing technology to consequential issues that can be resolved over this time scale. Clearly, systematics of recently extinct species, some of which may have deep phylogenetic roots such as the giant moa and the extinct ground sloth, help resolve larger evolutionary issues. Additionally, systematics of recently extinct species potentially provides an outgroup perspective that can greatly improve confidence in topologies of phylogenetic trees of extant taxa (29, 76, 98, 152).

The genetic record of temporal changes in populations is perhaps the most intriguing and least explored potential of ancient DNA research over millennia time scales. Prior to the discovery of ancient DNA, genetic changes within populations were reconstructed from current patterns of genetic diversity. This static view often assumed equilibrium conditions or specific demographic models of population growth and expansion (70, 117, 151). Given the dynamics of climate change and migration between and within continents, such assumptions are likely to be violated. Ancient DNA analysis can provide a direct record of the tempo and mode of genetic change within populations, and thus a means of testing and refining existing population models. The deposits with the most potential are those in Arctic permafrost, high altitude caves, and other cold, arid environments. In combination with stratigraphic information and radiocarbon dating, analysis of these remains provides an absolute temporal scale to view the mutational and demographic processes occurring within and among populations.

Conservation biology may be well served by ancient DNA analysis. Systematics of extinct species provides an important perspective on the remaining diversity of extant organisms. For example, research on avian remains from Polynesian islands suggests that over 2000 species (one quarter of all bird species) may have gone extinct in historic times (130). Molecular systematics of these extinct species would place their extinction in a phylogenetic context and better highlight the evolutionary diversity that was lost (144). Additionally, insights into island speciation and colonization could result. Population level analysis also has important conservation implications. Analysis of genetic diversity of past populations of endangered species establishes a baseline from which the loss of genetic variability can be measured. Additionally, geographic sampling from museum or archeological collections allows reconstruction of the past geographic limits of endangered species. These limits, in combination with the inferred genetic relationships among populations, provide a guide for reintroduction and augmentation (57). Genetic changes that accompany climatic or environmental shifts may also be uncovered by ancient DNA analysis. For example, the body size of the bushy-tailed woodrat (*Neotoma cinerea*), as estimated by the diameter of their feces, dramatically decreased since the last glaciation in association with global warming (125). DNA analysis of these feces may be possible with new techniques (111) and would allow the dissection of the causes of phenotypic change; do they reflect changes in species composition or evolutionary responses within populations?

Finally, DNA analysis of fecal remains and other tissue deposited by organisms may provide information about diet, movements, and past pathogens. Although some may despair over the unlikely prospect of Cretaceous DNA, most will recognize that the study of ancient DNA, if directed toward millennial scale changes, still has the potential to solve evolutionary mysteries. The initial excitement may be gone from the field, but with maturity of purpose, it will come to signify more than sound and fury.

ACKNOWLEDGMENTS

We thank Carles Vilà, Michael Kohn, Klaus Koepfli, Blaire Van Valkenburgh, Debra Pires, and Gary Shin for editing and commenting on the manuscript. This effort was supported by National Science Foundation grants to RKW and a Natural Environmental Research Council and Royal Society grants to AC.

Visit the Annual Reviews home page at http://www.AnnualReviews.org

LITERATURE CITED

1. Allard MW, Young D, Huyen Y. 1995. Detecting dinosaur DNA. *Science* 268:1192–93
2. Ambler R, Daniel M. 1991. Proteins and molecular palaeontology. *Philos. Trans. R. Soc. London Ser. B* 333:381–89
3. Audic S, Béraud-Colomb E. 1997. Ancient DNA is 13 years old. *Nat. Biotechnol.* 15:855–58
4. Austin JJ, Ross AJ, Smith AB, Fortey RA, Thomas RH. 1997. Problems of reproducibility—does geologically ancient DNA survive in amber-preserved insects? *Proc. R. Soc. London Ser. B* 264:467–74
5. Avise J. 1994. *Molecular Markers, Natural History, and Evolution*. New York: Chapman & Hall
6. Bailey J, Richards MB, Macaulay VA, Colson IB, James IT, et al. 1996. Ancient DNA suggests a recent expansion of European cattle from diverse wild progenitor species. *Proc. R. Soc. London Ser. B* 263:1467–73
7. Baker CS, Cipriano F, Palumbi SR. 1996. Molecular genetic identification of whale and dolphin products from commercial markets in Korea and Japan. *Mol. Ecol.* 5:671–85
8. Baron H, Hummel S, Herrmann B. 1996. *Mycobacterium tuberculosis* complex DNA in ancient human bones. *J. Archaeol. Sci.* 23:667–71
9. Best CH. 1994. Genetic analysis of ancient DNA from the hair of the woolly rhinoceros *Coelodonta antiquitatis*. *Proc. Arctic Sci. Conf., 45th, Anchorage*, pp. 37. Vladivostok: Dalnauka
10. Bouzat JL, Lewin HA, Paige KN. 1997. The ghost of genetic diversity past resurrected: ancient DNA analysis of the greater prairie chicken *Tympanuchus cupido*. *Bull. Ecol. Soc. Am.* 78:57
11. Braun M, Cook D, Pfeiffer S. 1998. DNA from *Mycobacterium tuberculosis* complex identified in north American, pre-Columbian human skeletal remains. *J. Archaeol. Sci.* 25:271–77
12. Brown T, Allaby R, Brown K, Jones M. 1993. Biomolecular archaeology of wheat: past, present and future. *World Archaeol.* 25:64–73
13. Cano R, Borucki MK, Higby-Shweitzer M, Poinar HN, Poinar GO, Pollard KJ. 1994. *Bacillus* DNA in fossil bees: an ancient symbiosis? *Appl. Environ. Microbiol.* 60:2164–67
14. Cano R, Poinar H, Pieniazek N, Acra A, Poinar G. 1993. Amplification and sequencing of DNA from a 120–135 million year old weevil. *Nature* 363:536–38

15. Cano R, Poinar H, Poinar G. 1992. Isolation and partial characterization of DNA from the bee *Proplebeia dominicana* (Apidae: Hymenoptera) in 25–40 million year old amber. *Med. Sci. Res.* 20:249–51
16. Cano R, Poinar H, Roubik D, Poinar G. 1992. Enzymatic amplification and nucleotide sequencing of portions of the 18s rRNA gene of the bee *Proplebeia dominicana* isolated from 25–40 million year old amber. *Med. Sci. Res.* 20:619–22
17. Cano RJ, Borucki M. 1995. Revival and identification of bacterial spores in 25- to 40-million-year-old Dominican amber. *Science* 268:1060–64
18. Cattaneo C, Gelstorpe K, Phillips P, Sokol R. 1995. Differential survival of albumin in ancient bone. *J. Archaeol. Sci.* 22:271–76
19. Cherafas J. 1991. Ancient DNA: still busy after death. *Science* 253:1354–56
20. Cherry MI. 1994. Ancient DNA and museums. *S. African J. Sci.* 90:437–38
21. Cipollaro M, Di Bernardo G, Galano G, Galderisi U, Guarino F, et al. 1998. Ancient DNA in human bone remains from Pompeii archaeological site. *Biochem. Biophys. Res. Commun.* 247:901–4
22. Clegg MT, Zurawski G. 1992. Chloroplast DNA and the study of plant phylogeny: present status and future prospects. In *Molecular Systematics of Plants*, ed. PS Soltis, JE Soltis, JJ Doyle, pp. 1–13. New York: Chapman & Hall
23. Cooper A, Mourer-Chauviré C, Chambers GK, von Haeseler A, Wilson A, Pääbo S. 1992. Independent origins of New Zealand moas and kiwis. *Proc. Natl. Acad. Sci. USA* 89:8741–44
24. Cooper A. 1997. Reply to Stoneking: ancient DNA–How do you really know when you have it? *Am. J. Hum. Genet.* 60:1001–2
25. Cooper A, Rhymer J, James HF, Olson SL, McIntosh CE, et al. 1996. Ancient DNA and island endemics. *Nature* 381:484
26. Curry G. 1987. Molecular palaeontology: new life for old molecules. *Trends Ecol. Evol.* 2:161–65
27. Del Pozzo G, Guardiola J. 1989. Mummy DNA fragment identified. *Nature* 339:431–32
28. DeSalle R, Barcia M, Wray C. 1993. PCR jumping in clones of 30 million year old DNA fragments from amber preserved termites (*Mastotermes electrodominicus*). *Experientia* 49:906–9
29. DeSalle R, Gatsey G, Wheeler W, Grimaldi D. 1992. DNA sequences from a fossil termite in Oligo-Miocene amber and their phylogenetic implications. *Science* 257: 1933–36
30. Doran GH, Dicke DN, Ballinger WE Jr, Agee OF, Laipis PJ, Hauswirth WW. 1986. Anatomical, cellular and molecular analysis of 8000-yr-old human brain tissue from the Windover archaeological site. *Nature* 323:803–6
31. Drancourt M, Aboudharam G, Signoli M, Dutour O, Raoult D. 1998. Detection of 400-year-old *Yersinia pestis* in human dental pulp: an approach to the diagnosis of ancient septicemia. *Proc. Natl. Acad. Sci. USA* 95:12637–40
32. Evershed R. 1993. Biomolecular archaeology and lipids. *World Archaeol.* 25:74–93
33. Fisher D, Holland M, Mitchell L, Sledzik P, Wilcox A, et al. 1993. Extraction, evaluation, and amplification of DNA from decalcified and uncalcified United States civil war bone. *J. Forensic Sci.* 38:60–68
34. Frankham R. 1996. Relationship of genetic variation to population size in wildlife. *Conserv. Biol.* 10:1500–8
35. Golenberg E, Giannasi D, Clegg M, Smiley C, Durbin M, et al. 1990. Chloroplast sequence from a Miocene Magnolia species. *Nature* 344:656–58
36. Goloubinoff P, Pääbo S, Wilson A. 1993. Evolution of maize inferred from sequence diversity of an *adh2* gene segment from archaeological specimens. *Proc. Natl. Acad. Sci. USA* 90:1997–2001

37. Graham RW, Lundelius EL Jr, Graham MA, Schroeder EK, Toomey RS III, et al. 1996. Spatial response of mammals to late Quaternary environmental fluctuations. *Science* 272:1601–6
38. Guthrie RD. 1990. *Frozen Fauna of the Mammoth Steppe: The Story of Blue Babe*. Chicago: Univ. Chicago Press
39. Guthrie RD. 1995. Mammalian evolution in response to the Pleistocene-Holocene transition and the break-up of the mammoth steppe: two case studies. *Acta Zool. Cracoviensis* 38:139–54
40. Hadly E, Kohn M, Leonard J, Wayne R. 1998. A genetic record of population isolation in pocket gophers during Holocene climate change. *Proc. Natl. Acad. Sci. USA* 95:6893–96
41. Hagelberg E. 1995. Mitochondrial DNA polymorphysms in modern and ancient Pacific Islanders. *Am. J. Phys. Anthropol.* 20:103
42. Hagelberg E. 1997. Ancient and modern mitochondrial DNA sequences and the colonization of the Pacific. *Electrophoresis* 18:1529–53
43. Hagelberg E, Clegg J. 1991. Isolation and characterization of DNA from archaeological bone. *Proc. R. Soc. London Ser. B* 244:45–50
44. Hagelberg E, Clegg J. 1993. Genetic polymorphisms in prehistoric Pacific islanders determined by analysis of ancient bone DNA. *Proc. R. Soc. London Ser. B* 252:163–70
45. Hagelberg E, Gray I, Jeffreys A. 1991. Identification of the skeletal remains of a murder victim by DNA analysis. *Nature* 352:427–29
46. Hagelberg E, Quevedo S, Turbón D, Clegg JB. 1994. DNA from ancient Easter islanders. *Nature* 369:25–26
47. Hagelberg E, Thomas MG, Cook CEJ, Sher AV, Baryshnikov GF, Lister AM. 1994. DNA from ancient mammoth bone. *Nature* 370:333–34
48. Handt O, Höss M, Krings M, Pääbo S. 1994. Ancient DNA: methodological challenges. *Experientia* 50:524–29
49. Handt O, Krings M, Ward RH, Pääbo S. 1996. The retrieval of ancient human DNA sequences. *Am. J. Hum. Genet.* 59:368–76
50. Handt O, Richards M, Trommsdorff M, Kilger C, Simanainen J, et al. 1994. Molecular genetic analysis of the Tyrolean Ice Man. *Science* 264:1775–78
51. Hänni C, Laudet V, Stehelin D, Taberlet P. 1994. Tracking the origins of the cave bear (*Ursus spelaeus*) by mitochondrial DNA sequencing. *Proc. Natl. Acad. Sci. USA* 91:12336–40
52. Hardy C, Callou C, Vigme J-D, Casane D, Dennebouy N, et al. 1995. Rabbit mitochondrial DNA diversity from prehistoric to modern times. *J. Mol. Evol.* 40:227–37
53. Hardy C, Casane D, Vigne JD, Callou C, Dennebouy N, et al. 1994. Ancient DNA from bronze age bones of European rabbit (*Oryctolagus cuniculus*). *Experientia* 22:487–89
54. Hardy C, Vigne J, Casan D, Dennebouy N, Mounolou J, et al. 1994. Origin of European rabbit (*Oryctolagus cuniculus*) in a Mediterranean island: zooarchaeology and ancient DNA examination. *J. Evol. Biol.* 7:217–26
55. Hauswirth WW, Dickel CD, Rowold DJ, Hauswirth MA. 1994. Inter-and intra population studies of ancient humans. *Experientia* 50:585–91
56. Hedges SB, Schweitzer MH. 1995. Detecting dinosaur DNA. *Science* 268:1191–92
57. Hedrick PW. 1995. Gene flow and genetic restoration: the Florida panther as a case study. *Conserv. Biol.* 9:996–1007
58. Henikoff S. 1995. Detecting dinosaur DNA. *Science* 268:1192
59. Higuchi R, Bowman B, Freiberger M, Ryder O, Wilson A. 1984. DNA sequence from the quagga, an extinct member of the horse family. *Nature* 312:282–84
60. Higuchi R, von Beroldingen C, Sensabagh

G, Erlich H. 1988. DNA typing from single hairs. *Nature* 332:543–46

61. Higuchi R, Wriscknik L, Oakes E, George M, Tong B, et al. 1987. Mitochondrial DNA of the extinct quagga: relatedness and extent of postmortem change. *J. Mol. Evol.* 25:283–87
62. Hillman G, Wales S, McLaren F, Evans J, Butler A. 1993. Identifying problematic remains of ancient plant foods: a comparison of the role of chemical, histological and morphological criteria. *World Archaeol.* 25:94–121
63. Horai S, Kondo R, Murayama K, Hayashi S. 1991. Mitochondrial DNA sequence in contemporary and ancient humans. *Am. J. Hum. Genet.* 49:561
64. Höss M, Dilling A, Currant A, Pääbo S. 1996. Molecular phylogeny of the extinct ground sloth *Mylodon darwinii*. *Proc. Natl. Acad. Sci. USA* 93:181–85
65. Höss M, Jaruga P, Zastawny T, Dizdaroglu M, Pääbo S. 1996. DNA damage and DNA sequence retrieval from ancient tissues. *Nucleic Acids Res.* 24:1304–7
66. Höss M, Kohn M, Pääbo S, Knauser S, Schroder W. 1992. Excremental analysis by PCR. *Nature* 359:199
67. Höss M, Pääbo S, Vereshchagin N. 1994. Mammoth DNA sequences. *Nature* 370: 333
68. Houde P, Braun M. 1988. Museum collections are a source of DNA for studies of avian phylogeny. *Auk* 105:773–76
69. Howell N, Kubacka I, Mackey DA. 1996. How rapidly does the human mitochondrial genome evolve? *Am. J. Hum. Genet.* 59:501 9
70. Hudson RR. 1990. Gene genealogies and the coalescent process. *Oxford Surv. Evol. Biol.* 7:1–44
71. Janczewski D, Yukhi N, Gilbert D, Jefferson G, O'Brien S. 1992. Molecular phylogenetic inference from saber-toothed cat fossils of Rancho La Brea. *Proc. Natl. Acad. Sci. USA* 89:9769–73
72. Jeffreys A, Allen M, Hagelberg E, Sonnberg A. 1992. Identification of the skeletal remains of Josef Mengele by DNA analysis. *Forensic Sci. Int.* 56:65–76
73. Kohn M, Wayne R. 1997. Facts from feces revisited. *Trends Ecol. Evol.* 6:223–27
74. Kohn MH, York EC, Kamradt DA, Haught G, Sauvajot RM, Wayne RK. 1999. Estimating population size by genotyping faeces. *Proc. R. Soc. London Ser. B* 266:657–63
75. Krajewski C, Driskell A, Baverstock P, Braun M. 1992. Phylogenetic relationships of the thylacine (Mammalia, Thylacinidae) among dasyuroid marsupials—evidence from *cytochrome b* DNA sequences. *Proc. R. Soc. London Ser. B* 50: 19–27
76. Krings M, Stone A, Schmitz RW, Krainitski H, Stoneking M, Pääbo S. 1997. Neanderthal DNA sequences and the origin of modern humans. *Cell* 90:19–30
77. Krings M, Geisert H, Schmitz RW, Krainitzki H, Pääbo S. 1999. DNA sequence of the mitochondrial hypervariable region II from the Neandertal type specimen. *Proc. Natl. Acad. Sci. USA* 96:5581–85
78. Kwok S, Higuchi R. 1989. Avoiding false positives with PCR. *Nature* 339:237–38
79. Lalueza C, Pérez-Pérez A, Prats E, Cornudella L, Turbón D. 1997. Lack of founding Amerindian mitochondrial DNA lineages in extinct aborigines from Tierra del Fuego-Patagonia. *Hum. Mol. Genet.* 6:41–46
80. Lawlor D, Dickel C, Hauswirth W, Parham P. 1991. Ancient HLA genes from 7500-year-old archaeological remains. *Nature* 349:785–88
81. Lehman N. 1998. Conservation biology: genes are not enough. *Curr. Biol.* 8:722–24
82. Lindahl T. 1993. Instability and decay of the primary structure of DNA. *Nature* 362:709–15
83. Lindahl T. 1993. Recovery of antediluvian DNA. *Nature* 365:700
84. Lindahl T. 1997. Facts and artifacts of ancient DNA. *Cell* 90:1–3

85. Logan GA, Boon JJ, Eglinton G. 1993. Structural biopolymer preservation in Miocene leaf fossils from the Clarkia site, northern Idaho. *Proc. Natl. Acad. Sci. USA* 90:2246–50
86. Lowenstein JM. 1985. Radioimmune assay of mammoth tissue. *Acta Zool. Fennica* 170:233–35
87. Matisoo-Smith E, Allen JS, Ladefoged TN, Roberts RM, Lambert DM. 1997. Ancient DNA from Polynesian rats: extraction, amplification and sequence from single small bones. *Electrophoresis* 18:1534–37
88. Matuschka F-R, Ohlenbusch A, Eiffert H, Richter D, Spielman A. 1996. Characteristics of Lyme disease spirochetes in archived European ticks. *J. Infect. Dis.* 174:424–26
89. McKusick V. 1991. The defect in Marfan syndrome. *Nature* 352:279–81
90. Merriwether DA, Rothhammer F, Ferrell RE. 1994. Genetic variation in the new world: ancient teeth, bone, and tissue as sources of DNA. *Experientia* 50:592–601
91. Monsalve M, Cardenas F, Guhl F, Delaney AD, Devine DV. 1996. Phylogenetic analysis of mtDNA lineages in South American mummies. *Ann. Hum. Genet.* 60:293–303
92. Morin P, Woodruff DS. 1996. Noninvasive genotyping for vertebrate conservation. In *Molecular Genetic Approaches in Conservation*, ed. TB Smith, RK Wayne, pp. 298–313. New York: Oxford Univ. Press
93. Morin PA, Moore JJ, Woodruff DS. 1992. Identification of chimpanzee subspecies with DNA from hair and allele-specific probes. *Proc. R. Soc. London Ser. B* 249: 293–97
94. Mullis KB, Faloona FA. 1987. Specific synthesis of DNA in vitro via a polymerase catalyzed chain reaction. *Meth. Enzymol.* 155:335–50
95. Mundy NI, Winchell CS, Burr T, Woodruff DS. 1997. Microsatellite variation and microevolution in the critically endangered San Clemente Island loggerhead shrike (*Lanius ludovicianus mearnsi*). *Proc. R. Soc. London Ser. B* 264:869–075
96. Noro M, Masuda R, Dubrovo IA, Yoshida MC, Kato M. 1998. Molecular phylogenetic inference of the woolly mammoth *Mammuthus primigenius*, based on complete sequences of mitochondrial *cytochrome b* and 12S ribosomal RNA genes. *J. Mol. Evol.* 46:314–26
97. O'Rourke DH, Carlyle SW, Parr RL. 1996. Ancient DNA: methods, progress and perspectives. *Am. J. Hum. Biol.* 8:557–71
98. Ozawa T, Hayashi S, Mikhelson VM. 1997. Phylogenetic position of mammoth and Steller's sea cow within Tethytheria demonstrated by mitochondrial DNA sequences. *J. Mol. Evol.* 44:406–13
99. Pääbo S. 1984. Uber den Nachweis von DNA in Altagyptischen Mumien. *Das Altertum* 30:213–18
100. Pääbo S. 1985. Molecular cloning of ancient Egyptian mummy DNA. *Nature* 314:644–45
101. Pääbo S. 1989. Ancient DNA: extraction, characterization, molecular cloning and enzymatic amplification. *Proc. Natl. Acad. Sci. USA* 86:1939–43
102. Pääbo S. 1993. Ancient DNA. *Sci. Am.* 269:86–92
103. Pääbo S, Higuchi R, Wilson A. 1989. Ancient DNA and the polymerase chain reaction. *J. Biol. Chem.* 264:9709–12
104. Pääbo S, Irwin D, Wilson A. 1990. DNA damage promotes jumping between templates during enzymatic amplification. *J. Biol. Chem.* 265:4718–21
105. Pääbo S, Wilson A. 1991. Miocene DNA sequences: a dream come true? *Curr. Biol.* 1:45–46
106. Palumbi SR, Cipriano F. 1998. Species identification using genetic tools: the value of nuclear and mitochondrial gene sequences in whale conservation. *J. Hered.* 89:459–64
107. Parr R, Carlyle S, O'Rourke D. 1996. Ancient DNA analysis of Fremont Amerindians of the Great Salt Lake wetlands. *Am. J. Phys. Anthropol.* 99:507–18

108. Persing D, Tedford III S, Rys PN, Dodge DE, White TJ, et al. 1990. Detection of *Borrelia burgdorferi* in DNA in museum specimens of *Ixodes dammini* ticks. *Science* 249:1420–23
109. Poinar G. 1994. The range of life in amber: significance and implications in DNA studies. *Experientia* 50:536–42
110. Poinar H, Cano R, Poinar G. 1993. DNA from an extinct plant. *Nature* 363:677
111. Poinar HN, Hofreiter M, Spaulding WG, Martin PS, Stankiewicz BA, et al. 1998. Molecular coproscopy: dung and diet of the extinct ground sloth *Nothrotheriops shastensis*. *Science* 281:402–6
112. Poinar HN, Höss M, Bada JL, Pääbo S. 1996. Amino acid racemization and the preservation of ancient DNA. *Science* 272:864–66
113. Rafi A, Spigelman M, Stanford J, Lemma E, Donoghue H, et al. 1994. *Mycobacterium leprae* DNA from ancient bone detected by PCR. *Lancet* 343:1360–61
114. Reed J, Tollit D, Thompson P, Amos W. 1997. Molecular scatology: the use of molecular genetic analysis to assign species, sex and individual identity to seal faeces. *Mol. Ecol.* 6:225–34
115. Reid AH, Fanning TG, Hultin JV, Taubenberger JK. 1999. Origin and evolution of the 1918 "Spanish" influenza virus hemagglutinin gene. *Proc. Natl. Acad. Sci. USA* 96:1651–56
116. Roelke-Parker ME, Munson L, Packer C, Kock R, Cleaveland S, et al. 1996. A canine distemper virus epidemic in Serengeti lions *Panthera leo*. *Nature* 379:441–45
117. Rogers AR, Harpending H. 1992. Population growth makes waves in the distribution of pairwise genetic differences. *Mol. Biol. Evol.* 9:552–69
118. Rollo F, Venanzi F, Amici A. 1991. Nucleic acids in mummified plant seeds: biochemistry and molecular genetics of pre-Columbian maize. *Genet. Res.* 58:193–201
119. Roy M, Girman D, Wayne R. 1994. The use of museum specimens to reconstruct the genetic variability and relationships of extinct populations. *Experientia* 50:551–63
120. Roy MS, Geffen E, Smith D, Wayne RK. 1996. Molecular genetics of pre-1940 red wolves. *Conserv. Biol.* 10:1413–24
121. Salo W, Aufderheide A, Buikstra J, Holcomb T. 1994. Identification of *Myco bacterium tuberculosis* DNA in a pre-Columbian Peruvian mummy. *Proc. Natl. Acad. Sci. USA* 91:2091–94
122. Salvo J, Allison M, Rogan P. 1989. Molecular genetics of pre-Columbian South American mummies. *Am. J. Phys. Anthropol.* 78:295
123. Shoshani J, Walz DA, Goodman M, Lowenstein JM, Prychodko W. 1985. Protein and anatomical evidence of the phylogenetic position of *Mammuthus primigenius* within the Elephantinae. *Acta Zool. Fennica* 170:237–40
124. Sidow A, Wilson A, Pääbo S. 1991. Bacterial DNA in Clarkia fossils. *Philos. Trans. R. Soc. London Ser. B* 333:429–33
125. Smith F, Betancourt J, Brown JH. 1995. Evolution of body size in the woodrat over the past 25,000 years of climate change. *Science* 270:2012–14
126. Soltis P, Soltis D, Smiley C. 1992. An rbcL sequence from a Miocene *Taxodium* (bald cypress). *Proc. Natl. Acad. Sci. USA* 89:499–501
127. Soltis PS. 1995. Fossil DNA: its potential for biosystematics. In *Experimental and Molecular Approaches to Plant Biosystematics*, ed. PC Hoch, AG Stephenson, pp. 1–13. St. Louis: Missouri Bot. Gard.
128. Soltis PS, Soltis DE. 1993. Ancient DNA: prospects and limitations. *New Zealand J. Bot.* 31:203–9
129. Stankiewicz BA, Poinar HN, Briggs DEG, Evershed RP, Poinar GO Jr. 1998. Chemical preservation of plants and insects in natural resins. *Proc. R. Soc. London Ser. B* 265:641–47

130. Steadman DW. 1995. Prehistoric extinctions of Pacific island birds: biodiversity meets zooarchaeology. *Science* 267: 1123–31
131. Stone AC, Milner GR, Pääbo S, Stoneking M. 1996. Sex determination of ancient human skeletons using DNA. *Am. J. Phys. Anthropol.* 99:231–38
132. Stone AC, Stoneking M. 1993. Ancient DNA from a pre-Columbian Amerindian population. *Am. J. Phys. Anthropol.* 92: 463–71
133. Stoneking M. 1995. Ancient DNA: How do you know when you have it and what can you do with it? *Am. J. Hum. Genet.* 57:1259–62
134. Swofford DL, Olsen GJ, Waddell PJ, Hillis DM. 1996. Phylogenetic inference. In *Molecular Systematics*, ed. DM Hillis, C Moritz, BK Mable, pp. 407–514. Sunderland, MA: Sinauer. 2nd ed.
135. Sykes B. 1991. Ancient DNA: the past comes alive. *Nature* 352:381–82
136. Sykes B. 1993. Ancient DNA: less cause for grave concern. *Nature* 366:513
137. Taberlet P, Bouvet J. 1992. Bear conservation genetics. *Nature* 358:197
138. Taberlet P, Fumagalli L. 1996. Owl pellets as a source of DNA for genetic studies of small mammals. *Mol. Ecol.* 5:301–5
139. Taylor G, Crossey M, Saldanha J, Waldron T. 1996. DNA from *Mycobacterium tuberculosis* identified in mediaeval human skeletal remains using polymerase chain reaction. *J. Archaeol. Sci.* 23:789–98
140. Taylor JW, Swann EC. 1994. DNA from herbarium specimens. In *Ancient DNA*, ed. B Hermann, S. Hummel, pp. 166–81. New York: Springer-Verlag
141. Templeton AR, Routman E, Phillips CA. 1995. Separating population structure from population history: a cladistic analysis of the geographical distribution of mitochondrial DNA haplotypes in the Tiger salamander, *Ambystoma tigrinum*. *Genetics* 140:619–33
142. Thomas R, Schaffner W, Wilson A, Pääbo S. 1989. DNA phylogeny of the extinct marsupial wolf. *Nature* 340:465–67
143. Thomas WK, Pääbo S, Villablanca FX, Wilson AC. 1990. Spatial and temporal continuity of kangaroo rat populations shown by sequencing mitochondrial DNA from museum specimens. *J. Mol. Evol.* 31:101–12
144. Vane-Wright RI. 1996. Systematics and the conservation of biological diversity. *Ann. Missouri Bot. Gard.* 83:47–57
145. Walden KKO, Robertson HM. 1997. Ancient DNA from amber fossil bees? *Mol. Biol. Evol.* 14:1075–77
146. Wang H-L, Yan Z-Y, Jin D-Y. 1997. Reanalysis of published DNA sequence amplified from Cretaceous dinosaur egg fossil. *Mol. Biol. Evol.* 14:589–91
147. Wayne R, Jenks S. 1991. Mitochondrial DNA analysis implies extensive hybridization of the endangered red wolf *Canis rufus*. *Nature* 351:565–68
148. Weiss KM, Buchanan AV, Daniel C, Stoneking M. 1994. Optimizing utilization of DNA from rare of archival anthropological samples. *Hum. Biol.* 66:789–804
149. Williams S, Longmire J, Beck L. 1990. Human DNA recovery from ancient bone. *Am. J. Phys. Anthropol.* 81:318
150. Woodward SR, Weyand NJ, Bunnell M. 1994. DNA sequence from Cretaceous period bone fragments. *Science* 266: 1229–32
151. Wright S. 1978. *Evolution and the Genetics of Populations*. Vol. 4: *Variability Within and Among Natural Populations*. Chicago: Univ. Chicago Press
152. Yang H, Golenberg E, Shoshani J. 1996. Phylogenetic resolution within the Elephantidae using fossil DNA sequence from the American mastodon (*Mammut americanum*) as an outgroup. *Proc. Natl. Acad. Sci. USA* 93:1190–94
153. Young DL, Huyen Y, Allard MW. 1995. Testing the validity of the *cytochrome b*

sequence from Cretaceous period bone fragments as dinosaur DNA. *Cladistics* 11:199–209

154. Yousten AA, Rippere KE. 1997. DNA similarity analysis of a putative ancient bacterial isolate obtained from amber. *FEMS Microbiol. Lett.* 152:345–47

155. Zischler H, Höss M, Handt O, von Haeseler A, van der Kuyl AC, et al. 1995. Detecting dinosaur DNA. *Science* 268:1192

156. Zischler H, Geisert H, von Haeseler A, Pääbo S. 1995. A nuclear 'fossil' of the mitochondrial D-loop and the origin of modern humans. *Nature* 378:489–92

Annu. Rev. Ecol. Syst. 1999. 30:479–513

Do Plant Populations Purge Their Genetic Load? Effects of Population Size and Mating History on Inbreeding Depression

D. L. Byers* and D. M. Waller+

*Department of Biological Sciences, Illinois State University, Campus Box 4120, Normal, Illinois 61790; e-mail: dlbyer2@ilstu.edu; +Department of Botany, University of Wisconsin-Madison, 430 Lincoln Drive, Madison, Wisconsin 53706; e-mail: dmwaller@facstaff.wisc.edu

Key Words mating system, genetic load, purging, self-fertilization, genetic effects of small population size

■ **Abstract** Inbreeding depression critically influences both mating system evolution and the persistence of small populations prone to accumulate mutations. Under some circumstances, however, inbreeding will tend to purge populations of enough deleterious recessive mutations to reduce inbreeding depression (ID). The extent of purging depends on many population and genetic factors, making it impossible to make universal predictions. We review 52 studies that compare levels of ID among species, populations, and lineages inferred to differ in inbreeding history. Fourteen of 34 studies comparing ID among populations and species found significant evidence for purging. Within populations, many studies report among-family variation in ID, and 6 of 18 studies found evidence for purging among lineages. Regression analyses suggest that purging is most likely to ameliorate ID for early traits (6 studies), but these declines are typically modest (5–10%). Meta-analyses of results from 45 populations in 11 studies reveal no significant overall evidence for purging, but rather the opposite tendency, for more selfing populations to experience higher ID for early traits. The likelihood of finding purging does not vary systematically with experimental design or whether early or late traits are considered. Perennials are somewhat less likely to show purging than annuals (2 of 10 vs. 7 of 14). We conclude that although these results doubtless reflect variation in population and genetic parameters, they also suggest that purging is an inconsistent force within populations. Such results also imply that attempts to deliberately reduce the load via inbreeding in captive rearing programs may be misguided. Future studies should examine male and female fitness traits over the entire life cycle, estimate mating histories at all levels (i.e. population and families within populations), report data necessary for meta-analysis, and statistically test for purging of genetic loads.

INTRODUCTION

> It often occurred to me that it would be advisable to try whether seedlings from cross-fertilised flowers were in any way superior to those from self-fertilised flowers.
>
> C. Darwin (36, p. 8)

In the course of working on inheritance, Darwin grew self- and cross-fertilized seedlings of *Linaria vulgaris* side by side in beds and was surprised to observe a clear difference between them. This led him to began systematic investigations into the effects of close inbreeding that extended over 52 taxa and 11 years (36). He was careful to pair selfed and outcrossed seeds for simultaneous germination and grew pairs side by side in the same pots. He soon realized progeny from related plants suffered constitutional changes and concluded "cross-fertilisation is generally beneficial, and self-fertilisation injurious." He further noted that selfed progeny suffered reductions in performance at almost every stage of growth even in taxa that were self-fertile. In *Ipomea purpurea*, Darwin found fewer seed capsules per self-pollination (69%), fewer seeds in selfed capsules (93%), and selfed seedlings that were shorter (76%). In the sixth generation of selfing, he found a plant with surprising vigor he christened Hero. Hero's descendants maintained unusual vigor, leading Darwin to note that the deleterious changes produced by selfing could be ameliorated under some circumstances. He concluded (36, p. 442) that "the advantages which follow from cross-fertilisation differ much in different plants" and "There does not seem to exist any close correspondence between the degree to which the flowers of species are adapted for cross-fertilisation and the degree to which their offspring profit by this process." Remarkably, he did all this work with no notion of Mendelian genetics.

Today, we refer to the reduced fitness we observe in progeny from matings among relatives as inbreeding depression. Inbreeding depression (ID) is indeed widespread among species and has been investigated thoroughly in domesticated plants and animals in this century (Ch. 2 in 182). We now interpret the occasional superior performance of selfed individuals like Hero and reductions in ID in inbred lines to be the result of purging, the preferential elimination of recessive deleterious alleles in inbred lines (26, 29, 31, 102, 151, 156, 182). While Darwin failed to find any correspondence between mating system and the degree of ID, several authors report reductions in ID in more inbred populations in both plants (7, 41, 72) and animals (12, 109, 113, 137). If purging is an effective process, we might expect it to progressively favor selfing in more inbred lines. This might account both for plant species that persistently self-fertilize (37, 77) and for the apparent tendency for plant species to either mostly self or mostly outcross (143). More selfing populations also tend to show reduced ID for early components of fitness as would be expected if selection preferentially purges lethal and semi-lethal mutations that act early (76). Thus, the evolutionary dynamics of inbreeding have become central in attempts to account for the remarkable diversity of plant mating systems (26, 65, 163).

Interest in ID dynamics has also expanded to include conservation biologists who wish to understand how inbreeding affects the persistence of small and/or inbred populations. Some researchers, impressed by the potential for purging, recommend intentional inbreeding as a strategy to reduce ID in captively bred populations (135, 152, 160). This faith in purging, however, may be misplaced if selection does not consistently reduce the load. Recent theoretical work suggests that purging may be ineffective in small populations in which mutations may accumulate to the point that they threaten population persistence (e.g., 110). Because theoretical predictions regarding ID dynamics hinge on assumptions and parameter values, there is now considerable interest in assessing how real populations respond to a history of inbreeding. This has spawned a renaissance of empirical work. Much of this work focuses on plants for the same reasons Darwin chose them for experimental work: Many are capable of the closest possible inbreeding (self-fertilization); they are relatively easy to control and cross; and they can be propagated in sufficient number to generate statistically meaningful results.

In this review, we first summarize the theory pertaining to ID and the accumulation of mutations in small populations. After considering mechanisms of ID, we explore how populations accumulate and reduce genetic load via mutation and selection. Theoretical work reveals that the efficiency of purging is strongly affected by the degree of dominance, the distribution of mutational effects, interactions among loci, and the breeding system. In addition, selection in small and/or inbred populations is often compromised, further affecting load dynamics. We next consider studies aimed at assessing how ID responds to varying histories of inbreeding. We concentrate on 52 empirical plant studies that compare ID among taxa, populations, or lineages that differ in their inferred histories of inbreeding. This burst of activity, mostly in the last 10 years, stands in contrast to the century after Darwin's initial work which produced few studies of ID outside agricultural settings. Because these studies differ considerably in design, scale, and execution, we first enumerate how frequently purging has been observed in various types of study. We then explore how the extent of purging varies in magnitude over studies, using regression and meta-analyses to assess how the type of experiment, plant characteristics, and the traits measured affect whether purging is observed. We conclude that although much evidence for purging exists, it is not consistent or substantial enough in most cases to favor fully inbred forms of mating or intentionally inbred schemes of mating for conserving small populations. To better understand purging and the genetic basis for ID, we recommend that future work of this kind incorporate a consistent set of methodological and analytical features.

THEORETICAL BACKGROUND

Mechanisms of ID

Deleterious alleles are often both recessive (38) and rare, meaning that their effects will usually be masked in heterozygous form in outbred populations. Thus, dominance is sufficient to account for the decline in fitness observed upon inbreeding

(35, 47, 182). Overdominance could also account for ID, however, in that fitness will also decline upon inbreeding as complementary alleles segregate as homozygotes (31, 32, 63, 184). Both dominance and overdominance predict similar declines in inbred fitness in proportion to the inbreeding coefficient, F (180a). If fitness effects interact independently across loci, the logarithm of fitness (or fitness components) should decline linearly with increased progeny inbreeding with a slope that measures the extent of the mutational load in terms of the number of lethal equivalents (104, 123, 180). Epistatic interactions among loci could also account for some ID (13, 90). Declines in log fitness that are less than linear (positive F^2 term) indicate diminishing epistasis, while fitness declines that are more than linear reflect synergistic or reinforcing epistasis (p. 79, 35).

Controversy over whether ID primarily reflects dominance or overdominance has persisted for much of this century (34). Much of the difficulty in resolving the question of mechanism stems from the fact that beneficial and dominant alleles may be linked to detrimental recessive alleles, causing pairs of such linked loci to mimic single overdominant loci (85, 130). Such associative (pseudo) overdominance could account for the positive correlations we often observe between fitness and heterozygosity (reviewed in 120) and empirical estimates of overdominance in natural populations (e.g., 53). While work in maize initially implicated overdominance, later work in maize and *Drosophila* observed declines in apparent dominance levels after further recombination (26, 33, 34, 79, 153). Associations between heterozygosity and fitness in *Drosophila* and pines appear attributable to associative overdominance (140, 159). Simulations confirm that associative overdominance arises in inbred lineages and that mutation to partially recessive deleterious alleles produces biologically realistic levels of ID (25, 29). A recent, well-designed study of two species of *Mimulus* concluded inbreeding depression was attributable to greater expression of recessive alleles (43).

Selection Against the Load

Classic population genetic models illustrate how recurrent mutation to deleterious alleles is balanced by selection against these alleles, leading to an equilibrium that depends on mutation rates, selective effects, the mating system and levels of dominance (35). If deleterious mutations occur at a locus at rate u and are selected against so that relative fitness is $1-s$ in homozygotes and $1-hs$ in heterozygotes, the equilibrium frequency of completely recessive mutations will be $\sqrt{}(u/s)$ in an infinite panmictic population. These alleles become much rarer (u/s) in a completely inbred population as these alleles are exposed more often to selection. Similarly, selection in heterozygotes becomes far more significant with even partial dominance, decreasing equilibrium frequency to approximately u/hs (or $2u/s$ with no dominance). In the absence of inbreeding $(F = 0)$ and dominance $(h = 0.5)$, the reduction in average fitness due to mutation–the mutational load (125) per locus–is independent of the selective effect and $L = 2u/(1 + u)$ (35). With partial dominance, $L \approx 2u$. If the mutant is completely recessive or if inbreeding causes most of them to be eliminated as homozygotes, the load is reduced ($L = u$). This implies that with complete recessivity, the maximum per locus decrease in the load due to purging is u. Over all loci, the load is equal to the sum of the mutation rates at each locus ($U = \Sigma u$).

The distribution of selective effects strongly affects the extent of purging. While lethal or semilethal mutations are efficiently purged (102), purging is far less efficient against mutations of minor effect, particularly if they affect heterozygote performance (26, 61). Classic work in *Drosophila* suggests that dominance of the wild type is rarely complete and most mutations have minor effects (89, 124). Assuming that inbreeding populations retain a substantial fraction of their load, the architecture of the load will shift with inbreeding to include fewer lethals and semilethals relative to mildly deleterious mutations. For example, five generations of selfing in *Mimulus guttatus* exposed many major mutations, leading to the loss of almost half of the lines, yet this did not appreciably diminish subsequent levels of ID in many of the surviving lines (44).

What mutation rates and levels of dominance characterize real populations? Total mutation rates are difficult to estimate, but it is clear that lethal mutations occur regularly in long-lived plants like ferns and mangroves (92, 93). Mukai's (124) classic experiments with chromosome II in *Drosophila* suggest a haploid genome mutation rate (U) of at least 0.17 with an average dominance (h) of 0.21 for minor viability genes at equilibrium (vs. 2–3% for lethals within a natural population; 33). However, recent work in Drosophila use improved methods have found lower mutation rates of 0.02 and suggest mutations have small effects (52a). Using these estimates of dominance and existing data on ID, Charlesworth (28) estimated the mutation rate to be $U = 1.3$–1.7 for *Leavenworthia crassa* and $U = 0.7$–0.9 in *L. uniflora*. In the diploid *Amsinckia spectabilis*, Johnston & Schoen (83) estimated U at 0.24–0.4 with h at 0.07–0.14, while the tetraploid congener, *A. gloriosa*, appeared to have roughly twice the mutation rate and load and higher levels of dominance. Other experiments suggest a genomic mutation rate of at least 0.25 (74), or 0.57 (112) or perhaps 1.0 or above (46, 97).

Selection is far less efficient in small and inbred populations as selection can act effectively only against mutations with selective effects greater than $1/(2\ N_e)$ (182). As N_e declines due to demographic factors or inbreeding, an increasing proportion of deleterious alleles becomes effectively neutral and invisible to selection, reducing the rate at which they are eliminated. Ultimately, we expect drift to result in these alleles being fixed at a probability equal to their starting frequencies. Once fixed, these mutations add to the load in a difficult-to-reverse process known as Muller's ratchet (62). In addition, strong selection against lethals and semilethals can incidentally fix mildly deleterious mutations via background selection or selective sweeps (24). This will cause many slightly deleterious mutations to be fixed in inbreeding populations even as genes with greater effects are being purged. Ironically, once fixed, these mutations decrease the fitness of both selfed and outcrossed progeny, reducing the difference between them and thus between any estimates of ID based on relative performance. Thus, studies that simply compare levels of

ID before and after inbreeding could observe reduced ID even in the absence of purging (see Discussion).

Mating System Evolution

Classically, ID has been considered the primary impediment to the evolution of increased selfing, with various flowering mechanisms (e.g., dioecy, gynodioecy, and monoecy) being interpreted as specifically evolved to avoid or reduce selfing (36). Fisher (51) first enunciated the substantial transmission advantage enjoyed by selfing and other forms of uniparental reproduction that avoid what Williams (176) termed the "cost of meiosis." Since then, the evolution of self-fertilization has usually been viewed as a balance between its transmission advantage (often assumed to be 50%) and the disadvantage of ID among selfed progeny (77, 173). If, however, increased selfing reduces opportunities to donate pollen, selfing may not enjoy the full transmission advantage commonly assumed (21, 69, 127). Selfing also brings the advantage of reproductive assurance following colonization (Baker's Law) (5, 36, 156), which may be more important than hitherto appreciated (147). Clearly, the mechanism by which selfing occurs also affects how selection can act on the mating system (i.e., prior, competing, or delayed selfing; 108). While most models place particular emphasis on genetic factors in mating system evolution, reproductive assurance, pollen discounting, and other ecological factors deserve similar emphasis in models of mating system evolution (65, 163).

Classic models of mating system evolution considered ID to be a property of the population and assumed its level to be fixed. ID is usually estimated as $\delta = 1 - W_s / W_o$, where W_s and W_o refer to the fitnesses of selfed and outcrossed progeny, respectively. This estimation of the parameter ID δ will be bounded from -1 to 1. Such models typically predict that selection (due to the transmission advantage) favors decreases in the rate of selfing if δ exceeds 0.5. In some situations, it may take two or more generations of selfing for the decline of fitness with F to exceed this threshold (116, 176). This might increasingly favor outcrossing in progressively more inbred populations, allowing a stable mixed (selfed and outcrossed) mating system (176). If δ is always less (or greater) than 0.5, however, simple models predict that plants should either completely self (or outcross) (102, 143).

Recent models that consider dynamic processes in mating system evolution make somewhat different predictions (163, 173). The most influential such model incorporated the capacity for populations to purge themselves of deleterious mutations and hence evolve lower rates of ID after a history of inbreeding (102). This model allowed ID and the selfing rate to coevolve by alternating selection on the mating system with selection against the load (due to recurrent major mutations at unlinked loci). As expected, reduced ID in more inbred lines can favor runaway selection for ever increasing levels of selfing in such models, reinforcing the binary prediction of classical static models. In contrast, models incorporating associations among load and mating system loci make a richer set of predictions.

In these simultaneous coevolutionary models, ID alone does not predict whether increased levels of selfing will evolve because genetic associations between load and modifier loci complicate selection (67, 68, 164). In addition, mixed mating systems can be evolutionarily stable under some circumstances (165). These models also predict that populations will retain among-family variation in ID and mating system characters as well as associations between these (23, 27, 29, 166). While some empirical studies detected such associations (169; N Takebayashi, LF Delph, in review; S-M Chang, MD Rausher, in review), others have not.

Our view of mating system dynamics is thus being extended to explore how levels of selfing, population structure, and ID all coevolve (173). This picture is complicated further by load dynamics in small inbred populations.

Genetic Hazards in Small and Inbred Populations

While conservation biologists have traditionally worried most immediately about demographic and environmental hazards, it is now clear that these are compounded by, and interact with, the genetic hazards faced by small populations (48, 52, 55). Lande (100) identifies three classes of genetic risk to small populations: immediate ID, the loss of possibly adaptive genetic variability, and the fixation of new deleterious mutations (accentuating ID). Early predictions based on the attrition of genetic variance in small populations suggested little genetic danger to populations above a few hundred—the so-called 50/500 rule (155). Recent theoretical work suggests that populations of several thousand may be necessary to maintain quantitative genetic variation and slow the accumulation of deleterious mutations (14, 101, 110). Empirical support is also emerging (e.g., small populations of *Silene regia* suffer reduced seed viability, 118).

Lynch and colleagues (54, 111) linked demographic and genetic models to demonstrate how the accumulation of deleterious mutations may reduce population persistence via a runaway process they term "mutational meltdown." These models have now been extended to include sexual populations; they conclude that populations with $N_e < 100$ are highly vulnerable to extinction on time scales of about 100 generations (110). In such populations, purging is "at best transient, as intentional inbreeding can only enhance the probability of fixation of deleterious alleles, and those alleles that are purged are rapidly replaced with new mutations." Mostly selfing plant populations appear to be at particular risk of mutational meltdown as they have small N_e and associations among alleles at different loci. Lande's (100) similar models led him to conclude that mildly deleterious mutations are "far more important in causing loss of fitness and eventual extinction than are lethal and semilethal mutations in populations with effective sizes, N_e, larger than a few individuals." Inbred populations also suffer from decreases in quantitative genetic variation that could further increase their risk of extinction (101).

Variable mutational effects, synergistic epistasis, and recurrent beneficial mutations all reduce the accumulation of mutations, but purging becomes impeded in small or inbred populations with N_e less than about 100 (148). Although linkage can increase the efficiency of selection in some circumstances (30), small and/or fluctuating population size also restricts selection for favorable mutations (131).

The strong selection that accompanies the purging of strongly deleterious alleles also causes allele frequencies at linked loci to change via background selection (24). This tends to increase population differentiation while further decreasing within-population genetic variation.

Paradoxes of Purging

If purging eliminates much of the load, we should expect inbred populations to show reduced levels of ID. The Charlesworth's earlier review (26) noted that several inbred plant and animal populations express relatively low levels of ID, but others retain substantial amounts. A more recent review of data from 79 populations in 54 species found that ID tends to decline with increases in the estimated selfing rate (rank correlation $r_s = -0.42$) (76). These authors found average δ levels of 0.23 in the selfers vs. 0.53 in predominantly outcrossed plants. This significant difference supports the purging hypothesis, but ID could also be depressed in selfing populations via fixation as noted above. Perhaps more tellingly, the ID remaining in most of the selfing species occurred later in life as expected if selection had effectively purged early-acting mutations of major effect.

Because purging appears effective, some suggest using intentional inbreeding as a conservation strategy in particular situations (135, 152, 160). In their work with Speke's Gazelle, Templeton & Read (161) reported that the mutational load decreased from 3.09 to 1.35 lethal equivalents for 30-day viability after a period of enforced inbreeding. These results, however, may reflect a statistical artifact (6, 178). The several genetic complications reviewed above suggest that purging will be relatively slow and constrained in its effects on ID for mutations of mild effect and partial dominance. At the same time, the inbred conditions necessary for purging to occur tend to accelerate the fixation of mildly deleterious alleles. Such equivocal results suggest that we should be cautious in promoting purging as a tactic to reduce the load.

Plants with persistently mixed mating systems pose an evolutionary enigma in that a history of inbreeding within a population is expected to purge populations of their genetic load and reduce ID. Thus, we expect species to undergo accelerating, or disruptive, selection for either complete outcrossing or ever-increasing levels of self-fertilization (102). While many plant species do appear to be either mostly selfing or mostly outcrossing (143), this partly reflects dominant modes of pollination (2). Exclusively selfing species are rare, and the many species with mixed mating systems do not appear to be undergoing strong directional selection (172).

Thus, theory leaves unresolved how effective purging will be in real populations. As both purging and mutation fixation plausibly occur in inbred populations, it is important to assess empirically how ID responds to inbreeding history.

APPROACH

Here, we review studies that test how levels of ID respond to a history of inbreeding. Our goal was to find studies that directly compare levels of ID among groups with divergent inbreeding histories. Most of this work is recent. In choosing studies for our analyses, we specifically sought studies that compare taxa, populations, or lineages that differ in their inferred levels of inbreeding. The studies differ considerably, however, in the taxa chosen, in how differences in inbreeding history were inferred (or manipulated), and in methods for estimating ID. After categorizing the studies according to their scale of comparison (taxon, population or lineage comparisons), we further categorize them according to how inbreeding history was inferred and how ID was measured. These divisions allow us to judge whether the estimated extent of purging varies across these groups.

We looked for studies via online literature searches and surveys of particular journals (*Evolution, Heredity*, and the *American Journal of Botany* since 1985), and we used personal contacts to find studies not yet in print. We chose not to include studies that only compare levels of ID in selfed lines to the 0.5 threshold used in classical models, as measurements of ID are sensitive to the trait chosen and assay conditions (42, 76). The comparisons presented here provide stronger and more quantitative assessments of the extent of purging.

Fifty-two studies provided usable results spanning 29 families and 52 species (full data online: www.bio.ilstu.edu/BEES/byers/). Ten studies involved *Mimulus*, reflecting its utility as a model genus due to its diverse mating systems and short life cycle. All studies found significant ID in at least one trait, but many did not explicitly test for purging.

Scale of Comparison

The studies we review address the degree of purging on one or more of the following scales:

1. *Among taxa*—In these 14 studies, authors compared levels of ID among related species within a genus that differ in flower morphology, self-compatibility, or some other indicator of the extent of inbreeding. The best such studies use molecular systematics methods to confirm that the species chosen are closely related (e.g., *Linanthus*—C Goodwillie, in review).
2. *Among populations within species*—These 21 studies compare the relative expression of ID among populations known or inferred to differ in their mating history (e.g., 72).
3. *Among lineages within a population*—In these 18 studies, levels of ID are compared among lines known or thought to differ with respect to their amount of inbreeding (e.g., 7). In some cases, inbreeding history is inferred

(e.g., from floral characters in Mimulus; 20) while in other cases it is experimentally manipulated (e.g., 19).

Selection has had progressively less time to act over these nested scales, suggesting that comparisons among them could indicate how effective purging is over various scales. Comparisons among related taxa are likely to reflect average differences in the extent of purging that occur over long periods (since divergence) and among several populations. Such comparisons lack strong controls, however, in that taxa could differ in many aspects of their inbreeding and ecological history. In contrast, studies that experimentally cross plants to compare lineages gain maximum control while losing the opportunity to assess how inbreeding affects purging over longer time periods and variable population circumstances. Comparisons among populations are somewhere in between. Comparisons at multiple levels in the same taxa might provide the most comprehensive information on how purging occurs in different contexts and its effects on mating system evolution.

Inferring Inbreeding History

Reliably inferring inbreeding history is critical for evaluating whether more inbred populations experience reduced levels of ID in accord with the purging hypothesis. Several distinct methods are used to infer the degree to which various taxa, populations, or lineages have experienced inbreeding in the past. Like Darwin (36), some use flower morphology or estimates of the degree of self-compatibility as measures of population inbreeding (e.g., 105). Others rely on population size, outcrossing rates, or inbreeding coefficients to infer inbreeding history. Each of these approaches has particular advantages and disadvantages summarized here.

1. Flower morphology—In many species, outcrossing rates increase with increasing flower size or stigma-anther distance, allowing one to infer historical patterns of mating from flower form (e.g., 133). These inferences are likely to be reliable if flower form is genetically based and stable, and if flower form and mating system are highly correlated. Such appears to be the case in recently derived selfing *Amsinckia* (146). Anther-stigma separation is also associated with outcrossing in populations of *Clarkia tembloriensis* (73), *Mimulus guttatus* (16a, 41), *Mimulus ringens* (88), *Turnera ulmifolia* (8), and *Ipomoea purpurea* (S-M Chang, MD Rausher, in review). Herkogamy and dichogamy are both correlated with the outcrossing rate (t) in *Aquilegia caerulea* (11). Flower form can also respond to microhabitat variation and seasonal changes, however (73).
2. Self-compatibility—Like flower form, levels of self-compatibility (SC) are often genetically based and remain stable long enough for populations with differing levels of SC to differ reliably in their inbreeding history. However, the expression of self-incompatibility may also vary with environmental conditions (157). We include only one study that used this approach in *Campanula rapunculoides* (169).

3. Population size—A few studies compare populations that differ in size, implicitly assuming population size to be stable and smaller populations to have experienced greater amounts of inbreeding (e.g., 168). Decreases in population size do increase selfing rates in some outcrossing species (e.g., 3). Such assumptions are unwarranted, however, in colonizing or other species where populations rapidly fluctuate in size. In addition, population substructuring influences mating (181) in that localized mating may cause biparental inbreeding even in large populations (15, 49, 50, 173). Few purging studies quantify local gene flow or the influence of population structure.
4. Outcrossing rate—Many studies use empirical estimates of the outcrossing rate (t) to infer historical levels of inbreeding in populations (e.g., 143, 174). Outcrossing rates are typically estimated using isozyme or microsatellite markers. Like SC and population size, the utility of t as an indicator of inbreeding history hinges on how stable it is over multiple generations. If outcrossing rates vary significantly among years in response to fluctuations in the availability of pollinators or other environmental conditions, t may be a poor predictor of inbreeding history. Such often appears to be the case (e.g., 8). Those who use this approach are advised to use multilocus estimators (e.g., 138) averaged over several years (e.g., 177).
5. Inbreeding coefficient—The inbreeding coefficient F estimates the probability of identity by descent between alleles at a locus and thus reflects inbreeding not only in the current generation but also in previous generations. This cumulative aspect of F makes it superior to t for inferring population history (172). Nevertheless, because a single generation of random mating resets F to 0, this approach could mischaracterize some historically inbred populations.

Ideally, studies should employ combinations of these approaches, and some do (see data online). Jain (78) compared several methods (population size, outcrossing rate, percent polymorphic loci, and percent heterozygosity) to determine their relationship to expression of inbreeding depression in seven populations. He did not find any relationship leading him to suggest the use of several methods to estimate genetic structure. Studies should also report how selfing is achieved in SC taxa (prior, competing, or delayed; 108).

Estimating Inbreeding Depression

Various techniques are used to estimate ID. Comparisons among taxa and populations typically rely on levels of ID as measured by comparing fitness traits between experimentally produced selfed and outcrossed progeny ($\delta = 1 - 6W_s/W_o$). We term these multiple comparisons (abbr.: M). Because ostensibly outcrossed progeny may actually be somewhat inbred, it behooves researchers to ensure that $F = 0$ in their parental group (26). Alternatively, when levels of inbreeding exceed

0 for outcrossed progeny or 0.5 for inbred progeny, one should plot log fitness directly against the inbreeding coefficient (173). The slope of this line (-b) reflects the number of lethal equivalents per gamete present in the population (123). Values as high as 16 have been reported in trees (154).

In some comparisons among populations, levels of ID are inferred indirectly from shifts in inbreeding levels over different life history stages and generations. We term these Ritland comparisons (abbr.: R) after K Ritland, who developed and applied this technique (139). If families differ in inbreeding history, however, they may also differ in levels of ID, adding to the variance and reducing the precision of comparisons among populations. In such cases, it may be preferable to experimentally manipulate the genetic background and levels of inbreeding. Studies that compare ID among lineages within populations often generate progeny at several levels of inbreeding via multiple generations of selfing (e.g., 7, 115). We term these S studies. Such studies clearly reduce the amount of unknown among-family variance, particularly if they start with a randomly outcrossed parental generation.

Measures of ID obtained from different environments do not always agree (see, e.g., 107, 119, 144, 170). Assays under greenhouse conditions may be less stressful than those in natural habitats, perhaps resulting in a decrease in the expression of ID (42, 122, 136, 142). Similarly, reducing opportunities for competition in greenhouse studies (e.g., by growing plants individually in pots) may reduce the expression of ID (179). Higher fertilizer levels increased apparent ID in *Schiedea* (129). In *Impatiens capensis*, high density enhanced ID in one experiment (144) but had little consistent effect in another (170). Field experiments, on the other hand, are prone to both disasters and environmental noise that may obscure even appreciable fitness differences (119, 170). Field experiments could also underestimate ID if field conditions restrict growth. Greenhouse studies that incorporate competition could provide reasonable estimates of ID.

The expression of ID may also differ among traits. Traits expressed early in the life cycle such as seed traits may be strongly influenced by maternal effects, clouding the effects of progeny inbreeding (15, 171). As a population's mating history may influence when ID is expressed (75), it is important for studies to report estimates of ID for multiple traits over the life cycle. Similarly, while most researchers report information on female components of fitness, male components are also sensitive to inbreeding (18, 19, 36, 80, 115) and should be reported.

Methods of Analysis and Meta-Analysis

We first tally the number of studies that find evidence of purging in relation to the type of study (multiple comparisons–M, successive selfing–S, Ritland's method–R) and level of comparison (taxa, population, or line; Table 1). Most studies involve multiple comparisons of populations or species (M-studies–see online Table (www.bio.ilstu.edu/BEES/byers/) for a fuller description). We consider these M studies in more detail as they appear to be the most straightforward and powerful for detecting evidence for purging. Our tallies of which studies find purging or not

TABLE 1 Number of studies that have or have not found evidence for purging of mutational load. Note study by Holtsford and Ellstrand (72) is listed twice since it consists of a greenhouse study and a separate field study.

Study type	Level of comparison	Evidence for purging (citations)		
		Yes	No	Maybe
Multiple Comparisons (M)	Taxa	4 (18, 105, 133, A)	6 (1, 28, 84, 87, 103, 141)	1 (81)
	Population	6 (60, 70, 72, 86, 115, 177)	10 (8, 45, 58, 72, 78, 114, 132, 168, 175, B)	2 (17, 91)
	Lineage	4 (129, 169, C, D)	3 (4, 20, 126)	—
	Total	14	19	3
Ritland Method (R)	Taxa	1 (41)	1 (139)	—
	Population	—	—	—
	Lineage	—	—	1 (96)
	Total	1	1	1
Successive inbreeding (S)	Taxa	1 (7)	—	—
	Population	2 (19, 117)	1 (107)	—
	Lineage	2 (44, 145)	8 (40, 59, 94, 106, 121, 128, 179, E)	—
	Total	5	9	0
TOTALS	Taxa	6	7	1
	Population	8	11	2
	Lineage	6	11	1
	Grand Total	20	29	4

A. C Goodwillie, in review.

B. DM Waller, unpub. data.

C. S-M Chang and MD Rausher, in review.

D. N Takebayashi and LF Delph, in review.

E. JH Willis, in review.

are generally based on the authors' judgement. For further analyses, we consider M type studies that compare lineages within a population separately, but exclude successive selfing to avoid bias as studies employing single seed descent allow only near lethal mutations to be purged (see Theory section). There were only three Ritland-type studies, making it difficult to further assess them as a group.

For studies that present results for a number of populations that differ in t or F, we plotted δ versus these predictors of population inbreeding and applied regression analysis (Figure 1, Table 2). This figure and table allow ready comparison among populations that naturally span a continuous range of inbreeding histories.

We also applied meta-analysis on a different but overlapping subset of studies to assess how evidence for purging varies with respect to plant life history, how inbreeding was inferred, and which traits were measured. This potentially powerful method weights each study for magnitude of effect and variance in estimates and sample size (which the tallies and regressions do not). Unfortunately, only a few studies (11) had sufficient information to be included in these analyses (a mean for each trait value, for sample sizes and standard deviations). Because this method requires assignment of populations to a category (either selfing or outcrossing), we included only studies with clearly distinguishable populations. If populations varied appreciably among years in F or t, the study was not included (e.g. 177). In one case, the populations spanned a wide range of outcrossing rates, so we included only the seven most selfed and outcrossed populations (8). These constraints limited our meta-analyses to 10 studies for early traits (4 among-taxa and 6 among-population comparisons), 9 for late traits (2 among-taxa and 7 among-populations), and 2 for cumulative fitness (both among taxa; see data online). We used the mixed model of Gurevitch & Hedges (56) to compare the magnitude of cumulative effects (of selfing relative to outcrossing) for early traits (seed weight or germination), late traits (number of flowers or fruits, seeds/fruit or plant, or biomass), and cumulative fitness (when provided). For each comparison, we present the cumulative effect size for populations (or taxa) with and without a history of inbreeding. The cumulative effect size is essentially the difference in means (selfed individuals–outcrossed) for a particular trait divided by a pooled standard deviation for the two groups. These values are then pooled as a weighted sum across studies with similar mating histories. The cumulative effect size values are evaluated for homogeneity within groups (Q_W, same mating history) and between groups (Q_B). Since we do not expect a particular value for either the self or outcross groups, we used the mixed model, and therefore the cumulative effect size was corrected by a constant (for further details see 56).

RESULTS

General Patterns

While many studies found evidence for purging, others did not (Table 1). Overall, 20 studies observed a significant decrease in ID (δ) with a greater history of selfing; 29 found no such decrease; and 4 were inconclusive. The type of study

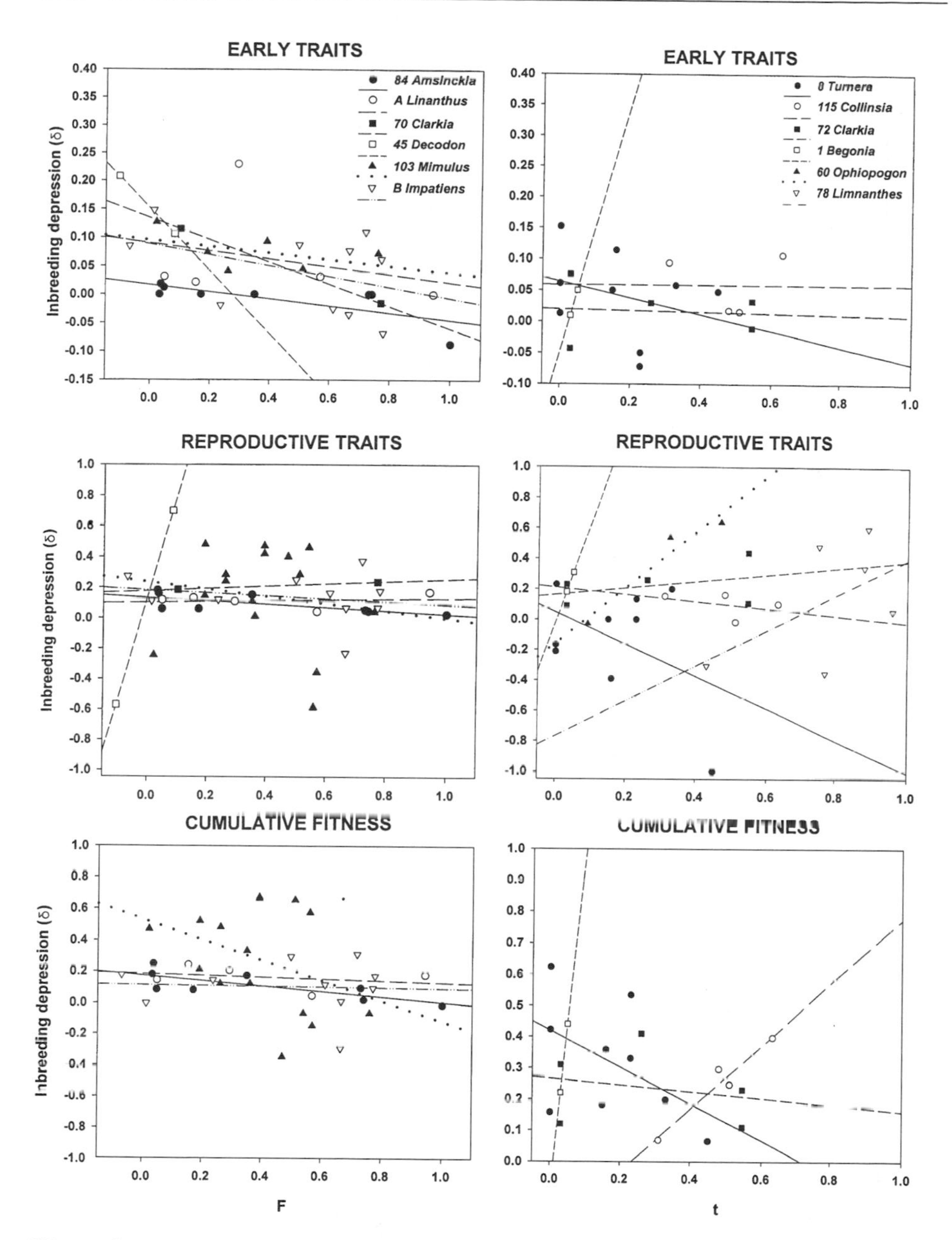

Figure 1 This series of graphs corresponds to the regression analysis in Table 2. Symbols represent populations within each study. Regression lines are shown for the individual studies. A negative slope with increasing values of F (left panels) indicates purging of mutational load, while a positive slope with increasing values of t (right panels) indicates purging. Legend indicates reference # except A = C Goodwillie, in review; B = DM Waller, unpub. data.

TABLE 2 Regression analyses: How the magnitude of inbreeding depression (ID) responds to mating history. Results are shown for each of the three ways to infer mating history. Regressions were performed on the $\log_{10}$ of population size. n = number of populations.

Reference		Early traits			Survivorship			Reproductive traits			Cumulative fitness		
number	**Genes**	**n**	**slope**	r^2	**n**	**slope**	r^2	**n**	**Slope**	r^2	**n**	**slope**	r^2
F *studies:*													
(45)	*Decodon*	2	−0.548	1.000	2	0.446	1.000	2	6.833	1.000			
(70)	*Clarkia*	2	−0.194	1.000	2	−0.187	1.000	2	0.075	1.000			
(84)	*Amsinckia*	8	-0.062^{*}	0.515				8	$-0.111^{0.06}$	0.466	8	-0.174^{*}	0.563
(103)	*Mimulus*	6	-0.057^{ns}	0.216				15	-0.240^{ns}	0.019	15	-0.653^{ns}	0.137
A	*Linanthus*	5	-0.069^{ns}	0.070	5	0.009^{ns}	0.054	5	0.023^{ns}	0.032	5	-0.060^{ns}	0.081
B	*Impatiens*	10	-0.095^{ns}	0.169				10	-0.099^{ns}	0.037	10	-0.025^{ns}	0.002
Overall		33	-0.102^{**}	0.222	9	-0.049^{ns}	0.097	42	-0.030^{ns}	0.001	38	-0.211^{ns}	0.063
t studies:													
(1)	*Begonia*	2	2.000	1.000	2	2.000	1.000	2	6.500	1.000	2	11.000	1.000
(8)	*Turnera*	9	-0.131^{ns}	0.084	9	-0.119^{ns}	0.322	9	-1.052^{ns}	0.186	9	-0.587^{ns}	0.247
(60)	*Ophiopogon*							3	1.840^{ns}	0.939			
(72)	*Clarkia*	5	-0.011^{ns}	0.004	5	-0.296^{ns}	0.223	5	0.219^{ns}	0.162	5	-0.105^{ns}	0.045
(78)	*Limnanthes*							5	1.161^{ns}	0.293			
(115)	*Collinsia*	4	-0.001^{ns}	0.000	2	−0.013	1.000	4	-0.230^{ns}	0.137	4	1.011^{*}	0.940
Overall		20	-0.013^{ns}	0.003	18	-0.074^{ns}	0.029	28	0.194^{ns}	0.029	20	-0.193^{ns}	0.073
Population size studies:													
(132)	*Salvia*	6	0.054^{ns}	0.159				6	0.011^{ns}	0.007			
(168)	*Scabiosa*				6	-0.013^{ns}	0.005	6	-0.073^{ns}	0.217	6	0.069^{ns}	0.092
Overall								12	-0.020^{ns}	0.014			

A. C Goodwillie, in review.

B. DM Waller, unpub. data.

only slightly affects the chance of finding declines in δ. Among the 36 multiple comparison studies (M), 14 produced evidence for purging, 19 did not, and 3 were inconclusive. Of the three studies that inferred ID from changes in F (R), one found a decline in genetic load, one did not, and one was inconclusive. The least frequent evidence for purging (5/14) occurred in studies involving successive generations of selfing within populations (as expected if small N_e limits purging in selfed lines). Those S studies with five generations of selfing (e.g., 7) usually found declines in δ, whereas similarly designed studies with 2–3 generations of selfing rarely did (e.g., 107, 121, 179).

The scale at which purging is examined only slightly affects the outcome. Of the 14 studies comparing species, 6 found support for purging, 7 did not, and 1 was inconclusive. Of the 21 among-population comparisons, 8 found reduced load in more inbred populations, 11 did not, and 2 were inconclusive. The least frequent evidence for purging (6/18) occurred in comparisons among lineages within a population, perhaps again reflecting diminished opportunities for purging nonlethal mutations in selfing lines.

Does Method of Inferring Mating History Matter?

Does the chance of finding evidence for purging vary with respect to what method is used to infer inbreeding history? The tallies here only include the M and R studies. One study used only population estimates of F, 6 studies used both F and outcrossing rate (t), 4 studies used just t, and 3 used Ritland's (139) method. Of these 14 studies, 9 found purging, 3 did not, and 2 were inconclusive. Studies that used both t and floral morphology or population size to determine mating history also did not find consistent purging (3 vs. 7). Those studies that used variation in floral morphology or breeding system to infer history also showed no particular pattern (5 vs. 4 and 1 inconclusive). None of the 5 studies that used population size to infer inbreeding history found evidence for purging. This may reflect inaccuracies in estimating N, variation in N over time, the inefficiency of purging, fixation of the load in small populations, or that N is simply a poor predictor of inbreeding.

Comparisons Among Taxa

Only 4 of 11 M studies comparing more and less inbred taxa found evidence for purging (18, 105, 133; C Goodwillie, in review). In *Phlox* and *Epilobium*, the cross × species interaction (or correlation of δ with mating history) was significant for early traits, indicating purging (105, 133). Similar studies of *Amphicarpaea* and *Linanthus* found evidence for the purging of mutations affecting later traits (reproduction or biomass) (134; C Goodwillie, in review). Three of the four studies that noted declines in δ for cumulative fitness in more inbred species found a significant cross × species effect (84; C Goodwillie, in review; 134). In one of these (84), however, the correlation of δ (for cumulative fitness) vs. $(1-t)$ was not significant, leading them to conclude that purging was not significant. Our regression analysis of these data (δ vs. F) found a significant relationship, however (Table 2). ID for biomass declined in more selfing taxa of *Mimulus*, although this was not statistically analyzed (18). In a similar study, δ for five fitness traits tended to decline in the more inbred of 15 populations of four species in the *Mimulus guttatus* complex, but this relationship was inconstant and only significant for height (103). The other studies did not directly analyze the relationship between ID and mating history and did not show any evidence of purging.

Comparisons Among Populations

In the first study of this kind, Jain (78) compared seven populations of *Limnanthes alba* that differed widely in outcrossing rate (0.43–0.97), heterozygosity (0.12–0.27), and percentage of polymorphic loci (0.29–0.57). Although ID was significant in 4 of the 7 populations for several traits, he observed no association between ID and these indicators of inbreeding history among populations. In our overall comparison, 6 of 18 M studies comparing populations found evidence for purging (Table 1). Of the 17 that examined early traits, two studies found a significant decrease in δ in populations with a greater history of selfing (70, 86). Several reported nonsignificant cross $\times$ population interactions or correlations of δ with mating history for early traits (8, 45, 57, 60, 114, 132, 175). The other studies provided no support for purging (but no inbreeding depression for early traits was observed in 5 studies).

Of 12 M studies comparing ID for late (reproductive or biomass) traits among populations, seven tested directly for purging and none were significant. Two studies found significant variation for δ among populations, but no clear association with mating history (8) or population size (168). In four populations of *Collinsia heterophylla* with significant ID, regression analysis showed ID to decrease in more selfing populations (115). ID for reproductive traits (seed weight, seed set, and fruit set) declined in more selfing populations, although this was not tested statistically (*Clarkia*; 72).

Regression Results

Fourteen of the 27 M studies of taxa and populations provided sufficient information to analyze linear relationships between δ and F, t, or estimated population size (Figure 1, Table 2). If purging is effective, we expect δ to decline in more inbred populations (higher F) and to increase in populations with higher outcrossing rates (t) or larger populations. Overall, these trends are modest, with only one F and one t study showing individual significance (Table 2). In *Amsinckia*, δ's for both early and cumulative fitness traits decline in more inbred species and populations (84). While this was the only F study to be individually significant, enough studies showed declines in δ for early traits for the overall trend to be significant ($p < 0.01$; Table 2). This result suggests that purging may be more effective against mutations affecting early traits (75). In contrast, decreases in δ observed across studies were not large or consistent enough to be significant for survivorship, reproductive traits, or cumulative fitness (Figure 1; Table 2).

Evidence for purging is less evident in studies exploiting variation in outcrossing rates. Only about half the individual regressions find δ to increase with t as expected under the purging hypothesis, and most overall regressions were in the wrong direction (Table 2). The only significant individual regression (for cumulative fitness in *Collinsia heterophylla*) was positive (115) supporting the purging hypothesis. Only two studies used population size to infer mating history, and they did not show any pattern.

Meta-Analysis Results

For the 11 studies that provided sufficient data (M studies of taxa and populations), we applied meta-analyses aimed at testing whether ID regularly declines in more selfing/inbred groups, as reflected by a significant value of Q_B^* (Table 3). Overall tests revealed no consistent difference between the more and less inbred groups ("Over all studies" lines; Table 3). In fact, there appeared to be more ID in the more inbred groups for early and late traits, contrary to the prediction of purging. The variation within groups, however, is too great to demonstrate significance. Likewise, with only 7 populations from 2 studies, we cannot draw conclusions regarding cumulative fitness.

Meta-analyses also reveal little effect of life history (Table 3) despite the fact that tallies suggest that herbaceous perennials are less likely to purge mutational load than annuals (2 of 10 vs. 7 of 14 studies, respectively). One of 4 studies involving woody perennials found purging.

Of the three methods used to infer inbreeding history (t, F, or Other), only studies comparing populations that differ in t showed a significant difference between selfing and outcrossing populations (Table 3). This difference, however, again showed inbreeding populations to have greater ID than outcrossing populations, contrary to predictions of the purging hypothesis. Tallies of the M (taxa and population) studies revealed no pattern except that studies using population size to infer mating history found no evidence of purging.

Lineages Within Populations

Most within-population studies compared lineages that differ in mating system characters known or expected to affect the selfing rate. Four of these seven studies found evidence for purging. Only one of three studies that compared δ between females and hermaphrodites within gynodioecious species found a reduction in the load in hermaphrodites as expected if purging occurs swiftly in response to selfing in these lines. Del Castillo (39) found higher ID for fruit set and seed production in females in *Phacelia dubia*. Mutikainen & Delph (126), however, examined maternal family by cross effects for four fitness traits and found δ was approximately twice as high (0.3) in hermaphrodites as in females, contrary to the purging hypothesis. Ashman (4) found no relation. Two of three studies that used

TABLE 3 Results of meta-analyses, n = number of populations and 'cumulative effect size' refers to estimated level of inbreeding depression (δ)/pooled standard deviation.

			For Selfing Populations		For Outcrossing Populations		Significant evidence of mating history on purging?		
		n	Cumulative effect size	95% conf. int	Cumulative effect size	95% conf. int	Q_w	Q^*_B	
Early Fitness Traits: OVER ALL STUDIES		45	−0.486	−0.899 to −0.072	−0.350	−0.753 to 0.053	436.70	0.213	n.s.
Broken down by groups according to:	Trait[a]								
Life History	Ann	24	−0.696	−1.232 to −0.160	−0.538	−1.123 to 0.046	404.28	0.151	n.s.
	Per	21	−0.153	−1.295 to 0.989	−0.161	−1.145 to 0.823	22.94	0.0001	n.s.
Method of inferring population history	*t*	13	−1.457	−2.212 to −0.702	−0.469	−1.162 to 0.224	308.66	3.57	P = 0.059
	F	18	−0.189	−1.205 to 0.827	−0.331	−1.471 to 0.810	10.52	1.52	n.s.
	Other	14	−0.067	−0.339 to 0.205	−0.261	−0.443 to −0.078	19.44	0.53[b]	n.s.
Late Fitness Traits: OVER ALL STUDIES		45	−0.808	−1.188 to −0.428	−0.540	−0.911 to −0.169	345.28	0.974	n.s.
Broken down by groups according to:	Trait								
Life History	Ann	24	−0.700	−1.202 to −0.198	−0.312	−0.855 to 0.230	183.15	1.057	n.s.
	Per	21	−0.987	−1.609 to −0.365	−0.764	−1.301 to −0.228	141.19	0.283	n.s.
Method of inferring population history	*t*	19	−1.607	−2.276 to −0.938	−0.869	−1.425 to −0.313	188.42	2.76	n.s.
	F	18	−0.455	−1.115 to 0.205	−0.196	−0.938 to 0.546	91.05	0.351	n.s.
	Other	8	−0.664	−1.242 to −0.087	−0.596	−1.176 to −0.016	9.61	0.027	n.s.
Cumulative Fitness: OVER ALL STUDIES		7	−0.550	−1.000 to −0.101	−0.615	−1.021 to −0.208	3.40	0.04[b]	n.s.

[a]Traits Ann = Annual; Per = Perennial.

[b]Not corrected for mixed model analysis due to small uncorrected Q_B value.

[c]Studies included: Early traits-Overall (8, 18, 87, C Goodwillie, in review, DM Waller unpub. data, 103, 115, 132, 133, 175), Life history–annuals (18, 103, 115, 133, C Goodwillie in review, DM Waller unpub. data) and perennials (8, 87, 132, 133, 175), Method of inferring population history-*t* (8, 18, 115), *F* (103, C Goodwillie in review, DM Waller unpub. data) and other (87, 132, 133, 175). Reproductive traits-Overall (8, 18, 103, 115, 132, C Goodwillie in review, DM Waller unpub. data, 133, 168), Life history-annuals (18, 103, C Goodwillie in review, DM Waller unpub. data, 115) and perennials (8, 132, 133, 168), Method on inferring population history-*t* (8, 18, 115, 168), *F* (103, C Goodwillie in review, DM Waller unpub. data), and other (132, 133). Cumulative fitness studies-Overall (133, C Goodwillie in review).

anther-stigma distance to infer inbreeding history found evidence for purging (e.g., *Ipomoea purpurea*. where δ values for late traits and cumulative fitness decrease in lineages with less anther-stigma separation; S-M Chang, MD Rausher, in review). Interestingly, male fitness responds more than female fitness to inbreeding in this species. Unfortunately, male traits are only rarely measured in other studies.

One study compared the variable expression of self-incompatibility with variation in ID. In *Campanula rapunculoides*, different *S* genotypes vary in their expression of self-incompatibility, causing variable rates of selfing that parallel slight decreases in δ for early traits (169).

Effects of lineage are also evident in the many taxa and population studies where maternal family by cross interactions are significant (1, 70, 132, 133, 141). Such results reflect the potential for greater selfing rates to coevolve with purging within populations. Unfortunately, such interactions are not often quantified or investigated in relation to known differences in the mating history of maternal families.

Experimental Conditions

We found no relationship between the conditions under which experiments were conducted and the likelihood of finding evidence for purging. Fewer than half (7 of 19) of the greenhouse studies found decreased ID in more inbred populations or lineages. Of the few studies that used common garden (3) or field (2) assays, two showed reduced load, two did not, and the fifth was inconclusive. Of the five studies that used greenhouse and field or garden assays, 1 found decreased ID, 3 did not, and one was inconclusive. Johnston's (81) study, for example, found limited ID in two species of *Lobelia*, but not any systematic difference in the expression of ID in the field vs. the greenhouse. Holtsford & Ellstrand (72) found greater expression of ID in the greenhouse, but Eckert & Barrett (45) found more ID in the field.

Several greenhouse studies varied growing conditions in order to examine how ID varied with respect to conditions. Three of 8 studies found ID to depend on conditions, but none of these found conditions to affect the detection of purging (8, 17; C Goodwillie, in review). Thus, while the greenhouse conditions may better control environmental variance, they do not appear to increase the probability of finding purging.

DISCUSSION

How Major and Consistent a Force Is Purging?

While some studies indicate that while purging of the mutational load occurs in individual cases, purging does not appear to act consistently as a major force in natural populations (Table 1, Figure 1). We chose and categorized studies according to their experimental design, favoring those that attempted to directly test how a history of inbreeding affects the expression of inbreeding depression. The strongest studies compared many populations or lineages that varied in estimated rates of selfing or inbreeding (e.g. 8, 20, 73, 84, 91, 103; C Goodwillie, in review; DM Waller unpublished data). However, only one of these (C Goodwillie, in review) found significant evidence for purging. In other studies, ID sometimes appears to decline in more selfing populations, but the relationship is not significant (e.g. 84). In an extensive study involving several *Mimulus* spp., δ appeared to decline with greater selfing (higher F) across species and populations, but the relationship was not significant for reproductive traits or cumulative fitness (103). These inconsistent results parallel Darwin's (36) conclusion cited in the Introduction and appear to undermine some recent assertions that purging can be a powerful force in inbred populations.

Parallel studies in animal populations also reveal variable results (e.g. 98, 99, 150, 167). Ballou (6) recently assessed the extent of purging in 25 captive mammalian populations by comparing levels of ID in the inbred progeny of parents with and without ancestral inbreeding. Inbreeding depression for neonatal survival declined in 15 of 17 species, but the decline was only statistically significant for the Sumatran tiger. While the combined data strongly support the existence of purging, the median effect was small (2% at $F = 0.25$) and similar to the slope observed in the F studies for early traits (=0.102) (Table 2). These data also suggest that purging affects early survival more than later traits.

What Limits Purging?

Why should reductions in the genetic load in more inbred groups be difficult to detect? Authors who did not find evidence for purging often volunteer explanations. For example, some suggest that experimental designs lack sensitivity for detecting changes in ID. Others suggest that their populations may be too similar in mating history or population size to provide good evidence for purging (e.g. 115, 132, 168). Others argue that long-lived species may be unlikely to shed their genetic load (e.g. 45). Finally, many argue that purging may be limited by its genetic basis (mildly deleterious or partially penetrant alleles, epistasis, etc—106, 126; JH Willis, in review; 128). Let us consider these in turn.

Experimental design—Does experimental design affect the likelihood of finding purging? Studies clearly differed greatly in design and setting in ways that could affect the detection of purging. Holtsford & Ellstrand (72), for example, found evidence that more selfed populations in the lathhouse purged some of their mutational load but did not in similar experiments in the field. We found no consistent patterns over all the studies, however.

The accuracy with which we can infer mating history may also influence our ability to test its effect on purging. Because outcrossing rates can vary widely among years in response to fluctuating ecological conditions (e.g. 45), we expected t to be a poor predictor of purging. Likewise, N may fluctuate widely and unpredictably over time, particularly in short-lived plants. The inbreeding coefficient

(F) appears better suited to comparative studies in that it often reflects the integrated effects of inbreeding history due to both small N and the mating system. This may explain why only the F studies provided overall evidence for purging (of early traits—Table 2). Such effects were not evident in the meta-analyses, however, suggesting few clear differences in the likelihood of finding purging among methods. Studies involving limited contrasts among mating histories are constrained in their ability to detect purging (e.g. 115), but even studies comparing populations across a broad range of mating histories did not consistently find evidence for purging. Of the studies that used t to infer inbreeding history, only one showed significant increases in δ in more outcrossed populations (Figure 1). While variable selfing rates likely affect individual studies, the lack of any overall pattern across studies suggests that purging is an inconstant force.

Life history traits influence the extent of purging—Perennials were less likely to decrease their genetic load than annuals (17% vs. 50%). If the purging of mildly deleterious mutations occurs slowly in mixed mating populations (22, 102), such results are not surprising.

Effectiveness of purging on early vs. later traits—Husband & Schemske (76) compared levels of ID in selfing and outcrossing species and found lower ID for early traits in the selfers. We therefore expected to find more purging for early traits. While the regression analyses reveal some purging of ID for early traits in the F studies (Table 2), no such pattern emerged from the meta-analysis. In some cases, purging appeared to occur for later traits (e.g. 8, 20, 72). Whereas species characteristics and experimental design obviously influence our ability to detect ID and thus purging, collectively these results suggest that purging may be limited in its extent and consistency.

The genetic basis of inbreeding depression—The extent to which purging occurs in more inbred populations strongly depends on genetic details including the distribution of selective effects, the degree of dominance, and interactions among loci (see Theoretical Background). Mutations that are even slightly penetrant in the heterozygote are far more likely to be eliminated as heterozygotes, greatly reducing the degree to which inbreeding can reduce the load (26). Similarly, although selection can act swiftly to eliminate recessive lethal mutations exposed upon inbreeding, selection acts far more slowly against mildly deleterious recessive mutations. Such selection also becomes ineffective in small populations for mutations below a threshold effect, limiting the extent of purging. Limitations may also exist in the form of selective interference among loci segregating for deleterious mutations.

Thus, variability in the extent of purging found among studies may reflect variation in dominance, mutational effects, and the history of selection. Recent empirical studies suggest that slightly deleterious mutations with intermediate degrees of dominance may contribute the most to ID (43, 45, 83; JH Willis, in review; 103). Many authors now explain failures to find purging as the likely consequence of slightly deleterious mutations contributing most of the genetic load (94; JH Willis; in review, 128). Mutation accumulation studies to document rates of mutation and the distribution of mutational effects are needed to bolster this conclusion.

Implications for Mating System Evolution

Initial dynamic models of mating system evolution suggested that periodic bottlenecks or pollinator failures could enforce selfing in a way that could purge mutational load due to recessive alleles of major effect (102). We expect such purging to cause runaway selection for ever-higher rates of selfing. Exclusively selfed populations and species are quite rare, however (36, 77, 172), suggesting that purging may not regularly act as an efficient force or that it may be countered in real populations by mutation accumulation in selfed or bottlenecked lines. This review reveals only limited evidence for purging and, in a few cases, evidence that load may even increase in more inbred populations or lines. If small or more inbred populations do not efficiently and regularly purge their mutational load, the potential for runaway selection for increased selfing is correspondingly limited. The low fraction of purging observed in woody perennials (1 of 4 studies) lends some support to the conjecture that there may not be sufficient time for purging to occur under some circumstances (102). However, the many populations that retain intermediate selfing rates (e.g. 71, 158) suggest evolutionary stability.

Ecological factors can also favor selfing for reasons unrelated to reduced ID (e.g., lower resource costs of selfing, pollinator limitation, etc–67, 163). Pollen discounting (reduced availability of pollen for outcrossing due to selfing) can favor mixed mating as well as pollen limitation (64, 163). Limited pollen and seed dispersal will also tend to structure local populations (49, 182) in a way that increases biparental inbreeding (66). Biparental inbreeding mimics selfing in its effects on both parent-offspring relatedness and purging and could help maintain mixed mating systems (162, 183).

These models indicate that the evolution of mixed mating systems depends on more than levels of ID. Johnston (82) and Schoen examined the joint effects of pollen discounting and ID on the mating system. Mixed mating is favored when an increase in selfing causes declines in the proportion of outcrossed ovules fertilized or a decrease in male fitness via pollen export. They also found evolutionarily stable intermediate selfing rates under conditions of low mutation rates and moderate dominance coefficients. Pollen discounting is apparently sufficient in *Ipomoea purpurea* to prevent more selfing lineages from becoming fixed in populations (S-M Chang, MD Rausher, in review). Fixation of deleterious mutations over successive generations of selfing will also counter ever-higher selfing rates (112).

Most studies that checked found evidence for significant family (or lineage) by cross-type interactions (e.g. 40, 129), indicating variation in how lines respond to inbreeding. Such variation could be maternal effects. Alternatively, any of several direct genetic effects could occur. First, the mutation of major effect will increase variation among lines; this could be increased further by drift. Second, variation in the amount of inbreeding among lineages implies that the recovery of fitness upon wide outcrossing will be greater for more inbred lines. This variation also contributes to the differences in heterozygosity among lines referred to as identity disequilibrium (10, 95). Third, variation in inbreeding history may cause purging in more inbred lines. Thus, the increased variance for δ among lines observed by Dudash et al (44) after five generations of selfing in *Mimulus guttatus* could reflect both differential purging and mutation fixation among lines. A recent model by Schultz & Willis (149) predicts that random genetic variation of mutations will contribute more to variation in the load among lines than mating history. This latter source of variation was predicted to be too small to reliably detect.

Note that these mechanisms make different predictions regarding possible associations between breeding history and ID. If random mutation and drift cause lines to differ, we expect little consistent relationship between ID and breeding history among lineages. If differential recovery of fitness is responsible for the differences among lines, we expect estimated levels of ID to be greater for more inbred lines (assuming parents are not randomly mated before testing ID). And if purging accounts for differences among lines, we expect ID to decrease in more inbred lines. Associations between load and mating system loci also influence mating system evolution (16, 163). ID declined significantly in more inbred lines in 6 of 18 lineage comparisons (Table 1), including two studies that inferred mating history using anther-stigma separation (S-M Chang, MD Rausher, in review; N Takebayashi, LF Delph, in review). This seems surprising in light of the Schultz and Willis prediction.

Implication for Small Population Conservation

Some have advocated deliberate inbreeding to purge the mutational load in captive populations (e.g. 161). Such programs appear risky, however, in that some mutations will be fixed as purging is occurring, particularly in small populations (110). The tendency for inbreeding to fix deleterious mutations has been noted in some empirical work (e.g. 44). Computer simulations of ID in large vs. small populations further indicate that there is increased variance in the expression of ID in small populations (57). Not detecting inbreeding depression in a small or inbred population does not necessarily demonstrate that purging has occurred. Fixation of the load can also cause ID to decline (by depressing outcrossed fitness rather than by raising selfed fitness). Similarly, selection occurring in multigeneration experiments can raise the fitness of both inbred and outbred progeny to similar extents, causing δ to decline even though no purging of mutations has occurred (as in corn—see 9). Thus, to accurately determine the extent to which small or inbred populations have shed or fixed mutations, one must compare absolute fitnesses (and, ideally, evaluate the progeny from crosses between populations). The studies reviewed here suggest that intentional inbreeding to promote purging could result primarily in temporary reductions in lethal and semilethal mutations at the cost of fixing many mildly deleterious mutations that could eventually erode population fitness.

Recommendations for Future Research

Future studies designed to assess purging would be most useful if they routinely incorporated a number of design and reporting features:

1. Studies of ID should examine multiple fitness traits that extend over the full life-cycle of the plant. This will ensure that one can judge the relative degree to which ID for both early and late traits may be purged.
2. Estimates of ID based on comparisons of progeny fitness should statistically control for maternal effects (e.g., differences in seed size) that can appreciably affect progeny performance and cloud genetic effects of inbreeding between populations.
3. Successive inbreeding studies should pay particular attention to male fitness components such as pollen number and viability. Most studies only quantify ID for female fitness, although male fitness can be significantly influenced by inbreeding (17, 80; JH Willis, in review). This lack of information is particularly critical if ID effects differ between the sexes (S-M Chang, MD Rausher, in review).
4. Comparative studies should incorporate appropriate controls and report more than simply the levels of inbreeding depression observed. In particular, we urge authors to report absolute selfed and outcrossed fitnesses and inbreeding load (the number of lethal equivalents).
5. Studies should routinely and clearly report sample sizes, variances (or S.D.), and means. Such data are needed to compile meta-analyses of the data.
6. Inbreeding studies should routinely test for family or line by cross-type interactions and whenever possible test for relations between levels of ID and any known differences in the breeding history of those lines.

CONCLUSIONS

Like Darwin, late twentieth century researchers usually take steps to make controlled and unbiased comparisons, measure several successive components of fitness, combine these to calculate overall estimates of inbreeding decline, and apply state-of-the-art statistical analyses. Some also follow Darwin in considering the effects of growing conditions on the expression of ID and documenting the cumulative effects of multiple generations of selfing. In doing such work, we should also attempt to emulate Darwin's work in terms of its careful execution, concern for male fitness traits, and attention to potentially confounding variables.

The efficiency with which selection can purge populations of their genetic load depends on the size of mutational effects, the degree of dominance, interactions among loci, the breeding system, and population size. The many theoretical impediments to efficient purging suggest that we should not expect it to occur universally in natural or captive populations. Thus, we should not be surprised that comparisons like Darwin's (36), Ballou's (6), and these 52 plant studies reveal only limited support for purging. Whereas these studies vary considerably in species examined, how inbreeding history was inferred (F, outcrossing rate, etc), how ID was evaluated, and the particular traits measured, these factors were of only limited utility in predicting when purging would occur. While other factors may yet better predict patterns of purging, these results suggest that purging may not occur even in consistently inbred lineages.

The variation we observe among studies probably reflects both variation in experimental design and conditions and real variation among populations and taxa in the extent to which purging occurs. Thus, although populations are capable of purging under some circumstances, purging appears neither consistent nor effective enough to reliably reduce ID in small and inbred populations.

ACKNOWLEDGMENTS

We are indebted to C Bristow for assistance in graphics, reading many versions of this paper and chasing down references in the library. We are grateful to the following authors that provided us with copies of unpublished manuscripts and papers: S Chang, L Delph, C Goodwillie, A Stephenson, and J Willis. We particularly wish to thank J Crow, J Dole, M Dudash, C Fenster, and D Charlesworth for their insightful comments on an earlier version of this paper. DB thanks D Andow for his encouragement to undertake this project.

LITERATURE CITED

1. Ågren J, Schemske DW. 1993. Outcrossing rate and inbreeding depression in two annual monoecious herbs, *Begonia hirsuta* and *B. semiovata*. *Evolution* 47:125–35
2. Aide TM. 1986. The influence of wind and animal pollination on variation in outcrossing rates. *Evolution* 40:434–35
3. Aldrich PR, Hamrick JL. 1998. Reproductive dominance of pasture trees in a fragmented tropical forest mosaic. *Science* 281: 103–5
4. Ashman TL. 1992. The relative importance of inbreeding and maternal sex in determining progeny fitness in *Sidalcea oregana* ssp. *spicata*, a gynodioecious plant. *Evolution* 46:1862–74
5. Baker HG. 1955. Self-compatibility and establishment after 'long-distance' dispersal. *Evolution* 9:347–48
6. Ballou JD. 1997. Ancestral inbreeding only minimally affects inbreeding depression in mammalian populations. *J. Hered.* 88: 169–78
7. Barrett SCH, Charlesworth D. 1991. Effects of a change in the level of inbreeding on the genetic load. *Nature* 352:522–24
8. Belaoussoff S, Shore JS. 1995. Floral correlates and fitness consequences of mating-system variation in *Turnera ulmifolia*. *Evolution* 49:545–56
9. Benson DL, Hallauer AR. 1994. Inbreeding depression rates in maize populations before and after recurrent selection. *J. Hered.* 85:122–28

10. Brown AHD, Feldman MW. 1981. Population structure of multilocus associations. *Proc. Natl. Acad. Sci. USA* 78: 5913–16
11. Brunet J, Eckert C. 1998. Effects of floral morphology and display on outcrossing in blue columbine, *Aquilegia caerulea* (Ranunculaceae). *Func. Ecol.* 12:596–606
12. Bryant EH, Meffert LM. 1990. Multivariate phenotypic differentiation among bottleneck lines of housefly. *Evolution* 44: 660–68
13. Bulmer MG. 1980. *The Mathematical Theory of Quantitative Genetics*. New York: Oxford Univ. Press. 254 pp.
14. Burger R, Lynch M. 1995. Evolution and extinction in a changing environment: a quantitative-genetic analysis. *Evolution* 49:151–63
15. Byers DL. 1998. Effect of cross proximity on progeny fitness in a rare and a common species of *Eupatorium* (Asteraceae). *Am. J. Bot.* 85:644–53
16. Campbell RB. 1986. The interdependence of mating structure and inbreeding depression. *Theor. Pop. Biol.* 30:232–44
16a. Carr DE, Fenster CB. 1994. Levels of genetic variation and covariation for *Mimulus* (Scrophulariaceae) floral traits. *Heredity* 72:606–18
17. Carr DE, Dudash MR. 1995. Inbreeding depression under a competitive regime in *Mimulus guttatus*: consequences for potential male and female function. *Heredity* 75:437–45
18. Carr DE, Dudash MR. 1996. Inbreeding depression in two species of *Mimulus* (Scrophulariaceae) with contrasting mating systems. *Am. J. Bot.* 83:586–93
19. Carr DE, Dudash MR. 1997. The effects of five generations of enforced selfing on potential male and female function in *Mimulus guttatus*. *Evolution* 51:1797–807
20. Carr DE, Fenster CB, Dudash MR. 1997. The relationship between mating-system characters and inbreeding depression in *Mimulus guttatus*. *Evolution* 51:363–72
21. Charlesworth B. 1980. The cost of sex in relation to mating system. *J. Theor. Biol.* 84:655–72
22. Charlesworth B, Charlesworth D. 1998. Some evolutionary consequences of deleterious mutations. *Genetica* 102/103:3–19
23. Charlesworth B, Morgan MT, Charlesworth D. 1991. Multilocus models of inbreeding depression with synergistic selection and partial self-fertilization. *Genet. Res. (Camb.)* 57:177–94
24. Charlesworth B, Nordborg M, Charlesworth D. 1997. The effects of local selection, balanced polymorphism and background selection on equilibrium patterns of genetic diversity in subdivided populations. *Genet. Res.* 70:155–74
25. Charlesworth D. 1991. The apparent selection on neutral gene marker loci in partially inbreeding populations. *Genet. Res.* 57:159–75
26. Charlesworth D, Charlesworth B. 1987. Inbreeding depression and its evolutionary consequences. *Annu. Rev. Ecol. Syst.* 18:237–68
27. Charlesworth D, Charlesworth B. 1990. Inbreeding depression with heterozygote advantage and its effect on selection for modifiers changing the outcrossing rate. *Evolution* 44:870–88
28. Charlesworth D, Lyons EE, Litchfield LB. 1994. Inbreeding depression in two highly inbreeding populations of *Leavenworthia*. *Proc. R. Soc. Lond. Ser. B* 258: 209–14
29. Charlesworth D, Morgan MT, Charlesworth B. 1990. Inbreeding depression, genetic load, and the evolution of outcrossing rates in a multilocus system with no linkage. *Evolution* 44:1469–89
30. Charlesworth D, Morgan MT, Charlesworth B. 1992. The effect of linkage and population size on inbreeding depression

due to mutational load. *Genet. Res.* 59: 49–61

31. Crow JF. 1948. Alternative hypotheses of hybrid vigor. *Genetics* 33:477–87
32. Crow JF. 1987. Muller, Dobzhansky and overdominance. *J. Hist. Biol.* 20:351–80
33. Crow JF. 1993. Mutation, mean fitness, and genetic load. *Oxford Ser. in Evol. Biol.* 9:3–42
34. Crow JF. 1999. The rise and fall of overdominance. *Plant Breed. Rev.* 17:225–57
35. Crow JF, Kimura M. 1970. *An Introduction to Population Genetics Theory*. Minneapolis, MN: Burgess. 591 pp.
36. Darwin C. 1876. *The Effects of Cross and Self Fertilization in the Vegetable Kingdom*. London: J. Murray
37. Darwin C. 1877. *Different Forms of Flowers of Plants of the Same Species*. London: J. Murray. 352 pp.
38. Davenport CB. 1908. Degeneration, albinism and inbreeding. *Science* 28:454–55
39. del Castillo RF. 1993. Consequences of male sterility in *Phacelia dubia*. *Evol. Trends Plants* 7:15–22
40. del Castillo RF. 1998. Fitness consequences of maternal and non-maternal components of inbreeding in the gynodioecious *Phacelia dubia*. *Evolution* 52:44–60
41. Dole J, Ritland K. 1993. Inbreeding depression in two *Mimulus* taxa measured by multigenerational changes in the inbreeding coefficient. *Evolution* 47:361–73
42. Dudash MR. 1990. Relative fitness of selfed and outcrossed progeny in a self-compatible, protandrous species, *Sabatia angularis* L. (Gentianaceae): a comparison in three environments. *Evolution* 44:1129–39
43. Dudash MR, Carr DE. 1998. Genetics underlying inbreeding depression in *Mimulus* with contrasting mating systems. *Nature* 393:682–84
44. Dudash MR, Carr DE, Fenster CB. 1997. Five generations of enforced selfing and outcrossing in *Mimulus guttatus*: inbreeding depression variation at the population and family level. *Evolution* 51:54–65
45. Eckert CG, Barrett SCH. 1994. Inbreeding depression in partially self-fertilizing *Decodon verticillatus* (Lythraceae): population-genetic and experimental analyses. *Evolution* 48:952–64
46. Eyre-Walker A, Keightley PD. 1999. High genomic deleterious mutation rates in hominids. *Nature* 397:344–47
47. Falconer DS, Mackay TFC. 1996. *Introduction to Quantitative Genetics*,. Essex, England: Longman Group. 464 pp. 4th ed
48. Falk DA, Holsinger KE, eds. 1991. *Genetics and Conservation of Rare Plants*, New York: Oxford Univ. Press. 283 pp.
49. Fenster CB. 1991. Gene flow in *Chamaecrista fasciculata* (Leguminosae) I. Gene dispersal. *Evolution* 45:398–409
50. Fenster CB. 1991. Gene flow in *Chamaecrista fasciculata* (Leguminosae) II. Gene establishment. *Evolution* 45:410–22
51. Fisher RA. 1941. Average excess and average effect of a gene substitution. *Ann. Eugen.* 11:53–63
52. Franklin IR. 1980. Evolutionary change in small populations. In *Conservation Biology: An Evolutionary-Ecological Perspective*, ed. ME Soule, BA Wilcox, pp. 135-49. Sunderland, MA: Sinauer

52a. Fry JD, Keightley PD, Heinsohn SL, Nuzhdin SV. 1999. New estimates of the rates and effects of mildly deleterious mutation in *Drosophila melanogaster*. *Proc. Natl. Acad. Sci. USA* 96:574–79

53. Fu YB, Ritland K. 1994. Marker-based inferences about fecundity genes contributing to inbreeding depression in *Mimulus guttatus*. *Genome* 37:1005–10
54. Gabriel W, Lynch M, Bürger R. 1993. Muller's ratchet and mutational meltdowns. *Evolution* 47:1744–57
55. Gilpin M, Hanski I, eds. 1991. *Metapopulation Dynamics: Empirical and Theoretical Investigations*, New York: Academic Press. 336 pp.
56. Gurevitch J, Hedges LV. 1993. Meta-

analysis: combining the results of independent experiments. In *Design and Analysis of Ecological Experiments*, eds. SM Scheiner, J Gurevitch, pp. 378–98. New York: Chapman & Hall

57. Hauser TP, Damgaard C, Loeschcke V. 1994. Effects of inbreeding in small plant populations: expectations and implications for conservation. In *Conservation Genetics*, eds. V Loeschcke, J Tomiuk, SK Jain, pp. 115–29. Basel, Switzerland: Birkhäuser Verlag
58. Hauser TP, Loeschcke V. 1994. Inbreeding depression and mating-distance dependent offspring fitness in large and small populations of *Lychnis flos-cuculi* (Caryophyllaceae). *J. Evol. Biol.* 7:609–22
59. Hauser TP, Loeschcke V. 1995. Inbreeding depression in *Lychnis flos-cuculi* (Caryophyllaceae): effects of different levels of inbreeding. *J. Evol. Biol.* 8:589–600
60. He TH, Rao GY, You RL, Ge S. 1998. Mating system of *Ophiopogon xylorrhizus* (Liliaceae), an endangered species in southwest China. *Int. J. Plant Sci.* 159: 440–45
61. Hedrick PW. 1994. Purging inbreeding depression and the probability of extinction: full-sib mating. *Heredity* 73:363–72
62. Heller R, Smith JM. 1978. Does Muller's ratchet work with selfing? *Genet. Res. (Camb.)* 32:289–93
63. Hill WG, Robertson A. 1968. The effects of inbreeding at loci with heterozygote advantage. *Genetics* 60:615–28
64. Holsinger K. 1991. Mass-action models of plant mating systems: the evolutionary stability of mixed mating systems. *Am. Nat.* 138:606–22
65. Holsinger K. 1992. Ecological models of plant mating systems and the evolutionary stability of mixed mating systems. In *Ecology and Evolution of Plant Reproduction*, ed. R Wyatt, pp. 169–91. New York: Chapman & Hall
66. Holsinger KE. 1986. Dispersal and plant mating systems: the evolution of self-fertilization in subdivided populations. *Evolution* 40:405–13
67. Holsinger KE. 1988. Inbreeding depression doesn't matter: the genetic basis of mating-system evolution. *Evolution* 42: 1235–44
68. Holsinger KE. 1991. Inbreeding depression and the evolution of plant mating systems. *Trends Evol. Ecol.* 6:307–8
69. Holsinger KE, Feldman MW, Christainsen FB. 1984. The evolution of self-fertilization in plants: a population genetic model. *Am. Nat.* 124:446–53
70. Holtsford TP. 1996. Variation in inbreeding depression among families and populations of *Clarkia tembloriensis* (Onagraceae). *Heredity* 76:83–91
71. Holtsford TP, Ellstrand NC. 1989. Variation in outcrossing rate and population genetic structure of *Clarkia tembloriensis* (Onagraceae). *Theoret. Appl. Genet.* 78: 480–88
72. Holtsford TP, Ellstrand NC. 1990. Inbreeding effects in *Clarkia tembloriensis* (Onagraceae) populations with different natural outcrossing rates. *Evolution* 44:2031–46
73. Holtsford TP, Ellstrand NC. 1992. Genetic and environmental variation in floral traits affecting outcrossing rate in *Clarkia tembloriensis* (Onagraceae). *Evolution* 46:216–25
74. Houle D, Hoffmaster DK, Assimacopoulos S, Charlesworth B. 1992. The genomic mutation rate for fitness in *Drosophila*. *Nature* 359:58–60
75. Husband BC, Schemske DW. 1995. Magnitude and timing of inbreeding depression in a diploid population of *Epilobium angustifolium* (Onagraceae). *Heredity* 75:206–15
76. Husband BC, Schemske DW. 1996. Evolution of the magnitude and timing of inbreeding depression in plants. *Evolution* 50:54–70
77. Jain SK. 1976. The evolution of inbreeding in plants. *Annu. Rev. Ecol. Syst.* 7:469–95
78. Jain SK. 1978. Breeding system in *Lim-*

nanthes alba: several alternative measures. *Am. J. Bot.* 65:272–75
79. Jinks JL. 1983. Biometrical genetics of heterosis. In *Heterosis, Reappraisal of Theory and Practice*, ed. R Frankel. New York: Springer Verlag
80. Jóhannsson MH, Gates MJ, Stephenson AG. 1998. Inbreeding depression affects pollen performance in *Cucurbita texana*. *J. Evol. Biol.* 11:579–88
81. Johnston MO. 1992. Effects of cross and self-fertilization on progeny fitness in *Lobelia cardinalis* and *L. siphilitica*. *Evolution* 46:688–702
82. Johnston MO. 1998. Evolution of intermediate selfing rates in plants: pollination ecology versus deleterious mutations. *Genetica* 102/103:267–78
83. Johnston MO, Schoen DJ. 1995. Mutation rates and dominance levels of genes affecting total fitness in two angiosperm species. *Science* 267:226–29
84. Johnston MO, Schoen DJ. 1996. Correlated evolution of self-fertilization and inbreeding depression: an experimental study of nine populations of *Amsinckia* (Boraginaceae). *Evolution* 50:1478–91
85. Jones DF. 1917. Dominance of linked factors as a means of accounting for heterosis. *Genetics* 2:466–79
86. Kärkkäinen K, Koski V, Savolainen O. 1996. Geographical variation in the inbreeding depression of scots pine. *Evolution* 50:111–19
87. Karron JD. 1989. Breeding systems and levels of inbreeding depression in geographically restricted and widespread species of *Astragalus* (Fabaceae). *Am. J. Bot.* 76:331–40
88. Karron JD, Jackson RT, Thumser NN, Schlicht SL. 1997. Outcrossing rates of individual *Mimulus ringens* genets are correlated with anther-stigma separation. *Heredity* 79:365–70
89. Keightley PD. 1994. The distribution of mutation effects on viability in *Drosophila melanogaster*. *Genetics* 138:1315–22
90. Kempthorne O. 1957. *An Introduction to Genetic Statistics*. New York: John Wiley
91. Kennington WJ, James SH. 1997. The effect of small population size on the mating system of a rare clonal mallee, *Eucalyptus argutifolia* (Myrtaceae). *Heredity* 78:252–60
92. Klekowski EJ. 1988. *Mutation, Developmental Selection, and Plant Evolution*. New York: Columbia Univ. Press. 373 pp.
93. Klekowski EJ, Lowenfeld R, Hepler PK. 1994. Mangrove genetics. II. Outcrossing and lower spontaneous mutation rates in Puerto Rican rhizophora. *Int. J. Plant Sci.* 155:373–81
94. Koelewijn HP. 1998. Effects of different levels of inbreeding on progeny fitness in *Plantago coronopus*. *Evolution* 52:692–702
95. Kohane I, Kidwell JF. 1983. Effect of selection, mutation, and linkage on the equilibrium structure of selfing systems. *J. Hered.* 74:175–80
96. Kohn JR, Biard JE. 1995. Outcrossing rates and inferred levels of inbreeding depression in gynodioecious *Cucurbita foetidissima* (Cucurbitaceae). *Heredity* 75:77–83
97. Kondrashov AS, Houle D. 1994. Genotype-environment interactions and the estimation of the genomic mutation rate in *Drosophila melanogaster*. *Proc. R. Soc. London Ser. B* 258:221–27
98. Lacy RC, Alaks G, Walsh A. 1996. Hierarchical analysis of inbreeding depression in *Peromyscus polionotus*. *Evolution* 50:2187–2200
99. Lacy RC, Ballou JD. 1998. Effectiveness of selection in reducing the genetic load in populations of *Peromyscus polionotus* during generations of inbreeding. *Evolution* 52:900–9
100. Lande R. 1994. Risk of population extinction from fixation of new deleterious mutations. *Evolution* 48:1460–69
101. Lande R. 1995. Mutation and conservation. *Conserv. Biol.* 9:782–91

102. Lande R, Schemske DW. 1985. The evolution of self-fertilization and inbreeding depression in plants. I. Genetic model. *Evolution* 39:24–40
103. Latta R, Ritland K. 1994. The relationship between inbreeding depression and prior inbreeding among populations of four *Mimulus* taxa. *Evolution* 48:806–17
104. Latter BDH, Robertson A. 1962. The effects of inbreeding and artificial selection on reproductive fitness. *Genet. Res.* 3:110–38
105. Levin DA. 1989. Inbreeding depression in partially self-fertilizing *Phlox*. *Evol.* 43: 1417–23
106. Levin DA. 1991. The effect of inbreeding on seed survivorship in *Phlox*. *Evol.* 45:1047–49
107. Levin DA, Bulinska-Radomska Z. 1988. Effects of hybridization and inbreeding on fitness in *Phlox*. *Am. J. Bot.* 75:1632–39
108. Lloyd DG. 1979. Some reproductive factors affecting the selection of self-fertilization in plants. *Am. Nat.* 113:67–79
109. Lorenc E. 1980. Analysis of fertility in inbred lines of mice derived from populations differing in their genetic load. *Zeierzeta Lab* 17:3–16
110. Lynch M, Conery J, Burger R. 1995. Mutation accumulation and the extinction of small populations. *Am. Nat.* 146:489–518
111. Lynch M, Gabriel W. 1990. Mutation load and the survival of small populations. *Evolution* 44:1725–37
112. Lynch M, Walsh B. 1998. *Genetics and Analysis of Quantitative Traits*. Sunderland, MA: Sinauer Assoc. 980 pp.
113. MacNeil MD, Kress DD, Flower AE, Blackwell RL. 1984. Effects of mating system in Japanese quail 2. Genetic parameters, response and correlated response to selection. *Theor. Appl. Genet.* 67:407–12
114. Mahy G, Jacquemart AL. 1998. Mating system of *Calluna vulgaris*: self-sterility and outcrossing estimates. *Can. J. Bot.* 76: 37–42
115. Mayer SS, Charlesworth D, Meyers B. 1996. Inbreeding depression in four populations of *Collinsia heterophylla* Nutt (Scrophulariaceae). *Evolution* 50:879–91
116. Maynard Smith J. 1978. *The Evolution of Sex*. New York: Cambridge Univ. Press
117. McCall C, Waller DM, Mitchell-Olds T. 1994. Effects of serial inbreeding on fitness components in *Impatiens capensis*. *Evolution* 48:818–27
118. Menges E. 1991. Seed germination percentage increases with population size in a fragmented prairie species. *Conserv. Biol.* 5:158–64
119. Mitchell-Olds ST, Waller DM. 1985. Relative performance of selfed and outcrossed progeny in *Impatiens capensis*. *Evolution* 39:533–44
120. Mitton JB, Grant MC. 1984. Associations among protein heterozygosity, growth rate, and developmental homeostasis. *Annu. Rev. Ecol. Syst.* 15:479–99
121. Molina-Freaner F, Jain SK. 1993. Inbreeding effects in a gynodioecious population of the colonizing species *Trifolium hirtum* All. *Evolution* 47:1472–79
122. Montalvo AM. 1994. Inbreeding depression and maternal effects in *Aquilegia caerulea*, a partially selfing plant. *Ecology* 75:2395–2409
123. Morton NE, Crow J.F. Muller H.J. 1956. An estimate of the mutational damage in man from data on consanguineous marriages. *Proc. Natl. Acad. Sci. USA* 42: 855–63
124. Mukai T, Chigusa SI, Mettler LE, Crow JF. 1972. Mutation rate and dominance of genes affecting viability in *Drosophila melanogaster*. *Genetics* 72:335–55
125. Muller HJ. 1950. Our load of mutations. *Am. J. Human Genet.* 2:111–76
126. Mutikainen P, Delph LF. 1998. Inbreeding depression in gynodioecious

Lobelia siphilitica: among-family differences override between-morph differences. *Evolution* 52:1572–82
127. Nagylaki T. 1976. A model for the evolution of self fertilization and vegetative reproduction. *J. Theor. Biol.* 58:55–58
128. Nason JD, Ellstrand NC. 1995. Lifetime estimates of biparental inbreeding depression in the self-incompatible annual plant *Raphanus sativus*. *Evolution* 49:307–16
129. Norman JK, Sakai AK, Weller SG, Dawson TE. 1995. Inbreeding depression in morphological and physiological traits of *Schiedea lydgatei* (Caryophyllaceae) in two environments. *Evolution* 49:297–306
130. Ohta T. 1971. Associative overdominance caused by linked detrimental mutations. *Genet. Res.* 18:277–86
131. Otto SP, Whitlock MC. 1997. The probability of fixation in populations of changing size. *Genetics* 146:723–33
132. Ouborg NJ, van Treuren R. 1994. The significance of genetic erosion in the process of extinction. IV. Inbreeding load and heterosis in relation to population size in the mint *Salvia pratensis*. *Evolution* 1994: 996–1008
133. Parker IM, Nakamura RR, Schemske DW. 1995. Reproductive allocation and the fitness consequences of selfing in two sympatric species of *Epilobium* (Onagraceae) with contrasting mating systems. *Am. J. Bot.* 82:1007–16
134. Parker MA. 1985. Local population differentiation for compatibility in an annual legume and its host specific fungal/pathogen. *Evolution* 39:713–23
135. Ralls K, Ballou JD. 1988. Estimates of lethal equivalents and the cost of inbreeding in mammals. *Conserv. Biol.* 2:185–93
136. Ramsey M, Vaughton G. 1998. Effect of environment on the magnitude of inbreeding depression in seed germination in a partially self-fertile perennial herb (*Blandfordia grandiflora*, Liliaceae). *Int. J. Plant Sci.* 159:98–104
137. Ribble DO, Miller JS. 1992. Inbreeding effects among inbred and outbred laboratory colonies of *Peromyscus maniculatus*. *Can. J. Zool.* 70:820–24
138. Ritland K, S. Jain. 1981. A model for the estimation of outcrossing rate and gene frequencies based on independent loci. *Heredity* 47:35–52
139. Ritland K. 1990. Inferences about inbreeding depression based on changes of the inbreeding coefficient. *Evolution* 44: 1230–41
140. Rumball W, Franklin IR, Frankham R, Sheldon BL. 1994. Decline in heterozygosity under full-sib and double first-cousin inbreeding in *Drosophila melanogaster*. *Genetics* 136:1039–49
141. Sakai AK, Karoly KK, Weller SG. 1989. Inbreeding depression in *Schiedea globosa* and *S. salicaria* (Caryophyllaceae), subdioecious and gynodioecious Hawaiian species. *Am. J. Bot.* 76:437–44
142. Schemske D. 1983. Breeding system and habitat effects on fitness components in three neotropical *Costus* (Zingiberaceae). *Evolution* 37:523–39
143. Schemske DW, Lande R. 1985. The evolution of self-fertilization and inbreeding depression in plants. II. Empirical observations. *Evolution* 39:41–52
144. Schmitt J, Ehrhardt DW. 1990. Enhancement of inbreeding depression by dominance and suppression in *Impatiens capensis*. *Evolution* 44:269–78
145. Schoen DJ. 1983. Relative fitnesses of selfed and outcrossed progeny in *Gilia achilleifolia* (Polemoniaceae). *Evolution* 37:292–301
146. Schoen DJ, Johnston MO, L'Heureux AM, Marsolais JV. 1997. Evolutionary history of the mating system in *Amsinckia* (Boraginaceae). *Evolution* 51:1090–99
147. Schoen DJ, Morgan MT, Bataillon T. 1996. How does self-pollination evolve? Inferences from floral ecology and molecular genetic variation. *Philos. Trans. R. Soc. Lond. Ser. B.* 351:1281–90
148. Schultz ST, Lynch M. 1997. Mutation

and extinction: the role of variable mutational effects, synergistic epistasis, beneficial mutations, and degree of outcrossing. *Evolution* 51:1363–71
149. Schultz ST, Willis JH. 1995. Individual variation in inbreeding depression: the roles of inbreeding history and mutation. *Genetics* 141:1209–23
150. Sharp PM. 1984. The effect of inbreeding on competitive male-mating ability in *Drosophila melanogaster*. *Genetics* 106:601–12
151. Shields WM. 1982. *Philopatry, Inbreeding, and the Evolution of Sex*. Albany, NY: State Univ. of New York Press
152. Simberloff D. 1988. The contribution of population and community biology to conservation science. *Annu. Rev. Ecol. Syst.* 19:473–511
153. Simmons MJ, Crow JF. 1977. Mutations affecting fitness in *Drosophila* populations. *Annu. Rev. Genetics* 11:49–78
154. Sorensen FC. 1969. Embryonic genetic load in coastal Douglas-fir, *Pseudotsuga menziesii*. *Am. Nat.* 103:389–98
155. Soulé ME. 1980. Thresholds for survival: maintaining fitness and evolutionary potential. In *Conservation Biology*, ed. ME Soulé, BA Wilcox, pp. 151–69. Sunderland, MA: Sinauer
156. Stebbins GL. 1957. Self fertilization and population variability in the higher plants. *Am. Midl. Nat.* 91:337–54
157. Stephenson AG, Winsor JA, Richardson TE, Singh A, Kao T-H. 1992. Effects of style age on the performance of self and cross pollen in *Campanula rapunculoides*. In *Angiosperm Pollen and Ovules: Basic and Applied Aspects*, eds. D Mulchay, G Bergamini-Mulchay, E Ottaviano, pp. 1-6. New York: Springer-Verlag
158. Stewart SC. 1994. Genetic constraints on mating system evolution in the cleistagomous annual *Impatiens pallida*: inbreeding in chasmogamous flowers. *Heredity* 73:265–74
159. Strauss SH. 1986. Heterosis at allozyme loci under inbreeding and crossbreeding in *Pinus attenuatus*. *Genetics* 113:115–34
160. Templeton AR, Hemmer H, Mace G, Seal US, Shields WM, et al. 1986. Local adaptation, coadaptation and population boundaries. *Zoo Biol.* 5:115–25
161. Templeton AR, Read B. 1984. Factors eliminating inbreeding depression in a captive herd of Speke's gazelle (*Gazella Spekei*). *Zoo Biol.* 3:177–99
162. Uyenoyama MK. 1986. Inbreeding and the cost of meiosis: the evolution of selfing in populations practicing biparental inbreeding. *Evolution* 40:388–404
163. Uyenoyama MK, Holsinger KE, Waller DM. 1993. Ecological and genetic factors directing the evolution of self-fertilization. *Oxford Surv. Evol. Biol.* 9: 327–81
164. Uyenoyama MK, Waller DM. 1991. Coevolution of self-fertilization and inbreeding depression I. Mutation-selection balance in haploids and diploids. *Theor. Pop. Biol.* 40:14–46
165. Uyenoyama MK, Waller DM. 1991. Coevolution of self-fertilization and inbreeding depression II. Symmetric overdominance in viability. *Theor. Pop. Biol.* 40: 47–77
166. Uyenoyama MK, Waller DM. 1991. Coevolution of self-fertilization and inbreeding depression III. Homozygous lethal mutations at multiple loci. *Theor. Pop. Biol.* 40:173–210
167. Van Noordwijk AJ, Scharloo W. 1981. Inbreeding in an island population of the great tit. *Evolution* 35:674–88
168. van Treuren R, Bijlsma R, Ouborg NJ, van Delden W. 1993. The significance of genetic erosion in the process of extinction. IV. Inbreeding depression and heterosis effects caused by selfing and outcrossing *Scabiosa columbaria*. *Evolution* 47:1669–80
169. Vogler DW, Filmore K, Stephenson AG. Inbreeding depression in *Campan-*

ula rapunculoides L. I: A comparison of inbreeding depression in plants derived from strong and weak self-incompatibility phenotypes. *J. Evol. Biol.* 12:483–94

170. Waller DM. 1984. Differences in fitness between seedlings derived from cleistogamous and chasmogamous flowers in *Impatiens capensis*. *Evolution* 38:427–40
171. Waller DM. 1985. The genesis of size hierarchies in seedling populations of *Impatiens capensis*. *New Phytol.* 100:243–60
172. Waller DM. 1986. Is there disruptive selection for self-fertilization? *Am. Nat.* 128:421–26
173. Waller DM. 1993. The statics and dynamics of mating system evolution. In *The Natural History of Inbreeding and Outbreeding*, ed. N Thornhill, pp. 97–117. Chicago, IL: Univ. Chicago Press
174. Whitlock MC, Fowler K. 1996. The distribution among populations in phenotypic variance with inbreeding. *Evolution* 50:1919–26
175. Widén B. 1993. Demographic and genetic effects on reproduction as related to population size in a rare, perennial herb, *Senecio integrifolius* (Asteraceae). *Biol. J. Linn. Soc.* 50:179–95
176. Williams GC. 1975. *Sex and Evolution*. Princeton, N.J.: Princeton Univ. Press
177. Willis JH. 1993. Partial self-fertilization and inbreeding depression in two populations of *Mimulus guttatus*. *Heredity* 71: 145–54
178. Willis K, Wiese RJ. 1997. Elimination of inbreeding depression from captive populations: Speke's gazelle revisited. *Zoo Biol.* 16:9–16
179. Wolfe LM. 1993. Inbreeding depression in *Hydrophyllum appendiculatum*: role of maternal effects, crowding, and parental mating history. *Evolution* 47:374–86
180. Wright S. 1922. Coefficients of inbreeding and relationship. *Am. Nat.* 56:330–38

180a. Wright S. 1922. The effects of inbreeding and crossbreeding on guines pigs. III. Crosses between highly inbred families. Bull. 1121 US Dep. Agric., pp. 1–60

181. Wright S. 1969. *Evolution the Genetics of Populations. The Theory of Gene Frequencies*. Chicago: Univ. Chicago Press
182. Wright S. 1977. *Evolution and the Genetics of Populations* Vol. 3: *Experimental Results and Evolutionary Deductions*. Chicago: Univ. Chicago Press. 611 pp.
183. Yahara T, Maki M. 1993. Effects of biparental inbreeding on the evolution of gynodioecy: a model and a case study in *Chionographis japonica* var. *kurohimensis*. *J. Plant Res.* 106:279–81
184. Ziehe M, Roberds JH. 1989. Inbreeding depression due to overdominance in partially self-fertilizing plant populations. *Genetics* 121:861–68

Annu. Rev. Ecol. Syst. 1999. 30:515–38

HISTORICAL EXTINCTIONS IN THE SEA

James T. Carlton
Maritime Studies Program, Williams College—Mystic Seaport, P.O. Box 6000, 75 Greenmanville Avenue, Mystic, Connecticut 06355; e-mail: jcarlton@williams.edu

Jonathan B. Geller
Moss Landing Marine Laboratories, P.O. Box 450, Moss Landing, California 95039; e-mail: geller@mlml.calstate.edu

Marjorie L. Reaka-Kudla
Department of Zoology, University of Maryland, College Park, Maryland 20742; e-mail: mr9@umail.umd.edu

Elliott A. Norse
Marine Conservation Biology Institute, 15806 NE 47th Court, Redmond, Washington 98052; e-mail: enorse@u.washington.edu

Key Words extinction, endangered, threatened

"The Frail Ocean"

—Wesley Marx, 1967 (79)

Monday, June 3, 1844
(*Date when the last Great Auks were seen alive, on the island of Eldey, Iceland*)

INTRODUCTION

Communities of organisms can change over historical (ecological) time in three ways: Species can be deleted (extinctions), added (invasions), or can change in relative abundance. In marine environments, while the latter two types of alterations are increasingly recognized (if not extensively studied), extinctions in historical time have received little recognition. This lack of attention to marine extinctions stands in striking contrast to the comparatively advanced recognition of the existence of extinctions, particularly of larger organisms, in terrestrial communities (7). Extinctions in historical time have been referred to as neoextinctions, and prehistoric extinctions as paleoextinctions (19), a distinction we follow here. Baillie & Groombridge (7) treat historical extinctions as those occurring in the past 400 years or less. Ehrlich et al (37) noted examples of birds that had become extinct "since 1776," noting that they had "chosen 1776 as our cutoff point somewhat arbitrarily, but reliable reports from before that point are few, museum specimens are rare, and documented extinctions are rarer still" (if this cutoff were to be extended beyond birds, it would not include the extinction of the Steller's Sea Cow, *Hydrodamalis gigas*, last observed in 1768).

We review the record of neoextinctions in the ocean and discuss these in terms of both temporal and spatial patterns. We further review the possible extent of underestimation of marine neoextinctions. Finally, we attempt to set the importance of what we do know about marine extinctions into a larger framework of the vulnerability of marine organisms to global deletion.

It is important to understand the diversity and number of extinctions in the oceans for a variety of reasons. At a general level, an understanding of marine extinctions provides a measure of the scale of susceptibility of the seas to human perturbations and alterations. More specifically, determining which species have become extinct can serve as a harbinger of further loss in particular habitats, providing both a rationale and an opportunity for increased protection of species guilds and habitats that may be most at risk. Knowledge of which species have regionally or globally disappeared is critical in understanding modern-day community and ecosystem structure and function. Energy flow, predator-prey networks, indirect interactions, and a host of other processes may change dramatically with the removal of a species—removals that have, by and large, preceded scientific study. Knowing which species were removed from communities in historical times is the sine qua non of understanding prealtered communities and how they evolved and functioned (20). Indeed, it is not impossible that some of our modern-day views and interpretations of the structure of many marine communities may be the result of species interactions that have been readjusted by means of unrecognized species deletions in ecological time.

Finally, there is a compelling value to knowing about extinct species in terms of evolutionary biology: Detecting species that have gone extinct but that are currently taxonomically buried and thus hidden in the synonymy of still-extant species may provide potentially important phylogenetic information.

SPATIAL AND FUNCTIONAL SCALES OF EXTINCTION

Extinction occurs at a variety of operative levels, in both spatial and functional terms:

Local extinction occurs when a population or populations of a species are displaced from a small area or habitat. This includes the local extirpation of a native species by an introduced species (that is, a reduction in the native species' fundamental niche). An example is the displacement of the native California mudsnail *Cerithidea californica* from open intertidal mudflats on San Francisco Bay, California, by the introduced American Atlantic mudsnail *Ilyanassa obsoleta*, resulting in *Cerithidea* being restricted to an upper intertidal refugium (105). We discuss the California abalone *Haliotis sorenseni* below. Local extinctions may result in loss of distinctive genetic stocks.

Regional extinction occurs when a species is removed from parts of its "fundamental range" (to parallel niche theory terminology) and reduced to a "realized range." Examples include the extirpation of the gray whale *Eschrichtius robustus* from the North Atlantic Ocean (87), restricting modern-day populations to the North Pacific Ocean, and the removal of the sea otter *Enhydra lutris* from large parts of its former range in the Northeastern Pacific Ocean (69, 110). We discuss below the extirpation of the native mussel *Mytilus trossulus* from southern California. As with local extinctions, regional extinctions may result in loss of distinctive genetic stocks (which, as discussed below, could later be found to represent distinct species).

Global extinction occurs when a species completely disappears. For smaller marine organisms, there is no consensus as to how long a species must remain undiscovered in order to be declared extinct.

In addition to these spatial scales, *functional* extinction occurs when a species is so reduced in abundance that it no longer plays a quantitatively important role in the energy flow or the structuring (bioengineering) of the community. An example is the large-scale removal of baleen whales from the southern oceans and the subsequent increase of their former prey, euphausiacean crustaceans (krill) (70, 92).

Commercial extinction occurs when a species is so reduced by hunting (fishing) that exploitation pressure is reduced or ceases. Many finfish and most whale populations have been so overfished that hunting them is no longer economically viable; this results in the pursuit of other species (115, 116). We focus here largely on global extinctions, although other spatial and functional scales of extinction provide critical lessons, as we discuss.

GLOBAL MARINE NEOEXTINCTIONS

Challenges in Establishing a List of Extinct Species

Which marine organisms have become extinct in historical time? We discuss below the absence of data that obscures the true number of extinct species. Here we address three phenomena pertinent to establishing an actual working list of extinctions: one, the level of taxonomic resolution of the species in question; two, the ecological boundaries placed on the habitat definition; and three, the subjectivity, noted above, as to when to declare a taxon extinct in the wild. In this review we take a conservative approach in addressing these challenges.

We treat the extinction of "subspecies" as regional extinctions of allopatric populations of the stem species: Genetic studies (which may be still possible on museum material) may establish the uniqueness of the taxon in question. For example, it remains unclear if the Jamaican Diablotin (a petrel), *Pterodroma hasitata caribbea*, the Bonin Night Heron, *Nycticorax caledonicus crassirostris*, and the Japanese sea lion, *Zalophus californianus japonicus*, were distinct subspecific

TABLE 1 Status of additional marine birds and mammals listed as extinct in Norse (93) and Vermeij (138)

Taxon	Common name	Comment
Class Mammalia		
Order CETACEA		
Family Eschrichtiidae		
Eschrichtius robustus	Atlantic Gray Whale	Former Atlantic populations are now considered to be the same species as Pacific populations (87)
Class Aves		
Order PROCELLARIIFORMES		
Family Procellariidae		
Pterodroma jugularis	Petrel (Hawaiian Islands)	Prehistoric extinction (138)
Pterodroma hasitata caribbea	Jamaican Diablotin	Subspecific status uncertain (40) and extinct status uncertain (25)
Family Hydrobatidae		
Oceanodroma macrodactyla	Guadalupe Storm Petrel	Extinct status uncertain (7, 40)
Order CICONIIFORMES		
Family Ardeidae		
Nycticorax caledonicus crassirostris	Bonin Night Heron	Subspecific status uncertain (7)
Order ANSERIFORMES		
Family Anatidae		
Tadorna cristata	Crested Shelduck	Not extinct (7, 25)
Chendytes lawi	Flightless Duck (California)	Prehistoric extinction (138)

(genetic) taxa or only clinal variants of the stem species (Tables 1 and 2). Baillie & Groombridge (7) did not list the first two and did list the Japanese sea lion as an extinct subspecies. In contrast, Vermeij (138) listed the Canary Islands Oystercatcher (as the "Canary Islands Black Oystercatcher") under the trinomial *Haematopus ostralegus meadewaldoi*; however, Baillie & Groombridge (7) treated it as a full species (*H. meadewaldoi*), a designation that we follow here.

As further examples, we have also not treated as global extinctions two marine birds whose taxonomic status as valid species remains unclear: the Auckland Island Shore Plover, *Thinornis rossi*, and Cooper's Sandpiper, *Pisobia (Tringa) cooperi*. Cooper's Sandpiper is based on a single specimen of a shorebird collected on Long Island, New York, in 1833 (40). Cooper's Sandpiper has long been rejected as a valid species, being interpreted as either a hybrid of the White-Rumped Sandpiper, *Calidris fusicollis*, and the Pectoral Sandpiper, *Calidris melanotos* (40)—although no valid hybrid shorebird has been recorded from North America (81)—or as an aberrant specimen of one or the other (despite being placed in *Pisobia*, a different genus). Molecular genetic analysis could be done on the extant type specimen if

TABLE 2 Additional marine birds and mammals that may be extinct

Taxon	Common name	Comment
Class Mammalia		
Order CARNIVORA		
Family Otariidae		
Zalophus californianus japonicus	Japanese Sea Lion	Subspecific status uncertain (110, 138)
Class Aves		
Order PROCELLARIIFORMES		
Family Procellariidae		
Pterodroma sp.	Petrel (Mauritius)	Listed as extinct (7) with no further data
Order CHARADRIIFORMES		
Family Charadriidae		
Thinornis rossi	Auckland Island Shore Plover	Specific status uncertain (40, 49, 52)
Family Scolopacidae		
Pisobia (Tringa) cooperi	Cooper's Sandpiper	Specific status uncertain (40, 81)
Prosobonia leucoptera	White-Winged Sandpiper	Marine, maritime, or inland (see text)

harvestable DNA is present. The Auckland Island Shore Plover, *Thinornis rossi*, is also known from a single specimen collected in 1840. This individual differs from all known specimens of its congener, the New Zealand Shore Plover, *Thinornis novaeseelandiae*, itself a threatened species (7, 40, 49). Greenway (49) suggested that *T. rossi* may have been a distinct sibling ("sympatric") species. As with Cooper's Sandpiper, the still-extant type specimen may be worthy of molecular examination.

Vermeij (138) noted four taxa as being "marginally marine," but included them in a table of "recently extinct marine species." For the purposes of this review, we define a "marine organism" as one that relies for some or all of its life on ocean resources (such as food, breeding sites, or habitat). We omit, however, maritime taxa, such as the Pallid Beach Mouse (*Peromyscus polionotus decoloratus*) or the Dusky Seaside Sparrow (*Ammodramus maritimus nigrescens*). Unfortunately, precise habitat data are lacking for many extinct taxa, leading to further potential omissions from the list of extinct marine taxa. An example is the extinct White-Winged or Tahitian Sandpiper, *Prosobonia leucoptera* (32, 49, 81). Its habitat is unknown ("near small streams"; 52) and thus their proximity to or use of the ocean is also unknown. However, given that its only congener, *Prosobonia cancellata*, is a marine shorebird (52), *P. leucoptera* may also have been marine.

World Conservation Union criteria up until 1996 indicated that a species was considered extinct if it had "not definitely been located in the wild during the past 50 years" (50). In the 1996 "Red Book" (7), the arbitrariness of 50 years (19) was replaced with the criterion of when "exhaustive surveys . . . over a time frame appropriate to the taxon's life cycle and life form." As discussed below, not finding a species may be due to inadequate exploration and incorrect identification. Thus, the Guadalupe Storm Petrel, *Oceanodroma macrodactyla*, returns to the "possibly alive" category (having been formerly considered extinct) because of the possibility of confusion with a related species and because all of its former island range has not been exhaustively surveyed (25, 40).

Marine Birds and Mammals Known to be Extinct

Treatments of marine vertebrate extinctions by Norse (93) and Vermeij (138) listed as extinct 10 and 13 species of birds and mammals, respectively. Setting aside two paleoextinctions (a flightless duck, *Chendytes lawi*, and a Hawaiian petrel, *Pterodroma jugularis*), Norse (93) and Vermeij (138) reported a total of 13 extinctions, with only a 62% overlap in their lists. The Guadalupe Storm Petrel, listed by both Norse (93) and Vermeij (138), may be extant, as noted above. The Korean Crested Shelduck, listed by Vermeij (138), is considered to be still living (Table 2). The Jamaican Diablotin, in Vermeij's (138) but not Norse's (93) list, is neither clearly a distinct taxon nor demonstrably extinct (Table 2). The Atlantic Gray Whale, listed by Norse but not by Vermeij, is not considered to be a distinct taxon (87).

We consider three mammals and five birds extinct, (Table 3). In Table 3, the date of record is the date the last individual(s) were actually seen (and usually killed), not necessarily when the species became extinct, but the last sighting can be considered an estimate of the date of *functional* extinction of the species. The status of seven other species of marine vertebrates discussed by Norse (93) and Vermeij (138) are summarized in Table 1, and five more species that may be extinct are listed in Table 2. We discuss below taxonomic, spatial, or temporal patterns in these extinctions.

Marine Invertebrates Known to be Extinct

There are, as discussed by Carlton et al (21) and Carlton (19), obstacles associated with assembling more than a rudimentary list of examples of extinct marine invertebrates. Comparison of pre-twentieth century accounts of marine invertebrates with museum collections (to detect sudden terminations in collections of specific taxa) or with modern faunal lists has not yet begun. In 1992, Carlton (19) suggested that four species of marine gastropods had become extinct, none of which has been found since then (Table 4). The data are too limited to resolve any patterns.

Clark (24, 25, and personal communication 1996) noted that the anaspidean seaslug *Phyllaplysia smaragda* was "possibly extinct." It was first described in 1977 and last collected in 1981, and only known from portions of the Indian River Lagoon system on the east coast of Florida. It may have specialized on the epiphytic algae growing on the basal stems of the seagrass *Syringodium*; this habitat was extirpated from the type locality of the slug, but remains widespread elsewhere in Florida and the Caribbean. Mikkelsen et al (89) reported specimens collected

TABLE 3 Marine birds and mammals extinct in historical time

Taxon	Common name	Geographic region (*)	Last known living (**)
CLASS MAMMALIA			
Order CARNIVORA			
Family Mustelidae			
Mustela macrodon	Sea Mink	NW Atlantic	1880
Family Phocidae			
Monachus tropicalis	West Indian Monk Seal	Caribbean, Gulf of Mexico	1952
Order SIRENIA			
Family Dugonidae			
Hydrodamalis gigas	Steller's Sea Cow	NW Pacific	1768
CLASS AVES			
Order PELECANIFORMES			
Family Phalacrocoracidae			
Phalacrocorax perspicillatus	Pallas's Cormorant	NW Pacific	ca. 1850
Order ANSERIFORMES			
Family Anatidae			
Mergus australis	Auckland Islands Merganser	SW Pacific	1902
Camptorhynchus labradorius	Labrador Duck	NW Atlantic	1875
Order CHARADRIIFORMES			
Family Charadriidae			
Alca impennis	Great Auk	NW/NE Atlantic	1844
Haematopus meadewaldoi	Canary Islands Oystercatcher	NE Atlantic	1913

*Region abbreviations: NE, Northeast; NW, Northwest, SW, Southwest.

**Final observations of living specimens:

Mustela macrodon: Day (32) stated that the last known specimen was taken in 1880 on an island in the Gulf of Maine; Campbell (16) stated it was taken at Campobello, New Brunswick, in 1894, but Waters & Ray (141) noted that the 1894 record is questionable. Vermeij (138) listed a date of "about 1900."

Monachus tropicalis: Knudtson (67) noted that the last authenticated sighting was in 1952 on Serranilla Bank in the western Caribbean. LeBoeuf et al (72) noted that surveys between 1973 and 1984 failed to discover it (thus Vermeij's (138) record of extinct "before 1973"). Solow (126) discussed the monk seal as an example of inferring extinction from sighting data (see also 82). In a prophetic statement, William T Hornaday wrote in 1913 (55) about West Indian monk seals, "... the Damocletian sword of destruction hangs over them suspended by a fine hair, and it is to be expected that in the future some roving sea adventurer will pounce upon the Remnant, and wipe it out of existence for whatever reason may to him seem good."

Hydrodamalis gigas: Vermeij (138) gave a date of 1750, and Day (32) a date of 1767; we follow Silverberg (121), Scheffer (118), and Rice (110) in using 1768.

Phalacrocorax perspicillatus: The date of "about 1850" is based on estimates made in 1882 of when the last birds were seen on a small island off the Komandorskiye (Commander) Islands (40, 49).

Mergus australis: Fuller (40) noted that the last pair of birds was collected in 1902. The dates of 1901 (49), 1905 (138), and about 1910 (32) appear to be either errors or speculations.

Camptorhynchus labradorius: A date of 1878 is occasionally cited, but Fuller (40) noted, this record cannot be verified, and added that "there is some doubt concerning the" date 1875 itself but did not elaborate.

Alca impennis (= Pinguinus impennis). The date of 1844 is widely agreed upon (32, 40, 49).

Haematopus meadewaldoi: Hayman et al (52) noted that the last "firm sighting" was 1913, but Fuller (1987) wrote the "black Oystercatcher was last seen on Tenerife in 1968." The 1968 date represents one of a series of sightings of "black oystercatchers" from regions such as the western Canaries (Tenerife) (in 1968 and 1981), and Senegal (in 1970 and 1975) where no oystercatchers, either the Canary Islands Oystercatcher or the African Black Oystercatcher (*H. moquini*) have ever been recorded. The temporal clustering of these records one decade could suggest a temporary expansion of the vagrant range of *H. moquini* (previously known only as far north as Angola, far to the south). Regardless, the long hiatus between 1913 and 1968, and the lack of reports since 1981 or 1975, compel us to use the date of 1913.

TABLE 4 Marine invertebrates extinct in historical time

Taxon	Common name	Geographic region	Last known living
PHYLUM MOLLUSCA			
Class GASTROPODA			
Lottia alveus alveus	Atlantic eelgrass Limpet	NW Atlantic	1929
"Collisella" edmitchelli	Limpet	NE Pacific	1861
Littoraria flammea	Periwinkle	NW Pacific	<1840s
Cerithidea fuscata	Horn Snail	NE Pacific	1935

in 1980 from the seagrass *Thalassia* in the Indian River Lagoon, but Clark, in Mikkelsen et al (89), felt that these may have been misidentified specimens of the related slug *P. engeli*. Clark (24, 25, and personal communication 1996) further noted that the sacoglossan seaslug *Stiliger vossi*, described in 1960 and known only from Biscayne Bay, in southeast Florida, has never been collected again, despite extensive searching, and he speculated that it was also extinct. Given the relatively recent discovery of these two seaslugs in the latter half of the twentieth century, and given that much of the Caribbean has not been thoroughly explored for rare opisthobranchs, we note these here as possible extinctions but do not formally admit them to Table 4 at this time.

Some other invertebrate species were thought possibly extinct:

- Wells et al (143) and Barnes (10) noted that the nudibranch seaslug *Doridella batava*, described from the Zuiderzee, Netherlands, may be extinct. Swennen & Dekker (130) demonstrated that this slug is *Corambe obscura*, described from North America and apparently introduced to Europe.
- Banks et al (8) speculated that the Kumamoto oyster, *Crassostrea sikamea*, "may be extinct in Japan" and thus survive only in the laboratory or in oyster culture in North America. Living specimens were recovered in 1996 in Japan (149).
- Runnegar (113) believed the Caribbean bivalve mollusk *Pholadomya candida* was extinct, but it was discovered living in Venezuela (45).
- Glynn & de Weerdt (47) reported that the hydrocoral *Millepora boschmai* had become extinct in its only known locality, the Gulf of Chiriqui, Panama, in the tropical Eastern Pacific Ocean. Glynn & Feingold (48) later reported the species had been rediscovered alive in 1992.
- Barnes (10) noted that the infaunal sea anemone *Edwardsia ivelli*, known only from a single lagoon in Sussex, England, "may. . .be extinct," as it had not been collected, despite searching, since 1983. However, as Barnes also notes, this is a small anemone of shallow soft mud bottoms, a habitat not

well explored for small invertebrates, and thus it "may well be living, unnoticed, in other localities."

- Edmondson (36) noted that the fiddler crab *Uca minor* was known only from and said to have been collected on the island of Oahu, Hawaiian Islands, in 1826–1827. It has never been found again in Hawaii (nor indeed were any species of fiddler crabs known from the islands). He speculated that it "may have become extinct here." Crane (28) believed the record, based upon the original description, represented mislabeled specimens of the widespread Indo-Pacific fiddler crab *Uca lactea*.

O'Clair & O'Clair (96) noted that the ectoparasitic cyamid amphipod crustacean (whale louse) "*Sirenocyamus rhytinae* was recorded from the Steller's Sea Cow, but that unfortunate marine mammal was driven to extinction in 1768 and its cyamid has not been seen since." No specimens were preserved, and cyamid amphipods are known only from whales and dolphins, and have never (since Steller) been recorded from sirenians. As Leung (73) noted, the species *Cyamus rhytinae* was resurrected by later workers based upon the assumption that a piece of dried skin discovered in St. Petersburg, Russia, was from the sea cow, and that the whale-lice attached to this skin were thus Steller's species. However, the specimens were *Cyamus ovalis*, a species well-known from the right whale *Balaena glacialis*. It was then assumed that the dried skin was that of a right whale. Leung (73) reported that *C. ovalis* is also known from the sperm whale, *Physeter catodon*, suggesting a third possibility for the skin's origin. As some cyamids are not host specific (73), it is possible that the cyamid found on Steller's sea cow was the same as a species found on North Pacific whales; arguing against this is a sirenian being an atypical habitat for a cyamid. Identifying this piece of dried skin by molecular means would thus be of interest.

Marine Fish, Marine Algae, and Marine Seagrass Neoextinctions

Although an increasing number of marine fish and marine plants are recognized as threatened and endangered, no fish, algae, or seagrasses are known to have become globally extinct in historical time. For seaweeds, this may reflect a taxonomic artifact, the difficulty of recognizing when poorly studied and systematically difficult taxa have disappeared.

Taxonomic, Geographic, and Temporal Patterns of Marine Vertebrate Neoextinctions

With only eight marine bird and mammal taxa unquestionably extinct, taxonomic, geographic, and temporal analyses are limited, but nonetheless offer some compelling insights. The eight extinct species represent five orders of mammals and birds, suggesting that extinction is not phylogenetically constrained. Three of the species occurred in the Pacific and five in the Atlantic-Caribbean. No Pacific marine vertebrate has gone extinct since 1902, but Atlantic extinctions continued to 1952.

Eight extinctions from 1768 to 1952 means an extinction every 23 years. Aside from the first and last (1768 and 1952), these extinctions occurred from 1844 to 1913. A marine bird went extinct approximately every 12 years in the latter half of the nineteenth century and into the first decade of the twentieth century. This period of human history is coincident with rapidly increasing global exploration, colonization, and industrialization, and thus rapidly increasing hunting and habitat destruction. The most recent extinction, that of the West Indies monk seal, occurred in the Caribbean Islands suggesting that in areas with complex geography, it may be more difficult to locate and kill every individual of a larger organism than it is along more open coastlines or on individual islands.

UNDERESTIMATION OF MARINE NEOEXTINCTIONS

The small number of invertebrate and vertebrate marine extinctions recorded to date suggests that the true number may be underestimated. We argue below that this limited record is not due to marine life being relatively immune to extinction.

Two phenomena are primarily responsible for the poor record of marine extinctions in historical time (19): one, the rich pre-twentieth century literature on marine organisms remains largely uninvestigated (see also 20), and two, a decreasing knowledge of marine biodiversity, especially in those coastal waters most susceptible to human-induced destruction and perturbation. The combination makes it difficult to detect losses, even of once-abundant species. Carlton et al (21) demonstrated that the disappearance from the New England shore of a marine snail (a limpet) that was recorded in 1929 as occurring by the thousands on the eelgrass *Zostera marina* had been overlooked for more than 50 years.

Issues in systematics also contribute to overlooking extinctions (19). Sibling species—those that are so similar that they are considered to belong to a single species without corroborating genetic data—abound in the sea (66). The discovery of sibling species within even well-studied genera (e.g., the mussel *Mytilus*, and the snails *Littorina* and *Nucella*) (80, 84, 97; see 66 for additional examples) suggests that many additional widespread species may be complexes of two or more species, with the conservation status of each unknown. Similarly, phylogenetically distinct populations of single species may be seen in the context of conservation biology as evolutionarily significant units (ESUs), each with its own evolutionary trajectory (140). Considered in this manner, local populations may be the appropriate units of concern (90, 139). There is no evidence that many geographically restricted marine sibling species or ESUs have become extinct or are even endangered, but the question has not been systematically addressed. We discuss below examples of the impact of undetected sibling species extinction and the difficulty of recognizing ESUs.

Inadequately described species (especially those not collected since the nineteenth century or earlier and for which no museum material exists) are usually assumed to be (*a*) completely unrecognizable taxa not assignable to a known species (but which, it is further assumed, could probably be so assigned if the description had been better), or (*b*) barely or questionably recognizable taxa that are placed in the synonymy of extant taxa based upon the best judgment of a systematist.

Extinct populations may have been of undescribed species; as suggested elsewhere in this review, this hypothesis is testable by molecular examination of museum material. The search should be for extinct populations that represented distant, peripheral, or end-of-range records or for extinct populations reported from habitats not typical of the species. As also noted by Carlton (19), undescribed parasites of extinct marine vertebrates and invertebrates can easily be overlooked. If undescribed species may go extinct before being collected or described, there is no way of knowing they ever existed.

We examine the question of extinction underestimation by reviewing the application of species-area relationship theory to coral reefs, a marine environment under global stress, and by reviewing molecular genetic analysis as a means of detecting now-extirpated sibling species and recognizable allopatric populations that, if rendered extinct, would result in genetic loss.

A Model System: Theoretical Estimations of Coral Reef Extinctions

It is possible to estimate species loss based on estimates of species diversity and documented range contraction and habitat loss. That species diversity is predictably related to the area of habitat in both continental and island faunas has been known empirically for over a century, and multiple regression studies in these environments have shown that area alone accounts for most of the variation in species number (30, 76).

Although early studies fitted the curve between species number and area in several ways, the most commonly accepted relationship now is that the logarithm of the number of species (S) is proportional to the logarithm of area (A) plus a constant (c), or $S = cA^z$ (11, 26, 76, 83, 145), where c depends upon the taxon, biogeographic region, and population density, and z is the rate of increase of species (log S) with area (log A). Of the z-values that have been determined empirically for beetles, ants, amphibians, reptiles, breeding birds, land vertebrates, land plants, freshwater diatoms, and crustaceans inhabiting coral heads, most cluster between 0.15 and 0.40 (76, 83, 122, 123, 147, 148). Although the reasons for slight variations in the parameters remain debatable, the fundamental relations between number of species and the area they occupy remains sound.

Particularly for marine environments, which are difficult to monitor for loss of species, the species-area curve can be useful in providing bracketed assessments of the numbers of species that should be present, and, given a documented or projected amount of habitat loss, how many species could go extinct. This approach has been used successfully, in consort with satellite imagery, to identify species loss associated with the massive destruction of rainforests (38, 147). Except for a few studies (3, 119, 122) however, species-area relationships have received relatively little attention in the oceans, in part because there have been few studies of the number of species present in marine habitats.

The total number of marine species in several major marine environments, with a focus on the number of species on global coral reefs, was estimated by Reaka-Kudla (107) to be about 220,000 described coastal species, of which about 195,000 were tropical. These estimates were based upon an empirically determined total number of described marine species (about 274,000) and the areas of global coastal zones and tropical coastal zones, and employed several empirically based assumptions about the rate of change in species numbers with latitude, coastal versus open ocean environments, and complexity of the habitat. Although estimates of global reef area vary (65, 107, 127), it is likely that for purposes of assessing species diversity, the relevant global reef area is somewhere between 250,000 and 650,000 km^2, and that global coral reefs and reef-associated habitats support about 75,000–95,000 described species (M Reaka, unpublished observations).

Many of these species could be lost if recent estimates of loss of reef habitat are correct. Wilkinson (144) classified the status of the world's coral reefs as "critical" (severely damaged and in imminent threat of extermination if current levels of anthropogenic stress are not reduced), "threatened" (currently show signs of stress and threatened with collapse if stresses continue to increase at current rates), and "stable" (should remain stable unless large-scale anthropogenic processes introduce unforeseen impacts), based upon his own and other monitoring programs. These programs included data on observed or quantified live coral cover, existing populations of fishes and fishing pressure, and pollution and sedimentation near expanding human populations. Wilkinson hypothesized that 10% of global coral reefs already have been degraded beyond recovery, 30% are in a critical state and could be lost in the next 10–20 years, 30% are threatened and could disappear in the next 20–40 years, and another 30% are stable.

Bryant et al (13) mapped locations of human activity (size and density of urban areas, type of land clearing and agriculture, rainfall and watershed topography, ports, oil facilities, mining activities, shipping lanes, destructive fishing practices) and summarized these impacts as four threat factors: coastal development, marine pollution, overexploitation and destructive fishing, and inland pollution and erosion. They mapped the coral reefs where these factors would pose high, medium, or low threats. Reefs were regarded as under "high threat" if they were highly threatened from at least one of the four threat factors, under "medium threat" if they received medium threats from at least one of the four threat factors, and under "low threat" if all four of the threat factors posed a low threat to the reef. Under this categorization, 58% of the world's reefs were considered to be at risk (27% under high and 31% under medium threat); outside of the Pacific, however, these data indicated that 70% of all reefs were at risk. More than 80% of the reefs in both Southeast Asia (56% under high and 26% under medium threat) and the Atlantic (excluding the Caribbean) were considered to be at risk (55% under high and 32% under medium threat). More than 60% of the reefs in the Caribbean were at risk (29% under high and 32% under medium threat). These figures are relatively consistent with Wilkinson's conclusions.

If one accepts the value that even 5% of the world's reef area has been degraded to a nonreef state, about 1,000 (1.3%) of the described species on global coral reefs would already have become extinct (984–1,200 if the original reefs occupied 2500,000 or 650,000 km^2 and $z = 0.25$; the latter figures are not intended to imply undue precision but simply to show that the calculations are robust relative to the estimated ranges of global reef area). If 30% of the world's reef area were to be lost in the next 10 to 20 years, almost 10% (approximately 6,000–8,000) of the world's described coral reef species could become extinct (M Reaka, unpublished observations).

Since the biota of the oceans is so poorly documented, however, estimates of the actual number of species (described plus undescribed), and their potential loss, on global coral reefs, would be of greater interest. The total number of global coral reef species (known plus unknown) has been estimated from reef area (about 6% that of global rainforests), assuming that coral reefs operate according to similar ecological principles as rainforests (107).

If rainforests contain 2 million species (an undoubtedly conservative number; 38, 147), global reefs would host between 750,000 and 950,000 total (known plus unknown) species (reef area = 250,000 to 650,000 km^2, $z = 0.25$). If 5% of the area of these coral reefs were destroyed, 10,000–12,000 species would become extinct; if 30% of the area of global reefs were degraded to nonfunctional states, reefs could lose 65,000–85,000 of their total (known plus unknown) species. Since emerging consensus favors a figure of about 14–18 million species on Earth, it is reasonable to expect that global rainforests may support as many as 10 million species. Comparable calculations (reef area = 250,000–650,000 km^2, $z = 0.25$) showed that global reefs then would contain 3–4 million total (known plus unknown) species (M Reaka, unpublished observations). Of these, 50,000–60,000 species could become extinct if only 5% of reef area is destroyed and 300,000–400,000 species would be lost if 30% of the area of global reefs were destroyed (M Reaka, unpublished observation).

Model Systems: Regional Mussel Extinction and Snail Demise in California

Mussels of the genus *Mytilus* are common on intertidal and shallow-water hard bottoms worldwide. Three species comprise a sibling species group: *M. edulis* is native to the North Atlantic Ocean, *M. galloprovincialis* is native in the Mediterranean Sea, and *M. trossulus* is native in the North Pacific Ocean (68, 84). In southern California, *Mytilus galloprovincialis*, introduced from southern Europe, is abundant, and the native *M. trossulus* is rare or absent (84, 117, 129). In contrast, *M. trossulus* is abundant north of San Francisco Bay (84, 117, 129).

Collections and reports of living mussels in southern and central California in the twentieth and nineteenth centuries indicate there have been temporally continuous populations of mussels (18, 60, 64, 78, 111). Thus, the modern domination of southern California by species thought to have been introduced in the twentieth century (18) suggests that a native species has been replaced. Geller (44) sequenced a portion of the mitochondrial 16S ribosomal RNA gene from dry mussels collected about 1900 from Santa Catalina Island, and in 1871 and 1884 from Monterey Bay. These three lots of mussels are *M. trossulus*, which proves that *M. trossulus* was present south of San Francisco Bay and later temporally overlapped *M. galloprovincialis*. This case illustrates how unrecognized species differences can obscure the regional extirpation of a member of a sibling species complex. It is not difficult to imagine the extinction of such a sibling species with a geographic distribution more restricted than in this example.

Could the decline of *M. trossulus* in southern California be a direct consequence of the invasion of *M. galloprovincialis*? Circumstantial evidence suggests interspecific competition as a contributing role: Environmental conditions in southern California may favor growth of *M. galloprovincialis* compared to *M. trossulus* (117). *Mytilus galloprovincialis* smothers another native mussel, *M. californianus*, in wave-protected areas in southern California (51, studying *M. galloprovincialis* under the name of *M. edulis*) and may exhibit similar behavior in patches of *M. trossulus*. *Mytilus galloprovincialis* is also a strong competitor in Europe, where it overlaps with *M. edulis*, having lower rates of predation by predatory snails, lower incidence of parasite infections, higher strength of attachment, and lower mortality rates (29, 42, 124, 146). In South Africa, invading *M. galloprovincialis* outcompetes a native mussel (*Aulacomya ater*) by overgrowth (54).

Cerithidea californica is a potamid mudsnail that is abundant on tidal flats in southern California. Populations also are known in northern California in San Francisco Bay, Bolinas Lagoon, Drake's Estero, and Tomales Bay, in marginal habitat at the upper fringes of salt marshes (17, 104, 105; J Byers, personal communication, JB Geller and JT Carlton, personal observation). Competition or egg-predation by invading snails (the Atlantic nassariid mudsnail *Ilyanassa obsoleta* and the Japanese potamid mudsnail *Batillaria attramentaria*) have had demonstrable impacts on the northern populations (104, 105; J Byers, personal communication). The northernmost population of *Cerithidea californica* known in the Eastern Pacific Ocean occurred in Bodega Harbor, 96 km north of San Francisco but went extinct in the late 1960s due to destruction of salt marsh habitat (17).

Emerging from these observations is a picture of large, robust populations in southern California and severely reduced and marginalized populations in northern California. Petryk (103) observed three mitochondrial haplotypes in a sample of 64 snails from southern California, in San Diego Bay, while only one haplotype (also found in San Diego) in a sample of 55 snails from Tomales Bay and San Francisco Bay. *Cerithidea* has no planktonic larva, and thus Petryk (103) concluded that gene flow between these two populations was absent or very low. Indeed, northern populations were once described as a separate species, *Cerithidea sacrata*, although the validity of the northern species is not widely recognized (1, but see 131). Regardless, Petryk's (103) work suggests that northern and southern populations are isolated. The smaller northern populations may thus be considered as a unit of conservation concern or as an ESU, pending the recognition of diagnostic characters.

THE SUSCEPTIBILITY OF MARINE ORGANISMS TO EXTINCTION

Numerous workers have reviewed geological, ecological, and biological attributes that make species susceptible to extinction (19, 39, 53, 71, 85, 106, 109, 135–137). Carlton (19) argued that restricted geographic distribution, restricted habitat, and limited dispersal abilities may prove to be major factors that render marine invertebrates most prone to extinction. Vermeij (138) and Norse (93) noted that marine vertebrate species with both small and large aboriginal ranges have become extinct, although complicating this interpretation is the difficulty in distinguishing three types of ranges among extinct animals: the breeding range, the range occupied by adults throughout their lifetime, and the vagrant range.

The eelgrass limpet of the Northwestern Atlantic Ocean disappeared after most of its habitat, the eelgrass *Zostera marina*, was lost to disease (21). The invertebrates that have likely disappeared from southern California and Florida were largely taxa linked to shallow embayments that are easily destroyed by human activities. In reviewing the probable demise of the apparently endemic seahare *Phyllaplysia smaragda* in a Florida lagoon, Clark (24) noted that "reconnection of mosquito impoundments may have increased nutrient levels and decreased oxygen levels." This seahare fed on epiphytic algae growing at the base of the seagrass *Syringodium*. High densities of drift algae apparently excluded the epiphytes; in addition, the slugs' egg masses were deposited near the grass bases, and the lowered oxygen levels (a result of releasing impounded marsh peat) may have lowered the reproductive success of the slugs.

Human predation (for food, traditional medicines, the aquarium trade, decoration, or other reasons) on marine mammals, fish, invertebrates, and algae can be overwhelming. In speculating about the extinction of Steller's Sea Cow, Anderson (5) has argued that human predation on one species could lead to a cascading effect that could lead to the extinction of another species. Anderson notes that human extirpation of sea otters along the Arctic and Pacific coasts of America would have resulted in increasing populations of sea urchins in shallow water, leading, in turn, to the disappearance of nontoxic seaweeds and their replacement (from deeper waters) by phenolic-rich species. Anderson suggests that if the sea cows evolved in shallow waters, in the presence of large sea otter populations and eating nontoxic seaweeds, their foraging grounds may have been destroyed by such algal replacements mediated by sea otter reductions. This, Anderson argues, could explain the early demise of sea cows from the mainland (where aboriginal hunters would have reduced otter populations) and their isolation by the eighteenth century to uninhabited offshore islands. Discovery by eighteenth century ship-borne hunters was the final blow to the reduced populations of the sea cow.

Few such complex relationships have been suggested in modern marine extinctions. Direct human hunting has severely affected edible species (9, 93, 99, 115, 116), and may imply that the twentyfirst century will witness vastly increased extinction rates. The number of species of marine mammals, turtles, and fish (7, 93, 116) that are now severely depressed appears to be unprecedented in human history. We review two case histories that illustrate how widespread marine organisms can quickly near extinction.

An Abundant Subtidal Mollusk Becomes Rare

The white abalone, *Haliotis sorenseni*, is a large herbivorous gastropod that ranged from Point Conception, California, to Isla Asuncion, central Baja California (86). White abalones occur mainly on low-relief rocky reefs at 26–65 m (31), deeper than the other seven Northeast Pacific species of *Haliotis*. Their populations were therefore largely ignored until other shallow-water species became rare from overexploitation.

A fishery for white abalone began around 1965 and collapsed after 1977 (31, 132). When Tutschulte (134) surveyed the Channel Islands in southern California in the early 1970s after heavy commercial fishing had begun, white abalone population density averaged 10,000 per hectare ($1/m^2$). By 1980–1981, the density had dropped three orders of magnitude, to $0.0021/m^2$, and by yet another order of magnitude to $0.0002/m^2$ in 1992–1993 (31). Surveys in 1996–1997 at greater depths using a research submarine showed densities of about 1 per hectare in suitable habitat; there thus appears to be no relict white abalone populations in the deeper part of its range (G Davis, personal communication, 1998). There appear to be no other factors contributing significantly to the population decline during this period, and there is evidence that white abalone populations in Mexico have declined in parallel (31, 132). If population densities in the historic population center of this species, California's Channel Islands, are typical of the entire range, only 500–600 white abalone exist today.

White abalone can spawn as many as 10–15 million eggs and have planktonic dispersal, attributes that suggest that it would be difficult to render this marine snail extinct by overfishing. The last recruitment event of the white abalone occurred in the late 1960s, at the beginning of the most intensive eight years of the fishery (31). Sporadic recruitment might have been more than sufficient to sustain this species if the fishery had not reduced population densities by three to four orders of magnitude in prime habitat. As population density fell far below $1/m^2$, it became impossible for fertilization to occur, resulting in complete reproductive failure (31, 132). Many marine invertebrates are broadcast spawners, and when their population densities decline beyond a certain threshold, fertilization and recruitment fall sharply, a positive feedback mechanism called the Allee Effect that leads to extinction.

Trawling the Sea Floor and the Near-Extinction of the Barn Door Skate

A large proportion of the world's fish catch comes from continental shelves (114, 116). Fishing methods such as dynamite fishing, muro ami, poisoning, and use of mobile fishing gear such as otter trawls, beam trawls, scallop dredges, clam dredges, and St. Andrews' crosses are harmful because, in addition to killing target species and those incidentally brought on deck (bykill), they severely disturb the seabed and organisms that provide food and hiding places (33, 142).

Dragging (trawling and dredging) has a long history on smooth, shallow bottoms near industrialized nations, but with technological advances, such as diesel engines on fishing boats (in the 1920s) (74) it has spread. Dragging extended into continental slope waters as deep as 2000 m (88) as more continental shelf species were overfished. Starting in the 1980s, the widening use of roller, rockhopper, and streetsweeper gear, global positioning systems, precision depth finders, and more powerful engines led to dragging over rough and steep bottoms. By the end of the twentieth century, mobile fishing gear can be used on any bottom type on the continental shelf, upper continental slope, and seamounts from subpolar to tropical waters, penetrating ecosystems that had once served as de facto marine refuges. The effects of mobile gear on seabed biota resemble those of clearcutting in forests, removing the complex structures that are hiding and feeding places for many species (142).

Bottom trawling affects a greater area than any other benthic disturbance: Watling & Norse (142) estimated that an area equivalent to 14.8×10^6 km^2 is trawled annually. Worldwide, an area equaling the world's continental shelf is trawled every two years. Trawling effort is uneven; some spots are hit repeatedly, while others are missed randomly or because they are uneconomical to fish. For longer-lived species, repeated removal and physical disturbance on so large a scale can make extinction all but inevitable. The long-lived, large (1.5 m long) barndoor skate (*Raja laevis*), a Northwest Atlantic shelf-dwelling fish that was once common in trawl bykill (12), is now nearing extinction (22). Risk et al (112), Butler (41), and Fuller & Cameron (14) included anecdotal accounts that dragging is eliminating very long-lived cold-water shelf-dwelling gorgonian corals (*Primnoa* spp. and others). Large skates and five-meter tall corals are among the most conspicuous benthic species, and their loss would be noticed long before small bryozoans, amphipods, or polychaetes were missed. As in tropical forests, the disappearance of the most observable species is likely a strong indication of greater extinction.

CONCLUSIONS

That the sea is not immune to extinctions is demonstrated by the ubiquitousness of paleoextinctions and neoextinctions in the sea (4, 39, 57, 58, 106, 128, 138). That marine organisms are "immune" to extinction because of some perceived attributes of a combination of having widespread populations, dispersal by ocean currents, or refugia from human predation is no longer a tenable position. Hudson (56) noted the "widespread and ingrained belief that some fishes are inherently resilient to extinction and will not disappear, however hard they are fished."

The unsurprising fact that more extinctions are not yet documented is a result of a number of factors. Few scientists work on marine extinction issues: data production on extinctions may be proportional to the number of investigators. As with the once abundant and widespread eelgrass limpet, the extinction of which went unremarked for more than five decades, the fact that these extinctions have not been documented is not evidence that they are not occurring.

The world ocean, covering 361×10^6 km^2 to an average depth of nearly 4 kilometers, constitutes more than 99% of the biosphere permanently inhabited by animals and plants (94). That most marine biota is not seen contributes to the belief that life in this vastness may be extinction-proof. In contrast, forests and woodlands cover 38×10^6 km^2 (102), or 7.5% of the Earth's surface. Because these terrestrial ecosystems are imperiled and are the best-studied wildlife habitat, much of the concern about biodiversity and habitat fragmentation (34, 133) came from studies in forests. Forest species are vulnerable, in part, because forest area is so limited that humans can eliminate, fragment, or simplify a very large proportion of it, eliminating species found only in forests and dramatically increasing extinction probabilities for species in forest remnants. It seems reasonable to assume that the sea's extent makes it less vulnerable than forests. However, the most productive marine ecosystems, hence the ones most heavily impacted by humans—the world's continental shelves—cover 28×10^6 km^2 (120), less than the area covered by forests. The most charismatic marine ecosystems, including coral reefs, kelp forests, seagrass beds, and mangrove forests, constitute a very small portion of the sea. Coral reefs, which are considered the marine equivalents of tropical forests, occupy only 0.6×10^6 km^2, or 0.1%, of the Earth's surface (107), a small fraction of the 12×10^6 km^2 of closed tropical forests (102). Widespread concern about the prospect of millions of extinctions in forests comes from calculations of how island biogeographic considerations of forest loss would affect numbers of species that forests could sustain (75), and we thus have reviewed above how comparable calculations might affect a portion of the sea.

In the two decades since Myers (91), Lovejoy (75), and Norse & McManus (95) announced that the loss of biological diversity is occurring on a scale without precedent in the last 65 million years, most concern about elimination of genetic, species, and ecosystem diversity has focused on the terrestrial realm. It is now clear that marine ecosystems are at equal risk, and that those marine ecosystems that receive the most sustained and unrelenting pressure from human activities—estuaries, coral reefs, intertidal shores, and continental shelves and slopes—are now at very serious risk (15). As the world human population grows and world marine populations decline, relying on a vast and deep ocean to be forever resilient will result in an ocean that in the twentyfirst century will see more extinctions than in all of human history.

ACKNOWLEDGMENTS

The concept for this paper grew from a workshop, "Endangerment and Extinction in the Sea," sponsored by the Center for Marine Conservation in Washington, D.C. in March 1996. Participants were BW Bowen, JT Carlton, KB Clark, RK Cowen, P Dayton, DG Fautin, J Geller, D Hedgecock, GR Huntsman, JBC Jackson, G Mace, RC Phillips, WF Ponder, K Ralls, M Reaka-Kudla, K Roy, and EA Norse and D Crouse (organizers). We thank R Cowie, G Davis, D Hedgecock, and M Wicksten for information on selected species. For assistance in securing critical literature, we thank B Holohan, B Butler, and I Stirling. This paper is dedicated to the memory of Kerry B. Clark, who died unexpectedly in January 1999, and who was one of the very rare workers who thought and wrote about the extinction of marine invertebrates in modern times.

LITERATURE CITED

1. Abbott RT. 1974. *American Seashells*. New York: Von Nostrand Reinhold. 663 pp.
2. Deleted in proof
3. Abele LG, Walters K. 1979. Marine benthic diversity: a critique and alternative explanation. *J. Biogeog.* 6:115–26
4. Allmon WD, Rosenberg R, Portell RW, Schindler KS. 1993. Diversity of Atlantic coastal plain mollusks since the Pliocene. *Science* 260:1626–9
5. Anderson PK. 1995. Competition, predation, and the evolution and extinction of Steller's sea cow, *Hydrodamalis gigas*. *Mar. Mamm. Sci.* 11:391–94
6. Deleted in proof
7. Baillie J, Groombridge B. 1996. *1996 IUCN Red List of Threatened Animals*. Gland, Switzerland: IUCN. 368 pp.
8. Banks MA, McGoldrick DJ, Borgeson W, Hedgecock D. 1994. Gametic incompatibility and genetic divergence of Pacific and Kumamoto oysters, *Crassostrea gigas* and *C. sikamea*. *Mar. Biol.* 121:127–35
9. Barlow J, Gerrodette T, Silber G. 1997. First estimates of vaquita abundance. *Mar. Mamm. Sci.* 13:44–58
10. Barnes RSK. 1994. *The Brackish-Water Fauna of Northwestern Europe*. Cambridge, UK: Cambridge Univ. Press. 287 pp.
11. Begon M, Harper JL, Townsend CR. 1990. *Ecology: Individuals, Populations, and Communities*. Boston: Blackwell Sci. Publ. 876 pp.
12. Bigelow HB, Schroeder WC. 1953. Fishes of the Gulf of Maine. *Fish. Bull.* 53:1–577
13. Bryant D, Burke L, McManus J, Spalding M. 1998. *Reefs at risk: A map-based indicator of threats to the world's coral reefs*. World Resource Inst. *Tech. Rep.* Washington, DC. 56 pp.
14. Butler M. 1998. Deep sea corals off Nova Scotia. In *Effects of Fishing Gear on the Sea Floor of New England*, ed. EM Dorsey, J Pederson. pp. 129–131. Boston: Conservation Law Found.
15. Butman CA, Carlton JT, eds. 1995. *Understanding Marine Biodiversity: A Research Agenda for the Nation*. Washington, DC: Nat. Acad. Press. 114 pp.
16. Campbell RR. 1988. Status of the sea mink, *Mustela macrodon*, in Canada. *Can. Field-Nat.* 102:304–6
17. Carlton JT. 1975. Extinct and endangered populations of the endemic mudsnail *Cerithidea californica* in northern Califor-

nia. *Bull. Am. Malacol. Union* 1975:65–66
18. Carlton JT. 1979. *History, biogeography, and ecology of the introduced marine and estuarine invertebrates of the Pacific coast of North America*. Ph.D. diss. Univ. California, Davis
19. Carlton JT. 1993. Neoextinctions of marine invertebrates. *Am. Zool.* 33:499–509
20. Carlton JT. 1998. Apostrophe to the ocean. *Conserv. Biol.* 12:1165–67
21. Carlton JT, Vermeij GJ, Lindberg DR, Carlton DA, Dudley EC. 1991. The first historical extinction of a marine invertebrate in an ocean basin: the demise of the eelgrass limpet *Lottia alveus*. *Bio. Bull.* 180:72–80
22. Casey JM, Myers RA. 1998. Near extinction of a large, widely distributed fish. *Science* 281:690–92
23. Clark KB. 1994. Ascoglossan (=Sacoglossa) molluscs in the Florida Keys: rare marine invertebrates at special risk. *Bull. Mar. Sci.* 54:900–16
24. Clark KB. 1995. Rheophilic/oligotrophic lagoonal communities through the eyes of slugs (Mollusca: Opisthobranchia). *Bull. Mar. Sci.* 57:242–51
25. Collar NJ, Crosby MJ, Stattersfield AJ. 1995. *Birds to Watch 2: The World List of Threatened Birds*. Birdlife Conservation, No. 4. Washington, DC: Smithsonian Inst. Press. 407 pp.
26. Deleted in proof
27. Deleted in proof
28. Crane J. 1975. *Fiddler Crabs of the World. Ocypodidae: Genus Uca*. Princeton NJ: Princeton Univ. Press. 736 pp.
29. Coustau C, Renaud F, Maillard C, Pasteur N, Delay B. 1991. Differential susceptibility to a trematode parasite among genotypes of the *Mytilus edulis/ galloprovincialis* complex. *Genetic. Res.* 57:207–12
30. Darlington PJ. 1957. *Zoogeography: the Geographical Distribution of Animals*. New York: Wiley. 675 pp.
31. Davis GE, Haaker PL, Richards DV. 1996. Status and trends of white abalone at the California Channel Islands. *Trans. Am. Fish. Soc.* 125:42–48
32. Day D. 1981. *The Doomsday Book of Animals. A Natural History of Vanished Species*. New York: Wiley. 288 pp.
33. Dayton PK, Thrush SF, Agardy MT, Hofman RJ. 1995. Environmental effects of fishing. *Aquatic Conserv. Mar. Freshwat. Ecosyst.* 5:205–32
34. Diamond JM. 1975. The island dilemma: lessons of modern biogeographic studies for the design of natural preserves. *Biol. Conserv.* 7:129–46
35. Deleted in proof
36. Edmondson CH. 1962. Hawaiian Crustacea: Goneplacidae, Pinnotheridae, Cymopoliidae, Ocypodidae, and Gecarcinidae. *Occas. Pap. B. P. Bishop Mus.* 33:1–27
37. Ehrlich PR, Dobkin DS, Wheye D. 1992. *Birds in Jeopardy*. Stanford, CA: Stanford Univ. Press. 259 pp.
38. Ehrlich PR, Wilson EO. 1991. Biodiversity studies: science and policy. *Science* 253:758–62
39. Eldredge N. 1991. *The Miner's Canary. Unraveling the Mysteries of Extinction*. Princeton NJ: Princeton Univ. Press. 246 pp.
40. Fuller E. 1987. *Extinct Birds*. London: Viking/Rainbird. 256 pp.
41. Fuller S, Cameron P. 1998. Marine benthic seascapes: fishermen's perspectives. *Marine Iss. Com. Special Publ. 3*, Ecology Action Centre, Halifax, Nova Scotia
42. Gardner JPA. 1994. The structure and dynamics of naturally occurring hybrid *Mytilus edulis* Linnaeus, 1758 and *Mytilus galloprovincialis* Lamarck, 1819 (Bivalvia: Mollusca) populations: review and interpretation. *Hydrobiologia* (suppl.) 99:37–71
43. Geller JB. 1998. Molecular studies of marine invertebrate biodiversity: status and prospects. In *Molecular Approaches to the Study of the Ocean*, ed. KE Cooksey, pp. 359–376. London: Chapman & Hall

44. Geller JB. 1999. Decline of a native mussel masked by sibling species invasion. *Conserv. Biol.* 13:661–664
45. Gibson-Smith J, Gibson-Smith W. 1981. The status of *Pholadomya candida* G. B. Sowerby, I, 1823. *Veliger* 23:355–56
46. Deleted in proof
47. Glynn PW, de Weerdt WH. 1991. Elimination of two reef-building hydrocorals following the 1982–83 El Nino warming event. *Science* 253:69–71
48. Glynn PW, Feingold JS. 1992. Hydrocoral species not extinct. *Science* 257:1845
49. Greenway JC. 1967. *Extinct and Vanishing Birds of the World.* 2nd rev. ed. New York: Dover. 520 pp.
50. Groombridge B. 1993. *1994 IUCN Red List of Threatened Animals.* Gland, Switzerland: IUCN. 286 pp.
51. Harger JRE. 1968. The role of behavioral traits in influencing the distribution of two species of sea mussel, *Mytilus edulis* and *Mytilus californianus. Veliger* 11:45–49
52. Hayman P, Marchant J, Prater T. 1986. *Shorebirds. An Identification Guide to the Waders of the World.* Boston: Houghton Mifflin. 412 pp.
52a. Hedgecock D, Li G, Banks MA, Kain Z. 1999. Occurrence of the Kumamoto oyster *Crassostrea sikamea* in the Ariake Sea, Japan. *Mar. Biol.* 133:65–68
53. Hiscock K. 1997. Conserving biodiversity in North-East Atlantic marine ecosystems. In *Marine Biodiversity. Patterns and Processes*, ed. RFG Ormond, JD Gage, MV Angel, pp. 415–27. Cambridge: Cambridge Univ. Press. 449 pp.
54. Hockey PAR, van Erkom Schurink C. 1992. The invasive biology of the mussel *Mytilus galloprovincialis* on the southern African coast. *Trans. R. Soc. S. Africa* 48:123–40
55. Hornaday WT. 1913. *Our Vanishing Wild Life. Its Extinction and Preservation.* New York: New York Zool. Soc. 411 pp.
56. Hudson E. 1996. Red listing marine fishes. *Species (Newsletter of the Species Survival Commission, IUCN, The World Conservation Union).* Nos. 26–7:31–32
57. Jablonski D. 1991. Extinctions: a paleontological perspective. *Science* 253:754–62
58. Jablonski D. 1994. Extinctions in the fossil record. *Philos. Trans. R. Soc. Lond.* B, 344:11–17
59. Deleted in proof
60. Johnson ME, Snook HJ. 1927. *Seashore Animals of the Pacific coast.* New York: Macmillan. 659 pp.
61. Deleted in proof
62. Deleted in proof
63. Deleted in proof
64. Keep J. 1935. *West Coast Shells.* Stanford CA: Stanford Univ. Press. 350 pp.
65. Kleypas JA. 1997. Modeled estimates of global reef habitat and carbonate production since the last glacial maximum. *Paleooceanography* 12:533–45
66. Knowlton N. 1993. Sibling species in the sea. *Annu. Rev. Ecol. Syst.* 24:189–216
67. Knudston PM. 1977. The case of the missing monk seal. *Nat. Hist.* 86:78–83
68. Koehn RK. 1991. The genetics and taxonomy of species in the genus *Mytilus. Aquaculture* 94:125–145
69. Lane JA. 1992. *Sea otters.* Golden, C. Fulcrum. 148 pp.
70. Laws RM. 1985. The ecology of the southern ocean. *Am. Sci.* 73:26–40
71. Leakey R, Lewin R. 1995. *The Sixth Extinction. Biodiversity and its Survival.* London: Weidenfeld & Nicolson. 271 pp.
72. LeBoeuf BJ, Kenyon KW, Villa-Ramirez B. 1986. The Caribbean monk seal is extinct. *Mar. Mamm. Sci.* 2:70–72
73. Leung Y-M. 1967. An illustrated key to the species of whale-lice (Amphipoda, Cyamidae), ectoparasites of Cetacea, with a guide to the literature. *Crustaceana* 12:279–91
74. Lindeboom HJ, de Groot SJ. eds. 1998. Impact II. The effects of different types of fisheries on North Sea and Irish Sea

benthic ecosystems. Texel, Netherlands: Netherlands Inst. Sea Res. (NIOZ)
75. Lovejoy T. 1980. Changes in biological diversity. In *The Global 2000 Report to the President, Vol. 2, The Technical Report. 327–332.* Washington, DC: Council on Environ. Qual. and US Dep. State
76. MacArthur RH, Wilson EO. 1967. *The Theory of Island Biogeography. Monographs in Population Biology.* Princeton NJ: Princeton Univ. Press. 203 pp.
77. Deleted in proof
78. MacGinitie GE, MacGinitie N. 1949. *Natural History of Marine Animals.* New York: McGraw-Hill. 473 pp.
79. Marx W. 1967. *The Frail Ocean.* New York: Ballantine Books. 274 pp.
80. Mastro E, Chow V, Hedgecock D. 1982. *Littorina scutulata* and *L. plena*: sibling species status of two prosobranch gastropod species confirmed by electrophoresis. *Veliger* 24:239–46
81. Matthiessen P. 1973. *The Wind Birds.* New York: Viking. 160 pp.
82. May M. 1993. Equations for extinction. *Am. Sci.* 81:331
83. May RM. 1975. Patterns of species abundance and diversity. In *Ecology and Evolution of Communities*, ed. ML Cody, JM Diamond, 81–120. Cambridge, MA: Harvard Univ. Press
84. McDonald JH, Koehn RT. 1988. The mussels *Mytilus galloprovincialis* and *M. trossulus* on the Pacific coast of North America. *Mar. Biol.* 99:111–18
85. McKinney ML. 1997. Extinction vulnerability and selectivity: combining ecological and paleontological views. *Annu. Rev. Ecol. Syst.* 28:495–516
86. McLean JH. 1978. *Marine Shells of Southern California.* Los Angeles County Mus. of Nat. Hist., Sci. Ser. 24, Rev. Ed. Los Angeles, CA. 104 pp.
87. Mead JG, Mitchell. 1984. Atlantic gray whales. In *The Gray Whale.* 33–53. New York: Academic
88. Merrett NR, Haedrich RL. 1997. *Deep-sea Demersal Fish and Fisheries.* London: Chapman & Hall. 282 pp.
89. Mikkelsen PM, Mikkelsen PS, Karlen DJ. 1995. Molluscan biodiversity in the Indian River Lagoon, Florida. *Bull. Mar. Sci.* 57:94–127
90. Moritz C. 1994. Applications of mitochondrial DNA analysis in conservation: a critical review. *Mol. Ecol.* 3:401–11
91. Myers N. 1979: *The Sinking Ark: A New Look at the Problem of Disappearing Species.* New York: Pergamon. 307 pp.
92. Nicol S and de la Mare W. 1993. Ecosystem management and the Antarctic krill. *Am. Sci.* 81:36–47
93. Norse EA ed. 1993. *Global Marine Biological Diversity: A Strategy for Building Conservation into Decision Making.* Washington, DC: Island. 383 pp.
94. Norse EA. 1994. Capsizing the cradle of life. *Global Biodiversity* 4:4–7
95. Norse EA, McManus RE. 1980. Ecology and living resources: biological diversity. In *Environmental Quality 1980: The Eleventh Annual Report of the Council on Environmental Quality.* 31–80. Washington, DC: Council on Environ. Qual.
96. O'Clair RM, O'Clair CE. 1998. *Southeast Alaska's Rocky Shores. Animals.* Auke Bay, Alaska: Plant Press. 564 pp.
97. Palmer AR, Gayron SD, Woodruff DS. 1990. Reproductive, morphological, and genetic evidence for two cryptic species of northeastern Pacific *Nucella. Veliger* 33:325–38
98. Palumbi SR. 1992. Marine speciation on a small planet. *Trends Ecol. Evol.* 7:114–18
99. Pauly D, Christensen V, Dalsgaard J, Froese R, Torres F. 1998. Fishing down marine food webs. *Science* 279:860–63
100. Deleted in proof
101. Deleted in proof
102. Perry DA. 1994. *Forest Ecosystems.* Baltimore: Johns Hopkins Univ. Press
103. Petryk M. 1996. Evolutionarily significant units and marine invertebrate extinctions in the California mudsnail *Cerithidea cal-*

ifornica. MS Thesis. Univ. North Carolina at Wilmington. 43 pp.

104. Race MS. 1981. Field ecology and natural history of *Cerithidea:californica* (Gastropoda: Prosobranchia) in San Francisco Bay. *Veliger* 24:181–27
105. Race MS. 1982. Competitive displacement and predation between introduced and native mud snails. *Oecologia* 54:337–47
106. Raup DM. 1991. *Extinction. Bad Genes or Bad Luck?* New York: Norton. 210 pp.
107. Reaka-Kudla ML. 1996. The global biodiversity of coral reefs: a comparison with rain forests. In *Biodiversity II: Understanding and Protecting Our Natural Resources*, ed. ML Reaka-Kudla, DE Wilson, EO Wilson, pp. 83–108. Washington, DC: Joseph Henry/Natl. Acad. Press
108. Deleted in proof
109. Rhymer JM, Simberloff D. 1996. Extinction by hybridization and introgression. *Annu. Rev. Ecol. Syst.* 27:83–109
110. Rice DW. 1977. A list of the marine mammals of the world. *NOAA Tech. Rep. NMFS* SSRF-711, 15 pp.
111. Ricketts EF Jr, Calvin J. 1939. *Between Pacific Tides*. Stanford CA: Stanford Univ. Press
112. Risk MJ, McAllister DE, Behnken L. 1998. Conservation of cold- and warm-water seafans: threatened ancient gorgonian groves. *Sea Wind* 12:2–21
113. Runnegar B. 1979. *Pholadomya candida* Sowerby: the last cadaver unearthed. *Veliger* 22:171–72
114. Ryther J. 1969. Photosynthesis and fish production in the sea. *Science* 166:72–76
115. Safina C. 1995. The world's imperiled fish. *Sci. Am.* 273:46–53
116. Safina C. 1998. *Song for the Blue Ocean*. New York: Henry Holt. 458 pp.
117. Sarver SK, Foltz DW. 1993. Genetic population structure of a species' complex of blue mussels (*Mytilus* spp.) *Mar. Biol.* 117:105–12
118. Scheffer VB. 1973. The last days of the sea cow. *Smithsonian* [Jan. 1973]: 64–67
119. Schopf TJM. 1974. Permo-Triassic extinctions: relation to sea-floor spreading. *J. Geol.* 82:129–43
120. Sharp GD. 1988. Fish populations and fisheries: their perturbations, natural and man-induced. In *Continental shelves. Ecosystems of the world, Vol. 27*, ed. H Postma, JJ Zijlstra, pp. 155–202. Amsterdam: Elsevier Sci.
121. Silverberg R. 1967. *The Dodo, the Auk and the Oryx. Vanished and Vanishing Creatures*. London: Puffin Books. 232 pp.
122. Simberloff DS. 1974. Equilibrium theory of island biogeography and ecology. *Annu. Rev. Ecol. Syst.* 5:161–82
123. Simberloff DS. 1976. Experimental zoogeography of islands: effects of island size. *Ecology* 57:629–48
124. Skibinski DOF, Roderick EE. 1991. Evidence of selective mortality in favour of the *Mytilus galloprovincialis* Lmk phenotype in British mussel populations. *Biol. J. Linn. Soc.* 42:351–66
125. Deleted in proof
126. Solow AR. 1993. Inferring extinction from sighting data. *Ecology* 74:962–64
127. Spalding M, Grenfell AM. 1997. New estimates of global and regional coral reef areas. *Coral Reefs* 16:225–30
128. Stanley SM, Campbell LD. 1981. Neogene mass extinction of Western Atlantic molluscs. *Nature* 293:457–59
129. Suchanek T, Geller JB, Kreiser B, Mitton JB. 1997. Zoogeographic distributions of the sibling species *Mytilus galloprovincialis* and *M. trossulus* (Bivalvia: Mytilidae) and their hybrids in the North Pacific. *Bio. Bull.* 193:87–94
130. Swennen C, Dekker R. 1995. *Corambe batava* Kerbert, 1886 (Gastropoda: Opisthobranchia), an immigrant in the Ne-

therlands, with a revision of the family Corambidae. *J. Moll. Stud.* 61:97–107
131. Taylor DW. 1981. Freshwater mollusks of California: a distributional checklist. *Cal. Fish Game* 67:140–63
132. Tegner MJ, Basch LV, Dayton PK. 1996. Near extinction of an exploited marine invertebrate. Trends in Ecology and Evolution 11:278–280
133. Terborgh J. 1974. Preservation of natural diversity: the problem of extinction-prone species. *Bioscience.* 24:715–22
134. Tutschulte TC. 1976. *The comparative ecology of three sympatric abalones.* Ph.D. dissertation, Univ. Calif., San Diego.
135. Vermeij GJ. 1987. *Evolution and Escalation. An Ecological History of Life.* Princeton, NJ: Princeton Univ. Press. 527 pp.
136. Vermeij GJ. 1989. Geographical restriction as a guide to the causes of extinction: the case of the cold northern oceans during the Neogene. *Paleobiology* 15:335–56
137. Vermeij GJ. 1991. Marine extinctions and their implications for conservation and biogeography. In Proc. Fourth Int. Congr. Syst. Evol. Biol. 143–148. *The Unity of Evolutionary Biology*, ed. EC Dudley, Portland OR: Dioscorides Press. 588 pp.
138. Vermeij GJ. 1993. Biogeography of recently extinct marine species: implications for conservation. *Conserv. Biol.* 7: 391–397
139. Vogler AP, DeSalle R. 1994. Diagnosing units of conservation management. *Conserv. Bio.* 8:354–63
140. Waples RS. 1995. Evolutionarily significant units and the conservation of biological diversity under the Endangered Species Act. *Am. Fish. Soc. Symp.* 17:8–27
141. Waters JH, Ray CE. 1961. Former range of the sea mink. *J. Mamm.* 42:380–83
142. Watling L, Norse EA. 1998. Disturbance of the seabed by mobile fishing gear: a comparison with forest clearcutting. *Conserv. Biol.* 12:1180–97
143. Wells SM, Pyle RM, Collins NM. 1983. *The IUCN Invertebrate Red Data Book.* Gland, Switzerland: IUCN. 632 pp.
144. Wilkinson CR. 1992. Coral reefs of the world are facing widespread devastation: Can we prevent this through sustainable management practices? *Proc. 7th Int. Coral Reef Symp.* 1:11–24
145. Williamson M. 1988. Relationship of species number to area, distance, and other variables. In *Analytical Biogeography*, ed. AA Myers, P.S. Gillers, chapter 4. London: Chapman & Hall
146. Willis GL, Skibinski DOF. 1992. Variation in strength of attachment to the substrate explains differential mortality in hybrid mussel (*Mytilus galloprovincialis* and *M. edulis*) populations. *Mar. Biol.* 112:403–8
147. Wilson EO. 1992. *The Diversity of Life.* Cambridge, MA: Harv. Univ. Press. 424 pp.
148. Wright SJ. 1981. Intra-archipelago vertebrate distributions: the slope of the species-area relation. *Am. Nat.* 118:726–48
149. Hedgecock D, Li G, Banks MA, Kain Z. 1999. Occurrence of the Kumamoto oyster *Crassostrea sikamea* in the Ariake Sea, Japan. *Mar. Biol.* 133:65–68

Annu. Rev. Ecol. Syst. 1999. 30:539–63

GENE FLOW AND INTROGRESSION FROM DOMESTICATED PLANTS INTO THEIR WILD RELATIVES

Norman C. Ellstrand[1], Honor C. Prentice[2], and James F. Hancock[3]

[1]*Department of Botany & Plant Sciences and Center for Conservation Biology, University of California, Riverside, California 92521-0124; e-mail: ellstrand@ucrac1.ucr.edu;* [2]*Department of Systematic Botany, Lund University. Ö. Vallgatan 14-20, Lund S-223 61, Sweden; e-mail: Honor.Prentice@sysbot.lu.se;* [3]*Department of Horticulture, Michigan State University, East Lansing, Michigan 48894; e-mail: hancock@pilot.msu.edu*

Key Words hybridization, introgression, plant conservation genetics, risks of transgenic crops, weed evolution

■ **Abstract** Domesticated plant taxa cannot be regarded as evolutionarily discrete from their wild relatives. Most domesticated plant taxa mate with wild relatives somewhere in the world, and gene flow from crop taxa may have a substantial impact on the evolution of wild populations. In a literature review of the world's 13 most important food crops, we show that 12 of these crops hybridize with wild relatives in some part of their agricultural distribution. We use population genetic theory to predict the evolutionary consequences of gene flow from crops to wild plants and discuss two applied consequences of crop-to-wild gene flow–the evolution of aggressive weeds and the extinction of rare species. We suggest ways of assessing the likelihood of hybridization, introgression, and the potential for undesirable gene flow from crops into weeds or rare species.

INTRODUCTION

Planting, growing, and harvesting plants for food, fiber, fodder, pharmaceuticals, and many other uses have profound effects on the ecology and evolution of organisms in the surrounding habitats. Some of these effects may be direct and obvious. For example, tilling affects the composition of the local plant and microbial communities; likewise, replacing pre-existing vegetation with sunflower fields alters nectar and pollen sources for local insect communities. Evolutionary effects may be less obvious and less immediate. For example, weeds may evolve crop mimicry under directional selection from constant mechanical weed control (14), and the continued use of herbicides may lead to evolution of herbicide resistance (63).

0066-4162/99/1120-0539$08.00

More subtle, but no less important, are evolutionary effects that arise from spontaneous mating of domesticated plants (those that have evolved under human selection and management) with their wild relatives. Such hybridization may lead to gene flow–"the incorporation of genes into the gene pool of one population from one or more populations" (52). If new or locally rare alleles from the domesticate persist in wild populations, gene flow may lead to significant evolutionary change in the recipient populations (e.g. 5). Crop-to-weed gene flow will have important practical and economic consequences if it promotes the evolution of more aggressive weeds (e.g. 5, 14). Hybridization with domesticated species has also been implicated in the extinction of certain wild crop relatives (e.g. 46, 112). Over the last decade, much attention has been focused on crop-to-weed hybridization as a potential avenue for the escape of crop transgenes into natural populations (e.g. 31, 32, 56). Although that narrow issue has been the topic of numerous theoretical, empirical, and synthetic publications, the general issue of the consequences of the flow of alleles from domesticated plants (whether genetically engineered or not) to their wild relatives has received scant attention (but see 41, 112). Indeed, a few scientists have questioned the validity of reports of spontaneous hybridization between crops and their wild relatives, asserting that "evidence for ... hybridization between crops and their wild relatives has often been only circumstantial" (89). Our review addresses the general question of gene flow from domesticated plants to their wild relatives; we find that evidence for this phenomenon is typically much more than "circumstantial." Instead, cases of gene flow from crops to their wild relatives provide examples of contemporary micro-evolution that also have important applied implications.

Our review addresses the following questions: "What do we know about spontaneous hybridization between domesticated plants and their wild relatives?" and "How and when does gene flow from domesticated plants to their wild relatives play a role in the evolution of wild populations?" Our specific focus is the flow of domesticated alleles into the populations of their wild relatives. The pollen parent in the initial cross may involve a plant in cultivation, a volunteer left from a previous planting or from seed spillage into a natural population, or a recent escape from cultivation. An alternate pathway may involve a wild plant that acts as the pollen parent in a cross with a plant in cultivation, giving rise to a hybrid that spontaneously backcrosses with wild plants.

While we recognize that gene flow in the reverse direction (into the crop) may have important evolutionary consequences for crop taxa, that topic is beyond the scope of our review. Also beyond our scope is the topic of artificial hybridization–the production of hybrids by hand-crossing or under laboratory conditions. Finally, although we discuss a few cases where hybridization has resulted in the evolution of new taxa, we do not address the evolutionary origins of wild crop relatives per se. We recognize that wild crop relatives may be the descendants of feral crops, plants with both wild and domesticated ancestors, or plants with no domesticated ancestors in their genealogy. We define "wild" plants as those that grow and reproduce without being deliberately planted. We define "weeds" as wild plants that interfere with human objectives.

First, we present an overview of gene flow by hybridization in plants. Next, we provide some examples of spontaneous hybridization between the world's most important crops and their wild relatives, and we use these to illustrate the generality of the phenomenon. We then examine the theoretical consequences of gene flow. We describe two applied implications–the evolution of aggressive weeds and the extinction of vulnerable species. Finally, we suggest ways of assessing the likelihood of hybridization, introgression, and the potential for undesirable gene flow from crops into weeds or rare species.

GENE FLOW BY HYBRIDIZATION IN PLANTS

"Hybridization is a frequent and important component of plant evolution and speciation" (101). More than 70% of plant species may be descended from hybrids (54). Natural interspecific and, in certain families, even intergeneric hybridization is not rare; well-studied examples number well over 1000 (10, 54). Nonetheless, even if hybridization is common, it is not ubiquitous. The incidence of natural hybridization varies substantially among plant genera and families (47). In certain cases, two taxa naturally form hybrid swarms in some areas of contact but not in others (e.g. 87).

Hybridization in plants depends on a variety of factors (54). Cross-pollination must occur. For this to happen, both taxa must be in flower at the same time. The plants must be close enough in space to allow a vector to carry pollen between them. They must be cross-compatible so that the pollen is able to germinate and effect fertilization. If the resulting embryos develop into viable seeds and germinate, the F_1 plants typically have some reduced fertility but are rarely fully sterile (cf. 116). In fact, many spontaneous plant hybrids do not suffer reduced fitness, and in certain cases, they do exhibit increased fitness relative to either or both of the parental taxa (10, 59, 71).

The frequent occurrence of fertile hybrids increases the chances of introgression–the incorporation of alleles from one taxon into another (adapted from 99). Note that this definition is essentially equivalent to that of gene flow at the inter-taxon level. Studies employing allozymes and DNA-based genetic markers have revealed dozens of instances of natural introgression in plants (101). In many cases, morphological intermediacy and molecular confirmation of introgression go hand in hand. But, in other cases, one or a few introgressed genetic markers may be found in otherwise morphologically pure individuals, even far beyond the morphologically defined limits of a hybrid zone of contact (cf. 102, 106).

GENE FLOW FROM DOMESTICATED PLANTS TO THEIR WILD RELATIVES

The hundreds of well-studied cases of natural hybridization and introgression involving wild plants suggest that most domesticated plants will hybridize naturally with their cross-compatible wild relatives when they come into contact. A growing number of both experimental and descriptive studies, using genetically based markers, have demonstrated that domesticated alleles can and do enter and persist in natural populations. The domesticated species involved are amazingly diverse, ranging from mushrooms (134) and raspberries (82) to ornamental shrubs (107) and forage crops (examples in 116). The accumulating evidence suggests these examples are probably the rule rather than the exception.

The risk of hybridization as an avenue for the escape of crop transgenes has stimulated researchers to evaluate opportunities for spontaneous gene flow from major cultivated species to wild plants. Some reviews have been regional in scope, addressing a country's major domesticated plants and their wild taxa in the local flora. For example, for the United Kingdom, about one third of the 31 domesticated species reviewed spontaneously hybridize with one or more elements of the local flora (96); for the Netherlands, that fraction was about one quarter of the 42 reviewed species (38). These proportions are surprisingly high, given that (*a*) the vast majority of the cultivated species of the United Kingdom and the Netherlands were domesticated elsewhere and (*b*) both countries are small in relation to the world-wide area covered by most of the cultivated species reviewed.

Other reviews are taxon-specific in scope, addressing whether a crop is known to hybridize with wild plants over all or part of its range. Examples of crops examined by such reviews include rapeseed (108), oat (25), and potato (81). These reviews typically report evidence for spontaneous mating between the crop in question and some subset of its wild relatives.

Rather than focusing on a single region or a single crop, we present the results of a literature review of a suite of different crops and ask whether these crops hybridize with wild taxa in any part of their range of cultivation. These case studies provide examples of the degree and extent of spontaneous hybridization between crops and their wild relatives. Our sample also illustrates the generality of the phenomenon; all but one of the reviewed crops hybridize naturally with wild relatives in some part of their agricultural distribution.

These results make sense in an evolutionary context. Domesticated plants represent lineages that diverged from their progenitors no more than a few thousand generations ago. There is no reason to assume that reproductive isolation should be absolute. Whether the evidence is reviewed on a regional basis or a crop-by-crop basis, it is clear that spontaneous hybridization and introgression of genes from domesticated plants to wild relatives is a common characteristic of domesticated plant taxa.

Case Studies: The World's 13 Most Important Food Crops

We chose the 13 most important crops (in terms of area harvested) grown for human consumption (including oil crops) for our case studies because their impact on human well-being makes them likely to be among the best-studied domesticated plants. We used the FAOSTAT Statistics Database (http://apps.fao.org/) to obtain the most recent (1997) estimates of area harvested for each crop. Some crops comprise more than one species; we reviewed as many species as were necessary to account for a substantial majority of the area harvested (Table 1). The number of crops reviewed was limited by considerations of time and space. The case studies represent a relatively heterogeneous group of 18 tropical, subtropical, and temperate species belonging to five different families.

We used the AGRICOLA bibliographic database as a point of departure for our literature search for information on whether the reviewed crops naturally hybridize with wild species. After compiling our conclusions for each crop, we consulted one or more experts (see Acknowledgments) to pre-review the conclusions, to suggest other experts, and to identify gaps in our treatment. We, however, take final responsibility for interpretation of the data.

For each crop below, we start with its name, its rank, and the name(s) of the taxa involved, following the nomenclature of Smartt & Simmonds (114). We then present evidence for natural hybridization and introgression (summarized in Table 1). In some cases, the only evidence for hybridization is the presence of morphologically intermediate plants in localities where the crop and wild relative are sympatric. Such inferred evidence of hybridization is relatively weak, because morphological intermediacy may result from phenotypic plasticity or from convergent evolution for crop mimicry rather than from gene flow (89). In many cases, however, sympatric intermediates have been genetically evaluated and compared with reference populations (putatively pure allopatric crop and wild populations). The presence of crop-specific alleles in morphologically intermediate populations provides strong evidence for a history of hybridization. Another approach is to examine experimentally whether a crop will mate spontaneously with a wild relative under field conditions. Progeny testing of seed set by wild plants grown adjacent to plantations of their crop relative can be used to identify whether the crop is the paternal father. Gene flow rates can be obtained directly by this method.

Wheat (Rank: No. 1) *Triticum aestivum* (L.) Thell. (bread wheat) and *T. turgidum* ssp. *turgidum* (L.) Thell. (durum wheat) (Poaceae)

Spontaneous intermediates between cultivated wheats and their wild relatives occur frequently on margins of wheat fields when wild *T. turgidum* subspecies or certain species of *Aegilops* are present (e.g. 70, 74, 95, 124). Hybrids are reported from the Middle East, Africa,Europe, Asia, and North America. At least 21 different natural hybrids, arising from 12 *Aegilops* species and involving both bread and durum wheat, have been reported (124). In addition to their morphological intermediacy, the presumed hybridity of these plants is supported by their sterility. Although breeders have produced fertile hybrids between wheat and its wild relatives (48, 64), "all natural hybrids... are highly sterile, although seeds may occasionally be found" (95, 124). This hybrid sterility may explain why hybridization generally appears to be restricted to F_1s with little evidence for subsequent introgression.

However, introgression from cultivated durum into wild emmer, *T. dicoccoides*, has been implicated in the evolution of a distinct race of wild emmer in the Upper

TABLE 1 Spontaneous hybridization between the world's most important food crops and their wild relatives

Rank	Crop	Scientific name	Kilohectares[a]	Evidence[b] for hybridization	Implicated in weed evolution	Implicated in extinction risk
1	Wheat	*Triticum aestivum*[c]	228131	+	no	no
		T. turgidum[c]		+	yes	no
2	Rice	*Oryza sativa*	149555	+	yes	yes
		O. glaberrima		+	no	no
3	Maize, including sweet and field corn	*Zea mays mays*	143633	+	no	no
4	Soybean	*Glycine max*	67450	+	yes	no
5	Barley	*Hordeum vulgare*	65310	m	no	no
6	Cotton seed	*Gossypium hirsutum*	51290	+	no	yes
		G. barbadense		+	no	no
7	Sorghum	*Sorghum bicolor*	45249	+	yes	no
8	Millet	*Eleusine coracana*[c]	38077	m	no	no
		Pennisetum glaucum[c]		+	yes	no
9	Beans, dry, green, and string	*Phaseolus vulgaris*[c]	28671	+	yes	no
10	Rapeseed	*Brassica napus*	24044	+	no	no
		B. rapa		+	no	no
11	Groundnut	*Arachis hypogaea*	23647	none	n/a	n/a
12	Sunflower seed	*Helianthus annuus*	19628	+	yes	no
13	Sugar cane	*Saccharum officinarum*[c]	19619	+	no	no

[a]estimated area of production for 1997 from the 1998 FAOSTAT website, 6/15/98.

[b]m, morphological intermediacy only; +, more substantial evidence for hybridization.

[c]Other taxa account for only a small portion of world production of this crop.

Jordan Valley. That race of wild emmer is native to a region with a history of durum cultivation. At present, wild emmer in the Upper Jordan Valley has a number of morphological, physiological, and isozyme traits that are common in durum and rare or absent in other wild emmer populations (19).

Rice (Rank: No. 2) *Oryza sativa* L. and *O. glaberrima* Steud. (Poaceae)
Both cultivated rice species are interfertile with certain close wild relatives; hand crosses are easily accomplished (e.g. 30, 88). F_1 fitnesses from crosses between wild and cultivated rice are generally high, but hybrids from certain crosses show reduced fertility (e.g. 29). In contrast, when the two cultivated species are intercrossed, the hybrids are highly sterile (26).

Spontaneous intermediates between cultivated rice species and their wild relatives occur frequently in and near rice fields when wild taxa are present. In the case of *O. glaberrima*, which is cultivated in west Africa, the sympatric wild relatives are weedy forms of the same species as well as *O. barthii* and *O. longistamina*. For the more widespread *O. sativa*, they are weedy conspecifics (*O. sativa* f. *spontanea*), as well as *O. nivara* and *O. rufipogon*. The intermediate plants usually appear as hybrid swarms. In the case of *O. sativa*, the intermediates may sometimes be relatively uniform and behave as stabilized races. Genetic analysis of the wild intermediates has been conducted using isozymes, RAPDs, and progeny segregation studies of morphological and physiological traits. Intermediate plants have a combination of alleles specific to both the pure crop and the pure wild taxon (e.g. 30, 90, 92, 118, 120), supporting the hypothesis that the intermediates are products of hybridization or introgression.

Natural rates of hybridization can be substantial. Allozyme progeny analysis of experimental mixed stands of cultivated *O. sativa* and the weed, red rice (*O. sativa* f. *spontanea*), in Louisiana revealed rates of natural hybridization ranging from 1% to 52%, depending on the cultivar acting as pollen parent (75). Hybridization rate tended to increase with phenological overlap. The hybrids demonstrated heterosis; they were generally taller and produced more tillers than either parental type (75).

A notable example of crop-to-wild gene flow involving rice occurred during the present century. To facilitate weeding, a red-pigmented rice cultivar was planted in India to distinguish its seedlings from those of the unpigmented weed. The strategy was thwarted by introgression. After a few seasons of gene flow and selection, the weed populations had accumulated the pigmentation allele at a high frequency (91).

Natural hybridization with cultivated rice has been implicated in the near extinction of the endemic Taiwanese taxon, *O. rufipogon* ssp. *formosana*. Collections of this wild rice over the last century show a progressive shift toward characters of the cultivated species and a coincidental decrease in fertility of seed and pollen (69). Indeed, throughout Asia, typical specimens of other subspecies of *O. rufipogon* and the wild *O. nivara* are now rarely found because of extensive hybridization with the crop (26).

Maize (Rank: No. 3) *Zea mays* ssp. *mays* L. (Poaceae)

Plants morphologically intermediate between maize and teosintes (various wild *Zea*) often occur spontaneously in and near Mexican maize fields when teosinte, particularly *Z.m.* ssp. *mexicana*, is abundant (132). At present, the genetic data do not offer a clear view of the extent of hybridization and introgression that have occurred. Allozyme analysis of accessions of the teosintes *Z. luxurians, Z. diploperennis*, and *Z.m.* ssp. *mexicana* revealed that alleles that are otherwise maize-specific occurred at extremely low frequencies, suggesting a very low level of introgression from maize into these teosinte taxa (42).

However, cytogenetic analyses "offer no evidence of... maize-teosinte introgression in either direction" (67). Allozyme comparisons of teosinte populations (involving *Z. diploperennis, Z. m.* ssp. *huehuetenangensis*, and *Z.m.* ssp. *mexicana*) that are both allopatric and sympatric with maize showed no evidence of introgression from the crop into the wild taxa (42). Such comparisons are not available for *Z. m.* ssp. *parviglumis* for which no crop-specific allozymes are available. To our knowledge, no one has genetically analyzed spontaneous, morphologically intermediate plants to test for hybridity in the same way as in the thorough rice studies discussed above. Such analyses would identify whether the maize-teosinte intermediates are true hybrids, introgressants, or crop mimics.

Strong reproductive barriers exist between maize and one teosinte taxon, *Z. perennis* (42). No evidence suggests that natural hybridization occurs between these taxa.

Soybean (Rank: No. 4) *Glycine max* (L.) Merr. (Fabaceae)

Wild plants morphologically intermediate to *G. max* and its wild relative *G. soja* often occur spontaneously near Chinese and Korean soybean fields when *G. soja* is present (72; R Palmer, personal communication). An apparently stabilized hybrid taxon from northeastern China, *G. gracilis* (= *G. max* forma *gracilis*) is intermediate to and interfertile with both *G. max* and *G. soja*. *Glycine gracilis* accessions in germplasm collections have been genetically characterized with morphological, chemical, and RLFP makers; all studies have shown the accessions have a mixture of alleles diagnostic for both putative parents (20, 21, 68). However, population-level analysis has not yet been conducted to determine the extent to which crop alleles have moved into natural populations. It is not yet clear whether *G. gracilis* represents a stabilized hybrid-derived taxon or a subset of the variation found within hybrid swarms between *G. max* and *G. soja*, maintained as self-fertilized accessions.

Barley (Rank: No. 5) *Hordeum vulgare* L. (Poaceae)

Wild plants morphologically intermediate to cultivated barley and its wild, weedy relative *H. spontaneum* (= *H. vulgare* ssp. *spontaneum*) often "occur where the two species are found together" (58) "in less intensively cultivated fields or areas adjacent to cultivated fields" in the Middle East (125). Despite morphological evidence for hybridization, we are not aware of any genetic analysis of the spontaneous intermediates. Barley can be crossed with *Hordeum* species other than *H. spontaneum*, but the resulting hybrids are highly sterile (58).

Cottonseed (Rank: No. 6) *Gossypium hirsutum* L. and *G. barbadense* L. (Malvaceae)

Allozyme and DNA analyses have demonstrated that limited interspecific introgression has occurred from the cultivated cotton species to certain wild relatives. In the New World tropics and subtropics, *G. barbadense*-specific alleles were found in low frequency in wild or feral populations of *G. hirsutum* that are sympatric with the crop. In the same regions, *G. hirsutum*-specific alleles occur in wild *G. barbadense* populations that are sympatric with cultivated *G. hirsutum* (22, 23, 128). Population-level analysis has not yet been conducted to determine the extent to which crop alleles have moved into natural populations. Intraspecific gene flow from these crops to their wild forms has apparently not been investigated.

Fryxell (51) suggested, on the basis of morphology, that two species, *G. darwinii* of the Galapagos Islands and *G. tomentosum* of the Hawaiian Islands, were at risk of extinction as a result of hybridization with *G. hirsutum*. The presence of *G. hirsutum*-specific allozyme alleles in wild populations has confirmed that *G. darwinii* has experienced substantial introgression from the crop (129). No evidence of interspecific hybridization between *G. hirsutum* and *G. tomentosum* has yet been detected; however, attempts to determine whether it occurs have been hampered by the lack of species-specific markers (36). Allozyme analysis has revealed limited introgression into the Brazilian endemic *G. mustelinum* (130).

Sorghum (Rank: No. 7) *Sorghum bicolor* (L.) Moench (Poaceae)

Spontaneous plants, morphologically intermediate between cultivated sorghum and its wild relatives (both conspecifics and congenerics), occur frequently in and near sorghum fields when wild taxa are present, in both the Old World and the New World (13, 43). Analyses of progeny segregation, allozymes, and RFLPs all reveal crop-specific alleles in wild *S. bicolor* when it co-occurs with the crop in Africa, suggesting that intraspecific hybridization and introgression are common (2, 3, 43).

Allozyme progeny analysis of the tetraploid weed *S. halepense*, planted around cultivated sorghum fields in California, detected spontaneous hybridization as far as 100 m from the crop (11). The resulting, presumably triploid, hybrids showed no fitness differences from the weed under field conditions (12).

Hybridization between crop sorghum and *S. propinquum* has been implicated in the origin of the weedy *S. almum*. Likewise, introgression from crop sorghum has been implicated in the evolution of enhanced weediness in one of the world's worst weeds, *S. halepense* (62). RFLP analysis has shown that both *S. almum* and *S. halepense* contain a combination of alleles specific to both putative parent species (93).

Millets (Rank: No. 8) *Eleusine coracana* (L.) Gaertn. (finger millet) and *Pennisetum glaucum* (L.) R. Br. (= *P. americanum* = *P. typhoides*) (pearl millet) (both in Poaceae). [Millets are a collection of unrelated, tropical small grain species; these two species account for most of the world's production (55).]

Plants with a range of morphologies between finger millet and the wild *E. coracana* ssp. *africana* are often found along roadsides as well as in, and at the edges of, finger millet fields where the two taxa co-occur in Africa (86, 94). These spontaneous intermediates sometimes behave as noxious weeds (39). Despite morphological evidence for hybridization, we are not aware of any genetic analysis of the spontaneous intermediates. Natural hybrids between finger millet and other wild relatives have not been reported.

Spontaneous intermediates between pearl millet and the wild species, *P. violaceum* (= *P. g.* ssp. *monodii*) and *P. sieberanum* (= *P. g.* ssp. *stenostachyum*), often occur in and near pearl millet fields in West Africa and northern Namibia when the wild taxa are present (24). Allozyme analysis of *P. violaceum* accessions revealed that populations sympatric with the crop were genetically more similar to the crop than allopatric populations (122), supporting the hypothesis of crop-to-wild introgression in areas of sympatry.

Hybridization between pearl millet and *P. violaceum* may have been involved in the evolution of *P. sieberanum*, "shibra," which is morphologically intermediate to the two species. "In Africa, shibras are found throughout much of the area of pearl millet cultivation" (24), where, as "obligate weeds of cultivation," they "do not persist for more than one generation after cultivated fields have been abandoned"(40).

Natural hybridization between wild and cultivated *Pennisetum* can occur at a substantial rate. Genetic markers (allozyme and morphological) were used to measure hybridization between naturally adjacent wild and cultivated populations of millet in Niger (84). The crop sired about 35% of the progeny of both *P. violaceum* and *P. sieberanum*. Also, a plot of interplanted wild and cultivated millet was genetically structured so that allozyme progeny analysis would identify hybrids (97). In this experiment, the crop sired 8% of the progeny of *P. violaceum* and 39% of the progeny of *P. sieberanum*.

Dry, String, and Green Beans (Rank: No. 9) *Phaseolus vulgaris* L. (Fabaceae) (This species, common bean, comprises most of the world's production of the crop.)

Spontaneous intermediates between cultivated bean and wild plants of the same species often occur at the margins of bean fields when the wild beans are present in South America (e.g., 15). These intermediates often persist for years (50). Spontaneous intermediates between common bean and its wild relatives, *P. aborigineus* and *P. mexicanus*, are known from Mexico (1, 37, 123). However, morphological intermediates are not always present where cultivated and wild taxa co-occur (35, 50).

Genetic analysis of the wild intermediates in South America has been conducted using phaseolin seed proteins and progeny segregation studies of morphological traits. Intermediate plants have a combination of alleles specific to both the pure crop and the pure wild taxon (15), supporting the hypothesis that the intermediates are products of hybridization or introgression. However, some of the weedy populations are fixed for both crop-specific and wild-specific characters, suggesting that they are genetically stabilized lineages arising from past hybridization. We are not aware of similar genetic analysis of the putative hybrids in Mexico.

Rapeseed (Canola) (Rank: No. 10) *Brassica napus* L. and *B. campestris* L. (= *B. rapa*) (Brassicaeae) (Other cultivars of *Brassica napus* and *B. campestris* are grown as vegetable and fodder crops.)

Spontaneous hybridization between a *B. campestris* vegetable cultivar and its wild conspecific has been experimentally measured (83). Wild plants homozygous for a recessive allele were planted at varying distances around stands of a cultivar homozygous for a dominant anthocyanin marker allele. Progeny testing from the wild plants revealed that *B. campestris* readily hybridizes with its weedy conspecific under field conditions. Hybridization rates involving oilseed cultivars probably will be similar to those for the vegetable cultivar. "It is uncertain whether or not *B. napus* exists in truly wild form," except for escapes from cultivation (85), but reproductive barriers that would prevent spontaneous hybridization between wild and cultivated *B. napus* are unlikely.

The hybrid between cultivated *B. napus*, a tetraploid, and wild *B. campestris*, a diploid, *(B. x harmsiana)* occurs sporadically in the British Isles in crops of *B. napus* that are adjacent to or sympatric with *B. campestris* (57). Jørgensen & Andersen (65) found *B. napus*-specific allozyme alleles in two wild *B. campestris*-like plants in Denmark–evidence of past hybridization and introgression.

Field-based experiments using genetically based morphological, cytogenetic, allozyme, and DNA markers have measured spontaneous hybridization rates between *B. napus* and wild *B. campestris*. *Brassica campestris* growing within stands of the crop produced anywhere from 9% to 93% hybrid progeny, depending on the experimental design (66). The same study showed that the *B. campestris x B. napus* hybrids spontaneously backcrossed to their wild parent. When these hybrids were grown with their parents under field conditions, the hybrids were significantly more fit than *B. campestris* (59). Both backcrosses and F_2s, however, have somewhat reduced fitness relative to their wild grandparent (60).

Field-based experiments using genetically based markers have also been used to measure spontaneous hybridization rates between *B. napus* and wild *B. juncea*. Experiments involving oilseed rape intermixed with *B. juncea* revealed that 3% of the seed set by the wild species were hybrids (18, 66). Likewise, field-based experiments using genetic markers have shown that *B. juncea x B. napus* hybrids spontaneously backcross to their wild parent (49).

Field-based experiments have also demonstrated that *B. napus* can act as a successful pollen parent in spontaneous intergeneric crosses, albeit at a very low rate. Herbicide resistant *B. napus* was interplanted with wild hoary mustard (*Hirschfeldia incana*) at low density (77). Progeny testing revealed that about 2% of seedlings from the wild plants were hybrids. The *Hirschfeldia-Brassica* hybrids produced almost no pollen grains and less than one seed per plant. In a similar experiment, herbicide resistant *B. napus* was interplanted with wild jointed charlock (*Raphanus raphanistrum*) under different densities in open field conditions (34). Progeny from the wild plants were tested for herbicide resistance. At low densities (1:600 charlock:rape) hybridization was detected at a very low rate–0.05% of the seedlings were hybrids. The *Raphanus-Brassica* hybrids have very low fertility under field conditions, averaging less than one seed per plant (27), but recovery of fertility was observed under repeated generations of backcrossing to the wild parent in the field (28). A third experiment involved *B. napus* interplanted with wild mustard (*Sinapis arvensis*) (76); despite progeny testing of millions of seeds from the wild species, no hybridization was detected.

Groundnut (Peanut) (Rank: No. 11) *Arachis hypogea* L. (Fabaceae)

Experimental crosses between groundnut and the wild South American species *A. monticola* readily produce fertile hybrids, but the crop does not cross freely with other wild relatives (110). We did not find any report suggesting that spontaneous hybridization occurs between groundnut and any wild relative. Although *A. hypogea* may show limited natural cross-pollination (S Hegde, personal communication), all *Arachis* species are geocarpic and are presumed to be predominantly self-pollinated (100), a situation that is likely to severely limit opportunities for spontaneous interspecific hybridization, even in sympatry.

Sunflower Seed (Rank: No. 12) *Helianthus annuus* L. (Asteraceae)

Cultivated sunflower is cross-compatible with wild *H. annuus* and other species in the genus (105). Both morphological and molecular evidence suggest that *wild H. annuus* naturally hybridizes with other wild sunflower species in areas of sympatry in western North America (e.g., 44, 61, 105), and that hybridization may lead to the formation of new species (e.g., 100). However, the existing data on hybridization between these wild taxa are of limited value in predicting gene flow dynamics between *cultivated H. annuus* and wild congeners.

Spontaneous hybridization between cultivated sunflower and wild *H. annuus* has been the subject of considerable research. Allozyme progeny analysis of wild *H. annuus*, planted around cultivated sunflower fields in Mexico, detected spontaneous hybridization at substantial rates and over distances of up to 1000 m from the crop (9). The fitness of these F_1s was compared to that of wild plants under field conditions; "in general, hybrid plants had fewer branches, flower heads, and seeds than wild plants, but in two crosses fecundity of the hybrids was not significantly different from that of purely wild plants" (113). Introgression of crop alleles into wild sunflower populations is likely to be impeded rather than prevented by the lowered fitness of the hybrids. Indeed, introgression from culitvated to wild *H. annuus* appears to occur freely.

Whitton et al (131) showed that crop-specific RAPD markers can persist in a wild population for up to five generations, following a single season of hybridization with a nearby crop plantation. Linder et al (80) conducted a similar analysis of wild *H. annuus* populations that had long-term (20 to 40 years) contact with the crop and found substantial introgression, with an "average overall frequency of cultivar markers greater than 35%" and "every individual... tested contained at least [one cultivar-specific allele]." Given the substantial gene flow from the crop to its wild form, it is not surprising that introgression from the crop has been implicated in the evolution of increased weediness in wild *H. annuus* (61).

Sugarcane (Rank: No. 13) *Saccharum officinarum* L. (This species, noble cane, comprises most of the world's production of the crop.)

Sugarcane does not need to flower to produce its crop. Nonetheless, spontaneous hybridization between cultivated and wild *Saccharum* in Australasia and islands of the Indian Ocean has been implicated in the origin of many cultivars (104). While most evidence for the hybrid origin of these cultivars comes from morphology, some supporting evidence also comes from genetically based traits. For example, certain wild cases from New Guinea that are morphologically classified as *S. spontaneum* "have leaf flavonoids and triterpenoids common to... *S. officinarum*" (33). Likewise, chromosome data indicate that certain wild canes from Java and Mauritius are hybrids between the cultivated *S. officinarum* and the wild *S. spontaneum* (117).

Spontaneous intergeneric hybridization has been implicated in the origin of particular accessions of wild sugarcane relatives (33). Chromosomal evidence supports the suggestion, based on morphology, that *Erianthus maximus* is derived from a natural cross of the cultivated *S. officinarum* with the wild genus, *Miscanthus* (33). Likewise, a *Miscanthus* clone was found to have the same chloroplast restriction fragment sites as cultivated *Saccharum* species, suggesting a history of introgression between the two genera in the lineage of that genotype (115). Population-level analysis has not yet been conducted to determine the extent to which crop alleles have moved into natural populations.

EVOLUTIONARY CONSEQUENCES OF GENE FLOW

Gene flow can be a potent evolutionary force. A small amount of gene flow is capable of counteracting the other evolutionary forces of mutation, drift, and selection (111). Gene flow's impact depends on its magnitude. The magnitude of gene flow among natural plant populations is idiosyncratic, varying among species, populations, individuals, and even years (45). In the same way, levels of gene flow from domesticated plants to their wild relatives are expected to be highly variable and to depend on a variety of spatiotemporal factors. Gene flow may be effectively zero for plants that are cross-incompatible, spatially isolated, or that do not overlap in flowering time. Even when compatible species with similar flowering phenologies grow in spatial proximity, levels of gene flow may vary. For example, differences in the relative sizes of the source and sink populations may result in different rates of hybridization and gene flow (reviewed by 46). The cultivation of one or a few plants in a backyard or dooryard garden may result in low levels of crop-to-weed gene flow. In contrast, a stand of an agronomic crop grown according to the practices of modern industrial agriculture may contain millions of plants. In such situations, weed populations hundreds of meters away may show high levels of hybridization and introgression.

Experimentally measured hybridization rates between crops and cross-compatible wild relatives typically exceed 1%, sometimes even over distances of 100 m or more (e.g. 9, 11, 75, 97). The rate of incorporation of foreign alleles under such levels of hybridization is likely to be orders of magnitude higher than typical mutation rates (53), and we would expect gene flow in these systems to be much more important than mutation.

The best-known evolutionary consequence of gene flow is its tendency to homogenize population structure (reviewed in 111). The conditions for homogenization will vary depending on whether immigrant alleles are neutral, detrimental, or beneficial in the ecological and genomic environment of the population that receives them.

Neutral Gene Flow

When gene flow is absent, the evolution of neutral alleles in a population depends on stochastic processes (genetic drift). Genetic drift will lead to genetic differentiation among populations, especially when effective population sizes are small (133). When gene flow is present, the rule of thumb is that one immigrant every other generation, or one interpopulation mating per generation, will be sufficient to prevent strong interpopulation differentiation of neutral alleles (111). Interestingly, this relationship is independent of the size of the recipient population. Conservation geneticists often conclude one migrant per generation will be sufficient to counteract the effects of drift, preventing divergence between populations and buffering against the loss of within-population variation (e.g., 4).

The homogenization of genetic variation by gene flow does not necessarily result in the enhancement of local variation. Ultimately, changes in local diversity will depend on levels of allelic richness and genetic diversity of the source population (in our case, the crop) relative to that of the sink population. Crop cultivars typically contain substantially less genetic variation than populations of their wild relatives (73). Thus, we can predict that levels of neutral variation in a wild population will often decrease under gene flow from a domesticated relative.

Detrimental Gene Flow

Immigration of disadvantageous alleles can counteract local directional selection and reduce local fitness (e.g. 6). Generally, detrimental alleles will swamp those that are locally adaptive when $m \geq s$, where m is the fraction of immigrants relative to the total population per generation, and s is the local selective coefficient against the immigrant alleles (111). That is, moderate rates of incoming gene flow (ca. 1%–5% per generation) are expected to be sufficient to introduce and maintain alleles when the local selective coefficient against them is of the same magnitude (i.e. 1%–5%).

Data from natural systems support this expectation. Reciprocal transplant studies often reveal local adaptive differentiation in plant populations (reviewed by 78, 126), but seldom at the microgeographic level at which substantial gene flow occurs (e.g. 8, 127). However, such differentiation becomes apparent if selection is very strong, $s > 0.99$ (e.g. 7). Adaptive differences between populations (in our case, adaptation to growth and reproduction under cultivation versus adaptation to growth and reproduction in unmanaged ecosystems) may lead to outbreeding depression, "a fitness reduction following hybridization" between populations (121). Outbreeding depression has been detected in many natural plant populations (126), and naturally occurring interspecific hybrids are often partially or completely sterile (e.g. 116). Similarly, attempts by plant breeders to introduce commercially desirable alleles to crops from distant relatives (the secondary gene pool) are often constrained by the reduced fertility and viability of hybrids resulting from wide crosses (109).

Beneficial Gene Flow

When gene flow and selection work in concert, gene flow will accelerate the increasing frequency of favorable alleles in a sink population. Wright (133) modeled several situations that involve gene flow in combination with favorable selection. His "Continent-Island" model, with one-way immigration from a large source population into a sink population, is the one most likely to be applicable to gene flow from crops to their wild relatives. Our primary concern is the movement of domesticated alleles into natural populations; these alleles are more-or-less fixed in the crop and absent from the wild population. With these assumptions, Wright's model can be simplified to predict that the fraction of crop-to-wild immigrants, m, will have the same effect on the speed of fixation of the favorable alleles as the local selective coefficient, s. That is, if the magnitudes of both migration and the local selective coefficient are equal, then the rate of allele frequency change per generation is doubled relative to the situation with selection alone. If m is twice as great as s, then the rate will be tripled compared to the situation without gene flow.

APPLIED IMPLICATIONS

Gene flow between domesticated plants and their wild relatives may have two potentially harmful consequences–the evolution of increased weediness and the increased likelihood of extinction of wild relatives. Applied plant scientists can collect data, from experimental or descriptive studies and (if available) from the pre-existing scientific literature, to evaluate the likelihood and potential importance of these gene flow risks. Such data should be used to guide decision-makers about the deployment of new cultivars or the introduction of a new crop into a region.

Guidelines have been created for evaluating the gene flow risks of transgenic crops (e.g. 103), but, to our knowledge, have never been extended to domesticated plants in general. Below, we examine two important risks of gene flow from a domesticated plant species to a wild taxon and address possible ways of evaluating those risks.

Weed Evolution

The first potential problem is the transfer of crop alleles to a weedy taxon to create a more aggressive weed. While some crop alleles may be disadvantageous in the wild, others may confer a selective advantage for weedy populations. As noted above, the spread of alleles under favorable selection is accelerated by gene flow. The issue of evolution of enhanced weediness has received considerable attention as a potential consequence of the field release of genetically engineered crops and their subsequent hybridization with wild relatives (e.g., 31, 103).

The problem is far from hypothetical. Increased weediness resulting from gene flow from traditionally improved crops to weedy relatives has been well documented. Crop-to-weed gene flow has been implicated in the evolution of enhanced weediness in wild relatives of 7 of the world's 13 most important crops (see Table 1). Another example is the recent evolution of a weedy rye derived from natural hybridization between cultivated rye, *Secale cereale*, and wild *S. montanum* in California. The weed's hybrid origin has been confirmed by genetic analysis (119). The weed has become such a serious problem that "farmers have abandoned efforts to grow cultivated rye for human consumption" (89).

Under what circumstances is hybridization with a crop most likely to result in the evolution of enhanced weediness? The evolution of enhanced weediness will depend on (*a*) whether hybridization can occur, (*b*) whether the hybrids can reproduce in the wild, and (*c*) whether one or more alleles from the domesticate confer an advantage on the weed.

Without hybridization, gene flow cannot occur. Problems with the evolution of enhanced weediness will only arise when a crop is grown in proximity to a cross-compatible relative. The appropriate way to test whether hybridization is likely to occur under field conditions is to create experimental stands of crop and weed under conditions that simulate those in which the crop and weed will co-occur after field release. Gene flow can then be measured by testing the progeny of the weed for crop-specific genetically based markers. Sample sizes must be large enough to detect gene flow at biologically important levels (sensitive enough to detect hybridization rates of at least 1%). Our case studies include several experimental studies of this type that measure crop-to-wild hybridization rates (e.g., 9, 11, 75). Hand pollination and embryo rescue are inappropriate techniques for the investigation of crop-to-weed gene escape because they do not simulate natural vectors or natural seed production.

How these data on hybridization are interpreted is as important as the data themselves. Phrases such as "gene flow drops off rapidly with distance" are misleading.

If any hybridization is detected within distances that are typical of those occurring between the crop and the weed, the appropriate interpretation is "crop genes will move into weedy populations."

The next question is "Will hybrids reproduce in the wild?" If hybrids reproduce, and if the weedy populations are exposed to a constant rain of compatible pollen from the crop, favorable crop alleles will persist, even if the hybrids do not show heterosis. Appropriate data on hybrid persistence are those that indicate whether first generation hybrids will reproduce (by seed and/or by vegetative reproduction) under field conditions. We are aware of only a few studies that have measured the fitness of weed-crop hybrids compared to that of their weed progenitors under field conditions (e.g. 12, 59, 71). Interestingly, in the majority of such studies, the fitness of the hybrid tends to be the same as or even higher than that of its wild parent.

An alternative to experimental approaches to the first two questions is to conduct descriptive population genetic studies that ask whether crop-specific genetic markers occur in natural populations that grow in the vicinity of the crop in question. The presence of and the frequency of crop-specific markers can then be used to determine levels of past introgression. Such studies must be based on sufficiently large population samples. If crop-specific alleles occur at biologically important frequencies (>1%) in adjacent weedy populations, then it is clear that domesticated alleles can become incorporated into natural populations via hybridization and introgression. Numerous studies have already demonstrated introgression of crop alleles into natural populations; some of these are mentioned in our case studies (e.g. 15, 64, 80).

The final question is "Will the crop genes enhance weediness?" Unlike the preceding two questions, this question cannot be addressed without direct consideration of the biology of the new cultivar. Certain novel traits (e.g., herbicide tolerance, disease resistance, insect resistance) will enhance particular fitness components of a weed in particular environments but may (or may not) have attendant costs (16). Pleiotropic fitness effects may also occur with alleles in a novel genetic context (e.g, 17). Ideally, the fitness impact of the novel crop alleles should be examined experimentallly in the context of the wild genome and the environment of the weed. The fitness of hybrids containing the novel allele could be compared with the fitness of those without the allele under field conditions. Alternatively, the field-measured fitness of first generation backcrosses to the weed could be compared with that of pure wild plants. Selective factors–herbivores, diseases, and herbicides–should not be excluded from the experiments if they are common features of the field environment. Few such experiments have been reported (e.g. 60)–largely motivated by the desire to estimate the fitness impact of a transgene in a hybrid genome. The interpretation of the fitness data is critical. We can expect that a fitness boost of 5% will enhance weediness to a point that would have important practical consequences.

In summary, bringing a novel cultivar or a new crop taxon into contact with a cross-compatible weed may lead to evolution of enhanced weediness if hybridization occurs under field conditions, the hybrids persist to reproduce, and one or more crop alleles give a fitness boost to the weed populations. The economic and environmental impact of that enhanced weediness will depend on how much fitness increases and under what environmental conditions.

Extinction of Wild Relatives

The other potential problem with natural hybridization between domesticated plants and their wild relatives is the increased risk of extinction of wild taxa. Largely neglected as a general conservation issue, extinction by hybridization has recently begun to attract attention (e.g., 46, 79, 98). Rare taxa may be threatened by hybridization in two ways.

As noted above, outbreeding depression from detrimental gene flow will reduce the fitness of a locally rare species that is mating with a locally common one. An alternate route to extinction is by swamping, which occurs when a locally rare species loses its genetic integrity and becomes assimilated into a locally common species as a result of repeated bouts of hybridization and introgression (46). We would expect swamping to result from gene flow that is largely neutral or beneficial.

Both outbreeding depression and swamping are frequency-dependent phenomena and show positive feedback. With each succeeding generation of hybridization and backcrossing, genetically pure individuals of the locally rare species become increasingly rare until extinction occurs (46). Both phenomena can lead to extinction rapidly. If 900 individuals of a locally common species mate randomly with 100 individuals of the locally rare one, extinction by outbreeding depression and/or swamping can occur in two generations (N Ellstrand, unpublished data).

In contrast to the problem of enhanced weediness, the problem of extinction by hybridization does not depend on relative fitness, but only on patterns of mating. Specifically, the risk decreases as intertaxon mating rates depart from random mating (that is, as assortative mating increases). Consider the example above, with 900 individuals of a locally common species mating with 100 individuals of the locally rare one. If we impose assortative mating such that the rare species mates with its own kind five times more frequently than the rate expected under random mating, the time to extinction by outbreeding depression and/or swamping is expected to double relative to random mating (N Ellstrand, unpublished data).

Extinction by hybridization with domesticated species has been implicated in the extinction or increased risk of extinction of several wild species, including wild relatives of two of the world's 13 most important crops (Table 1; also cf. 112). A risk of extinction by hybridization will occur when a previously allopatric rare taxon becomes sympatric or peripatric with a recently introduced crop. That risk will increase as hybridization and introgression proceed over generations and will also increase with intertaxon hybridization rates.

Measuring hybridization rates is critical for the assessment of the risk of extinction by hybridization. The appropriate way to assess hybridization rates under field conditions is to create experimental stands of the crop and wild taxon under conditions comparable to those in which the crop and the wild taxon will co-exist when field release occurs. Progeny testing of the wild taxon for crop-specific genetic markers can then be used to measure gene flow. Sample sizes must be large enough to detect gene flow at biologically significant levels (sensitive enough to detect hybridization rates of at least 10%). We are not aware of any experiments measuring hybridization rates between a common plant species and a rare relative. However, the experimental studies mentioned above that measure crop-to-weed hybridization rates are appropriate models.

After assessing hybridization rates, the sizes of the populations of the rare taxon can be estimated by direct counts and adjusted to allow for future disturbance and fragmentation. These numbers, combined with the experimentally assessed hybridization rates, can then be used to develop crude estimates of the numbers of individuals in each wild population that will be replaced by hybrids and introgressants with each generation of hybridization. The speed of replacement will indicate whether the risk of extinction is biologically important.

ACKNOWLEDGMENTS

This project grew out of projects funded with USDA grant support to NCE and SJFR (the Swedish Forestry and Agricultural Research Council) grant support to NCE and HCP. Helena Parrow helped with the initial compilation of case studies. The following scientists prereviewed our case studies: J Antonovics, P Arriola, L Blancas, AHD Brown, E Buckler, J Doebley, D Garvin, P Gepts, W Hanna, S Hegde, K Hilu, T Holtsford, F Ibarra-Perez, R Jørgensen, P Keim, J Kohn, H Morishima, H Oka, R Palmer, L Rieseberg, Y Sano, K Schertz, A Scholz, B Sobral, H Suh, JG Waines, and J Wendel. This paper improved as a result of thoughtful suggestions from N Barton, D Bartsch, S Baughman, JMJ deWet, J Endler, S Frank, D Gessler, J Hamrick, and A Rankin.

LITERATURE CITED

1. Acosta J, Gepts P, Debouck DG. 1994. Observations on wild and weedy accessions of common beans in Oaxaca, Mexico. *Annu. Rep. Bean Improv. Coop* 37:137–38
2. Aldrich PR, Doebley J. 1992. Restriction fragment variation in the nuclear and chloroplast genomes of cultivated and wild *Sorghum bicolor*. *Theor. Appl. Genet.* 85: 293–302
3. Aldrich PR, Doebley J, Schertz KF, Stec A. 1992. Patterns of allozyme variation in cultivated and wild *Sorghum bicolor*. *Theor. Appl. Genet.* 85:451–60
4. Allendorf FW. 1983. Isolation, gene flow, and genetic differentiation among populations. In *Genetics and Conservation: A Reference for Managing Wild Animal and Plant Populations,* ed. CM Schonewald-Cox, SM

Chambers, B MacBryde, WL Thomas, pp. 51–65. Menlo Park, CA: Benjamin-Cummings

5. Anderson E. 1949. *Introgressive Hybridization.* New York: Wiley & Sons
6. Antonovics J. 1976. The nature of limits to natural selection. *Ann. Mo. Bot. Gard.* 63: 224–47
7. Antonovics J, Bradshaw AD. 1970. Evolution in closely adjacent populations. VIII. Clinal patterns at a mine boundary. *Heredity* 25:349–62
8. Antonovics J, Ellstrand NC, Brandon RN. 1988. Genetic variation and environmental variation: expectations and experiments. In *Plant Evolutionary Biology: A Symposium Honoring G. Ledyard Stebbins,* ed. L. Gottlieb, SK Jain, pp. 275–303. London: Chapman & Hall
9. Arias DM, Rieseberg LH. 1994. Gene flow between cultivated and wild sunflowers. *Theor. Appl. Genet.* 89:655–60
10. Arnold ML. 1997. *Natural Hybridization and Evolution.* New York: Oxford Univ. Press
11. Arriola PE, Ellstrand NC. 1996. Crop-to-weed gene flow in the genus *Sorghum* (Poaceae): spontaneous interspecific hybridization between johnsongrass, *Sorghum halepense,* and crop sorghum, *S. bicolor. Am. J. Bot.* 83:1153–60
12. Arriola PE, Ellstrand NC. 1997. Fitness of interspecific hybrids in the genus *Sorghum:* persistence of crop genes in wild populations. *Ecol. Appl.* 7:512–18
13. Baker HG. 1972. Human influences on plant evolution. *Econ. Bot.* 26:32–43
14. Barrett SCH. 1983. Crop mimicry in weeds. *Econ. Bot.* 37:255–82
15. Beebe S, Toro O, González AV, Chac6n MI, Debouck DG. 1997. Wild-weed-crop complexes of common bean (*Phaseolus vulgaris* L., Fabaceae) in the Andes of Peru and Columbia, and their implications for conservation and breeding. *Genet. Resourc. Crop. Evol.* 44:73–91
16. Bergelson J, Purrington CB. 1996. Surveying patterns in the cost of resistance in plants. *Am. Nat.* 148:536–58
17. Bergelson J, Purrington CB, Wichmann G. 1998. Promiscuity in transgenic plants. *Nature* 395:25
18. Bing DJ. 1991. *Potential of gene transfer among oilseed Brassica and their weedy relatives.* MS thesis, Univ. Saskatchewan, Saskatoon
19. Blumler MA. 1999. Introgression of durum into wild emmer and the agricultural origin question. In *The Origins of Agriculture and the Domestication of Crop Plants in the Near East.* ed. AB Damania, J Valkoun, Aleppo: ICARDA. In press
20. Broich SL, Palmer RG. 1980. A cluster analysis of wild and domesticated soybean phenotypes. *Euphytica* 29:23–32
21. Broich SL, Palmer RG. 1981. Evolutionary studies of the soybean: the frequency and distribution of alleles among collections of *Glycine max* and *G. soja* of various origin. *Euphytica* 30:55–64
22. Brubaker CL, Koontz JA, Wendel JF. 1993. Bidirectional cytoplasmic and nuclear introgression in the New World cottons, *Gossypium barbadense* and *G. hirsutum* (Malvaceae). *Am. J. Bot.* 80:1203–8
23. Brubaker CL, Wendel JF. 1994. Reevaluating the origin of domesticated cotton (*Gossypium hirsutum;* (Malvaceae)) using nu clear restriction fragment length polymorphisms (RFLPs). *Am. J. Bot.* 81:1309–26
24. Brunken J, de Wet JMJ, Harlan JR. 1977. The morphology and domestication of pearl millet. *Econ. Bot.* 31:163–74
25. Burdon JJ, Marshall DR, Oates JD. 1992. Interactions between wild and cultivated oats. In *Wild Oats in World Agriculture,* ed. AR Barr, RW Medd, *Proc. 4th Int. Oat Conf.* 2:82–87
26. Chang TT. 1995. Rice. See Ref. 114, pp. 147–55
27. Chèvre AM, Eber F, Baranger A, Hureau G, Barret P, et al. 1998. Characterization of backcross generation obtained under field conditions from oilseed rape–wild radish

F1 interspecific hybrids: an assessment of transgene dispersal. *Theor. Appl. Genet.* 97:90–98
28. Chèvre AM, Eber F, Baranger A, Renard MI. 1997. Gene flow from transgenic crops. *Theor. Appl. Genet.* 97:90–98
29. Chu YE, Morishima H, Oka HI. 1969. Reproductive barriers distributed in cultivated rice species and their wild relatives. *Jpn. J. Genet.* 4:207–23
30. Chu YE, Oka HI. 1970. Introgression across isolating barriers in wild and cultivated *Oryza* species. *Evolution* 24:344–55
31. Colwell RK, Norse EA, Pimentel D, Sharples FE, Simberloff D. 1985. Genetic engineering in agriculture. *Science* 229:111–12
32. Dale PJ. 1994. The impact of hybrids between genetically modified crop plants and their related species: general considerations. *Mol. Ecol.* 3:31–36
33. Daniels J, Roach BT. 1987. Taxonomy and evolution. In *Sugarcane Improvement through Breeding,* ed. DJ Heinz, pp. 7–83. Amsterdam: Elsevier
34. Darmency H, Lefol E, Fleury A. 1998. Spontaneous hybridizations between oilseed rapes and wild radish. *Mol. Ecol.* 7: 1467–73
35. Debouck DG, Toro O, Paredes OM, Johnson WC, Gepts P. 1993. Genetic diversity and ecological distribution of *Phaseolus vulgaris* in northwestern South America. *Econ. Bot.* 47:408–23
36. DeJoode DR, Wendel JF. 1992. Genetic diversity and origin of the Hawaiian Islands cotton, *Gossypium tomentosum. Am. J. Bot.* 79:1311–19
37. Delgado Salinas A, Bonet A, Gepts P. 1988. The wild relative of *Phaseolus vulgaris* in Middle America. In *Genetic Resources of Phaseolus Beans,* ed. P Gepts, pp. 163–84. Dordrecht: Kluwer
38. deVries FT, van der Meijden R, Brandenburg WA. 1992. Botanical files: a study of the real chances for spontaneous gene flow from cultivated plants to the wild flora of the Netherlands. *Gorteria* (suppl.) 1:1–100
39. deWet JMJ. 1995. Finger millet. See Ref. 114, pp. 137–40
40. deWet JMJ. 1995. Pearl millet. See Ref. 114, pp. 137–40
41. deWet JMJ, Harlan JR. 1975. Weeds and domesticates: evolution in the man-made habitat. *Econ. Bot.* 99–107
42. Doebley J. 1990. Molecular evidence for gene flow among Zea species. *BioScience* 40:443–48
43. Doggett H, Majisu BN. 1968. Disruptive selection in crop development. *Heredity* 23:1–23
44. Dorado O, Rieseberg LH, Arias DM. 1992. Chloroplast DNA introgression in southern California sunflowers. *Evolution* 46:566–72
45. Ellstrand NC. 1992. Gene flow among seed plant populations. *New For.* 6:241–56
46. Ellstrand NC, Elam DR. 1993. Population genetic consequences of small population size: implications for plant conservation. *Annu. Rev. Ecol. Syst.* 24:217–42
47. Ellstrand NC, Whitkus RW, Rieseberg LH. 1996. Distribution of spontaneous plant hybrids. *Proc. Natl. Acad. Sci. USA* 93:5090–93
48. Feldman M, Lipton FGH, Miller TE. 1995. Wheats. See Ref. 114, pp. 184–92
49. Frello S, Hansen KR, Jensen J, Jørgensen RB. 1995. Inheritance of rapeseed (*Brassica napus*) specific RAPD markers and a transgene in the *cross B. juncea x (B. juncea X B. napus). Theor. Appl. Genet.* 91: 236–41
50. Freyre R, Ríos R, Guzmán L, DeBouck DG, Gepts P. 1996. Ecogeographic distribution of *Phaseolus* ssp. (Fabaceae) in Bolivia. *Econ. Bot.* 50:195–215
51. Fryxell PA. 1979. *The Natural History of the Cotton Tribe (Malvaceae, Tribe Gossypieae).* College Station: Texas A&M Univ. Press
52. Futuyma DJ. 1998. *Evolutionary Biology.* Sunderland: Sinauer. 3rd ed.

53. Grant V. 1975. *Genetics of Flowering Plants*. New York: Columbia Univ. Press
54. Grant V. 1981. *Plant Speciation*. New York: Columbia Univ. Press. 2nd ed.
55. Hancock JF. 1992. *Plant Evolution and the Origin of Crop Species*. Englewood Cliffs, NJ: Prentice Hall
56. Hancock JF, Grumet R, Hokanson SC. 1996. The opportunity for escape of engineered genes from transgenic crops. *HortScience* pp. 31:1080–85
57. Harberd DJ. 1975. *Brassica* L. See Ref. 116, pp. 137–39
58. Harlan J. R. 1995. *Barley*. See Ref. 114, pp. 140–47
59. Hauser TP, Shaw RG, Østergård H. 1998. Fitness of F1 hybrids between weedy *Brassica rapa* and oilseed rape (*B. napus*). *Heredity* pp. 81:429–35
60. Hauser TP, Jørgensen RB, Østergård H. 1998. Fitness of backcross and F2 hybrids between weedy *Brassica rapa* and oilseed rape (*B. napus*). *Heredity* pp. 81:436–43
61. Heiser CB. 1978. Taxonomy of *Helianthus* and the origin of domesticated sunflower. In *Sunflower Science and Technology,* ed. JF Carter, pp. 31–54. Madison, WI: Am. Soc. Agron., Crop Sci. Soc. & Soil Sci. Soc. Am.
62. Holm LG, Plucknett DL, Pancho JV, Herberger JP. 1977. *The World's Worst Weeds: Distribution and Biology*. Honolulu: Univ. Press Hawaii
63. Jasieniuk M, Brûlé-Babel AL, Morrison IN. 1996. The evolution and genetics of herbicide resistance in weeds. *Weed Sci.* pp. 44:176–93
64. Jiang J, Freibe B, Gill BS. 1994. Recent advances in alien gene transfer in wheat. *Euphytica* pp. 73:199–212
65. Jørgensen RB, Andersen B. 1994. Spontaneous hybridization between oilseed rape *(Brassica napus)* and weedy *Brassica campestris* (Brassicaceae): a risk of growing genetically modified oilseed rape. *Am. J. Bot.* pp. 81:1620–26
66. Jørgensen RB, Andersen B, Landbo L, Mikkelsen T. 1996. Spontaneous hybridization between oilseed rape *(Brassica napus)* and weedy relatives. In *Proc. Int. Symp. on Brassicas/Ninth Crucifer Genetics Workshop,* ed. JS Dias, I Crute, AA Monteiro, pp. 193–97. Lisbon: ISHS
67. Kato TA. 1997. Review of introgression between maize and teosinte. In *Gene Flow among Maize Landraces, Improved Maize Varieties, and Teosinte: Implications for Transgenic Maize,* ed. JA Serratos, MC Willcox, F Castillo, pp. 44–53. Mexico: DF: CIMMYT
68. Keim P, Shoemaker RC, Palmer RG. 1989. Restriction fragment length polymorphism diversity in soybean. *Theor. Appl. Genet.* pp. 77:786–92
69. Kiang YT, Antonovics J, Wu L. 1979. The extinction of wild rice *(Oryza perennis formosana)* in Taiwan. *J. Asian Ecol.* 1:1–9
70. Kimber G, Feldman M. 1987. *Wild Wheat. Special Rep. 353.* Columbia: Univ. Missouri Press
71. Klinger T, Ellstrand NC. 1994. Engineered genes in wild populations: fitness of weed-crop hybrids of radish, *Raphanus sativus L. Ecol. Appl.* pp. 4:117–120
72. Kwon SH. 1972. Studies on diversity of seed weight in the Korean soybean landraces and wild soybean. *Korean J. Breeding* pp. 4:70–74
73. Ladizinsky G. 1985. Founder effect in crop-plant evolution. *Econ. Bot.* 39:191–99
74. Ladizinsky G. 1992. Crossibility relations. In *Distant Hybridization of Crop Plants,* ed. G Kalloo, JB Chowdhury, pp. 15–31. Berlin: Springer-Verlag
75. Langevin S, Clay K, Grace JB. 1990. The incidence and effects of hybridization between cultivated rice and its related weed red rice *(Oryza sativa* L.) *Evolution* 44:1000–08
76. Lefol E, Danielou V, Darmency H. 1996. Predicting hybridization between transgenic oilseed rape and mustard. *Field Crops Res.* 45:153–61

77. Lefol E, Fleury A, Darmency H. 1996. Gene dispersal from transgenic crops. II. Hybridization between oilseed rape and the wild hoary mustard. *Sex. Plant Reprod.* 9:189–96
78. Levin DA. 1984. Immigration in plants: an exercise in the subjunctive. In *Perspectives on Plant Population Ecology*, ed. R Dirzo, J Sarukhán, pp. 242–60. Sunderland, MA: Sinauer
79. Levin DA, Francis-Oretega J, Jansen RK. 1996. Hybridization and the extinction of rare plant species. *Conserv. Biol.* 10:10–16
80. Linder CR, Taha I, Seiler GJ, Snow AA, Rieseberg LH. 1998. Long-term introgression of crop genes into wild sunflower populations. *Theor. Appl. Genet.* 96:339–47
81. Love SL. 1994. Ecological risk of growing transgenic potatoes in the United States and Canada. *Am. Potato J.* 71:647–58
82. Luby JJ, McNichol RJ. 1995. Gene flow from culitvated to wild raspberries in Scotland: developing a basis for risk assessment for testing and deployment of transgenic cultivars. *Theor. Appl. Genet.* 90.1133–37
83. Manasse RS. 1992. Ecological risks of transgenic plants: effects of spatial dispersion of gene flow. *Ecol. Appl.* 2:431–38
84. Marchais L. 1994. Wild pearl millet population (*Pennisetum glaucum*, Poaceae) integrity in agricultural Sahelian areas. An example from Keita (Niger). *Pl. Syst. Evol.* 189:233–45
85. McNaughton IM. 1995. Swedes and rapes. See Ref. 114, pp. 68–75
86. Mehra KL. 1962. Natural hybridization between *Eleusine coracana* and *E. africana* in Uganda. *J. Indian Bot. Soc.* 41:531–39
87. Meyn O, Emboden WA. 1987. Parameters and consequences of introgression of *Salvia apiana X S. mellifera* (Lamiaceae). *Syst. Bot.* 12:390–99
88. Morishima H, Sano Y, Oka HI. 1992. Evolutionary studies in rice and its wild relatives. *Oxford Surveys in Evol. Biol.* 8:135–84
89. National Academy of Sciences. 1989. *Field Testing Genetically Modified Organisms: Framework for Decisions.* Washington, DC: Natl. Acad. Sci.
90. Oka HI. 1988. *Origin of Cultivated Rice.* Tokyo: JSSP/Elsevier
91. Oka HI, Chang WT. 1959. The impact of cultivation on populations of wild rice, *Oryza sativa* f. *spontanea. Phyton* 13:105–17
92. Oka HI, Chang WT. 1961. Hybrid swarms between wild and cultivated rice species, *Oryza perennis* and *O. sativa. Evolution* 15:418–30
93. Paterson AH, Schertz KF, Lin YR, Liu SC, Chang YL. 1995. The weediness of wild plants: molecular analysis of genes influencing dispersal and persistence of johnsongrass, *Sorghum halepense* (L.) Pers. *Proc. Natl. Acad. Sci. USA* 92:6127–31
94. Phillips SM. 1972. A survey of the genus *Eleusine Gaertn.* (Gramineae) in Africa. *Kew Bull.* 27:251–70
95. Popova G. 1923. Wild species of *Aegilops* and their mass hybridisation with wheat in Turkestan. *Bull. Appl. Bot.* 13:475–82
96. Raybould AF, Gray AJ. 1993. Genetically modified crops and hybridization with wild relatives: a UK perspective. *J. Appl. Ecol.* 30:199–219
97. Renno JF, Winkel T, Bonnefous F, Benzançon G. 1997. Experimental study of gene flow between wild and cultivated *Pennisetum glaucum. Can. J. Bot.* 75:925–31
98. Rhymer JM, Simberloff D. 1996. Extinction by hybridization and introgression *Annu. Rev. Ecol. Syst.* 27:83–109
99. Richards AJ. 1986. *Plant Breeding Systems.* Hemel Hempstead, UK: Allen & Unwin
100. Rieseberg LH, Carter R, Zona S. 1990. Molecular tests of the hypothesized hybrid origin of two diploid *Helianthus* species. *Evolution* 44:1498–1511

101. Rieseberg LH, Ellstrand NC. 1993. What can molecular and morphological markers tell us about plant hybridization? *Crit. Rev. Pl. Sci.* 12:213–41
102. Rieseberg LH, Wendel JF. 1993. Introgression and its evolutionary consequences in plants. In *Hybrid Zones and the Evolutionary Process*, ed. R Harrison. pp. 70–109. New York: Oxford Univ. Press
103. Rissler J, Mellon M. 1996. *The Ecological Risks of Engineered Crops.* Cambridge, MA: MIT Press
104. Roach BT. 1995. Sugar canes. See Ref. 114, pp.160–66
105. Rogers CE, Thompson TE, Seiler GJ. 1982. *Sunflower Species of the United States.* Bismarck, ND: Natl. Sunflower Assoc.
106. Runyeon H, Prentice HC. 1997. Genetic differentiation in the bladder campions, *Silene vulgaris* and *S. uniflora* (Caryophyllaceae), in *Sweden. Biol. J. Linn. Soc.* 61:559–84
107. Sanders RW. 1987. Identity of *Lantana depressa* and *L. ovatifolia* (Verbenaceae) of Florida and the Bahamas. *Syst. Bot.* 12:44–60.
108. Scheffler JA, Dale PJ. 1994. Opportunities for gene transfer from transgenic oilseed rape (*Brassica napus*) to related species. *Transgenic Res.* 3:263–78
109. Simmonds NW. 1981. *Principles of Crop Improvement*, London: Longman
110. Singh AK. 1995. Groundnut. See Ref. 114, pp. 246–50
111. Slatkin M. 1987. Gene flow and the geographic structure of natural populations. *Science* 236:787–92
112. Small E. 1984. Hybridization in the domesticated-weed-wild complex. In *Plant Biosystematics*, ed. WF Grant, pp. 195–210. Toronto: Academic
113. Snow AA, Moran-Palma P, Rieseberg LH, Wszelaki A, Seiler GJ. 1998. Fecundity, phenology, and seed dormancy of F1 wild-crop hybrids in sunflower (*Helianthus annuus*, Asteraceae). *Am. J. Bot.* 794–801
114. Smartt J, Simmonds NW. 1995. *Evolution of Crop Plants,* Harlow, UK: Longman. 2nd ed.
115. Sobral BWS, Braga DPV, LaHood ES, Keim P. 1994. Phylogenetic analysis of chloroplast restriction enzyme site mutations in the *Saccharinae* Griesb. subtribe of the *Andropogoneae* Dumort. tribe. *Theor. Appl. Genet.* 87:843–53
116. Stace CA. 1975. *Hybridization and the Flora of the British Isles.* London: Academic
117. Stevenson GC. 1965. *Genetics and Breeding of Sugar Cane.* London: Longmans
118. Suh HS, Sato YI, Morishima H. 1997. Genetic characterization of weedy rice (*Oryza sativa* L.) based on morphophysiology, isozymes and RAPD markers. *Theor. Appl. Genet.* 94:316–21
119. Sun M, Corke H. 1992. Population genetics of colonizing success of weedy rye in northern California. *Theor. Appl. Genet.* 83:321–29
120. Tang LH, Morishima H. 1988. Characteristics of weed rice strains. *Rice Genet. Newsl.* 5:70–72
121. Templeton AR. 1986. Coadaptation and outbreeding depression. In *Conservation Biology: The Science of Scarcity and Diversity,* ed. M Soulé, pp. 105–16. Sunderland, MA: Sinauer
122. Tostain S. 1992. Enzyme diversity in pearl millet (*Pennisetum glaucum* L.). 3. Wild millet. *Theor. Appl. Genet.* 83:733–42
123. Vanderborght T. 1983. Evaluation of *Phaseolus vulgaris* wild and weedy forms. *Plant Genet. Res. Newslett.* 54:18–25
124. van Slageren MW. 1994. *Wild Wheats: a Monograph of* Aegilops *L. and* Amblyopyrum *(Jaub. & Spach) Eig (Poaceae).* Wageningen: Veenman Druckers
125. von Bothmer R, Jacobsen N, Baden C, Jørgensen RB, Linde–Laursen I. 1991.

An Ecogeographical Study of the Genus. Hordeum. Rome: IBPGR

126. Waser NM. 1993. Population structure, optimal outbreeding, and assortative mating in angiosperms. In *The Natural History of Inbreeding and Outbreeding, Theoretical and Empirical Perspectives*, ed. NW Thornhill, pp. 173–199. Chicago: Univ. Chicago Press
127. Waser NM, Price MV. 1985. Reciprocal transplants with *Delphinium nelsonii* (Ranunculaceae): evidence for local adaptation. *Am. J. Bot.* 72:1726–32
128. Wendel JF, Brubaker CL, Percival AE. 1992. Genetic diversity in *Gossypium hirsutum* and the origin of upland cotton. *Am. J. Bot.* 79:1291–1310
129. Wendel JF, Percy RG. 1990. Allozyme diversity and introgression in the Galapagos Islands endemic *Gossypium darwinii* and its relationship to continental *G. barbadense*. *Biochem. Ecol. Syst.* 18:517–28
130. Wendel JF, Rowley R, Stewart JMcD. 1994. Genetic diversity in and phylogenetic relationships of the Brazilian endemic cotton *Gossypium mustelinum*. *Pl. Syst. Evol.* 192:49–59
131. Whitton J, Wolf DE, Arias DM, Snow AA, Rieseberg LH. 1997. The persistence of cultivar alleles in wild populations of sunflowers five generations after hybridization. *Theor. Appl. Genet.* 95:33–40
132. Wilkes HG. 1977. Hybridization of maize and teosinte, in Mexico and Guatemala and the improvement of maize. *Econ. Bot.* 31:254–93
133. Wright S. 1969. *Evolution and the Genetics of Populations*. Vol. 2. *The Theory of Gene Frequencies,* Chicago: Univ. Chicago Press
134. Xu J, Kerrigan RW, Callac P, Horgen PA, Anderson JB. 1997. Genetic structure of natural populations of *Agaricus bisporus,* the commercial mushroom. *J. Hered.* 88:482–88

Annu. Rev. Ecol. Syst. 1999. 30:565–91

RESISTANCE OF HYBRID PLANTS AND ANIMALS TO HERBIVORES, PATHOGENS, AND PARASITES

Robert S. Fritz
Department of Biology, Vassar College, Poughkeepsie, New York 12604-0133; email: fritz@vassar.edu

Catherine Moulia
Laboratoire Genome, Populations, Interactions; UPR 9060 CNRS, CC105, UM II Place Eugene Bataillon, Montpellier Cedex O5, France

George Newcombe
Washington State University, Puyallup Research & Extension Center, Puyallup, Washington 98371-4998 USA

Key Words fitness, fungi, introgression, herbivory, hybrid zones

■ **Abstract** Interspecific hybridization can disrupt normal resistance of plant and animal species to their parasites. Resistance to parasites is affected by hybridization in the following ways: no difference between hybrids and parentals, additivity, hybrid susceptibility, and dominance to susceptibility. Similar patterns were seen across host taxa. Responses of different parasite species vary widely to the same hybrid host, which indicates diverse genetic effects of interspecific hybridization on resistance. Differences between field and common garden or laboratory studies suggest that environmental factors in hybrid zones influence the patterns seen in the field. Based on recent studies of hybrid-parasite interactions, three avenues of future research will provide a more complete understanding of the roles of hybrids and the roles of parasites in host evolution. First, the relationship between inheritance of putative resistance mechanisms of hosts and responses of parasites needs study using analyses of recombinant progenies. Second, the interaction among environmental variation in hybrid zones, resistance mechanisms, responses of parasites, and the impact of parasites on host fitness needs experimental analysis using reciprocal transplant experiments in hybrid zones. Finally, the role of hybrids in the community structure and interactions of parasites needs study.

INTRODUCTION

Studies of hybridization of plant and animal species have traditionally focused on the maintenance of hybrid zones, the mechanisms and evolution of reproductive isolation, and genetic exchange between hybridizing populations. Except for

0066-4162/99/1120-0565$08.00

interactions with pollinators (33), the importance of the biotic environment has been mostly ignored; instead attention has focused on the interactions between hybrids and abiotic factors in hybrid zones. Evolutionary ecologists have recently begun to investigate the interactions of hybrid plants and animals with their parasites. We use the generic sense of "parasite" to refer to most herbivores of plants, including insects and mites, internal and external parasites of animals, and fungal pathogens of plants. Parasites are thought to be a potent selective agent for the evolution of plant and animal taxa (e.g., 11), and not surprisingly, hybridization between two species with different histories of interactions with parasites can modify the mechanisms of resistance of each taxon to its parasites.

Biologists have taken several perspectives in studying the interactions of parasites and pathogens with hybrids. First, parasites and pathogens could reach high levels on hybrid animals and plants (75, 84), reducing fitness of hybrids (4), limiting the abundance of hybrid individuals, limiting the extent and location of the hybrid zone (59, 60), and causing selection among hybrid genotypes (12). Second, the population dynamics and epidemiology of parasites and pathogens on nearby parental populations of hosts could be influenced by their abundance on hybrid individuals (e.g., 85). Third, hybrid-parasite interactions can be useful for investigating mechanisms of resistance in the parental populations (29). Parasite-hybrid interactions can be model systems for studying basic ecological and evolutionary interactions between parasites and hosts. Here, we review recent studies of herbivores and fungal parasites of plant hybrids and internal and external parasites of animal hybrids; we consider interspecific and intersubspecific hybrids that are known to occur in nature or have a natural equivalent. We first examine the models of resistance of hybridizing organisms, then compare field studies and laboratory or common garden experiments, and finally consider the ecological and evolutionary consequences of these interactions on both host and parasite populations.

MODELS OF HYBRID RESISTANCE

Quantitative Models

Several hypotheses concerning the resistance of hybrid organisms relative to parental species assume that resistance is polygenic and that quantitative genetics models describe the patterns of parasite response (1, 8, 25, 28, 30, 56, 78). Some models specify that hybrids are F_1s (25, 30), but other models do not specify the genotypes of hybrids (1, 8, 9). Tests for resistance of hybrids and parental species may employ post hoc comparisons (85), but a priori contrasts provide more explicit tests of alternative hypotheses and are preferable (8, 25, 28).

If parental species and hybrids do not differ in parasite abundance, the null hypothesis of no difference is supported (1, 8, 30). If the parental species differ in their parasite resistance and hybrids do not differ from the midparent value, resistance is additive (or intermediate) (1, 8, 28). Hybrid resistance may resemble that of a parental species, in which case the dominance hypothesis is supported (28). If hybrids resemble the susceptible parent, the pattern is dominance of susceptibility, whereas if hybrids resemble the resistant parent, the pattern is dominance of resistance. The prevalence of each of these two dominance patterns is indicative of the genetics of resistance mechanisms and influences the relative fitness of hybrid plants (25, 54). If hybrids have intermediate resistance that differs from exact additivity but tends toward one of the parents, partial dominance can be inferred (30). When susceptibility of hybrids is significantly greater than that of both parents, hybrids are considered to have overdominance for susceptibility, outbreeding depression (8), or hybrid susceptibility (28, 30); however, when resistance of hybrids is greater than that of both parents, heterosis (8), overdominance for resistance, or hybrid resistance (28, 30) exists. Whitham (84) coined the term "hybrids-as-sinks" to describe the situation where hybrids have greater herbivore abundances than do parental species, but the term implies more than hybrid susceptibility. It implies a population process whereby herbivore abundances are maintained at lower levels on parental species (because they are drawn away from the parents) than would be the case if hybrids were absent (78). Thus, support for hybrid susceptibility does not provide support for the population dynamics implied by the "hybrids-as-sinks" hypothesis.

Due to genetic recombination, F_2 hybrids are predicted to demonstrate hybrid breakdown (25, 61, 75, 84). Such patterns imply the breakup of coadapted gene complexes, which may be one cause of hybrid susceptibility. With backcross hybrids, specifically backcrosses of F_1 to pure parents, resistance is often predicted to be intermediate (25, 30, 78, 85), but breakdown of resistance in backcrosses or dominance may also occur (25, 30, 57, 78).

Hybrid swarms are populations consisting of a range of hybrid genotypes resulting from extensive crossing between hybrid genotypes and parental species. Regression analyses are best suited for examining the patterns of resistance in hybrid swarms, where continuous variation in hybrid genotypes exists. Boecklen & Spellenberg (9) presented methods to test hybrid intermediacy (additive), hybrid susceptibility, and dominance hypotheses using linear, quadratic, and cubic regressions, respectively, when diverse hybrid genotypes are present. These regressions are performed by regressing herbivore load against hybrid genotype class or a hybrid index based on morphology or molecular markers.

Mendelian Models

Resistance that is due to a single gene of major effect may show segregation of discrete phenotypes rather than the continuous variation in resistance expected from quantitative models. Flor's (24) development of the gene-for-gene theory of the disease resistance of plants was based on the recognition of discrete infection types in the flax rust pathosystem. The cloning of genes for resistance and avirulence has confirmed the validity of the gene-for-gene model (13, 41). Discrete resistance phenotypes, undergirded by a gene-for-gene relationship, also characterize plant parasites other than rust fungi (13, 41). This model does not appear to be limited in its applicability to domesticated agricultural plants. Natural plant pathosystems also operate in a gene-for-gene manner (63, 66, 81).

When distinct phenotypic classes of resistance can be recognized, Mendelian hypotheses of simple inheritance can be tested. A controlled-cross pedigree, in which resistance to a particular pathogen segregates, is necessary for testing. Ideally the controlled cross will mimic, or reproduce, natural hybrids seen in the field. The F_1 progeny of a controlled cross may be either all susceptible or all resistant, matching the dominance-to-susceptible and dominance-to-resistance hypotheses, respectively. However, segregation for resistance to *Venturia populina* and to *Melampsora medusae* in F_1 progenies of *Populus trichocarpa* x *P. deltoides* has been reported (64). The *P. deltoides* parents of F_1 progenies apparently can be heterozygous for resistance to *Venturia populina* (64). Similarly, *P. trichocarpa* can be heterozygous for resistance to *Melampsora medusae* (63, 66).

The segregation of recessive resistance in an F_2 progeny of two susceptible parents can be especially striking. An example of this can be found in the resistance of *Populus trichocarpa* x *P. deltoides* to stem canker caused by *Septoria musiva.* Although *P. deltoides* is resistant its F_1 progeny are always susceptible. However, when two susceptible F_1 clones are crossed, resistant F_2 individuals are seen (G Newcombe, ME Ostry, unpublished data). Conversely, resistance to *Melampsora occidentalis* is dominant, so that interspecific hybrid F_1s are all resistant. Yet the latter, when crossed, produce some susceptible progeny (65, 66).

PATTERNS OF HYBRID RESISTANCE

Comparison of Studies

We have compiled results of 162 tests of the expected pattern of hybrid resistance from 47 studies that compare responses of parasites to hybrid and parental plants and animals (Table 1). Parasite and host species cover a wide range of taxa, but studies of herbivorous insects on plants predominate. A range of parasite responses have been evaluated, including density, abundance, prevalence, intensity, damage, growth rate, development time, mortality, and fecundity. Abundances of several parasites are often evaluated on the same hybrid and parental hosts. Frequently, the same species or guild of parasites has been studied in different experiments, or separate measures have been made of the responses of a parasite to hybrids (e.g., feeding preference and larval growth rate), so not all observations are independent of each other. The genetic status of hybrids is known to varying degrees in these studies; historical, morphological, or molecular methods are used to determine hybrid status. For these reasons, statistical analyses of the data are not appropriate. Despite these limitations, the comparison of results between plant and animal hybrids, among the taxa of parasites, and among types of hybrid responses, reveals some general patterns. In evaluating these studies, we categorize the results of the study according to the hypotheses described above and use means and statistics presented in each paper to determine the fit of results to the hypotheses.

TABLE 1 Summary of studies of parasites on hybrid plants and animals

Plants/Herbivores						
Taxa	Genotype	Method	Type of data	Parasite	Measure of response	Hypothesis/ Reference
Quercus dumosa x Q. engelmanni	MX	HI	FC	*Andricus californicus*	Incidence	DS 56
				Andricus californicus	Density	S
Quercus depressipes x Q. rugosa	MX	HI	FC	*Nepticula* spp.	Density	R 8
				Camararia spp.	Density	R
				Phyllonorycter spp.	Density	R
				Leaf gallers	Density	DR
				Stem gallers	Density	ND
Quercus coccolobifolia x Q. emoryi	MX	HI	FC	*Nepticula* spp.	Density	R
				Camararia spp.	Density	ND
				Phyllonorycter spp.	Density	ND
				Leaf gallers	Density	ND
				Stem gallers	Density	ND
Quercus grisea x Q. gambelii	MX	HI	FC	*Nepticula* spp.	Density	DR 1
				Camararia spp.	Density	ND
				spp.	Density	ND
				Leaf gallers	Density	ND
				Stem gallers	Density	ND
Populus fremontii x P. angustifolia	MX	RFLP	FC	*Pemphigus betae*	Density	S 84
			CG	*Pemphigus betae*	Density	S

(Continued)

TABLE 1 (*Continued*)

Plants/Herbivores						
Taxa	Genotype	Method	Type of data	Parasite	Measure of response	Hypothesis/ Reference
Populus fremontii x P. angustifolia	MX	RFLP	FC	*Pemphigus betae*	Density	S 70
			CG	*Pemphigus betae*	Density	S
Populus fremontii x P. angustifolia	MX	RFLP	CG	*Chrysomela confluens*	Devel. time	DS 23
				Chrysomela confluens	Larval mortality	ND
				Chrysomela confluens	Male mass	ND
				Chrysomela confluens	Female mass	ND
				Chrysomela confluens	Fecundity	A
Populus deltoides x P. angustifolia x P. balsamifera	MX	HI	FC	*Aceria parpopuli*	Density	S 43
				Aceria parpopuli	Density	ND
Salix sericea x S. eriocephala	MX	RAPD	FC†	*Phyllonorycter salicifoliella*	Density	S 30
				Phyllocnistis sp.	Density	A
				Aculops tetanothrix	Density	DR
				Phyllocolpa nigrita	Density	A
				Phyllocolpa eleanorae	Density	DR
				Phyllocolpa terminalis	Density	S
				Leaf folder-LF	Density	S
				Leaf folder-V	Density	A
Salix sericea x S. eriocephala	MX	RAPD	CG	*Phyllonorycter salicifoliella*	Density	S 29
				Phyllocnistis sp.	Density	A
				Aculops tetanothrix	Density	ND
				Phyllocolpa nigrita	Density	ND

				Phyllocolpa eleanorae	Density	ND	
				Phyllocolpa terminalis	Density	A	
				Leaf folder-LF	Density	A	
				Leaf folder-V	Density	DS	
				Caloptilia sp.	Density	S	
Salix sericea x S. eriocephala	F1	RAPD	CG	*Phyllonorycter salicifoliella*	Density	ND	
				Phyllocnistis sp.	Density	A	
				Aculops tetanothrix	Density	ND	
				Phyllocolpa nigrita	Density	ND	
				Phyllocolpa eleanorae	Density	A	
				Phyllocolpa terminalis	Density	A	
				Leaf folder-LF	Density	A	
				Leaf folder-V	Density	DS	
				Caloptilia sp.	Density	ND	
Salix sericea x S. eriocephala	MX	HI	LE	*Chrysomela scripta*	Preference	DS	69
				Chrysomela knabi	Preference	DS	
				Calligrapha multi. bigsby.	Preference	S	
				Plagiodera versicolora	Preference	S	
			FC	*Popilla japonica*	Damage	DR	
Salix caprea x S. phylicifolia	F1	CR	CG	*Lochmaea caprea*	Damage 1994	ND	38
				Lochmaea caprea	Damage 1995	S	
Salix caprea x S. repens	F1	CR	CG	*Pontania pedunculi*	Density	DS	39
				Pontania brigmanii	Density	DS	

(*Continued*)

TABLE 1 (*Continued*)

Plants/Herbivores						
Taxa	Genotype	Method	Type of data	Parasite	Measure of response	Hypothesis/ Reference
				Iteomyia capreae	Density	DS
				Crepidodera fulvicornis	Density	ND
				chrysomelid larvae	Density	ND
				Dasineura rosaria	Density	ND
Ulmus wilsoniana x 'Urban' elm 3	F1	CR	LE	*Xanthogaleruca luteola*	Longevity-Male	A 35
				Xanthogaleruca luteola	Longevity-Female	A
				Xanthogaleruca luteola	Eggs/female	A
				Xanthogaleruca luteola	Oviposition percent	A
				Xanthogaleruca luteola	Preoviposition	A
				Xanthogaleruca luteola	Egg/Ovipositing Female	A
Cercidium floridum x C. microphyllum	MX	HI, CH	FC	*Mimosestes amicus*	Emergence/egg	A 76
				Mimosestes amicus	Survival 1987	ND
				Mimosestes amicus	Survival 1991	DS
				Mimosestes amicus	Larval mort	DS
				Mimosestes amicus	Oviposition	ND
				Stator limbatus	Survival 1987	ND
				Stator limbatus	Larval mort	DS
			LE	*Stator limbatus*	Larval mort	DS
Alnus glutinosus x A. incana	F1	HS	FC	*Pterocallis alni*	Density	PD 31

			FE	*Pterocallis alni*	Mortality	A	
				Pterocallis alni	Development time	ND	
				Pterocallis alni	Growth rate	ND	
				Pterocallis alni	Fecundity	DS	
Pinus thunbergiana x P. densiflora	F1	CR	CG	*Matsucoccus matsumurae*	Density	DS	53
	F1	CR	CG	*Matsucoccus matsumurae*	Survival	DS	
Pinus thunbergiana x P. massoniana	F1	CR	CG	*Matsucoccus matsumurae*	Density	DS	
	F1	CR	CG	*Matsucoccus matsumurae*	Survival	DS	
Pinus densiflora x P. massoniana	F1	CR	CG	*Matsucoccos matsumurae*	Density	DS	
	F1	CR	CG	*Matsucoccus matsumurae*	Survival	DS	
Pinus edulis x P. californiarum	MX	HI	FC	*Dioryctria albovittella*	Density	S	10
				Pitch moth	Density	DS	
Picea rubens x P. mariana	MX	HI	FC	*Choristoneura fumiferana*	Damage	A	43
Artemesia tridentata tridentata x A. tridentata vaseyana	MX	HI, CE	FC	*Odocoileus hemionus*	Damage	DS	32
				Grasshoppers	Damage	S	
				Total galls	Density	S	
				Aphids	Frequency	S	
				Cercopeus artemisiae	Density	DR	
	F2	CR	CG	*Rhopalomyia obovata*	Occurrence	DS	54
				Apterona helix	Occurrence	DS	
				Obtusicauda coweni	Density	DS	
				Melanoplus sanguinipes	Survival	S	
				Melanoplus sanguinipes	Growth	S	

(*Continued*)

TABLE 1 (*Continued*)

Plants/Herbivores						
Taxa	Genotype	Method	Type of data	Parasite	Measure of response	Hypothesis/ Reference
				Trirhabda pilosa	Growth	S
				Apterona helix	Occurrence	DS
				Obtusicauda fillifoliae	Occurrence	ND
				Calstoptera brunnea	Occurrence	ND
				Rhopalomyia ampullaria	Occurrence	ND
Betula pubescens tortuosa x B. nana	F1	CA	LE	*Epirrata autumata*	RGR	DS 36
				Amauronematus amplus	RGR	A
				Amauronematus fallax	RGR	ND
				Nematus umbratus	RGR	A
				Nematus brevivalvis	RGR	A
				Nematus pravus	RGR	A
				Nematus viridis	RGR	ND
				Arge fuscinervis	RGR	A
				Priophorus pallipes	RGR	A
				Dineura viridridorsata	RGR	ND
Typha latifolia x T. angustifolia	F1	HI	FC	*Lymnaecia phragmitella*	Density	R 19
				Dycimolomia julianalis	Density	A
Eucalyptus obliqua x E. baxteri	MX	HI	FC	Leaf gall	Density	ND 57
				Glycaspis cameloides	Density	ND
				Schedotrioza serrata	Density	DR
				Stem gall	Density	A

Plants/ Fungal Pathogens							
Silene alba x S. dioica	F1	CR	LE	*Ustilago violacea*	Infection success	DS	6
				Ustilago violacea	Infection success	A	
Populus trichocarpa x P. deltoides	F1	CR	FE	*Septoria populicola*	Necrotic leaf lesions	DR	65
				Septoria musiva	Stem canker	DS[a]	
				Melampsora occidentalis	Leaf rust	DR	63
				Linospora tetraspora	Necrotic leaf lesions	DS[b]	
Carex canescens x C. mackenziei	F1	HI	FC	*Anthracoidea fischeri*	Number	S	21
Lilium longiflorum x L. dauricum	F1	CR	GH	*Fusarium oxysporum*	Basal rot	DR	47
Eucalyptus nitens x E. globulus	F1	CR	FE	*Mycosphaerella cryptica*	Necrotic lesions—juvenile leaves	S	18
				Mycosphaerella cryptica	Necrotic lesions—adult leaves	DS	
				Mycosphaerella cryptica	Necrotic lesions—adult leaves	A	
Eucalyptus bicostata x E. globulus	F1	CR	FE	*Mycosphaerella cryptica*	Necrotic lesions—adult leaves	S	
				Mycosphaerella cryptica	Necrotic lesions—juvenile leaves	DS	
Nicotiana glutinosa x N. debneyi	F1	CR	GH	*Cercospora nicotianae, Phytophthora parasitica var. nicotianae, Pseudomonas*	Various	R	2

(*Continued*)

TABLE 1 *(Continued)*

Taxa	Plants/Herbivores Genotype	Method	Type of data	Parasite	Measure of response	Hypothesis/	Reference
				syringae pvs. tabaci and syringae, and tobacco mosaic virus			
Eucalyptus amygdalina x E. risdonii	F1	HI	FC	*Yellow leaf spot*	Necrotic leaf lesions	DS	85
	B1	HI	FC	*Yellow leaf spot*	Necrotic leaf lesions	S	
Castanea dentata x C. mollissima	F2	CR	FE	*Endothia parasitica*	Canker expansion	DR	44
Salix sericea x S. eriocephala	MX	CR	FC	*Melampsora* sp.	Uredinia score	S	28
			CG	*Melampsora* sp.	Uredinia score	S	74
	F1	CR	CG	*Melampsora* sp.	Uredinia score	S	
Salix viminalis x S. dasyclados	F1	CR	LE	*Melampsora epitea*—86 VIM	Incidence	A	24
				Melampsora epitea—77 DAS	Incidence	DS	
				Melampsora epitea—164 VXT	Incidence	DS	
				Melampsora epitea—86 VIM	Uredinia/Leaf	A	
				Melampsora epitea—77 DAS	Uredinia/Leaf	A	
				Melampsora epitea—86 VIM	Spores/Uredinia	DS	
				Melampsora epitea—77 DAS	Spores/Uredinia	DS	
Salix caprea x S. repens	F1	CR	CG	*Melampsora* sp.	Uredinia score	S	39
Animals							
Mytilus edulis x M. galloprvinialis	MX	PE	FC, FE	*Prosorhynchus squamatus*	Prevalence	A	12
Barbus meridionalis x B. hassi	MX	PE	FC	*Dactylogyrux* sp.	Prevalence	A	20
Barbus meridionalis x B. barbus	MX	PE	FC	*Diplozoon gracile*	Prevalence	A	45

Alburnus alburnus x Rutilio rutilio	F1	HI, PE	FC	*Diplozoon* sp.	Prevalance	S	17
				Bolbophorus confusus	Prevalence	S	
				Pomphorhyncus bosniacus	Abundance	S	
Anas platyrhyncos x A. rubripes	MX	HI?	FC	*Sarcocystis* sp.	Prevalence	S	50
Geomys bursarius x G. lutescens	MX	HI, CA, PE	FC	*Geomydoecus geomydis*	Identity	DR	37
				Geomydoecus nebrathkensis	Identiy	DR	
Mus musculus musculus x M. musculus domesticus	MX	PE	FC	*Syphacia* sp., *Aspiculuris* sp., *Protospitura* sp., *Trichocephalus* sp., *Trichuris* sp., *Hymenolepis* sp., *Catenotaenia* sp., *Taenia* sp.	Abundance	S	75
	MX	PE	FC	*Syphacia obvelata, Aspiculuris tetraptera*	Abundance	S	59
	MX	PE	LE	*Aspiculuris tetraptera*	Intensity	S	60
	F1	CR	LE	*Aspiculuris tetraptera*	Intensity	R	61
Mus m. domesticus Scottish chromosomal races	MX	CA	FC	*Syphacia obvelata, Aspiculuris tetraptera*	Abundance	R	72

[a]Newcombe and Ostry, unpublished.

[b]Newcombe, unpublished.

Genotype—F1, F2, BC, MX—Mixed genotypes; Method—HI-morphology hybrid index, CA—chromosomal analysis, CR—Cross, HS—history, CH—chemical traits, PE—protein electrophoresis, RFLP, RAPD.

Type of Data—FC—Field censuses, FE—Field experiment, CG—Common garden, RT—Reciprocal transplant, LE—Lab experiment, GH—Greenhouse.

Hypothesis supported—A—Additive; DS—Dominance to susceptible parent, DR—Dominance to resistant parent, PD—Partial Dominance, S—Hybrid susceptibility, R—Hybrid resistance, ND—No difference among parents and hybrids response—Prevalence: presence/absence of parasites in a host sample; Intensity: parasite loads of parasitized animals in the host sample; Abundance: parasite loads in the host sample (including the unparasitized host) (49).

The four most commonly reported responses of herbivorous insects to hybrid and parental plants across all studies are (a) no difference, (b) additive, (c) hybrid susceptibility, and (d) dominance of the susceptible parent (Table 2). Field studies and common garden studies report about equal percentages of no difference, additive, and hybrid susceptibility patterns (Table 2). Hybrid resistance and partial dominance are less frequently reported in field censuses, but dominance to the susceptible parent and dominance of resistance were equally frequent. The low occurrence of partial dominance, either of susceptibility or of resistant parents, is probably due to lack of specific tests of the hypothesis to distinguish it from dominance. The commonness of dominance of susceptibility and the absence of dominance of resistance or hybrid resistance in common garden and laboratory studies suggest that resistance is typically recessive (e.g., 53). These differences suggest that environmental variation, not controlled in field studies, influences the responses of herbivores to hybrid plants. The role of interaction between genetic and environmental components of hybrid resistance needs careful study (25) and is a central issue in hybrid zone theory (3).

Studies of plant pathogens on hybrid plants are fewer in number but show many of the same patterns noted with herbivores (Table 1, Table 2). Dominance of susceptible parent (10 of 28 studies) and susceptibility (8 of 28 studies) are the most common patterns, followed by additivity (6 of 28 studies). Hybrid resistance occurred in 1 of 28 studies and dominance of resistant species occurred in 4 of 28 studies. Most of these studies were performed on known F_1 genotypes in common garden or in laboratory experiments, indicating that the patterns are largely due to genetic differences in susceptibility among the taxa.

Only 14 tests of parasite loads on animal hybrids have been performed. Three support the additive hypothesis (12, 20, 45); 7 support the hybrid susceptibility

TABLE 2 Number and percentage of tests that support hybrid resistance hypotheses from plant-parasite studies presented in Table 1

	Plants/Herbivores									
Hypothesis	**Field census**		**Common garden**		**Laboratory**		**Plants/Fungi**		**Animals**	
No difference	17	30.4%	17	34.7%	3	14.3%	0	0%	0	0%
Additive	8	14.3%	8	16.3%	12	57.1%	5	17.9%	3	21.4%
Dominance										
Susceptible	7	12.5%	16	32.7%	4	19.0%	10	35.7%	0	0%
Resistant	7	12.5%	0	0%	0	0%	4	14.3%	2	14.3%
Partial dominance	1	1.8%	0	0%	0	0%	0	0%	0	0%
Susceptible	11	19.6%	8	16.3%	2	9.5%	8	28.6%	7	50%
Resistant	5	8.9%	0	0%	0	0%	1	3.6%	2	14.3%
Total	56		49		21		28		14	

hypothesis (17, 50, 59–61, 75); 2 may be regarded as a case of dominance of resistance (37), and 2 suggest hybrid resistance (61, 72). The apparent commonness of hybrid resistance and dominance of resistance among these animal studies warrant attention to determine their possible causes and whether there may be a fundamental difference from plant hybrid and herbivore/pathogen studies in expression of resistance in some hybrids.

Common Garden and Laboratory Studies

The effects of environmental variation in field studies of hybrid zones confound the interpretation of genetic causes of variation in hybrid resistance. A good example comes from the *Mus musculus musculus/domesticus* hybrid zone. From in situ studies, a genetic basis of the hybrid susceptibility to helminth parasites was proposed (59, 75). Meiotic recombination was suggested to have disrupted the coadaptation of resistance genes in each subspecies' genome. However, the results of these studies could not allow rejection of the alternative hypothesis that eco-ethological differences between hybrid and parental populations of mice could explain their differential level of infestation. Only experimental work in standardized environmental conditions could definitively establish the origin of the hybrid susceptibility. Thus, two experimental approaches were undertaken with *Aspicularis tetraptera* and hybrid mice. The first study compared the intrinsic susceptibility of experimentally infested mice that originated from the hybrid and the parental populations of the hybrid zone. This study showed that the hybrid mice are more heavily infested than parental ones and clearly established the genetic origin of the hybrid overinfestation (60). The hybrid breakdown hypothesis (59, 75) was tested by producing experimental F_1 hybrids (absent in the hybrid zone). In the genome of F_1 hybrids, the sets of genes of the two taxa are combined without recombination. Laboratory infestations of these mice showed they have a higher level of resistance to pinworms than do the parental mice. Thus, not only hybrid vigor in F_1 hybrids was demonstrated, but also the role of recombination in the hybrid susceptibility was strongly supported (61).

A common garden study conducted on F_2 hybrids (54) of subspecies of *Artemmisid tridentata tredentata* and *A. t. vaseyana* (Table 1) showed the hybrid susceptibility and dominance of the susceptible subspecies for seven parasite species, documenting the genetic effects of hybridization on resistance; no difference was found for three parasite species, showing the absence of genetic effects in those cases. The susceptibility of F_2 plants was not compared to F_1 plants in this study because F_1's were not included in the garden.

Fritz et al (29) compared the susceptibility of clones of parent willows and of hybrids of unknown pedigree, taken from the field, and also compared parental and F_1 hybrids from crosses of pure parents in common garden experiments. These studies confirmed that genetic effects of hybridization on resistance were present for several herbivore species, but they also showed the absence of genetic effects for several species. Comparison to field studies on the same herbivores showed that in several cases hypotheses supported in the common garden differed with those found in the field, implicating environmental effects on field resistance patterns.

The results from these studies argue for the significance of genetic effects of hybridization on resistance to parasites, for the coadaptation of resistance traits in species, and for the importance of environmental variation within and across hybrid zones in affecting resistance. There is a need for careful comparisons of parental species and F_1 and F_2 progeny in common garden or laboratory studies to test the coadaptation hypothesis. The involvement of genetic and environmental variation in hybrid resistance presents important research opportunities for studies of genotype by environment interactions on resistance and for testing models of hybrid fitness across hybrid zones (3, 4, 25).

Variable Parasite Responses to Hybrids

A common feature of nearly all plant hybrid-herbivore studies, whether conducted in the field, common garden, or laboratory, is that different herbivore species respond differently to the same hybrid individuals (25, 29, 30, 78). Studies of multispecies assemblages find that several hypothesized herbivore responses are supported (Table 1) (7, 25, 28, 30, 54, 57, 85). Even related parasite species can support different hypotheses on the same hybrids (7, 30). This apparently idiosyncratic response of herbivores to hybrids suggests two conclusions. First, resistance is specific to the parasite being considered. Different parasites are affected by different resistance traits or attractants or respond differently to variation in the same trait. Second, mechanisms that condition resistance to parasites in the parental species are affected differently by hybridization. Traits may be additive, dominant, or recessive in hybrids. Because of the variable results shown above, studies of fitness in these same systems could suggest different roles of parasites in selecting against hybrids or in favoring hybrids and promoting introgression.

Unlike the plant hybrid studies, the studies of several intestinal parasite species on intersubspecific hybrids of mice show a common pattern of hybrid susceptibility (75). Moreover, F_1 cyprinids were found to have a pattern of susceptibility similar to that of parasites located in different tissues of their hosts (17). In these two cases, studies of the mechanisms of resistance could reveal if these parasites are responding to the same resistance traits.

MECHANISMS OF HYBRID RESISTANCE

Hybrids inherit the resistance traits of both parental species, but it is the expression and interaction of the resistance traits or attractants that determine their effect on parasitism. Hybrids can be intermediate in expression of potential resistance mechanisms such as trichome density (31, 77) or chemical defenses (52, 68, 77). Some chemical traits in hybrids are inherited as dominant traits (e.g. 40, 46, 67, 73). Other resistance mechanisms are recessive in hybrids. Pod abscission, a defensive trait of palo verde (*Cercidium microphyllum*), appears to be recessive in hybrids with *C. floridum*, whereas seed coats of *C. floridum* are chemically defended against bruchid weevil attack, and this trait is also recessive in hybrids (76). Hybrids may show novel resistance traits not found in their parental species (73, 83), and some chemical defenses show heterosis in hybrids (52).

Few studies have made specific connections between inheritance of defensive or attractant traits of hybrids and the responses of parasites, but these could be especially valuable at uncovering mechanisms of resistance in pure species and in explaining susceptibility of hybrids. Parasites may show additive, dominant, threshold, or overdominant responses to hybrid traits. Additive responses suggest parallels between herbivore response and variation in a chemical or physical defensive or attractant trait (e.g., 77), whereas dominant responses suggest inverse response of parasites to dominant resistance traits (69) and positive responses of parasites to dominant attractants (56).

The threshold model of hybrid resistance seems to operate in some systems. Trichome density appeared to be a resistance trait of *Alnus incana*, perhaps leading to dislodgment of aphids on these leaves. In the F_1 hybrid with *A. glutinosa* (named *A. pubescens*) trichome density was intermediate between the two parents, but aphid susceptibility showed a dominance deviation toward *A. glutinosa*, suggesting a threshold density of trichomes for resistance that exceeds the levels found in the F_1 hybrid (31). A similar threshold model could explain why hybrid willows are susceptible to the leaf beetle *Plagiodera versicolora* (69) when concentrations of two possible chemical defenses (phenolic glycosides and tannins) are inherited additively (68). Painting experiments with phenolic glycosides indicate a defensive role of these chemicals in resistance of hybrid willows to Japanese beetles (*Popillia japonica*) (69), suggesting that hybrid concentrations of phenolic glycosides exceed a threshold level to confer resistance. A threshold of attractants was suggested in the susceptibility of plant hybrids to herbivores (54, 56).

Hybrid susceptibility and hybrid resistance could occur by several mechanisms. Hybrids with novel defenses or with heterotic quantities of defenses could be more resistant to parasites. In contrast, hybrids might be more susceptible to parasites if the different resistance mechanisms from the parental species are additively inherited but are below a threshold to confer resistance or if the expression of resistance mechanisms is disrupted by recombination.

Experimental studies of *Mus musculus musculus/domesticus* hybrids in the laboratory suggested that heterosis of F_1 resistance was due to complementation of resistance systems (61), and they further suggested that hybrid breakdown present in field populations of advanced generation hybrid mice is due to breakup of coadapted sets of resistance genes. Higher growth rate in a grasshopper (*Melanoplus sanguinipes*) and a leaf-feeding beetle (*Trirhabda pilosa*) on F_2 hybrids of basin and mountain big sagebrush suggests breakdown in resistance. Without comparisons to F_1s, however, it is not possible to determine if breakdown occurred in F_1s as well, or only in F_2s, the latter of which would suggest the presence of coadapted resistance genes (54). More experimental studies that compare parasite abundance and performance on both parents, F_1s, and F_2s, are needed to determine if or in what genotypes hybrid breakdown occurs.

No direct connection may exist between susceptibility and specific resistance traits. Hybridization may affect behavior, phenology, or physiology in ways that lead to altered hybrid resistance. For example, behavioral changes in hybrid fish in rivers in France brought hybrids into contact with parasites (45), but no intrinsic differences existed in the susceptibility of hybrid and parental fish to the parasites. Earlier flowering phenology of wild x crop sunflower hybrids was partially, but not completely, responsible for the much greater parasitism of achenes compared to wild sunflowers (*Helianthus annuus*) (14). The role of environmental variation in influencing the expression of resistance or attractant traits and interactions between environmental and behavior, phenology, or physiology is largely unexplored.

CONSEQUENCES FOR HYBRIDS

Parasite Load on Naturally Occurring Hybrids

The overall load of parasites on hybrids in hybrid zones when multispecies assemblages have been evaluated is useful in estimating the relative fitness of hybrids, testing models of hybrid zone theory, and in assessing the role of parasites in selecting for resistance in hybrid populations. Several studies have shown that overall abundance of herbivores on hybrids is intermediate between parental species (1, 8, 28, 30). Lower herbivore densities were found on hybrids in one oak hybrid zone, but densities were equal to densities on one of the parental species in another oak hybrid zone (8). Lack of difference between hybrid and parental species has also been found (10). In animal hybrids, overall susceptibility has been found (58–61, 75). Two studies found higher abundances of herbivores on hybrid plants (30, 57), but in another hybrid zone, relative abundances of herbivores were greater on probable backcrosses to each parental species than on either intermediate hybrids or parents (57). Thus, nearly the same range of variation in individual species seen in field and common garden studies is found when total herbivore abundances are considered. Despite the evident importance of resistance in determining plant fitness, no studies so far have experimentally linked the total herbivore or parasite load of hybrid plants with hybrid fitness.

Fitness of Hybrids

The pattern of hybrid susceptibility would suggest that parasites might have a detrimental effect on hybrid fitness. If so, then parasites could be instrumental in limiting the size of hybrid zones and in limiting the extent of introgression. Hybrid susceptibility was found 10–20% of the time in studies for insect herbivores, 27% of the time for fungal parasites on hybrid plants, and 50% of the time in animal studies (Table 2). Parasites will not lead to lower fitness of hybrids relative to parents if resistance is no different, additive, partially dominant, or dominant to resistance. Although specific studies of hybrid fitness due to parasites have not been conducted in most cases, the high frequency of these patterns does suggest that hybrid fitness will not usually be lower than that of parental species. However, the low frequency of hybrid resistance also does not suggest that hybrids will have higher fitness than both parents, at least regarding parasite impact.

A few studies have investigated the fitness of hybrid plants and the role that parasites play in determining hybrid fitness. Percentage of fungal infection (*Ramularia* sp.) and seed predation was higher on hybrids between *Eucalyptus melanophloia* and *E. crebra* but was intermediate on hybrids of *E. populnea* and *E. crebra* (16). In the former case, fitness of hybrids was much lower than that of parent species. Capsule production of hybrids between *Eucalyptus obliqua* and *E. baxteri* was lowest on the intermediate hybrids in the hybrid zone and was greatest on the parents and apparent backcrosses to parents in the hybrid zone (57). In a fourth *Eucalyptus* hybrid zone, capsule production was found to be equal on hybrid classes and on one parental species, but significantly higher on the other parental species in the hybrid zone in 1990 and 1992, although in 1980 capsule production was lower on hybrids than on parental species (85).

Models of hybrid zone dynamics distinguish between endogenous and exogenous selection acting on hybrids. Endogenous selection is postulated in dynamic equilibrium models where hybridization disrupts coadapted gene complexes leading to lower fitness, regardless of environment (5). Exogenous selection, predicting an important role of environmental variation, is postulated in the bounded hybrid superiority model (55) where hybrids are predicted to have higher fitness in specific habitats, different from those of parental species. There are no tests of the effects of parental and hybrid zone environments on hybrid fitness due to parasite impact. Tests of the alternative hypotheses will require, preferably parental, F_1, and F_2 progeny reciprocally transplanted to parental and hybrid habitats (e.g., 4, 82), along with measurement of parasite load and fitness.

If parasites affect host fitness and if hybridization with a resistant host species produces resistant hybrids, then hybrid genotypes can be favored in an introgressed population. The mussel *Mytilus edulis* and hybrids with *M. galloprovincialis* were highly parasitized by the trematode *Prosorhynchus squamatus*, which caused total castration and possibly death (12). Cousteau et al concluded that the parasite produced intense selection, favoring the spread of *M. galloprovincialis* genes. Such an interaction can cause movement of a hybrid zone, favoring the resistant species over the susceptible species. At some point where other factors limit the spread of *M. galloprovincialis*, introgression with *M. edulis* would be favored since genes that confer resistance to the trematode will confer higher fitness of hybrids. Grazers are postulated to have favored introgression between two shrubs in western North America, *Cowania stansburyana* and *Purshia tridentata* (79). Stutz & Thomas suggested that selection by grazers favored introgression of defensive traits of *Cowania stansburyana*, which is avoided by grazers, into some populations of *Purshia tridentata* (bitterbrush), a heavily grazed plant in western North America. In other examples, introgression of jack pine resistance traits appears to explain geographical variation in resistance of lodgepole pine to several insects and diseases (87), and introgression of red spruce into black spruce resulted in greater resistance to spruce budworm (49). The role of parasites in selecting for introgression of resistance genes is virtually unstudied, but it could be a creative process in the evolution of plant defenses.

CONSEQUENCES FOR PARENTAL SPECIES

Do Hybrids Maintain Higher Parasite Populations on Parental Species?

The observation that most of the population of a gall-forming aphid was found on hybrids rather than on pure parental species suggested the possibility that susceptible hybrids could limit the abundance of a parasite on parental species (sink hypothesis) (84). An alternative hypothesis is that hybrids could be a reservoir for parasites, which could disperse and colonize parental species at higher rates than would occur in the absence of hybrids (source hypothesis) (78). These alternative population consequences of hybrid hosts in populations could profoundly affect the dynamics and epidemiology of natural parasite populations in hybrid zones. Where parental and hybrid individuals coexist, unless specialized adaptation of the parasite to the hybrid host has occurred, parasite populations ought to be maintained at higher levels on parents than would occur in pure single parent populations or in contact zones.

Several field studies of hybrid and pure parental zones compare herbivore abundances on parental species. No difference was found between parents in pure zones and parents in hybrid zones in 18 of 20 comparisons in *Eucalyptus* systems, but in 2 of 20 comparisons lower densities occurred on parents in pure zones than in the hybrid zone (57, 85). Environmental variation between pure and hybrid zones confounds the interpretation of these results but does not suggest broad support for either the sink or the source hypothesis. Experimental studies are needed where mixed hybrid and pure one or two parent populations are established and monitored for herbivore abundances or parasite infection rates. Such experiments would be very informative about the population dynamic consequences of hybridization if the effects of local colonization versus regional parasite dispersal could be controlled.

COMMUNITY CONSEQUENCES OF HYBRID-PARASITE INTERACTIONS

Do Generalist and Specialist Parasites Differ in Hybrid and Parent Host Use?

Among insect herbivores on plants there are both specialists and generalists, and likewise, fungal parasites of plants usually are highly specialized but may be generalists. We define specialists as those species that attack only the genus of hybridizing plant, which may mean only one of the parental species. Generalists are defined as those that feed on more than one genus of plants, including the hybridizing genus. Specialists and generalists are predicted to differ in their responses to hybrids: Generalists are predicted to show higher abundances on hybrids of all types because of the breakdown of resistance mechanisms (85). Whitham et al (85) found greater abundance of generalists on hybrid than on parental *Eucalyptus*. In other studies, dominance for susceptibility was found for a highly generalist bagworm (54), dominance of resistance was found for the Japanese beetle (69), and an additive pattern was found for the spruce budworm on spruce (48). At present there appears to be no consistency among the results.

The predicted differences between herbivores or pathogens that are single-species specialists and those that utilize both parental species may prove to be more consistent. For example, specialists on only one of the parental species are predicted to attack F_1 hybrids, providing resistance is additive, but to show higher abundance on backcrosses to the susceptible parent. Among the 28 specialists of the *E. amygdalina* x *E. risdonii* hybrid zone, 68% (19) were most abundant on all hybrid classes, and 4% (1) were most abundant on the least similar hybrid category (85). However, gall-forming cynipids occurred equally on parental and all hybrids, but not on a nonhost oak (56).

Biodiversity on Hybrids

Hybrids often have morphology (e.g., 8, 28, 76, 85) and chemistry (e.g., 68) that is intermediate between their parents. Occasionally, hybrids may have transgressive or unique morphology or chemical traits (73). This leads to three predictions. First, this uniqueness of ethological or physiological traits of hybrids could explain why hybrids could be unsuitable for the development of any parasite species (37). Second, hybrids may also be attractive to and suitable for parasite species that do not utilize either parental species (17). In spruce, 12% to 15% of insect species were unique to spruce hybrids (51). Third, hybrids share characteristics of both parental species and therefore should share the herbivore species of both parental species (51, 84, 86). Thus, there should be higher species diversity on hybrids than on any pure parent species, and some evidence supports this hypothesis. Species richness was higher on intermediate hybrids and putative backcrosses than on parental species in *Populus* (15) and *Eucalyptus* (57, 85, 86) hybrid systems. Consequences of the last two patterns may be that unique interspecific interactions and evolution may occur on hybrids, creating greater interaction diversity (80).

Tritrophic Effects on Hybrid Plants

A few studies have considered the tritrophic effects of parasitoids or predators on herbivores of hybrid and parental plants. No studies have considered the effects of higher trophic levels on animal parasites or pathogens of plants, but there are known fungal parasites of plant pathogens that could provide comparable studies (64). The impact of the third trophic level might explain some of the variation in abundance and fitness of herbivores on hybrid plants. For example, lower parasitism or predation could explain the higher abundance of some herbivores on hybrids compared to parental species. No difference in parasitism by chalcid parasitoids was found on multichambered galls of *Andricus californicus* on *Quercus dumosa* and a wide range of hybrid genotypes (56). Gange (31) found that parasitism of aphids on hybrids of *Alnus* increased with aphid densities across parents and hybrids but could not be attributed to independent hybrid traits. Eisenbach (19) found the parasitism rates of a seed-eating herbivore of two cattails and their F_1 hybrid were significantly greater on the hybrid, even though densities of the herbivore were significantly lower on the hybrids. This result suggests a direct effect of hybridization on susceptibility to parasites. Parasitism of *Phyllonorycter* on oak hybrids was lower than on parental species after densities were considered (71). Less parasitism of the leaf-mining moth *Phyllonorycter salicifoliella* on hybrids of willows was found in the field in one year, although this pattern was not repeated in the next year nor on F_1 hybrid plants in a common garden experiment (27). Predation on beetle larvae on hybrid hosts was intermediate between that on both parental species (34). These few studies show that a wide range of effects of parasitoids and predators on herbivores occurs on hybrids and parental plant species. Parasitoids may have a lower impact on herbivores on hybrid hosts in some cases, even when herbivore densities are intermediate or higher on hybrids (27, 71), but in other systems parasitoids have a greater impact on herbivores of hybrids (19). The former effect could favor herbivores that choose hybrid plants, but the latter case would select against herbivores that utilize hybrids. The outcome of these interactions will influence the population dynamics of parasites of hybrids and their effects on hybrid plant fitness.

CONCLUSIONS

We are struck by the similarity of patterns of responses of a diverse array of parasite taxa to plant and animal hybrids. Common features of parasite susceptibility of plant and animal hybrids are, one, the diversity of responses of parasites to hybrid hosts, two, the different responses of different parasites to the same host taxa, an especially important parameter to evaluate the evolutionary outcome of the hybridization process, and three, the commonness of genetic variation in resistance in most studies. One important difference between most studies of herbivore insects and those of plant pathogens and animal parasites is the role of complex life cycles. Populations of plant pathogens, animal parasites, and aphids on hybrid hosts and in hybrid zones may be affected by the availability of alternative hosts. The absence of alternate hosts could lead to erroneous conclusions about hybrid resistance. No studies have yet investigated availability of alternate hosts as a factor in parasite response to hybrids. One difference between studies of fungal pathogens of hybrids and the other studies of plant and animal hybrids is that fungi may be highly specific in their virulence to parental and hybrid hosts. In the case of highly specific rust fungi, host and parasite taxa may be matched so that the congeneric pair of host parental species are parasitized by a congeneric pair of rust species. Pathotypes of the same parasite species specific to each parental species may also exist in a single population. Since pathotypes are defined by differential interactions with hosts, a local population comprising mixed pathotypes will make the resistance pattern, for any given pathotype, uninterpretable.

Progress in this research indicates that experimental studies are needed to lead to further progress in understanding: 1. the genetic causes of parasite resistance in hybrids and the mechanisms of its breakdown, 2. the interaction between environmental variation in hybrid zones and mechanisms of resistance, 3. fitness of hybrids and its relationship to hybrid zone theory, 4. selection by parasites on introgression of traits in hybrids and across hybrid zones, and 5. the role of hybrids in the population dynamics of parasites, in biodiversity of parasites, and in tritrophic interactions of parasites.

ACKNOWLEDGMENTS

We thank CG Hochwender and Jean-Marc Derothe for comments on the manu script. A Shuckett assisted in preparing the manuscript. RSF was supported by NSF DEB 96-15038. GN was supported by the US Department of Energy, Biomass Energy Technology Division (supplement to grant 19X-43382C).

Visit the Annual Reviews home page at http://www.AnnualReviews.org

LITERATURE CITED

1. Aguilar JM, Boecklen WJ. 1992. Patterns of herbivory in the *Quercus grisea* x *Quercus gambelii* species complex. *Oikos* 64:498–504
2. Ahl Goy P, Felix G, Métraux JP, Meins F. 1992. Resistance to diseases in the hybrid *Nicotiana glutinosa x Nicotiana debneyi* is associated with high levels of chitinase, β-1.3-glucanase, peroxidase and polyphenoloxidase. *Physiol. Mol. Plant Pathol.* 41:11–21
3. Arnold ML. 1997. *Natural Hybridization and Evolution.* New York: Oxford Univ. Press
4. Arnold ML, Hodges SA. 1995. Are natural hybrids fit or unfit relative to their parents. *Trends Ecol. Evol.* 10:67–71
5. Barton NH, Hewitt GM. 1985. Analysis of hybrid zones. *Annu. Rev. Ecol. Syst.* 16:113–48
6. Biere A, Honders S. 1996. Host adaptation in the anther smut fungus *Ustilago violacea* (*Microbotryum violaceum*): infection success, spore production and alteration floral traits on two host species and the F1-hybrid. *Oecologia* 107:307–20
7. Boecklen WJ, Larson KC. 1994. Gall-forming wasps (Hymenoptera: Cynipidae) in an oak hybrid zone: testing hypotheses about hybrid susceptibility to herbivores. In *The Ecology and Evolution of Gall-Forming Insects,* ed. P Price, W Mattson, Y 8. Baranchikov, pp. 110–20. St. Paul: North Central For. Exp., Station, For. Serv., USDA
8. Boecklen WJ, Spellenberg R. 1990. Structure of herbivore communities in two oak (*Quercus* spp.) hybrid zones. *Oecologia* 85:92–100
9. Boecklen WJ, Spellenberg R. 1998. Tests

of hypotheses regarding hybrid resistance in the *Quercus coccolobifolia* x *Quercus viminea* species complex. In *The Biology of Gall-Inducing Arthropods*, ed. G Csoka, W Mattson, G Stone, pp. 295–304. St. Paul: North Central For. Exp. Station, US For. Serv.
10. Christensen KM, Whitham TG, Keim P. 1995. Herbivory and tree mortality across a pinyon pine hybrid zone. *Oecologia* 101:29–36
11. Combes C. 1996. Parasites, biodiversity and ecosystem stability. *Biodiversity Conserv.* pp. 5:953–62
12. Coustau C, Renaud F, Maillard C, Pasteur N, Delay B. 1991. Differential susceptibility to a trematode parasite among genotypes of the *Mytilus edulis/galloprovincialis* complex. *Genet. Res.* 57:207–12
13. Crute IR. 1998. The elucidation and exploitation of gene-for-gene recognition. *Plant Pathol.* 47:107–13
14. Cummings CL, Alexander HM, Snow AA. 1999. Increased pre-dispersal seed predation in sunflower crop-wild hybrids. *Oecologia.* In press
15. Dickson LL, Whitham TG. 1996. Genetically-based plant resistance traits affect arthropods, fungi, and birds. *Oecologia* 106:400–6
16. Drake DW. 1981. Reproductive success of two *Eucalyptus* hybrid populations. II. Comparison of predispersal seed parameters. *Austr. J. Bot.* 29:37–48
17. Dupont F, Crivelli AJ. 1988. Do parasites confer a disadvantage to hybrids? *Oecologia* 75:587–92
18. Dungey HS, Potts BM, Carnegie AJ, Ades PK. 1997. *Mycosphaerella* leaf disease: genetic variation in damage to *Eucalyptus nitens*, *Eucalyptus globulus*, and their F_1 hybrid. *Can. J. For. Res.* 27:750–59
19. Eisenbach J. 1996. Three-trophic-level interactions in cattail hybrid zones. *Oecologia* 105:258–65
20. El Gharbi S, Renaud F, Lambert A. 1992. Dactylogirids (Platyhelminthes: Monogenea) of *Barbus* spp. (Teleostei, Cyprinidae) from the Iberian Peninsula. *Res. Rev. Parasitol.* 52:103–16
21. Ericson L, Burdon JJ, Wennström A. 1993. Interspecific host hybrids and phalacrid beetles implicated in the local survival of smut pathogens. *Oikos* 68:393–400
22. Floate KD, Martinsen GD, Whitham TG. 1997. Cottonwood hybrid zones as centres of abundance for gall aphids in western North America: importance of relative habitat size. *J. Anim. Ecol.* 66:179–88
23. Floate KD, Kearsley MJC, Whitham TG. 1993. Elevated herbivory in plant hybrid zones: *Chrysomela confluens*, *Populus* and phenological sinks. *Ecology* 74:2056–65
24. Flor HH. 1971. Current status of the gene-for-gene concept. *Annu. Rev. Phytopathol.* 9:275–96
25. Fritz RS. 1999. Resistance of hybrid plants to herbivores: genes, environment, or both? *Ecology* 80:382–91
26. Fritz RS, Johansson L, Åstrom B, Ramstedt M, Häggström H. 1996. Susceptibility of pure and hybrid willows to isolates of *Melampsora epitea* rust. *Eur. J. Plant Pathol.* 102:875–81
27. Fritz RS, McDonough SE, Rhoads AG. 1997. Effects of plant hybridization on herbivore-parasitoid interactions. *Oecologia* 110:360–67
28. Fritz RS, Nichols-Orians CM, Brunsfeld SJ. 1994. Interspecific hybridization of plants and resistance to herbivores: hypotheses, genetics, and variable responses in a diverse community. *Oecologia* 97:106–17
29. Fritz, RS, Roche BM, Brunsfeld SJ. 1998. Genetic variation in herbivore resistance of hybrid willows. *Oikos* 83:117–28
30. Fritz RS, Roche BM, Brunsfeld SJ, Orians CM. 1996. Interspecific and temporal variation in herbivore responses to hybrid willows. *Oecologia* 108:121–29
31. Gange AC. 1995. Aphid performance in

an alder (*Alnus*) hybrid zone. *Ecology* 76:2074–83
32. Graham JH, Freeman DC, McArthur ED. 1995. Narrow hybrid zone between two subspecies of big sagebrush (*Artemisia tridentata*: Asteraceae). II. Selection gradients and hybrid fitness. *Am. J. Bot.* 82:709–16
33. Grant V. 1981. *Plant Speciation*. New York: Columbia Univ. Press. 2nd ed.
34. Häggström H, Larsson S. 1995. Slow larval growth on a suboptimal willow results in high predation mortality in the leaf beetles *Galerucella lineola*. *Oecologia* 104:308–15
35. Hall RW, Townsend AM. 1987. Suitability of *Ulmus wilsoniana*, the 'Urban' elm, and their hybrids for the elm leaf beetle, *Xanthogaleruca luteola* (Müller) (Coleoptera: Chrysomelidae). *Environ. Entomol.* 16:1042–44
36. Hanhimäki S, Senn J, Haukioja E. 1994. Performance of insect herbivores on hybridising trees: the case of the subarctic birches. *J. Anim. Ecol.* 63:163–75
37. Heaney LR, Timm RM. 1985. Morphology, genetics, and ecology of pocket gophers (genus *Geomys*) in a narrow hybrid zone. *Biol. J. Linn. Soc.* 25:301–17
38. Hjältén J. 1997. Willow hybrids and herbivory: a test of hypotheses of phytophage response to hybrid plants using a generalist leaf-feeder *Lochmaea caprea* (Chrysomelidae). *Oecologia* 109:571–74
39. Hjältén J. 1998. An experimental test of hybrid resistance to insects and pathogens using *Salix caprea*, *S. repens* and their F_1 hybrids. *Oecologia* 117:127–32
40. Huesing J, Jones D, Deveerna J, Myers J, Collins G, et al. 1989. Biochemical investigations of antibiosis material in leaf exudate of wild *Nicotiana* species and interspecific hybrids. *J. Chem. Ecol.* 15:1203–17
41. Hutcheson SW. 1998. Current concepts of active defense in plants. *Annu. Rev. Phytopathol.* 36:59–90
42. Deleted in proof
43. Kalisichuk AR, Gom LA, Floate KD, Rood SB. 1997. Intersectional cottonwood hybrids are particularly susceptible to the poplar bud gall mite. *Can. J. Bot.* 75:1349–55
44. Kubisiak TL, Hebard FV, Nelson DD, Zhan J, Bernatzky R, et al. 1997. Molecular mapping of resistance to blight in an interspecific cross in the genus *Castanea*. *Phytopathology* 87:751–59
45. Le Brun N, Renaud F, Berrebi P, Lambert A. 1992. Hybrid zones and host-parasite relationships: effect on the evolution of parasitic specificity. *Evolution* 46:56–61
46. Levy A, Milo J. 1991. Inheritance of morphological and chemical characters in interspecific hybrids between *Papaver bracteatum* and *Papaver pseudo-orientale*. *Theor. Appl. Genet.* 81:537–40
47. Löffler HJM, Meijer H, Straathof ThP, van Tuyl JM. 1996. Segregation of *Fusarium* resistance in an interspecific cross between *Lilium longiflorum* and *Lilium dauricum*. *Acta Hortic.* 414:203–7
48. Manley SAM, Fowler DP. 1969. Spruce budworm defoliation in relation to introgression in red and black spruce. *For. Sci.* 15:365–66
49. Margolis L, Esch GW, Holmes JC, Kuris AM, Schad G. 1982. The use of ecological terms in parasitology. *J. Parasitol.* 68:131–33
50. Mason JR, Clark L. 1990. Sarcosporidiosis observed more frequently in hybrids of Mallards and American Black Ducks. *Wilson Bull.* 102:160–62
51. Mattson WJ, Haack RK, Birr BA. 1996. F_1 hybrid spruces inherit the phytophagous insects of their parents. In *Dynamics of Forest Herbivory: Quest for Pattern and Principle* ed. W Mattson, P Niemela, M Rousi, 183:142–49. St. Paul: North Central For. Exp. Station, For. Serv., USDA
52. McArthur ED, Welch BL, Sanderson SC. 1988. Natural and artificial hybridization between big sagebrush (*Artemisia triden-*

tata) subspecies. *J. Hered.* 79:268–76
53. McClure MS. 1985. Susceptibility of pure and hybrid stands of *Pinus* to attack by *Matsucoccus matsumurae* in Japan (Homoptera: Coccoidea: Margarodidae). *Environ. Entomol.* 14:535–38
54. Messina FJ, Richards JH, McArthur ED. 1996. Variable responses of insects to hybrid versus parental sagebrush in common gardens. *Oecologia* 107:513–21
55. Moore WS. 1977. An evaluation of narrow hybrid zones in vertebrates. *Q. Rev. Biol.* 52:263–77
56. Moorehead JR, Taper ML, Case TJ. 1993. Utilization of hybrid oak hosts by a monophagous gall wasp: How little host character is sufficient? *Oecologia* 95:385–92
57. Morrow PA, Whitham TG, Potts BM, Ladiges P, Ashton DH, Williams JB. 1994. Gall-forming insects concentrate on hybrid phenotypes of *Eucalyptus*. In *The Ecology and Evolution of Gall-Forming Insects,* ed. P Price, W Mattson, Y Baranchikov, pp. 121–34. St. Paul: North Central For. Exp. Station, For. Serv., USDA
58. Moulia, C. 1999. Parasitism in plant and animal hybrids: Are facts and fates the same? *Ecology* 80:392–406
59. Moulia C, Aussel JP, Bonhomme F, Neilsen JT, Renaud F. 1991. Wormy mice in a hybrid zone: a genetic control of susceptibility to parasite infection. *J. Evol. Biol.* 4:679–87
60. Moulia C, Le Brun N, Dallas J, Orth A, Renaud F. 1993. Experimental evidence of genetic determinism in high susceptibility to intestinal pinworm infection in mice: a hybrid zone model. *Parasitology* 106:387–93
61. Moulia C, Le Brun N, Loubes C, Marin R, Renaud F. 1995. Hybrid vigour against parasites in interspecific crosses between two mice species. *Heredity* 74:48–52
62. Newcombe G. 1996. The specificity of fungal pathogens of *Populus.* In *Biology of Populus and its Implications for Management and Conservation,* ed. R Stettler, H Bradshaw Jr, P Heilman, T Hinckley. pp. 223–246. Ottawa: NRC Res. Press
63. Newcombe G. 1998. Association of *Mmd1,* major gene for resistance to *Melampsora medusae* f. sp. *deltoidae,* with quantitative traits in poplar rust. *Phytopathology* 88:114–21
64. Newcombe G. 1998. A review of exapted resistance to diseases of *Populus. Eur. J. For. Pathol.* 28:209–16
65. Newcombe G, Bradshaw HD. 1996. Quantitative trait loci conferring resistance in hybrid poplar to *Septoria populicola,* the cause of leaf spot. *Can. J. For. Res.* 26:1943–50
66. Newcombe G, Bradshaw HD Jr, Chastagner GA, Stettler RF. 1996. A major gene for resistance to *Melampsora medusae* f. sp. *deltoidae* in a hybrid poplar pedigree. *Phytopathology* 86:87–94
67. O'Donoughue LS, Raelson JV, Grant WF. 1990. A morphological study of interspecific hybrids in the genus *Lotus* (Fabaceae). *Can. J. Bot.* 68:803–12
68. Orians CM, Fritz RS. 1995. Secondary chemistry of hybrid and parental willows: phenolic glycosides and condensed tannins in *Salix sericea, S. eriocephala,* and their hybrids. *J. Chem. Ecol.* 21:1245–53
69. Orians CM, Huang CH, Wild A, Dorfman KA, Zee P, et al. 1997. Willow hybridization differentially affects preference and performance of herbivorous beetles. *Entomol. Exp. Appl.* 83:285–94
70. Paige KN, Capman WC. 1993. The effects of host-plant genotype, hybridization and environment on gall aphid attack and survival in cottonwood: the importance of genetic studies and the utility of RFLP's. *Evolution* 47:36–45
71. Preszler RW, Boecklen WJ. 1994. A three-trophic level analysis of the effects of plant hybridization on a leaf-mining moth. *Oecologia* 100:66–73
72. Ressouche L, Ganem G, Derothe J-M, Searle J-B, Renaud F, Moulia C. 1998. Host

chromosomal evolution and parasites of the house mouse *Mus musculus domesticus* in Scotland. *Zeigtschrift für Säugertierkunde* 63:52–57

73. Rieseberg LH, Ellstrand NC. 1993. What can molecular and morphological markers tell us about plant hybridization? *Crit. Rev. Plant Sci.* 12:213–41
74. Roche BM, Fritz RS. 1998. Effects of host plant hybridization on resistance to willow leaf rust caused by *Melampsora* sp. *Eur. J. For. Pathol.* 28:259–70
75. Sage RD, Heyman D, Lim KC, Wilson AC. 1986. Wormy mice in a hybrid zone. *Nature* 324:60–63
76. Siemens DH, Ralston BE, Johnson CD. 1994. Alternative seed defence mechanisms in a palo verde (Fabaceae) hybrid zone: effects on bruchid beetle abundance. *Ecol. Entomol.* 19:381–90
77. Soetens P, Rowell-Rahier M, Pasteels JM. 1991. Influence of phenolglucosides and trichome density on the distribution of insects herbivores on willows. *Entomol. Exp. Appl.* 59:175–87
78. Strauss SY. 1994. Levels of herbivory and parasitism in host hybrid zones. *Trends Ecol. Evol.* 9:209–14
79. Stutz HC, Thomas LK. 1964. Hybridization and introgression in *Cowania* and *Purshia. Evolution* 18:183–95
80. Thompson JN. 1994. *The Coevolutionary Process*. Chicago: Univ. Chicago Press
81. Thompson JN, Burdon JJ. 1992. Gene-for-gene coevolution between plants and parasites. *Nature* 360:121–25
82. Wang H, McArthur ED, Sanderson SC, Graham JH, Freeman DC. 1997. Narrow hybrid zone between two subspecies of big sagebrush (*Artemisia tridentata*: Asteraceae). IV. Reciprocal transplant experiments. *Evolution* 51:95–102
83. Weber DJ, Wang DR, Halls SC, Smith BN, McArthur ED. 1994. Inheritance of hydrocarbons in subspecific big sagebrush (*Artemisia tridentata*) hybrids. *Biochem. Syst. Ecol.* 22:689–97
84. Whitham TG. 1989. Plant hybrid zones as sinks for pests. *Science* 244:1490–93
85. Whitham TG, Morrow PA, Potts BM. 1994. Plant hybrid zones as centers of biodiversity: the herbivore community of two endemic Tasmanian eucalypts. *Oecologia* 97:481–90
86. Whitham TG, Martinsen GD, Floate KD, Dungey HS, Potts BM, Keim P. 1999. Plant hybrid zones affect biodiversity: tools for genetic-based understanding of community structure. *Ecology* 80:416–28
87. Wu HX, Ying CC, Muir JA. 1996. Effect of geographic variation and jack pine introgression on disease and insect resistance in lodgepole pine. *Can. J. For. Res.* 26:711–26

Annu. Rev. Ecol. Syst. 1999. 30:593–616

EVOLUTIONARY COMPUTATION: An Overview

Melanie Mitchell* and Charles E. Taylor**
**Santa Fe Institute, 1399 Hyde Park Road, Santa Fe, New Mexico 87501; e-mail: mm@santafe.edu, **Department of Organismic Biology, Ecology, and Evolution, University of California, Los Angeles, California 90095; e-mail: taylor@biology.ucla.edu*

Key Words artificial life, computational modeling

INTRODUCTION

Evolutionary computation is an area of computer science that uses ideas from biological evolution to solve computational problems. Many such problems require searching through a huge space of possibilities for solutions, such as among a vast number of possible hardware circuit layouts for a configuration that produces desired behavior, for a set of equations that will predict the ups and downs of a financial market, or for a collection of rules that will control a robot as it navigates its environment. Such computational problems often require a system to be adaptive—that is, to continue to perform well in a changing environment.

Problems like these require complex solutions that are usually difficult for human programmers to devise. Artificial intelligence practitioners once believed that it would be straightforward to encode the rules that would confer intelligence on a program; expert systems were one result of this early optimism. Nowadays, however, many researchers believe that the "rules" underlying intelligence are too complex for scientists to encode by hand in a top-down fashion. Instead they believe that the best route to artificial intelligence and other difficult computational problems is through a bottom-up paradigm in which humans write only very simple rules and provide a means for the system to adapt. Complex behaviors such as intelligence will emerge from the parallel application and interaction of these rules. Neural networks are one example of this philosophy; evolutionary computation is another.

Biological evolution is an appealing source of inspiration for addressing difficult computational problems. Evolution is, in effect, a method of searching among an enormous number of possibilities—e.g., the set of possible gene sequences—for "solutions" that allow organisms to survive and reproduce in their environments. Evolution can also be seen as a method for adapting to changing environments. And, viewed from a high level, the "rules" of evolution are remarkably simple: Species evolve by means of random variation (via mutation, recombination, and other operators), followed by natural selection in which the fittest tend to survive and reproduce, thus propagating their genetic material to future generations. Yet these simple rules are thought to be responsible for the extraordinary variety and complexity we see in the biosphere.

There are several approaches that have been followed in the field of evolutionary computation. The general term for such approaches is *evolutionary algorithms*. The most widely used form of evolutionary algorithms are *genetic algorithms* (GAs), which will be the main focus of this review. Other common forms of evolutionary algorithms will be described in the third section.

A SIMPLE GENETIC ALGORITHM

The simplest version of a genetic algorithm consists of the following components:

1. A population of candidate solutions to a given problem (e.g., candidate circuit layouts), each encoded according to a chosen representation scheme (e.g., a bit string encoding a spatial ordering of circuit components). The encoded candidate solutions in the population are referred to metaphorically as chromosomes, and units of the encoding (e.g., bits) are referred to as genes. The candidate solutions are typically haploid rather than diploid.
2. A fitness function that assigns a numerical value to each chromosome in the population measuring its quality as a candidate solution to the problem at hand.
3. A set of genetic operators to be applied to the chromosomes to create a new population. These typically include selection, in which the fittest chromosomes are chosen to produce offspring; crossover, in which two parent chromosomes recombine their genes to produce one or more offspring chromosomes; and mutation, in which one or more genes in an offspring are modified in some random fashion.

A typical GA carries out the following steps:

1. Start with a randomly generated population of n chromosomes.
2. Calculate the fitness $f(x)$ of each chromosome x in the population.
3. Repeat the following steps until n offspring have been created:
   (*a*) Select a pair of parent chromosomes from the current population, the probability of selection increasing as a function of fitness.
   (*b*) With probability p_c (the crossover probability), cross over the pair by taking part of the chromosome from one parent and the other part from the other parent. This forms a single offspring.
   (*c*) Mutate the resulting offspring at each locus with probability p_m (the mutation probability) and place the resulting chromosome in the new population. Mutation typically replaces the current value of a locus (e.g., 0) with another value (e.g., 1).

4. Replace the current population with the new population.
5. Go to step 2.

Each iteration of this process is called a generation. A genetic algorithm is typically iterated for anywhere from 50 to 500 or more generations. The entire set of generations is called a run. At the end of a run, there are typically one or more highly fit chromosomes in the population. Since randomness plays a large role in each run, two runs with different random-number seeds will generally produce different detailed behaviors.

The simple procedure just described is the basis for most applications of GAs. There are a number of details to fill in, such as how the candidate solutions are encoded, the size of the population, the details and probabilities of the selection, crossover, and mutation operators, and the maximum number of generations allowed. The success of the algorithm depends greatly on these details.

A BRIEF HISTORY OF EVOLUTIONARY COMPUTATION

In the 1950s and the 1960s several computer scientists independently invented different evolutionary computation methods. In the 1960s, Rechenberg introduced evolution strategies (61), a method he used to optimize real-valued parameters for devices such as airfoils. A "population" consisted of two individuals—a parent and a child mutated from the parent—each encoding a set of real-valued parameters to be optimized. The fitter of the two was selected to be the parent for the next generation. Mutation consisted of incrementing or decrementing a real value according to a given distribution. The parameters of this distribution were themselves encoded as part of each individual and thus coevolved with the parameters to be optimized. This idea was further developed by Schwefel (65, 66), and the theory and application of evolution strategies has remained an active area of research (see, e.g., 6).

Fogel, Owens, & Walsh developed *evolutionary programming* (23), a technique in which candidate solutions to given tasks were represented as finite-state machines and were evolved by randomly mutating their state-transition diagrams and selecting the fittest. As in evolution strategies, random mutation was the only source of variation. A somewhat broader formulation of evolutionary programming also remains an area of active research (see, e.g., 21).

The techniques called genetic algorithms (GAs) were first invented by Holland in the 1960s (31). GAs are population-based algorithms in which mutation and crossover are sources of random variation. GAs typically worked on strings of bits rather than real-valued parameters. The simple GA given above is close to the algorithm proposed by Holland. Holland's original proposal also included an "inversion" operator for reordering of bits on a chromosome. In contrast with evolution strategies and evolutionary programming, Holland's goal was not to design algorithms to solve specific problems, but rather to formally study the phenomenon of adaptation as it occurs in nature and to develop ways in which the mechanisms of natural adaptation might be imported into computer systems. Several other people working in the 1950s and 1960s developed evolution-inspired algorithms for optimization and machine learning; see (22, 26, 49) for discussions of this history.

In the last several years there has been widespread interaction among researchers studying various evolutionary computation methods, and the boundaries between evolution strategies, evolutionary programming, genetic algorithms, and other evolutionary computation methods have broken down to some extent. In this review most of our examples involve what researchers have called genetic algorithms, though in many cases these will be of a somewhat different form than Holland's original proposal.

In this review we first survey some applications of evolutionary computation in business, science, and education, and we conclude with a discussion of evolutionary computation research most relevant to problems in evolutionary biology. Due to space limitations, we do not survey the extensive work that has been done on the theoretical foundations of evolutionary computation; much work in this area can be found in the various *Foundations of Genetic Algorithms* proceedings volumes (11, 58, 76, 77) and in recent reviews (6, 21, 26, 49).

COMMERCIAL APPLICATIONS OF EVOLUTIONARY ALGORITHMS

We suggested in the introduction that evolution can be viewed as a method for searching through enormous numbers of possibilities in parallel, in order to find better solutions to computational problems. It is a way to find solutions that, if not necessarily optimal, are still good—what Simon (68) has termed "satisficing." Problems where it is sufficient to have satisficing solutions arise frequently in business and engineering. For many of these applications the traditional methods of search (e.g., hill-climbing and conjugate-gradient methods) perform well (55). It is well established that no single method of optimization is best for all applications—the best method will depend on the problem and computational devices at hand. We are aware of no good theory that will generally predict when one method or another is best for any particular problem. It has been speculated that evolutionary algorithms perform relatively well when there is a large number of parameters to be determined and when there is a high degree of epistasis so the adaptive surface of solutions is complex, having many intermediate optima. Lewontin (44) has stated that evolutionary computation has not solved any problems that could not be solved by traditional means. This might be true or not—we know of no tests of this statement. Nevertheless, evolutionary computation is finding use in a variety of commercial and scientific applications, and that number is certainly growing. Some examples include integrated circuit design, factory scheduling, robot sensor design, image processing, financial market prediction, and drug design.

For example, a common problem in drug design is to learn how well an arbitrary ligand will bind to some given enzyme. This problem is known to be NP-complete (52). A particular problem of great significance is to know how well inhibitors will bind to HIV protease enzymes. When an HIV-1 virus matures, it must cleave proteins manufactured by the viral genetic material into the individual proteins required by HIV-1. Such cleavage takes place at the active site in the protease. Protease inhibitors bind tightly into this active site and disrupt the life cycle of the virus. Such inhibitors are finding widespread use in treating AIDS. The market for protease inhibitors is thus huge. Companies would like to screen candidate molecules and determine whether they will fit into the active site and how well they will bind there. Natural Selection Inc. provided software to Aguron Pharmaceuticals that combines models of ligand-protein interactions for molecular recognition with evolutionary programming to search the space of all possible configurations of the ligand-protein complex (24). Each candidate solution of the evolving population is a vector with rigid-body coordinates and the angles about its rotatable bonds. Those individuals with the lowest calculated energies are then used as parents for the next generation. Mutations occur as random changes in the bond angles in the offspring candidates. In practice it is useful to have the magnitudes of the mutations themselves evolve as well.

Supply-chain management provides examples of a very different sort. For instance, Volvo trucks are built to order and have dozens of options for each tractor cab, giving millions of configurations that must be scheduled and inventory checked, after which tools and time must be provided at the appropriate place on the plant floor and then delivered there. Starting from a collection of average-quality schedules, the scheduling program provided by I2 Technologies evolves a satisficing schedule for plant production each week (57). Deere & Company was probably the first to use such methods (54), also provided by I2 Technologies, for making their John Deere tractors, and now employs the methods in six of their assembly plants.

Evolution in such cases is based on an optimizing-scheduling procedure that was developed and employed at the US Navy's Point Magu Naval Airbase. Each chromosome in the evolving population encodes a schedule—a permutation of the tasks to be performed—whose fitness is the cost of the schedule after it is further optimized by an additional scheduling program.

An altogether different problem is to predict stock market prices. Several large corporations, including Citibank, Midland Bank, and Swiss Bank (through their partner Prediction Company), have been evolving programs that attempt such predictions (38). Typically such methods involve backcasting—withdrawing the most recent data from the evaluators, then determining how well each program in the population predicts that data. Not surprisingly, the details of how such programs work, including their performance, are trade secrets, though, for reasons discussed in the section on the Baldwin effect, it seems that such programs are likely to contain some other sorts of learning mechanisms in addition to evolution.

Asahi Microsystems is building an evolvable hardware (EHW) chip that is part of an integrated circuit for cellular telephones. When computer chips are made, there is slight variation from chip to chip, due to differences in capacitance, resistance, etc. A percentage of such chips do not perform up to specification and so must be discarded. In general, as chips are made smaller they are less likely to perform to specification. In the case of analog cellular telephones, where size is a major issue, certain filters must perform to within 1% of the central frequencies. The laboratory of T Higuchi at Tsukuba (50) has shown how to build chips that are tunable with 38 parameters as they come off the assembly line. Their EHW microchip will alter these parameters using evolutionary computation and then set the appropriate switches on a field programmable gate array (FPGA). The EHW chip is leading to improved yield rates of the filter chip and also will lead to smaller circuits and power consumption. The chip is to appear in January 1999 with a target production of 400,000 chips per month.

Descriptions of various other commercial applications can be found in journals such as *Evolutionary Computation* or *IEEE Transactions on Evolutionary Computing*, and in various conference proceedings (e.g., 3, 7).

SCIENTIFIC APPLICATIONS OF EVOLUTIONARY ALGORITHMS

In addition to their commercial applications, evolutionary algorithms have been used extensively in science, both as search methods for solving scientific problems and as scientific models of evolutionary systems including population genetics, clonal selection in immune systems, innovation and the evolution of strategies in economic systems, and the evolution of collective behavior in social systems. In this section we review two scientific applications, one in ecology and one in computer science.

Using Genetic Programming to Evolve Optimal Foraging Strategies

Koza, Rice, & Roughgarden used a type of evolutionary algorithm (called genetic programming) to investigate two ecological questions: What makes for an optimal foraging strategy, and how can an evolutionary process assemble strategies that require complex calculations from simple components? (41). They addressed this by building on Roughgarden's work on foraging strategies of Anolis lizards. These lizards wait in trees for insects to appear in their field of vision and then pursue and eat those insects they deem most desirable.

The model strategies involved four variables: the abundance a of insects (the probability of an insect appearing per square meter per second, assumed to be uniform over space and time), the sprint velocity v of the lizard (assumed to be constant), and the coordinates x, y of the insect in the lizard's view, assumed in this case to be two dimensional. (It is also assumed that only one insect at a time is viewed, all insects are identical, and that each chase consists of the lizard leaving and returning to its perch before a new insect appears). A *strategy* is a function of these four variables that returns 1 if the insect is to be chased, -1 otherwise. The goal is to devise a function that maximizes food capture per unit time. Clearly not every insect is worth chasing; if an insect is too small or too far away it will take too much time to catch it and might even be gone by the time the lizard reaches it.

In Koza, Rice, & Roughgarden's model, a single simulation of a lizard's behavior consisted of assigning values to the variables a and v, and then allowing 300 simulated seconds in which insects appear at different x, y coordinates (with uniform probability over x and y); the lizard uses its strategy to decide which ones to chase. In one of Koza, Rice, & Roughgarden's experiments, the lizard's 10×20 meter viewing area was divided into three regions, as shown in Figure 1(a). Insects appearing in region I always escaped when chased; those appearing in region II never escaped; and those appearing in region III escaped with probability zero on the x axis and linearly increasing with angle to a maximum of 0.5 on the y axis. The optimal strategy is for a lizard to always ignore insects in region 1, chase those in region 2 that are sufficiently close, and chase those in region 3 that are

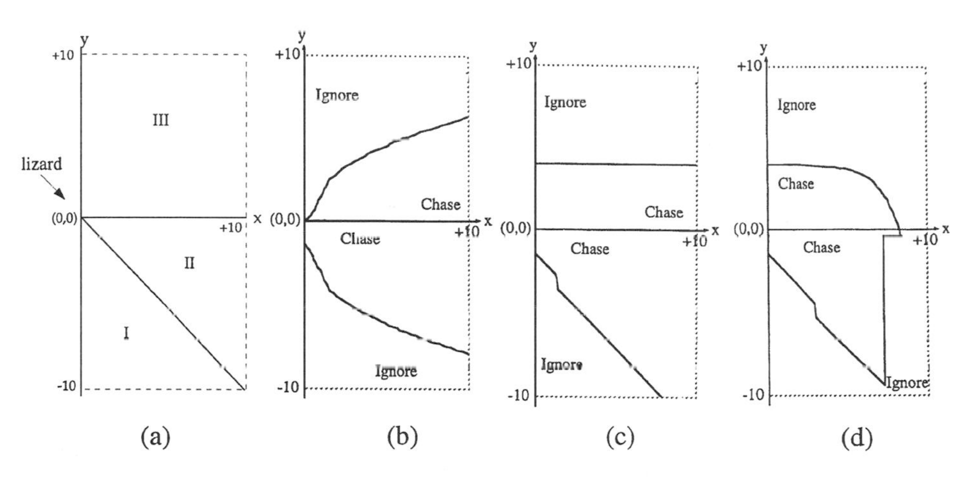

Figure 1 (a) The lizard's viewing area, divided into three regions, each with a different escape probability for insects. (b)–(d) Switching curves to illustrate the behavior of the best program in the population at generations 0 (b), 12 (c), and 46 (d) from one run of the genetic programming algorithm. The curves divide the lizard's viewing area into regions in which it will chase insects and regions in which it will ignore insects. (Adapted from 41.)

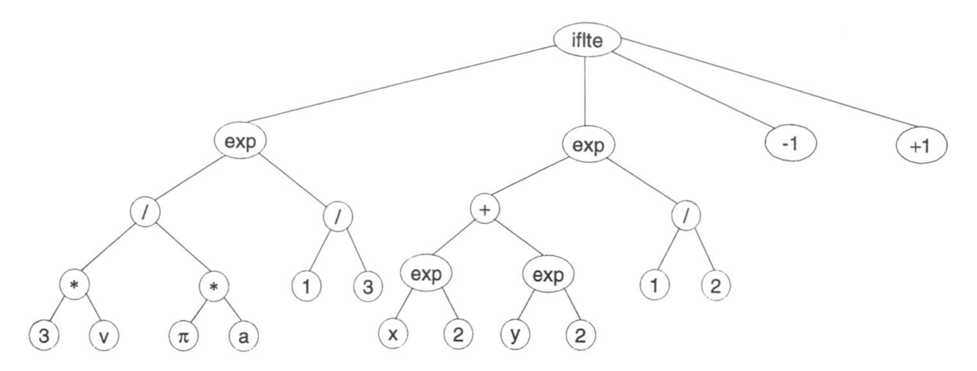

Figure 2 The optimal strategy for a simpler version of the foraging problem, encoded as a parse tree.

sufficiently close and have a high-enough probability of not escaping. No analytic form for the optimal strategy in this model was known.

The genetic programming algorithm is a type of genetic algorithm devised by Koza (40) in which the individuals undergoing evolution are computer programs encoded as parse trees. For example, Koza, Rice, & Roughgarden derived by hand the following formula for an optimal strategy in a simpler version of the problem in which insects never escape in any region:

$$Sig\left[\left(\frac{3v}{\pi a}\right)^{\frac{1}{3}} - (x^2 + y^2)^{\frac{1}{2}}\right], \qquad 1.$$

where $Sig(z)$ returns $+1$ if z is positive and -1 otherwise. Figure 2 gives this same strategy in the form of a parse tree (similar to those formed by compilers when parsing computer programs), in which nodes contain either functions (e.g., $+$, $-$, exponentiation) or the terminals on which the functions are to be performed (e.g., v, a, π, 1, 2, 3). In Figure 2, *exp* is an exponentiation operator that takes the absolute value of its first argument and raises that to its second argument, and *iflte* is the "if less than or equal to" operator: If its first argument is less than or equal to its second argument, it returns its third argument; otherwise it returns its fourth argument.

The objective is to discover a structure like that in Figure 2 that encodes an optimal (or at least good) strategy for the foraging task. It starts by forming a population of M randomly generated programs in parse-tree format. The fitness of each population member is then calculated by simulating the behavior of the corresponding strategy for several values of the variables (fitness cases). The highest fitness individuals then reproduce—either by direct copying or by crossover (no mutation was used)—to produce a new population; this whole evolutionary process is iterated for many generations.

The initial random programs are generated using a predefined set of functions and terminals. One of the arts of genetic programming is figuring out what should

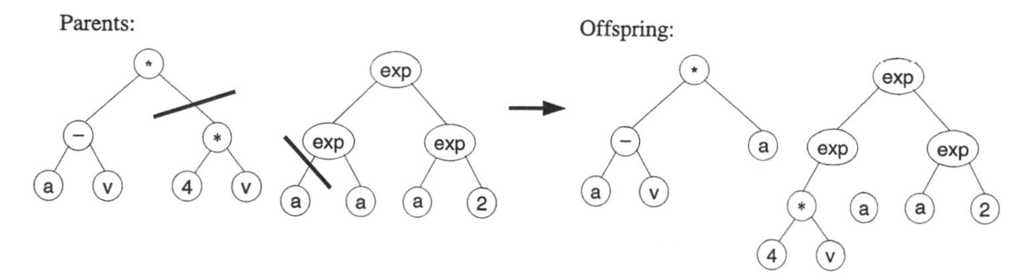

Figure 3 Illustration of crossover between two parse trees to produce two offspring. The (randomly chosen) point of crossover on each tree is marked by a dark line.

be in this set. Koza, Rice, & Roughgarden's set of terminals was $\{a, v, x, y, \mathcal{R}\}$, where $\mathcal{R}$ produces, each time it appears in an initial program, a random floating point number between -1 and $+1$. Their set of functions was $\{+, -, *, /,$ *exp, iflte*$\}$. Koza, Rice, & Roughgarden presumably constructed this set of functions and terminals via intelligent guesswork. For more details about how the initial population is generated, see (41).

The fitness of each program in the population was calculated by simulating the program with several different values for a, v, x, y and measuring the total number of insects that were eaten over the different simulations. Once the fitness of each individual program has been calculated, some fraction of the highest fitness individuals form a new population via copying themselves directly or by crossing over to create offspring. A crossover between two programs is illustrated in Figure 3. Two parents are selected to cross over, a random point is chosen in each program, and the subtrees at that point are exchanged to form two offspring, which are added to the new population. The copying and crossover procedures are repeated until a new population of M individuals has been formed. This whole process is repeated for some number G of generations. In Koza, Rice, & Roughgarden's algorithm, $M = 1000$ and $G = 61$.

Although the evolved strategies themselves were often hard to interpret, nonetheless, runs exhibited a sequence of progressively improved strategies. Each program's behavior can be visualized on a number of cases via "switching curves"—curves that illustrate in what regions respectively the lizard will chase or ignore insects. Figures 1(b)–(d) give switching curves for the best individuals in the population at generations 0, 12, and 46. It can be seen that genetic programming produced individuals with increasingly fit behavior over the course of evolution. For example, the best individual at generation 46 will avoid insects in an area that approximates region I, chase insects in a region that approximates region II, and in region III the distance the lizard is willing to travel is greatest on the x-axis and decreases with angular distance on the y axis.

In short, Koza, Rice, & Roughgarden's work showed that genetic programming can evolve increasingly fit foraging behavior in this particular simplified model. The evolved strategies can be considered hypotheses about real-life foraging strategies—possibly difficult for humans to devise—and experiments in the real world can test their validity.

Hosts and Parasites: Using GAs to Evolve Sorting Networks

Our second example of a scientific application is work by Hillis on "host-parasite" coevolution as applied to genetic algorithms (29). Hillis, like many other GA researchers, found that in genetic algorithms, adaptation in a static environment results in both the loss of diversity in a population of candidate solutions and evolved solutions that are "overfit" to that static environment—that is, solutions that do not generalize well when placed in new environments. His solution was to have the environment itself—in the form of "parasites"—evolve to be increasingly challenging for the evolving candidate solutions.

The problem Hillis tackled was that of evolving minimal sorting networks. Sorting is a much studied problem in computer science whose goal is to place the elements in a data structure in some specified order (e.g., numerical or alphabetic) in minimal time. One particular approach to sorting is the sorting network, a parallelizable device for sorting lists with a fixed number n of elements. In a simplified form, a sorting network consists of an ordered list of comparisons to be made between elements in the given list; the compared elements are to be swapped if they are out of order. For example, the sorting network

$$(2, 5), (4, 2), (7, 14) \ldots$$

specifies that the second and fifth elements are to be compared (and possibly swapped), then the fourth and second elements are to be compared, followed by the seventh and fourteenth, and so on. A correct sorting network will take any list of a fixed length n and, after performing the specified comparisons and swaps, return the list in correctly sorted order.

In the 1960s several researchers had worked on the problem of finding correct sorting networks for $n = 16$ with a minimal number of comparisons. It was first believed that the minimum was 65 comparisons, but then smaller and smaller networks were discovered, culminating in a 60-comparison sorter. No proof of its minimality was found, but no smaller network was discovered. (See 39 for a discussion of this history).

Hillis used a form of the genetic algorithm to search for minimal sorting networks for $n = 16$. There were two criteria for networks in the population: correctness and small size. Small size was rewarded implicitly due to a diploid encoding scheme in which networks with fewer comparisons were encoded as chromosomes with more homozygous sites; smaller networks were more robust to crossovers and thus tended to be implicitly favored. Correctness was rewarded explicitly via the fitness function: Each network was tested on a sample of fitness cases (lists to be sorted). There were too many possible input cases to test each network exhaustively, so at each generation each network was tested on a set of cases chosen at random. The fitness of a network was equal to the percentage of cases it sorted correctly.

Hillis's GA was a considerably modified version of the simple GA described above. The individuals in the population were placed on a two-dimensional lattice; thus, unlike in the simple GA, there was a notion of spatial distance between two strings. Hillis hoped that this scheme would foster "speciation"—that different types of networks would arise at different spatial locations—rather than having the whole population converge to a set of very similar networks.

Nonetheless, the GA got stuck at local optima. The GA found a number of moderately good (65-comparison) solutions but could not proceed to correct smaller solutions. One reason was that after early generations the randomly generated test cases used to compute the fitness of each individual were not challenging enough. The GA had evolved strategies that worked well on the test cases they were presented with, and the difficulty of the test cases remained roughly the same. Thus, after the early generations there was no pressure on the evolving population to change the current suboptimal sorting strategies.

To solve this problem, Hillis used a form of host-parasite (or predator-prey) coevolution, in which the sorting networks were viewed as hosts and the test cases (lists of 16 numbers) as parasites. Hillis modified the system so that a population of networks coevolved on the same grid as a population of parasites, where a parasite consisted of a set of 10–20 test cases. Both populations evolved under a GA. The fitness of a network was now determined by the parasite located at the network's grid location. The network's fitness was the percentage of test cases in the parasite that it sorted correctly. The fitness of the parasite was the percentage of its test cases that the network sorted incorrectly.

The evolving population of test cases was thought to provide increasing challenges to the evolving population of networks. As the networks got better and better at sorting the test cases, the test cases presumably got harder and harder, evolving to specifically target weaknesses in the networks. This forced the population of networks to keep changing—to keep discovering new sorting strategies—rather than staying stuck at the same suboptimal strategies. With coevolution, the GA discovered correct networks with only 61 comparisons—a real improvement over the best networks discovered without coevolution, though a single comparison away from rivaling the smallest known network for $n = 16$.

Hillis's work is important because it introduced a new, potentially very useful GA technique inspired by coevolution in biology, and his results are a convincing example of the potential power of such biological inspiration. Additional work on coevolution in genetic algorithms has been done by a number of people; see, e.g., (35, 53, 62).

Educational Applications of Evolutionary Computation

It is sobering to reflect that polls show 40% of all Americans do not believe in Darwinian evolution and that this is true for 25% of college-educated Americans as well. Several factors contribute to this deficiency, including religious convictions and a lack of understanding by students and teachers alike. The mode of

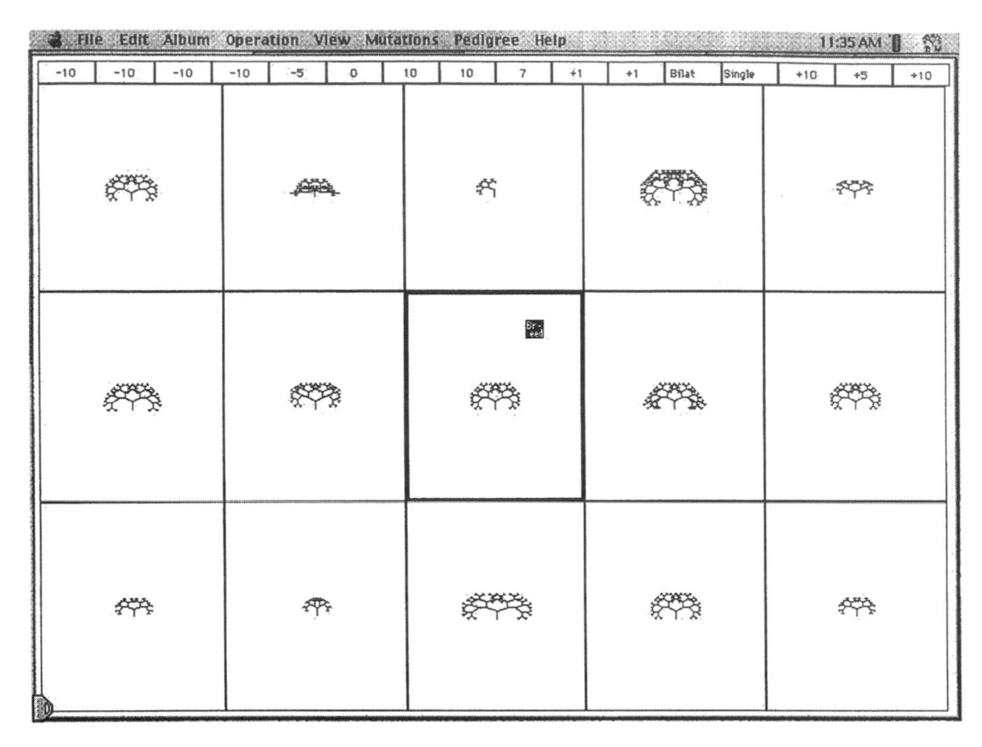

Figure 4 "Breeding screen" of the Blind Watchmaker program. The parent biomorph is present at the center of the screen and offspring biomorphs are surrounding it. Figures in the line immediately above the screen signify the genotype of the parent. Offspring biomorphs differ from the parent by mutations in the genotype string occurring with a probability determined by a parameter in the parent. Modified from (17).

presentation is also important. It is well-established that students are less likely to understand when they are passively presented with facts than when they are engaged in experimentation and construction. Several educational computer programs use evolutionary computation to address this challenge.

One such program, The Blind Watchmaker (17), is an exemplary means to teach students how evolution works. Figure 4 shows a typical screen when running the program. Near the middle is an individual "biomorph" that was chosen from the prior generation and was the "seed" for the current generation. It is surrounded by variant biomorphs that were generated by mutations. They may involve mutations of morphological features such as branch length, symmetry, gradients of length, number of branches, or even of the mutation rates and size themselves. The values for these parameters are shown in the line of figures just above the gallery of biomorphs. The figure illustrates what sorts of changes are possible even within one generation. The user can chose any of the variants shown by clicking on it, then a new generation is generated, based on variants of that biomorph, and displayed around it. It becomes an engaging challenge to evolve forms that are exotic in one way or another, and while doing so the user acquires a real feeling for mutation, selection, and the enormous variety of phenotypes that evolution can produce. The program is keyed to Dawkins' popular book with the same title (18), which uses the program to illustrate many key features of evolution.

Like the biomorphs, the Blind Watchmaker program has itself been the seed for several interesting variants. One such variant, where the focus is on developing interesting pictures, was developed by Karl Sims and displayed at the Centre Georges Pompidou in Paris. Additional variants of this evolution have contributed to his stunning videos, such as Liquid Selves, some of which are described in (69). Besides the programs described here, there is a large variety of other popular software that employs and illustrates the creativity of evolution, including SimLife (37) and Creatures (15), nurturing hope that evolutionary computation will provide a useful teaching experience and that the next generation of American students might show a better understanding of evolution than this one has.

RELEVANCE TO PROBLEMS IN EVOLUTIONARY BIOLOGY

There is a major intellectual divide in the field of evolutionary computation, much as there is in the field of artificial intelligence. On the one hand there is a desire to make systems as efficient as possible for solving engineering problems, irrespective of any relevance to the biological world. The commercial applications discussed above, for example, frequently involve populations of 100 or so candidate solutions per generation with mutation rates on the order of 10^{-2}. Such conditions are rarely if ever met in the biological world. On the other hand, a great deal of work is now directed toward understanding general features of complex adaptive systems, including organic and cultural evolution, learning, immune reactions, and so on. The hope has been that by studying simple systems that can be dissected and rerun, it will be possible to identify rules governing adaptation that might otherwise be difficult to analyze. This pertains especially to systems with several levels of organization or that involve collective action giving rise to emergent behavior. Most of this research has been directed toward understanding the trade-offs among two or more modes of adaptation, such as learning versus evolution.

Here we can give only a sample of the biologically relevant problems that have been addressed, but we hope it gives a feel for the variety of problems, approaches, and findings that the field encompasses. We begin by discussing evolution in systems that probably do not exist in the biological world right now but that exhibit adaptation nonetheless—systems with Lamarckian evolution. (They are surprisingly effective at solving some problems.) We next discuss an interaction between two modes of adaptation—development and evolution—variously described as "genetic assimilation" or the "Baldwin effect." The Baldwin effect is thought to be important for adaptation in many artificial systems, though it has received only scant attention by biologists. This is followed by a discussion of cultural and social evolution. These have proven difficult to study by traditional means, and the findings from evolutionary computation are often quite different from those in more traditional evolutionary biology. Finally we address some of the findings from attempts to optimize the response to evolution itself, such as optimizing the mutation or recombination rates that typically fall under the heading "evolution of genetic systems."

Lamarckian Evolution

Lamarckian Evolution refers to the evolution of traits that are modified through experience and are passed on, in their modified form, to the genotype of the next generation (42, 43). While this is consistent with certain pre-Mendelian theories of inheritance, including that which Darwin himself used, it is now recognized never to occur due to the lack of a mechanism for accomplishing it in natural systems. Artificial organisms are, of course, not subject to such constraints, and the study of Lamarckian evolution in such systems sheds some light on issues of general evolvability. For Lamarckian evolution to occur requires both a means of adapting within a generation (e.g., via development or learning) and a means of passing those gains to the genotype of the subsequent generation. Models of learning studied in this context include neural networks (63, 64), Hopfield networks (34) and production systems (28).

Hopfield networks have the ability to learn associations and, most remarkably, exhibit content addressable memory. The mere smell of a cookie, for example, might evoke all sorts of memories that have nothing to do with cookies themselves; sensing just a few properties of an object can recover a whole host of other properties. Starting from random configurations of a Hopfield network, the number of memories reliably learned and stored is approximately 0.15 times the number of nodes in a completely connected system (32). This result depends on the starting conditions, and some configurations can lead to much greater ability to remember. Imada & Araki (34) presented a set of inputs to a population of Hopfield networks capable of learning connection weights (encoded as real-valued vectors) via Hebbian (reinforcement) learning to perform a pattern-recognition task. Each generation there were learning trials, where the vectors in the evolving population were modified via a supervised-learning method. At the end of several learning trials, the weights modified via learning replaced those that had started the generation. Thus evolution was Lamarckian. Then all possible inputs were presented, and it was observed how many stable fixed points (memories) were reached by the system. If exactly those states corresponding to the inputs were obtained, then the fitness was set to its maximum value. If different vectors or more vectors were observed as fixed points, then fitnesses were diminished accordingly. After this fitness evaluation, mutation and recombination occurred, and a next generation was formed, followed by another round of learning, fitness evaluation, and selection. Nearly twice as many memories could be reliably acquired with Lamarckian evolution as could be acquired without the Hebbian learning phase. Further, this larger number of memories could be learned even more rapidly than the much smaller number of memories acquired by networks evolving in a purely Darwinian fashion. Clearly the ability to learn and to pass this experience on through the genotype accelerated evolution.

Exploring this in more depth, Tokoro and coworkers (63, 64) found that adaptation by Lamarckian evolution was indeed much faster for neural networks than Darwinian evolution when the vectors to be learned were the same from generation to generation, that is, when the environment was constant. But when the environment changed randomly from generation to generation, then Darwinian evolution was superior. Further, when modifier genes that determined the amount of Lamarckian abilities the networks possess were themselves allowed to evolve, the Lamarckian abilities were lost completely in a randomly changing environment.

Apparently the relative advantages of Lamarckian versus Darwinian evolution alone must depend on the degree of correlation in the environment from one generation to the next, in much the way that modifiers of recombination and sexuality do. In view of the very large differences in adaptability observed here, we must expect that these differences will be likely to be exploited in practical applications of evolutionary computation.

Baldwin Effect

Lamarckian evolution is often more effective than Darwinian evolution because the space of phenotypes can be searched more extensively when each individual can try out not just one phenotype but a whole suite of new possibilities serially within their lifetime, perhaps even guided by learning. For example, in the experiments on the evolution of pattern recognition by Hopfield networks just described, each individual instantiated a genotype that generation. Under Darwinian evolution alone, a total of *number_of_agents* $*$ *number_of_generations* networks can be explored, maximum. But with learning, each trial during the Hebbian learning phase could explore yet another network, so the maximum now is *number_of_agents* $*$ *number_of_generations* $*$ *number_of_trials*, which is potentially much greater. The problem is how to pass on successful discoveries to the next generation. As is well-known, the lack of a suitable mechanism prevents biological organisms from exploiting Larmarckian evolution. There is, however, a reasonable alternative that is both possible and well-suited for evolution. This is "genetic assimilation" or the "Baldwin effect."

Many years ago C. Waddington (74) observed that *Drosophila melanogaster* normally produce posterior cross-veins in their wings. But when exposed to heat shock as pupae, they occasionally fail to develop the cross-veins in their wings when they become adults. Waddington started selection from a base population in which all of the adults had cross-veins. Each generation he heat-shocked the offspring and selected from those who were later cross-veinless as adults. After 14 generations, 80% of those who were heat-shocked were cross-veinless, and a few began to be cross-veinless even without the shock. With subsequent selection he obtained lines with as many as 95% cross-veinless in the absence of shock. He recognized that this was not Lamarckian evolution, but that it rather resulted simply from changing the thresholds for expression of the cross-vein trait; Waddington termed the phenomenon "genetic assimilation" (74). It also happened that a similar phenomenon had been described earlier by JM Baldwin and is sometimes called the "Baldwin effect" (10). In textbooks of evolution this phenomenon is occasionally mentioned but seldom receives more than a brief note.

The Baldwin effect has been observed in evolutionary computation studies (see, e.g., 1, 30). In Waddinton's study the problem was to select for a trait (cross-veinlessness) that is almost never expressed. The importance for evolutionary computation is slightly different; it sometimes occurs that a trait is enormously useful if it is fully developed or expressed, but it is of no use otherwise. The problem is to hit upon the right (and rare) configuration of alleles, then preserve it for further selection and elaboration. In an asexual population, the right ensemble of alleles might never (or almost never) arise. In a sexual population it might arise but would tend to be broken up immediately by recombination. However, if learning or other forms of adaptation during individuals' lifetime are available, the desired configuration can arise via these mechanisms; while the trait itself will not be passed on to offspring, the genetic background producing it will be favored. Thus, according to Baldwin, learning and other forms of within-lifetime adaptation can lead to increased survival, which can eventually lead to genetic variation that produces the trait genetically.

This effect has been demonstrated in simple evolutionary computation settings. For example, Hinton & Nowlan (30) considered neural networks that evolved via GAs. At each generation every individual in the population had a "lifetime" during which its weights were learned. Each weight was coded by a different locus in the network's genome. The alleles at each locus were 0, 1, or ?, where "?" signified that the value varied with learning, and where "learning" consisted of a series of trials in which the ? values were guessed to be 0 or 1. A final weight of value 1 came either from having the "1" allele in one's genome or from having adopted it in a guessing trial. Populations of networks evolved under a fitness function that highly rewarded networks when all connections were set to 1 sometime during the network's lifetime but not otherwise. If the 1 state was adopted early in a network's lifetime, then the fitness was higher than if it was adopted later. With this combination of evolution and learning, Hinton & Nowlan observed that correct settings for all loci were achieved after about 20 generations of the GA, whereas they never occurred under evolution alone. Hinton & Nowlan interpreted this result as an (extremely simple) example of the Baldwin Effect. Maynard Smith (46) calculated that if phenotypes were strictly determined by genotype, without opportunity for learning, then about 1000 generations would have been required in an asexual population and would probably never evolve in a strictly sexual population. As described by Hinton & Nowlan (30), learning makes the difference between finding a needle in a haystack and finding a needle in the haystack when someone tells you when you are getting closer. It is evident from these studies, both with Lamarckian evolution and the Baldwin effect, that learning often allows organisms to evolve *much* faster than their non learning equivalents. Hence its importance for tracking the stock market by adapting agents, and quite possibly for evolution in the natural world.

Cultural and Social Evolution

In view of the effect of learning on evolution, it is natural to ask how culture affects evolution. It would seem that artificial systems like those used for studying Lamarckian evolution and the Baldwin effect would be natural vehicles to explore the elements of cultural transmission. But in fact, there have been relatively few such studies.

Studies of the evolution of cooperation in the Prisoner's dilemma, begun by Axelrod (5), have stimulated a great deal of investigation. These typically do not involve cultural evolution, though in the real world such traits would have a very strong cultural component. There have been a few studies on the evolution of communication: acquiring the ability to communicate and agree on common lexical items have been modeled with some success (4, 70, 75). In addition, a few studies have addressed the very difficult problems concerned with how actual languages are learned and evolve (see 51).

Learning human languages presents serious theoretical problems for complex adaptive systems. For example, Gold's problem (25) is concerned with how, after hearing only a finite number of sentences (many of which may have errors), each of us learns a grammar that can generate an infinite number of grammatically correct sentences. A second problem is to account for the many changes that occur through time. Speakers of modern English can typically read Shakespeare (early Modern English from 400 years ago) without much difficulty. We can read Chaucer (Middle English from 800 years ago) with a moderate amount of lexical help. Only scholars can read Bede (Old English from 1300 years ago). Spoken Chaucer would be incomprehensible because of changes in vowel sounds that have occurred since it was spoken, but the spelling has remained similar. The challenge is to describe and, possibly predict, the course of language evolution in cases such as this. Learnability is a major issue here, and it is generally felt that the evolution of languages is largely driven by how easy it is to learn. Niyogi & Berwick (51) have used evolutionary computation methods to model how populations of agents can acquire language from hearing samples from the prior generation and then themselves provide examples for the next generation to learn. Using Chomsky's principles and parameter model of language (12), they found that some parameters were more stable than others. Further, they found that learnability alone was an inadequate explanation for some of the changes in grammatical form known to have occurred in the evolution of English, such as verb order, where Old English resembles present-day German. For a review of other attempts to model the coevolution of genes, cultures, and societies see Epstein & Axtell (20).

Evolution of Computational Ecologies

A novel approach to computational studies of ecologies was pioneered by T Ray (59). His goal was to evolve self-replicating, cooperating bits of computer code that "live" in a virtual computer world. Most computer instructions are too brittle to permit random mutations and recombinations that will provide fragments of code that are both meaningful and capable of being strung together with other such fragments. Koza's genetic programming paradigm, described in the section on using Genetic Programming to Evolve Optimal Foraging Strategies, is one method to get around this brittleness; neural networks are another. Ray's "Tierra" program uses a different method involving a specially designed assembly language to construct self-replicating programs. The resulting "ecology" provides interesting parallels to natural life—including competition for (memory) resources, trophic structures, and so on (47). Ray's current efforts are directed toward the evolution of self-contained but cooperating programs that emerge through evolutionary computation and arc analogous to multicellular organisms. Success has so far been limited, but Ray does observe differentiation into something akin to somatic and reproductive code. (60).

Building on Tierra, C Adami (2) has developed a computational world, "Avida," with spatial structure that Tierra lacks. J Chu (14) has further developed Avida to run on the massively parallel Intel Paragon computer, so that very large numbers of simulations can be run in fairly large environments, e.g. 100×100 units, with as many as 10,000 competing bits of code. Chu observed what seem to be invariant power laws, where the log of the number of copies of a program that have "lived" can be plotted against how frequently such sequences were observed in the evolutionary sequence. When selection was strong he found a $-3/2$ slope for this, just as is observed for the number of families in the fossil record (67) and is observed for avalanche size in the higher-dimension sandpiles of Bak's models of self-organized criticality (9). Chu developed arguments based on the theory of branching process to explain why this should be true. Such relationships, if found to be general, might point to a radically different theory of evolution than we now have, based on principles of self-organizing systems that are both more general and also more capable, in that they can capture phenomena that have so far resisted adequate explanation (19).

Adaptive Surfaces and Evolution of Genetic Systems

Until recently, the field of evolutionary computation has made surprisingly little use of quantitative genetics or population genetics theory. One reason for this is the belief by computer scientists that most difficult problems are highly nonlinear—that is there exists widespread interaction among the parts of a candidate solution, so that attempts to minimize interactions (as by attempts to linearize in quantitative genetics) or to treat genes as individuals (in single-locus models of population genetics) are bound to be self-limiting. A second reason for this lack of reciprocity in theory is that the conditions for evolution are often different in evolutionary computation settings (e.g., population sizes of a few hundred and mutation rates of 10^{-2}) than in biology. Further, while most evolutionary computation systems include recombination, the life cycle of individuals is like that of a moss, with a short diploid and a long haploid phase—not at all what most genetic theory addresses.

This is not to say that population genetics is inconsistent or inapplicable. Christiansen & Feldman (13) showed how to derive parts of Holland's GA theory (31) from principles of population genetics. Further, theoretical predictions from population genetics do help explain certain observations of evolutionary computation: e.g. in evolutionary computation applications where mutation rates and magnitudes are allowed to evolve, it is typically observed that they evolve downwards after sufficient time (reviewed in 8), as expected from equilibrium theory in population genetics.

One of the most challenging problems in population genetics has concerned the manner that populations traverse their adaptive landscape. Does evolution carry out a gradual hill-climbing, leading to some sort of optimization, as RA Fisher argued, or does it proceed by jumps and starts, with chance playing a significant role, as argued by Sewall Wright? In spite of the centrality of this issue for many questions in evolutionary theory, it has proven extremely difficult to test different proposals (16). Evolutionary computation has addressed this problem from a purely practical standpoint and has typically found that population subdivision ("island models") significantly speeds evolution (e.g., 8, 27, 71). From a different vantage, theoretical approaches to evolutionary computation, such as those proposing mechanisms underlying metastability in evolution (72, 73), may provide new theoretical bases for describing many of these phenomena.

One feature of adaptive landscapes, in both evolutionary computation and biological settings, is that broad plateaus of fitness seem common, and chance plays a major role in moving about on them. Where the population can move next seems to depend critically upon where it has drifted on the plateau. For example, Huynen, Stadler & Fontana (33) used computational models for predicting molecular structures to describe the 3D structure of tRNA. While 30% of nucleotide substitutions seemed to be neutral, the high dimensionality made it possible to traverse sequence space along a connected path, changing every nucleotide present, without ever changing the structure. It is no surprise then, that when a population is begun with all sequences identical, but with a small amount of mutation, the initial point in sequence space diffuses outward into a cloud, to a limit in accord with theoretical expectations, then drifts along the plateau. Different subpopulations can reach very different parts of the sequence space before dramatic improvement results from finding one improvement or another. Fitness assumes a step-like improvement, coinciding with Wright's expectation that "Changes in wholly nonfunctional parts of the molecule would be the most frequent ones but would be unimportant, unless they occasionally give a basis for later changes which improve function in the species in question, which would then become established by selection" (56), p. 474.

Moving from plateau to plateau of fitness is frequently observed in evolutionary computation and is typically associated with changes in complexity. Taking just one example, Miglino, Nafisi, & Taylor (48) used genetic algorithms to evolve controllers for small robots. The controllers were neural networks that could be described by equivalent finite state automata, the complexity of which can be readily observed and measured. The task presented was to traverse as many squares as possible in a grid placed on the floor in a limited amount of time. Starting from random networks it was initially sufficient merely to move forward and turn left when encountering a corner. Many neural networks prescribed this behavior, but some of these made it easier to make jumps to radically more sophisticated behavior, with correspondingly more complex programs. There was much variation from run to run, with chance largely determining which populations were able to find one improved solution or another.

While these studies showed quite clearly that plateaus on the adaptive surface are common, with stepwise improvement in fitness, it must be stressed that this is not always the case—especially when fitnesses are frequency-dependent. Very complex dynamics are sometimes observed, including plateaus interspersed with periods of chaos (45). An interesting example of this is provided by competition among bit strings in a series of studies by K Kaneko and co-workers (summarized in 36). In this system strings are assumed to compete to the extent that they are similar (measured by their Hamming distance)—more similar strings compete more strongly, so fitness is frequency-dependent. But strings too far apart have less success in reproduction. Mutation among the strings is allowed. It is also possible to include predator/prey interactions in this system, where the strength of predator-prey interactions depends on the Hamming distance. Such systems are high-dimensional and highly nonlinear. In a way, their interactions resemble logistic maps, which are known to be chaotic over much of their parameter space, except that here they are high-dimensional and can escape from having their fitness reduced by competition, as it were, through mutation to a less frequent form. In a series of papers Kaneko and co-workers analyzed the dynamics of this system, numerically calculating the Lyapunov exponents, and observed high-dimensional, weakly chaotic dynamics in the evolution of this system that often led to dynamic stability and robustness against external perturbations. He termed this situation "homeochaos" and suggested that such system dynamics may be very general features of evolution, both in computational evolution and in the real world.

CONCLUSIONS

There are many parallels between biological evolution searching through a space of gene sequences and computer evolution searching through a space of computer programs or other data structures. Several approaches to exploit these similarities have developed independently and are collectively termed evolutionary algorithms or evolutionary computation. Such methods of computation can be used to search through the space of candidate solutions to problems and are now finding application in an increasing number of industrial and business problems.

While there is no general theory that will identify the best method to find optima, it appears that evolutionary computation is best suited for problems that involve nonlinear interactions among many elements, where many intermediate optima exist, and where solutions that are satisficing—merely very good without necessarily being the absolute optimum—will do. Such problems are common in business and in biological studies, such as cooperation, foraging, and coevolution.

Evolutionary computation can sometimes serve as a useful model for biological evolution. It allows dissection and repetition in ways that biological evolution does not. Computational evolution can be a useful tool for education and is beginning to provide new ways to view patterns in evolution, such as power laws and descriptions of non-equilibrium systems. Evolutionary theory, as developed by biologists, typically tries to linearize systems, for ease of analysis with differential equations, or to treat units in isolation, as in single-locus selection. While evolutionary computation is not inconsistent with such theory, it tends to be outside it, in that real difference in capacity and complexity are often observed and are not really describable by stable equilibria or simple changes in gene frequencies, at least in ways that are interesting. There is reason to believe that theories of evolutionary computation might extend the language of biological evolutionary theory and contribute to new kinds of generalizations and analyses that have not been available up to now.

ACKNOWLEDGMENTS

MM acknowledges the Santa Fe Institute and the National Science Foundation (grant NSF-IRI-9705830) for support. CT acknowledges NSF grant #5BR9720410.

Visit the Annual Reviews home page at http://www.AnnualReviews.org

LITERATURE CITED

1. Ackley D, Littman M. 1992. Interactions between learning and evolution. In *Artificial Life II*, ed. CG Langton, C Taylor, JD Farmer, S Rasmussen, pp. 487–509. Reading, MA: Addison-Wesley
2. Adami C. 1998. *Introduction to Artificial Life*. New York: Springer-Verlag
3. Angeline PJ, ed. 1997. *Evolutionary Programming VI: 6th Int. Conf. EP97*. New York: Springer
4. Arita T, Koyama Y. 1998. Evolution of linguistic diversity in a simple communication system. In *Artificial Life VI*, ed. C Adami, RK Belew, H Kitano, CE Taylor, pp. 9–17. Cambridge, MA: MIT Press
5. Axelrod R. 1984. *The Evolution of Cooperation*. New York: Basic
6. Bäck T. 1996. *Evolutionary Algorithms in Theory and Practice: Evolution Strategies, Evolutionary Programming, Genetic Algorithms*. Oxford: Oxford Univ. Press
7. Bäck T, ed. 1997. *Proceedings of the Seventh International Conference on Genetic Algorithms,* San Francisco, CA: M. Kaufmann
8. Baeck T, Hammel U, Schwefel HP. 1997. Evolutionary computation: comments on

the history and current state. *IEEE Trans. Evol. Computation* 1:3–17
9. Bak P. 1996. *How Nature Works: The Science of Self-Organized Criticality.* New York: Springer-Verlag
10. Belew RK, Mitchell M, eds. 1996. *Adaptive Individuals in Evolving Populations: Models and Algorithms.* Reading, MA: Addison Wesley
11. Belew RK, Vose MD, eds. 1997. *Foundations of Genetic Algorithms 4.* San Francisco, CA: M. Kaufmann
12. Chomsky N. 1995. *The Minimalist Program.* Cambridge, MA: MIT Press
13. Christiansen FB, Feldman MW. 1998. Algorithms, genetics, and populations: the schemata theorem revisited. *Complexity* 3(3):57–64
14. Chu J. 1999. *Computational explorations of life.* PhD thesis. Calif. Inst. Technol., Pasadena, CA
15. Cliff D, Grand S. 1999. The 'Creatures' global digital ecosystem. *Artificial Life.* In press
16. Coyne JA, Barton N, Turelli M. 1997. Perspective: a critique of Sewall Wright's shifting balance theory of evolution. *Evolution* 51:643–71
17. Dawkins R. 1989. The evolution of evolvability. In *Artificial Life*, ed. CG Langton, 201–220. Reading, MA: Addison-Wesley
18. Dawkins R. 1996. *The Blind Watchmaker: Why the Evidence of Evolution Reveals a Universe Without Design.* New York: Norton. 2nd ed.
19. Depew DJ, Weber BH. 1995. *Darwinism Evolving.* Cambridge, MA: MIT Press
20. Epstein J, Axtell R. 1996. *Growing Artificial Societies.* Cambridge, MA: MIT Press
21. Fogel DB. 1995. *Evolutionary Computation: Toward a New Philosophy of Machine Intelligence.* New York: IEEE Press
22. Fogel DB, ed. 1998. *Evolutionary Computation: The Fossil Record.* New York: IEEE Press
23. Fogel LJ, Owens AJ, Walsh MJ. 1966. *Artificial Intelligence Through Simulated Evolution.* New York: John Wiley
24. Gehlhaar D, Verkhivker G, Rejto P, Sherman C, Fogel D, et al. 1995. Molecular recognition of the inhibitor AG-1343 by HIV-1 protease: conformationally flexible docking by evolutionary programming. *Chem. Biol.* 2:317–24
25. Gold EM. 1967. Language identification in the limit. *Inform. Control* 10:447–74
26. Goldberg DE. 1989. *Genetic Algorithms in Search, Optimization, and Machine Learning.* Reading, MA: Addison-Wesley
27. Gordon VS, Whitley D. 1993. Serial and parallel genetic algorithms as function optimizers. In *Proc. Fifth Int. Conf. Genetic Algorithms*, ed. T Bäck, pp. 177–183. San Mateo, CA: M. Kaufmann
28. Grefenstette JJ. 1991. Lamarckian learning in multi-agent environments. In *Proc. 4th Int. Conf. on Genetic Algorithms and Their Applications*, ed. RK Belew, L Booker, pp. 303–10. San Mateo, CA: M. Kaufmann
29. Hillis WD. 1990. Co-evolving parasites improve simulated evolution as an optimization procedure. *Physica D* 42:228–34
30. Hinton GE, Nowlan SJ. 1987. How learning can guide evolution. *Complex Systems* 1:495–502
31. Holland JH. 1975. *Adaptation in Natural and Artificial Systems.* Ann Arbor, MI: Univ. Michi. Press
32. Hopfield J. 1982. Neural networks and physical systems with emergent collective computational abilities. *Proc. Nat. Acad. Sci. USA* 79:2554–58
33. Huynen MA, Stadler F, Fontana W. 1996. Smoothness within a rugged landscape: The role of neutrality in evolution. *Proc. Natl. Acad. Sci. USA* 93:397–401
34. Imada A, Araki K. 1996. Lamarckian evolution and associative memory. In *Proc. 1996 IEEE Third Int. Conf. Evol. Computation (ICES-96)*:676–80
35. Juillé H, Pollack JB. 1998. Coevolutionary learning: a case study. In *ICML '98-Proc.*

Int. Conf. Machine Learning. San Francisco, CA: M. Kaufmann
36. Kaneko K. 1994. Chaos as a source of complexity and diversity in evolution. *Artificial Life* 1:163–77
37. Karakotsios K, Bremer M. 1993. *SimLife: The Official Strategy Guide*. Rocklin, CA: Prima
38. Kelly K. 1994. *Out of Control: The Rise of Neo. Biological Civilization*. Reading, MA: Addison-Wesley
39. Knuth DE. 1973. *The Art of Computer Programming*. Vol. 3: *Sorting and Searching*. Reading, MA: Addison-Wesley
40. Koza JR. 1992. *Genetic Programming: On the Programming of Computers by Means of Natural Selection*. Cambridge, MA: MIT Press
41. Koza JR, Rice JP, Roughgarden J. 1992. Evolution of food foraging strategies for the Caribbean anolis lizard using genetic programming. *Adaptive Behav.* 1(2):47–74
42. Lamarck JB. 1809. *Philosophie Zoologique, ou Exposition des Considérations Relatives a l'Histoire Naturelle de Animaux*. Paris: Chez Dentu et L'Auteur
43. Lamarck JB. 1996. Of the influence of the environment on the activities and habits of animals, and the influence of the activities and habits of these living bodies in modifying their organization and structure. See Ref. 10, pp. 39–57
44. Lewontin R. 1998. Survival of the nicest. *NY Rev. Books*. 22 Oct. 1998, 59–63
45. Lindgren K. 1992. Evolutionary phenomena in simple dynamics. In *Artificial Life II*, ed. CG Langton, C Taylor, JD Farmer, S Rasmussen, pp. 295–312. Reading, MA: Addison-Wesley
46. Maynard Smith J. 1987. Natural selection: when learning guides evolution. *Nature* 329:761–62
47. Maynard Smith J. 1992. Byte-sized evolution. *Nature* 355:772–73
48. Miglino O, Nafasi K, Taylor CE. 1994. Selection for wandering behavior in a small robot. *Artificial Life* 2:101–16
49. Mitchell M. 1996. *An Introduction to Genetic Algorithms*. Cambridge, MA: MIT Press
50. Murakawa M, Yoshizawa S, Adachi T, Suzuki S, Takasuka K, et al. 1998. Analogue EHW chip for intermediate frequency filters. In *Evolvable Systems: From Biology to Hardware*, ed. M Sipper, D Mange, pp. 134–43. New York: Springer
51. Niyogi P, Berwick RC. 1995. *The Logical Problem of Language Change. Tech. Rep. A. I. Memo No. 1516*. MIT Artificial Intelligence Lab. Cambridge, MA
52. Papadimtriou CH, Sideri M. 1998. On the evolution of easy instances. *Unpublished manuscript,* Computer Science Dept., University of California, Berkeley, CA
53. Paredis J. 1997. Coevolving cellular automata: Be aware of the red queen! In *Proc. Seventh Int. Conf. Genetic Algorithms*, ed. T Bäck, pp. 393–400. San Francisco, CA: Morgan Kaufmann
54. Petzinger Jr. T. 1995. At Deere they know a mad scientist may be the firm's biggest asset. *Wall Street J.* 14 July 1995, p. A1
55. Press WH, A. Teukolsky S, Vetterling WT, Flannery BP. 1992. *Numerical Recipes in C*, New York: Cambridge Univ. Press
56. Provine WB. 1986. *Sewall Wright and Evolutionary Biology*. Chicago, IL: Univ. Chicago Press
57. Rao SS. 1998. Evolution at warp speed. *Forbes Mag.*
58. Rawlins G, ed. 1991. *Foundations of Genetic Algorithms*. San Mateo, CA: M. Kaufmann
59. Ray TS. 1991. An approach to the synthesis of life. In *Artifical Life II*, ed. CG Langton, C Taylor, J Farmer, S Rasmussen, pp. 371–408. Reading, MA: Addison–Wesley
60. Ray TS, Hart J. 1998. Evolution of differentiated multi-threaded digital organisms. In *Artificial Life VI*, ed. C Adami, RK Belew, H Kitano, CE Taylor, pp. 295–306. Cambridge, MA: MIT Press
61. Rechenberg I. 1973. *Evolutionsstrategie*. Stuttgart: Frommann-Holzboog

62. Rosin CD, Belew RK. 1995. Methods for competitive coevolution: finding opponents worth beating. In *Proc. Sixth Int. Conf. Genetic Algorithms*, ed. LJ Eshelman, pp. San Francisco, CA: M. Kaufmann
63. Sasaki T, Tokoro M. 1997. Adaptation toward changing environments: Why Darwinian in nature? In *Proc. Fourth Eur. Conf. on Artificial Life*, 145–53. Cambridge, MA: MIT Press
64. Sasaki T, Tokoro M. 1999. Evolvable learnable neural networks under changing environments with various rates of inheritance of acquired characters: comparison between Darwinian and Lamarckian evolution. *Artificial Life*. In prcss
65. Schwefel HP. 1975. *Evolutionsstrategie und Numerische Optimierung*. PhD thesis, Technische Univ. Berlin, Berlin
66. Schwefel HP. 1995. *Evolution and Optimum Seeking*. New York: Wiley
67. Sepkowski JJ. 1992. *A Compendium of Fossil Marine Animal Families*. Milwaukee, WI: Milwaukee Public Mus. 2nd ed.
68. Simon H. 1969. *The Sciences of the Artificial*. Cambridge, MA: MIT Press
69. Sims K. 1994. Evolving 3D morphology and behavior by competition. In *Artificial Life IV*, ed. RA Brooks, P Maes, pp. 28–39. Cambridge, MA: MIT Press
70. Steels L, Kaplan F. 1998. Stochasticity as a source of innovation in language games. In *Artificial Life VI*, ed. C Adami, RK Belew, H Kitano, CE Taylor, pp. 368–78. Cambridge, MA: MIT Press
71. Tanese R. 1989. Distributed genetic algorithms. In *Proc. Third Int. Conf. on Genetic Algorithms*, ed. JD Schaffer, pp. 434–39. San Mateo, CA: M. Kaufmann
72. van Nimwegen E, Crutchfield JP, Mitchell M. 1999. Statistical dynamics of the Royal Road genetic algorithm. *Theoret. Computer Sci.* To appear
73. van Nimwegen E, Crutchfield JP, Mitchell M. 1997. Finite populations induce metastability in evolutionary search. *Phys. Lett. A*, 229(2):144–50
74. Waddington CH. 1953. Genetic assimilation of an acquired character. *Evolution* 7:118–26
75. Werner GM, Dyer MG. 1991. Evolution of communication in artificial organisms. In *Artificial Life II*, ed. CG Langton, C Taylor, J Farmer, S Rasmussen, pp. 659–87. Reading, MA: Addison Wesley
76. Whitley LD, ed. 1993. *Foundations of Genetic Algorithms 2*. San Mateo, CA: M. Kaufmann
77. Whitley LD, Vose MD, eds. 1995. *Foundations of Genetic Algorithms 3*. San Francisco, CA: M. Kaufmann

SUBJECT INDEX

I

Cumulative Indexes

CONTRIBUTING AUTHORS, VOLUMES 26–30

CHAPTER TITLES, VOLUMES 26–30

Volume 26 (1995)

Special Section on Sustainability

Volume 27 (1996)

Volume 28 (1997)

Volume 30 (1999)

NAME

ADDRESS

Place Stamp Here

ANNUAL REVIEWS
4139 EL CAMINO WAY
PO BOX 10139
PALO ALTO CA 94303-0139